# TRANSACTIONS OF THE INTERNATIONAL ASTRONOMICAL UNION

## VOLUME XVIII B – PROCEEDINGS

10.8.1927 - 19.8.1982

PRESIDENT DE L'UNION ASTRONOMIQUE INTERNATIONALE

MANALI KALLAT VAINU BAPPU

PRESIDENT OF THE INTERNATIONAL ASTRONOMICAL UNION

1979–1982

INTERNATIONAL ASTRONOMICAL UNION

UNION ASTRONOMIQUE INTERNATIONALE

TRANSACTIONS

OF THE

INTERNATIONAL ASTRONOMICAL UNION

VOLUME XVIII B

# PROCEEDINGS OF THE EIGHTEENTH

# GENERAL ASSEMBLY

## PATRAS 1982

*Edited by*

RICHARD M. WEST

*General Secretary of the Union*

D. REIDEL PUBLISHING COMPANY

DORDRECHT : HOLLAND / BOSTON : U.S.A. / LONDON : ENGLAND

Library of Congress Cataloging in Publication Data

CIP

ISBN-13: 978-90-277-1569-2 e-ISBN-13: 978-94-009-7108-0
DOI: 10.1007/978-94-009-7108-0
LC-Number 30-10103

*Published on behalf of*
*the International Astronomical Union*
*by*
*D. Reidel Publishing Company, P. O. Box 17, 3300 AA Dordrecht, Holland*


*Softcover reprint of the hardcover 1st edition 1983*

*Sold and distributed in the U.S.A. and Canada*
*by Kluwer Boston, Inc.,*
*190 Old Derby Street, Hingham, MA 02043, U.S.A.*

*In all other countries, sold and distributed*
*by Kluwer Academic Publishers Group,*
*P. O. Box 322, 3300 AH Dordrecht, Holland*

*D. Reidel Publishing Company is a member of the Kluwer Group*

# Preface

The XVIIIth General Assembly of the International Astronomical Union was held in Patras, Greece, from 17-26 August 1982. It was marked by the tragic death of the President of the IAU, Professor M.K.V. Bappu, on August 19, 1982. This sad event, without precedent in the history of the Union, posed serious problems to the organization of the General Assembly, which could only be overcome by the full collaboration of all members, the organizers, and the Executive Committee.

A tribute to the memory of Prof. Bappu was paid during a plenary meeting on 23 August 1982. The full texts of the speeches are published in Chapter I of this volume.

The excellent scientific programme in Patras was organized by the Presidents of the 40 IAU Commissions and coordinated by the IAU General Secretary (1979-1982), Professor P.A. Wayman. The local arrangements were taken care of by Professor C. Goudas and his collaborators from the Patras University. Due to the unexpected withdrawal in 1979 of another invitation to host the 1982 IAU General Assembly, the organizers in Greece had less than two years available for the extensive preparations, and our hearty thanks are due to them for their persistent efforts, which made this General Assembly an outstanding success.

Although the usefulness of IAU General Assemblies, as compared to more specialized meetings, has been questioned at various times, there is little doubt that the XVIIIth General Assembly was a clear demonstration of their value. First of all, it became a most fertile meeting ground for scientists from all over the world, providing a highly efficient forum for evaluation of on-going international collaborative projects and also for the initiation of new ones. Secondly, and not of less importance, it was for many young astronomers the first massive contact with more experienced colleagues from other countries and a unique opportunity to make the acquaintance with authors of scientific papers whom they had only known by name before. For them, as well as for many astronomers working in countries where our science has not yet been firmly established, the meetings in Greece were highly incentive and acted as catalysts for future research projects.

The present volume, IAU Transactions XVIII B, summarizes the work of the 18th General Assembly. The Inaugural Ceremony and the first session of the General Assembly were held in the evening of 17 August 1982, in the Ancient Odeon of the City of Patras under the open sky. The inaugural speeches are published in Chapter II. The proceedings of the General Assembly will be found in Chapter III, which includes the Resolutions and other aspects of the administration of the Union. They provide the official record of the business of the General Assembly, and, together with the Executive Committee Report (Chapter V), the permanent record for the Union in the period 1979-1982.

In addition, this volume contains the Commission Reports from Patras, compiled by the Presidents of Commissions (Chapter IV). Pending re-edition of the complete Astronomer's Handbook, which is foreseen within the next few years, Chapter VI gives updated information about the various activities of the Union. Finally, Chapter VII contains the members lists for IAU Commissions and also an alphabetical list of the more than 5000 individual members of the IAU.

In order to make the information contained in this volume available at the earliest possible date, a tight production schedule was imposed. This would not have been possible without the efficient cooperation of the Presidents of Commissions who provided the camera-ready manuscripts for Chapter IV. Similarly, I am most grateful to Mrs. B. Manning and Ms. P. Smiley (IAU Secretariat, Paris), and Mrs. E. Völk (ESO, Munich) for their expert help in typing and assembling the manuscript for publication. I am also much indebted to my predecessor, Professor P.A. Wayman and Dr I. Elliott of the Dunsink Observatory, Dublin, Ireland, who kindly provided the computer print-outs with the lists of Commission members. Finally, it is a pleasure to acknowledge the friendly, professional support by the IAU Publisher, D. Reidel Publishing Co., and in particular Mrs. N.M. Pols-v.d. Heijden.

IAU Secretariat
61, Ave. de l'Observatoire
F-75014 Paris
France

Richard M. West
General Secretary, IAU
4 December 1982

# Contents

# I.
# Tribute to the Memory of Professor M. K. V. Bappu

Professor M.K.V. Bappu, President of the International Astronomical Union (1979-1982) died in hospital in Munich, Federal Republic of Germany, at the time of the XVIIIth General Assembly. He had undergone a major heart operation on 9 August and although the surgical intervention was apparently successful, serious complications set in a few days later. After a heroic struggle, assisted by the foremost medical expertise, Professor Bappu expired in the early evening of Thursday, 19 August 1982. Word about this tragic event was immediately received in Patras and the IAU Executive Committee in an emergency meeting the same evening decided that a special tribute should be paid to the memory of their esteemed colleague, friend and President. This took the form of a meeting held at the University of Patras, on 23 August 1982 at 13:00 hours, with speeches by representatives of the IAU, of the Indian National Science Academy, of the Greek National Committee for Astronomy, of the Local Organising Committee and of Professor Bappu's wide circle of friends. The meeting brought together the majority of General Assembly participants and was a sad farewell to a highly respected and admired scientist, a brilliant organisor and a beloved friend.

The texts of the speeches, which were also printed in the General Assembly Newspaper, "Astrocosmos", on 24 August, are reproduced below.

## PROFESSOR E.K. KHARADZE - ON BEHALF OF THE IAU

Dear Colleagues, Friends, Ladies and Gentlemen,

On behalf of the International Astronomical Union I express the most profound sorrow caused by the most untimely death of the current President of our Union, the distinguished scientist Professor Manali Kallat Vainu Bappu, 55 years of age.

We have all gathered here today at this mournful moment to pay a worthy tribute to his memory.

During Professor Bappu's triennial Presidency of our Union, we have witnessed his devotion to the interests of the Union, his very fruitful activity aimed at its consolidation and strengthening of its scientific effectiveness for the further development of our science - astronomy, all over the planet. The members of our Union will also remember Professor Bappu's contribution at the earlier time of his office in the Union as Vice-President in 1967-1973.

Professor Bappu's prominent activity in the Union and in the world-wide astronomical community in general carries the features not only of the skill of an experienced eminent scientist, but also of personal tact, kindness and benevolence to everybody, that has always been so highly appreciated by his colleagues and that has made collaboration with him so pleasant and easy.

We are all sharing Mrs Bappu's inconsolable grief.

Every effort was made in Munich to save Professor Bappu. The medical care was on the highest possible level. But, unfortunately, it was not Professor Bappu's fate to recover.

With his death, not only we - the Union, suffer a great loss, but so does the entire astronomical world. Moreover, the astronomical community in India is feeling it badly. They have lost their leading astronomer, who was full of plans and ideas concerning further development of astronomy in India.

Last year Professor Bappu visited several observatories in the USSR. We have a vivid memory of that visit, and of the enthusiasm with which he spoke on the development of astronomy in India. Many of his plans are left unfinished. But however profound the sorrow may be, however great the loss, our Indian colleagues should be encouraged to go on with their endeavours. That would be the best tribute paid to Professor Bappu's memory by his compatriots.

Being a member of the USSR delegation to this General Assembly, I permit myself to express on behalf of the USSR astronomers also, most hearty brotherly condolence and sympathy to the Indian delegation.

DR J.C. BHATTACHARYYA - ON BEHALF OF THE INDIAN NATIONAL SCIENCE ACADEMY

I was with a profound shock that we learnt about the sudden, unexpected passing away of Professor M.K. Vainu Bappu at Munich on the 19th of this month. A few days ago, we had the vision of a grand reunion of old friends here; we did not even dream of such a calamity of cruel fate snatching away our beloved, illustrious compatriot at this juncture. It is difficult to believe that Vainu Bappu, our charming friend and leader who was so full of life, will never be among us again.

Bappu completed his fifty-fifth year only a few days ago; we have been looking forward to many more years of his company and guidance. The great upsurge of astronomical activities in India, particularly in the Institute of Astrophysics, which he created, has suddenly lost its main guiding force. It is like a nightmare of a group of sailors in mid-ocean waking up to find their captain missing. He leaves behind not only the team of scientists he has guided and brought up, but also his wife Yemuna and his aged mother to suffer this unbearable loss.

Bappu's contribution to astronomy cannot be summed up easily. He has discovered new objects, new relations, created and guided new schools of astronomers inspired active societies and taken part in a wide range of international activities. He had a deep understanding of many problems; whether it was in instrumentation, or stellar processes or problems concerning tenuous gas clouds, he was able to suggest ways most likely to lead to tangible answers. A gem of a teacher, as anybody who has listened to his lectures would testify, he could attract and hold audiences spellbound. A brillant organiser who could foresee the minutest details of a projected venture years in advance; in short, Bappu represented a personality of an ideal world, so rare in real life.

We have been fortunate to have this magnificent personality so close to us, and that makes our present loss overwhelming. His circle of friends covered the entire world; many of them are present here, and I am confident that all will agree how his adorable personality charmed everyone who came in contact with him.

One great aspect of Bappu was his love of nature which he worshipped and to which he tendered his creative ideas as offerings; the Institute premises at Bangalore, Kavalur and Kodaikanal bear mute testimony to this enchanting quality.

He implored others to develop a sincere love for nature. Speaking on one occasion he remarked: "To a person who enjoys a glorious sunset or marvels at the beauty of a cobweb glistening with the morning dew, nature unfolds her unending stream of magical charms."

Bappu is no more; he now belongs to eternity. His mortal remains have returned to mother nature he loved so much. Only a few days ago, as I was talking to him by telephone from India, he appeared so full of vitality and confidence. The thought uppermost in his mind was of this Assembly. "Do not worry about me", he advised, "go and join the General Assembly which must be a success. Nothing - nothing should mar the proceedings." That was his wish; we all owe him our sincere efforts to see that it is fulfilled.

ACADEMICIAN J. XANTHAKIS - ON BEHALF OF THE GREEK NATIONAL COMMITTEE FOR ASTRONOMY

Le Comité National d'Organisation exprime ses sincères condoléances pour la mort de notre président, le professeur Bappu.

Avec vive émotion, je me rapelle qu'il y a huit mois, le professeur Bappu nous a rendu visite avec le Secrétaire Général de l'UAI, Monsieur Wayman, pour discuter des affaires concernant la préparation du 18ème congrès.

A cette époque, il était en pleine santé et a manifesté un grand intérêt et beaucoup de dynamisme pour l'organisation de notre congrès.

L'annonce de sa mort subite, nous a laissés dans une profonde consternation et les drapeaux en berne sont le reflet de notre grande peine.

PROFESSOR C.L. GOUDAS - ON BEHALF OF THE LOCAL ORGANISING COMMITTEE of the XVIIIth General Assembly

Members of the Assembly, Ladies and Gentlemen,

We are gathered here to mourn for the loss of an excellent scientist and a noble man. To mourn for Vainu Bappu, the President of our Union, whose personality and high qualities are known, to a larger or smaller extent, by all.

The death of a President during the General Assembly he was supposed to preside over is an event without precedence and most likely never to be repeated.

As Chairman of the Local Organising Committee of the General Assembly that will be marked and remembered not only for its scientific works but also the loss of Vainu Bappu, I wish, in retrospect, to remind you of the story of Diagoras. Being a veteran of past Olympic games, and while lifted on the shoulders of his two olympic laureate sons inside the stadium of Olympia, Diagoras was cheered by the crowds with the unusual wish "Die now Diagoras". The crowds wished him to die at the moment of the peak of his glory.

Unable as we were, and are, to influence the decisions of fate, we cannot but observe that, since Vainu could not but leave us, the choice of moment for the inevitable was for him a grace - a moment of glory, recognition and appreciation.

Of course, such thoughts can hardly now be mentioned to the immediate family of the deceased. To them we express our deepest condolences.

Long live the memory of Vainu Bappu.

## PROFESSOR A. BLAAUW - AS PAST PRESIDENT OF THE IAU

Dear friends and colleagues of Vainu Bappu,

Three years ago, at the closing of the General Assembly at Montreal, when Vainu assumed the presidency, he conveyed to us his feelings of gratitude and joy in anticipation of the task ahead. "As each member," he said, " individual and country, contributes stone by stone to the vast edifice of astronomical knowledge, we rejoice without reservation and in all friendliness at the triumph of the human mind over a facet of Nature. In encouraging opportunities and providing that little extra fillip needed to cross the threshold into the domain of creative productivity, the Union can play an important role."

It is in this spirit that Vainu Bappu, during these past years has served the Union, with his many talents as a scientist and as an organiser. Three years during which he helped guide the affairs of the Union with determination and with the experience and foresight that had already made him so distinguished a renovator of astronomy in his home country. Members of the Executive Committee and of course even more his close collaborators, the Union officers, will long remember his deep concern for all that was in the interest of the IAU.

How immensely would he have enjoyed being among us these two weeks, seeing the accomplishments of tasks undertaken, the realization of plans designed - and above all, how would he have rejoiced in the company of so many dear friends from all over the world, engaged with them in scientific discourse in the subject of astronomy that had fascinated him from his early childhood.

It has not been given to him and to us, for him to be with us during these days. But we know how much he gave his thoughts to this Assembly, long before its official beginnings and with how much care he devoted himself to contributing to its successful proceedings; the several weeks spent in Munich in quiet contemplation of the many affairs to be dealt with and we can only surmise the deep disappointment he must have felt when it became clear that he would be unable to participate himself.

But we have had the privilege of sharing his thoughts of those last weeks by the message he conveyed to us at the opening of this Assembly. May I remind you of the words Vainu passed on to us in his presidential address and let me quote a few sentences from it:

> "While innovation in technique is a factor of much significance in the discoveries that will yet be made, it is certain that an even greater role will undoubtedly be of the human intellect. Time and again we have seen how an individual has appeared on the scene and transformed a picture of gathering confusion into one of logical rigour and aesthetic simplicity. In the final reckoning, it is this aspect of Man that is responsible for the culminating triumph. It is, therefore, a responsibility shared by each one of us in our individual roles, be it of teacher or senior colleague, as well as of the astronomical community and the Union as a whole, to nurture such possibility. In its longterm view of contributing to the development of Astronomy, the Union must necessarily give considerable attention to the very important role of the individual in Astronomy."

As we take up our tasks for the Union and for research and teaching, let us keep these words of Vainu Bappu in mind.

## DR HARLAN G. SMITH - AS A CLOSE FRIEND

Vainu Bappu was probably my closest friend. For 33 years the strands of our lives have interwoven, even to the extent of his being best man at my wedding, and my just missing (because of academic schedules) being something of an equivalent at his wedding in India. I mention this because so few now in the IAU have had the privilege of really knowing him, and it is my grevious task to say farewell to him as a friend, on behalf of us all.

When the IAU was founded most astronomers knew each other, often rather well. Vainu was proud of the growth of our science, but regretted that he could only know so few of us, especially the young astronomers who are coming up so fast. For those who did not know him, I want to share several memories. There is no way I can truly bring out the richness of Vainu's life and mind and personality, but these vignettes - like flashes of lightning revealing fragments of landscape - may help you to glimpse the wonderful human being behind the formal degrees, the papers, and the honours he so well deserved.

In January 1949 we entered the Harvard Graduate School of Astronomy - entered with trepidation yet excitement to be studying in halls peopled with memories of the likes of Bond and Pickering, and the very real presence of Shapley, Bok, Payne-Gaposchkin, Menzel and Whipple. In the so-called liberal arts there is now a cult - I believe unfortunate and destructive - of the anti-hero, with its emphasis on sickness and psychosis, on feet of clay. But Vainu belonged to an older and finer school. His spirit was generous, kind, non-cynical - even hero-worshipping in the best sense of the word. He deeply admired if not even venerated his teachers and those, living or dead, who had really accomplished something. He well knew and understood human frailties, but to Vainu the important things were those qualities and works of great men and women which should be admired and emulated as far as possible in his own life. Then, as now, it was fashionable for graduate students to work very hard, and we did - sometimes tired and discouraged. I recall an occasion walking home with Vainu late one winter night, our conversation turning to whether we might ever get jobs or amount to anything in astronomy. Nearly always cheerful, he brightened up and proceeded to quote from his memory, well-stocked with classics, the entirety of his favorite poem, Kipling's "If". Though I cannot quote them exactly, most of us are familiar with a few of its famous lines, to the effect

> If you can keep your head,
>   when all about are losing theirs,
> ...
> Then you'll be a man, my son.
>
> If you can hazard all upon a toss,
>   and lose, and rise to try again,
> ...
> Then you'll be a man, my son.

The spirit of this poem was truly a beacon to his life - encouragement always to try to answer those "Ifs" with "I can, and I will".

Yet Vainu was anything but solemn. He laughed, and joined in most of the games and jokes. He found the necessary paraphernalia, and organised at the Observatory what must have been the only cricket team within a few hundred miles, teaching us the mysteries of sticky wickets and googlie-balls. On one occasion he gaily commented on some virtues of yoga, to the amused condescension of some of the other graduate students. So he inveigled me into a scientific test. We both stood on our heads (one form of yoga) for ten or fifteen minutes before each major exam. And, that year, we two were the only students to pass them all. QED! (but Vainu was too gentle ever to rub it in).

In those innocent days we had never even dreamed of being given money to travel to meetings. Yet Vainu never wanted to miss a session of the American Astronomical Society. So we went to all of them in our part of the country, doing what was necessary to make it possible, such as driving in my $75 Model A Ford, literally spending nights on the lawns of host observatories in our war-surplus sleeping bags. Hot or cold, hungry, mosquitoes - Vainu never complained, and usually managed to bring the conversation around to astronomy.

He finished his PhD in near-record time. Then, after a couple of brilliantly successful years as a post-doc at Cal Tech, he could easily have obtained a comfortable job in the US. But I never sensed any doubt in his mind. His dream was to build and to do astronomy in India. I think it is hard for us now to realize the courage of this decision - to leave the centre of world astronomy, the 200-inch, and return to an India which then, just after Independence, lacked any of the structure of modern astronomy, with essentially no institutes, telescopes, colleagues...not even a job. That first period effectively of unemployment, in the midst of such desperate poverty, might indeed have made him wonder whether - having risked his all upon a toss - he might have lost, and could he rise again?

Then came the chance, largely through his own persuasion of a reluctant minister, to build a U.P. State Observatory at Naini Tal. And shortly thereafter came Yemuna, a pillar of love and strength throughout his life. From then on the path was up. You know at least the outline of the rest, but again I think, cannot easily imagine the problems to be overcome, the incredible effort involved, in an India trying to build nearly everything at once from scratch, to in effect create three astronomical institutions, a national society, a national journal, optical and machine shops, engineering and computer contacts and facilities adequate to design and build entirely within India a 2.34-meter telescope. Yet during all this time he was constantly looking for, and then looking out for, dozens of students who are now putting India on the astronomical map.

He did these things, and more, and reached the very top. This meeting in Patras would have been the peak of his career. I believe it meant more to him than words can convey, something he had been building toward for 40 years and more. Yet, when the urgency of the operation became clear, he adjusted with remarkable speed and calm philosophical acceptance both to missing this peak, and to the very real risk of the operation.

My wife and I spent a day and a night with Vainu and Yemuna just before the operation. He was vigorously planning at least three major things for the future, finishing the Kavalu telescope, getting the 1985 IAU plans off to a good start, and, finally, contemplating the course of his life after mandatory retirement around 1985 as director. In general he was planning to travel to many institutes, to renew old friendships and make new ones, and to concentrate on science.

Now these travels will be in our memories, and the work done by others. Meanwhile our hearts are with Yemuna and with Vainu's mother. I want them to know that for us Kipling's poem has come true. Each of the "Ifs" has become a "Because", or "Since", for Vainu really did live up to all of them. And I hope his mother might take comfort in adding a final couplet:

Since you ran the course, against all odds, and won,

You did indeed become a man, my son.

# II.
# Inaugural Ceremony
# Tuesday, 17th August 1982, 20$^h$

The Inaugural Ceremony was held in the Ancient Odeon, near the centre of the city of Patras, in the presence of His Excellency, Mr. Euripides Kafantaris, Governor of the Prefecture of Achaia and representatives of the City and University of Patras.

The chair was taken by Professor Constantine Goudas, Chairman of the Local Organizing Committee for the XVIIIth General Assembly of the IAU.

Two musical interludes were provided by the OTE Mixed Choir, Athens, directed by Mr. A. Papayannopoulos and the Patras Mixed Choir, directed by Mr. S. Solomos.

Professor Goudas introduced Academician J. Xanthakis, Chairman of the Greek National Organizing Committee.

## Allocution de M.J. Xanthakis, Président du Comité National d'Organisation

"C'est un grand événement pour la Grèce et un grand honneur pour l'Université de Patras d'accueillir le 18eme Congrès International d'Astronomie. Des savants et des jeunes astronomes venus des différents pays du monde sont ici réunis pour présenter les résultats de leurs recherches récentes dans les différentes disciplines d'astronomie et d'astrophysique. Ils vont aussi discuter et exposer leurs idées nouvelles et leurs opinions pour les recherches de l'avenir.

Chers collègues, nous sommes ici, dans la région où a vu pour la première fois la lumière du jour, ce nouveau né, qui était la Science du Ciel, avec les observations des peuples anciens du Proche et d'Extrême Orient. Ce nouveau né, dans ce pays des Grecs anciens, avec le génie de Thalès de Millet, de Pythagore, de Philolaos, d'Héraclite et d'Aristarque de Samos est devenu un adulte.

C'est dans l'âge de la puberté que ce jeune homme a émigré quelques siècles après dans l'Occident pour se former en un jeune savant, sous l'impulsion des découvertes de Copernic, de Galilée, de Képler, de Newton et d'autres grands savants de la Renaissance et des années récentes.

C'était donc une nécessité pour les successeurs des astronomes des peuples anciens, de vouloir visiter, après 25 siècles, le lieu où a vu la lumière et a fait ses premiers pas solides, la science du ciel. Nous vous accueillons ici cordialement, dans ce petit coin de l'Europe avec sa longue histoire et sa civilisation ancienne. Mais nous regrettons que nos moyens financiers ne nous permettent pas d'être à la hauteur de notre tradition hospitalière, de "Ξενίος Διαζ", de dieu Jupiter de nos ancêtres. Les Comités d'Organisation National et Local ont essayé de faire tout ce qu'ils pouvaient pour faciliter autant que possible votre travail ici, et pour rendre votre séjour sur cette côte pittoresque du Péloponnèse aussi agréable que possible.

Professor Goudas regretted that the President of the Hellenic Republic, Mr. Constantine Karamanlis, due to urgent duties elsewhere, had been unable to come to Patras. Professor Goudas delivered the President's speech in Greek; it was thereafter given in French and English translation.

## ΜΗΝΥΜΑ ΠΡΟΣ ΤΗΝ 18Η ΓΕΝΙΚΗ ΣΥΝΕΛΕΥΣΗ

'Αποσταλέν ἀπό τόν Πρόεδρο τῆς Δημοκρατίας κ. Κ. ΚΑΡΑΜΑΝΛΗ

Μέ ἰδιαίτερη ἱκανοποίηση χαιρετίζω τήν σύγκλιση τῆς 18ης Γενικῆς Συνελεύσεως τῆς Διεθνοῦς 'Αστρονομικῆς 'Ενώσεως εἰς τήν 'Ελλάδα καί καλωσορίζω τά μέλη της, εἰς τήν χώρα αὐτή, ἡ ὁποία ὑπῆρξε μία ἀπό τίς ἀρχαῖες κοιτίδες τῆς 'Αστρονομίας. Καί λυποῦμαι διότι, λόγω ἐκτάκτου ἀπασχολήσεως, δέν μοῦ εἶναι δυνατόν, παρά τήν μεγάλη ἐπιθυμία μου, νά κηρύξω αὐτοπροσώπως τήν ἔναρξη τῶν ἐργασιῶν τοῦ Συνεδρίου σας.

Εἶναι δι'ἡμᾶς ἰδιαιτέρως σημαντικό τό γεγονός ὅτι εἰς τό Συνέδριό σας μετέχουν ἐπιστήμονες ἀπ'ὅλο τόν κόσμο. 'Η ἑνότης αὐτή μεταξύ τῶν σπουδαστῶν τοῦ διαστήματος ἀποτελεῖ ὑπόδειγμα εἰρηνικῆς συνεργασίας καί παράδειγμα, τό ὁποῖο θά ἦταν εὐχῆς ἔργο νά ἀκολουθήσει ὁ ταραγμένος κόσμος τῆς ἐποχῆς μας, γιά νά ἀνακαλύψει νέους δρόμους προόδου, χάριν ὅλων τῶν λαῶν τῆς Γῆς.

Τό πρόγραμμα τῶν ἐργασιῶν σας προδίδει τήν ἐπιθυμία τῶν συνέδρων νά καταστήσουν τήν συνάντηση αὐτή σταθμό εἰς τόν τομέα τῆς ἀνταλλαγῆς πληροφοριῶν μεταξύ τῶν μελετητῶν τοῦ διαστήματος.

Εὔχομαι ἡ προσπάθειά σας αὐτή νά ἐπιτύχει, ὄχι μόνο διά τό καλό τῆς ἐπιστήμης, ἀλλά καί διά τό καλό τῆς ἀνθρωπότητος, γενικώτερον·

## MESSAGE TO THE ASSEMBLY sent by the President of the Republic Mr. CONSTANTINE KARAMANLIS

"It is a great pleasure for me to salute the convocation of the 18th General Assembly of the International Astronomical Union in Greece, and to welcome its members to this country, which is one of the ancient cradles of Astronomy. I am very sorry, indeed, that due to unforeseen duties, it has become impossible for me inaugurate personally the works of your Assembly.

I deeply appreciate the fact that in your Assembly participate scientists from all over the world. This unity among students of space constitutes a model of peaceful cooperation and an example, which would be most desirable if the troubled world of our times could follow, in order to find new ways towards progress, for the sake of all peoples of the Earth.

The program of this Convention manifests the wish of the participants to make this meeting a landmark in the field of exchanging information among students of space.

I wish full success to your efforts, not only for the benefit of science, but also for the good of humanity in general."

Allocution de bienvenue par M. G. Maniatis,

Recteur de l'Université de Patras

"Dear Participants of the Eighteenth General Assembly of the International Astronomical Union, Ladies and Gentlemen:

It gives me great pleasure to welcome you on behalf of the University of Patras. It is an honour for our young University to be the host of the Eighteenth General Assembly of the IAU.

Your meeting is the largest international gathering ever held on our campus, but I trust that, thanks to the hard work of the Organizing Committee, we will succeed in meeting our responsibilities.

I am not in a position to speak about the particular scientific significance of your meeting, since I am not a specialist in astronomy. As a biologist, however, I feel a kinship with astronomy: In a way astronomy and biology are both concerned, although from opposite ends, with the same fundamental problem: the understanding of the Universe and of man's position in it.

Meetings such as this are extremely important for the progress of science. The personal contacts and exchanges afforded by these meetings can not be substituted by the conventional means of scientific communication. They are essential for the development of new ideas.

But your meeting is especially important for our University, which although young and relatively small, has a sizable and active community of astronomers. We hope that your presence here will provide an impetus for the further development of this group.

In closing, I wish you a productive meeting and a pleasant stay in Greece."

In the absence of the President of IAU, Professor M.K.V. Bappu, his written speech was read by Vice-President, Academician E.K. Kharadze.

Allocution de M. M.K.V. Bappu, Président de l'UAI

"Your Excellency, Mr. President, Mr. Chancellor, Professor Xanthakis, Professor Goudas, Ladies and Gentlemen:

It is with great pleasure that I wish to convey to you the sincere gratitude of the International Astronomical Union, for your very kind invitation to hold our General Assembly in Greece, and for the very warm welcome you have extended to us. This delightful seaside venue with its ancient castle, the several attractions for the visitor, and its place of learning, provides a setting for our deliberations of the forthcoming week, that is both popular and practical. Sitting here today, in this ancient Odeon of Patras, whose walls have held the rapturous silence of deep concentration or the thunderous ovation of approval of many a Greek classic, here beneath a glowing canopy of stars, a present generation of astronomers seeks to keep a tryst as it were, with their predecessors of an ancient past. The Greek philosophers of antiquity have provided us much of the basis of our present concept of man, state and society. From the point of view of those gathered here, the Greek mind has demonstrated the power of abstract thought and mathematical ingenuity. Its picture of the world was intertwinned with reasoning and apparent harmony.

It is therefore very gratifying to find that Astronomy in the Greece of today has not just rested on these very creditable laurels that assure it a place in the history of Mankind. There are several flourishing schools at many centres of learning whose contributions in such fields as mechanics, galactic dynamics and solar physics, to mention just a few, have been highly rated by the astronomical fraternity This awareness of its role in human history and willingness to participate in international cooperation is seen in the fact that Greece has been a member country of our Union since its inception. Its individual members have contributed very effectively to the progress of our science. We can offer no better testimony of your interest in the affairs of the Union, than the fact that our General Secretary of a few years ago, has been a distinguished colleague from this country.

We are happy to be here today. There is an air of excitement akin to homecoming. For it is a privilege to be invited as astronomers to this cradle of human endeavour from where Man first speculated on the nature of the Universe and his environment."

Finally, Prof. Goudas introduced Professor Pericles Theocaris, President of the Academy of Athens, who spoke as follows:

Allocution de M. Pericles Theocaris,

Président de l'Académie d'Athènes

"Mr. Chairman, Ladies and Gentlemen:

It is an honour and pleasure to me, as President of the National Academy of Athens, to welcome you in Patras, the city of my origin, and to declare open the Eighteenth General Assembly of the International Astronomical Union.

The organization in Greece of this activity, as befittingly mentioned by the General Secretary of your Society, looks like pilgrimage of Astronomy to the place of its birth.

It is true that the Greek contribution to Astronomy, the most ancient of all sciences, is particularly marked in the following three spheres: (a) the improvement of astronomic measurements; (b) the development of geometrical models for the explanation of stellar movements; (c) the calculation of cosmic dimensions. In the first case the Greeks merely carried on, from the point reached by the Egyptians and Babylonians. But, in the other two, they opened new chapters in the history of astronomy, which resulted in far-reaching advances.

The best criterion of the improvement made in astronomical observations during the Greek period is the degree of accuracy attained in determining the length of the year. The solar year of 365 days was known to the Greeks at the end of the sixth century B.C. from Egyptian sources, just as they got their knowledge of the monthly period from Babylonia.

The Greek astronomers of the fifth century B.C. narrowed down the earlier approximation by observations of the solstices on the shortest and longest days of the year. In 432 B.C. Meton of Athens estimated the length of the year as 365 days, 6 hours, 18 minutes, 56 seconds, a figure which is only 30 minutes, 10 seconds more than the correct value. In the next three hundred years the margin of error was steadily reduced, until Hipparchus, whose main observations were completed by about 130 B.C., arrived at the figure of 365 days, 5 hours, 55 minutes, 12 seconds - an error of only 6 minutes, 26 seconds. Thus, the accuracy was of the order of approximately one part in one hundred thousand, and still more accurate was the mean length of the month, as known in the time of Hipparchus.

This degree of accuracy, which so remarkably demonstrates the patient perseverance of the Greek astronomers in their observations of the heavens, set the standard for medieval and modern workers in the same field.

Bearing in mind the paucity of precise measurements, made by the Ancient Greeks in all the other physical sciences, we cannot help asking why astronomy was an exception. It is usual to stress the practical, and particularly the economic, factors which led to the development of astronomical research. The importance of the stars to navigation is frequently mentioned in ancient poetry; and there is literary evidence from very early times for the connection between agriculture and the knowledge of the heavens.

But more important that these practical considerations were irrational factors, the roots of which go down to still earlier times and which are connected with the history of astrology. Especially noteworthy in this connection is the effect produced on man by the cyclical character of heavenly phenomena. The changes in the phases of the moon and their periodity, the progress of the sun along the zodiacal belt with its attendant changes of season, the complicated movements of the planets, which also have a cyclical regularity, and above all the twenty-four hours periodicity of the whole heavenly dome with the accompanying alternations of day and night, all this vast array of eternally recurring cycles awoke the consciousness of ancient man to the great contrast between the firm-set certainty of the heavens and the uncertainty of human life on the earth.

Here there is a twofold contrast: Man's life is unique and can be lived only once, whereas in the heavens there is endless recurrance; the individual is the plaything of chance, whereas the stars obey a law of marvellous constancy. This contrast is the source of that religious fascination, which from the earliest times held man's mind in thrall to the heavenly bodies, expressing itself as simple-star worship, or in a rationalized form as an urge to follow the minute details of those manifestations of precision and regularity.

To this irrational element there was added the practical aim of fixing the religious festivals by the calendar.

This union of science and religion in the approach to the phenomena of the heavens was disturbed by the new, intensely rational, scientific method introduced by the Milesian School, in the very early days of Greek science. We know that Anaximander explained the revolutions of the sun and the moon with the aid of mechanical models; moreover, Anaxagoras said that the sun and the stars were flaming stones, and similar opinions are attributed to Leucippus and Democritus, the founders of the atomic school. Some of Empedocles' conjectures about heavenly phenomena are also purely physical in character. Still, even at that time, the religious notion was not entirely excluded from scientific thought; indeed, the scientific sublimation of star-worship appears still more strongly in the Pythagorean School, where religious and mystic tendencies prevailed and combined with the scientific approach.

The cyclical motion of the heavenly bodies displays this combination of immortality and continuity in its purest form, thus giving dynamic proof of the divinity of the stars.

This linking together of the soul and the stars had a characteristic sequel in Plato's teaching. His famous proof of the immortality of the soul begins with the following sentence: "Every soul is immortal - for whatever is in perpetual motion is immortal"; and his view about the divinity of the stars found its classic expression in the Epinomis. In this culmination of the scientific sublimation of star-worship along Pythagorean lines we may also see some elements of the foundation, on which Aristotle's dynamics was built: "Men ought to have regarded the possession of intelligence by the stars and all their movements as proved by the uniformity of their action, and by the fact that they continue to carry out the counsels formed long ago, and do not wander about with varying revolutions, altering their counsels this way and that, and doing first one thing and then another. The majority of us have adopted exactly the opposite view, thinking, because of the uniformity of their actions, that these beings have no soul; the multitude thus follows the lead of fools, and considers that the human race shows intelligence and life by its mutability, whereas that of the gods is devoid of intelligence, because it remains in the same orbits. Man might, however, have adopted a view nobler, better, and acceptable - the view that which acts always in the same uniform way and under the influence of the same causes, should on the very account be regarded as possessed of intelligence; and this specially applies to the nature of stars, which form so glorious a spectacle, and while performing the movements of their dance, of all dances the most lovely and magnificent, discharge their duty to all things that live. And, as a proof that we are right in ascribing life to them, let us first consider their size. For they are not small, as they appear to the eye, but each of them is immense in bulk; this we are bound to believe, for it is established by sufficient demonstrations. The sun as a whole may be rightly regarded as larger than the earth as a whole; and all the stars in their courses are of wonderful size. Let us consider how it could be that anything could cause so great a bulk to move in its orbit in exactly the same time in which it now performs its course" (Aristotle's De caelo).

We, of the machine age, have grown accustomed to an entirely different association of ideas. The essence of every machine is that it repeats the same movement exactly; so that we use the expression "automatic" to indicate precisely a movement that is devoid of reason, a "soulless" movement. But in the age of arts and handicrafts, the exact reproduction of a model or form was regarded as a sign of the artist's divine inspiration. It was Plato's educational influence that decided in favour of the view that the stars are divine and against the purely physical tendency of the pre-Socratic period.

In the twelfth book Plato sums up his views about the existence of the gods. He finds two proofs for this: the immortality of the soul which results from its perpetual motion, and the movement of the heavens. He asks if it is conceivable that "...those who study these objects in astronomy and the other necessary allied arts become atheists through observing, as they suppose, that all things come into being by necessary forces and not by the mental energy of the will, aiming at the fulfilment of good". His answer is that: "The position at present is exactly the opposite of what it was when those who considered these objects considered them to be soulless. Yet even then they were objects of admiration, and the conviction which is now actually held was suspected by all who studied them accurately - namely, that if they were soulless, and consequently devoid of reason, they could never have employed with such precision calculations so marvellous".

This polemic, too, is directed against Anaxagoras and the physical way of looking at the heavens, as is shown by the next passage, in which it is reiterated that "all that moves in the heavens appeared to them to be full of stones, earth and many other soulless bodies which dispense the causes of the whole cosmos". At the end there appears the Pythagorean elements in Plato's philosophy: "An in addition to this, as we have often affirmed, he must also grasp the reason which controls what exists among the stars...and he must observe also the connection therewith of musical theory...and he must be able to give a rational explanation of all that admits of rational explanation".

We may quote at length from Plato, showing that it was his opinion that decided this important dispute. Once Aristotle had accepted it and given it a broader physical basis, the fate of Greek science was sealed and the division between heaven and earth became an integral part of ancient physics and the Greek cosmos.

This position remained unchanged till Galileo. Even Epicurus, who tried to follow in the steps of Anaxagoras and Democritus and therefore contested Plato's views about the divinity of the stars, maintained this separation, though in exactly the opposite sense. The accepted view was that the soul and intelligence of the stars was revealed in the conformity to law and absolute regularity of their movements. Epicurus therefore tried to undermine the Platonic theory by directly and indirectly raising doubts about the existence of such conformity to law in the phenomena of the heavens, while maintaining that terrestrial phenomena were strictly obeying the causal laws.

Modern science was born in the seventeenth century, when it was shown that the laws of terrestrial mechanics hold good also for the movements of the planets; in other words, with the breaking down of the barrier which separated heaven from earth.

In closing this very rapid and therefore incomplete introduction we may state that the Egyptian civilization created technology in the prescientific era. On its decline, the civilization of Greece gave birth to science without its technical application. Then, after a thousand years of paralysis, the civilization of Europe inaugurated the era of integration of science and technology. As participators in this era, we are in danger of suffering from a distorted perspective. Still, there can be no gainsaying the fundamental fact that this integration has become the mainspring of a creativeness and rapid progress, alike in the theoretical and practical spheres, which are unparalleled in earlier cultures.

The fructification of technology by science is plain for all to see; the converse effect is no less profound and many sided. Not only does technology give fresh impetus to pure science, but technical achievements have been harnessed to the service of fundamental science. In this connection, it is sufficient to mention the tremendous service rendered by the development of scientific instruments and scientific machines in extending man's knowledge of nature beyond the limits of his five senses, thus enabling him to overcome that "weakness of the senses", which Anaxagoras regarded as the chief obstacle to ascertaining the truth.

If the intellectual adventure of modern science is perhaps the greatest of all the adventures inaugurated by the modern era, this is due to the development of mathematics as the key to nature's laws. It is true that our cosmos has been drained of all the "human" content, which it contained in the Greek period; it is true that the naive world of the senses is separated from the world of science by an eye-widening chasm; it is true that understanding this world of science calls for enormous powers of abstraction and a professional and intellectual training, which is becoming ever more rigorous. But, on the other hand, this cosmos - from the nucleus of the atom to the distant galaxies- is being filled more and more with new and marvellous contents, which make the experience of those who share in this development, certainly not less rich than the cosmic experience of the first natural philosophers of Ancient Greece.

Ladies and Gentlemen:

Of particular importance is the fact that we are today gathered, scientists from all over the world, to participate in this meeting at the country where astronomy was born. The union and collaboration of all space-scientists and astronomers constitutes a hopeful message and a precept how humanity must handle all its problems for the benefit of peace in the world, for the development of understanding of the various peoples, and for the abolition of poverty and misery from so many parts of our world.

Considering the wide spectrum of subjects which will be discussed and the emphasis given to various subjects of a heavy agenda, the ideas expressed and the results derived in this congress will constitute a useful knowledge of the actual status of astronomy, as well as the targets to which is aiming its future development.

I hope that the relaxing surroundings of the University of Patras will contribute to a successful meeting. Finally, I hope that you will find some time to enjoy also the country and especially the close-by site of Olympia and the temple of the Epicurian Apollo. I am sure that the University of Patras, the Local Organizing Committee and the Local Committees will do their best to make your stay both productive and pleasant."

# III.
# Report on the XVIIIth General Assembly of the International Astronomical Union

AGENDA

(including Motions Proposed)

First Session: 17 August 1982, 21.15 in the Ancient Odeon, Patras

1. Opening of the General Assembly by the President

2. Appointment of Official Interpreters

3. Report of the Executive Committee, 1979-1981

4. Report of the Executive Committee January-August 1982
   (a) IAU Membership nominations
   (b) Report of the General Secretary
   (c) Motion of the Executive Committee to ratify changes in the adhering bodies for China

The Executive Committee proposes that the General Assembly of the Union,

*noting* that progress has been achieved in restoring full adherence to the IAU of China, as was anticipated at the XVIIth General Assembly of 1979 at Montreal, and that agreement was obtained by 1 May 1980 on the mode of listing, in the official list of member countries, two adhering bodies for China as a temporary measure,

*ratifies* the arrangements made by the Executive Committee for the adherence of China to the Union during the period elapsed since the XVIIth General Assembly.

5. Report by the President on the work of the Special Nominating Committee 1979-1982

The Special Nominating Committee, proposed to the XVIIIth General Assembly at Patras as members of the IAU Executive Committee for the term 1982-85:

President: Professor R. Hanbury Brown (Australia)

Vice Presidents continuing from the former period:

Dr M.W. Feast (South Africa)

Dr L. Kresak (Czechoslovakia)

Professor R. Wilson (UK)

Vice Presidents to be newly proposed:

Professor R.P. Kraft (USA)

Dr M. Peimbert (Mexico)

Dr Ya. S. Yatskiv (USSR)

General Secretary: Dr R.M. West (Denmark)

Assistant General Secretary: Dr J.P. Swings (Belgium)

Advisers to the Executive Committee:

Professor M.K.V. Bappu (India)

Professor P.A. Wayman (Ireland)

6. Announcement of:

(a) The names of Official Representatives of Adhering Organisations

(b) The names of representatives to serve on the Nominating Committee

(c) The names of Acting Presidents of Commissions

7. Appointment of the Finance Committee

8. Appointment of the Resolutions Committee

9. Revisions of the Statutes and By-laws (no revisions are proposed)

10. Resolutions submitted by Adhering Bodies (no motions have been submitted)

11. Resolutions submitted by Commissions or by associated Inter-Union Commissions

The following resolutions have been proposed:

1. Resolution proposed on behalf of Commission 5:

Commission 5

*considering* the present unsatisfactory situation of the transfer of astronomical data between astronomical institutions

*recommends* that all astronomical computer facilities recognise and support the Flexible Image Transport System (FITS) for the interchange of binary data on magnetic tape, as described in Astronomy and Astrophysics Supplement, vol. 44, p.363 and 371.

2. Resolution proposed on behalf of Commission 15:

The International Astronomical Union

*recognising* that it is particularly desirable that pre-selected Comet Halley Days for co-ordinated observation over a limited time be supported

*recommends* that observatory directors and observing program committees give high priority to Comet Halley observation during the interval 1985 - 1987.

3. Resolution proposed on behalf of Commission 15:

The International Astronomical Union

*noting* that, in order to organise and marshall ground observations of Comet Halley throughout its 1986 perihelion passage and to co-ordinate them with space missions, an international program, the International Halley Watch, has been established

*and wishing* to avoid duplication of effort at the international level and to encourage participation in this program

*endorses* the International Halley Watch as the international co-ordinating agency for Comet Halley observations.

4. Resolution proposed on behalf of Commission 46

*Whereas* astronomy is an integral part of the physical sciences, both in education and in research,

*and whereas* some countries now without astronomy may desire to introduce astronomy in their education or scientific institutions,

*be it resolved* that the International Astronomical Union supports the introduction of astronomy in such countries and encourages activities by its Commissions toward that purpose.

---

Second Session: 26 August 1982, 10:00 in the University of Patras

12. Report of the Finance Committee

13. Financial Resolutions of the Executive Committee

(a) Budget Estimate for 1982

(b) Unit of Contribution to the IAU, 1983-1985

The Executive Committee proposes that the unit of contribution (1983-1985) be fixed at Swiss Francs 1760.-.

(c) Budget Estimate for 1983-85

The Executive Committee proposes a balanced budget of Swiss Francs 1.433.700 for this triennium.

(d) Other financial resolutions, not included in the Budget Estimate proposed

1. The Executive Committee,

*having examined* the proposal of the President of Commission 22 to establish a Meteor Data Center of the Commission at the Observatory of Lund in Sweden,

*proposes* the allocation of SwFr 1 800 per annum as a financial contribution to the costs of such a center, so established, for the period 1983-85.

2. The Executive Committee,

*noting* the expressed wish of the official representative of the Swedish National Committee of Astronomy at the XVIIth General Assembly of 1979 at Montreal that fellowships for students from developing countries be encouraged,

*and recognizing* the valuable discussions carried out in the interim by the Working Group of Commissions 46 and 38 on the Education of Astronomers from Developing Countries,

*proposes* that a sum of SwFr 35 000 be allowed during the period 1983-85 to support the carrying out of a Contract Scheme for Visiting Lecturers with a sponsoring institution in a country newly entering into astronomical work, to be selected before June 1983, as the result of active enquiry within IAU member countries and other selected countries.

14. Resolutions submitted by the Executive Committee

15. Recommendations of the Resolutions Committee on the resolutions submitted by Commissions

16. Appointment of the Special Nominating Committee

17. Nomination of new Members of the Union

18. Changes in Commissions

(a) Commissions discontinued

(b) new Commissions

The Executive Committee proposes to the General Assembly that a new Commission of the Union be established in respect of the search for life in the Universe, the name of the Commission to be Commission 51, Search for Extraterrestrial Life .

(c) names of Commissions

The following changes of name have been proposed for existing Commissions of the Union:

Commission 26 (Double Stars) to be renamed "Double and Multiple Stars"

Commission 34 (Interstellar Matter and Planetary Nebulae) to be renamed "Interstellar Matter"

(d) Election of Presidents and Vice-Presidents of Commissions for 1982-1985

(e) Organising Committees of Commissions

19. The place and date of the XIXth General Assembly

20. Election to the Union of a President, six Vice-Presidents, a General Secretary and an Assistant General Secretary

21. Address by the President 1979-1982

22. Address by the President 1982-1985

23. Closing Ceremonies

# Proceedings

## FIRST SESSION

Vice-President, Academician E.K. Kharadze, in the Chair as Acting President

### 1. Formal Opening

In the absence of the President, Professor M.K.V. Bappu, the Acting President welcomed the Members of the Union, Invited Participants, Registered Guests, Representatives of Adhering Countries and National Committees of Astronomy, and the representatives of other Unions and Organisations, and formally opened the first session of the 18th General Assembly of the International Astronomical Union in the Ancient Odeon of Patras, on Tuesday, 17 August 1982, at 21.30, with the following words:

"It is with the greatest disappointment that I have to inform participants at this General Assembly that our President, Professor M.K.V. Bappu, required a major heart operation in Munich on 9 August, and is therefore unable to be with us. Following a successful operation, we have been distressed to learn that serious complications have occured, and I cannot do other than report to you that Professor Bappu's condition at present gives cause for serious anxiety. I ask you all to join us in our eager desire for Professor Bappu's recovery."

The Acting President then asked Vice-President of the IAU, Dr M.W. Feast, to read the President's address prepared by Professor Bappu to the opening of the 18th General Assembly:

Members of the Union, Ladies and Gentlemen:

Circumstances, totally unexpected, and to some extent beyond my control prevent me from being with you during this General Assembly. Like many amongst you, I had looked forward to a feast of scientific fare with a gourmet's anticipation and especially the opportunity to meet both a young generation of IAU members, as well as those already established in the profession, when I was a newly initiated member. Herein lies the strength of our Union and at the same time its principal charm that keeps us all together. Besides the aspects of international cooperation, the Union has a prime responsibility of promoting the study and development of Astronomy in all its branches. It does so by fostering an awareness of accomplishment, and providing some of the sources of stimuli that can result in creative intellectual productivity. On this occasion, the joint discussions, the invited discourses, the joint commission meetings, to mention a few, cover a wide variety of astronomical enterprise; the cream of such effort and its implications, discussed and absorbed, before we return homewards for a fresh look at the problems that interest each of us.

The years ahead hold forth the promise of striking developments on a magnitude never witnessed before. We are on the threshold of an era when we shall have the opportunities of using telescopes from space that will have apertures, considered large, even from ground-based standards. With the veil of limitations imposed by the earth's atmosphere, behind us, and with the variety of new technological triumphs at our disposal, we approach in our quest, the ultimate in resolution and limiting brightness. This has always been the astronomer's fondest dream. And the nature of physical discovery that we have witnessed over the ages gives us the feeling of assurance that a major technological leap of this kind, will in its wake, stabilize some of our conjectures, bring out the totally unforeseen characteristics in some of the objects we study, that will revolutionize our thinking, and in general produce a multitude of data that needs to be sifted through with care and judgement. This avalanche of information will undoubtedly need the coordinated effort in analysis of so many dedicated to astronomy, that it will need to be global in nature. As a Union of such likeminded individuals, we see ourselves in an important role of the task of assimilation of these new results; it would be a perfect setting for international cooperation where the entire fraternity is engrossed in the adventure of exploring the vistas of the unknown. Should we not explore this important angle in great detail before we are engulfed in the actuality of subsequent events?

At this time and stage of development of our science, I cannot rule out the feeling that our approach to the unknown and the unexplored has been overcautious and without the touch of daring, bordering on rashness, that has characterized the bold explorer of the past. Are we enmeshed in the fetters of inhibition originating from a conservatism of outlook? Could we have had discoveries, ahead of the times they were actually made, if only a freshness of approach was not lacking? I am of the opinion that it has been so.

Assuming that my surmise is likely to be valid, I fear that we have been slow to respond to the lessons of experience of the recent past. In our search for objects that display peculiarities of a kind beyond our normal familiarity, we could with advantage follow the pattern adapted by that illustrious pioneer, Kapteyn, of sampling to the limits of brightness all aspects of the radiation received from objects in a restricted area of sky. With the era of the large space telescopes round the corner, perhaps this philosophy of approach would be a dire necessity.

While innovation in technique is a factor of much significance in the discoveries that will yet be made, it is certain that an even greater role will undoubtedly be of the human intellect. Time and again we have seen how an individual has appeared on the scene and transformed a picture of gathering confusion into one of logical rigour and aesthetic simplicity. In the final reckoning, it is this aspect of Man that is responsible for the culminating triumph. It is, therefore, a responsibility shared by each one of us in our individual roles, be it of teacher or senior colleague, as well as of the astronomical community and the Union as a whole, to nurture such possibility. The history of the human race provides the evidence that the spark of enthusiasm which fires the genius in Man has no regional preferences for its origin. To fan this streak into the flame of intellectual achievement is at once an obligation and an assurance of a dynamic future. In its longterm view of contributing to the development of Astronomy, the Union must necessarily give considerable attention to the very important role of the individual in Astronomy.

I shall dwell very briefly on one further aspect to which the Union must find the resources necessary for effective implementation. Since the dawn of human civilization, we have seen how emulation has spread the zone of achievement. The facility of travel has been the prime means of doing so. The examples witnessed and experienced first hand, the contacts made, the discussions, the arguments and the listening which update ones awareness of the problem, all apply to any form of enterprise, astronomical or otherwise. In research and in the propagation of learning, it is well known that a brief period of contact with the great masters, whichever part of the globe they be in, has always provided the stimulus for achievement; the confidence of ones capability coupled with a spark of ambition inevitably brings on a measure of success. Our efforts within the Union, to promote such possibilities, have been of modest dimensions, for the resources we usually provide Commission 38 are meagre. This is an important area which deserves considerable attention. In the years ahead, it might indeed be an important lifeline of existence.

I send each of you my greeting and good wishes. I hope you will all have a very useful and enjoyable stay at Patras.

---

The Acting President added his own thanks to Professor John Xanthakis and his colleagues on the Greek National Organizing Committee and to Professor Constantine Goudas and the Local Organising Committee in Patras, for all the work done. He also thanked Professor George Maniatis, Rector of the Patras University, on the resources of which the influx of almost 2000 participants has placed considerable strain. He further expressed a deep appreciation of the privilege to meet in the Ancient Odeon of the City of Patras, under the ceiling provided by the night sky: indeed a unique occasion in the history of the Union.

It was agreed to send a message of sympathy to Professor and Mrs Bappu.

A special word of welcome was extended to the past Presidents and General Secretaries of IAU, present at the General Assembly: Professors V. Ambartsumian, A. Blaauw, L. Goldberg, G. Contopoulos, C. de Jager, E.A. Müller and J.-C. Pecker.

Furthermore, the Acting President wished a particularly hearty welcome to the official representatives of ICSU and Sister Unions as follows:

Representation of Other Bodies

| | |
|---|---|
| COSPAR | C. de Jager |
| IUGG | P.E.G. Pâquet |

| | |
|---|---|
| URSI | G. Westerhout |
| IUGS | H. Masursky; G. Wetherill |
| FAGS and BIH | B. Guinot |
| SCOSTEP | M.R. Kundu |
| CODATA | B. Hauck |
| IUTAM | P.S. Theocaris |
| IAF | M.D. Moutsoulas |
| ESA | R. Reinhard, B. Serene, and A. Heck |
| IUHPS | M.A. Hoskin |

Good wishes had been received from UNESCO and COSTED, which regretted being unable to send representation.

The Acting President asked the members to stand while the General Secretary read the names of members who had died since the 17th General Assembly, or whose death had not been known at the time of that meeting. The General Secretary read the following list:

Deceased Members

| | | |
|---|---|---|
| J. Ashbrook | P. Herget | H.H. Plaskett |
| U. Baehr | H. Hirose | J.G. Porter |
| A. Beer | T. Ikeda | N.B. Richter |
| A. Boury | H.L. Johnson | A. Romana |
| E. Buchar | K. Kordylewski | J. Schilt |
| E. Bullard | N.I. Kozhevnikov | D.W. Schuerman |
| G.R. Carrasco | D. Lauterborn | K. Serkowski |
| G. Chis | K. Lubeck | M.B. Shapley |
| P. Couderc | G.D. Mamedbejli | F.M. Strauss |
| I.F. da Silva | N. Matsunami | C.H. Smiley |
| P.M. Djurkovic | O.A. Melnikov | H.H. Swope |
| A.E. Douglas | H. Michel | S.I. Syrovatskij |
| J.D. Fleckenstein | J.M. Mohr | K.N. Tavastsherna |
| K.P. Florensky | F. Moreau | B.M. Tinsley |
| L.I. Gama | V.P. Myerscough | H.C. Urey |
| V.M. Grigorevsky | J. Neyman | J. Wempe |
| U. Guntzel-Lingner | W. Norlind | J.A.J. Whelan |
| V. Guth | N.E. Nørlund | S.P. Wyatt |
| W.R. Hartner | L. Pajdusakova | I.D. Zhongolovich |
| N. Heidmann | C.H. Payne-Gaposchkin | |

## 2. Appointment of Official Interpreters

The General Assembly appointed Professor M. de Groot as official interpreter from French to English, and Professor A.G. Velghe as official interpreter from English to French.

3. Report of the Executive Committee (1979-1981)

The Acting President invited discussion about the report of the Executive Committee for the three-year interval 1979-81, as presented in draft form in IAU Information Bulletin 48 of June 1983. He mentioned that it had been given consideration by official representatives of Adhering Organisations, at their meeting earlier in the day, and that, after incorporation of some minor corrections and with the important reservation that the financial section of the report would come under scrutiny of the Finance Committee of the General Assembly, the official representatives were prepared to recommend adoption of the report, subject to further discussion here.

There were no points raised from the audience, and the General Assembly unanimously approved the report of the Executive Committee (1979-81), subject to a relevant report by the Finance Committee (The full report is printed in this volume on pp. 347-381.

4. Report of the Executive Committee (January-August 1982)

(a) IAU Membership Nominations

The Acting President asked the General Secretary to inform the General Assembly about proposals of individuals to IAU membership.

The General Secretary indicated that, as a result of his requests to Adhering Organisations for proposals of individuals to IAU membership, he had received almost 750 nominations. A further 35 were received from Presidents of Commissions and members of the Executive Committee. At the meeting of the Nominating Committee, earlier in the same day, it was decided to refer these nominations to the Executive Committee, whose report will be taken under Item 17 of the agenda during the second session on August 26, 1982.

(b) Report of the General Secretary

The General Secretary then gave the following report to the General Assembly about the affairs of the Union in the period from January 1 - August 17, 1982:

Let me begin by expressing on my own behalf and on behalf of all participants in this General Assembly the thanks that are due to you, Academician Kharadze, for your generous consent to undertake the function of our President in his regrettable enforced absence from our Assembly. This has been a real disappointment to us all but especially to Professor and Mrs Vainu Bappu themselves. For me, there have been three years of preparation for personal cooperation with Vainu on this occasion that are only fulfilled in his absence. I have been in Patras twice in his company and he has contributed a great deal towards overcoming the difficulties that the holding of an IAU General Assembly necessarily presents. He has also looked forward constructively to the possibility of meeting in his own country of India in 1985.

I also wish to acknowledge the support of my colleagues on the Executive Committee of the Union who have given every assistance to the work of the Secretariat since the Montreal General Assembly, both by correspondence and in person at the Executive Committee Meetings held in Amsterdam in 1980 and in Abastumani in 1981.

The record of the IAU during 1979, 1980 and 1981 stands for all to read in the Executive Committee Report for those years which has been published in the Information Bulletin of the IAU, and which should have reached every individual IAU member. My main task now is to report briefly on the eight months of the current year that have elapsed, months that have been largely occupied in preparing for this General Assembly. My chief task has been to produce the scientific programme for the Assembly in such a way as to avoid occasions when astronomers wish to be in two places at the same time. Taking into account such problems as finding that one

Commission represented more time than the whole programme provided and expressed at the same time a desire not to have any meeting scheduled for Saturday morning, I hope to discover through your comments that a satisfactory compromise has been made. Small readjustments of the programme are being made, so please watch the notice boards and the daily programme sheet carefully for any alterations that are found to be necessary. Please regard the printed Daily Diary as a guide only to your selection of meetings.

One piece of work that has been undertaken in preparation for this General Assembly is to draw up the requirements for a new Commission of the IAU to be concerned with the detection of life processes elsewhere than on the Earth. Brief mention has been made in the IAU Information Bulletin but members will wish to know that the Executive Committee will propose at the Second Session of this General Assembly, has been a great pleasure to me and to the staff of the IAU terrestrial life". This is an important and interesting new direction for our Union to prepare for.

During the present year of 1982, the Union has arranged one event of a new type - an International Astronomy Seminar that presented astronomical science to a particular audience. This was in conjunction with the United Nations Conference of this Year - UNISPACE 82 - which has taken place in Vienna in Austria in the past two weeks. Several of our distinguished colleagues have given their time towards ensuring that the international work of our science is represented on that occasion. It took place on Thursday, 12th August, under the Chairmanship of Professor Bengt Strömgren. Unfortunately, due to a number of factors, including UN security precautions, the attendance was somewhat poor. My thanks go to Dr Richard West for taking over the detailed preparations of this particular event during the past six months, and to Professor Alla Massevich of Moscow for arranging the necessary facilities.

The work of Presidents of Commissions, and in certain cases the Vice-Presidents acting on behalf of Presidents in preparing their scientific meetings for this General Assembly has been a great pleasure to me and to the staff of the IAU Secretariat. Seven Joint Discussions and a total of over 150 other meetings have been arranged by the forty Presidents and by some fifty other participating Chairmen. The Union draws heavily on this work voluntarily undertaken by so many dedicated individual IAU members.

I draw your attention to the professional handling of the affairs of the Paris IAU Secretariat by our permanent staff of two - Mme Brigitte Manning and Ms Patricia Smiley. I have depended very heavily on them in my commuting between Dublin and Paris, leaving behind me handwritten material which had to be dealt with in my absence. They have both expended their particular skills in the service of the IAU.

Other aspects of our preparations have depended upon many persons, too numerous to mention. but four have been outstanding - the two Advisers to the Executive Committee, Professor Adriaan Blaauw of Leiden and Groningen in the Netherlands and Professor Edith Müller of Geneva in Switzerland. These two have ceaselessly responded to my requests for helpful discussions with unfailing patience and courtesy. In Patras itself, the work of Professor Goudas in preparing his University and his City for this event has been outstanding. He has shouldered personally far more than his fair share of both responsibility and detailed effort. In Athens, the efforts of our Greek colleagues have been led by Academician John N. Xanthakis, Chairman of the Greek National Committee for the IAU General Assembly and of the Greek National Committee of Astronomy.

This, Mr Vice-President, concludes my brief Report; I look forward to having many opportunities in the next ten days of amplifying this sketchy information on the current position of our Union in private conversations and at the remaining administrative meetings held on behalf of the General Assembly.

The General Secretary's report was received with acclamation. There were no questions from the audience.

## (c) Ratification of the IAU Membership of China

The Acting President then asked the General Secretary to read a recommendation to the General Assembly, from the Executive Committee, concerning ratification of the full adherence of China to the Union. The General Secretary gave the following explanation to the proposed motion:

A resolution of the 1979 XVIIth General Assembly (IAU Transactions, Vol. XVIIB, 1980, p.49) authorizes the Executive Committee to implement the conclusion of negotiations on Chinese Membership of the Union, subject to the ratification of such actions at the next Meeting of the General Assembly.

The documents referred to in the 1979 resolution are reproduced on pages 526 and 527 of IAU Transactions Vol. XVIIB. These documents left only one question undetermined (as described below) and all the requirements of the Chinese Astronomical Society in Nanjing as an adhering body to the Union have been met since May 1980. That is, individual IAU members in the Peoples' Republic of China have been listed for acceptance by the Executive Committee and names have been assigned for Commission membership; IAU travel monies have been offered to astronomers from China; correspondence has taken place in respect of present and future IAU cooperation for the General Assembly and for scientific meetings, etc; and China has participated in the work of some of the IAU (etc) agencies such as Bureau International de l'Heure.

The style of reference in the official IAU list, however, was left undetermined for China in 1979. By 1 May 1980 this question was agreed to the satisfaction of the Chinese Astronomical Society and with the compliance of the adhering body in Taipei, Taiwan, China. The entry as published in IAU Transactions Vol. XVIIB (1980) was agreed early in 1980.

Official Representatives, following the Executive Committee resolution given in the Agenda of the XVIIth General Assembly, are now requested to accept these actions of the Executive Committee as full confirmation of IAU membership for the whole of China by their ratification in the resolution under Item 4(c) of that Agenda.

The text of the proposed resolution was read in English and French. Then the General Assembly indicated its unanimous acceptance by acclamation:

### RESOLUTION A1:

### Ratification of IAU Membership of China

The General Assembly

*noting* — that progress has been achieved in restoring full adherence to the IAU of China, as was anticipated at the XVIIth General Assembly of 1979 at Montreal, and that agreement was obtained by 1 May 1980 on the mode of listing, in the official list of member countries, two adhering bodies for China as a temporary measure,

*ratifies* — the arrangements made by the Executive Committee for the adherence of China to the Union during the period elapsed since the XVIIth General Assembly.

Ratification de l'adhésion de la Chine à l'UAI

L'Assemblée Générale de l'Union

*constatant* le progrès accompli pour le rétablissement de la pleine adhésion de la Chine à l'UAI, tel qu'il avait été anticipé à la XVIIème Assemblée Générale de 1979 à Montréal, et l'accord obtenu au 1er Mai 1980 pour la dénomination, dans la liste des pays membres, de deux organisations adhérentes pour la Chine à titre provisoire,

*ratifie* les accords passés par le Comité Exécutif pour l'adhésion de la Chine à l'Union pendant la période écoulée depuis la XVIIème Assemblée Générale.

The Acting President then declared the ratification of the de facto membership of China to the IAU to be accepted by the General Assembly and invited Professor Wang Shou-Guan to say a few words on behalf of the Chinese Astronomical Society. Professor Wang spoke as follows:

Dear friends & colleagues,

The Chinese Astronomical Society celebrates its 60th anniversary this year. Its reunion with this international community today is an event that is highly appreciated by all its 900 members. I and my colleagues here are very glad to have this opportunity of speaking on behalf of our Society and its members to express our most cordial greetings and most sincere thanks to you all. Thank you!

The General Assembly cordially applauded Professor Wang's address.

5. Report on the Work of the Special Nominating Committee

The Acting President invited Professor A. Blaauw, former Chairman of the Special Nominating Committee, and former President of the Union, to speak on behalf of the Special Nominating Committee:

The Special Nominating Committee, convened during 1980 and 1981 by the President of the Union, has selected the following IAU Members to be proposed to the XVIIIth General Assembly at Patras as members of the IAU Executive Committee from 26 August 1982:

As President - Professor R. Hanbury Brown (Australia)

As Vice-Presidents continuing from the former period -

Dr M.W. Feast (South Africa)
Dr L. Kresak (Czechoslovakia)
Professor R. Wilson (UK)

As Vice-Presidents to be newly proposed -

Professor R.P. Kraft (USA)
Dr M. Peimbert (Mexico)
Dr Ya.S. Yatskiv (USSR)

As General Secretary - Dr R.M. West (Denmark)

As Assistant General Secretary - Dr J.P. Swings (Belgium)

As Advisers to the Executive Committee -

Professor M.K.V. Bappu (India)
Professor P.A. Wayman (Ireland)

The French text was then ready by Professor Velghe, and the Acting President informed the General Assembly that, according to the by-laws, the formal election will take place at the second session of the General Assembly on Thursday, August 26, 1982.

## 6. Announcement of (a) Official Representatives of Adhering Organisations and (b) Representatives to serve on the Nominating Committee

At the request of the Acting President, the General Secretary announced the following names:

| | Official Representative | Nominating Committee |
|---|---|---|
| Argentina | E. Bajaja | E. Bajaja |
| Australia | B. Melrose | R.N. Manchester |
| Austria | H.F. Haupt | H.F. Haupt |
| Belgium | A.G. Velghe,<br>L. Houziaux | L. Houziaux |
| Brazil | E. da Rocha Vieira | S. Ferraz-Mello |
| Bulgaria | M. Popova | B. Kovatchev |
| Canada | W.H. Wehlau | J.R. Percy |
| Chile | H. Moreno,<br>F. Noël | A. Gutierrez |
| China, Nanjing | Wang Shou-guan | Qu Qin-yue |
| China, Taipei | C.S. Shen | W.T. Ni |
| Colombia | J. Arias-de-Greiff | E. Brieva |
| Cuba | | |
| Czechoslovakia | V. Bumba | Z. Ceplecha |
| Denmark | A. Reiz | A. Reiz |
| Egypt | A.S. Asaad | A.M.I. Osman |
| Finland | J. Tuominen | T. Markkanen |
| France | Ch. Fehrenbach | Y. Andrillat |
| Germany, D.R. | G. Ruben | G. Ruben |
| Germany, F.R. | R. Kippenhahn | W. Priester |
| Greece | J.N. Xanthakis | L. Mavridis |
| Hungary | I. Almar | I. Almar |
| India | J.C. Bhattacharya | V.R. Venugopal |
| Indonesia | B. Hidayat | W. Sutantyo |
| Iran | | |
| Iraq | Hamid M.K. Al-Naimy | Hamid M.K. Al-Naimy |
| Ireland | I. Elliott | M. de Groot |
| Israel | R. Ramaty | R. Ramaty |
| Italy | L. Rosino | F. Bertola |

| | Official Representative | Nominating Committee |
|---|---|---|
| Japan | Y. Kozai | Y. Uchida |
| Korea, Dem. P.R. | Kim Yong-hek | Choi Bon Chol |
| Korea, Republic | Y.K. Minn | Y.K. Minn |
| Mexico | M. Peimbert | S. Torres-Peimbert |
| Netherlands | C. de Jager | H. van Woerden |
| New Zealand | P.J. Edwards | J.B. Hearnshaw |
| Norway | E. Jensen | R. Stabell |
| Poland | W. Iwanowska | W. Iwanowska |
| Portugal | J. Pereira Osorio | J. Pereira Osorio |
| Roumania | E. Marcus | E. Marcus |
| South Africa | E. Baart | A.H. Jarrett |
| Spain | A. Orte | F. Sanchez |
| Sweden | A. Elvius | B. Höglund |
| Switzerland | E.A. Müller | G.A. Tammann |
| Turkey | Z. Aslan | E. Hamzaoglu |
| USSR | E.R. Mustel | A.V. Tutukov |
| UK | A. Hewish | R.J. Tayler |
| USA | O. Gingerich, A.V. Landolt | A.A. Hoag |
| Uruguay | | |
| Vatican | M.F. McCarthy | M.F. McCarthy |
| Venezuela | F. Fuenmayor | J. Stock |
| Yugoslavia | Z. Dadic | L. Randic |

(c) Acting Presidents of Commissions

The General Secretary announced that the Executive Committee had appointed the following Vice-Presidents of IAU Commissions to act for Presidents of Commissions unable to attend the General Assembly:

| Commission | Acting President |
|---|---|
| 4 | T. Lederle (for A.M. Sinzi) |
| 20 | E. Roemer (for G. Sitarski) |
| 33 | R. Wielen (for G. Kuzmin) |
| 44 | H. Oda/Y. Kondo (for R.J. van Duinen) |

The General Assembly unanimously agreed that these lists were in accord with the requirements of the Union.

## 7. Appointment of the Finance Committee

In accord with Statute 18(a), the General Assembly appointed the following Finance Committee consisting of one representative from each Adhering Body:

| | | | |
|---|---|---|---|
| *Argentina* | E. Bajaja | *Ireland* | I. Elliott |
| *Australia* | R.D. Brown | *Israel* | G. Shaviv |
| *Austria* | H.F. Haupt | *Italy* | G. Godoli |
| *Belgium* | A.G. Velghe | *Japan* | Y. Kozai |
| *Brazil* | P. Kaufmann | *Korea, Dem. P.R.* | Kim Iul |
| *Bulgaria* | M. Kalinkov | *Korea, Republic* | Y.K. Minn |
| *Canada* | V. Gaizauskas | *Mexico* | L. Rodriguez |
| *Chile* | F. Nöel | *Netherlands* | W.B. Burton |
| *China, Nanjing* | Ye Shu-hua | *New Zealand* | J.B. Hearnshaw |
| *China, Taipei* | W.T. Ni | *Norway* | O. Elgarøy |
| *Colombia* | E. Brieva | *Poland* | S. Piotrowski |
| *Cuba* | | *Portugal* | E.M.L. Cabrita |
| *Czechoslovakia* | Z. Ceplecha | *Roumania* | E. Marcus |
| *Denmark* | S. Laustsen | *South Africa* | E.E. Baart |
| *Egypt* | N.H. Youssef | *Spain* | L. Quijano |
| *Finland* | V. Piirola | *Sweden* | B. Gustafsson |
| *France* | F. Spite | *Switzerland* | P. Wild |
| *Germany, D.R.* | G. Ruben | *Turkey* | A. Marsoglu |
| *Germany, F.R.* | W. Fricke | *USSR* | Y.S. Yatskiv |
| *Greece* | L.N. Mavridis | *UK* | D. Walsh, Chairman |
| *Hungary* | B.A. Balazs | *USA* | P.B. Boyce |
| *India* | M.B.K. Sarma | *Uruguay* | |
| *Indonesia* | B. Hidayat | *Vatican* | J. Casanovas |
| *Iran* | | *Venezuela* | D.J. McConnell |
| *Iraq* | Hamid M.K. Al-Naimy | *Yugoslavia* | I. Pakvor |

## 8. Appointment of the Resolutions Committee

The Acting President informed the Assembly that the Executive Committee proposes a Resolutions Committee to be established under the chairmanship of Professor M. McCarthy, and with Professors A.H. Batten and B. Hidayat as members. The General Secretary and Assistant General Secretary will attend the meetings of the Resolutions Committee in an advisory capacity.

The General Assembly unanimously agreed to this composition of the Resolutions Committee.

## 9. Revisions of the Statutes and By-laws

No revisions of Statutes and By-laws were proposed for the 18th General Assembly by the due date for incorporation in the agenda.

## 10. Resolutions Submitted by Adhering Bodies

No resolutions were proposed to the 18th General Assembly by Adhering Bodies by the due date for statutory inclusion in the agenda, and none had subsequently been received for incorporation.

11. Resolutions Submitted by Commissions or by Associated Inter-Union Commissions

The Acting President mentioned that four resolutions proposed by Commissions have been notified to Adhering Organisations for inclusion in the agenda. However, the Executive Committee proposed that these four resolutions, together with other Commission resolutions, and resolutions of the Executive Committee, be referred to the Executive Committee and be taken under Item 15 of the agenda at the second session of the 18th General Assembly.

The Acting President then formally adjourned the meeting to Thursday, August 26, 1982, 10 a.m. and closed the session with a word of thanks to the participants.

## SECOND SESSION

held in the Main Auditorium of
the Patras University on

Thursday, August 26, 1982, from 10.10-13.00

Vice-President Dr D. Heeschen in the Chair as Acting President

The Acting President declared the session open and informed that it was his sad duty to announce the death of the oldest member of the Union, Professor Georgio Abetti, a founding member of the IAU. A message of condolence had been sent to Mrs Abetti.

Messages of good wishes were sent to absent former IAU Presidents and General Secretaries: Professor O. Heckmann, J.H. Oort, B. Strömgren, P. Swings, L. Perek and D.H. Sadler.

The Acting President mentioned that discussions in the Executive Committee on a subject connected with Human Rights had resulted in letters being written to the Secretary of the ICSU Committee on the Safeguard of Pursuit of Science and to the President of the Soviet Academy of Sciences.

Before passing to the agenda, the General Secretary, called upon by the Acting President, established the quorum and found eleven adhering countries not present. Two countries, which had not paid their dues for 1981 and were thus not permitted to vote, were among those absent.

Professors J.C. Pecker and M. Schmidt were appointed tellers.

12. Report of the Finance Committee

The General Secretary mentioned that copies of the financial report were in the hands of the official representatives and invited the Chairman of the Finance Committee, Professor D. Walsh, to read the report:

Report of the Finance Committee to the XVIIIth General Assembly

1. At the First Session of the XVIIIth General Assembly on 17 August 1982, the Finance Committee, consisting of the persons nominated by the Adhering Organisations, was duly appointed. The Committee met on that date, and the General Secretary outlined its function. The Finance Committee appointed D. Walsh (UK) as its Chairman. It elected a Sub-Committee to make a detailed examination of the accounts for the period 1979-1981 and of the budgetary proposals of the Executive Committee. The Sub-Committee consisted of the

Chairman *ex officio*, P. Boyce (USA), D. MacConnell (Venezuela), G. Ruben (GDR), A. Velghe (Belgium), and P. Wild (Switzerland). The Sub-Committee met with the General Secretary, the Assistant General Secretary and the Secretary of the Union on the 19, 20 and 21 August 1982 and reported to the Finance Committee at its second meeting on 21st August 1982.

2. The Chairman reported to the Finance Committee that the Sub-Committee had examined in detail the accounts for the 1979-1981 triennium. The Secretariat had supplied to the Sub-Committee verification of the accounts by the Auditor to the Union, M. Roger Bacle of Paris.

The Sub-Committee compared the accounts for 1979, which was the final year of the previous triennium 1977-1979, with the Budget for 1979 adopted by the XVIIth General Assembly. It concluded that the accounts were consistent with the intentions of that General Assembly. The Sub-Committee then examined the accounts for 1980 and 1981, the first two years of the current (1980-1982) triennium, bearing in mind the budget adopted for this period by the XVIIth General Assembly. It was satisfied that the accounts were within the intent of that budget. The Sub-Committee noted with satisfaction that the proportion of cash held in the form of interest-bearing savings accounts had increased in the past three years from 59% to 80%. The Sub-Committee concluded that the accounts were in order and complimented the Secretariat on the high standard of presentation.

Having considered the report of the Sub-Committee, the Finance Committee recommends to the General Assembly that the accounts for 1979-80 be accepted.

3. The Sub-Committee reported to the Finance Committee that it had examined the budgetary proposal for 1982 as published in the Report of the Executive Committee together with the amendment put by the Executive Committee to the General Assembly. The Sub-Committee noted that both the Receipts and the Payments anticipated in 1982 were considerably higher than those in the Residual Budget for 1982 as published in the Report of the Executive Committee, but that the excess of Payments over Receipts was almost exactly equal to that in the Residual Budget. Some adjustments of payments between headings had been made, the major one being the large increase in cost of the XVIIIth General Assembly over that anticipated in 1979. In view of the special circumstances of this General Assembly, this increase was considered acceptable. It was noted that, over the 1980-82 triennium as a whole, a balanced budget will be achieved.

The Finance Committee, after discussion of the report of the Sub-Committee, concluded that the proposed budget for 1982 was consistent with the policies of the 1980-82 budget as approved by the XVIIth General Assembly, and recommends that it be adopted.

4. The Sub-Committee reported the Finance Committee that it had discussed in detail the budgetary provisions for 1983-85 proposed by the Executive Committee. It examined first the payments envisaged for this period and noted that:

(a) the largest single budget item is the operational costs of the Paris Secretariat. It agreed that the Secretariat at present operates in an economical manner. Based on the operating costs anticipated in the 1982 budget proposals and allowing for a modest rate of inflation of costs, together with limited purchases of furniture and equipment, the budgeted figure of SwFr 500,000 appears to be appropriate.

(b) A substantial increase is proposed to support IAU Symposia and Colloquia. This is in order to provide increased support for each Symposium and Colloquium rather than to increase their number. Presidents of

Scientific Organising Committees will thus have greater freedom in organising their event; they have recently felt unreasonable financial restriction. Scientific meetings represent one of the most important activities of the Union, and if they are to retain a truly international character, the Sub-Committee felt that the proposed budget figure is justified.

(c) Some individual budget items have been restricted to compensate increases in necessary areas, thereby limiting the overall budget figure to a modest increase over the 1980-82 triennium.

In considering the anticipated income of the Union, the Sub-Committee noted that:

(a) The figure assumed for Interest on Accounts is dependent on interest rates which cannot be predicted with confidence; however, any decrease in such interest rates is likely to be accompanied by a reduction in the rate of inflation costs.

(b) The figure assumed for UNESCO and ICSU contributions is thought to be a minimum figure and may be exceeded.

The Sub-Committee concluded that the expenditure itemized in the proposed budget was appropriate to the proper functioning of the Union. In order to provide for this expenditure, the proposal of the Executive Committee to increase the Unit of Contribution to SwFr 1,760 is necessary. The Sub-Committee noted that the Unit of Contribution had not been increased at the previous General Assembly, thus the proposed Unit represents an increase of 22% over a total period of six years, equivalent to 3.3% p.a.

The Finance Committee discussed the Sub-Committee's report at length. On the basis of a vote of 26 for, 0 against, with 11 abstentions, the Finance Committee recommends to the General Assembly adoption of the annexed budget.

5. The Sub-Committee recommends that the Executive Committee examine the possibility of preparing the Budget on an annual basis. As a minimum, the Budget proposed to the General Assembly should be detailed for each of the years that it covers; this should be reviewed by the Executive Committee annually and Adhering Organisations should be informed of revisions. The Sub-Committee also recommends that the headings used in the Budget and accounts be revised to provide more details, possibly using a form similar to that of the Report supplied by the IAU to ICSU.

6. The Sub-Committee wishes to draw attention to the fact that the IAU has benefitted greatly from the support of the home institutions of the General Secretary and the Assistant General Secretary. This cannot be counted on indefinitely.

D. Walsh

Chairman, Finance Committee

The General Assembly approved the Report with acclamation and the Acting President expressed the Union's appreciation of the work done by the members of the Finance Committee and the Finance Sub-committee. He especially thanked the Chairman, Professor D. Walsh.

## 13. Financial Resolutions of the Executive Committee

### (a) Budget Proposal for 1982

The Acting President asked the General Secretary to present the budget estimate for 1982, as contained in the Executive Committee Report in

IAU Information Bulletin No. 48, pp.36-40. He proposed, on behalf of the Executive Committee, that an amendment should be adopted in the sense that the budget figure for International Schools for Young Astronomers (item 15: Swiss Francs 22 000) should be decreased to Swiss Francs 5 000 (since no School would be held in 1982). The difference should be deducted from the excess of payments over income, which would now read Swiss Francs 69 086. The Acting President put the amended budget proposal 1982 to a vote: It was adopted with one abstention.

(b) Unit of Contribution to the IAU 1983-85

The General Secretary explained the background for the proposed increase in the unit of contribution to the IAU from Swiss Francs 1465.- to Swiss Francs 1760.- for the 1983-85 period. An extensive discussion in the Executive Committee had resulted in this proposal, in particular in connection with the stated intention to increase IAU support of scientific acitivities, e.g. a 63% increase was proposed for travel grants to Symposia and Colloquia. At the same time, the Executive Committee felt that there was no way of diminishing the administrative expenses without jeopardizing the proper functioning of the Union, and with the current rate of inflation, some increase had to be made for this budgetary item. This opinion was shared by the Financial Sub-committee.

The General Secretary finally mentioned that the proposed increase corresponds to a 3.3% p.a. increase in the 6-year interval since the unit of contribution was last adjusted in 1976, i.e. well below the inflation rate in most countries.

The proposed increase was adopted by the General Assembly with 109 votes for, 11 votes against, and 30 votes abstaining.

(c) Budget Proposal for 1983-1985

The General Secretary informed that this budget estimate had been distributed in January 1982 to all Adhering Bodies, together with the draft agenda of the 18th General Assembly:

BUDGET FOR 1983 - 1985

Unit of Contribution: 1 760.00 Swiss Francs

| RECEIPTS | Swiss Francs |
|---|---|
| 1. Contribution from Adhering Bodies (225 units per year) | 1 188 000 |
| 2. Revenue from IAU Publications | 51 000 |
| 3. Interest on Accounts | 125 000 |
| 4. UNESCO and ICSU Contributions | 68 000 |
| Total Receipts | 1 433 700 |
| Excess of Payments over Receipts | 1 700 |
| | 1 433 700 |

| PAYMENTS | | Swiss Francs |
|---|---|---|
| 1. | Administrative Office | |
| | 1.1 Paris Secretariat (including furniture, equipment, etc.) | 500 000 |
| | 1.2 Secretarial help to GS and AGS | 10 000 |
| 2. | Subscription to ICSU | 29 700 |
| 3. | Commission Expenses | 15 000 |
| 4. | Projects of Commissions | |
| | 4.1 Exchange of Astronomers (Commission 38) | 58 000 |
| | 4.2 Teaching of Astronomy (Commission 46) | 40 000 |
| | 4.3 IAU Telegram Bureau (Commission 6) | 10 000 |
| | 4.4 Minor Planet Centre (Commission 20) | 10 000 |
| | 4.5 Variable Star Catalogue (Commission 27) | 10 000 |
| 5. | General Assembly | 200 000 |
| 6. | Publications (IAU Information Bulletin) | 60 000 |
| 7. | Distribution of IAU Publications to Developing Countries and Executive Committee Members | 20 000 |
| 8. | Executive Committee Meetings | 70 000 |
| 9. | Officers' Meetings | 20 000 |
| 10. | Symposia and Colloquia | 225 000 |
| 11. | Inter-Union Commissions | 40 000 |
| 12. | Projects of the Executive Committee | 14 000 |
| 13. | Representation | 30 000 |
| 14. | Bank Charges | 1 000 |
| 15. | Young Astronomers' Schools | 36 000 |
| 16. | Regional Meetings | 30 000 |
| 17. | Expenses of the Special Nominating Committee | 5 000 |
| | Total Payments | 1 433 700 |

There being no discussion, the Acting President put the budget to a vote: It was adopted with 137 votes for, none against, and 13 abstentions.

(d) Other Financial Resolutions not included in the Budget Estimate Proposal

The General Secretary presented the Executive Committee proposal concerning the support of a Meteor Data Center in Lund, Sweden, as stated in the agenda. The French text was read by Professor Velghe. Upon a question from the floor about the duration of this project, the General Secretary answered that the project will be reviewed from time to time by the Executive Committee. In the ensuing vote, the resolution was adopted with 147 votes for, and 3 abstentions.

## RESOLUTION A2:

### Support of the Meteor Data Center

The General Assembly, following the recommendation of the Executive Committee,

*having examined* the proposal of the President of Commission 22 to establish a Meteor Data Center of the Commission at the Observatory of Lund in Sweden,

*agrees to* the allocation of SwFr 1800 per annum as a financial contribution to the costs of such a center, so established, for the period 1983-85.

### Aide Financière au Centre de Données sur les Météores

L'Assemblée Générale, sur recommandation du Comité Exécutif,

*ayant examiné* la proposition du Président de la Commission 22 d'établir un Centre de Données sur les Météores à l'Observatoire de Lund en Suède,

*décide* d'allouer une somme de 1.800 Francs Suisses par an comme contribution aux frais de ce centre, pour la période 1983-85.

Next, the General Secretary presented the proposal by the Executive Committee for a Contract Scheme for Visiting Lecturers. The French text was read by Professor Velghe.

There was no discussion and this resolution was also adopted with 147 votes for and 3 abstentions.

## RESOLUTION A3:

### Contract Scheme for Visiting Lecturers

The General Assembly, following the recommendation of the Executive Committee,

*noting* the expressed wish of the official representative of the Swedish National Committee of Astronomy at the XVIIth General Assembly of 1979 at Montreal that fellowships for students from developing countries be encouraged,

*and recognizing* the valuable discussions carried out in the interim by the Working Group of Commissions 46 and 38 on the Education of Astronomers from Developing Countries,

*agrees* that a sum of SwFr 35000 be allowed during the period 1983-85 to support the carrying out of a Contract Scheme for Visiting Lecturers with a sponsoring institution in a country newly entering into astronomical work, to be selected before June 1983, as the result of active enquiry within IAU member countries and other selected countries.

Un Plan de Contrat pour Conférenciers Visiteurs

L'Assemblée Générale, sur recommandation du Comité Exécutif,

*notant* le voeu exprimé par le représentant officiel du Comité National Suédois d'Astronomie à la XVIIème Assemblée Générale de 1979 à Montréal que soient encouragées les bourses pour les étudiants des pays en voie de développement,

*et reconnaissant* les discussions valables engagées depuis par le Groupe de Travail des Commissions 46 et 38 sur la formation des Astronomes des pays en voie de développement,

*décide* d'allouer une somme de 35.000 Francs Suisses pour la période 1983-85 destinée à financer un Plan de Contrat pour Conférenciers Visiteurs avec un institut de tutelle dans un pays nouvellement acquis à la recherche astronomique, dont le choix sera arrêté avant Juin 1983 et dépendra d'une enquête approfondie dans les pays membres de l'UAI et dans d'autres pays sélectionnés.

## 14. Resolutions Submitted by the Executive Committee

A motion concerning membership of Syria had been withdrawn by the Executive Committee.

A motion on the formation of a new Commission 51 (cf. Item 18(b) of the agenda) was put forward by the General Secretary, on behalf of the Executive Committee and carried unanimously.

**RESOLUTION A4:**

Creation of Commission 51: "Search for Extraterrestrial Life"

The General Assembly, following the recommendation of the Executive Committee,

*resolves* that a new Commission of the Union be established in respect of the search for life in the Universe, the name of the Commission to be Commission 51 (Search for Extraterrestrial Life).

Création de la Commission 51: "Recherche de la Vie dans l'Univers"

L'Assemblée Générale, sur recommandation du Comité Exécutif,

*décide* la création d'une nouvelle Commission de l'Union, relative à la recherche de la vie dans l'Univers, le nom de la Commission étant Commission 51 (Recherche de la Vie dans l'Univers).

Similarly, the proposed changes of names of Commissions 26 and 34 were also agreed to unanimously.

RESOLUTION A5:

Changes of Names of Commissions 26 and 34

The General Assembly, following the recommendation of the Executive Committee,

*resolves that* — Commission 26 (Double Stars) be renamed "Double and Multiple Stars"

Commission 34 (Interstellar Matter and Planetary Nebulae) be renamed "Interstellar Matter"

Changement des Noms des Commissions 26 et 34

L'Assemblée Générale, sur recommandation du Comité Exécutif,

*décide que* — La Commission 26 (Etoiles Doubles) deviendra "Etoiles Doubles et Multiples"

La Commission 34 (Matière Interstellaire et Nébuleuses Planétaires) deviendra "Matière Interstellaire"

15. Recommendations of the Resolutions Committee on the Resolutions submitted by Commissions

At the request of the Acting President, Dr M. McCarthy, Chairman of the Resolutions Committee, presented the report of this Committee. More than 30 resolutions had been received and discussed in detail during the past few days and Dr McCarthy expressed his appreciation of the help offered to the Committee by the IAU Officers and Secretariat. He then presented a total of 11 resolutions to the General Assembly; for all of these, the text had been published in the August 25th issue of the General Assembly newspaper "Astrocosmos". Professor B. Donn spoke briefly to the General Assembly about Resolution R1 and R2. A minor change was introduced into the wording of Resolution R6 and R9 was not read in *extenso*, due to an administrative mistake. Resolution R11 confirms and endorses the Resolutions, made by individual Commissions of the Union. The General Assembly unanimously adopted all resolutions.

The final text of resolutions R1 to R11 is as follows (French translations are also included):

RESOLUTION R1:

Comet Halley Days

The International Astronomical Union

*recognizing* — that it is particularly desirable that pre-selected Comet Halley Days for co-ordinated observation over a limited time be supported

*recommends* — that observatory directors and observing program committees give high priority to Comet Halley observation during the interval 1985-87.

Journées "Comète de Halley"

L'Union Astronomique Internationale

*estimant* qu'il est particulièrement souhaitable que l'on consacre des Journées "Comète de Halley", choisies à l'avance, à des observations coordonnées pendant une durée limitée

*recommande* que les directeurs d'observatoires et les comités de programme attribuent une forte priorité aux observations de la Comète de Halley dans l'intervalle 1985-87.

## RESOLUTION R2:

International Halley Watch (IHW)

The International Astronomical Union

*noting* that, in order to organise and marshall ground-based observations of Comet Halley throughout its 1986 perihelion passage and to co-ordinate them with space missions, an international program, the International Halley Watch, has been established

*and wishing* to avoid duplication of effort at the international level and to encourage participation in this program

*endorses* the International Halley Watch as the international co-ordinating agency for Comet Halley observations.

L'Union Astronomique Internationale

*constatant* qu'un programme international de surveillance de la Comète Halley dit "International Halley Watch" a été mis en place afin de préparer et d'orchestrer les observations depuis la Terre de la Comète de Halley tout au long de son passage au périhélie en 1986 et de les coordonner avec les missions spatiales

*et désirant* éviter la duplication des efforts au niveau international et encourager la participation à cette campagne

*reconnaît* l'International Halley Watch comme l'organisation internationale coordonnant les observations de la Comète de Halley.

## RESOLUTION R3:

1980 IAU Theory of Nutation

The International Astronomical Union

*recognizing* that since the Seventeenth General Assembly, Commission 4, 7, 8, 19 and 31 have adopted the 1980 IAU Theory of Nutation in place of the 1979 IAU Theory of Nutation *now endorses* — the adoption of the 1980 IAU Theory of Nutation.

Théorie de la Nutation UAI 1980

L'Union Astronomique Internationale

*considérant* — que depuis la XVIIème Assemblée Générale, les Commissions 4, 7, 8, 19 et 31 ont adopté la Théorie de la Nutation UAI 1980 en remplacement de la Théorie de la Nutation UAI 1979,

*souscrit* — désormais à l'adoption de la Théorie de la Nutation UAI 1980.

## RESOLUTION R4:

Synoptic Solar Observations

The International Astronomical Union

*recognizing* — that continuing long-term synoptic observations of solar activity are vitally necessary, and that some long-term programs have been lost or are threatened in the present world-wide economic situation,

*recommends* — that all countries endeavour to plan well balanced programs of synoptic solar observations for the sake of future generations.

Observations Synoptiques Solaires

L'Union Astronomique Internationale

*reconnaissant* — que les observations synoptiques continues et de longue durée de l'activité solaire sont d'une nécessité vitale et que certains programmes à long terme ont été perdus ou sont menacés par suite de la situation économique mondiale actuelle,

*recommande* — que tous les pays s'efforcent d'organiser des programmes rationnels d'observation synoptique solaire pour les générations futures.

## RESOLUTION R5:

MERIT Campaign

The International Astronomical Union

*noting* — the success of the MERIT short campaign

*and recognizing* — that the results to be obtained during the MERIT main campaign will be of short and long-term benefit for the studies of the planet Earth

*supports* — the following two resolutions adopted by the International Association of Geodesy (IAG) at its General Meeting of Tokyo (May 1982):

IAG Resolution 1

The International Association of Geodesy

*noting* that the results to be obtained during the MERIT main campaign will be of long-term benefit to geodesy and its applications

*strongly endorses* the proposals of the COTES and MERIT Working Groups that during the campaign

(a) very long baseline radio interferometric and satellite and lunar laser ranging systems be used for co-located observations of high precision at the recommended sites, and

(b) observations be made intensively for a limited period to detect any short-period variations in the derived earth-rotation parameters

*and urges* the appropriate resources and facilities be made available for these activities by the countries involved.

IAG Resolution 2

The International Association of Geodesy

*considers* that it is important that the new terrestrial reference frame to be derived from high-precision observations during the MERIT Main Campaign should be extended and related to existing services as accurately and quickly as possible,

*and urges* that co-ordinated precise positioning observations be made during the campaign by satellite radio tracking systems at the Very Long Baseline Interferometric, Lunar and Satellite Laser Ranging sites as well as at a larger number of well distributed sites around the world.

*recommends* (a) that the co-ordinates of the stations of the International Service of Latitude be determined in the Reference System (NW1OF) adopted for the analysis of observations, by the Doppler method of artificial satellite of TRANSIT or NOVA type.

(b) That special support be given to BIH acting as the co-ordinating center for the MERIT international campaign.

Programme MERIT

L'Union Astronomique Internationale

*prenant note* du succès de la campagne courte du programme MERIT et

*reconnaissant* que les résultats à obtenir au cours de la Campagne Principale de MERIT seront bénéfiques à long et court terme pour les études de la planète Terre

*souscrit* aux deux résolutions suivantes adoptées par l'Association Internationale de Géodésie (AIG) lors de sa réunion générale de Tokyo (Mai 1982):

AIG Résolution 1

La Réunion Générale de l'Association Internationale de Géodésie

*notant* que les résultats que fournira la Campagne MERIT seront d'un profit durable pour la géodésie et ses applications,

*approuve fortement* les propositions des Groupes de travail COTES et MERIT que, lors de cette Campagne,

(a) les systèmes d'interférométrie à très longue base et de télémétrie laser sur les satellites et la Lune soient utilisés à des observations de haute précision en colocation sur les sites recommandés, et que

(b) des observations intensives soient faites pendant une période limitée afin de détecter des variations à courte période dans les paramètres de rotation de la Terre ainsi déterminés,

*et insiste* pour que les moyens appropriés soient donnés à ces activités par les pays concernés.

AIG Résolution 2

La Réunion Générale de l'Association Internationale de Géodésie

*considérant* qu'il est important que les programmes d'observation associés au nouveau repère terrestre qui doit se déduire des mesures de haute précision pendant la Campagne Principale MERIT soient étendus et reliés aux services existants aussi exactement et rapidement que possible,

*insiste* pour que des observations coordonnées de positionnement précis soient faites pendant cette Campagne par des systèmes de poursuite radioéléctrique de satellites aux sites d'interférométrie à très longue base et de télémétrie laser sur la Lune et les satellites, ainsi qu'à un grand nombre de sites bien répartis autour du globe.

*recommande* (a) que les coordonnées des stations du Service International des Latitudes soient déterminées dans le système de référence (NW1OF) adopté pour l'analyse des observations par la méthode Doppler des satellites artificiels du type Transit ou Nova.

(b) qu'une aide spéciale soit octroyée au Bureau International de l'Heure (BIH) agissant en qualité de centre coordinateur pour la campagne internationale MERIT.

## RESOLUTION R6:

AAVSO Data on Cataclysmic Variables

The International Astronomical Union

*recognizing* the recent increased interest in cataclysmic variables and the need for long-term light-curves to make possible the correlation of theoretical and observational research

*noting* that data collected by AAVSO observers and other groups throughout the world are available on magnetic tape

*supports* the search for means to publish this valuable reference material.

Données AAVSO sur les Variables Cataclysmiques

L'Union Astronomique Internationale

*reconnaissant* l'intérêt croissant que présentent les variables cataclysmiques et la nécessité de disposer de courbes de lumière à long terme afin de permettre la corrélation entre la recherche théorique et observationnelle

*notant* que les données rassemblées par les observateurs AAVSO et par d'autres groupes dans le monde sont disponibles sur bandes magnétiques,

*recommande* la recherche de moyens en vue de la publication de ces précieuses données de référence.

## RESOLUTION R7:

Astronomy and Astrophysics Abstracts

The International Astronomical Union

*confirming* Resolution No. 3 adopted by the XVIth General Assembly in 1976,

*considers* that Astronomy and Astrophysics Abstracts (AAA) fulfil excellently the specialized needs for an abstracting service in Astronomy and Astrophysics,

*and recommends* strongly that AAA continued to be produced by the Astronomisches Rechen-Institut, Heidelberg, Fed. Rep. Germany, under the auspices of the IAU.

L'Union Astronomique Internationale

*reconfirmant* la Résolution No. 3 adoptée par la XVIème Assemblée Générale en 1976

*considère* que les "Astronomy & Astrophysics Abstracts" (AAA) satisfont pleinement aux besoins spécifiques d'un service chargé des notes abrégées dans le domaine de l'astronomie et de l'astrophysique,

*et recommande* vivement que les AAA continuent à être publiés par l'Astronomisches Rechen-Institut à Heidelberg (Rép. Féd. d'Allemagne) sous les auspices de l'UAI.

## RESOLUTION R8:

### Very Long Baseline Interferometry

The International Astronomical Union

*recognizing* the importance of interdisciplinary scientific use of large steerable antennas for astrophysical, astrometric and geodetic research in Very Long Baseline Interferometry (VLBI),

*endorses* the formation of a working group under Commission 40 to collect and disseminate information concerning plans of VLBI experiments in astronomy and geodesy, and to encourage cooperation between observatories internationally.

### Interférométrie à Très Longue Base

L'Union Astronomique Internationale

*reconnaissant* l'importance de l'utilisation scientifique interdisciplinaire de grandes antennes orientables pour la recherche astrophysique, astrométrique et géodésique dans l'Interférométrie à Très Longue Base (VLBI),

*souscrit* à la création d'un groupe de travail dépendant de la Commission 40 en vue de rassembler et de diffuser l'information concernant les projets d'expériences VLBI en astronomie et en géodésie, et d'encourager la coopération internationale entre les observatoires.

## RESOLUTION R9:

### Protection of Radio Frequency Bands

The International Astronomical Union

*recalling* the considerations (a) to (d) of IAU Resolution No. 3, passed at the XVIIth General Assembly in 1979 concerning harmful interference to radio astronomy observations *and considering* — the pioneering use by radio astronomers of the radio spectrum at frequencies above 275 GHz,

*recommends*

1. The provision by national administrations of frequency bands for radio astronomy continuum and polarization measurements at nearly octave intervals throughout the radio spectrum,

2. The provision of bands at the frequencies of the astrophysically most important spectral lines tabulated in the 1982 report of IAU Commission 40, and

3. The protection of these frequency bands from harmful interference from in-band, band-edge and sub-harmonic emissions, especially from space-borne transmitters.

Protection de Bandes de Fréquences Radio

L'Union Astronomique Internationale

*rappelant* — les considérations (a) à (d) de la Résolution UAI No. 3 passée à la XVIIème Assemblée Générale en 1979, relative aux interférences nuisibles aux observations radioastronomiques,

*et considérant* — l'utilisation d'avant-garde faite par les radioastronomes du spectre radio à des fréquences supérieures à 275 GHz,

*recommande*

1. l'attribution par les administrations nationales de bandes de fréquences destinées aux mesures radioastronomiques du spectre continu et de la polarisation à des intervalles de près d'une octave dans tout le spectre radio,

2. l'attribution de bandes à des fréquences correspondant aux raies spectrales les plus importantes pour l'astrophysique telles qu'elles figurent dans le rapport 1982 de la Commission 40 de l'UAI,

3. la protection de ces bandes de fréquences contre toute interférence nuisible dans la bande, en bord de bande et dans les bandes sous-harmoniques, et en particulier contre toute émission en provenance d'émetteurs embarqués.

## RESOLUTION R10:

Sydney Observatory Instruments

The International Astronomical Union

*noting* — the decision to terminate scientific work of the Sydney Observatory, and

*recognizing* — the role of the Sydney Observatory in international campaigns for the improvement of the astrometric reference frame in the Southern Hemisphere, *recommends* that the instruments now at Sydney continue to be made available for astronomical observations.

Les instruments de l'Observatoire de Sydney

L'Union Astronomique Internationale

*notant* la décision de mettre fin aux travaux scientifiques de l'Observatoire de Sydney et,

*reconnaissant* le rôle de l'Observatoire de Sydney dans les campagnes internationales visant à améliorer le système de référence astrométrique dans l'hémisphère Sud,

*recommande* que les instruments se trouvant actuellement à Sydney continuent à être utilisables pour les observations astronomiques.

**RESOLUTION R11:**

Endorsement of Commission Resolutions

The XVIIIth General Assembly of the International Astronomical Union

*having* full confidence in its Commissions,

*endorses* other resolutions submitted by them to the Resolutions Committee. These will be published in the Official Languages of the Union, French and English, in Transactions IAU XVIIIB.

Adoption des Résolutions des Commissions

La XVIIIème Assemblée Générale de l'UAI

*accordant* toute sa confiance dans ses Commissions,

*souscrit* aux autres résolutions qu'elles ont soumises au Comité des Résolutions. Ces résolutions seront publiées dans les deux langues officielles de l'Union, le français et l'anglais, dans les Transactions de l'UAI XVIIIB.

---

The full texts of the 16 Resolutions by IAU Commissions (C1-C16) are as follows:

**RESOLUTION C1:**

Flexible Image Transport System

Commission 5

*considering* the present unsatisfactory situation of the transfer of astronomical data between astronomical institutions *recommends* — that all astronomical computer facilities recognize and support the Flexible Image Transport System (FITS) for the interchange of binary data on magnetic tape, as described in Astronomy and Astrophysics Supplement, vol. 44, pp.363 and 371.

Système de Transport d'Image Souple

La Commission 5

*considérant* — la situation actuelle peu satisfaisante de l'échange de données entre instituts d'astronomie

*recommande* — que tous les ordinateurs utilisés en astronomie reconnaissent et adoptent le Système de Transport d'Image Souple pour l'échange de données binaires sur bande magnétique, tel qu'il est décrit dans "Astronomy & Astrophysics Supplement", vol. 44, pp. 363 et 371.

## RESOLUTION C2:

Introduction of Astronomy

Commission 46

*considering* — that astronomy is an integral part of physical sciences, both in education and in research,

*and whereas* — some countries now without astronomy may desire to introduce astronomy in their educational or scientific institutions,

*recommends* — that the International Astronomical Union supports the introduction of astronomy in such countries and encourages activities by its Commissions towards that purpose.

Promotion de l'Astronomie

La Commission 46

*considérant* — que l'astronomie fait partie intégrante des sciences physiques tant dans l'enseignement que dans la recherche

*et alors que* — certains pays dépourvus d'astronomie peuvent souhaiter promouvoir cette science dans leurs institutions scientifiques ou d'enseignement,

*recommande* — que l'Union Astronomique Internationale soutienne la promotion de l'astronomie dans ces pays et encourage les activités s'y rapportant par l'intermédiaire de ses Commissions.

## RESOLUTION C3:

### UT1 in Air and Nautical Almanacs

Commissions 4 and 31

*noting* that the present method of keeping UTC within $0^s.9$ of UT1 by means of leap seconds both provides the second of SI and meets the need for safe celestial navigation,

*recommend* that the Air and Nautical Almanacs continue to be published with UT1 as the argument.

### UT1 dans les Annuaires Aéronautiques et Nautiques

Les Commissions 4 et 31

*notant* que la méthode actuelle consistant à conserver l'écart entre le UTC et le UT1, dans les limites de $0^s.9$, au moyen de secondes intercalaires, fournit à la fois la seconde de SI et satisfait aux besoins de la securité de la navigation par les méthodes astronomiques,

*recommandent* que les annuaires aéronautiques et nautiques continuent d'être publiés avec le UT1 comme argument.

## RESOLUTION C4:

### Sign of Terrestrial Longitude

Commissions 4, 19, 31

*noting* that the International Meridian Conference held in Washington in October 1884 adopted a resolution that from the meridian of the Observatory of Greenwich "longitude shall be counted in two directions up to 180 degrees, east longitude being plus and west longitude minus",

*and noting* that there is an increasing interaction between astronomy and geodesy, a field in which the sign convention "east longitude is positive" is in common use,

*recommend* that as soon as practicable all national ephemerides and other astronomical publications adopt the convention that terrestrial longitude be measured positively to the east.

### Signe de Longitude Terrestre

Les Commissions 4, 19 et 31

*notant* que la Conférence du Méridien international tenue à Washington en Octobre 1884 a adopté une résolution selon laquelle à partir du méridien de l'Observatoire de Greenwich "la longitude sera comptée dans deux directions jusqu'à 180°, la longitude Est étant positive et la longitude Ouest étant négative", *et notant* qu'il existe une interaction croissante entre l'astronomie et la géodesie dans lequel la convention de signe "longitude positive vers l'Est" est d'un usage courant,

*recommandent* que toutes les éphémérides nationales et autres publications astronomiques adoptent dès que possible la convention selon laquelle la longitude terrestre soit comptée positivement vers l'Est.

## RESOLUTION C5:

### UT1 and Greenwich Mean Sidereal Time

Commissions 4, 19 and 31

*considering* Commission resolution (4) of the XVIIth General Assembly and that it is planned to introduce the IAU (1976) System of Astronomical Constants, the 1980 IAU Theory of Nutation, and the equinox of the FK5 on 1984 January 1,

*recommend* that (a) the relationship between mean sidereal time and UT1 be modified so that there is no change in either value or rate of UT1, as determined from stellar observations, due to a correction to the zero point of right ascensions of the FK4 or to a correction for the motion of the zero point, both to be introduced in FK5 (b) the new expression for Greenwich mean sidereal time of $0^m$UT1 be

$$\text{GMST of } 0^h\text{UT1} = 6^h41^m50\overset{s}{.}54841 + 8640184\overset{s}{.}812866\, T_u + 0\overset{s}{.}093104\, T_u^2 - 6\overset{s}{.}2 \times 10^{-6}\, T_u^3$$

where $T_u$ is the number of Julian centuries of 36525 days of universal time elapsed since 2000 January 1, $12^h$ UT1 (JD 2451545.0).

### UT1 et Temps Moyen Sidéral de Greenwich

Les Commissions 4, 19 et 31

*considérant* la Résolution 3 de la Commission 4 prise lors de la XVIIème Assemblée Générale et qu'il est prévu d'introduire le Système des Constantes Astronomiques de l'UAI 1976, la Théorie de la Nutation UAI 1980 et l'équinoxe du FK5 au 1er Janvier 1984.

*recommandent* (a) de modifier la relation existant entre le temps sidéral moyen et le UT1 de façon à ne changer ni la valeur ni le taux de UT1, déterminés à partir d'observations stellaires et introduits, par suite d'une correction à l'origine des ascensions droites du FK4 et du mouvement de cette origine, toutes deux étant apportées au FK5.

(b) que la nouvelle relation du temps sidéral moyen de Greenwich à 0h UT1 soit de

$$GMST\ \text{à}\ 0hUT1 = 6h41mn50s.54841 + 8640184s.812866\ TU + 0s093104\ TU^2 - 6s.2 \times 10^{-6} TU^3$$

où TU est le nombre de siècles juliens de 36525 jours de temps universel écoulés depuis le 1er Janvier 2000, à 12h UT1 (JD 2451545, 0).

## RESOLUTION C6:

### Positional Astronomy and Fundamental Reference System

Commission 8

*considering* that the determination of the equinox and equator will remain an important task of positional astronomy

*recommends* that
(a) the Sun, major and minor planets be included in observational programs with transit circles.
(b) the Sun, and major planets be included in programs with astrolabes.
(c) minor planets be included in programs of photographic astrometry,
so that from differential and/or absolute observations the positions can be rigorously referred to the fundamental reference system and thus serve for further improvements of the zero points of this system.

### Astronomie de Position et Système de Référence Fondamental

La Commission 8

*considérant* que la détermination de l'équinoxe et de l'équateur resteront une tâche importante pour l'astronomie de position,

*recommande* que
(a) le Soleil, les grosses et les petites planètes soient incluses dans les programmes d'observation au moyen d'instruments de passage
(b) le Soleil et les grosses planètes soient incluses dans les programmes d'astrolabe et
(c) les petites planètes soient incluses dans les programmes d'astronomie photographiques
en sorte qu'à partir d'observations différentielles et/ou absolues, les positions puissent être rigoureusement rattachées au système de référence fondamental et servir à améliorer les origines de ce système.

## RESOLUTION C7:

Solar Seismology

Commission 12

*recognizing* the extreme importance of the observation of solar seismology

*strongly supports* international cooperation in establishing a world-wide network of observing stations.

Sismologie Solaire

La Commission 12

*reconnaissant* l'extrème importance de l'observation de sismologie solaire,

*encourage* vivement la coopération internationale par la mise en place d'un réseau mondial de stations d'observation.

## RESOLUTION C8:

Dangers to Astronomy from Space Techniques

Commission 50

*viewing* with deep concern the proposals for a Satellite Power System which would place very large structures into Earth orbit with the potential of destroying the conditions for a major part of astronomical research in many regions of the electromagnetic spectrum,

*recommends* national representatives to bring this concern to the notice of space agencies in their countries, and to ensure that the IAU is kept fully informed of proposed developments in space techniques which would involve dangers to astronomy.

Dangers que font courir à l'Astronomie certains projets spatiaux

La Commission 50

*considérant* avec une profonde inquiétude les propositions relatives à un système de centrale d'énergie sur satellite qui placerait en orbite autour de la Terre d'importantes structures ayant la capacité de détruire les conditions d'observation pour une grande partie des recherches astronomiques, et ceci dans de nombreuses régions du spectre électromagnétique,

*recommande* aux représentants nationaux d'attirer l'attention des agences spatiales de leur pays sur ce problème, et de s'assurer que l'UAI est tenue informée des projets proposés dans le domaine spatial qui impliqueraient des dangers pour la recherche astronomique.

## RESOLUTION C9:

Zonal Tides (Table on p. 52)

Commissions 19 and 31

*considering* the need, in the processing and publication of the universal time, to clarify the role of the zonal tide,

*recommend*

1. that when the effect of the short period zonal tides, i.e. periods less than 35 days, is computed for the above purposes, the attached tabulation based on Yoder, Williams and Parke (J. Geophys. Res. 86, 881, 1981) be used, and
2. that the letter R be added to the notation of the relevant quantities to indicate that the correction for the short period zonal terms has been made (for instance UT1R).

Marées Zonales

Les Commissions 19 et 31

*considérant* le besoin de représenter clairement le rôle des marées zonales dans le calcul et la publication du temps universel,

*recommandent*

1. que lorsque l'effet des marées zonales à court terme (période inférieure à 35 jours) est calculé dans les buts précités, on utilise le tableau ci-joint, tiré des travaux de Yoder, Williams et Parke (J. Geophys. Res. 86, 881, 1981), et
2. que le suffixe R soit ajouté à la notation des quantités concernées pour signifier que la correction des effets à court terme a été effectuée (exemple: UT1R).

## RESOLUTION C10:

Value of k

Commission 4

*considering* that it is planned to introduce the IAU (1976) System of Astronomical Constants, the 1980 IAU Theory of Nutation, and the equinox of the FK5 on 1984 January 1,

*recommends* that in the calculation of lunar and solar eclipses, there be a single value for k, the ratio of the radius of the lunar profile to the Earth's radius, and that it correspond to the mean radius of Watt's datum as determined by observation of occultations and to the adopted radius of the Earth; the value of k = 0.2725076.

## Terms Due to Zonal Tides, with Periods up to 35 Days

Based on Yoder, Williams, and Parke: 1981, J. of Geophys. Res., Vol. 86, 881 with K/C = 0.94

UT1R, DR, $\omega$R represent the corrected forms of UT1, of the duration of the day D and of the angular velocity of the Earth $\omega$.

The units are $10^{-4}$s for UT, $10^{-5}$s for D, and $10^{-14}$Rad/s for $\omega$

| | ARGUMENT | | | | | PERIOD | UT1-UT1R | D-DR | $\omega$-$\omega$R |
|---|---|---|---|---|---|---|---|---|---|
| N | $l$ | $l'$ | F | $D$ | $\Omega$ | DAYS | COEFFICIENT OF SIN (ARGUMENT) | COEFFICIENT OF COS (ARGUMENT) | |
| 1 | 1 | 0 | 2 | 2 | 2 | 5.64 | -0.02 | 0.3 | -0.2 |
| 2 | 2 | 0 | 2 | 0 | 1 | 6.85 | -0.04 | 0.4 | -0.3 |
| 3 | 2 | 0 | 2 | 0 | 2 | 6.86 | -0.10 | 0.9 | -0.8 |
| 4 | 0 | 0 | 2 | 2 | 1 | 7.09 | -0.05 | 0.4 | -0.4 |
| 5 | 0 | 0 | 2 | 2 | 2 | 7.10 | -0.12 | 1.1 | -0.9 |
| 6 | 1 | 0 | 2 | 0 | 0 | 9.11 | -0.04 | 0.3 | -0.2 |
| 7 | 1 | 0 | 2 | 0 | 1 | 9.12 | -0.41 | 2.8 | -2.4 |
| 8 | 1 | 0 | 2 | 0 | 2 | 9.13 | -0.99 | 6.8 | -5.8 |
| 9 | 3 | 0 | 0 | 0 | 0 | 9.18 | -0.02 | 0.1 | -0.1 |
| 10 | -1 | 0 | 2 | 2 | 1 | 9.54 | -0.08 | 0.5 | -0.5 |
| 11 | -1 | 0 | 2 | 2 | 2 | 9.56 | -0.20 | 1.3 | -1.1 |
| 12 | 1 | 0 | 0 | 2 | 0 | 9.61 | -0.08 | 0.5 | -0.4 |
| 13 | 2 | 0 | 2 | -2 | 2 | 12.81 | 0.02 | -0.1 | 0.1 |
| 14 | 0 | 1 | 2 | 0 | 2 | 13.17 | 0.03 | -0.1 | 0.1 |
| 15 | 0 | 0 | 2 | 0 | 0 | 13.61 | -0.30 | 1.4 | -1.2 |
| 16 | 0 | 0 | 2 | 0 | 1 | 13.63 | -3.21 | 14.8 | -12.5 |
| 17 | 0 | 0 | 2 | 0 | 2 | 13.66 | -7.76 | 35.7 | -30.1 |
| 18 | 2 | 0 | 0 | 0 | -1 | 13.75 | 0.02 | -0.1 | 0.1 |
| 19 | 2 | 0 | 0 | 0 | 0 | 13.78 | -0.34 | 1.5 | -1.3 |
| 20 | 2 | 0 | 0 | 0 | 1 | 13.81 | 0.02 | -0.1 | 0.1 |
| 21 | 0 | -1 | 2 | 0 | 2 | 14.19 | -0.02 | 0.1 | -0.1 |
| 22 | 0 | 0 | 0 | 2 | -1 | 14.73 | 0.05 | -0.2 | 0.2 |
| 23 | 0 | 0 | 0 | 2 | 0 | 14.77 | -0.73 | 3.1 | -2.6 |
| 24 | 0 | 0 | 0 | 2 | 1 | 14.80 | -0.05 | 0.2 | -0.2 |
| 25 | 0 | -1 | 0 | 2 | 0 | 15.39 | -0.05 | 0.2 | -0.2 |
| 26 | 1 | 0 | 2 | -2 | 1 | 23.86 | 0.05 | -0.1 | 0.1 |
| 27 | 1 | 0 | 2 | -2 | 2 | 23.94 | 0.10 | -0.3 | 0.2 |
| 28 | 1 | 1 | 0 | 0 | 0 | 25.62 | 0.04 | -0.1 | 0.1 |
| 29 | -1 | 0 | 2 | 0 | 0 | 26.88 | 0.05 | -0.1 | 0.1 |
| 30 | -1 | 0 | 2 | 0 | 1 | 26.98 | 0.18 | -0.4 | 0.3 |
| 31 | -1 | 0 | 2 | 0 | 2 | 27.09 | 0.44 | -1.0 | 0.9 |
| 32 | 1 | 0 | 0 | 0 | -1 | 27.44 | 0.53 | -1.2 | 1.0 |
| 33 | 1 | 0 | 0 | 0 | 0 | 27.56 | -8.26 | 18.8 | -15.9 |
| 34 | 1 | 0 | 0 | 0 | 1 | 27.67 | 0.54 | -1.2 | 1.0 |
| 35 | 0 | 0 | 0 | 1 | 0 | 29.53 | 0.05 | -0.1 | 0.1 |
| 36 | 1 | -1 | 0 | 0 | 0 | 29.80 | -0.06 | 0.1 | -0.1 |
| 37 | -1 | 0 | 0 | 2 | -1 | 31.66 | 0.12 | -0.2 | 0.2 |
| 38 | -1 | 0 | 0 | 2 | 0 | 31.81 | -1.82 | 3.6 | -3.0 |
| 39 | -1 | 0 | 0 | 2 | 1 | 31.96 | 0.13 | -0.3 | 0.2 |
| 40 | 1 | 0 | -2 | 2 | -1 | 32.61 | 0.02 | -0.0 | 0.0 |
| 41 | -1 | -1 | 0 | 2 | 0 | 34.85 | -0.09 | 0.2 | -0.1 |

$l = 134^\circ.96 + 13^\circ.064993$(MJD-51544.5) MEAN ANOMALY OF THE MOON

$l' = 357^\circ.53 + 0^\circ.985600$(MJD-51544.5) MEAN ANOMALY OF THE SUN

$F = 93^\circ.27 + 13^\circ.229350$(MJD-51544.5) $L - \Omega$: L : MEAN LONGITUDE OF THE MOON

$D = 297^\circ.85 + 12^\circ.190749$(MJD-51544.5) MEAN ELONGATION OF THE MOON FROM THE SUN

$\Omega = 125^\circ.04 - 0^\circ.052954$(MJD-51544.5) MEAN LONGITUDE OF THE ASCENDING NODE OF THE MOON

Valeur de k

La Commission 4

*considérant* qu'il est prévu d'introduire le Système des Constantes Astronomiques UAI 1976, la Théorie de la Nutation UAI 1980 et l'équinoxe du FK5 au ler Janvier 1984,

*recommande* qu'il y ait, dans le calcul des éclipses lunaires et solaires, une seule valeur pour k, rapport du rayon du profil lunaire au rayon de la Terre, et que ce rapport corresponde au rayon moyen de la donnée de Watts déterminée par observation d'occultations et au rayon qui a été adopté pour la Terre; cette valeur est k = 0,2725076.

RESOLUTION C11:

Proper Motions of High Luminosity Stars

Commission 8

*in view* of the importance of accurate positions and proper motions of High Luminosity Stars for galactic research

*reaffirms* its previous recommendation that such stars be included in transit circle programmes.

Mouvements propres d'étoiles très lumineuses

La Commission 8

*étant donné* l'importance des positions et mouvements propres précis d'étoiles très lumineuses pour la recherche galactique,

*reconfirme* sa recommandation antérieure selon laquelle ces étoiles soient incluses dans des programmes d'instruments méridiens.

RESOLUTION C12:

Transit Circle Observations of Subset of HIPPARCOS Stars

Commission 8

*recommends* the observation by transit circle of the subset of stars requested by the HIPPARCOS Input Catalogue Consortium. The positions should be available before 1985.5. A moderate accuracy of ± 1.0 arcsec at that epoch is required; however ± 0.3 arcsec (in the FK4 system) is desirable.

Observations par instruments méridiens du sous-ensemble d'étoiles HIPPARCOS

La Commission 8

*recommande* d'observer par instruments méridiens le sous-ensemble d'étoiles demandé par le HIPPARCOS Input Catalogue Consortium. Les positions devraient être disponibles avant 1985.5.

Une précision modérée de ± 1,0 arcsec est requise à cette époque; toutefois, ± 0,3 arcsec (dans le système FK4) est souhaitable.

## RESOLUTION C13:

### Meridian Positions of Subsets of HIPPARCOS Stars

Commission 8

*noting* — that the HIPPARCOS stars will be selected for their astrophysical and/or astrometric interest and that such stars would traditionally be included in meridian observing programmes, and that very accurate meridian positions of suitable subsets of the HIPPARCOS stars would give checks of the space observations and calibration of the ground-based instruments,

*recommends* — that such subsets, e.g. faint fundamental stars and International Reference Stars be observed.

### Positions méridiennes des sous-ensembles d'étoiles HIPPARCOS

La Commission 8

*notant* — que les étoiles HIPPARCOS seront sélectionnées pour l'intérêt qu'elles présentent du point de vue astrophysique et/ou astrométrique et que ces étoiles devraient, selon l'usage, être incluses dans les programmes d'observation méridienne, et que les positions méridiennes très précises des sous-ensembles appropriés d'étoiles HIPPARCOS permettraient d'effectuer des sondages sur les observations effectuées dans l'espace et sur la calibration d'instruments au sol, il est

*recommandé* — d'observer de tels sous-ensembles, c'est-à-dire les étoiles fondamentales faibles et les Etoiles de Référence Internationale.

## RESOLUTION C14:

### International Heliospheric Study

Commission 49

*noting* — decision 5/82 adopted at the XXIVth COSPAR meeting in May 1982,

*recommends* — the introduction of the International Heliospheric Study during January 1988 - December 1990 as an IAU activity conducted under the auspices of COSPAR in conjunction with other appropriate ICSU bodies.

L'Etude Héliosphérique Internationale

La Commission 49

*notant* — la décision 5/82 adoptée à la XXIVème réunion du COSPAR tenue en Mai 1982,

*recommande* — d'introduire l'Etude Héliosphérique Internationale de Janvier 1988 à Décembre 1990 parmi les activités de l'UAI, sous les auspices du COSPAR, conjointement aux autres organismes appropriés de l'ICSU.

RESOLUTION C15:

Fireball Networks

Commission 22

*noting* — the importance of the recent contributions from the Canadian and European Fireball Networks

*recommends* — that continued support be provided for these observational programs in order that a more extensive body of reliable data may be gathered for critical analysis

*note*: — The recovery of meteorites from these two major camera networks is likely to remain limited to rare events; progress in the interpretation of fireball data is now leading to significant studies of both the astronomical properties of meteorites and the physics of their atmospheric interaction.

Réseaux de Bolides

La Commission 22

*notant* — l'importance des récentes réalisations des Réseaux de Bolides Canadiens et Européens,

*recommande* — de poursuivre sans discontinuer l'aide apportée à ces programmes d'observation afin qu'un plus vaste ensemble de données fiables puisse être rassemblé en vue d'une analyse critique,

*note* — que la récupération des météorites au moyen de ces deux principaux réseaux restera vraisemblablement limitée à quelques cas rares; les progrès realisés dans l'interprétation des données relatives aux bolides permet actuellement d'entreprendre des études sérieuses à la fois sur les propriétés astronomiques des météorites et sur la physique de leur interaction dans l'atmosphère.

RESOLUTION C16:

Austrian Stations of the European Fireball Networks

Commission 22

*recognizing* the key position of the two Austrian stations of the European Fireball Network and their essential role in the location and recovery of meteorites which fall on Austrian soil,

*recommends* that the University Observatory of Vienna give its fullest support to the continued operation of these stations.

Stations Autrichiennes du Réseau Européen des Bolides

La Commission 22

*reconnaissant* la position clé des deux stations autrichiennes du Réseau Européen des Bolides et leur rôle fondamental dans la localisation et la récupération des météorites tombant sur le sol autrichien,

*recommande* à l'Observatoire de l'Université de Vienne d'accorder son soutien le plus total à la poursuite des activités de ces deux stations.

## 16. Appointment of Special Nominating Committee

The Chairman asked the General Secretary to report on the names of the members proposed for appointment by the General Assembly to the Special Nominating Committee 1982-85. These persons will be convened by the President of the IAU for the purpose of proposing names to the 19th General Assembly (1985) for the IAU Executive Committee membership (1985-88). The six persons appointed to the Special Nominating Committee were: Professors C. Hayashi (Japan), H. van der Hulst (Netherlands), R. Kippenhahn (FRG), D. Osterbrock (USA), I.S. Shklovsky (USSR), and J. Smak (Poland). The appointment was unanimously confirmed by the General Assembly.

## 17. Nomination of New Members of the Union

The General Secretary announced that the Executive Committee had, on the proposal of the Adhering Bodies and with the advice of the Nominating Committee, admitted around 780 new members to the Union. The names of the new members had been displayed in a prominent place in the course of the General Assembly. The General Secretary informed that the names will be incorporated in the alphabetical list of IAU members to appear in Transactions Vol. XVIIIB.

## 18. Changes in Commissions

At the request of the Chairman, the General Secretary informed the General Assembly that

(a) no commissions are discontinued,

(b) One new commission has been established (No. 51: "Search for Extraterrestrial Life", cf. Resolution A4), and

(c) the names of Commissions 26 and 34 have been changed (cf. Resolution A5)

(d) Election of Presidents and Vice-Presidents of Commissions for 1982-85

The General Secretary then read the proposals of the Executive Committee for Commission Presidents and Vice-Presidents for the period 1982-85:

| No. | Commission Name | President | Vice-President |
|---|---|---|---|
| 4. | Ephemerides | T. Lederle | B.L. Morando |
| 5. | Documentation and Astronomical Data | W.D. Heintz | G.A. Wilkins |
| 6. | Astronomical Telegrams | M. Candy | A. Mrkos |
| 7. | Celestial Mechanics | J. Kovalevsky | V.A. Brumberg |
| 8. | Positional Astronomy | J.A. Hughes | Y. Requieme |
| 9. | Astronomical Instruments | W.C. Livingston | C.M. Humphries |
| 10. | Solar Activity | E.A. Tandberg-Hanssen | M. Pick |
| 12. | Radiation and Structure of the Solar Atmosphere | R.W. Noyes | M. Kuperus |
| 14. | Atomic and Molecular Data | A.H. Gabriel | R.W. Nicholls |
| 15. | Physical Study of Comets, Minor Planets and Meteorites | C.R. Chapman | L. Kresak |
| 16. | Physical Study of Planets and Satellites | V.G. Tejfel | G.E. Hunt |
| 19. | Rotation of the Earth | Ya.S. Yatskiv | W.J. Klepczynski |
| 20. | Positions and Motions of Minor Planets, Comets, Satellites | E. Roemer | C.J. van Houten |
| 21. | The Light of the Night Sky | R.H. Giese | K. Mattila |
| 22. | Meteors and Interplanetary Dust | O.I. Belkovich | D.E. Brownlee |
| 24. | Photographic Astrometry | W. Gliese | A.R. Upgren |
| 25. | Stellar Photometry and Polarimetry | J. Tinbergen | F. Rufener |
| 26. | Double and Multiple Stars | M.G. Fracastoro | K.D. Rakos |
| 27. | Variable Stars | N.H. Baker | B. Szeidl |
| 28. | Galaxies | V.C. Rubin | P.C. van der Kruit |
| 29. | Stellar Spectra | J. Jugaku | G. Cayrel-de-Strobel |
| 30. | Stellar Radial Velocities | A.G. Davis Philip | J. Andersen |
| 31. | Time | G. Hemmleb | D.D. McCarthy |
| 33. | The Structure and Dynamics of the Galactic System | R. Wielen | W.B. Burton |
| 34. | Interstellar Matter | M. Peimbert | J. Lequeux |
| 35. | Stellar Constitution | A.N. Cox | D. Sugimoto |
| 36 | Theory of Stellar Atmosphere | B. Gustafsson | K. Kodaira |
| 37. | Star Clusters and Associations | K.C. Freeman | D.C. Heggie |
| 38. | Exchange of Astronomers | F.B. Wood | E.A. Müller |
| 40. | Radio Astronomy | K.I. Kellermann | J.E. Baldwin |

| No. | Commission Name | President | Vice-President |
|---|---|---|---|
| 41. | History of Astronomy | O. Pedersen | J.A. Eddy |
| 42. | Close Binary Stars | A.H. Batten | J. Smak |
| 44. | Astronomy from Space | M. Oda | Y. Kondo |
| 45. | Stellar Classification | V. Straizys | R.F. Garrison |
| 46. | Teaching of Astronomy | L.N. Houziaux | M. Rigutti<br>C. Iwaniszewska |
| 47. | Cosmology | J. Audouze | G. Setti |
| 48. | High Energy Astrophysics | R. Giacconi | C. Cesarsky |
| 49. | The Interplanetary Plasma and the Heliosphere | I.W. Roxburgh | S. Grzedzielski |
| 50. | Protection of Existing and Potential Observatory Sites | A.A. Hoag | S. van den Bergh |
| 51. | Search for Extraterrestrial Life | M.D. Papagiannis | N.S. Kardashev<br>F.D. Drake |

This proposal was received by acclamation by the General Assembly.

e) Organizing Committees of Commissions

The General Secretary informed that, in order to save time, it had been decided not to present the lists of members in Organizing Committees of Commissions, but that the lists were available at the IAU Secretariat for inspection. They will be printed in Transactions 18 B.

## 19. Place and Date of the 19th General Assembly

The Acting President called upon Prof. J.C. Bhattacharyya, official representative of the Indian delegation to the 18th General Assembly. Prof. Bhattacharyya spoke as follows:

"Mr. Chairman, members of the Executive Committee, colleagues, friends, Ladies and Gentlemen:

On behalf of the Indian National Science Academy, I have the honour to invite the International Astronomical Union to hold its next General Assembly in India. The Government of India has assured that all invited participants of the Assembly will be welcome to take part in the first ever large international gathering of astronomers in the country.

Some preliminary organisational work has already been done. The venue will be our capital city of New Delhi, where a conference complex with adequate facilities exists. All necessary arrangements for smooth functioning of the General Assembly will be possible here. We have made arrangements for the IAU to hold meetings between 19th and 28th of November 1985.

This is the first time that the IAU will hold its triennial General Assembly late; on this point, India as the venue should represent a very happy choice; November is the most pleasant time in India. Our autumn holds the promise of the mild Indian winter and is the traditional season for festivities. The participants will find the climate most congenial both for the meetings and for out-door activities. The romantic charm of India, a world in itself with a glorious past and a dynamic present, will greet every one of you.

India has nurtured many civilizations; it is one of the four areas in the world where man stepped out from the mists of antiquity to the light of civilization. Millenia ago, the land witnesses the blossoming of several fundamental ideas in mathematics and astronomy; in the present age of the rapid advancement of knowledge, her share in the common human effort is also significant. We request the International Astronomical Union to accept our cordial invitation and hold its next Assembly in the land, where ancient traditions of the quest for the truth are being blended with the efforts of a new generation of young scientists seeking clearer ideas about the Universe."

The General Assembly accepted this invitation with acclamation, and the Acting President expressed the gratitude of the Union to the Indian National Science Academy to meet in 1983 in India, a country with such long-standing astronomical traditions.

## 20. Election of the New Executive Committee

The General Assembly approved by acclamation the proposal of the Acting President that Professor R. Hanbury Brown be elected the new President of the Union, for the term 1982-1985.

The Acting President then moved that Profs. R. Kraft, M. Peimbert and Ya. S. Yatskiv be elected Vice-Presidents for the term 1982-85. This motion was also approved by acclamation.

The Acting President finally proposed that Dr R.M. West be elected General Secretary of the Union, and Dr J.P. Swings Assistant General Secretary of the Union, for the term 1982-85. This proposal was also approved by acclamation.

The Acting President then invited Prof. Hanbury Brown, Profs. Kraft, Peimbert and Yatskiv, and Dr Swings to join the Executive Committee on the platform.

## 21. Address by the Acting President

In the absence of the President of the IAU, the Acting President, Dr D.S. Heeschen, spoke as follows:

"Normally, the retiring President would at this point make some remarks. Although I have been delegated by the Executive Committee to preside at this meeting, there is obviously no way I can substitute for Professor Bappu in this respect, nor should I try. On the other hand, I am filled with admiration and respect for the work of my colleagues on the Executive Committee, and their efforts should not go unremarked. Thus, in spite of my difficulty about speaking in a role that is not rightfully mine, permit me to make a few very short comments.

The retiring Vice-Presidents, Profs. Kharadze and van den Bergh, and the retiring advisors, Profs. Blaauw and Müller, have labored hard and effectively on behalf of the Union for many years. The value of their wisdom and experience has been especially evident in the trying circumstances occasioned by the death of Prof. Bappu.

The General Secretary bears the brunt of administrative responsibility of the IAU, and even in normal times this is a formidable task. But in the last three years, Patrick Wayman has had more than the normal load. He has relocated and reorganized the Secretariat, he has had to cope, along with our Greek colleagues, with the difficult task of organizing this General Assembly in a 2-year time interval rather than the more normal 4 or 5 years, and finally he was burdened with still additional duties and problems resulting from the loss of our President. He has done all this, and the usual tasks of the General Secretary, with astonishing efficiency and good grace. We are all of us the fortunate beneficiaries of his tribute and hard work.

Patrick will be succeeded as General Secretary by Richard West, whose performance as Assistant General Secretary these past 3 years assures us that the affairs of the Union will remain in good hands.

I hope the General Assembly will join me in expressing its appreciation of the IAU Secretariat."

22. Address by the President 1982-85

Prof. R. Hanbury Brown addressed the General Assembly as follows:

"First let me say that I am honoured to be President of this Union. I welcome the appointment to Australia; for many people Australia is almost off the map, but that, I believe, is not true for astronomers.

My first reaction to being nominated as President was one of alarm at the thought that I should be responsible for the affairs of such a large and complicated organization, particularly as I have not served an apprenticeship as a Vice-President. However, the late Professor Bappu, being a considerate man, invited me to attend the meetings of the Executive Committee here in Greece so that I could see how the Union is run. My fears have been largely allayed; it is in capable hands. Patrick Wayman is a hard act to follow, but I am confident that Richard West can do it. I look forward to working with him and with our new Assistant General Secretary, Jean Pierre Swings. I had, sad to say, been looking forward to learning the job of President from Professor Bappu; instead I have had to learn it from the Vice-Presidents who have done a good job, for which we are all grateful.

I count it as a privilege to serve a Union which has contributed so much to the development of Astronomy. To meet in Greece is to be reminded of that long silver thread which runs through history, the growth of our understanding of the world around us. To read the programme of this meeting, with all its many ways of looking at the sky, is to be reminded that if our work is to lead to a coherent understanding of the world we must work together. I have seen enough of the development of Radio-Astronomy since its early days to realize how much its integration with Astronomy owes to the international Unions URSI and IAU. Anyone who attended, for example, the joint URSI/IAU Symposium on Radio-Astronomy in Paris in 1958 will know what I mean. That meeting was a real milestone in the subject. Much the same thing has happened and will, I hope, happen in the many other new branches of Astronomy. Astronomy is not only of basic importance to the progress of Science but, as we were reminded in the ancient Odeon, to the continued health of our whole culture. At a time when, in many parts of the world, the support for basic science is either diminishing or seriously in question, it is our responsibility not only to encourage support for astronomy, but also to ensure that what we already have is used to the best advantage; that necessarily implies that we must work together. You may be assured that as your new President I shall do everything I can to foster the ideal which inspired the formation of this Union, the advancement of Astronomy through international cooperation."

23. Closing Ceremonies

The Acting President then invited Prof. Sidney van den Bergh to propose a vote of thanks on behalf of all participants.

"It is my privilege to have a chance to thank all of those who have made this 18th General Assembly of the International Astronomical Union such a success. Three years ago, when the Union met in Canada, I had the opportunity to see at first hand how much work is really required to make a meeting such as this one a success. An IAU General Assembly is a bit like an iceberg - most of it is below the surface. Only those few things that do not work smoothly stick up above the water where they are noticed by all. Well, the temperature here in Patras certainly showed that not much of a cooling iceberg stuck up above the sea!

Je voudrai d'abord remercier l'Académicien Xanthakis pour l'assistance et l'hospitalité qu'il nous a donné de la part de l'Académie des Sciences de la Grèce.

Next we should thank Professor Goudas, both for his splendid hospitality here in Patras and for overseeing all of the local arrangements. To him we all say from the bottom of our hearts Efcharisto Poli - thank you very much! Prof. Goudas was, of course, assisted by the Members of the Local Organizing Committee, of whom I can only mention a few:

| | | |
|---|---|---|
| S. Kaplanis | - | transportation |
| C. Macris | - | registration |
| I. Georgopoulou | - | hotels |
| J. Papaioannou | - | social events |
| V. Goudas | - | tours |
| Dr. Roy and V. Markellos | - | published 'Astrocosmos' |
| G. Pande | - | ticket sales |
| S. Katsiaris | - | excursions |
| G. Karagorgopoulos | - | technical assistance |
| G. Katsiaris | - | finance, and finally |
| P. Pavlopoulou | - | typing |

And, of course, we should not forget the IAU secretaries, Brigitte Manning and Patricia Smiley, who often worked deep into the night. And we are also particularly thankful to Mrs. Bergevin who came from Montreal to share her experience and to tell us which mistakes not to make again!

I think we also owe a special debt of gratitude to our departing General Secretary, Patrick Wayman, whose task was made incomparably more difficult by the grievous illness of our past President, Vainu Bappu.

Finally, we thank all of the people of Greece who have made our stay here not only profitable but also very enjoyable. Thank you."

On behalf of the registered guests, Mrs. Heather Hanbury Brown expressed her thanks as follows:

"Mr. Chairman, Ladies and Gentlemen,

I speak for those of us who stand on the fringes of this conference - fringes which are perhaps the more visible for being less absorbed by meetings and discussions.

As guests, we come from singular places, accompanying singular persons, but we also are a coherent community with much in common. Furthermore, we have our own patterns of behaviour planned and prepared for us by Mrs. Goudas and Mrs. Katsiaris and their committees, which only interfere with the visibility of your fringes on the nicest possible occasions.

We have been introduced to Greek heroes and to each other. We have imbibed Greek Culture with Greek wine. Daily, we have been cheerfully encouraged to plunder Patras for anything from St. Andrew's blessing to the latest bargain. We have been driven skilfully and safely through your beautiful countryside to your sacred sites. Magnificent performances of Ancient Drama and Modern Music have kept us glued to our marble seats - it's all been a splendid way to savour the character of our host country, to form new friendships and rediscover old ones.

I know from experience how much agony lies behind an IAU Guest's Programme - how much imagination you need to reach decisions and how much persistence you need to carry them out. I also know that you, Mrs. Goudas, and the others, have been working under special difficulties for many months on our behalf.

We much appreciate it. It is good to be wanted and welcomed, and you have made us feel just that. As a token of our gratitude we have some small gifts for you and your ladies. Please accept them with our warmest thanks. EFCHARISTO POLI!"

The Acting President then called upon the retiring General Secretary, Professor Patrick A. Wayman.

"Mr. Chairman and Vice-Presidents, Members of the Union, Colleagues, Ladies and Gentlemen:

I wish to express my gratitude to all those who have supported me in my attempts to have the business of the XVIIIth General Assembly of the IAU carried out effectively, especially to you, Dr Heeschen and your fellow Vice-President, Professor Kharadze; also to those who are not with us at Patras who have furthered the affairs of the Union since the time of the Montreal General Assembly.

Foremost among those who are not with us is our former President, the late Professor Vainu Bappu. It was his enthusiasm during a visit of July 1980 that enabled a decision in favour of Patras as a venue to be made. From the beginning, it was clear to us that Patras could not offer what other meeting places had offered; but it provided other attractions that were paramount. It possessed no ancient centre of learning such as a long-established university or major research academy, but it did have a fully adequate modern university campus. It did not have a large congress hall that would suit the inaugural ceremony, but it possessed an auditorium unique in IAU experience, the Ancient Odeon of Patras, which has served the Union with supreme dignity and elegance, skilfully adapted to our special requirements. Then again, there was not in Patras a group of astronomers deeply associated with IAU affairs; there was, however, among several IAU members here, they key person we required. Professor Constantine Goudas, formerly known at the University of Manchester in England, and with many Mancunian connections would, we estimated, by his standing in Patras and in the University, be able to clear the path for the many helpers that he would need. These helpers came from other cities as well, and our senior colleague of the Academy of Athens, Academician John Xanthakis, an IAU member of some years standing, Chairman of the former Greek National Committee for Astronomy, bravely bore the load of leadership of the project in its formal responsibilities. From Canada came Mlle Madeleine Bergevin to bring her special expertise in an advisory and executive capacity with extraordinary intensity.

If the achievements of the General Assembly have fallen short of what was desired, and I can report that very few opinions in this direction have come to me, rather the reverse, it could be remarked that the bizarre sequence of events that we have witnessed, which culminated in the tragic death of Vainu Bappu on 19th August, the third day of our Assembly, made procedures difficult to sustain. In fact I believe this was not so; the shadow of our untimely loss, for which we could not have been prepared, served, I believe, to curb excesses in any controversial matters in which we might have been engaged. The formalities of our First Session on 17th August perhaps lacked spontaneity, but it was necessary to carry out business at that unaccustomed hou without undue delay and to avoid any ill-prepared debate at that stage.

In my experience as General Secretary, since 1979, I have been gratified in many IAU matters. The degree of co-operation of my colleagues who are Presidents of Commissions and who lead for a few years the scientific work of our Union in a particular field has resulted in their responding to my requests with unfailing patience and courtesy; I have been blessed with a vintage crop of Commission Presidents. Once I did have the comment: "I have not answered all your letters - they were too many!"

Also I have been honoured to enjoy unfailing loyalty and happy personal relationships with our IAU Secretary, Mme Brigitte Manning. To take two persons from different countries and put them in proximity to carry out work new to them both could have been perilous from the beginning. In fact we agreed well from the start. I recall that her very first reaction to the complexity of the IAU financial accounts was one of dismay, which I attempted to smooth away by saying that for the first few weeks we can assume that the banks do not make any mistakes! Since then she has paid special care ir re-designing the IAU financial records so that the General Secretary's responsibility can be more readily exercised.

Then I must give a special word of thanks to the two advisers to the Executi Committee who now relinquish their responsibilities. Adriaan Blaauw and Edith Müller were always available to me and I depended upon them countless times for consultation and advice, particularly since telephone conversations between Ireland and India were impractical. I was more than usually fortunate in my advisers and to them, as much as anyone, must I attribute my ability to carry through the affairs of the IAU since 1979.

I wish to say a few words about the responsibilities of IAU members. Our late President spoke of the cultivation of a sense of individual responsibility, and I believe that if the IAU has something special to offer it arises from a wide application of the principle that Presidents of Commissions are encouraged themselves, and they in turn encourage others, to devote special care in carrying out their programme of work for the IAU through their three years of office. If there is any criticism that I make of IAU members it is that some National Committees are not served with the same degree of personal attention as is shown in Commission activities. Many Adhering Organisations have responded well to the apparently endless stream of documents that are necessary to establish a General Assembly and I must specially thank the US National Committee for the IAU for exercising an extremely helpful attitude during the uncertain period that followed the withdrawal of the invitation for Sofia. Generally, I ask Chairmen of National Committees to ensure that IAU documents are properly dealt with in their home countries, even if this means extra work in penetrating the bureaucracy of their particular Academy or other adhering body. This will make the load of the IAU Secretariat, and hence the administrative cost to the Union, that much lighter.

This has been a General Assembly in which a most extraordinary sequence of events has taken place. It will remain always in my memory. I look forward to a change in my routines that will be accompanied by continued contact with my able successor, Richard West; to him and to Mrs. West I offer my very best wishes for what I believe will be a demanding and rewarding experience."

Finally, the Acting President invited the newly elected General Secretary, Dr R.M. West, to address the General Assembly.

"Ladies and Gentlemen, Friends and Colleagues,

A little more than four years ago, the Chairman of the IAU Special Nominating Committee, Prof. A. Blaauw, asked me whether I would consider to be nominated as Asssistant General Secretary. It is difficult for me to explain to you exactly how I felt - on the one side this was of course much too great an honour, but at the same time I had a thrilling feeling of a sudden, unexpected possibility to contribute to the furthering of our science in a way quite different from what I had known before.

I am now about to take upon me the even more complex task of General Secretary. Strangely enough, I today feel more confident that I did that time. In the intervening period I have had the privilege to participate in the affairs of the Union and the members of the Executive Committee have done their best to teach me and to show me, by their own illustrous example, how all kinds of matters can best be handled.

Above all, I am deeply indebted to our late President, Prof. Vainu Bappu, who is for me the outstanding example of the wonderful combination of a great scientist and a delightful person. It is with great sorrow in my heart that I must now carry on without his clear and friendly advice. No words can express the loss I feel.

Those of you who have seen my immediate predecessor, Prof. Patrick Wayman, in action during the past years and days, will have realized how ably he has carried the heavy responsibility of the daily affairs. Patrick, I admire you greatly and I shall do my best to carry on in your spirit. I shall always be thankful for your support.

Among the many tasks which confront us, I feel that two are of particular importance: to help the propagation of astronomy in countries where our science has not yet gained a firm foothold and to promote the all too scarce opportunities available for young astronomers, in the present world of ever-growing financial restrictions. I shall never forget the help I received from my home institute in Copenhagen and also in several other countries. I am convinced of the importance to help young people in every way possible, within the Union's limited resources. Similarly, I hope that it shall also be possible to intensify our support to countries where astronomy is still in an embryonic state. We must never forget our moral obligations to less favoured colleagues, even if we are ourselves in safe positions.

There are of course many other activities within the IAU, not easily summarized. A Danish philosopher of the 19th century, Søren Kirkekård, once said that the main dilemma of life is that "Experience is gained by looking backwards, but life must be lived forwards!" This maxim is indeed valid in a business as complex as the IAU. I am therefore particularly fortunate that my term will coincide with the presidency of Professor Hanbury Brown and I also appreciate having an old friend, Dr J.P. Swings, at my side as Assistant General Secretary. Furthermore, I sincerely hope that you, the members of the Union, will not hesitate to let me know, without delay, about all matters you feel that I ought to attend to. I am there to serve you, not just for my pleasure.

J'ai aussi eu la chance que mon prédécesseur, le Prof. Wayman, a déjà tout arrangé à Paris. Grâce à son travail, aidé par l'efficace collaboration de la secrétaire de l'UAI, Mme Manning, ainsi que Mlle Smiley, les affaires sont maintenant, trois ans après le début dans les nouveaux locaux, complètement en ordre. Je suis sure qeu peu de Secrétaires Généraux, ont trouvé au debut de leur mandat autant des facilités à leur disposition.

Finalement, j'aimerais, encore une fois, exprimer ma profonde reconnaissance pour la confiance que vous m'avez montré en m'élisant aujourd'hui. Restez assurés que je ferai mon mieux, et permettez-moi d'ajouter que, avec la compréhension et l'aide morale de ma famille et surtout celle de ma femme, Tamriko, je me sens beaucoup mieux preparé pour cet honorable devoir. Merci beaucoup!"

The Acting President then closed the session, thanking all participants for their attention and wishing them a happy return home.

# IV.

# Reports of Meetings of Commissions
# Comptes Rendus des Seances des Commissions

## COMMISSION 4: EPHEMERIDES (ÉPHÉMÉRIDES)

### Report of Meeting 1982 August 20

ACTING PRESIDENT: T Lederle

SECRETARIES: H Schwan, B D Yallop

### 1. Organisation and Membership

The Vice-President T Lederle reminded members that about a month ago he had informed members and consultants of the Commission that he had taken over as Acting President since the President, M A Sinzi was unable to carry out his duties. He then put forward proposals for the officers and membership of the Commission; these were agreed as follows:

New Officers of the Commission for the next three years:

President: T Lederle

Vice-President: B L Morando

Organizing Committee: V K Abalakin, S Aoki, R L Duncombe, J H Lieske, A Orte, P K Seidelmann, M A Sinzi, G A Wilkins, B D Yallop

New members of the Commission: F Chollet, Y Kubo, N-y Li

New Consultants: H-J Felber, K G Steinert

Resignations from the Commission: J Kovalevsky, H G Walter

### 2. Working Group on Nutation

The Chairman said that the future of the Working Group on Nutation should be considered since its recommendations had now been adopted.

J H Lieske drew attention to J Vondrák's view that some significant planetary terms had been omitted. P K Seidelmann, the Chairman of the Working Group, said that the uncertainties in the nutation are insignificant in comparison with the relativistic effects which should be included in the reduction. E M Standish suggested that it might be more useful to disband the Working Group on Nutation and create a new one on astronomical constants. R L Duncombe agreed that the Working Group on Nutation had served a useful purpose and proposed that it should be discontinued; this was adopted with a vote of thanks to the retiring Chairman, P K Seidelmann.

### 3. Bureau on Astronomical Ephemerides

B L Morando, the Director of the International Information Bureau on Astronomical Ephemerides (BIIEA), asked the meeting to consider the possibility of closing down the Bureau. The BIIEA is sponsored by COSPAR and supported by the IAU, but now the IAU is no longer supplying money, the Bureau has stopped supplying information and the users no longer enquire about the service. Summarizing the situation R L Duncombe said that the Bureau once met a real need but had now served its purpose; he proposed a vote of thanks to the retiring Chairman.

4. Transits of Mercury

A paper on the results of observations on the transits of Mercury made at Belgrade in 1970 and 1973 was read by V Protitch-Benishek. The observations were recorded photographically and did not use the method of timing by visual contacts as in earlier transits. The method yielded a much better value for the solar parallax than before.

5. Motion of Uranus

A paper on the deviations of astrolabe observations of Uranus taken in Chile since 1975 was read by F P Noël, who compared the observed right ascensions and declinations of Uranus directly with the Astronomical Ephemeris, the Astronomical Almanac and DE118. R L Duncombe stressed the importance of continuing to make astrolabe observations of planets because they are subject to different types of error from transit circle observations.

6. Radius of the Lunar Profile

P K Seidelmann proposed that a new value should be adopted for k, the ratio of the radius of the lunar profile to the Earth's radius (see Resolution 1), in calculations of lunar and solar eclipses; this was necessary to prevent different almanacs using different values. The term "lunar radius" in the original draft was modified to "radius of the lunar profile" and the resolution was then adopted without objection.

## Report of Joint Meeting of Commission 4, 7, 8, 19 and 31 on 1982 August 20

Chairman: T Lederle

HOMOGENIZATION OF COORDINATE SYSTEMS

1. Nutation

P K Seidelmann, Chairman of the Working Group on Nutation, discussed the developments after 1979 that led to the resolution on the 1980 IAU Theory of Nutation (see Resolution 2, and also Report for Commission 4, Section III and Appendix in Trans. IAU XVIIIA). He asked the meeting to endorse the adoption of the 1980 IAU Theory of Nutation which had been agreed by the majority of members of Commissions 4, 7, 8, 19, 24 and 31 in separate postal votes. Resolution 2 was adopted unanimously.

2. Report on the FK5

W Fricke gave a report on the FK5. The specification of the FK5 was fixed by the 1976 IAU resolution on astronomical constants. The determination of equator and equinox of the FK5 was completed in 1981. The systematic corrections to the FK5 will be finished by December 1983, and individual star corrections for the FK5 will be ready by December 1985. Fricke said that about 250 catalogues had been investigated involving 3 400 000 individual observations, and he gave a list of names of those who had made significant contributions to the determination of the FK5 equinox.

J Kovalevsky drew attention to the overlap between the MERIT campaign and the date of the introduction of the FK5 system and asked on what system will the astronomical observations be reduced during the campaign. G A Wilkins replied that the procedure for the computation of star positions during the campaign will be decided by the MERIT Working Group on Standards. He hoped that the report will be distributed before the end of the year and final decisions will be taken at the second MERIT Workshop in May 1983.

P D Hemenway asked Fricke about the completeness of the final selection of faint FK5 stars. Fricke replied that there are about 50 000 possible candidates with sufficient observational history and a good distribution in the sky. At most 5 000 of these stars will be included in the FK5.

3. Terrestrial Reference Frame

I I Mueller read a paper by S-y Zhu and himself on the effects of adopting new definitive precession, nutation and equinox corrections on the terrestrial reference frame. It was shown that the effect on polar motion is a diurnal periodic term with an amplitude increasing linearly with time, whilst the effect on UT1 is a linear term. General principles were given to determine the effects of small rotations (such as precession, nutation or equinox corrections) of the frame of a Conventional Inertial Reference System (CIS) on the frame of the Conventional Terrestrial Reference System (CTS) in both stellar and non-stellar observations. Seven options to change the CTS to accommodate these small rotations (corrections) were produced, one of which will have to be chosen. The last option, which requires no change in the origin of terrestrial longitudes or in UT1, is advocated. This option would be maintained by eventually referencing Greenwich Mean Sidereal Time to a fixed point on the equator, instead of the mean equinox of date, as is the current practice. Accommodating possible future changes in the astronomical nutation was also considered. The effects of differences which may exist between the various CTS's and CIS's (inherent in the various observational techniques) on Earth rotation parameters (ERP) and how these differences may be determined were mentioned. CTS differences are determined from observations made at the same site, whilst CIS differences are determined by comparing the ERP's obtained from the different techniques during the same period. The paper will be published in Bulletin Géodésique, 1982.

J H Lieske disagreed with Mueller on the size of the effect on longitude for a non-stellar type of observation in his first option which was being investigated at JPL. The large $E_0$ term will not appear whilst the precession and equinox drift effects will largely cancel each other out. It is expected that the total effect from VLBI and radio tracking observations between the current system and the new system will be less than one metre. Mueller replied that geodetic applications require better than one metre, and suggested that JPL is solving for the rotation of the frame and so is absorbing some of the effect.

4. Relationship between GMST and UT1

P K Seidelmann introduced a resolution on the expression for GMST at $0^h$ UT1 (see Resolution 3) by drawing attention to the small changes from the resolution adopted in 1979. S Aoki suggested that the definition should give the relationship between Greenwich sidereal date and any instant of UT1, not merely $0^h$ UT1. Seidelmann said that this could be regarded as a direct extension of the original fundamental relation. J H Lieske opposed the resolution on the grounds that the $T^2$ and $T^3$ terms served no physical purpose. Furthermore if UT1 is deduced from a transit observation its value will depend upon whether the reduction is made in the FK4 or the FK5 system. Seidelmann replied that the resolution adopted in 1979 included numbers which were based on preliminary values of E and $\dot{E}$, therefore the equation had to be changed. The correction for the quadratic and higher terms were introduced to correct the effect of Newcomb's error of 1890 and so the equation has a sound basis for future investigations. S-h Ye read a paper by B-x Xu, J-p Ren and Z-s Sun on a new definition of Universal Time where UT1 is directly proportional to the sidereal time, which is a more natural and a much easier concept. T Lederle commented that others have thought on similar lines. B Guinot said he had not changed his mind about the use of the non-rotating origin, but it is not possible to introduce a completely new concept in so short a time; we must adopt the proposed GMST/UT1 relation now.

Seidelmann agreed to Mueller's suggestion that the words "with respect to the stellar reference frame" be added to the resolution in the first sentence after "there is no change in either value or rate of UT1", and pointed out that Mueller's comments indicated that there was still room for study in these areas.

The resolution was adopted with some abstentions.

5. Sign Convention for Terrestrial Longitude

G A Wilkins proposed a resolution that the sign convention east longitude is positive should be used for terrestrial longitude (see Resolution 4). He drew attention to the many circumstances in which this convention is already in use and to the difficulties caused by the use of the opposite convention in some astronomical almanacs and publications. S V Débarbat asked if it was also necessary to adopt a sign convention for azimuth. Wilkins replied that this should be considered as a separate issue and that the resolution should not be extended. T van Flandern pointed out the inconsistancy that will still exist between planetographic and planetocentric longitudes on the Earth. The resolution was adopted by an overwhelming majority.

## Report of Joint Meeting of Commissions 4, 19 and 31 on 1982 August 20

Chairman: T Lederle

ASTRONOMICAL CONSTANTS AND EPHEMERIDES

1. New Ephemerides

E M Standish described the work done at JPL in collaboration with members of USNO in producing the DE200/LE200 ephemerides. They will form the basis of the *Astronomical Almanac* from 1984 onwards. The observations used to construct the ephemerides were lunar laser ranging to 30 cm precision, Earth-Mars ranging (10m), and planetary radar (1-2 km), together with optical transits and astrometry (350 km, or 100 km at 1 au) to tie the ephemeris to the FK4 equinox. JPL has produced software for distribution; one routine will provide the ephemeris of a body or bodies and the other will provide a complete list of the constants used. The fundamental ephemeris was produced by fitting a simultaneous integration of all bodies, including the Moon to the observations; for nutation the 1980 IAU Theory of Nutation was used. The origin of the DE200/LE200 reference frame is the mean equator and dynamical equinox of J2000.0 which is almost identical to the origin of the FK5. The accuracy was 0.″05 or better for the inner planets and 0.″2 to 1.″0 for the outer planets.

P K Seidelmann discussed the work done at USNO on the new ephemerides and showed comparisons between transit observations at USNO and DE200/LE200. There were large waves in declination for Saturn and the fit for the outer planets ran off. It was found impossible to fit Uranus observations to the ephemeris before 1900 without the current ephemeris going off. It was also found necessary to modify some of the 1976 IAU astronomical constants in order to obtain the best fit.

B L Morando described the work done at the Bureau des Longitudes on new theories for the Moon and planets. New analytical theories had been developed and had been fitted to DE200/LE200. They have used the IAU (1976) System of Astronomical Constants. Tapes are available containing the analytical series. The basis of the method will be published in Astron. & Astrophys. The Connaissance des Temps will use the ephemerides based on the new theories and will give the elements of the orbits for all the bodies.

## 2. Discussion on Resolutions

P K Seidelmann proposed a resolution that considering that it is planned to introduce the IAU (1976) System of Astronomical Constants, the 1980 IAU Theory of Nutation, the equinox of the FK5 and Dynamical Time, in 1984 January 1, it is recommended that the ephemerides published in the international and national ephemerides be based on the above resolutions and the available observational data to the extent possible to provide the best prediction ephemerides possible.

B L Morando and others raised objections to the resolution, particularly on the grounds that its meaning was not sufficiently precise, and so Seidelmann withdrew it.

Seidelmann then put forward a second resolution which recommended that some of the values of the IAU (1976) System of Astronomical Constants should be revised to form the IAU (1982) System of Astronomical Constants. He pointed out that it had been necessary to use the new values in order to obtain an optimum fit of the new fundamental ephemerides DE200/LE200 to new high precision observational data. Morando considered that the IAU system should not be changed until a working group had considered in detail the evidence and the arguments for change; the 1976 values had been used in the new ephemerides for the Connaissance des Temps. Aoki pointed out that the 1976 values would also be used for the Japanese Ephemeris. Lederle considered that such a resolution should be submitted for discussion in advance of the Assembly. H F Fliegel drew attention to the analogy with the change in the adopted nutation theory; the Commissions had refused to adopt a new theory in 1979 but had later found it desirable to do so. Duncombe supported the view of Lederle and proposed that the discussion of the resolution should be suspended. This was agreed by a vote of 12 in favour, with 2 against.

Seidelmann then proposed a third resolution which was intended to confirm the general practice that the dynamical equinox be defined as the intersection of the mean celestial equator with the smoothly moving ecliptic plane at which the Sun's declination is changing from South to North. He pointed out that it is also possible to determine the position of the equinox by treating the ecliptic as a fixed plane and that leads to a difference of $0''.1$ in position. After some discussion it was agreed not to vote on the resolution, even though there were no objections to the wording, since it was not a new definition or practice.

## RESOLUTIONS

### Resolution 1 of Commission 4 on the value of the ratio of the radius of the lunar profile to the Earth's radius.

Commission 4 considering

that it is planned to introduce the IAU (1976) System of Astronomical Constants, the 1980 IAU Theory of Nutation, and the equinox of the FK5 on 1984 January 1,

recommends

that in the calculation of lunar and solar eclipses, there be a single value for k, the ratio of the radius of the lunar profile to the Earth's radius, and that it correspond to the mean radius of Watts' datum as determined by observation of occultations and to the adopted radius of the Earth. The value is k = 0.2725076.

Resolution 2 of Commissions 4, 7, 8, 19 and 31 on the 1980 IAU Theory of Nutation.

Commissions 4, 7, 8, 19 and 31 recognizing

that since the Seventeenth General Assembly of the IAU, Commissions 4, 7, 8, 19 and 31 have adopted the 1980 IAU Theory of Nutation in place of the 1979 IAU Theory of Nutation,

endorse

the adoption of the 1980 IAU Theory of Nutation.

Resolution 3 of Commissions 4, 19 and 31 on the expression for GMST at $0^h$ UT1.

This resolution is an amendment to the provisional part of Resolution 3 of Commissions 4, 19 and 31 made at the Seventeenth General Assembly of the IAU.

Commissions 4, 19 and 31 considering

that it is planned to introduce the IAU (1976) System of Astronomical Constants, the 1980 IAU Theory of Nutation, and the equinox of the FK5 on 1984 January 1,

recommend

that (a) the relationship between mean sidereal time and UT1 be modified so that there is no change in either value or rate of UT1 with respect to the stellar reference system, due to a correction to the zero point of right ascensions of the FK4 and to a correction for the motion of the zero point to be introduced in FK5,

(b) the new expression for Greenwich mean sidereal time of $0^h$ UT1 be

$$\text{GMST of } 0^h \text{ UT1} = 6^h\ 41^m\ 50\overset{s}{.}548\ 41 + 8\ 640\ 184\overset{s}{.}812\ 866\ T_u + 0\overset{s}{.}093\ 104\ T_u^2 - 6\overset{s}{.}2 \times 10^{-6}\ T_u^3$$

where $T_u$ is the number of Julian centuries of 36525 days of Universal Time elapsed since 2000 January 1, $12^h$ UT1 (JD 245 1545.0).

Resolution 4 of Commissions 4, 19 and 31 on the sign convention for terrestrial longitude.

Commissions 4, 19 and 31 noting

that the International Meridian Conference held in Washington in October 1884 adopted a resolution that from the meridian of the Observatory of Greenwich "longitude shall be counted in two directions up to 180 degrees, east longitude being plus and west longitude minus",

noting

that there is an increasing interaction between astronomy and geodesy in which field the sign convention "east longitude is positive" is in common use,

recommend

that as soon as practicable all national ephemerides and other astronomical publications adopt the convention that terrestrial longitude be measured positively to the east.

# COMMISSION 5: DOCUMENTATION AND ASTRONOMICAL DATA
# DOCUMENTATION ET DONNEES ASTRONOMIQUES

## Report of Meetings, 18, 21, and 23 August 1982

### I. WORKING GROUP ON ASTRONOMICAL DATA

CHAIRMAN: B. Hauck.

The first part of the meeting was devoted to business. The Chairman reported on the activity of the Working Group during the period 1979-1982. The main activity was concentrated on trying to centralize information concerning astronomical plate vaults (see IAU Inf.Bull.48, June 1982). The Working Group also prepared a resolution concerning FITS. The Chairman attended the assembly of CODATA at Kyoto in 1980 as IAU Delegate.

Drs. Mead and Terashita were elected as new members of the Organizing Committee, the membership of which is otherwise unchanged. Chairman and Vice-Chairman will remain in office. It was decided to modify the structure of the W.G. by removing the notion of subgroup, and by no longer considering "Nomenclature" to be relevant to this Working Group. G.Wilkins then presented the finalized "Guide to the Presentation of Astronomical Data" (CODATA Bulletin 46, 1982). This Guide contains general recommendations to editors, referees, and authors on the reporting of numerical data obtained from astronomical observations. Copies may be ordered from the CODATA Secretariat, 51 Bd de Montmorency, 75016 Paris. B.Hauck reported on the plate vaults and was able to announce that two observatories had already sent the copies of their files to the Stellar Data Centre at Strasbourg.

The scientific part of the meeting was devoted to "Transfer of Astronomical Data". P.Grosbol reported on the Flexible Image Transport System (FITS), the purpose of which is to simplify the problems involved in the transfer of arrays of "image" data between different observatories, and different computer systems in particular. Data networks are also important for the transfer of data. Presentations on such networks in the United Kingdom, Italy, and France were prepared, respectively, by K.Tritton (read by G.Wilkins), G.Sedmak (read by S.Catalano), and F.Ochsenbein.

The composition of the Working Group as of 1982/83 is:
Chairman: B.Hauck Vice-Chairman: G.Westerhout
Organizing Committee: W.P.Bidelman, O.G.Dluzhnevskaya, R.Duncombe, W.Heintz, C.Jaschek, T.Lederle, J.Mead, Y.Terashita, G.Wilkins.
Members: J.F.Dolan, R.S.Dixon, P.Grosbol, A.Heck, P.Lantos, G.Lynga, J.Nishimura, F.Ochsenbein, L.E.Pasinetti, A.G.D.Philip, J.Shakeshaft, F.Spite, K.Tritton, P.T.Wallace, W.Warren.

### II. WORKING GROUP ON EDITORIAL POLICIES

CHAIRMAN: S. Mitton.

The Chairman presented the draft of the Style Manual for the IAU Astronomers' Handbook, and it was discussed by the participants of the meeting. Some amendments were suggested, and the proposed changes were agreed on after discussion.

The document will be circulated, prior to its final publication, to editors of astronomical periodicals, in order to invite their comments also. The Chairman agreed to present a copy to the IAU EC immediately for comments. The Working Group believes that its continuing existence is not necessary once it has discharged the duty of revising the Handbook.

## III. WORKING GROUP ON CLASSIFICATION SYSTEMS AND INFORMATION RETRIEVAL

CHAIRMAN: P. Lantos.

While this W.G. presented its report at the Full Commission Meeting, a special meeting was convened to discuss controlled vocabularies, their compatibility and desirable extent, and the feasibility of allowing composite, derivative, or free key words. L.Schmadel and P.Lantos will be in contact in order to examine currently used or proposed vocabularies in these respects. (From notes by W.Heintz.)

## IV. SPECIAL WORKING GROUP ON DESIGNATIONS

CHAIRMAN: C. Jaschek. SECRETARY: W. Heintz.

W.P.Bidelman has chaired the W.G. during the past term, and has completed his work with a comprehensive report; he was unable to attend the Patras conference. The business now at hand of the W.G. will be to implement the projects approved in Montreal in 1979, and to reach more impact of its recommendations, particularly with the help of scientific editors of journals. In pursuance of previous efforts, object designations should reach a degree of uniformity that permits reliable retrieval of material from all modes of data storage.

A related matter are the catalog codes, on which F.Spite reported. The draft of the standard list (Fernandez, Lortet, and Spite) was presented. (Note: A revised version was issued after the conference, in September 1982.) Completion of the work and its publication (about 120 pp.) are expected shortly. The W.G. calls attention of regional organisations to this catalog, and invites its publicizing in local newsletters.

Designation practices in various object categories will be the subject of a booklet to be published by J.Mead; in her absence this was reported on by Parsons. In view of new methods of discovery, Pecker advised that first publications should reference earlier catalogs relating to the objects in question. The Index of Double Stars (IDS) was cited as an example of a desirable compilation format. The recommendations on designations from past IAU Assemblies - a substantial though not always consistent list that has accrued - should be screened in order to arrive at standards which the periodicals will accept. Commission 37 reported its agreement to use the equator 1950 for cluster center coordinates.

Grosbol discussed the presentation of catalogs, especially in view of their exchange by magnetic tape. He advised the creation of an ad-hoc group (with Harten, Ochsenbein, Tritton, and Warren) which will study the matter, and expects to formulate appropriate recommendations.

Jaschek proposed to Commission 5 the following actions: That the mandate of the Working Group on Designations be extended for another period; that the W.G. contact journals in order to emphasize conformity with the recommendations approved at Montreal 1979; and that further input be solicited from IAU Commission presidents concerning unified practices of designation.

The 1980 Report of the W.G. as written by Bidelman, and published in Bull.CDS 18, 41 will go to Commission 5 and to the IAU Executive Committee for adoption. Further work on designations is expected to be published in the Bulletins of CDS (Strasbourg, France) and of NSSDC at Goddard Center (Greenbelt, USA).

## V. FULL COMMISSION MEETING (23 August 1982)

PRESIDENT: W. D. Heintz. SECRETARY: R. J. Davis.

Attendance: 23. Agenda: (1) Reports, (2) Business, (3) Resolutions and recommendations, (4) Papers, (5) Announcements.

(1) The Chairman recalled the recent history of expanding coordinative responsibilities of the Commission, which includes the secondary processing of scientific material from the time it leaves the authors' desks, to its retrieval by librarians, abstracting and data services, and their clientele. The restructuring of the Commission and its Working Groups, accompanied by a doubling of membership, has taken place largely under the leadership of Past President J.-C.Pecker. A vote of thanks to him was applauded.

Some concern had arisen that the activities might be too broad and diversified, and better served by two commissions. However, the interfacing nature of coordinating activities (related to computers, standardisation, retrieval procedures, and the corresponding contacts with other Commissions) had already in Montreal supported the general consensus that this administrative unity should remain intact.

One successful result of this integration was the IAU Colloquium 64 (in Strasbourg, July 1981). C.Jaschek is credited with initiating and organizing the conference (together with his colleagues), and also with efficient editing work so that the volume of the Proceedings appeared already in May 1982.

The Chairman then reported on the ICSU Abstracting Board, which is in a process of reorganisation toward its expected goals in the 1980's. Relevant recommendations will be formulated (with IAU participation) at the Executive Committee meeting in December 1982. It is recommended to the Commission and to the IAU EC that the membership in AB be maintained at least until its new structure has emerged. The volume of abstracting work, the technology employed, and the sales revenues required to maintain good abstracting services, all of this renders AB a meeting of experienced specialists, working in cooperation with - and advised by - ICSU Unions but not under their exclusive control. The 60-year-old Commission 5 of IAU always has been a forum of interfacing on bibliographical matters; most other Unions do not have a similar body. One concern is the growing habit to treat astronomy and astrophysics as part of physics. This results in a bibliographic processing which simply does not meet the needs of the astronomical research community.

Some amusing stir has been provided by private "agencies" trying to sell star names for profit. Within ICSU, and thus under UNESCO auspices, the IAU and its authorized committees are in charge of nomenclature of celestial objects. Although not subject to copyright, the names have to comply with documentation standards to serve as useful identifiers (the Spite et al master list of codes, for example). Having exposed the "star registries" as rip-offs without scientific backing, any further pressing-the-point by IAU would be a waste of time.

G.Wilkins, Vice-President, spoke about the documents on standards for the publication of data. He suggested a recommendation advocating compliance of astronomers with the "Guide" (CODATA Bull. 46), after its pending endorsement by the IAU Executive Committee.

Following a suggestion by Jaschek at the Strasbourg meeting, a small Working Group on a glossary of terms was set up. One portion, that on Computing, has so far been completed. Wilkins suggests that the work of this group continue.

The present document on subject classification for astronomy was issued in 1975 and is now in need of revision. A group working on the revision is chaired by Wilkins. Unfortunately for the input, this document is not widely known to astronomers. It can be bought from the British Standards Institute, 2 Park St., London W1A 2B5, and has the catalog number BS1000(52):1977. Of interest to astronomers is also the FID publication No. 483.

B.Hauck, IAU delegate to CODATA, referred to the reports in CDS Bull.23 (1982) and in CODATA Bull.36, chap. 3 of which is devoted to astronomy. Astronomers may benefit greatly by keeping track of CODATA activity. Jaschek suggested that CODATA sponsor a school for young astronomers as an introduction to data techniques.

P.Lantos, for the W.G. on Classification Systems, had a progress report on the IAU vocabulary and the key-word list. The vocabulary, using a number of standard sources, was originally created in English; a French version has now been evaluated by a number of French scientists. About 1000 entries are in the key-word list presently. Pecker observed the importance of uniform documentation, and of the responsibility of the authors in selecting key words; he suggested a deadline for completion of the vocabulary project. Heintz felt that the key-words should be agreed on and finalized first, and that IAU be asked to maintain the offer of publication expenses for the vocabulary.

S. Mitton summarized the revision of the Style Manual for the Astronomers Handbook, which is undergoing review now. Publication in IAU Trans. XVIII B is expected. Copies will be sent to the editors of the main journals. It is to be hoped that the IAU Manual will receive more attention.

(2) The business section of the agenda was opened by J.-C.Pecker in the chair. He recalled the Rules of the Commission, and its status as Committee of the IAU Executive Committee. For that reason the officers normally serve two terms. Pecker proposed the re-election of Heintz as President, Wilkins as Vice-President, and Schmadel as continuing member of the Organising Committee. This was approved by the vote of the Commission. Resuming the chair, Heintz led a discussion on the Working-Group structure of the Commission.

Hauck and Lantos were confirmed as continuing Chairmen of their respective Working Groups, and as ex officio members of the Organising Committee. Mitton's group was discontinued, following completion of work on the new Manual, and the Commission appreciated Mitton's effort with applause.

The Special Working Group (Task Force) on Designations had been chaired by W. Bidelman, who indicated by letter to the Commission that he thinks that the bulk of the work has been finished. However, from the tasks assigned to that Group some incomplete segments of work ensue, viz.: To finalize the reports in discussion with other IAU Commissions (which may require some more time on part of the Commissions), in order to unify the practices of object designations, and to get the results to the status of IAU recommendations, published, and put into practice. It would be pointless to make resolutions unless there were sufficient inclination to enforce them. The Commission resolved: (a) To adopt the Bidelman Report as printed in CDS Bull.18 and to express great appreciation to Bidelman for his work; (b) To continue and to promote the group to a full Working Group (Title: Designations); To install Jaschek as Chairman of the W.G., with H.Dickel, M.Lortet, J.Mead, J.-C.Mermilliod, and F. Spite as other nominated members.

There were requests for the appointment of two new Task Forces.

The Special Group on Standardisation of Exchange of Astronomical Data, to be chaired by P.Grosbol, was approved as proposed in the preceding meeting on designations. Following a discussion at the Abstracting Board meeting in Pine Mountain in 1981, a Special Group to explore guidelines for Abstracts in Astronomy was suggested. This was approved, and the Cairman asked L.Schmadel to temporarily chair this Group and to prepare a slate of proposed members.

The Working-Group structure for 1982-85 is thus: Astronomical Data (Hauck); Classification Systems (Lantos); Designations (Jaschek); Standardisation of Data Exchange (Special Group; Grosbol); Guidelines for Abstracts (Sp.Group; Schmadel).

For two vacancies on the Organising Committee, three nominations were at hand. C.Jaschek having already become a new ex officio member of the O.C., the Commission voted O.B.Dluzhnevskaya and C.E.Worley into office.

The Chairman paid tribute to three deceased members: J.O.Fleckenstein, J.Wempe and H.Hirose. Two members and several consulting members have resigned.

Six new members were admitted: S.Nishimura, K.P.Tritton, P.T.Wallace, P.Wayman, I.Heinrich, F.Ochsenbein (the latter two pending approval of proposed IAU membership). Reconfirmed consultant members: Bérardini, Berthélot, Collins, Dudley, Grassi, Guidoni, Laforge, Laloe, Martin, Schiminovich; new consultants: Corbin and de Narbonne. The Commission file now lists 59 members and 12 consultants.

(3) The Commission acted affirmatively on two Resolutions (see Report of General Assembly, 26 August) and endorsed several papers to be forwarded to the IAU General Secretary, including Wilkins's Guide and Mitton's Manual. The Chairman reported that no motion concerning the Rules of the Commission had been received.

(4) B.Corbin (U.S.Naval Observatory Librarian, and member of the Special Libraries' Association) reported on a compilation list of astronomical serials from their beginning and of astronomical books before 1800, including data on where available. In the Amer.Astr.Society a section has been established for the preservation of documents and instruments; microfilming the old papers is intended. In order to avoid duplication, information is requested if work of this kind is already obtainable elsewhere. Librarians deem the Astr.Astrophys.Abstracts a superb and important publication, and they would much appreciate to have it also in machine-readable form. Schmadel responded that the matter is under negotiation. Wilkins mentioned the archive preservation program of the Royal Greenwich Observatory. He and Heintz recommended to reactivate the initiative and input by librarians in the Commission and the IAU, encouraging Dudley (Greenwich) and Corbin to further contacts.

G.Grassi Conti (Rome Observatory Library) had a report (read by M.Smyth) on the catalog of 17th-century European books; the list will perhaps be merged with the 15-16th-century catalog currently being updated. Since response to the request for information was not quite complete, support by the Commission was desired. Libraries with pre-1700 holdings are urged to aid in the extensive efforts on an important compilation by contacting Grassi.

(5) The new edition of the Multilingual Dictionary by J.Kleczek is finished, except for some final checking. Since it was found not feasible to include Chinese and Japanese presently, and the work should go to press shortly, these languages should be incorporated in a later supplement. - Other announcements concerned new databases and reference papers.

The Commission voted to co-sponsor a symposium contemplated for 1984 by the President of Commission 42, with support by Comm.29, under the title "Calibration in Astronomy". In view of the (intentionally) broad scope of the subject, the participation by Commission 5 was invited. - The Commission meeting then adjourned to 1985.

# COMMISSION 6: ASTRONOMICAL TELEGRAMS (TELEGRAMMES ASTRONOMIQUES)

## Report of Meeting, 20 August 1982

PRESIDENT: J. Hers SECRETARY: B.G. Marsden

After having been welcomed by the President, members and guests were asked to stand in memory of Dr. E. Buchar, a former President of the Commission, who died in 1979. This was followed by the confirmation of nominations for officers of the Commission for the period 1982-1985: President, M.P. Candy; Vice President, A. Mrkos. No changes were proposed in the general membership of the Commission.

The President reported that, in view of constantly rising costs, it had again been necessary to approach the Executive Committee regarding an increase in the IAU subvention for the Central Bureau for Astronomical Telegrams. This resulted in an increased amount of S Fr. 10000 for the period 1982-1985 being proposed in the Draft Budget, which was subsequently approved by the Finance Committee. While this represents less than 5 percent of the actual running costs of the Bureau, it nevertheless remains tangible proof of the high regard in which the Bureau is held by the Executive Committee, and a letter to this effect received from the General Secretary was noted with appreciation.

Dr. Marsden, in his capacity as Director of the Central Bureau for Astronomical Telegrams, briefly discussed the report on the Bureau's activities during the period 1979-1982, as published in Transactions XVIIIA. He noted that all Commission members present at the Meeting were, without exception, also members of Commission 20, which might give the impression that Commission 6 was mainly a subsection of Commission 20. This was by no means the case, and to stress this fact a special effort had been made some years previously to include representatives from other Commissions. However, these had taken little or no part in Commission 6 activities. The reason might be that rapid dissemination of astronomical data still remained, as in the past, mainly of interest to observers of minor planets, comets and related objects, while it was evident that workers in other fields were entirely satisfied with the operation of the Bureau.

Dr. Marsden explained that the IAU Circulars were currently being produced by him with the aid of two IBM electric typewriters, one in the office and one at home, which made it possible for urgent Circulars to be sent out at very short notice. A more modern approach might make use of a microcomputer plus word processor, enabling the work to be done more quickly, and maintaining a uniformly high standard. This would have the added advantage that the material would be in machine readable form, and might thus be made directly available to users in that form. Postal services had deteriorated tremendously during the preceding few years (a brief enquiry among members present showed that the Circulars seldom took less than 5 days, and not infrequently took more than 20 days to reach their destination) and any method to speed up the process deserved serious consideration.

Questions raised in the ensuing discussion included the probable frequency with which users might dial for information (daily, say, or weekly?) and how this service was to be paid for. It was pointed out that a large part of the income of the Bureau was derived from subscriptions to the Circulars, and this was an income which the Bureau could not afford to lose. On the other hand, users outside the U.S.A. might find it difficult and expensive, perhaps even impossible, to make use of such a service, and the Circulars would doubtless have to continue to exist in their present form for many years to come. Nevertheless, it was agreed by the meeting that the modern way would be electronic rather than by means of pieces of paper, and it was agreed that investigations should continue.

This was followed by a report by Dr. Marsden on a meeting held on August 18 between representatives of Commission 6 (Hers and Marsden), Commission 20, and the Working Group on Planetary System Nomenclature (WGPSN). This meeting had been called, and Commission 6 had become involved in it, because it had become evident that there was a tendency to allocate designations and names to objects in the Solar System without due regard to existing well-established practices. These practices, based on long experience, and designed to reduce the possibility of errors and misidentifications to a minimum, had never received the formal endorsement of the IAU, and the Bureau was experiencing increasing difficulty in having to deal with incorrectly designated objects. The meeting had therefore drawn up a table, agreed to by all concerned, which allocated responsibility for all the various steps, from initial temporary designation to the final naming of planets, satellites, rings and comets. According to this, Commission 6, through the Central Bureau for Astronomical Telegrams, remains responsible for the temporary designation of principal planets, satellites, rings and comets (but not minor planets), and for the naming of comets.

The Meeting ended by approving the proposed allocation of responsibility, which was to be submitted to the Executive Committee for formal IAU approval.

## COMMISSION 7 : CELESTIAL MECHANICS(MÉCANIQUE CÉLESTE)

Report of Meetings on August 19, 1982

PRESIDENT: Y. Kozai SECRETARY: H. Kinoshita

### Business Session

OFFICERS OF THE COMMISSION

Recommendation for the following list was moved and unanimously approved:
President: J. Kovalevsky, the retiring Vice-President,
Vice-President: V.A. Brumberg
Organizing Committee: E.P. Aksenov, J. Chapront, S. Ferraz-Mello, J. Hadjidemetriou, J. Henrard, W. Jefferys, Y. Kozai, H. Scholl, P.K. Seidelmann, A. Sinclair

MEMBERSHIP OF THE COMMISSION

The president reported that when China became a member country of the IAU the following three Chinese astronomers were approved as new members of the Commission: Tong Fu, Yi Zhao Hua and Zhao Xian Zi.

The following members of the IAU were proposed as new members of the Commission 7 and approved:
R. Meire and M.B.Moons(Belgium), W. Sessin(Brasil), B.B. Baghos, Y.E. Helali and M.Y. Tawadrous(Egypt), Z. Sima and M. Šidlichovský(Czechoslovakia), D. Benest and F. Mignard(France), A.G. Mavraganis and Ch. Varvoglis(Greece), B. Érdi(Hungary), K.B Bhatnagar(India), A. Milani(Italy), J. de Orus (Spain), D.B. Taylor(UK), and P.E. Elyasberg, N.V. Emelianov, V.A. Izvekov, B.N. Noskov, B.S. Novselov and S.K. Tatevian(USSR).

As S. Böhme left the commission the total number of the members is now 148.

CONSULTANTS

The following 6 persons were approved as consultants of the commission for the coming three years and approved:
V.A. Arnold, R.R. Allan, V.R. Bond, C.J. Brookes, J.K. Moser and J. Boels.

### Scientific Session

The five review papers were presented between 10:00 and 13:00 with much discussions at the scientific session.

1) Planetary Theories( J. Kovalevsky)

In the present days, no group has endeavoured to cope simultaneously with the three tasks that have to be performed in order to construct a planetary theory: collect and reduce observations, obtain a theoretical representation of a set of solutions in the vicinity of the actual motions and compute the best fitting orbits. Two approaches have permitted to skip at least one of these tasks. Numerical integrations - the most recent being D.E. 200 by E.M. Standish(JPL) - skip the second task. In the approach by Bureau des Longitudes(P. Bretagnon, J. Chapront and J.L. Simon), an analytical theory has been derived for the 8 major planets, and the fit is made with D.E. 200 that is used as a kind of a collection of normal points.

Two methods have been applied. In the iterative method, the solution of order $n$-1 is substituted in the right hand members of the equations and after integration and fitting of the pre-established constants of integrations, one obtains the $n$-th approximation. In the ordered solution, equations are developed in powers of the variations of the elements, so that there is no subsitution, but it is necessary to compute a great number of partial derivatives. The final solution in Bureau des Longitudes was obtained in Poisson series of time to the third order of masses for the 4 inner planets by the ordered method and to the sixth order of masses for the outer planets by a combination of both methods - checked by the iterative method. After fitting to D.E. 200, maximum errors for 100 years are ± 0."002 for Mercury, ± 0."005 for Venus and Earth, ± 0."02 for Mars and Uranus and ± 0."01 for Jupiter, Saturn and Neptune.

But it is very necessary to improve also the general theories where the time is not allowed to appear outside trigonometric arguments. It is shown as an example, that the number of arguments must be larger by at least 2 or 3 orders of magnitude. The best available general theories for all planets present discrepancies of the order of 1/1000 with Cohen and Oesterwinter numerical integration over 700 000 years. It is believed that one can do much better and that more work should be done in improving general planetary theories.

2) Analytical Lunar Theories (H. Kinoshita)

The accuracy of recent analytical theories has increased considerably during the last decade due to the requirement of lunar laser ranging observations and due to the rapid increase of capability of handling mathematical formulas by computer. Recent lunar theories are throughly investigated by different authors and various perturbation methods and have superseded Brown's theory. Recent analytical lunar solutions are divided into the following four categories according to the treatment of coefficients of Fourier series, by which solutions are expressed: analytical (or literal) solution, partly analytical solution, semi-numerical solution, and semi-analytical solution. Advantages and disadvantages of each type of solution were briefly discussed. The precision of each solution was also discussed with use of comparison with numerical integration. ELP developed by M. Chapront-Touze and J. Chapront, one of these solutions, is a complete solution in the sense that it includes all perturbations acting on the Moon except the effect due to non-rigidity of the Moon. The accuracy of solution of main problem in ELP is about a few centimeters and the accuracy of planetary perturbation in ELP is about a few meters over one hundred years. In order for an analytical solution to be compatible with the accuracy of lunar laser observations, we have to 1) increase precision of planetary perturbations(especially direct perturbations due to Venus and complete second-order planetary perturbations) and 2) investigate coupling effect among various perturbations(especially coupling between $J_2$ perturbation and planetary perturbation) and 3) increase precision of external theories(planetary theory, physical libration of the Moon, etc).

3) Planetary Rings (J.A. Burns)

The discovering of the rings about Jupiter and Uranus, plus the vast growth in our knowledge of Saturn's rings, has allowed substantial progress to be made in our understanding of their dynamical evolution. This talk summarizes the physical properties of the three ring systems and discusses the gross evolution(flattening, collisional spreading, location within the Roche limit) of unbound rings. It then describes how the lifetimes of small grains can be limited by dynamical processes (plasma drag and Poynting-Robertson effect) or by erosion(spattering and micrometeoroid collision); micron-sized grains, such as found in Jupiter's ring and Saturn's E-ring, must be continually regenerated. Satellites have profound effects on ring structure: according to the mechanism proposed by Goldreich and Tremaine, satellites can repel disks of ring material and initiate spiral density waves at resonant locations. About twenty-five spiral density waves have been identified in Saturn's rings with two of them bending waves.

Satellites also confine Saturn's F-ring and probably the Uranian rings; local short-time perturbations may produce the kinks and clumps in the F-ring while its bending is as yet unexplained. While our understanding of planetary rings has developed substantially over the last dcades, many questions remain unanswered.

4) Periodic Orbits in Dynamical Systems and Their Stability (G. Contopoulos)

When a family of stable orbits becomes unstable, it "inherits" its stability to a new family bifurcating from it. In general we have an infinity of successive "pichfork bifurcations" leading to an infinity of unstable families and a large degree of stochasticity. The ratio of successive intervals between bifurcations tends to the universal number 8.72(Feigenbaum, Eckman etc). But we may have infinite bifurcations of the same family, which becomes successively stable and unstable. The corresponding ratio depends on the dynamical system and is not universal. We have also found a case of a Feigenbaum sequence followed by an inverse Feigenbaum sequence, leading back to a single stable family.

Another mechanism leading to large stochasticity is the breaking of a family to an infinity of families of periodic orbits separated by gaps. The extent of trapping around each family gives the non-linear stability of this family. We noticed that beyond each gap the central family of periodic orbits "inherits" its associated tube orbits to the next family.

5) Dynamical Evolutions of Satellite Orbits (Y. Kozai)

In the solar system there are many pairs of satellites, for which commensurable relations of the mean motions hold. Recently Voyager spacecrafts discovered several other commensurable pairs in Saturnian system. Why such commensurable relations, namely resonant configurations, are dominant in satellite systems? It is hardly believed that such relations have been dominant from the beginning of the solar system, and, therefore, some kinds of dynamical evolution processes should have slowly driven the systems in such resonant configurations. The most important factor to have driven the orbits is believed to be, for most of the cases, due to tidal dissipations in planets and satellites.

The tidal dissipations in the satellites tend to reduce the eccentricities even after the synchronization of the orbital motion with the rotational motion is realized. However, there are satellites, for which the eccentricity cannot be reduced to zero because of the forced eccentricity due to the resonant perturbations. For such pairs the tidal dissipation never stops and has produced volcanic activities in Io, one of Galilean satellites.

Martian satellites are likely to be composed of carbonaceous chondrites, and the secular accelerations observed in the mean longitude of Phobos are due to the tidal dissipations. It is believed that Phobos was captured by Mars and come to the present orbit due to the tidal dissipation.

# COMMISSION 8: POSITIONAL ASTRONOMY
(ASTRONOMIE DE POSITION)

## Report of Meetings of 19,20,21,23, and 25 August 1982

PRESIDENT: E Høg SECRETARIES: B L Klock, T E Corbin

CONTENT



## I. COMMISSION BUSINESS (19 August)

### a) Constitution of the Commission

President: J A Hughes Vice-President: Y Requieme

Organizing Committee: G Billaud, G Carrasco, W Fricke, V S Gubanov, E Høg, B L Klock, J A Lopez, Luo Ding-jiang, M Miyamoto, L V Morrison, I Nikoloff, G Teleki, Y S Yatskiv

Deceased Member: K N Tavastsherna

Retired: none

New Members: L Helmer (DK), L Stange (GDR), R Bien, S Röser (GFR), L N Mavridis (Greece), M Yoshizawa (Japan), L Lindegren (Sweden), R L Branham, P D Hemenway, R Stone (USA), V A Fomin, Yu L Kokurin (USSR)

Consultants: C Fabricius (DK), K C Blackwell (UK), J L Fanselow (USA), V I Orelskaya (USSR), M Dachich, I Pakvor (Yug)

Numerical Strength (with previous values in brackets) 15 (13) Officers, 103 (99) Members, 6 (9) Consultants, Total 124 (121)

b) Working Groups

Study Group on Astronomical Refraction

Chairman: G Teleki. Members: W J Altenhoff (Bonn), F K Brunner (Kensington), B Garfinkel (New Haven), J A Hughes (Washington), I G Kolchinskij (Kiev), A I Nefed'eva (Kazan), K Ramsayer (Stuttgart), J Saastamoinen (Ottawa), C Sugawa (Tokyo), E Tengström (Uppsala), and H Yasuda (Tokyo), L N Marridis (Thessaloniki).

Study Group on Horizontal Meridian Circles

Chairman: E Høg. Members: R d'E Atkinson (Bloomington), J W Gietzen (RGO), G van Herk (Leiden), B L Klock (USNO), J Osorio (Oporto), G J Pinigin and G M Timashkova (Pulkovo), Xu Tong-qi (Shanghai), and Hu Ning-sheng (Nanjing).

Working Group on Astrolabes

Chairman: G Billaud. Members: S Debarbat, W Gliese, V S Gubanov, F Noel.

A Joint Working Group for coordination of work on the Stellar Reference Frame was established by Commissions 8 and 24 in 1979. The continuation of the WG was approved by Commission 8, but the question of continuation was not treated by Commission 24 so that, formally, the WG has ceased to exist. Moreover, it seems difficult to have a WG on a subject which is the main subject of the whole Commission 8. This difficulty was perhaps the reason that the report given by the Chairman of the WG on 23 August contained considerations on the stellar reference frame rather than a report from the WG.

Should the President of the Commission like to restablish the WG he may of course do so.

SRS Committee

Chairman: W Fricke. Members: T Lederle, J A Hughes, C Smith, VA Fomin, DD Polozhentsev.

c) Resolutions

Proposed resolutions on the observation of planets, of high luminosity stars and of HIPPARCOS stars were presented to the Commission and adopted in principle. A committee of W Fricke, J A Hughes, E Høg and Y Requieme was given the task of formulation. After a meeting in the afternoon the following resolutions were written and were later endorsed by the IAU.

1. In considering that the determination of the equinox and equator will remain to be an important task of positional astronomy, it is recommended that
   (a) the Sun and major and minor planets are included in observational programs with transit circles,
   (b) the Sun and major planets are included in programs with astrolabes, and
   (c) minor planets are included in programs of photographic astrometry,

   such that from differential and/or absolute observations the positions can be rigorously referred to the fundamental reference system and serve for further improvements of the zero points of this system.

2. In view of the importance of accurate positions and proper motions of High Luminosity Stars for galactic research the Commission reaffirms its previous recommendation that such stars be included in transit circle programmes.

3. It is recommended to observe the subset of stars requested by the HIPPARCOS Input Catalogue Consortium. The positions should be available before 1985.5. A moderate accuracy of ±1.0 arcsec at that epoch is required, however ±0.3 arcsec (in the FK4 system) is desirable.

4. Noting that the HIPPARCOS stars will be selected for their astrophysical and/or astrometric interest and that such stars would traditionally be included in meridian observing programmes, and
noting that very accurate meridian positions of suitable subsets of the HIPPARCOS stars would give checks of the space observations and calibration of the ground-based instruments
it is recommended that such subsets, e.g. faint fundamental stars and International Reference Stars be observed.

5. A fifth resolution in support of the Sydney Observatory was formulated and later passed in agreement with Commissions 24 and 40.

d) Commission Rules

New rules of the Commission has recently been approved by mail vote.

e) 1980 IAU Theory of Nutation

This theory was adopted by mail vote of the Commission.

f) SRS

A meeting on SRS was held at Brorfelde, Denmark, on 3 March 1982. The full report has been distributed to members of the Organizing Committee and the conclusions have been accepted for Pulkovo by K N Tavastsherna.

The following contains the main points of the meeting:

Participants:
E Høg, President of IAU Commission 8, (Chairman),
T Lederle and H Schwan, Astronomisches Rechen-Institut (ARI)
J Hughes and C Smith (Secretary), U.S. Naval Observatory (USNO)

The meeting was first proposed by Dr. Tavastsherna in about July 1980 as a meeting on SRS of the president of Commission 8 and representatives from Pulkovo, ARI and USNO. It was therefore now particularly regretted that no representatives from Pulkovo were present, even though the meeting had already been postponed once (from October 1981) in order to make their presence possible.

It was recommended to the President to appoint W Fricke as Chairman of the SRS Committee. The president has approved this, and it has been accepted by Fricke.

Exchange of Data and Completion of Catalog

It was agreed that ARI would act as the point of exchange of data between USNO and Pulkovo.

Recently USNO has deposited a magnetic tape at ARI of averaged Leoncito RA results of SRS stars referred to FK4 for immediate transmittal to Pulkovo.

A second tape giving Dec in addition to RA was deposited at ARI with the understanding that this second tape would be transmitted to Pulkovo whenever they shall also have deposited a tape of their SRS results with Dec at ARI. A third tape of the Washington Six-inch SRS results referred to the FK4 system will be prepared and deposited with ARI for transmittal to Pulkovo under the same conditions.

It is noted that in accordance with the original SRS agreement only averaged results of SRS stars have been exchanged and no results about FK4 stars. A meeting in Heidelberg 1975 recommended the exchange of results from individual nights and of FK4 stars. This is still considered highly desirable. The Perth 70 results of individual nights and of FK4 stars have however been distributed to all parties in 1976.

The meeting agreed to recommend to Commission 8 that both Pulkovo and USNO should be free to complete the compilation of the SRS catalog with all data which may be available to them as of the end of 1982, and that all data whose exchange is now pending will have been deposited with ARI by 1 Nov 1982 (i.e. the third USNO tape mentioned above and the Pulkovo Dec).

It must be anticipated that there will regrettably be two different SRS catalogues, one from USNO referred to FK4 and another one from Pulkovo referred to an improved SRS system. The USNO catalog is expected by Hughes to be available by the end of 1983.

The participants in the meeting agreed that the results of the compiled SRS catalogs, once they are completed, will be made available for open distribution to all interested parties.

Meetings on SRS in Patras

It is recommended to the Chairman of the SRS Committee that he should select members for that Committee in correspondence with Pulkovo and the USNO. It is further recommended to the Chairman of the SRS Committee that he schedule a meeting of the Committee early in the General Assembly so that reports may be given to Commission 8 before the close of the General Assembly.

(E Høg 3 April 1982)

Members of the SRS Committee are listed in Ib).

g) Symposium on Astrometric Techniques

The proposed IAU Symposium was supported by the Commission to be held in Venezuela January 1984. Chairman of the Scientific Organizing Committee: H Eichhorn.

## II. INSTRUMENTATION

Meeting of Commission 8 (19 and 25 August)
Chairmen: E Høg on 19 Aug., J A Hughes 25 Aug.
On 23 Aug:

G Billaud: The Photoelectric Astrolabe of CERGA.

The fully automatic instrument employs a full pupil angle of zero expansion material and a Maksutov-Cassegrain reflector. A photoelectric slit micrometer measures the two stellar images, one at a time.

Standard deviations of 0.0038s and 0".12 have been obtained for the determination of UTO-UTC and latitude, respectively, from each group of 32 stars.

Y Requieme, J M Mazurier: Bordeaux AMC, Application to Faint Stars.

The automatic meridian circle is used for routine differential observations since 1980 and about 18000 observations have been obtained. The tracking photoelectric micrometer with rotating knife-edge sensor is used and mean errors of 0".11 and 0".17 are obtained in RA and Dec respectively. With a diaphragm of 20" diameter a few hundred stars up to $m_v$ = 12 were observed and no significant increase of mean error was noticed. The quasar 3C273B fainter than $12^m.5$ has been observed three times. Average observing time is 70s per star.

E Høg: Observation of Faint Stars with a Slit Micrometer.

A theoretical study has shown that stars of $m_v$ = 12 to 13 will be observable in 70s provided a photomultiplier with low dark current about 5 Hz is used.

J A Hughes: Progress Report on the Instrumental Development of the Seven-Inch TC.

The development involves the following major components: New objective lens with a solar attachment, a new glass circle, a photoelectric circle scanning system, an automatic telescope setting system, a photoelectric micrometer utilizing an image dissector, and on line computer control.

R Stone: Processing of Images from USNO's Transit Circle Image Dissector.

L V Morrison: Results from the Carlsberg Automatic Transit Circle.

H Yasuda, M Yoshizawa: Tokyo Photoelectric Meridian Circle.

The instrument is being installed this year. It uses a slit micrometer and photon counting techniques. Nadir, zenith mirror, azimuth marks, Sun, Moon and planets will be observable in an automatic mode controlled by on-line computer.

M Miyamoto, Ch Kühne: Derivation of Division Corrections from Unbiased Microscope Scales.

B L Klock, T J Rafforty: An Investigation of US Naval Observatory Glass Circles.

Thick transparent glass circles have been used elsewhere, thick enough to carry themselves. The USNO circle is a thin glass annulus cemented onto the conventional steel wheel that was previously directly engraved. High quality division lines and uniform background being stable for many years have been obtained.

E. Høg: Report of Study Group on Horizontal Meridian Circles, by its Chairman.

Since the report of Commission 8 published this year new information on the Pulkovo Horizontal Meridian Circle and the Glass Meridian Circle has been given:

Declination Observations with the Pulkovo Horizontal Meridian Circle

Kirian, Pinigin, Sukharev and Timashkova have (May 1982) given preliminary results of declinations obtained during 3 months:

(1) The photoelectric circle reading gives an mse of ±0".02, and the zero point is determined by a pendulum mirror to ±0".04 or 0".06. A photoelectric ocular micrometer registers the star position in the field of view to ±0".10 near zenith.
(2) The ventilation between inner and outer tubes is effective and has reduced the vertical temperature gradient of air inside the tube from a previous 0.5° C to 0.1° C per tube diameter.
(3) Observations of star declinations show that the equator point does not change with temperature, that the equators obtained from the two tubes coincide and that the final HMC system does not show variations in declination within ±0".3. It is therefore concluded that the instrumental flexure of the HMC is small, if at all existing.

Further observations will be obtained and the great potential possibilities of this instrument is expected to be realized in the nearest future.

Workshop on the GMC held in China

Dr. Hu Ningsheng arranged lecture and discussion about meridian astrometry and the Glass Meridian Circle with E Høg on 3 and 4 June 1982 at Nanjing. Shanxi Observatory organized a workshop on the GMC at Lintong (near Xian) on 1 to 9 June with 25 participants from Institutes in Lintong, Xian, Beijing, Wuhan, Tianjin, Nanjing and Shanghai, and working papers by T B Andersen, Hu Ningsheng and E Høg on the GMC have appeared in Chinese translation in Scientific Version of Shanxi Observatory B Series, 1982, Vol. 1, pp. 1-79. Høg participated in the Workshop during the last three days, and further cooperation is planned and awaiting approval with Shanxi Observatory, Purple Mountain Observatory, and Nanjing Astronomical Instruments Factory on design study, manufacturing and operation of the GMC.

On 25 August:

G. Teleki: Report of Study Group on Astronomical Refraction.

A brief review was given on papers written recently that relate to refraction. A summary of the meeting of the working group at Dubrovnik, Yugoslavia. Papers presented there were listed. Particular mention was made of anomalous refraction caused by meridional tilt of atmospheric layers and the possibility of obtaining refractional values directly from astrometric observations.

I S Guseva, E G Zhilinsky, G S Kosin: On the Fifth Edition of the Pulkovo Refraction Tables.

The tables will appear in logarithmic form, as before; however, algorithms in Fortran-4 for the direct computation of values will now appear. These may be used for zenith distances of 0° to 80°.

I S Guseva: Calculation of the Astronomical Refraction for Different Aerological Data.

The calculations are based on atmospheric physics and optics. An ellipsoidal Earth is used in the computations, which are done by direct numerical integration. This procedure results in a $\Delta R < 0''.05$ at 85° zenith distance.

A S Kharin: On the Refraction Anomalies from Meridian Observations.
A study of refraction anomalies was presented. A vertical circle has been used to make the necessary measures.

R Fukaya, M Yoshizawa: Numerical Calculation of the Atmospheric Refraction at a Given Wavelength Region.
Numerical calculations based on meteorological data and a spherical, concentric model of the atmosphere. Currently investigating atmospheric tilt. Comparisons with the Pulkovo tables show agreement to within 0.″02 from $0^\circ$ to $75^\circ$ zenith distance. The work illustrates that refraction, in its first order theory, can be determined by the meteorological data in the region of the telescope objective.

## III. STELLAR REFERENCE FRAME

Joint Meeting of Commissions 8 + 24 (23 and 25 August)
Chairmen: E Høg on 23 Aug., J A Hughes 25 Aug.

On 23 August:

A resolution was proposed by C A Murray requesting the General Assembly to note with concern the decision to terminate the astronomical work at the Sydney Observatory and to urge that the Sky Survey continues and that the instruments at the Observatory remain available for observations. The resolution was endorsed by Commission 40 and passed by a vote of Commissions 8 and 24.

H K Eichhorn: Report of IAU Commission 8 Working Group on the Stellar Reference Frame.
A review was given of present and possible future needs of the accuracies of various sets of stars that define the stellar reference frame at various magnitude intervals (FK4, GC, AG and AC). Future extensions of the fundamental catalog, beyond FK5, were discussed and it was recommended that these involve improvement of the positions and motions of the GC stars. The need for an improved SAO catalog was also emphasized.

W Fricke: Report of the SRS Committee.
This report was deferred since C Smith was to report on the compilation of the SRS.

W Fricke: Status of Work on the FK5.
A review was given of the resolutions that began the work and led to the adoption of the 1976 system of astronomical constants. 260 catalogs observed after 1900 contribute 55,000 positions to the new system. There are 30 new absolute catalogs being used for the improvement of the fundamental system. In addition, there are 100 new differential catalogs as well as 130 catalogs, observed 1900 to 1950, that will contribute to the faint fundamental extension. Corrections to the FK4 were shown. The $\Delta\alpha_\delta$ exhibit the largest corrections in the zone $-60^\circ$ to $-80^\circ$ while the $\Delta\delta_\delta$ show large oscillations in this region. The current schedule for the completion of the work is:
1. By 1 January 1984 the system will be complete. 2. After January 1984 tables giving the conversions from FK4 to FK5 will be available. 3. By the end of 1985 all of the FK5 will be completed.

C A Murray: Measurement of the Cape Survey, CPC2.
Progress in measuring the Cape plates on the GALAXY machine was

reported. 3000 of the 5000 plates, $-40^{\circ}$ to $-90^{\circ}$, have been measured. The internal precision was reported as $\pm 0''.1$, and the limiting magnitude is $10^{m}.5$.

C Smith (read by J A Hughes): Progress Report on a Revision of the IRS.
The distribution of the International Reference Stars (AGK3R & SRS) has been studied to identify regions of low density. It is proposed that about 1200 stars be added to the IRS, mostly in the zone $0^{\circ}$ to $-30^{\circ}$, to give the list a more uniform density.

C Smith (read by J A Hughes): Progress Report on the Compilation of the SRS.
A preliminary SRS catalog has been formed from the results of those observatories that reported both FK4 and SRS observations. The mean errors of the preliminary positions are $\pm 0^{s}.007$ in right ascension and $\pm 0''.12$ in declination. The final SRS will be available by the end of 1983.

T E Corbin: The Proper Motions of the Southern IRS (SRS) Stars.
A preliminary system of 4968 positions and proper motions on the system of the FK4 has been computed from 18 catalogs that were observed using screens. This system will now be used to study magnitude equations in other southern catalogs and reduce them to the FK4 system.

Y Yatskiv: Review of Astrometric Work in the USSR.
The catalog of Pulkovo observations in the zone $-47^{\circ}$ to $-90^{\circ}$ has been completed. The catalog contains 5511 SRS, 829 BS and 280 FK4 stars. The mean error of a single observation is $\pm 0^{s}.014 \sec\delta$ in right ascension and $\pm 0''.39$ in declination. Kiev has begun compilation of the Bright Star Catalog. The contributing catalogs have been listed and the work of combining them will begin at the end of this year. A list of high-luminosity stars has been proposed for observation. Pulkovo has prepared the list which contains about 2000 stars, mostly in the 6th to 9th magnitude range. The Ratan 600 radio telescope has made observations of Mercury and Venus. The mean errors of a single observation in right ascension are $\pm 0^{s}.1 \sec\delta$ for Mercury and $\pm 0^{s}.06 \sec\delta$ for Venus.

J A Hughes: Progress Report on the Proposed New Zealand Program.
The 7" Transit Circle at latitude $-41^{\circ}.8$ in New Zealand and the 6" Transit Circle at $+38^{\circ}.9$ in Washington will simultaneously observe the FK5 and IRS (AGK3R & SRS). The program is to begin in 1983, and the observations of the two instruments will overlap by $60^{\circ}$ and will be made fundamentally.

Y Requieme: Ground-based Measurements for the HIPPARCOS Input Catalog.
The Input Catalogue will require epoch 1988.0 positions of 40,000 bright stars ($B < 10$ mag.) with accuracies of $\pm 1''.0$ and of 10,000 faint stars ($10 < B < 13$ mag.) with accuracies of $\pm 1''.5$. Some 20,000 to 30,000 may need additional observations to obtain the required accuracy. The Cape survey should provide many of the southern positions, and the Bordeaux automatic meridian circle has been proposed to make the necessary observations in the north.

E Høg: The TYCHO Reference System of 400000 Stars.
An exploitation of the HIPPARCOS star mapper has been approved by ESA as the TYCHO project. This project will give positions of at least 400000 stars brighter than $B = 11$ with an accuracy at $B = 10$ of

0.03 arcsec. An accurate stellar reference frame with 10 stars per square degree can thereby be established.

Presented by Xu Tong-qi: A Preliminary Report of the General Catalogue of Stars Observed with Photoelectric Astrolabes in China, and: Some Observations of Bright Radio Stars by Means of Astrolabes in China.

The results of eight series of astrolabe observations have been combined to form the General Catalogue (GCPA). This has resulted in mean errors of ±0$^s$.0033 in right ascension and ±0$''$.058 in declination in the GCPA positions. The GCPA is used to investigate errors in the FK4 and comparisons were given in the form of several figures. Of the 1579 GCPA stars, 642 are from the FK4, 516 are from FK4 Sup and 421 are from the GC. Work has begun on the problem of the radio-optical connection. Since 1979 15 radio stars have been observed with the astrolabe at Shanghai (from the list endorsed by Commission 24). Observations of Algol have been given special attention.

H Yasuda, K. Hurukawa: Northern PZT Star Catalog.

Results from ten instruments have been combined to form the NPZT with a mean epoch of 1974. The catalog contains 1719 stars with catalog positions whose average mean errors are ±0$^s$.0033 sec $\delta$ in right ascension and ±0$''$.079 in declination. Comparisons were made with AGK3R and AGK3 to look for magnitude effects, but none were found.

E Marcus: Main Features of the Bucharest NPZT Reference Star Results.

A program of 951 NPZT stars was observed using the Bucharest Meridian Circle. Using AGK3 proper motions the Bucharest results were compared with the Tokyo $NPZT_{74}$. The computed internal precision is ±0$^s$.016 in right ascension and ±0$''$.29 in declination. The comparison with Tokyo gave external errors of ±0$^s$.019 in right ascension and ±0$''$.21 in declination.

F Laclare, F Chollot: Observations du Soleil avec Astrolabe.

Astrolabes at Sao Paulo and CERGA began observing the Sun in 1974 and 1975, respectively. They are now involved in a joint effort to regularly observe the Sun. Utilizing prisms that allow observations at zenith distances of 30$^o$, 45$^o$ and 60$^o$ each can cover 70% of the orbit.

G Westerhout: Concluding Remarks. (These had not been received at the time of writing).

On 25 August:

S N Sadzakov, M D Dacic, V A Fomin: An Investigation of the Systematic Errors of the Belgrade NPZT and AGK3 Catalogues on the Basis of Modern PZT Observations.

The NPZT catalog was observed at Belgrade from 1972 to 1980. The accuracy was reported to be ±0$^s$.012 sec $\delta$ in right ascension and ±0$''$.10 in declination. Comparisons were made for several different stations.

S Sadzakov, V A Fomin, M Dachich, D Saletich: Systematic Errors in Latitude Observations of Various Observatories.

Work by Yasuda and the AGK3 were used to estimate the external accuracy of the NPZT. Curves of $\Delta\alpha_\alpha$ and $\Delta\delta_\alpha$ were shown. The systematic component was shown to be due to the systematic errors in the AGK3 proper motions.

J Phan Van: Observations Planetaires a l'Astrolabe du CERGA.

The results of observations of Mars and Jupiter made in February and March of 1980 at CERGA. A 45° prism was used. Comparisons were made with Washington 6" observations for the same period that show systematic differences between the two series of observations. The differences are in the 1" to 2" range in both coordinates. However, subsequent discussions revealed that the two sets of observations had been treated differently with respect to $\Delta T$.

G Vigouroux: Catalogues Astrolabes a 2 Distances Zenithales.

G Carrasco, P Loyola: RA Observations with the MC of Cerro Calan 1978 to 1981.

The Repsold Meridian Circle at Cerro Calan started observing the FKSZ stars in December 1979. In addition, FK4 stars have been observed over a wide arc, from +42° to -90° and from -90° to -68° lower culmination. Comparisons of Instrument - FK4 in right ascension with similar differences from other observatoris show very good agreement.

## IV. EXTRAGALACTIC REFERENCE FRAME

Joint Meeting of Commissions 7,8,24,33,40 and 44 (21 August)
Chairman: A N Argue

An extensive report will be given elsewhere of papers by: Argue; Johnston; Eichhorn; Kovalevsky and Froeschle; Lindegren; Chollet and Debarbat; Fanselow, Purcell, Sovers and Thomas; Hemenway; de Vegt; Quijano.

E. Høg
President of the Commission

COMMISSION 9: INSTRUMENTS AND TECHNIQUES (INSTRUMENTS ET TECHNIQUES)

Report of Meetings: 18, 20, 21, 24 and 25th August 1982

PRESIDENT: E.H. Richardson

## Report of President

Since the 17th General Assembly in Montreal, the major activity organized by Commission 9 was IAU Colloquium 67, held 8-10 September, 1981, on the observing floor of the 6 metre telescope of the Special Astrophysical Observatory, USSR. The Proceedings have been published by D. Reidel Company, entitled "Instrumentation for Astronomy with Large Optical Telescopes". Commission 9 also participated in 'ICO-12', the Congress and Twelfth Assembly of the International Commission for Optics held at Graz, Austria, 31 August to 5 September 1981. The papers presented at that meeting have been published in Optica Acta, 1982, vol. 29.

The scientific and technical reports of the Commission 9 meetings at the 1982 General Assembly are arranged below according to the Working Group indicated, with the exception of the general meeting on 20th August which combined the interests of the Working Group on Large Telescopes and the Working Group on Infrared and General Techniques.

## Business Meeting, 21st August 1982

In anticipation of a poor attendance at the meeting scheduled for 21st August (which proved to be so), the election of new officers was started at the end of the well-attended general meeting of 20th August and the results were confirmed at the meeting on 21st August.

Dr W.C. Livingston was elected President, there being no other nominations. For the office of Vice-President, two nominations were received and a ballot was held. Nominations for membership of the Organising Committee were made by the outgoing President and Vice-President in consultation with the Commission members who were present. This resulted in the election of the following:- Vice-President: C.M. Humphries; Committee members: J.C. Bhattacharyya, J. Davis, J.L. Heudier, I.M. Kopylov, G. Lelièvre, J.P. Maillard, A.B. Meinel, N.N. Mikhel'son, E.H. Richardson (Retiring President), J. Ring, J. Rösch, N. Steshenko, R.G. Tull, M.F. Walker, W.L. Wilcock, G. Wlérick.

Dr Richardson reported that in early 1982 all members of Commission 9 who could be reached by mail were asked to reply if they wished to remain members. Those who did not reply were removed from the membership list unless another member spoke on their behalf. It was agreed that the names of all new applicants should be added to the membership list.

## General Meeting, 20th August: TELESCOPES AND INSTRUMENTS (Chairman: E.H. Richardson)

The following communications were presented:

LEST - A SUGGESTED DESIGN FOR A NEXT GENERATION SOLAR TELESCOPE (A. Wyller; Stockholm Observatory)

Design of the proposed Large European Solar Telescope (LEST), initiated by the Joint Organisation for Solar Observations, is proceeding on the basis of a specification for an instrument of large aperture combined with high spatial resolution (< 0.3 arcsec, preferably 0.1 arcsec), tracking at this accuracy for at least 30 min, low intrinsic polarization properties, and an adequate field of view (2 arcmin minimum, 30 arcmin preferably). The Ritchey-Chrétien optical design is based upon that of the NASA Space Telescope with a 2.4m honeycomb primary (f/2.3) and a 0.84m diameter secondary. To minimise thermal problems, evacuated and He filled tube structures have been considered (see next paper). The telescope will be built on a tower of minimum height 30m with a cap dome similar to that used by Rösch at Pic-du-Midi, and is likely to be located either at La Palma or at Tenerife.

TESTS OF A HELIUM FILLED SOLAR TELESCOPE (R.B. Dunn, O. Engvold, W. Livingstone and R. Smartt; Kitt Peak National Observatory). With additional presentation by J. Rösch, Pic-du-Midi Observatory.

Since helium has 1/8th of the refractive index (in terms of n-1) of air and almost 10 times greater heat conductivity, it should be capable of reducing thermal problems in a sealed telescope tube structure without the need for evacuation. In order to test this, the internal seeing of the KPNO 'vacuum' telescope was evaluated from video recorded interferograms obtained with the solar image present, at various pressures of air and of helium. At evacuated air pressures up to 25 Torr, motion of the fringe pattern could just be detected whereas at higher pressures the fringes became more and more chaotic until, at atmospheric pressure, the fringes were barely visible. When the tube was filled with helium, no image motion occurred until the pressure had increased to 650 Torr. The conclusion is that a helium filled solar telescope at normal atmospheric pressure will hardly differ from an evacuated one. Furthermore, the helium removes heat effectively from the optical surfaces, reducing thermal effects such as distortion and change of focus.

MICRO-ARCSECOND ASTROMETRIC ORBITER CONCEPT (S.M. Rathjen, D.C. Black, J.P. Murphy, L.W. Bandermann and W.H. Alff; NASA Ames Research Center and Lockheed Palo Alto Research Laboratories)

A shuttle launched free-flying satellite characterised by the ability to make astrometric measurements with exceptionally high angular resolution is being proposed. The mission objective is the discovery and characterisation of other planetary systems which in turn would lead to a better understanding of the formation and evolution of our own solar system. The instrumentation consists of a telescope which feeds an optical system utilising a Ronchi grating, an apodized pupil and a split-field detector giving an angular resolution capability of $10^{-5}$-$10^{-6}$ arcsec. With a 1m diameter telescope aperture and 1 hour integrations, such an instrument would allow dozens of stars to be searched for Earth-mass planets and thousands of stars for Jupiter-mass planets. The concept is based on a 5 year mission, refurbishable, with 10 acquisitions per year per target star.

ULTRA-THIN MIRRORS (E.J. Hysom; A.E. Optics Ltd., Luton, England)

For some time now, A.E. Optics has been making mirrors of diameter 23cm and thickness 1cm which were figured to a moderate astronomical quality. Recently we received a request for mirror substrates of diameter 80cm and thickness 3mm (d/t = 270/1) with a focal ratio $\sim$ f/6; 81 such mirrors were required by the Rutherford Appleton Laboratory for use in a tessellated mirror Cerenkov detector. The method of manufacture consisted of preforming to the required radius by slumping the substrates onto a steel mould. At the early testing stage, difficulties were encountered with an edge effect due to the extreme thinness. To ease manufacturing problems it was agreed to increase the thickness to 4mm, and work the substrates at 80cm diameter before edge cutting to hexagons of circumscribed diameter 70cm. It is considered that increasing the thickness still further to ∿ 10-12mm would permit figuring to give an image size which would be acceptable for certain types of astronomy, whilst still retaining the cost advantage that results from their low weight. It is estimated that a 1cm thick 70cm diameter substrate could be worked relatively quickly to give an image size of a small fraction of a millimetre and a number of such mirrors driven by stepping motors could form the basis of a cheap multi-aperture telescope.

ACTIVE OPTICS INSTRUMENT FOR USE AT MAUNA KEA (Laird A. Thompson; Univ. of Hawaii).

An active optics image stabilising instrument is currently under construction for use at Mauna Kea Observatory. It is designed to remove image jitter with frequencies < 70Hz and with amplitudes < 13 arcsec. There is a special need for such an instrument at Mauna Kea because stellar images, when viewed from this site, shrink to diameters < 0.5 arcsec. When this happens the image core almost invariably shows a residual motion with a frequency ∿ 20Hz and an amplitude ∿ 0.5 arcsec. Until such image jitter is removed, full advantage cannot be made of the superb Mauna Kea conditions.

The primary optical element in the image stabilizing system is an active mirror which folds the incoming converging Cassegrain beam by $45° \pm \Delta\theta$, where a range of $\Delta\theta < 16$ arcmin provides image mobility of ± 13 arcsec in the image plane. The active mirror, 12.5cm in diameter, gives an unvignetted field of view of 8 arcmin x 11 arcmin. The mirror is controlled by four linear motors (Ling drivers) which are linked in pairs to provide two axes of motion. Image position can thereby be controlled to within ± 0.01 arcsec in the image plane. Following the active mirror the converging beam intercepts a beam splitter (or alternatively a small central mirror) which provides two optical paths: one to the Cassegrain focal plane and the second to the guide probe. Freezing image motion at the guide probe (by feedback to the active mirror) simultaneously freezes image motion at the instrument focal plane.

As a guide probe, we use a set of 5 closely packed optical fibres linked individually to 5 photoelectric cells. The guide probe gives not only an error signal for controlling image motion, but also provides a real-time measure of the image quality. The image quality information is used to control an active shutter placed immediately above the Cassegrain focal plane. When the image quality deteriorates momentarily, the active shutter closes to maintain excellent image quality.

PRIME FOCUS CORRECTOR LENS FOR A 300-INCH (7.6M) TELESCOPE (E.H. Richardson, C.F.W. Harmer and W.A. Grundmann; Dominion Astrophysical Observatory)

A new design of three-element prime focus corrector lenses for Ritchey-Chrétien (R-C) telescopes has superior resolution and a greater back focal distance than previous designs. In the older design where the size is restricted, the resolution would decrease if the lenses were enlarged.

The proposed 300-inch Texas telescope made severe demands on the design of a corrector: the primary focus is very fast, F/2, and a resolution of 0.5 arcsec was sought over a very wide spectral region. It has been found from this work that exceptionally fast Ritchey-Chrétien telescopes can now be well corrected at the prime focus with three-element corrector lenses covering a wide field and a broad spectral region. Although larger than previous refractive lens designs, they are feasible and still have much smaller central obstruction than corrector mirrors.

UKIRT INSTRUMENTATION (W.J. Zealey; Royal Observatory, Edinburgh)

Improvements in the mechanical performance of the United Kingdom Infrared Telescope have continued with the installation of new RA and Dec position encoders (giving 3 arcsec RMS pointing accuracy), the provision of secondary mirror chopping (max. amplitude ± 100 arcsec at 10Hz, the re-writing of all control software in Fortran (instead of the less flexible assembly language), and an improvement in the magnitude limit (V = 18.5) for which autoguiding can be used. Fully commissioned instrumentation comprises InSb photometers for use with J, H and K filters or CVF (R = 130, 1.4 to 2.8μm); InSb photometers for use at J, H, K, L and M (optimised at L and M) or with CVF (R = 100, 2.3 to 4.6μm); a Ge bolometer (L to 35μm); and a cooled grating spectrometer (single channel, 1 to 5μm, R = 500 at 2.0μm in first order). Instrumentation under construction includes a 7 channel InSb array spectrometer with option of a 2 to 5.3μm Fabry-Perot system; an 8 channel bolometer with 4 spatial channels at any two of 10, 20 or 34μm; a 16 channel SiAs array photometer (10, 20μm); and a multichannel data acquisition system. Future instrumentation includes a 7-14μm SiAs array spectrometer; a 32 element InSb array spectrometer, a polarimeter (1 to 5μm); a $^3$He cooled submillimetre photometer (350μm to 1mm); and heterodyne receivers for use at submillimetre and millimetre wavelengths. At the Cassegrain focus (f/9 or f/35), up to six instruments can be mounted simultaneously. Effort has also been directed towards achieving remote operation of the telescope from the sea-level base at Hilo and from the Royal Observatory, Edinburgh (note added after the meeting: the first successful link giving remote control and direct data transfer between UKIRT and Edinburgh using the TELENET satellite and Packet Switching Service occurred on 6 September 1982).

CFHT INFRARED INSTRUMENTATION (J.P. Maillard)

Infrared instrumentation at the Canada-France-Hawaii Telescope has been tested and commissioned recently (October 1981 to April 1982). The telescope is now equipped with an infrared top end holding a chopping f/35.5 secondary mirror (maximum throw 6 arcmin, frequency 1 to 50Hz). Each instrument using this focus is mounted on a Cassegrain adaptor providing acquisition of the field and guiding. Two photometers (InSb and Ge bolometer) provide complete coverage from 1.25 to 33μm by discrete filters and CVFs. A high resolution 1-5μm interferometer (60cm path difference) is also part of the Cassegrain instrumentation and is completely computer controlled. Preliminary photometric tests at the telescope show that a magnitude limit of 19.3 at waveband K can be reached in 1 hour with S/N = 1. First results with CVF spectra in the 3-5μm region of some young objects provide evidence that their envelopes are composed of different grains of silicate.

FOURIER TRANSFORM SPECTROMETER FOR THE IRTF ON MAUNA KEA (Reinhard Beer; Jet Propulsion Laboratory, Pasadena)

This is a dual input, dual output, step-and-lock spectrometer for the region 1.1 to 5.5μm, extendable to 20μm. Resolution is 0.01 $cm^{-1}$ with a maximum scan rate of 400 steps/sec and a path difference accuracy < 0.5Å peak to peak in 1Hz bandwidth (< 10Å in 3KHz bandwidth). The reflective optics are sapphire overcoated silver mirrors giving an instrument transmittance of > 75% (beam size 25mm), the beam splitters are either Si on $SiO_2$ or Si on $CaF_2$, and the detectors are InSb housed in I.R.L. dewars with overcoated ZnSe relay and field lenses. The instrument is equipped with a wide range of filters and has built-in facilities for laser alignment and test-lamp calibration. The data handling system utilises two lock-in amplifiers with programmable gain control, digitised to 12 bits, a 24 bit integrator/ratiometer, and a DEC PDP/8E control computer with 66M byte disk for interferogram co-adding, and tape facilities for off-line interferogram transformation. Individuals with appropriate observing facilities who might be interested in using this instrument for their own programs are urged to contact the author.

RESULTS WITH THE HYDROGEN FLUORIDE PRECISION RADIAL VELOCITY TECHNIQUE
(B. Campbell; CFHT Corpn., and A. Irwin; Univ. of Victoria)

This method utilises a hydrogen fluoride absorption cell to superimpose reference wavelength lines upon coudé spectra. A series of high S/N spectra of Arcturus obtained on two nights gave radial velocities with internal errors of 6 m $s^{-1}$. There is, however, a change in velocity of about 100 m $s^{-1}$ between the two nights. This is likely to be real since similar series of spectra of Procyon on the same two nights show no velocity variations. Additional observations of Procyon on nine other nights suggest that any systematic errors in this velocity technique are less than 20 m $s^{-1}$.

THE NEW DIFFERENTIAL PHOTOMETRY SYSTEM OF THE UNIVERSITY OF CALGARY
(E.F. Milone)

The Rapid Alternate Detection System developed at the Rothney Astrophysical Observatory over the past decade for differential variable star photometry uses a single telescope, photometer and detector, and consists of gated, pulse-counting electronics synchronized to a chopping secondary. The mirror is successively driven (Ling driver, ± 10 arcmin maximum throw) to four positions permitting rapid alternate detection of a variable star, comparison star and two sky background regions. The system compensates adequately for the effects of transparency, sky brightness, and detector sensitivity variations occurring on timescales longer than the system's duty cycle, and has been used successfully with as much as three magnitudes of cloud cover and attendant sky brightness variations. Information about this system is also being communicated to Commission 25.

SOME RECENT CHINESE INNOVATIONS IN INSTRUMENT DESIGN (T. Kiang; Dunsink Observatory)

A "photoelectric astrolabe" of novel design was completed in China in 1974. It differs from the Danjon Astrolabe in the following respects: (1) its optical components are placed in an evacuated chamber and the light enters through two horizontal windows on the upper side. In this way, all refraction effects that can be attributed to a plane-parallel atmosphere are eliminated. The vacuum also keeps the mercury horizon clean for a very long time; (2) the Danjon 60°-prism is replaced by a side-by-side split mirror receiving the direct and reflected stellar beams; (3) reflecting surfaces are used exclusively. Thus, there are no chromatic effects and larger apertures are possible (actual aperture = 20cm); (4) a Wollaston prism is not required. The direct and reflected images traverse a reflecting-transmitting grid and the phase of the cut-off signals produced by the grid is registered photoelectrically. In this way, all the light from a star is utilized. A detailed description in English can be found in the translation journal Chinese Astronomy 1977, 1, pp.79-87. The following are some test results, compared with results from parallel observations using a Danjon instrument (in brackets): precision of a single star position: ± 0".14 (± 0".21); closing error in the instrumental almucantar: -0".13 (+1".67); thermal correction -0".025/°C (+0".160/°C). Routine observations with this Chinese photoelectric astrolabe have been in operation in three observatories (Shanghai, Beijing and Yunnan) and the results obtained are generally accessible in Chin. Astron.

Beginning with Vol. 5 (1981), this translation journal was re-titled Chinese Astronomy and Astrophysics. A recent issue (C.A.A. 6, No. 2, 1982) contains reports on two minor innovations. One refers to a three-channel photometer capable of uninterrupted monitoring of a program star, a comparison star and a control star. Three photoelectric multipliers are used and are mounted on three of the four arms of a cross, rotatable in its own plane. The other concerns the chopping arrangement in an IR detector. Unlike the design in general use, the rotating chopper is used to feed in the sky background only, while the fixed mirror feeds in star + sky. This simple exchange of roles reduces excessive demand on the setting and alignment of the chopper.

A SCANNING MICROPHOTOMETER FOR ON-LINE DATA REDUCTION (H. Lorenz; Zentral-institut für Astrophysik, Potsdam)

The measurement of astronomical plates using a scanning microphotometer gives information of high redundancy together with considerable noise. On the basis of HAAR-transform a processing is described which separates the noise optimally by a further variable digital filter (KALMAN filter). The system fits its own bandwidth to the incoming signal and reduces the number of data by a factor of up to 100. For the measurement of astronomical photograms, an evaluation system (ASAG) is planned in cooperation with VEB Carl-Zeiss-Jena. The system will include software packages for different applications and ensure high mechanical accuracy and stability (air bearings, laser metering). The method used allows an improvement in sensitivity for the evaluation of faint objects in the background by 1-3 mag.

DESIGN FOR A U.K. 18M TELESCOPE (A. Boksenberg; Royal Greenwich Observatory)

Design studies carried out at the RGO for an optical/IR telescope of effective aperture $\sim$ 18m favour a multi-mirror concept. The configuration being studied comprises six close-packed f/2 paraboloids (diameter 7.5m) which form an integrated, filled aperture in a cell supported by two sets of relatively short but stiff trusses of an altazimuth mounting; each element of the aperture is a two-mirror system equipped with 0.5° field corrector. The particular advantages of this arrangement are: (1) on-axis optics can be produced more easily than off-axis; (2) the use of two-mirror systems removes the requirement for excessively strict tolerances on focal lengths; (3) each component of the aperture is optically clean and suitable for IR work; (4) it can be used as an aligned, coherent array; (5) flexibility for the efficient matching of instrumentation to the telescope beam by using either multiple instruments at the individual foci or, for large instruments, the combined focal station.

THE NTT AND VLT PROJECTS OF ESO (J.P. Swings; Institut d'Astrophysique, University of Liège)

The New Technology Telescope (NTT) being built by ESO for completion in 1986/87 is intended as a test for several new design concepts before embarking on the much larger VLT project. These include: weight reduction; unique type of focus; compact rotating building; closed loop active control of thin primary mirror (aspect ratio 15/1); reduced auxiliary instrumentation; remote control. With an aperture of 3.5m, its approved budget is DM 24 million (1982) which is less than a third of that for the existing 3.6m telescope at La Silla.

The VLT (Very Large Telescope) is conceived as a powerful new instrument of large aperture with a capability for interferometry. The proposed design consists of four separate telescopes, each with aperture in the range 8-10m, arranged as an array in which the beams can be combined by coherent or incoherent addition. With an equivalent aperture of 16m and diffraction limited imaging in the infrared, the VLT is likely to be used particularly for high resolution spectroscopy of faint objects and interferometry at optical and IR wavelengths.

## Working Group on Detectors (Chairman: W.C. Livingston)

At a meeting of the Working Group held on 24th August the following papers were presented:-

A) Electronography

PROGRESS OF THE ELECTRONIC CAMERAS AND PHOTOCATHODES PRODUCED AT PARIS OBSERVATORY (G. Wlérick, M. Duchesne, B. Servan, F. Gex and J.M. Munier)

Production and research: The photocathode laboratory, which has been transferred to a new building and includes clean room facilities, produces about 200 photocathodes each year: S11 cathodes (20 to 81mm), S20 cathodes (20 to 30mm). Recent developments include S20 cathodes of 81mm diameter and S1 infrared photocathodes. Two cameras for the C.F.H. telescope have been described at the Zelenchuk meeting (IAU Colloquium 67):- (1) the 40mm camera with electrostatic focusing, developed by M. Duchesne, (2) the 81mm "caméra électronique grand champ" which, since the retirement of A. Lallemand, has been operated by B. Servan and G. Wlérick; these cameras are due to be shipped to Hawaii before March 1983.

Observations at Haute Provence: (i) Infrared: In January 1982, M. Duchesne and Y. Andrillat made observations with a camera equipped with an S1 cathode that was both sensitive and homogeneous. They took plates of emission line regions in Orion: [AIII] 775.1nm., [SIII] 906nm. and [SII] 1.033μm. Electronographic cameras are probably the best receivers for monochromatic studies in the 1 to 1.1μm region since they are free from interference fringes; (ii) Caméra grand champ: This camera has been operated mostly by G. Wlérick, B. Servan, V. Cayatte and L. Renard. Different programmes have been carried out: studies of radio sources including the separation of the components of BL Lac type objects (in collaboration with A. Bijaoui); stellar photometry (open cluster M37, studied by A. Robin); Saturn's rings (Sicardy et al.), gravitational lenses (0957 + 561, Vanderriest et al.).

Studies of electronographic emulsions: To choose the most convenient emulsion for a particular observation, G. Wlérick, V. Cayatte, A. Sellier and D. Michet have been engaged in a systematic photometric programme to determine parameters such as characteristic curve, speed, signal/noise ratio, and homogeneity. Whereas the emulsions show very different speeds, the signal/noise ratio is not very sensitive to the emulsion type and to the choice of developer.

ELECTRONOGRAPHIE AVEC LE TELESCOPE CFH (G. Lelièvre)

D'ici un an, les trois types de caméras électroniques seront opérationnelles au T.C.F.H.: (1) La caméra électrostatique de 40mm dévelopée par M. Duchesne est destinée à la spectroscopie avec le spectrographe à fente longue ainsi qu'avec le spectrographe multifentes. M. Duchesne a dû résoudre des problèmes difficiles sur les flexions et l'influence du champ magnétique terrestre afin de respecter les tolérances imposées: les images ne doivent pas bouger de plus de 4μm pour des poses supérieures à 1 H à 45°; (2) La caméra magnétique "grand champ" est décrite par G. Wlérick; (3) La caméra électrostatique à vanne a été presentée à Zelenchuk en septembre 1981. Elle est régulièrement utilisée au foyer primaire du TCFH. Les résultats les plus marquants concernent les mesures UBV de la galaxie responsable du quasar double 0957+561 ainsi que les sondages profonds pour donner une limite de la magnitude limite du noyau de la comète Halley: $m_V > 25.8$ à 3 sigmas.

Pour faciliter les conditions d'observation, le support d'instrumentation au foyer primaire a été reconstruit afin de permettre d'empiler les instruments suivants:- correcteur grand champ; réseaux objectif; bonnette; caméra électronique. Avec des images de 0.7", les magnitudes limites attendues sont: B = 26.5 en photométrie et V = 23 pour la spectroscopie des raies d'émission des quasars. Une nouvelle caméra, avec une photocathode de diamètre 40mm est en cours de réalisation pour remplacer le modèle actuel de 30mm.

ELECTROGRAPHIC CAMERAS AT BEIJING OBSERVATORY (Fu Delian)

A Kron camera which has been developed has been tested and proven to be reliable. The working field is only 12mm in diameter owing to simplified electron optics. Details of a novel gate valve are given. Cs-Sb photocathodes having a sensitivity of 40μa/lum are achieved. This camera is presently being evaluated at the telescope. An improved version having better electron optics and hence a larger field (∿40mm) is being designed. Finally, a magnetic focus camera similar to that built at Meudon is under development. A format of 100mm is the goal.

UBV PHOTOMETRY OF STARS IN M3 (H.D. Ables, C.C. Dahn and A.V. Hewitt; US Naval Observatory, Flagstaff Station)

As part of a continuing program to establish UBV sequences for use with small format panoramic detectors, we have measured photoelectrically and/or electrographically 130 stars in an 8 arc min diameter region centered approximately 13 arcmin east of the nucleus of M3. The stars range in brightness from V = 12.85 to V = 21.37 and almost all appear to be members of M3. The details of the photometry are discussed and the resulting color-magnitude and color-color diagrams are presented.

STELLAR PHOTOMETRY WITH THE 1.5m DANISH TELESCOPE AND 90mm McMULLAN ELECTRONOGRAPHIC CAMERA (A. Blecha, M. Walker and J. Andersen)

A comparison of B-V photometric results for the HODGE-11 LMC cluster has been made from independent analyses of the same data (two B and two V 3 hour exposures), using (1) a Joyce-Loebl microdensitometer at Lick and (2) the Geneva microdensitometer and HP 1000 computer system. Much better plate-to-plate agreement and a more comprehensible colour-magnitude diagram are achieved with the Geneva system working with the sophisticated multiple profile fitting software. In the magnitude range 16-22, a precision of ± 0.04 RMS is obtained. This program is part of a larger photometric study of LMC and SMC globular clusters.

ELECTROGRAPHY AT PIC-DU-MIDI OBSERVATORY (J.L. Vidal)

A short review is given of the main results obtained in recent years at Pic-du-Midi using Lallemand-Duchesne cameras. These include (1) studies of Saturn (discovery of the new satellite, Dione B; measurement of the thickness of Saturn's rings and their scattering properties; detection of E-ring), (2) galactic globular clusters and HII regions (map of the core of M15; discovery of a new breed of nebular condensations in the vicinity of the Trapezium in Orion), (3) extragalactic studies (ionised gaseous content of NGC 1068, supernova in NGC 2903 and in NGC 1199; high resolution plates of the jet in M87).

## B) Solid State Detectors

A WINDOWLESS INTENSIFIED CCD IMAGE SENSOR FOR A FAR-UV ECHELLE SPECTROGRAPH ON A SOUNDING ROCKET (E.B. Jenkins; Princeton University Observatory)

A photon-counting camera to record high resolution spectra ($\lambda/\Delta\lambda = 2 \times 10^5$) of bright stars has been developed at Princeton. Since the spectral range to be recorded spans the Lyman limit and 1100Å, the image must be recorded with a windowless detector. The camera design for this mission uses an opaque photocathode and has an image section with an 18kV potential difference to accelerate the photoelectrons towards a thinned, back-illuminated RCA CCD (320 x 256 active pixels, each 30μm square). A 150 gauss magnetic field directs the electrons at a small angle from the incoming optical beam and focuses them onto the CCD. Each high-energy photoelectron generates several thousand charges, with a lateral spread of about one pixel, within the CCD.

With the CCD running at 15 frames per second, we obtain a readout noise of approximately 200 electrons rms per pixel, which is about a factor of ten below the charge of a single photoevent. Rapid readout permits us (1) to count relatively high photon fluxes, (2) to run the CCD uncooled, and (3) to bypass the requirement for fine image stablization by allowing frame-by-frame post-flight corrections of small translations of the spectrum. In recognising individual photoevents in the data reduction, we can identify the much larger pulses produced by ions and reject them. This is crucial for sounding rocket applications where the vacuum conditions are poor ($P > 10^{-5}$ Torr). Laboratory tests of the resolution of the image section demonstrate that the electron focus is good over the whole field of view, confirming an earlier calculation of electron trajectories for the oblique magnetic focusing.

The flight test of this detector is expected to occur in November 1982 when the spectrograph will be flown for the first time.

IMPROVED INTERFEROMETRIC TESTING FOR TELESCOPE MIRRORS (M.J. Smyth; Dept of Astronomy, University of Edinburgh)

The production of large precise optical components is a cyclical process of measurement and optical working. A new interferometric measurement system, developed at the University of Edinburgh in collaboration with NEI Grubb Parsons Limited, replaces the conventional photographic method and improves the process by shortening the cycle time and improving accuracy and spatial resolution. It comprises a Reticon silicon diode array detector attached to a wavefront shearing interferometer and microcomputer.

The system images the fringe pattern on the detector array which senses the intensities at 512 points across the mirror diameter. These are stored and the measurements are repeated for a total of 5 stepped phase values. Within 10-15 minutes of measurement and computing time, the microcomputer displays a plot of relative phase across the mirror diameter. Interactive graphics allow 360° phase discontinuities to be removed, and the display can be folded to allow comparison and superposition of points equidistant from a selected centre of symmetry. For routine measurements of uncoated mirror surfaces, the relative phase precision is $\sim 0.01$ wavelength. The phase data are stored on a floppy disc and converted into wavefront errors using a summation procedure, giving a precision for the measured wavefront errors of $\sim 0.03$ wavelength (depending on the amount of shear used). For a shear value of 100 mm, the phase precision is equivalent to an image resolution of 0.01 arcsec.

SOLID STATE DETECTOR DEVELOPMENTS AT THE ROYAL GREENWICH OBSERVATORY (P.R. Jorden)

The RGO is concentrating on developing CCD cameras to be used on the telescopes of the new La Palma Observatory. In 1981 a prototype camera was built for the Anglo-Australian Telescope and a description of this and some performance characteristics of the RCA 53612 CCD can be found in Proc. SPIE 331, Tucson 1982 (Jorden, Thorne and Van Breda). A camera for the 2.5m (INT) prime focus is complete and further cameras for the 2.5m Cassegrain and the 1m telescope are under construction. A revised multi-detector system is being designed for the 4.2m telescope along with a CCD autoguider which will be designed in 1983/84. Problems due to Fabry-Perot interference fringe effects in the RCA thinned CCD are being investigated. We have shown that the patterns generated by night-sky emission lines can be effectively removed.

In the U.K., the General Electric Company are making a "speculative" development of a large CCD for the AAT. RGO, together with the Institute of Astronomy, Cambridge, will be testing this device. The format is 1501 x 1501 pixels of size 26 microns square and the noise is expected to be 10 electrons rms. Production is due at the end of this year and we plan to test them early in 1983. At present at RGO we are examining the GEC P8600 CCD in comparison with the RCA 53612 device. Other detector applications, including a micro-channel plate intensified system, are also being studied.

KAVALUR FAINT IMAGE DETECTION DEVICE (J.C. Bhattacharyya; Indian Institute of Astrophysics, Bangalore)

A system for producing digitally recorded faint spectra at the coudé spectrograph focus of the 1m telescope at Kavalur is described. It consists of a 4-stage image intensifier coupled to a CCD detection device by a relay lens arrangement. An Intel 8086 based microprocessor generates the stream of pulses for CCD operation and controls the digitization and storage of the data in an auxiliary dynamic RAM. The system permits reading of selected pixels and can handle a variety of two dimensional detector matrices.

SOLID STATE AREA SCANNER PHOTOMETER (K.D. Rakos; University of Vienna) - presented at General Meeting, 20th August.

An alternative to the use of a photomultiplier and moving slit aperture for extended field photometry is a solid state one-dimensional multidiode detector. A linear self-scanned photodiode array has been tested for this purpose. The detector is a linear array of 128 silicon photodiodes, each 2.5mm in length and arranged on 0.025mm centers with zero aperture effective dead space between them. The peak quantum efficiency is about 80% and the dark current is typically about 5 pA per diode at 25°C. Using a 1m telescope, integration time 1 s and thermoelectric cooling, the difference in brightness of double stars to visual magnitude 10 and as close as 0.5 arcsec can be measured with very high accuracy. Results of UBV and uvby photometry, position angles and separations obtained with this equipment for close visual binaries have been published (Rakos et al., Astron. Astrophys. Suppl. Ser. 1982, 47, 221).

FIBRE-COUPLED CCD'S FOR COUDE SPECTROSCOPY (B. Fort; Toulouse Observatory)

CCD image detectors for use in astronomy have been developed in France through work performed at Paris Observatory (T. Fauconnier), Toulouse Observatory (B. Fort), and the Nuclear Astrophysics Center of Saclay (L. Vigroux). The program started in 1979 with the construction of a prototype using a Fairchild 202 (100 x 100 pixels) and continued with the production of several cameras using thin back-illuminated RCA chips (see SPIE, volume 290, Solid State Imagers for Astronomy). A review of this work is given. The use of an optical fibre dissector with a CCD for coudé spectroscopy is described with particular reference to performance and limitations. Measurements of the shift of the quantum efficiency bandpass at low temperatures for thick and thin CCD's are also presented, and an optimum temperature which results from a trade-off with dark current shot noise is derived for far red observations.

## C) Image Photon Counting Systems

A PULSE COUNTING ARRAY DEVELOPED AT MT. STROMLO OBSERVATORY (Bruce A. Peterson)

Stapinski, Rodgers and Ellis (Publ. Astron. Soc. Pacific 1981, 93, p.242; Advances in Electronics and Electron Physics 1979, 52, p.389) have developed a two-dimensional pulse counting array (2D-PCA) which has been used for photometry and spectrophotometry. It has also been used by Peterson and Savage for stellar calibration photometry (15-19 mag.) at the 1m telescope at Siding Spring Observatory. A large-field version of the 2D-PCA consisting of two CCD's coupled by fiber optics to a microchannel-plate image tube is planned for operational use before the end of 1982.

A TWO-DIMENSIONAL PHOTON COUNTER (S. Shectman; Mt. Wilson and Las Campanas Observatories)

Based on many of the same principles as the Reticon photon counters, this instrument detects photon event scintillations in a multi-stage image intensifier, using a Fairchild CCD 221 diode array. A diode readout rate of 15 MHz, and the ability to inspect as little as one-eighth of the vertical format, make possible frame repetition times as short as 1 millisecond. Photon event centers are located to 0.25 diode horizontally and 0.5 diode vertically.

PHOTON COUNTING INTERFEROMETRY AT THE 6m TELESCOPE (Michel Marcelin; Marseilles Observatory)

The Marseilles Observatory Photon Counting System has worked successfully on several large telescopes (2m O.H.P. telescope, 3.6m E.S.O. telescope, 6m Soviet telescope). Interesting new results have been obtained with the 6m telescope, mainly with Perot-Fabry interferometry, for the study of nearby galaxies (Boulesteix et al., I.A.U. Colloquium 67, Zelenchuk, 1981). In January 1982 we added to our I.P.C.S. a new automatically processed focal reducer equipped with a scanning Perot-Fabry interferometer. This sophisticated set of equipment is called CIGALE ("Cinématique des Galaxies") and will be used mainly at the Cassegrain focus of the 3.6m C.F.H. telescope (first observing run in December 1982). CIGALE will be a powerful instrument for the detailed observation of the velocity field of galaxies in Hα light. It should help in solving problems such as the density wave structure in spiral galaxies through the observation of arm-disk velocity discontinuities.

SPECTROSCOPIC OBSERVATIONS WITH MULTI-ANODE MICROCHANNEL ARRAY DETECTOR SYSTEMS (Bruce Bohannan and J. Gethyn Timothy; Sommers-Bausch Observatory, Colorado)

Multi-anode Microchannel Array (MAMA) detector systems with formats as large as 256 x 1024 pixels are now being prepared for use in a series of ground-based, balloon, and sounding rocket observing programs. The modes-of-operation and the formats of the different detector systems are described briefly and high signal-to-noise spectroscopic data recorded with a visible-light (1 x 1024)-pixel MAMA detector at a number of different ground-based observatories are presented. An outline of our plans for stellar spectroscopy using the 60cm telescope on the University of Colorado campus is also given.

Working Group on Photographic Problems
(Chairman: J.L. Heudier)

The Working Group has been very active since the Montreal meeting in 1979. The mailing list is now about 450 people, indicating the degree of interest worldwide in astronomical photography. A three-day meeting of the Working Group, held at Nice in April 1981, was attended by 80 people from 20 countries. The Proceedings of that meeting have been published (Astronomical Photography 1981, eds. Heudier & Sim) and nearly 600 copies distributed. The Working Group has also encouraged publication of papers in the AAS Photobulletin (more than half the papers published in the Photobulletin are now from non-American contributors). Excellent contacts have been maintained with the AAS Working Group and with Eastman Kodak Company. In the period under review we have made some improvements in methods of ordering, shipping and storing plate material, and a new emulsion has become available. Kodak Technical Pan Film type 2415 is a fine grain panchromatic emulsion, similar in spectral response to the IIIa-F, but with higher resolution. The film version responds well to hypersensitisation but the same emulsion on glass (Kodak Special Plate Type 153-01, backed, or 153-02, unbacked) does not respond so well, at least in its initial tests.

During the General Assembly a meeting of the Organising Committee on 18th August 1982 preceeded a business session of the Working Group and two sessions of short contributed papers. The membership of the next organising committee was adopted as follows: Chairman: Elizabeth Sim (Royal Observatory, Edinburgh, UK); Secretary: K. Ishida, (Tokyo, Japan); Members: O. Dokuchaeva (USSR), J.L. Heudier (France), H. Maehara (Japan), D.F. Malin (Australia), A.G. Millikan (USA), W.E. Schoening (USA), K.R. Sivaraman (India), A.G. Smith (USA), R.M. West (ESO, Munich) and O. Zichova (Czechoslovakia).

The meeting noted with regret the death of W.C. Miller in October 1981, in Pasadena. Bill was a consultant to the IAU Working Group and a former Chairman of the AAS Working Group. His contributions to astronomical photography were fundamental and outstanding. The Working Group expressed its wish to be associated with the tribute to Bill which is being prepared for publication in the Photobulletin.

Seventeen papers were offered for inclusion in the working sessions. Most were presented as short contributions, and included the following: the AAS Photobulletin (A.G. Smith); hypersensitisation of 2415 film (A.G. Smith), and of IIIa-J plates (Takase); the importance of drying hypersensitised plates (McCarthy); automatic processing of large plates (Takase); the incidence of 'gold spots' on IIIa plates (Sim); processing and storage methods to prevent gold spot formation (A. Lecart, Kodak Pathe). Descriptions were given of a new Schmidt Plate Library in Edinburgh (Sim) and the new photographic laboratories at ESO, Munich (West). Applications and research reports included an evaluation of noise on UKSTU plates digitised with COSMOS (Okamura, Tokyo); a study of granularity and resolution, grain size and pixels (Everhart, Denver); two reports on photometry with electronographic emulsions (H. Smith; Wlérick); a report of astrometry of open clusters using the Danish 1.5 metre telescope (Reiz); colour photographs of the zodiacal cloud and of Comet West (Koutchmy & Lamy).

There is a proposal to hold another meeting of the Working Group at the Royal Observatory, Edinburgh, UK, probably in April 1984.

## Working Group on High Angular Resolution Interferometry
(Chairman: J. Davis)

The following papers were presented at meetings of the Working Group held on 21st and 25th August:-

STELLAR INTERFEROMETRY AT CERGA: A REVIEW OF OBSERVATIONAL METHODS, TECHNICAL PROBLEMS AND IMPROVEMENTS (M. Faucherre)

The two telescope interferometer (I2T) at CERGA continues to obtain astrophysical data. Michelson's approach has been adopted to obtain measurements of angular diameters of giant stars: the method involves attempting to see fringes at the limit of detectability for the longest baseline so that the position of the zero of the visibility function can be determined more accurately. A simulation of I2T and atmospheric turbulence has allowed low fringe contrasts to be calibrated and observers to be trained to give improved visibility estimates. Good agreement has been found for observations of $\alpha$ Lyr. The angular diameters of five giant stars have been measured. The paths for I2T can only be matched initially to within ± 1mm and this range has to be searched for fringes since the coherence length is only 30μm at $\lambda = 0.55\mu m$. A new spectrograph is being built which will increase the coherence length to 300μm. When this is completed a photon counting camera will be used in conjunction with electronics to compute the Fourier transform of each frame. This will give the amplitude at the fringe spatial frequency and hence the first objective measure of fringe visibility.

THE LARGE STELLAR INTERFEROMETER AT CERGA (A. Labeyrie and G. Schumacher)

The aim of this project is to build an array of optical telescopes having 1.5m apertures and working in coherence. Currently the first spherical mount telescope works and follows stars with ± 1 arc second accuracy. It is able to send an afocal beam to the central station. The second telescope and the optical table where the beams will be combined are under construction and it is hoped to obtain the first fringes early in 1983.

A PROTOTYPE FOR A LONG BASELINE MICHELSON STELLAR INTERFEROMETER (J. Davis)

Since the completion of the programme with the Narrabri Stellar Intensity Interferometer, the Chatterton Astronomy Department of the University of Sydney, Australia, has devoted its efforts to the development of an interferometer with $\sim$100x the sensitivity and $\sim$100x the resolution of the Narrabri instrument. An amplitude interferometer, although technically more difficult than a very large intensity interferometer, promises greater sensitivity and lower cost and we are building a prototype of a modern form of Michelson's stellar interferometer. This prototype instrument is currently being installed in the grounds of Australia's National Measurement Laboratory near Sydney. It features coelostats at each end of an 11 metre baseline to direct starlight into the instrument, internal path compensation, active wave-front tilt correction and rapid signal sampling to overcome the problem of phase fluctuations. The prototype interferometer will be used to establish whether fringe visibility can be measured through the atmosphere with the reliability and accuracy required to make useful astrophysical measurements. If successful, an interferometer with baselines extending to the order of a kilometre or more is planned.

A TWO-COLOR FULLY-COHERENT INTERFEROMETER FOR LARGE ANGLE AND DIFFERENTIAL ASTRONOMETRY (M. Shao)

We report initial measurements with the NRL-MIT stellar interferometer. The instrument is a phase coherent interferometer that maintains the extreme positioning of its optical elements by the extensive use of active optics technology. The first star observed was Vega for which the angle tracking servo maintained wavefront parallelism to 0.2 arcsec R.M.S. The fringe tracking servo maintained path length equality to better than 0.1 microns. The fringe tracker occasionally lost the fringe but the automatic fringe reacqusition algorithm was successful in finding the fringe every time (in less than 1 sec) over a period of an hour and a half without intervention by the observer. The system can be considered operational in a routine way. Data are currently being analysed to identify instrumental and atmospheric effects which must be calibrated for milliarcsecond astrometry. It is noted that there are some operational advantages of using active optics for both angular diameter measurements and astrometry. Advantages include shorter integration times, lack of ambiguities such as $n2\pi$ in phase measurements and no loss of temporal coherence in visibility measurements.

A MULTIPLE SPACECRAFT MICHELSON SPATIAL INTERFEROMETER (R.V. Stachnik)

A design for a multiple spacecraft Michelson spatial interferometer has been considered. The design includes two one-metre telescopes in orbits which are identical except for slightly differing inclinations. The telescopes achieve separations as large as 10 kilometers and relay starlight to a central station which has a one-meter optical delay line in one interferometer arm. Our four key findings are as follows: 1) a 1000 kilometer altitude, zero mean inclination orbit affords natural scanning of the 10km baseline with departures from optical pathlength equality which are well within the corrective capacity of the optical delay line; 2) electric propulsion is completely adequate to provide the required spacecraft motions (principally those needed for repointing); 3) all necessary technology is already in a high state of development; and 4) resolution and magnitude limits of $10^{-5}$ arcsecond and $m_v$ = 15 to 20 are achievable.

TWO APERTURE INFRARED HETERODYNE INTERFEROMETER AT CERGA (J. Gay and Y. Rabbia)

The infrared correlation interferometer at CERGA uses heterodyne detection, two 1m aperture telescopes and operates at a wavelength of 11 microns. The instrument is still under development. Preliminary observations have revealed that the current sensitivity falls short of the estimated theoretical performance but development continues.

EXTENSION OF PHASE CLOSURE IN OPTICAL ASTRONOMY (N. Arnot and A.H. Greenaway)

Redundant aperture systems may be re-distributed prior to recording images in such a way that they behave as non-redundant apertures and thus allow one to use 'phase closure' to reconstruct diffraction-limited images from single turbulence-degraded frames of data. The re-distribution involves the manipulation, within an interferometer, of three sub-apertures within the redundant aperture system. The scheme will work in the high photon flux limits, extensions to lower fluxes may be possible by summing several solutions. The best signal to noise is obtained by using a T shaped aperture. Preliminary reconstructions using computer simulations indicate that satisfactory solutions may still be obtained with up to 5% multiplicative noise on the data. Detailed simulations with satisfactory atmospheric models and path length correction errors etc., have yet to be done.

IMAGE RECONSTRUCTION FROM PHASELESS FOURIER DATA BY MAXIMUM ENTROPY (J. Skilling)

In general, a large number of images can be compatible with a given dataset. This difficulty is exacerbated with phaseless data, where there is a circular degree of freedom corresponding to each measured Fourier amplitude.

Maximum entropy selects from the compatible images that one with greatest configurational entropy $S = -\Sigma p \log p$ (sum over pixels). This criterion corresponds to seeking a maximally non-committal answer to the question "Where would the next photon come from?", considering the image as a radiating pattern. The resulting image is uniquely free of artifacts, (and maximum entropy has been used successfully in conventional interferometry, optical deconvolution, medical tomography and elsewhere).

A program has now been written for phaseless data from 2-D images of arbitrary size [and tested up to 120 x 120 pixels]. Even with incomplete and noisy amplitude data, sparse fields of point sources appear to reproduce accurately. Complex objects, though, yield reconstructions which often differ considerably from the original. There can also be subsidiary local maxima of the entropy: amplitude data do not suffice to recover complex objects uniquely.

Future work will upgrade the program to 3-D for X-ray crystallography, and will attempt to use maximum entropy in speckle interferometry.

Working Group on Computer Processing of Astronomical Data (Chairman: R. Albrecht)

At the meeting of the Working Group on 24th August the Chairman's Report was read by P. Boyce.

Chairman's Report

The Working Group was originally formed as an ad hoc Working Group during the 1976 IAU Meeting in Grenoble, France. Tom Bolton of David Dunlap Observatory, Toronto, became the first chairman of the Working Group. The "Circular of the Working Group" was initiated to serve as a communication vehicle for the members of the group. The immediate reason for forming the Working Group was the increasing popularity of powerful minicomputers, which for many astronomers meant sudden exposure to the complex problems of software development. Another strong reason was the availability to many astronomers of the PDS measuring engines and their assorted hardware and software problems.

In 1979 the Working Group was incorporated formally as a Working Group of Commission 9. The name of the "Circular of the Working Group", was changed to "Astronomical Image Processing Circular", both to streamline it a bit and to take note of the fact that the growing field of astronomical image processing was now the driving force of all non-theoretical software development. Reacting to widespread complaints that there was no adequate mechanism for the publication of software and of software-related papers, efforts were made to upgrade the "Image Processing Circular" to make it attractive for the publication of such technical papers. I am sharing the editorship of the "Circular" with M. Capaccioli of Padova Observatory. A quarterly publication schedule was adopted and continued throughout 1979 and 1980. There followed a pause in the publication schedule during 1981 for two reasons: contributions were not forthcoming in expected numbers; and the change of affiliation from the University of Vienna to the Space Telescope Science Institute absorbed my productivity, especially during the early days of the Institute when total manpower and support were low. However, the "Circular" has been started up again and will eventually return to the quarterly publication schedule.

As already indicated, it is the purpose of the "Circular" to disseminate information related to software, both SW application and development, and especially information on a very technical level which is usually not publishable in traditional astronomical journals. In fact, even source code listings of moderate length can be published in the "Circular". The intention behind this is to allow the authors of software to get credit for it in the form of a publication, at the same time encouraging better software documentation and making the software available to the community fast. Clearly, the goal has to be more than that. Ideally there should be a software library where software and software-related information is sorted. Using state-of-the-art retrieval systems, software development can be made much more efficient through coordination and sharing. I will go into that in more detail towards the end of this report.

There are no formal rules for membership of the Working Group. Anybody with a legitimate interest in astronomical software can be a member of the Group. The membership list is identical with the mailing list of the "Circular". Since the members of the group are spread all over the world it would be very difficult for the group to engage in more technical activities. This has been taken over by local groups which have been formed recently. In Europe the "Ad Hoc Working Group for the Coordination of Astronomical Software" was formed. Chairman of this group is Phil Crane of the European Southern Observatory. In the United States the "Astronomical Software Working Group" has been formed under the auspices of the American Astronomical Society. Chairman of that group is Don Wells of the National Radio Astronomy Observatory. In cooperation with both chairmen I will now briefly report on the activities of these two groups.

The European group has now been in existence for two years and has held several meetings hosted by major European astronomical institutes. Building on efforts by the European Space Agency to find the best approach for the development of software for the Faint Object Camera of the Space Telescope, the group extended those efforts to the global problem of astronomical software development. The major emphasis is on guidelines for software standardization and software documentation. Software design issues like transportability, maintainability, and image processing techniques have been discussed, as were networks and remote processing. The "Flexible Image Transport System" (FITS), developed and maintained by Eric Greisen and Don Wells, was adopted as a data exchange standard and a FITS-compatible format has been recommended for the exchange of source code. An effort was made by myself to start up a software library but, after a promising start, this also fell victim to my move to Baltimore. I hope that I will be able to revive the project at the ST ScI although at this point I have to be careful with promises.

The main result of these activities is the awareness that software is important, crucial in fact, and that efforts to improve the quality of software and software development more than pay for themselves in the long run. Most important, people know who is doing what, and where they can get information they require. A good example is the Italian computer network ASTRONET which is currently being built up. In order to spend their money as efficiently as possible Giorgio Sedmak and his people use, wherever applicable, available solutions that have been developed elsewhere. And this, after all, is what software coordination is all about.

The AAS working group for Astronomical Software was formed later than the European Group. There have been no meetings so far but the first is scheduled for the January 1983 meeting of the AAS in Boston. Activity is expected to pick up considerably after this meeting. The objectives of the Group are to identify and propagate mechanisms and standards by which coordinated development and exchange of software will be possible. This includes coding, documentation, and data format standards. In particular, FITS will be further developed and improved. Currently, a Microprocessor Special Interest Group is being formed within the Working Group. Also, it was suggested that a "Catalog of Sites and People" be developed, and Dan Klinglesmith of Goddard Space Flight Center has volunteered to collect the data.

The first approximation to such a data base is, of course, the mailing list of the "Image Processing Circular". In October 1980 a one-page questionnaire was sent to all persons and institutes on that list requesting information on type and size of their computers, peripheral devices, software, and software systems. About 70% of the questionnaires came back and it was possible to compile some statistics. I do not want to present any numbers here because they must certainly have changed by now, but the global conclusions are undoubtedly still correct. The computer most widely used throughout the community is the DEC PDP-11. This is not as surprising as the fact that the VAX was already second in 1980. Data General Novas and HPs come next. A wide variety of other machines is used in insignificant numbers.

In the area of software it was found that all but two installations supported, and were actually using, FORTRAN, in conjunction with a vendor-supplied operating system. Several installations use non-standard operating systems like UNIX but so far the numbers are small. A remarkable result was the almost total lack of dedicated software support manpower. With the exception of some large visitor-oriented institutes, all software development is done part-time by scientists, particularly graduate students. Everybody agreed that cooperation and software sharing was desirable even though some did not think it was possible.

Image processing was cited by the majority as the area where such collaboration was most important. This is not surprising since digital image processing is also the principal motivation for the formation of all three working groups. Microdensitometers, digital television cameras, and aperture synthesis radio telescopes are sprouting like mushrooms. The research budgets of nations and of institutes have been, and still are being, strained to provide the computing resources to process and analyze the digital imagery which these new data acquisition systems are creating. The new computer hardware demands new software, and the programming resources of institutes are also being strained to keep up with the demands made by astronomers. The people responsible for managing these programming resources naturally wish to minimize the degree to which they are 're-inventing the wheel', and so they want to coordinate the software development efforts occurring in the various institutes. There is a growing need to be able to transport software and data from institute to institute since astronomers do a lot of travelling and collaborating. Reports of experiences with various hardware and software components used in digital image processing systems are of great value to people who are establishing new installations. These various motivations have all contributed to the formation of the three working groups.

Despite such widespread general agreement there is not much that people seem willing to do about it. Sometimes I get the impression that software is like sex; everybody has it, needs it, but nobody talks about it. And doing it for its own sake, and maybe even improving it, is actually considered dirty. It is a fact that an awful lot of "re-inventing the wheel" is going on out there. The most scarce and most valuable resources, manpower and time, are being wasted. Also, there are some contradictions between the way in which astronomical software development is treated and the treatment of other areas of astronomy; for example, although it is accepted that scientists have an obligation to publish new data and the conclusions derived from them, the same approach is not always encouraged for advances in software.

In the area of hardware development, off-the-shelf components are used widely; home-building some piece of equipment that can be obtained much easier and faster commercially is frowned upon. Yet in the area of software development we are committing all these sins. For instance, nobody publishes software and quite often it is not even communicated to colleagues working in the same institute. This is a waste of resources and should be of interest to the funding organizations. It might be worthwhile to work out some mechanism of active persuasion by which such information can be made accessible to the community.

It seems to me that what we need in the near future is some kind of software library, maybe organized in ways similar to the software libraries maintained by the user groups of some of the computer vendors. A practical solution would be to attach such a library to one of the large national institutes. This would only require adding some storage devices to existing computers and to add one, maybe two, operator-level type of people to the staff.

In Europe this is being done already by STARLINK, the British network, although the software library is intended mainly for the participants of the net. It will also be done to a certain extent by the European Coordinating Facility, the organization managing the European end of Space Telescope Science operations.

In the U.S. a similar exercise will be carried out by the Space Telescope Science Institute, this being the institute which in the long run will be the center of the biggest community of sophisticated users. Of course in both cases the emphasis will be on software useful for ST data analysis.

This concludes my report on the activities of the IAU Working Group for Computer Processing of Astronomical Data. I would like to thank the chairmen of the two regional working groups, Phil Crane and Don Wells, for contributing material to this report.

BABSY - AN IMAGE PROCESSING SYSTEM BASED ON SOFTWARE EXCHANGE (P. Steffen, C. Crezelius, P.T. Rayner, R. Steube; Radioastronomisches Institut der Universität Bonn)

In late 1981 we started to build up the Bonn Astronomisches Bildrechner System (BABSY), based on a VAX 11-780 and an ARGS 7000 image display. In addition to our own software we have installed the following packages: AIPS (VLA), ASPIC and PICPRO (STARLINK) and GIPSY (Groningen). This work was done with the active cooperation of members of the IAU Working Group on co-ordination of astronomical software. Current work includes standards for documentation and "Online Help" and a command language structure for multiprocess control.

A NEW MICROPROCESSOR-BASED CONTROL SYSTEM FOR THE PDS-1010 MICRODENSITOMETER (H. Jenkner, M. Stoll and J. Hron; Institute for Astronomy, Vienna) - presented at General Meeting, 20th August.

An ageing DEC PDP-12 computer, used for control of a PDS microdensitometer and for data acquisition, had to be replaced due to limited functionability, increased unreliability, and maintenance problems. The new system uses a compact Motorola Exorset-30 development system, featuring an M6809 microprocessor, 2 mini-floppy disk drives, keyboard and CRT with graphics capabilities, as well as serial and parallel interfaces. The Exorset-30 is configured to execute all control, data acquisition, and user communication tasks, whereas data storage in general formats is implemented on a DEC PDP-11/34 host computer via a serial RS-232 link. The VIPS (Vienna PDS System) software, written in Basic and assembler, provides an efficient and self-explanatory user interface (continuous display and individual input of parameters and coordinates, single key commands, argument prompting) and increased functionability (parameter/coordinate storage and retrieval, continuous scanning, real-time graphic display during scanning, optional data transmission to host computer).

## COMMISSION 10: SOLAR ACTIVITY (ACTIVITÉ SOLARE)

### Report of Business Meeting

PRESIDENT: V. Bumba SECRETARY: R. Howard

A business meeting was held on August 19. The slate of new officers was approved, and the Commission elected the following Organizing Committee:

President: E. Tandberg-Hanssen

Vice-President: M. Pick

Organizing Committee: A. Bhatnagar, V. Gaizauskas, E. A. Gurtuvenko, J. Machado, D. Melrose, E. Priest, K. Tanaka, and H. Zirin.

The names of new Commission members were read and approved. The Commission approved the following nominations of IAU representatives to international organizations: IUWDS, H. Coffey; QBSA, F. Moriyama; SCOSTEP, M. Kundu.

A report of the SCOSTEP committee on Long-Range Planning for Future Projects was presented by M. Dryer. The International Heliospheric Study and Transfer of Energy within the Solar System were planned, and other suggestions were solicited.

The President read a report by L. Dezso on the efforts in Hungary to continue the Greenwich Photoheliographic Results. The first portion of this work will be published before the end of this year. There will be a study of an overlapping interval to ensure continuity. Spot polarities will be used to decide spot groupings.

A. Koeckelenberg was appointed to replace P. Simon as Chairman of the Working Group on International Programs.

SCIENTIFIC SESSIONS

1. Solar Luminosity Variations (J. A. Eddy) August 18.
   - H. S. Hudson: Observations of Short-Term Solar Irradiance Variations from Spacecraft.
   - P. V. Foukal: Interpretation of Short-Term Solar Irradiance Variations.
   - J. A. Eddy: Historical Reconstruction of Solar Constant Variations.
   - G. Frohlich: Radiometry of Solar Irradiance Variations over Long Time Scales.
   - J. Christiensen-Dalsgaard: Theory of Evolutionary Changes in Solar Luminosity.
   - W. C. Livingston: Indirect Diagnostics of Solar Irradiance Changes.
   - G. W. Lockwood: Luminosity Variations in Late-Type Stars.

2. Solar Maximum Year Special Session (E. Tandberg-Hanssen) August 20.
   - E. Tandberg-Hanssen: Opening Remarks.
   - C. de Jager: The SMY Idea-An Overview.
   - Z. Svestka: Report of FBS Activities.
   - D. M. Rust: Report of SERF Activities.
   - M. Dryer: Report of STIP Activities.
   - J. Smolkov/E. Stepanov: SMY Activities in the USSR.
   - H. Tanaka: Report on Hinotori.
   - K. J. Frost: Report of SMM.
   - M. Pick: Radio Astronomical Studies During SMY.

3. Recent Solar Results (V. Bumba) August 21.

H. Snodgrass and R. Howard: Differential Rotation of Solar Magnetic Fields.

R. W. Noyes: Magnetically Sensitive Emission Lines in the Far Infrared.

D. Ventkatesan: Solar Coronal Holes.

M. Pognerusse, C. Caroubalos, and J. Z. Steinberg: Stereo 5 Measurement of Type III and Type IIIb Radiation Patterns.

A. O. Benz: Solar Maximum Mission and Decimetric Radio Observations.

A. Dollfus: A New Solar Filter at the Meudon Observatory.

L. E. Alissandrakis: Preliminary Results from the Clark Lake Radioheliograph.

L. E. Alissandrakis: Ring Structure of an Active Region at 6 cm Wavelength.

T. Takakura: Sub-Second Fine Structure in a Solar Burst Observed Simultaneously at Hard X-Rays and mm Waves.

S. Martin and P. Kaufmann: Mm-Microwave Source Outside the Solar Limb Detected After a Major Prominence Disruption.

O. F. Smith: Microwave Signature of Thick-Target Electron Beams.

E. Becklin: Recent 30, 50, 100 and 200 μm Determination of the Solar Limb Extension.

K. V. Sheridan: A New Frequency at the Culgoora Radiospectroheliograph.

4. Magnetic Field Geometry in Active Phenomena.

This meeting, organized by Dr. R. G. Giovanelli, consisted of several papers on sunspot magnetic field configurations, including penumbral filaments, the heights of the horizontal field component (magnetic canopies), radio observations of quiet and active coronal fields, and airborne eclipse observations bearing on the increase of the filling factor with height above the active regions.

COMMISSION 12: RADIATION AND STRUCTURE OF THE SOLAR ATMOSPHERE
(RADIATION ET STRUCTURE DE L'ATMOSPHERE SOLAIRE)

Report of the Meetings held in conjunction with the 18th General Assembly

President : Y. Uchida

## Business Meeting of Commission 12

Business Session on August 19, 1982 was chaired by Y. Uchida with the secretary P. Foukal, and the following items were decided :

### I. Organizing Committee

The Commission elected the following new Executives and Organizing Committee Members of Commission 12 for the term 1982 - 1985.

President : R. W. Noyes
Vice President : M. Kuperus
Organizing Committee Members : Chen Biao, F. Deubner, E. Fossat, J. W. Harvey, V. A. Kotov, J. Leibacher, K. R. Sivaraman, J. O. Stenflo, P. R. Wilson, Y. Uchida ( ex-president ).

### II. New Members of the Commission

New general members of the Commission were endorsed as follows :
C. E. Alissandrakis, D. L. Book, J. B. Breckenridge, V. F. Chistyakov, E. Chvojkova, J. W. Cook, L. Dunkelman, P. Foukal, V. Gaizauskas, V. P. Gaur, M. V. Goldman, S. S. Hassan, R. M. E. Illing, P. Kaufmann, S. L. Keil, M. R. Kundu, B. J. La Bonte, J. T. Mariska, M. R. Nicolas, R. E. Rees, V. Rusin, M. Rybansky, J. Sakai, T. Sakurai, S. J. Schwartz, R. Stebbins, A. H. Vaughan, J. K. Vial, D. Wentzel, L-X. Wu,

### III. Working Group on Eclipses

Report from the Working Group on Eclipses ( Drs. Leroy and Fiala ) on the past and future eclipses was read by the chairman. It was decided that the Working Group is maintained with the same member in the term 1982 - 1985. Further meeting for the Indonesian eclipse in 1983 was organized separately by M. Aizenman.

### IV. Resolution

The Commission supported the following resolution :
"Commission 12, recognizing the importance of the observation of solar seismology, strongly supports international cooperation in establishing a world-wide network of observing stations.

Scientific Sessions and Symposium Cosponsored by Commission 12

I. IAU Symposium No 102, "Solar and Stellar Magnetic Field : Origins and Coronal Effects"

This symposium sponsored by Commission 12 with the cosponsorship of Commissions 10, 27, 35, 36, and 44, was held in Zurich on August 2 - 6 in conjunction with the 18th General Assembly. It was organized by J. O. Stenflo under the support of Swiss Federal Institute of Technology. Scientific Organizing Committee consisted of J. O. Stenflo ( chairman ), D. Mihalas, R. W. Noyes, A. Severny, M. Stix, Y. Uchida, G. Vaiana, N. Weiss, O. C. Wilson, and C. Zwaan, and Local Organizing Committee was headed by M. C. E. Huber. 17 invited talks and 34 contributed talks were given on the problems of the origin and the effects of the solar and stellar magnetisms. Detailed report will be published as the Proceedings of IAU Symposium No 102, separately.

II. Joint-Discussion "Solar Luminosity Variations"

This Joint-Discussion cosponsored by Commission 12 jointly with Commissions 10, 27, and 35, was organized by the Scientific Organizing Committee consisted of J. Eddy ( chairman ), P. V. Foukal, D. O. Gough, G. A. Newkirk and G. W. Lockwood and was held on August 18, 1982. The program consisted of seven invited talks and six contributed talks. Details will be found in the Highlights of Astronomy.

III. Special Session "Solar Maximum Year"

The special session cosponsored by Commission 12 jointly with Commissions 10, 40, and 44, was organized by E. Tandberg-Hanssen, and was held on August 20, 1982. A whole day was devoted to the special session which consisted of ten talks summarizing the organized efforts of Solar Maximum Year Campaign including the results obtained by satellites SMM and Hinotori. Details will be found in the Highlights of Astronomy.

IV. Joint-Meeting "Recent Solar Results"

This joint-meeting cosponsored by Commission 12 jointly with Commissions 10 and 44, was organized by V. Bumba and was held on August 21, 1982. The description of this half-day session will be found in the report of Commission 10.

V. Joint-Meeting "Magnetic Field Geometry in Active Phenomena"

This joint-meeting, cosponsored by Commission 12 jointly with Commission 10, and 49, was organized by R. Giovanelli and was held on August 23, 1982. The description of this whole-day session will be found in the report of Commission 10.

VI. Joint-Meeting "Solar and Stellar Spots and Activity Cycles"

This joint-meeting, sponsored by Commission 12 with the cosponsorship of Commission 10, was organized by R. W. Noyes and was held on August 24, 1982. R. W. Noyes opened the discussion by pointing out that recent data on spots and activity cycles of lower main sequence stars other than the Sun is now providing a rich new source of information on the general phenomenon of starspots and stellar activity cycles. The purpose of the discussion meeting was to compare and contrast related phenomena of magnetic activity on the Sun and on similar stars. ( It was noted that the topics covered relate closely to the more in-depth proceeings of the recent IAU Symposium No 102 "Solar and Stellar Magnetic Fields : Origins and Coronal Effects", and IAU Colloquium No. 71 "Magnetic Activity on Red Dwarf Stars" ).

The major content of this discussion meeting consisted of three invited review papers. First, M. Rodono reviewed "The Starspot Phenomenon", comparing physical properties of spots on BY Dra stars and RS CVn stars with those on the Sun. The speaker reviewed the observations of photometric waves on spotted stars and implications for physical properties of stellar spots and their rotation and differential rotation. Second, A. Vaughan discussed "Solar and Stellar Activity Cycles", reviewing the accumulation of data from Mt. Wilson Observatory on activity cycles of lower main sequence stars. Two principal conclusions were that solar-like cycles appear clearly only in stars with rotation periods of 20 days or more, and that for stars showing activity cycles there is at present no evidence for a dependence of cycle period upon rotation rate or on spectral type. In the third invited paper, Stix reviewed "Theories of Stellar Spots and Activity Cycles". This discussion covered the theory of concentration of magnetic flux into flux knots, and spots, as well as dynamo theories of cycles and ways in which they can be reconciled with observations of stellar cycles.

Contributed papers include : (1) H. Hudson and H. Jones ; Solar Magnetism and Rapid Variations of Solar Irradiance. (2) L. Doherty : Mg II h, k Emission in F and G Dwarfs. (3) A. Charles and L. Booth : The X-ray Transient 3A 1431-409. (4) P. Wilson and P. McIntosh ; A New Model for the Solar Magnetic Field and its Relation to the Solar Cycle. (5) R. Howard and B. LaBonte ; The Observed Relationships between some Solar Rotation Parameters and the Activity Cycle. (6) R. Giovanelli ; The Mechanism of the Sunspot and Solar Magnetic Cycles as inferred from Observations. Poster presentations made during the session include : (1) A. Boesgaard and T. Simon ; Observations of Chromospheric Activity through Rotation Cycles of the Young Solar-type Star $\chi^1$ Ori . (2) D. Dravins, P. Linde, K. Fredga, and G. Gahm ; Beta Hydri ; Chromospheric Activity in a Very Old Solar-type Star. (3) J. Arsenijevic ; A Possible Eruptive Event on the Cool Supergiant $\mu$ Cephei.

In concluding, I would like to acknowledge with the heartiest appreciation the valuable cooperation of Dr. R. W. Noyes, the Vice President, and of all the Members of the Organizing Committee, Drs. P. Delache, F. Deubner, A. Gabriel, S. Gopasyuk, J. Harvey, R. Howard, P. Wilson, and C. Zwaan, throughout the term 1979 - 1982. Finally, representing Commission 12, I would like to express our deepest sorrow for the passing away of Prof. M. K. V. Bappu, the President of IAU as of the 18th General Assembly, and the former President of Commission 12 in 1976 - 1979, and would like to deliver our heartiest condolences to his family.

Yutaka UCHIDA
Presidnet, Commission 12.

## COMMISSION 14: ATOMIC AND MOLECULAR DATA (DONNEES ATOMIQUES ET MOLECULAIRES)

Report of Meeting, 19 August 1982

PRESIDENT: John G. Phillips SECRETARY: Ralph W. Nicholls

The meeting of the Commission was opened by a brief business session, followed by a number of communications in a scientific session. About 30 members of the commission attended as the meeting progressed.

### I. Business Session

A. The President called attention to reprints of the triennial Draft Report, that had just been received. In referring to the truncated form of some of the Working Group reports, he reminded the meeting of the plan to distribute the full reports in mimeographed form.

B. Three years ago, at the 1979 Montreal meeting, a major item of discussion was the efficacy of our current Working Groups. The only substantive results of that discussion was the change in the name of the Commission, from "Fundamental Spectroscopic Data," to "Atomic and Molecular Data," thus opening up the possibility that the Commission might include Working Groups charged with the monitoring of important non-spectroscopic material, if such were deemed desirable. However, no changes or additions were made at the time. Numerous opinions expressed at Montreal were repeated and added to. They included:

1. While historically, the current Working Groups evolved as the needs of astrophysics for data developed, and were scientifically useful in their time, the present distribution of Working Groups may not any longer reflect the current needs of astronomy for atomic and molecular data. Some Working Groups may have served their purpose and should be retired. A possible example of the latter that was discussed at Montreal was Working Group 1 (Wavelength Standards); at that time it was the consensus that it should be continued, since to astronomers good secondary standards are of vital interest, especially when new spectral regions become available to astronomical observation.
2. Perhaps some Working Groups now cover too great a field of current research and should be further subdivided (e.g., Collision Cross Sections and Line Broadening; Molecular Spectroscopy).
3. Perhaps other Working Groups should be added to reflect other data needed by astronomers (e.g., Thermonuclear Processes; Elementary Particle Physics).

During a wide-ranging discussion of these questions, the following points were made:

1. In their present form the reports of the Commission are not as valuable to astronmers as they might be. Many do not know of the existence of these reports, yet astronomers are always anxious to have available atomic and molecular data from critically filtered data bases without the need for them to address the principal literature.
2. Beyond the compilation of bibliographies of original results, an essential feature that would be welcome by astronomers would be a critical review of data.
3. The production of such reviews is a lot of work. Probably the Trans. IAU is not the best vehicle for publication of critical reviews. The Commission should encourage publication of such reviews in, say, the Annual Reviews of Astronomy and Astrophysics, etc.
4. When the Commission considers other fields (e.g., Elementary Particle

Physics) to be monitored by additional Working Groups, it should be sure that some other Commission is not already doing so.

After consideration of the points of the discussion the meeting adopted the proposal of the chairman that:

"A subcommittee of three persons be established to consider, after consultation with colleagues, which fields our Working Groups should cover, and how they should be organized. The subcommittee will report back before the 1985 General Assembly of the IAU so that their recommendations can be fully discussed at the business meeting of the Commission at that General Assembly."

A. H. Gabriel, R. W. Nicholls, and K. Takayanagi were unanimously adopted as the members of this committee.

C. In a broader discussion of the service of the Commission to the astronomical community, the chairman reported on informal discussions he had carried out with a number of influential astronomers. What emerged was a general consensus well-expressed by Dalgarno, that our current role of data compilation in triennial reports might be outdated, in view of the evolution of Data Centers in Europe, the US and Japan. The discussion at the meeting raised the following points:

1. The production of a loose-leaf Commission 14 data handbook might help to keep our information current.

2. While the triennial reports already in part refer to the various data bases, they might in the future be restructured to lead astronomers more specifically to data bases of value in the various fields.

3. Data bases are all very well, but what is also needed is a series of critical compilations of reliable data. Existing data bases are of varying quality and may have been oversold.

4. What is needed is a practical and usable index of critical data bases for working astronomers.

5. Critical compilation is a hard and thankless task viewed by some as secondary scholarship, and not research. Thus few would do it as a service to the community.

6. A data base for species of interest in stellar atmospheres is being set up at University College London. Reference was made to the CODATA Bulletin on Atomic and Molecular Data. Few members of the Commission present at the meeting were aware of it. Reference was also made to the Berkeley Newsletter as a continually updated springboard to the literature fo molecular spectra. Phillips reported that it accessed over 30 journals, had a mailing list of nearly 500 workers and had been running since 1958.

7. There is a lack of uniformity of information in, and updating of, the various data bases and mode of access to them. What are urgently needed by the user are internationally agreed protocols for atomic and molecular data bases.

8. There is a great need for all developers and users of data banks to establish a dialogue and reach some sort of consensus before investing a lot of time and money in such endeavors.

In the light of these comments, the chairman suggested that the Triple Commission of Spectroscopy (IAU/IUPAC/UIPAP) be informed of our concerns and be asked to help coordinate needs. Accordingly, it was moved and adopted that:

"There be a charge to the IAU representatives on the Triple Commission on Spectroscopy to raise the profile of data banks on atomic and molecular data with their colleagues in the Triple Commission and the respective Unions to seek to achieve some sense of uniformity in data gathering and acquisition."

The current IAU members on the Triple Commission are Edlen, Phillips and Seaton. Initial informal discussions via the Triple Commission could eventually lead to more formal action by the three Unions and the involvement of the International Council of Scientific Unions (ICSU). It was emphasized that users should be consulted extensively on data needs and format. Nicholls referred to his questionnaire to be distributed to stellar spectroscopists on their needs for stellar spectroscopic molecular data.

D. In connection with possible cooperation with other Unions, the chairman reported on his discussion with Professor A. Schawlow of Stanford, the chairman of Commission 15 (Atomic and Molecular Physics and Spectroscopy) of IUPAC on possible collaboration and cooperation. Prof. Schawlow has subsequently reported general receptivity of possible collaboration from members of his Organizing Committee. They will be waiting to see what the astronomers suggest as an outcome of this General Assembly. Their advice and collaboration will be of great value in connection with the charge to the Triple Commission mentioned above.

E. Appointment of President, Vice-President, Organizing Committee and Chairmen of Working Groups for the 1982-85 period, and other committee appointments.

President: A. H. Gabriel
Vice-President: R. W. Nicholls

Organizing Committee: E. Trefftz had asked to be retired from the Organizing Committee for personal reasons. The President was instructed to send to Prof. Trefftz a letter of thanks for her service to the Commission. H. Nussbaumer was appointed to succeed her, so that the 1982-85 members on the Organizing Committee are:

D. R. Johnson
S. L. Mandel'shtam
H. Nussbaumer
J. G. Phillips
S. Sahal
K. Takayanagi
W. Wiese

It was suggested by the incoming President, A. H. Gabriel, that the new Organizing Committee should co-opt some astronomers onto the Organizing Committee.

Chairmen of Working Groups:

| | | |
|---|---|---|
| 1. | Wavelength Standards | K. M. Baird |
| 2. | Atomic Transition Probabilities | W. Wiese |
| 3. | Collision Cross Sections and Line Broadening | A. Dalgarno |
| 4. | Atomic Spectra | W. C. Martin |
| 5. | Molecular Spectra | R. W. Nicholls |

Other Memberships:
1. Consultative Committee on Definition of the Metre: A. H. Cook
2. Triple Commission on Spectroscopy: B. Edlen, J. G. Phillips, M. J. Seaton

Commission Memberships: The Commission approved the addition of the following to the membership of the Commission:

Alan W. Irwin
Keith Adrian Berrington
D. A. Varshalovich
V. S. Strelnitsky

## II. Scientific Session

### 1. ON A NEW DEFINITION OF THE METRE

A. H. Cook
The Cavendish Laboratory
Cambridge, ENGLAND

Professor A. H. Cook who is the representative of the Union on the Consultative

Committee for the Definition of the Metre of the International Committee for Weights and Measures, described recent developments in the stabilisation and measurement of frequencies of laser radiation and the way in which they have led to the now accepted principle that the standard of length, the Metre, should be re-defined by the distance travelled by light in a specified time, the speed of light being taken to be the invariable, conventional value of 299 792 458 m/s. The choice of that value for the speed of light ensures that any new definition of the Metre according to those principles will be consistent with the present (less precisely realisable) definition in terms of the wavelength of a line in the spectrum of Kr.86. The Consultative Committee met in 1982 June in Sevres and adopted a resolution (of which the English text is appended) recommending a form of definition of the Metre. It is expected that the proposed new definition will be considered at the next General Conference of Weights and Measures. The definition as proposed by the Consultative Committee corresponds with geophysical and astronomical practice whereby terrestrial and astronomical distances are determined by the times of passage of electromagnetic signals.

## Appendix

On a New Definition of the Metre
Recommendation M 1 (1982)

The Comité Consultatif pour la Définition du Mètre, recalling its Recommendation M 2 (1979) and considering:

1 - that the present definition does not allow a sufficiently precise realization of the metre for all requirements;
2 - that progress made in the stabilization of lasers allows radiations to be obtained that are more reproducible and easier to use than the radiation emitted by a krypton-86 lamp;
3 - that progress made in the measurement of the frequency and wavelength of these radiations has resulted in concordant determinations of the speed of light of which the accuracy is limited principally by the practical realization of the present definition of the metre;
4 - that wavelengths determined from frequency measurements and a given value for the speed of light have a reproducibility superior to that which can be obtained by comparison with the wavelength of the standard radiation of krypton-86;
5 - that there is an advantage, notably for astronomy and geodesy, in maintaining unchanged the value of the speed of light recommended in 1975 by the 15th Conférence Générale des Poids et Mesures in its Resolution 2 ($c$ = 299 792 458 m/s);
6 - that a new definition of the metre has been envisaged in various forms all of which have the effect of giving the speed of light an exact value, equal to the recommended value, and that this introduces no appreciable discontinuity in the unit of length, taking into account the indetermination of $\pm 4 \times 10^{-9}$ of the best realizations of the present definition of the metre;
7 - that these various forms, making reference either to the distance travelled by light in a specified time interval or to the wavelength of a radiation the frequency of which is measured or specified, have been the object of consultations and deep discussions, have been recognized as being equivalent and that a consensus has emerged in favour of the first type;
8 - that the Comité Consultatif pour la Définition du Mètre is now in a position to give instructions for the practical realization of such a definition, instructions which could include the use of the orange radiation of krypton-86 used as standard up to now, and which may in due course be extended or revised.

recommends:

- that the metre be defined as follows:

"the metre is the length of the path travelled by light in vacuum during the fraction (1/299 792 458) of a second";

- that the definition of the metre in force since 1960, based upon the transition between the levels $2p_{10}$ and $5d_5$ of the atom of krypton-86 be abrogated.

## 2. ATOMIC PROCESSES IN PLANETARY NEBULAE

G. Peach
Department of Physics and Astronomy
University College London, England

A detailed knowledge of many different types of atomic process is required in order to successfully interpret the complex features observed in the spectra of Planetary Nebulae. Much progress has been made in the last few years in the development of methods that can give accurate atomic data for this purpose, and in this paper recent advances are summarised. This report is based on the talks given by Dr. C. Mendoza, Dr. P. Storey and Professor R. McCarroll at the symposium on Planetary Nebulae, held at University College London just before the General Assumbly of the IAU.

The types of atomic data required may be listed as follows:

1) Atomic Transition Probabilities
2) Rate Coefficients for Electron Excitation
3) Photoionization Cross Sections
4) Radiative Recombination
5) Charge Exchange Reactions.

### Electron-Atom Processes

If $A^{+m}(SL)$ denotes an m times ionized atom A in a state SL, the following processes are of interest:

1) $A^{+m}(SL) + h\nu \rightleftarrows A^{+m}(S'L')$
2) $A^{+m}(SL) + h\nu \rightleftarrows A^{+m+1}(S'L') + e$
3) $A^{+m}(SL) + e \rightleftarrows A^{+m}(S'L') + e$
4) $A^{+m}(SL) + e \rightleftarrows A^{+m-1}(S''L'') \rightleftarrows A^{+m}(S'L') + e$

New computers, for example Cray-1, enable us to get accurate data for atoms of the 1st row and to get less accurate but nevertheless consistent data for 2nd row atoms. Calculations for heavier systems such as $Fe^{+m}$ can now be attempted. A compilation of transition probabilities, electron excitation coefficients and photoionization cross sections of interest in Planetary Nebulae has been made and critically evaluated by Dr. C. Mendoza (University College London).

### Radiative Recombination

Radiative recombination of an atomic system may take place in two ways, i.e.:

1) $A^{+m+1}(S'L') + e \rightleftarrows A^{+m}(SL) + h\nu$
2) $A^{+m+1}(S'L') + e \rightleftarrows A^{+m}(S''L'') \rightleftarrows A^{+m}(SL) + h\nu$

Process 1) represents direct radiative recombination, whereas process 2) is that of dielectronic recombination, i.e. the recombination proceeds through an intemediate doubly excited state S"L" of the atomic ion. It has been shown that the results for processes of type 1) as given by Aldrovandi and Péquignot (1973) can be wrong by as much as a factor of two or three at temperatures of $10^4$ K (Peach 1982, unpublished). Storey and Nussbaumer (1982, unpublished) have studied ions of carbon, nitrogen and oxygen and have found that the Burgess formula for dielectronic recombination grossly underestimates the effects for temperatures less than $10^4$ K, because the formula neglects the effects of low-lying resonances. More calculations are in progress at University College London (Peach and Storey).

### Charge Exchange Reactions

The charge exchange reactions that are important are:

1) $A^{+m} + H \rightarrow A^{+m-1} + H^{+}$

2) $A^{+m} + He \rightarrow A^{+m-1} + He^{+}$

where the atom A denotes C, N, O, Ne or S. Collisions of the type 1) and 2) at thermal energies can be the dominant mechanism for determining the relative concentrations of the ions. The inverse processes are not usually important, as $A^{+m-1}$ is usually formed in an excited state, which rapidly decays radiatively. So far, the conclusions that have been reached are that firstly the Landau-Zener formula can be very inaccurate, and secondly that good results can be obtained for processes in which no rearrangement of the core $A^{+m}$ takes place. The accuracy of the data of Butler et al. (1980) on the two-electron charge-exchange processes cannot yet be assessed (McCarroll, 1982).

### References

Aldrovandi, S.M.V. and Péquignot, D.: 1973, Astron. and Ap. 25, p. 137.
Aldrovandi, S.M.V. and Péquignot, D.: 1976, Astron. and Ap. 47, p. 321 (erratum).
Butler, S.E., Heil, T.G., and Dalgarno, A.: 1980, Astron. and Ap. 89, p. 379.
Mendoza, C.: 1982, "Recent Advances in Atomic Calculations and Experiments of Interest in the Study of Planetary Nebulae," invited talk at the IAU symposium on Planetary Nebulae, University College London.
McCarroll, M: 1982, "Charge Exchange Reactions in Astrophysical Plasmas," invited talk at the IAU symposium on Planetary Nebulae, University College London.
Nussbaumer, H. and Storey, P.: 1982, contributed paper to the IAU symposium on Planetary Nebulae.
Storey, P.: 1982, "Radiative Recombination," invited talk at the IAU symposium on Planetary Nebulae, University College London.

## 3. RECENT RESULTS ON THE SPECTRUM OF Cr II WITH ASTROPHYSICAL APPLICATIONS

Sveneric Johansson
Department of Physics
University of Lund

### I. Introduction

An extended analysis of Cr II was initiated for special reasons besides the general purpose of increasing the knowledge of the atomic structure of neutral and ionized atoms. The following four points are discussed in more detail:

(1) The structure of Cr II is similar to that of Fe II, previously analysed by the present author (1978), due to exchange of electrons to holes in the 3d- and 4s shells.

(2) Iso-ionic studies of the singly ionized 3d elements (Johansson et al. 1980) revealed gaps in the knowledge of Cr II, particularly for high configurations.

(3) In an article by Cowley and Arnold (1978) it was shown that many lines in the optical spectra of Ap stars could be identified as predicted lines of Cr II, calculated from known levels, but not observed in the laboratory.

(4) There is an urgent need of Cr II data in the short wavelength region of IUE. In the previous analysis of Cr II (Kiess 1951) no lines were reported below 1780 Å.

#### A. THE ATOMIC STRUCTURE OF Cr II

The fact that 3d- and 4s electrons have almost the same binding energy means that the configurations $3d^{k+1}$, $3d^{k}4s$ and $3d^{k-1}4s^{2}$ have nearly the same energy. This set of configurations in Cr II, with k=4, and in Fe II, with k=6, is given in Table

1. These two groups of low even configurations generate the same set of LS terms, as $3d^7$ and $3d^6$ are equivalent to $3d^3$ and $3d^4$ respectively, while the presence of $4s^2$ does not affect the possible values of L and S. It can also be visualized if the three Fe II configurations are written in the number of holes instead of electrons for the 3d- and 4s shells. We then get $3d^3 4s^2$ (for $3d^7$), $3d^4 4s$ (for $3d^6 4s$) and $3d^5$ (for $3d^5 4s^2$), which is equivalent with the three electron configurations in Cr II.

| | Low Even Configurations | Low Odd Configurations |
|---|---|---|
| | $3d^{k+1}$, $3d^k 4s$, $3d^{k-1} 4s^2$ | $3d^k 4p$, $3d^{k-1} 4s4p$, $(3d^{k-2} 4s^2 4p)$ |
| CrII, k=4, | $3d^5$, $3d^4 4s$, $3d^3 4s^2$ | $3d^4 4p$, $3d^3 4s4p$ |
| FeII, k=6, | $3d^7$, $3d^6 4s$, $3d^5 4s^2$ | $3d^6 4p$, $3d^5 4s4p$ |

Allowed Transitions

$3d^{k+1} \nearrow 3d^k 4p \nwarrow 3d^k 4s \nearrow 3d^{k-1} 4s4p \nwarrow 3d^{k-1} 4s^2$

$\nearrow$ 3d-4p transition, $\nwarrow$ 4s-4p transition

Table 1: Synoptic table of the low configurations in the spectra of neutral and singly ionized iron group elements. The strongest transition arrays are displayed.

The overlapping of these three low configurations of the same parity is considerable through the whole iso-ionic sequence of singly ionized iron group elements. This is illustrated in Figure 1 where the relative positions of the configurations are represented by their lowest LS term. Together these configurations generate a great number of metastable levels, 73 in Cr II and 61 in Fe II, and thereby a large set of possible forbidden transitions. The increasing binding energy with k of the 3d electron relative to the 4s electron in the two halves of the 3d shell allows $3d^{k+1}$ to constitute the ground state in five cases and $3d^k 4s$ in four cases, while $3d^{k-1} 4s^2$ never gets this status in second spectra. When a 3d- or 4s electron is excited to a 4p orbital we get three odd configurations, as illustrated in Table I, but the third one, $3d^{k-2} 4s^2 4p$, is predicted to appear high up in the level system and is generally not observed. The strongest absorption lines in the second spectra of the iron group elements are consequently due to 3d-4p and 4s-4p transitions, which normally appear below 3000 Å (see Figure 5).

B. ISO-IONIC REGULARITIES

The iso-ionic regularities through the sequence of singly ionized iron group atoms provide possibilities to roughly predict the positions of the lowest terms in different configurations. In Figure 2 the term value, i.e. the absolute value of the binding energy, is plotted for the lowest level of the observed configurations. From this graph the position of the completely unknown configuration $3d^3 4s4p$ in Cr II was interpolated and the strongest transitions down to $3d^4 4s$ were predicted to appear around 1430 Å. This is the most prominent group of lines below 2000 Å and it will be discussed further in Section III.

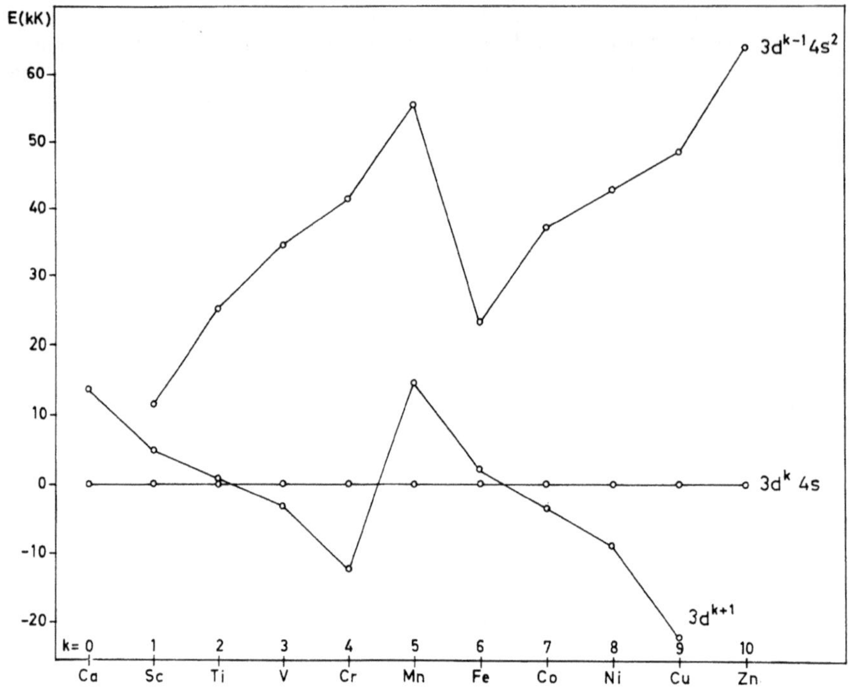


Fig. 1: Positions of the lowest terms of the $3d^{k+1}$ and $3d^{k-1}4s^2$ configurations relative to the lowest term of $3d^k4s$ for singly ionized iron group elements.

C. PREDICTED Cr II LINES IN STELLAR SPECTRA

The reinvestigation of neutral and singly ionized iron group elements by means of hollow cathode lamps have offered better possibilities to separate and analyse the spectra. The additions to the term structure have often been extracted from new observations in the vacuum ultraviolet and the infrared regions in the later studies. In some cases the line rich optical region (3000-5000 Å) has not been reexamined, which means that the old line lists are still valid for this region. Most of the old analyses were based on recordings of the spectrum from an arc in air with serious blends with molecular bands as well as other impurity lines. However, in most cases the high quality of the analyses is impressive, and the very few mistakes in the previous work on e.g. Fe II and Cr II give evidence for that. These analyses were carried out without the support of parametric calculations of the term structure, which is an important tool in modern spectral analysis. It should be mentioned that the data compiled in the Revised Multiplet Tables (Moore 1945) for the neutral and singly ionized iron group elements are extracted from these analyses from the 1920s and 1930s. From this background it is not surprising that many lines, calculated from later level compilations give rise to new identifications in stellar spectra. Thus, the important suggestion by Cowley and Arnold (1978) that doubtful identifications may be replaced by permitted transitions in iron group spectra illustrate the continuous need of updating all useful compilations and tables. Cowley will illustrate this identification technique by some specific examples at the Joint Commission Meeting and his article together with Arnold as well as his talk elucidates the need of more laboratory spectroscopy of elements with high cosmic abundances.

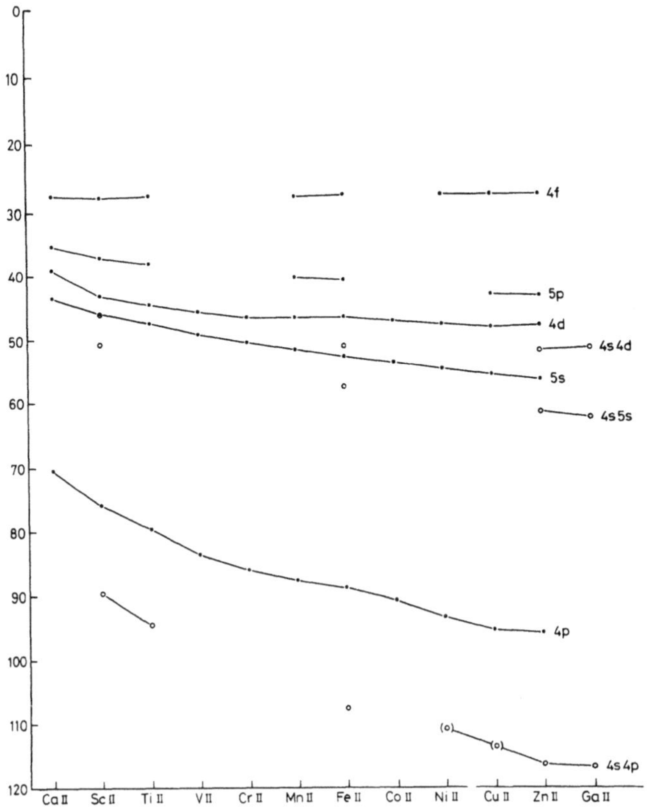


Fig. 2: Term values of the lowest-limit levels with highest possible J-quantum number in some excited configurations of singly ionized iron-group elements (from Johansson et al. 1980).

D. Cr II DATA IN THE IUE REGION

The work by Kiess on Cr II (1951) does not include any lines below 1780 Å, which explains why the $3d^3 4s4p$ configuration was unknown. In the present work about 1000 lines have been identified below this earlier limit. The self-evident fact that lines at lower wavelengths and higher energies are assigned to transitions from deep levels in the energy diagram give them high probability to contribute to the opacity in stellar spectra.

## II. Experiments and Results

A. RECORDINGS OF DATA

The Cr II spectrum has been recorded in the vacuum ultraviolet region with the 10.7 m normal incidence spectrograph at the National Bureau of Standards, Washington D.C., whch has an inverse dispersion of 0.78 Åmm$^{-1}$. The advantage of using that instrument has become evident during the analysis of the spectrum. As the $3d^5$ terms theoretically (to the first approximation) should have no fine structure the observed splittings of the terms were expected to be very small. That means a need of high dispersion in order to resolve the components of a LS multiplet. Take a $^4P$ in Cr II as an example, where the two lowest levels are separated by only 0.3 cm$^{-1}$, which corresponds to a wavelength difference of 0.01 Å at 2000 Å and 0.03 Å at 3000 Å. Another example is a very narrow LS multiplet at 2330 Å, UV 47. The small fine structures of the combining terms, two quartets, gather all the ten lines in the multiplet within 1.4 Å. Plates in the air region have been taken at Lund and

Madrid. In Figure 3 we can see how the separation of different ionization stages is possible on hollow-cathode spectrograms recorded with a stigmatic instrument. The figure contains two spectrograms of Fe II taken under different conditions of the light source, the top one is from a continously burning hollow cathode lamp and the bottom one from a pulsed discharge. The distribution of the light inside the cathode is projected on the slit and the stigmatic image of the illuminated slit on the plate preserves the distribution in the shape of the vertical line.

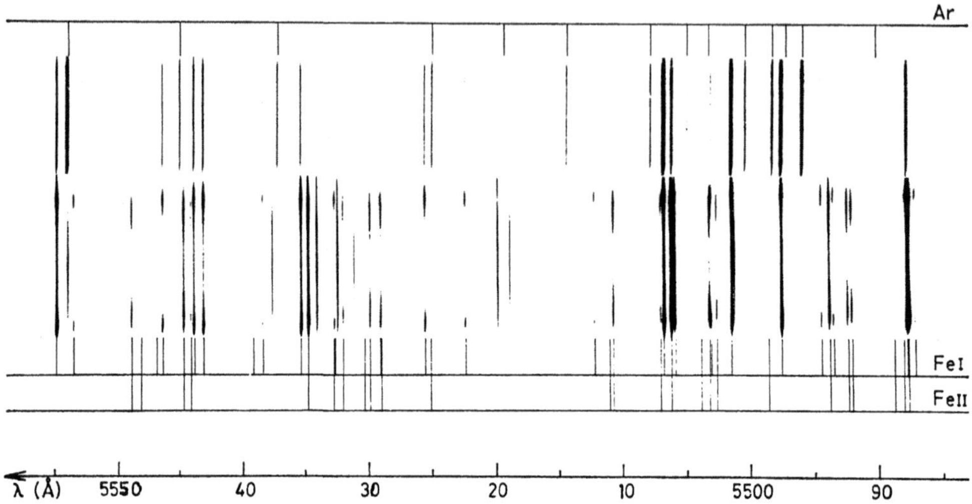


Fig. 3: Spectrograms demonstrating the methods for sorting out Fe II lines. The upper spectrogram was taken with a contininuous and the lower with a pulsed discharge.

B. RESULTS

The results of this investigation will eventually be presented in long tables of wavelengths and level values. To give an idea of the quantity of data some statistics are given in Table 2. Even if the number of known levels is now about 600 there is only one completely known configuration, namely $3d^5$ with 37 levels.

Table 2

Condensed Data for Cr II

| | |
|---|---|
| Number of measured spectral lines | ≃ 15,500 |
| Number of lines classified as Cr II | ≃ 6,000 |
| Number of identified Cr II lines | ≃ 4,900 |
| Number of unidentified Cr II lines with an intensity of > 1 | ≃ 300 |
| Number of known Cr II levels | ≃ 600 |

In Figure 4 is given the distribution of known levels of Cr II, where the hatched areas represents the additions from the present investigation. The sharp cut-off at 110 kK may be due to the excitation conditions in the light source or to the decreasing transition probabilities for lines from highly excited levels. The level system is fairly complete up to 100 kK as only some 30 levels are missing below that value. In the energy range 100–110 kK about 150 levels are still unknown but predictable from parametric calculations. The configurations $3d^3 4s4p$ and $3d^4 4d,5d,4f,5p$ account for the major part of the hatched area in the histogram.

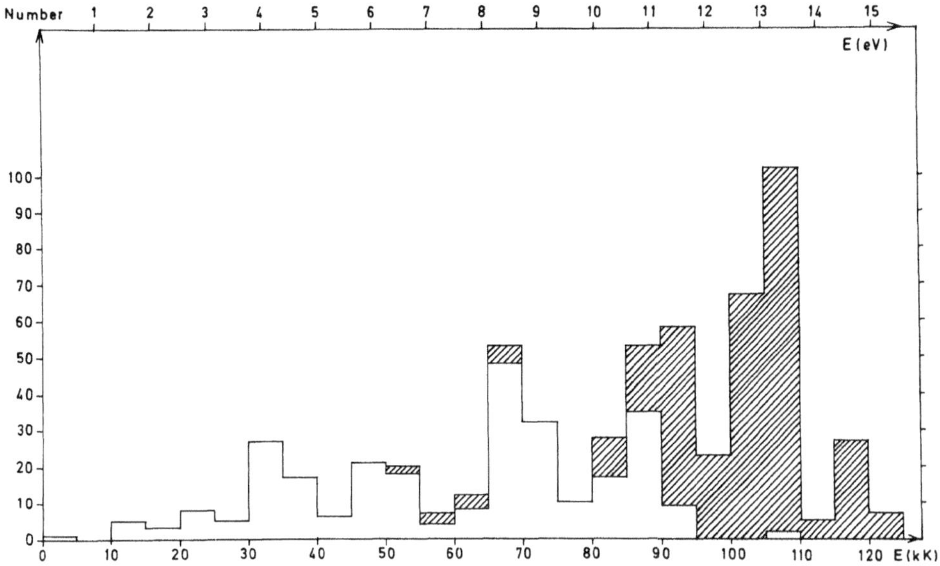


Fig. 4: The distribution of known levels in Cr II as a function of energy. The hatched area represents the additions from the present investigation.

The spectral distribution of the observed lines is illustrated in Figure 5, where the number of identified lines in certain wavelength intervals is represented by the area of a box. The upper limit of wavelength is set by the investigation itself, while the lower limit is set by the radiating atoms in the light source. We also infer from this histogram that the major part of the transitions appear in the IUE region. The regions where the strongest radiation occurs from different transition arrays are indicated below the histogram.

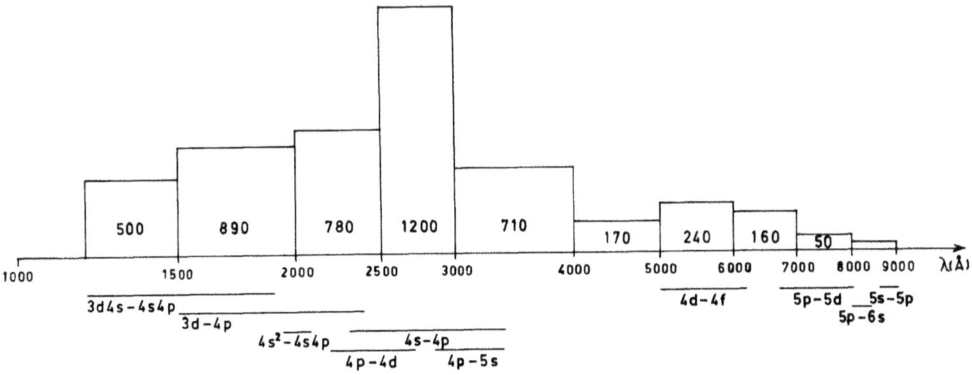


Fig. 5: The spectral distribution of the identified lines in Cr II. The number of lines in each box is represented by the area. Below is indicated the regions, where the different transitions occur.

C. SELECTIVE POPULATION

When the light source is run under different conditions the excitation and ionization rates are drastically changed. This is illustrated in Figure 6, which shows a part of the spectrum around 2000 Å, produced under different discharge modes. The upper one is from a pulsed discharge with a peak current of 80 A and a repetition of 100 $s^{-1}$ and the lower one is from a continuously burning discharge with a current of 0.8 A. Normally the Cr II spectrum is much stronger in a pulsed discharge, while the continuous discharge favours the spectrum of the neutral atom. However, on these spectrograms almost all lines are due to Cr II, even those on the low current spectrogram. The strong multiplet, which is indicated on the lower spectrogram, is generated from selectively excited Cr II levels. The excitation potential of these levels is about 14.5 eV and they are populated through a charge transfer reaction between $Ne^+$ and Cr. When the ionization energy of $Ne^+$ is transferred in a collision the neutral Cr atom will be ionized and left in a highly excited state. This process will consequently affect the spectrum from the low-energetic light source, which normally contains lines only from the neutral atom and probably the resonance lines from the singly ionized spectrum. This can be seen in Figure 6 as well, where the lines marked R are resonance transitions in Cr II and they appear much weaker than the high level transitions in the low current discharge.

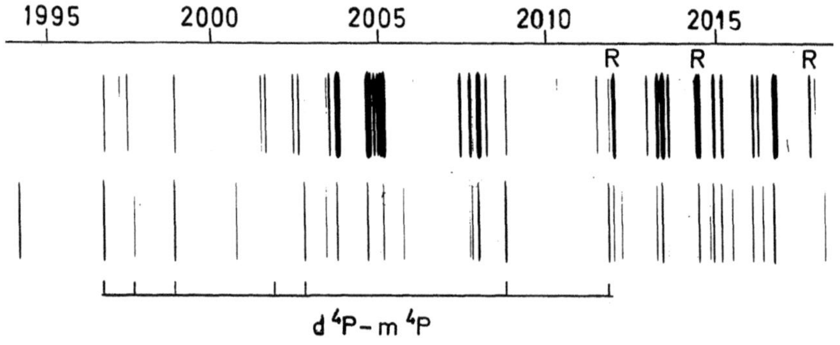


Fig. 6: Spectrograms demonstrating selective excitation of high levels of Cr II due to charge transfer. The upper spectrogram (pulsed discharge) contains three resonance lines (R), which are weak on the lower spectrogram (continuous discharge). The upper term in the indicated multiplet, $m^4P$, with an E.P. of 14.5 eV is selectively populated in the continous discharge.

## III. Astrophysical Applications

The high abundance of Cr II in some CP stars has made it meaningful to compare the laboratory spectrum with stellar spectra. For instance, Cowley has supplied me with a line list of HR 4816, arranged by Bidelman, and in the region 3700 – 4800 Å all identified laboratory Cr II lines are present in the spectrum of this star. There are also traces in the stellar spectrum which might be identified as 5p-6d transitions with a lower EP of about 11 eV. If such lines noticeably contribute to the opacity, then the corresponding transitions in Fe II would appear due to the normally higher abundance of iron. However, these lines are not known in Fe II and cannot be predicted since the 6d configuration is still unknown. This is the case in most singly ionized iron group spectra and the occurence of such high level transitions would imply a need of extensions of most existing analyses. In this last section we will come back to the prominent group of lines around 1430 Å mentioned in section IB. Recently "An Atlas of Solar Spectra in the VUV Region Recorded on SKYLAB" was issued by Cohen (1981). In Figure 7 is shown a part of the spectrum from a solar active region, where a group of strong absorption lines appear. On the bottom of the figure the strong Cr II group is displayed on the same wavelength scale. There is a perfect agreement between the absorption features in the solar spectrum and the two Cr II multiplets. It should further be mentioned that all identified Fe II and Ni II lines in the spectrum of this active region appear in emission and that some more Cr II lines show up as weak absorption features. A more detailed description of this coincidence is described in a paper submitted to Monthly Notices (Johansson 1982).

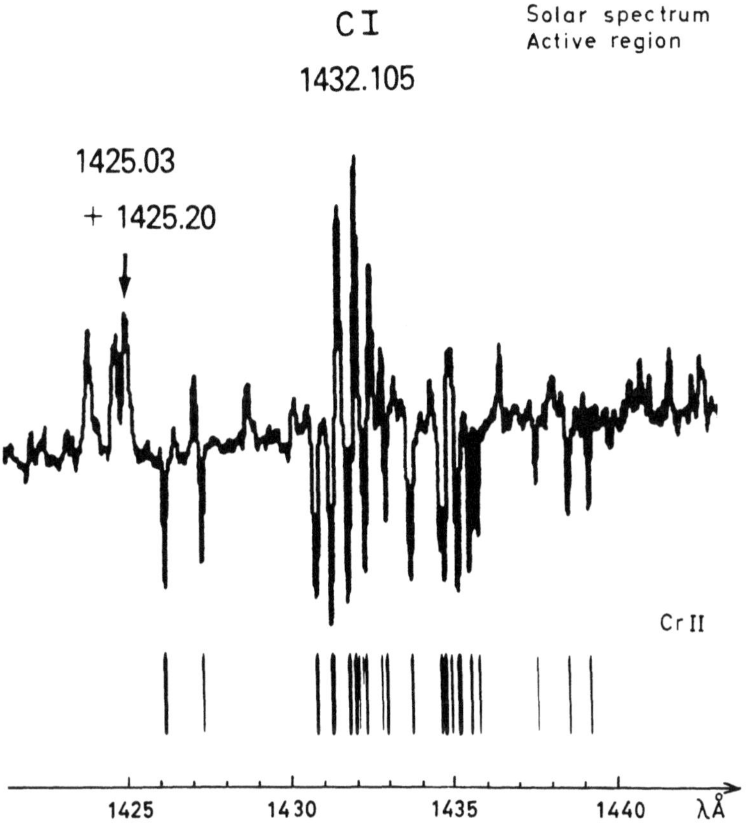


Fig. 7: A part of the spectrum of an active region on the sun, extracted from an atlas of solar spectra (Cohen 1981), compared to the laboratory spectrum of Cr II.

## References

Cohen, L.: 1981, NASA Reference Publication 1069.
Cowley, C.R. and Arnold, C.N.: 1978, Ap. J. 226, p. 420.
Johansson, S.: 1978, Phys. Scr. 18, p. 217.
Johansson, S., Litzén, U., Sinzelle, J., and Wyart, J.-F.: 1980, Phys. Scr. 21, p. 40.
Johansson, S.: 1982, Mon. Not. Roy. Astron. Soc., in press.
Kiess, C.C.: 1951, J. Res. Nat. Bur. Stand. 47, p. 385.
Moore, C.E.: 1945, Contrib. Princeton Obs. No. 20.

## 4. DATA BASES FOR ASTROPHYSICAL MOLECULAR SPECIES

R. W. Nicholls
York University, Canada

### I. Introduction

The impacts of laser phenomena on research in chemical physics together with increasing demands on spectroscopy for diagnostic tools in atmospheric remote sensing have led in recent years to greatly increased research activity in molecular spectroscopy. This activity has been in both of the major research areas of a) determination of structure constants, and b) line profile parameters. Relatively few molecular spectroscopy research laboratories devote much of their attention to spectra solely of astrophysical importance.

It is impossible to do justice in this report or the draft report, to the very vigorous research activity of the past three years, on many aspects of the spectroscopy of diatomic molecules. The vigour of the field is clear from examination of the many literature citations in each issue of the well known and invaluable Berkeley Newsletter [1].

Even a cursory survey of the recent critical compilation "Spectroscopic Constants" by Huber and Herzberg [2] reveals many gaps in the structure constant data for electronic states of astrophysically important diatomic molecules.

Astrophysical and other users cannot always wait until these data gaps are filled. Two tasks should thus be undertaken, perhaps at the instigation of Commission 14. They are:

a) To develop approximate methods for the interpretation of those molecular astrophysical spectra for which reliable data are sparse.
b) To develop internationally accessible and easy-to-use data bases of critically compiled constants of astrophysical molecules. The compilations should be presented in formats most useful to the astrophysical community. It is anticipated that thermochemical as well as molecular structure and transition probability constants would be included in such data bases. They would allow for the straightforward computation of synthetic molecular spectra of astrophysical importance.

The difficulties of implementing these tasks, particularly the second, should not be underestimated.

The first task presents some interesting theoretical challenges, which can be individually tackled [3].

Progress on the second task is made more difficult by the perhaps mistaken perception of many individual research workers that the development of critical compilations of data, no matter how useful and badly needed such compilations are, is an activity of "secondary scholarship," and not therefore as attractive as primary research. The tasks involved in critical assembly of continually updated data compilations are enormous and very time consuming. To undertake them in an orderly fashion calls for well-coordinated distribution of effort among many collaborators. The total effort will be best divided among a number of easily identified (and publishable) sub-tasks, each of which is complete in itself.

The more experienced the research workers who are involved in making the compilations, the more reliable and authoritative they will be.

### II. Compilations and Data Bases

Numerous compilations do of course exist to aid the astrophysical user of diatomic molecular spectroscopy. They include the regular listing of literature references in the Berkeley Newsletter [1], the listings of principal features of many band systems in Pearse and Gaydon's indispensible Identifications of Molecular Spectra [4], supplemented by the CNRS Donnees Spectroscopique edited by Rosen [5], Suchard's Spectroscopic Data [6], the various collections by Krupenie et al. of spectroscopic data for individual molecules [7], the compilation of Huber and Herzberg [2], the line lists on CN and $C_2$ spectra of Phillips et al. [8,9], and the York University series of "Identification Atlases of Molecular Spectra."

None of these absolves the astronomical user from making an individual assessment of the primary literature.

Except for the work of Phillips et al. [8,9], no ordered compilations of rotational line identifications, and line intensities appear to be available for astrophysical diatomic molecular spectra comparable to, for example, the AFGL lists by McClatchey et al. for $CO_2$ and other atmospheric molecules [10]. This continually updated data base is also available on computer-compatible tape for those who need to use the data for truly realistic studies of atmospheric transmission.

We work in an environment in which computer-accessible data bases of many kinds are becoming increasingly more available tools for research. Typical examples are:

a) literature search data bases
b) the JANAF Thermochemical data base
c) the AFGL Atmospheric Molecules data base
d) the Univ. Sussex Molecular Potentials data base

Some of these (e.g. (a)) are accessible on a "user-pay" basis. Others, (e.g. (b)) are published in hard cover and provide for many species a continuously updated set of thermochemical data tables. Yet others, (e.g. (c)) are available, by subscription, on computer tape. And others, (e.g. (d)) may be accessed through remote terminals.

Of these, item (c), the AFGL data base on positions and strengths of lines of atmospheric polyatomic molecules, such as $CO_2$, which contribute to atmospheric opacity in the infrared and microwave regions of the spectrum, is a good model for a data base of astrophysical molecular spectra. More spectral information should undoubtedly be contained in the astrophysical data base.

## III. An Astrophysical Molecular Spectrum Data Base

In the context of the above comments, it is proposed to limit this discussion to data base needs for electronic spectra of diatomic molecules of astrophysical interest in general, and in particular to those of importance to stellar spectrocopy.

To identify the optimum contents and format of a data base on molecular contributions to stellar spectra, the user community, and molecular spectroscopists interested in stellar spectra should be invited for comments on:

a) What wavelength range (waveband)?
b) Which molecular species?
c) Which spectral features of these species?
d) Which molecular (and other) data on these species?
e) What data accuracy would be acceptable?
f) What data presentation format data is most useful?
g) What presentation medium is most useful?
h) What are the priorities comments on each of these?

Astronomers have access to different instruments and will thus be intrested in different spectral wavebands. They work on different astronomical objects and thus are interested in different molecular species, and different spectral features of the species. They also work at different resolutions. Some need line information. Some would be content with band data. Different problems call for different molecular data. Some problems call for line positions. Some call for line and band strengths and related transition probability data. Others call for line profile information. Different problems also call for fundamental data at different precisions.

There is probably a wide spectrum of opinions on optimum data display formats. Similarly because of the many media on which data are commonly displayed (hard copy, computer tape, microcomputer diskette, etc.) there is likely to be a wide spectrum of opnions on the best media on which to make parts of the data base available.

Answers to the above questions by a representative number of astrophysical users of molecular data would provide a most valuable indication of the optimum content of an Astrophysical Molecular Spectra Data Base. It is hoped that during the coming year, Working Group 5 (Molecular Spectra) can gather this information from at least Commission 27 (Stellar Spectra). Later, a wider group of potential users might be polled. On the basis of this information, it is hoped that Commission 14 might be able to encourage the collaborative and cumulative provision of an Astrophysical Molecular Spectrum data base, perhaps one band system at a time.

The minimum data which it is anticipated would be provided for each band system are:

a) line list (band and branch number identifications)
b) associated wavelengths and wave numbers
c) line strength (Honl-London Number)
d) line widths under typical stellar atmosphere conditions
e) band Franck-Condon Factors and r-centroids
f) band system electronic transition moment
g) band system molecular structure constants
h) typical synthetic spectra
i) principal references

In astronomical applications, where spectral features of a number of molecular species are interlaced in a particular observational waveband, parts of the band-system data bases could be appropriately merged by users if display formats and media are chosen with care.

It is hoped that this proposal might be considered by interested members of the astronomical and spectroscopic community before the next meeting of Commission 14.

## References

[1] Phillips, J.G., Davis, S.P., and Eakin, D.M.: Berkeley Newsletter, Departments of Astronomy and Physics, University of California, Berkeley, California.

[2] Huber, K.P. and Herzberg, G.: 1979, Constants of Diatomic Molecules, Van Nostrand Reinhold, New York.

[3] Nicholls, R.W.: 1982(a), Ap. J. Suppl. 47, pp. 279-290. 1982(b), J. Quant. Spect. Rad. Transf. (in press). 1982(c), J. Chem. Phys. 77, pp. 1614-1616.

[4] Pearse, R.W.B. and Gaydon, A.G.: 1976, The Identification of Molecular Spectra, (4th Edition), Chapman and Hall, London.

[5] Rosen, B. (Ed.): 1970, Spectroscopic Data Relative to Diatomic Molecules, Pergamon Press, Oxford.

[6] Suchard, S.: 1975(a), Spectroscopic Data: Volume I - Heteronuclear Diatomic Molecules, (Part A), Plenum, New York. 1975(b), Spectrocopic Data: Volume I - Heteronuclear Diatomic Molecules, (Part B), Plenum, New York. Suchard, S. and Meltzer, J.E.: 1976, Spectroscopic Data: Volume II - Homonuclear Diatomic Molecules, Plenum, New York.

[7] Krupenie, P.H.: 1972, J. Phys. Chem. Ref. Data 1, No. 2.
[8] Davis, S.P. and Phillips, J.G.: 1963, The Red System of the CN Molecule, University of California Press, Berkeley, CA.
[9] Phillips, J.G. and Davis, S.P.: 1968, The Swan System of the $C_2$ Molecule and the Spectrum of the HgH Molecule, University of California Press, Berkeley, CA.
[10] McClatchey, R.A. et al.: 1973, AFCRL Atmospheric Absorption Line Parameters Compilation, AFCRL-TN-73-0096.

## Appendix

The following questionnaire was distributed at the scientific session. It is hoped to distribute an updated version of it by mail to interested parties during 1982 and 1983.

COMMENTS ON NEEDS FOR AN ASTROPHYSICAL MOLECULAR SPECTROSCOPIC DATA BASE

1) Which molecular species are most important to you?
2) Which band systems, bands, or lines of these species are most important?
3) Which wavelength ranges or wavebands are of principal and subsidiary importance?
4) What types of molecular data are most needed for your work?
5) What is the preferred format and medium (hard copy, tape, diskette, other) for these data?
6) Comments you wish to make?

Please send comments by air mail to:

Professor R. W. Nicholls
Centre for Research in Experimental Space Science
York University
4700 Keele Street
Downsview
Ontario, CANADA M3J 1P3

## 5. THE MOLECULAR SPECTROSCOPY FILE AT JPL

D. D. Locanthi
Jet Propulsion Laboratory

About 14 years ago the Planetary Atmospheres Section at JPL (Jet Propulsion Laboratory), at the instigation of Dr. C. B. Farmer, began maintaining in a systematic manner a file of papers about molecular spectra. Only those molecules, radicals, and ions that are found (or expected to be found) in the atmospheres of the planets or comets of the solar system are included in the list, which grew from about 115 to 345. More recently molecules found in interstellar space were added.

For each paper two information cards have been typed and kept in one section of the M.S.F. (Molecular Spectroscopy File) one giving the author's name on the top line, the other having the accession number at the top. A coding system on the cards provides access to information described in the articles: molecular constants, processes, experimental conditions, etc. The papers themselves are arranged in order of an accession number. About thirty articles are added per month. There are now about 6000 papers in the file.

To facilitate searching for papers about given molecules, the information on the typed cards has been transferred to standard punched cards. For each paper there is first a title card, which shows accession number, name of first author, abbreviated title, year of publication, volume number, page, and journal of publication. In the beginning, the list of journals was alphabetic, but, as new journals were added, alphabetization could no longer be maintained. A second punched card for each paper gives the code numbers that correspond to information in the article. A program for searching and listing all papers that our file contains for a given molecule was written by H. Aumann; but it is seldom used because of the expense involved in computer search. Instead a small file is kept with a card for each molecule/radical/ion. Whenever a new paper is added to the system its accession number is typed on the appropriate card. Thus, if one wants a list of all papers on HCl, for example, one may simply pull the punched cards for all papers noted on the HCl card, and have the computer list them.

The future of the M.S.F. is a little uncertain, owing to drastic decrease in funds allocated to NASA. Though our section's concerns are now necessarily limited to molecules found only in the Earth's stratosphere, we are attempting to continue including papers about all molecules in the solar system and interstellar space.

## 6. ADDITIONS TO REPORT OF WORKING GROUP 2 (ATOMIC TRANSITION PROBABILITIES)

John Black
Center for Astrophysics
Harvard College Observatory

The Atomic and Molecular Physics Division of the Harvard-Smithsonian Center for Astrophysics has constructed an ion trap and is using it for the first laboratory measurements of A-values for the intersystem lines of light ions that are used in the diagnosis of astrophysical plasmas. Many of these lines have A-values of the order of $10^2$ sec$^{-1}$ to $10^4$ sec$^{-1}$ and none have been measured. As a consequence, astronomers have used unconfirmed calculated values until now.

Preliminary results for the Si III $3s^2\ ^1S_0 - 3s3p\ ^3P_1^\circ$ line at 1892 Å give $A = (1.63 \pm 0.26) \times 10^4$ sec$^{-1}$. This value is in good agreement with the calculations of Cowan et al. (1982) and Laughlin and Victor (1979), but is smaller than the results of Nicolas (1977), $2.25 \times 10^4$ sec$^{-1}$. Our measurements are being described at IAU Colloquium 73 by Parkinson, Kwong, Smith and Johnson (1982).

Similar measurements for the $^4D - ^2P^\circ$ multiplets of C II and Si II at ~2320 Å, for the $^1S - ^3P^\circ$ line of C III at 1909 Å, and for the $^3P - ^5S^\circ$ multiplet of O III at 1660 Å are underway.

Electron impact excitation cross sections are being measured with inclined electron and ion beams at this laboratory by G. P. Lafyatis and J. L. Kohl. The near threshold cross section for excitation to the upper state of the C II λ1335 Å spectral line was measured from threshold to 14.5 eV and measurements of the well-established cross section for C IV λ1550 Å have verified the experimental methods being used (Lafyatis, 1982). A measurement of dielectronic recombination in C IV is currently in progress.

### References

Cowan, R.D., Hobbs, L.M., and York, D.G.: 1982, Ap. J. 257, p. 373.

Lafyatis, G.P.: 1982, "An Experimental Study of Electron Impact Excitation of Positive Ions," Ph.D. Thesis, Harvard U.

Laughlin, C. and Victor, G.A.: 1979, Ap. J. 234, p. 407.

Nicolas, K.R.: 1977, "The Application of Si III Line Intensity Ratios to Determine the Solar Transition Zone Density and Pressure," Ph.D. Thesis, U. of Maryland. See also, Nicolas et al.: 1979, Ap. J. 233, p. 741.

Parkinson, W.H., Kwong, H.S, Smith, P.L., and Johnson, B.C.: 1982, "Measurements of A-values for Intersystem Lines Used in Diagnosis of Astrophysical Plasmas," to be presented at IAU Colloquium No. 73, Dublin, Ireland, August, 1982.

## 7. FRANCK-CONDON FACTORS AND SYNTHETIC PROFILES FOR MOLECULES OF ASTROPHYSICAL INTEREST

Basile Petropoulos
Research Center for Astronomy and Applied Mathematics
Athens Academy, Greece

Plasma spectroscopy is a good diagnostic tool to study the chemical reactions in the plasma of the sun of comets, atmospheres of planets and in interstellar space.

This method is consisting of the comparison of the computed synthetic profiles with the measured, absolute intensities obtained by spectrophotometric methods. For this comparison it is possible to determine plasma parameters, such as the concentration and the temperature of the molecular plasma. The following data are needed for the computation of synthetic profiles: 1) Hönl London factors; 2) Franck-Condon (F.C.) factors; 3) electronic transition moments; 4) measurements in laboratory as Spectroscopical constants, dissociation energies, lifetimes of electronic states, oscillator strengths.

Franck-Condon factors have been computed used a Morse potential to resolve the one dimension Schrödinger equation, for the most diatomic molecules of astrophysical interest by McCallum _et al._ [1].

Different authors have used the more correct R.K.R. Cooley method to compute F.C. of diatomic molecules of astrophysical interest, Petropoulos [2], Splinder [3], and Petropoulos _et al._ [4,5]. A rotational effect has to be taken into account for the computation of F.C. factors, at high temperatures. This effect is more important for molecules of type XH. Consequently F.C. factors are needed for the following transitions: $CH^+$(A-X), MgH(A-X), $MgH^+(^1\Sigma-^1\Sigma)$, AlH(A-X), $AlH^+$(ZΠ-ZΣ), $SiH^+$(A-X), $PH(^3\Pi-^3\Sigma)$, HS(A-X), CaH(A,B,C-X), FeH(A-X), CoH(A-X), NiH(A,B,C-X), BaH(A-X), TiH(A-X). As also for the following transitions of molecules: MgO(B-X), ScO(A,B-X), YO(A-X), ZrO(A,B,C-X), TiO(B-X,b-d), VO(A-X), SiO(A-X) and CN (Red and Ballik Ramsay transitions).

For polyatomic molecules the computation of F.C. factors is more difficult. Different methods have been proposed to integrate the multidimensional Schrödinger equation. The method of generated functions has been used by Sharp and Rosenstock [6], which have computed F.C. factors for $CO_2$, $C_2H_2$, $C_2D_2$, $CS_2$. An harmonic potential has been admitted by these authors to resolve the Schrödinger equation. The normal coordinates $Q^1$ of the upper electronic state of the molecule, can be obtained by the following orthogonal transformation of the normal coordinates Q of the lower electronic state for each symmetric vibration:

$$Q^1 = /J/\ Q + /K/ \qquad (1)$$

where /J/ is a rotation matrix and /K/ a transition matrix. The rotation effect was pointed out by Duchinsky [7] (Duchinsky effect). Lucas [8] has observed that this effect is only a first order approximation, because for the Eckart's [9] axes system the transformation of coordinates is not orthogonal.

In Sharp-Rosenstock [6] computations the Duchinsky effect has been taken into account, in its first approximation given by relation (1). The method has been applied to compute F.C. for $H_2O$, $H_2O^+$ and $NH_3$, $NH_3^+$, by Botter _et al._ [10] for small or large molecules of hydrocarbons by Warshel _et al._ [11,12] and for Benzen by Metz [13].

Coon [14], Craig [15] and other authors have used the one dimensional method to compute F.C., but they have not taken into account the Duchinsky effect.

Doktorov [16] introduced another method to compute F.C. of Benzen, as also Faulkner [17], where Duchinsky effect is also taken into account in its first approximation. Petropoulos [2,18] proposed a general theory which takes into account the Lucas [18] transformations of coordinates and the one dimension theory to resolve the Schrödinger equation. Morse functions or R.K.R. potentials can be used for computing F.C. factor for the symmetric vibrations of every molecule. This method has been applied to compute F.C. factors of the electronic transitions:

$CO_2(X^1\Sigma) \rightarrow CO_2(X^2\Pi u)$ Petropoulos [18]

$C_2H_2(X^1\Sigma_g) \rightarrow C_2H_2(X^1\Pi u)$ Petropoulos [18]

$CO_2^+(X^2\Pi u) \rightarrow CO_2^+(X^2\Pi g)$ Petropoulos [18]

and also for

$H_2O(\tilde{X}^1A_1) \rightarrow H_2O^+(\tilde{X}^2A_1)$ Petropoulos [2]

F.C. factors computed by this method are in good agreement with the measured values of transition probabilities, and the Computed F.C. factors by Botter et al. [10]. McCallum [19] have also computed F.C. factors for $CO_2$ and $CO_2^+$ by used Morse potential. For these linear molecules Duchinsky rotation effect can be neglected, Petropoulos [2].

Polyatomic molecules have been observed in spectra of interstellar medium and comets as $H_2O$, HCO, $H_3^+$, $HCO^+$, $C_2H$, HCN, HNC, $N_2H$, $H_2S$, OCS, $SO_2$, HNO, $CH_2$, $CO_2$, ($CO_2^+$, $H_2O^+$, $NH_2$, $C_3$: comets only) $H_2CO$, $NH_3$, HNCO, $H_2CS$, $C_3N$, HCOOH, $HC_3N$, $CH_2NH$ (comets only), $NH_2CN$, $H_2CCO$, $HCH_2OH$, $NH_2CHO$, $CH_3CN$, $CH_3C_2H$, $CH_3CHO$, $HC_5N$, $CH_3NH_2$, $CH_2CHCN$, $CH_3COOH$, $(CH_3)_2O$, $CH_3CH_2OH$, $CH_3CH_2CN$, $HC_2CN$. And in the atmospheres of planets: $N_2O$, $NO_2$, $CO_2$, $H_2O$, $H_2S$, COS, $SO_2$, $O_3$, $S_3$, $CO_2^+$. For the above molecules Franck-Condon factors are needed in order to compute synthetic spectra by the Whiting et al. [20] program which has been applied for diatomic molecules, by Danylewych et al. [21] for the Swan bands of $C_2$. These authors by comparing their results with absolute spectrophotometric measurements of the Swan bands of the comets (1973) (XII) Kohoutek and 1975 IX (Kabayaski-Berger-Milon), have determined one relationship between the molecular column density and the electronic excitation temperature. Macris and Petropoulos [22] have also used the above program to compute the synthetic profile of the CO $(A^1\Pi - X^1\Sigma)$ transition in low and high temperatures. The computed intensities for low temperatures have been compared to absolute intensities measured in the Mars atmosphere by the Mariner's 9 spectrometer. From this comparison, the distribution of CO has been obtained, for different altitudes.

The computed profiles for high temperatures and concentrations, have been compared to the measured absolute intensities, of the sunspots, observed by the telescope of Sky-Laboratory. From this comparison they have obtained the minimum solar temperature.

## References

[1] McCallum, J.C., Nicholls, R.W. and Jarmain, W.R.: 1970, "Franck-Condon Factors and Related Quantities for Diatomic Molecular Spectra," Spectrosc. Rep. Cent. Res. Exp. Space Sciences, York University.
[2] Petropoulos, B.: 1974, These doctorat of Sciences, Paris VI, No. CNRS 3216.
[3] Splinder, R.J.: 1965, J. Quant. Spectr. Rad. Transfer 5, p. 165.
[4] Petropoulos, B., Dessaux, O., Chaffiol, D., Goudmand P.: 1967, Com. Rend. Acad. Sci., 265, pp. 355-358.
[5] Petropoulos, B. and Herman, L.: 1967, Com. Rend. Acad. Sci. 264, p. 250.
[6] Sharp, T.E. and Rosenstock, H.M.: 1964, J. Chem. Phys. 41, p. 3453.

[7] Duchinsky, F.: 1937, Acta Physicochim. U.R.S.S. 7, p. 551.
[8] Lucas, N.J.D.: 1973, J. Phys. (B) 6, p. 155.
[9] Eckart, C.: 1935, Phys. Rev. 47, p. 553.
[10] Botter, R. and Rosenstock, H.M.: 1969, J. Res. Nat. Bur. Stand. Sect. A73, p 313.
[10a]Botter R., Dibeler, V.H., Walker, J.A., and Rosenstock, H.M.: 1966, J. Chem. Phys. 44 (3), p. 1275.
[11] Warshel, A. and Karplus, M.: 1975, J. Chem. Phys. 62, p. 214.
[12] Warshel, A.: 1975, J. Chem. Phys. 62, p. 214.
[13] Metz, F., Robey, M.J., Schlay, E.W. and Dorr, F.: 1977, Chem. Phys. Lett. 51, p. 8.
[14] Coon, J.B. and De Wanes, Loyd C.M.: 1962, J. Mol. Spectr. 8, p. 825.
[15] Craig, D.P.: 1950, J. Chem. Soc. 2146.
[16] Doktorov, E.V., Malkin, I.A., and Marka, V.I.: 1975, J. Mol. Spectr. 56, p. 1; 1977, 64, p. 302.
[17] Faukner, T.R. and Richardson, F.S.: 1979, J. Chem. Phys. 70, p. 1201.
[17a]Faukner, T.R.: 1979, J. Chem. Phys. 71, p. 2737.
[18] Petropoulos, B. and Botter, R.: 1968, Com. Rend. Acad. Sci. 266, p. 104; 266, p. 276; 266, p. 933.
[19] McCallum, J.C. and Nicholls, R.W.: 1972, J. Phys. Atom. Mol. 4, p. 1096; 5, p. 1417.
[20] Whiting, E.E.: 1972, Ph.D. thesis, York University, Toronto.
[20a]Whiting, E.E., Paterson, J.A., Kovaks, I., and Nicholls, R.W.: 1973, J. Mol. Spectrosc. 47, p. 84.
[20b]Whiting, E.E.: NASA Tech. Note D7268.
[21] Danylewych, L., Nicholls, R.W., Neff, J.S., and Tatum, J.: 1978, Icarus 35, p. 112.
[22] Macris, C. and Petropoulos, B.: 1982 (in press), Com. Rend. Acad. Sci., Proc. of the 6th European Congress of I.A.U., Belgrade, 1981.

## COMMISSION 15: PHYSICAL STUDY OF COMETS, MINOR PLANETS AND METEORITES (L'ETUDE PHYSIQUE DES COMETES, DES PETITES PLANETES ET DES METEORITES)

Report of Meetings, 20, 21, 24 and 25 August 1982

PRESIDENT: B. Donn SECRETARY: J. Rahe

20 August 1982

BUSINESS MEETING

The meeting of the Commission opened with noting the deaths of Past President N. Richter and A. E. Douglas and holding a brief period of silence in respect. The serious illness of Vice-President B. Levin was also noted. Because of this illness, Dr. Levin could not accept the nomination as President and with regret his name was withdrawn.

The Trienniel Report of the Commission was distributed. Because of the late receipt of the Soviet contribution by O. V. Dobrovolsky it was distributed separately.

1. Election of Officers, New Members and Consultants

| | |
|---|---|
| President: | Clark R. Chapman |
| V. President | L. Kresak |
| Organizing Committee: | B. Donn, F. Dossin, H. Fechtig, M. Festou, D. L. Matson, Z. Sekanina, H. V. Schmidt, L. M. Shul'man, J. Wasson, G. W. Wetherill, L. Wilkening |
| New Members: | E. Barker, W. Cochran, B. Cosmovici, J. Crovisier, J. C. Degewij, P. Farinella, J. A. Fernandez, M. Gaffey, D. M. Gibson, J. Gradie, E. Grun, D. W. Hughes, R. Hutchinson, W. H. Ip, W. Irvine, S. Isobe, K. Jockers, A. C. Levasseur-Regourd, D. F. Lupishka, S. M. P. McKenna-Lawlor, R. Newburn, P. Pellas, D. Revelle, M. Shimizu, H. Takeda, E. F. Tedesco, P. R. Weissman, H. Tanabe, T. Nakamura |
| Consultants: | F. Begemann, G. Briggs, V. A. Bronshten, K. Churyumov, M. Combi, P. Eberhard, H. Hoopes, Z. Knezevic, G. Kurat, S. Larson, M. Oppenheimer, K. Saito, P. Signer, G. Turner, R. Zerull |

2. Resolutions

Two resolutions were approved by the Commission membership for submission to the General Assembly. One called for IAU recognition of the International Halley Watch and the other requested observatories to give high priority to selected Comet Halley observations. These were adopted by the General Assembly and appear under resolutions of the General Assembly.

3. Working Group Reports

Filter standards. M. A'Hearn described the status of the program to develop and make available standard sets of filters for cometary photometry. To obtain a set of these filters contact Dr. M. F. A'Hearn, Astronomy Program, University of Maryland, College Park, Maryland, 20742, USA.

Comet Filter Specifications

diameter - 2.54 cm ± 0.05, (1.000 in), thickness < 0.95 cm (0.375 in)

| Species | $\lambda_o$ | FWHM | Peak Trans. |
|---|---|---|---|
| $C_2$ | 5139 | 90 ± 10 | 0.50 |
| Cont. | 4845 ± 10 | 65 ± 10 | 0.50 |
| $C_3$ | 4060 ± 5 | 70 (+ 10, -5) | 0.40 |
| CN | 3871 | 50 (+ 0, -10) | >0.20 |
| Cont. | 3650 ± 10 | 80 (+ 10, -5) | >0.25 |

Additional Filters

| | | | |
|---|---|---|---|
| OH | 3085 | | |
| $CO^+$ | 4623 | 45 | |
| Cont. | 7500 | 80 | |
| $H_2O^+$ | | | |

Fifteen sets have been distributed, another twenty-five have been assigned, five are available and additional sets are on order. There is a charge of $335 per set of five. This will provide funds for expanding the set. The initial purchase was made with a grant from the U. S. National Science Foundation. This enables the cost to be subsidized for observers who do not have funds to purchase the filters.

Cometary Archives. The Notebooks of N. Bobrovonikouv are being prepared for publication by A. H. Delsemme. These are expected to be ready before much longer.

The preparation of an Atlas of Comet Halley, 1910, by J. Rahe, B. Donn and J. Brandt is well underway. PHotographic collections from the Cordaba, Harvard, Heidelberg, Helwan, Kodaikanal, Lick, Lowell, M. Wilson and Yerkes observatories were made available. These included exposures made in Hawaii. The set for inclusion in the Atlas have been selected and prepared.

Workshop on Modern Observational Techniques for Comets. The Workshop was organized by J. Brandt, B. Donn. J. M. Greenberg and J. Rahe and held at Goddard Space Flight Center, October 22-24, 1980. The proceedings of the meeting have been published as JPL Publication 81-68. Copies may be obtained from Dr. J. C. Brandt, Code 680, Goddard Space Flight Center, Greenbelt, Maryland 20771, USA.

Atlas of Cometary Spectra. Work on the updated version of the Swings and Haser "Atlas of Representative Cometary Spectra" by C. Arpigny, B. Donn, F. Dossin, J. Rahe and S. Wyckoff is continuing. Enlarged reproductions of essentially all spectra with dispersions greater than 20 Å/mm have been prepared. Medium resolution spectra of several of these comets will be included to facilitate the interpretation of such spectra by comparison with high dispersion observations. In addition to photographic reproductions, tracings of spectra will be presented. With the increasing trend to digitized spectra using electronic detectors, example of such data will be shown. Examples of ultraviolet and near infrared spectra are planned. Tables of identified species and wavelengths and unidentified lines will be included.

Coordinated Comet Observing. Plans under development in this area have effectively been taken over by the comprehensive International Halley Watch. Detailed accounts of the IHW were distributed at Patras. Anyone wishing descriptions of the IHW or copies of the Newsletter should write to:

Ray L. Newburn, Jr.
MST-1166
Jet Propulsion Laboratory
4800 Oak Grove Drive
Pasadena, CA 91109 USA

Dr. Jurgen Rahe
Director-Remeis Observatory
Sternwartstrasse 7
D-8600 Bamberg
FEDERAL REPUBLIC OF GERMANY

Other Items. Sample copies of the Lunar and Planetary Information Bulletin prepared and distributed by the Lunar and Planetary Institute, 3303 NASA Road One, Houston, TX 77058, USA were sent to all Commission 15 members before the General Assembly. It was decided at Patras that anyone wishing to continue receiving them should inform the Bulletin editor at LPI, Frances B. Waranius, phone (713) 486-2135.

Dr. D. Morrison described the composition and activities of NASA's Planetary Exploration Committee. One very relevent action was their support for modifying the mission of the International Sun-Earth Explorer satellite, ICEE-3 to convert it into a comet tail probe. This change is now in effect and the probe will intercept the tail of Comet Giacobini-Zinner in September 1985.

21 August 1982

The first session briefly reviewed the IHW and allowed IHW Discipline Specialists and interested observers in each area to meet.

Session 2 was a Joint Meeting of Commissions 15 and 34 "Spectroscopic Investigations of Grains". The program consisted of three invited papers by R. Knacke, W. Kratschmer and B. Donn and contributed papers by V. Pirronello and L. d'Hendecourt. These are summarized in the Report of Commission 34.

24 August 1982

Comet Halley Apparition, 1986

J. Rahe - The International Halley Watch

D. Yeoman - Tied together orbits from 1911 back to 241 B.C. and integrated motion back to 1404 B.C. There is no evidence for significant change in brightness. Uniformity and characteristics of non-gravitational forces indicate direct rotation, no spin axis precession and no significant change in outgassing rate over this interval.

R. Newburn devised a Halley coma model from gas production/luminosity ratios of recent comets and gas dust ratio of A'Hearn.

B. McIntosh described a new model for the Halley meteor stream. He concludes meteors not perturbed since $\sim$ 1400. They lie in a shell 0.44 AU wide and 0.04 AU thick.

H. Fechtig described the ESA Giotto mission to Halley. Launch will be July 7, 1985 and intercept, past perihelion will be March 13, 1986. A camera will obtain 140 images of the nucleus and a impact mass-spectrometer dust analyzer will seek the dust composition. Other experiments were not described.

Bertaux reported on the French-Soviet Vega mission with two spacecraft. After dropping probes into the Venusian atmosphere the main S/C continued on to intercept Halley at $10^4$ km. The first intercept will be March 8, 1986, the second a few days later. Experiments include a camera, IR and UV spectrometers, plasma wave detectors, magnetometer, and dust detector.

Oda described the Japanese Planet A Mission. Two S/C will be launched in January and August 1985, and intercept at 0.1 AU. Experiments consist of a CCD Ly-alpha imaging device, solar wind and plasma detectors.

J. Brandt described the U.S. Space Telescope capabilities for observing Halley. Launch is planned for the first quarter of 1985. There are also plans for the Space Lab Optical Science mission to observe Halley.

E. Gerard reported a search for Comet Halley on December 2-3, 1981 with the French-Canadian telescope in Hawaii to a limiting mag of 25.2.

Norris briefly presented a proposal for an interactive experiment on Comet Halley using an OH maser beam to turn off the cometary OH emission and watch its decay.

Dunham reported on the ISEE-3 mission to Giacobinni-Zinner.

The second Commission 15 session was on "Coma Models and Physical Processes". Invited papers by E. Barker, D. A. Mendis and M. Wallis appear in "Highlights of Astronomy".

D. Malaise gave an invited paper "The Nuclear region of Comets". Other papers were: Z. Sekanina - Two Dust Populations in Cometary Coma. V. Vanysek - Solar Corona After Impact of Coemt 1979XI; Rickman - Thermal Model of Comet Halley Nucleus; K. Jockers - In Production in the Coma via Solar Wind Charge Exchange.

## 24 August 1982

### Minor Planets - Physical Properties and Future Research Plans

The four scheduled minor planet sessions were combined into two on August 24. The following papers were presented:

E. Helin - Apollo 1982DB: The Earth's Most Accessible Neighbor

E. Tedesco: Results from Recent 8-Color JHK and Radiometric Observations of Asteroids

E. Bowell and K. Lumme: A Possible New Magnitude System for Atmosphereless Solar System Objects.

A. Harris: Statistics of Asteroid Rotation Rates.

D. Watson: Physical Properties of Asteroids from IRAS.

A. Brahic: Asterex and European Plans for an Asteroid Mission.

A. Harris: Recent Asteroid Mission Planning Efforts in the United States.

## 25 August 1982

### Condensation, Accretion and Early Parent Body Processes

G. Morfill - Some Cosmochemical Consequences of a Turbulent Protoplanetary Cloud.

H. J. Volk - Grain Growth in Turbulent Protostellar Accetion Disks.

B. Donn - Experiments on the Formation, Properties and Evolution of Primitive

Grains.

J. T. Wasson - Origin of Meteoritic Chondrules and Mechanical Fractionation During the Accretion Process.

G. Wetherill - Time Scale of Planet and Planetesimal Formation.

P. Pellas - Early Thermal History of Meteorites.

L. L. Wilkening - Possible Chemical Relationships Between Comets and Asteroids Based on Their Starting Compositions.

With turbulence, the time scale for disk formation and evolution are 10 and $10^6$ years, respectively. Grains in disk vaporize in inner zones, vapor flows out and recondenses. This leads to element separation with distance from center. Grains limited to fraction of mm. Solid bodies form by particle accretion but large bodies not possible unless turbulence varies in intensity. Experiments indicate amorphous, non-stoichiometric refractory grains form and current nucleation theories are not applicable. Composition of chondrites suggests everything vaporized where they formed by high temperature equilibrium condensation. There appears to be a conflict between condensation experiments and interpretation of chondrite formation conclusions. Chondrules seem to require several distinct precursors. Occurrence of cm size isotopic anomalous grains requires some massive particle infall from interstellar medium. Numberical treatment of accretion of planets after decay of turbulence would require $10^6$-$10^7$ years. If Jupiter stopped asteroid growth, it must have formed more rapidly. Formation of giant gaseous protoplanets or planetesimal formation in turbulent nebula may provide rapid formation but the answer is not clear. Comets and asteroids are smallest, therefore least changed in composition. C1 meteorites show best correlation with solar abundances but show evidence of aqueous processing. C3 may be most primitive. Low and high temperature material need to be mixed. Turbulence will do this.

## COMMISSION 16: PHYSICAL STUDY OF PLANETS AND SATELLITES (ETUDE PHYSIQUE DES PLANETES ET DES SATELLITES)

Report of Meetings 18, 19, 20, 24 August 1982

PRESIDENT: Bradford A. Smith

### 18 August 1982

COMMISSION BUSINESS

Presiding: Bradford A. Smith, President of Commission 16
Recording: Garry E. Hunt, Executive Secretary of Commission 16

1. Election of Officers for 1982-1985

The following slate was presented and elected unanimously by the members present:

President: Victor G. Tejfel'
Vice President: Garry E. Hunt
Executive Secretary: Andre Brahic
Organizing Committee: R.F. Beebe, J-L Berteaux, J. Elliot, D. Gautier, L.V. Ksanfumaliti, M.Ya. Marov, H. Masursky, G. Neukum, T. Owen, V.V. Shevshenko, B.A. Smith

2. New Members of the Commission

The following list of new members of the Commission was unanimously endorsed by the members present:

| | | |
|---|---|---|
| R.F. Beebe | J.A. Burns | J.J. Caldwell |
| J. Degewij | L.W. Esposito | P. Farinella |
| K. Fox | G.P. Horedt | R.F. Jurgens |
| A.L. Lane | J.B. Lewis | M.Ya. Marov |
| Y. Nakagawa | G. Neukum | K. Pauy |
| E.L. Ruskol | L.A. Soderblom | E.C. Stone |
| D.F. Strobel | R.J. Terrile | T.W. Thompson |
| L.H. Wasserman | C.F. Yoder | |

3. Meetings of Interest to Commission Members

Attention was called to the following meetings that will be held prior to the next General Assembly:

A. Planetary Spectroscopy
Org: Dr. R.W. Saunders
Rutherford Appleton Laboratory
Chilton, DIDCOT
Oxon, OX11 0QX, UK

Place: Oxfordshire, England

Time: 5-9 April 1983

B. Natural Satellites
Org: Dr. Joseph A. Burns, Cornell University, Thurston Hall, Ithaca, NY 14853, USA
Place: Ithaca, New York
Time: 5-9 July 1983

C. IUGG (including Syemour Hess Memorial Symposium)
Place: Hamburg, FRG
Time: 15-22 August 1983

D. COSPAR
Org: Mr. Harold Masursky, U.S. Geological Survey, 2255 N. Gemini Drive, Flagstaff, Arizona 85701, USA
Place: Graz, Austria
Time: 25 June-7 July 1984

This is not intended to be a complete list. Members who wish to bring additional meetings to the attention of the Commission 16 membership should contact the incoming Executive Secretary, Dr. A. Brahic.

4. COSPAR Decisions of interest to Commission 16 Membership

a) COSPAR has set up a group concerned with a reference atmosphere for Venus. Members are Klione, Moroz, Bauer, Biswas, Hunten, Marov, Seiff and Taylor.

b) COSPAR has issued a statement to the national space agencies urging them to try to find sufficient funds to ensure that more regular space missions are carried out in planetary exploration. COSPAR is deeply concerned by the reduction in solar sytem space missions, as has occured during the last decade.

5. Planetary Nomenclature

H. Masursky reported that meetings of the IAU Working Group for Planetary System Nomenclature have been held on 14 and 15 August in Patras to discuss outstanding issues relating to nomenclature of solar system objects. Unresolved issues included the naming of newly discovered satellites and the question of how to treat the numerous rings and gaps in the Saturn ring system.

6. Other Commission Business

There was a brief discussion of problems with archival storage of digital data. B. Smith expressed concern over the current status of large amounts of digital data from planetary spacecraft which, through poor archiving and loss of related documentation, are losing some of their value. He indicated that Mariner 9 and Viking data were good examples of large data sets that could be lost to future investigators. The problem has become more acute as increasing numbers of research groups have access to digital image processing systems and prefer to work with the original digital data rather than hard copy photographs. Smith noted that some Mariner 9 and Viking data were already irretrievably lost. Following discussion of this matter, the members present suggested that a small group draft a proposal for subsequent discussion as a possible means of getting an IAU resolution supporting improved archiving and documentation procedures for digital data at appropriate national data storage facilities. The President appointed G. Hunt, T. Johnson and H. Masursky to prepare a draft proposal.

Note: Because of inadequate time to prepare this proposal, the action has been deferred until the next General Assembly.

19 August 1982

I. SCIENTIFIC SESSION

Chairman: B.A. Smith

| | |
|---|---|
| S.K. Runcorn | Lunar Magnetism and Its Implications |
| F. Meyer | Creation of a Data File for Lunar Profiles from a New Set of Measurements |
| E.A. Whitaker | The Lunar Procellarum Basin |
| P. Moore | Recent Results of Lunar Transient Phenomena |
| R.H. Brown et al. | Diameters and Albedos of the Satellites of Uranus and Neptune |
| C.F. Capen | International Mars Observation Computer Program Studies of Seasonal Meteorological Phenomena and Current Changes in the Martian Arctic |
| G.P. Mantas | Comparison of Theoretical and Observed Photoelectron Energy Spectra in the Ionosphere of Mars |
| I.P. Williams | The Breakup of Comets to Form Meteor Streams |
| M.A. Janssen et al. | High Resolution Microwave Imaging of the Venus Atmosphere |
| R. Beer and M.A. Janseen | High Resolution Spectroscopy and Narrow-Band Imaging of Venus Around 4.8 m |
| B. Petropoulos et al. | A Model for the Upper Atmosphere of Jupiter after Voyager's Measurements |
| G.E. Hunt | Voyager Observations of Vortex Streets in the Atmosphere of Saturn |
| I. de Pater | New High Resolution Radio Observation of Jupiter and Saturn |
| Y. Leblanc | Jovian and Saturnian Radio Emissions Observed with the Voyager Spacecraft Radio Experiment |
| A. Dollfus | First Planetary Results from the 2-m Telescope at Pic-du-Midi |
| T.V. Johonson | Galileo Mission to Jupiter |
| M.Ya. Marov | Future Soviet Planetary Missions |

20 August 1982

I. JOINT DISCUSSION WITH COMMISSION 44: EXPLORATION OF THE SOLAR SYSTEM
Organizing Committee: R. Von Eshelman, D.M. Hunt, H. Masursky, T.C. Owen, ~~G.H.~~ Pettengill, B.A. Smith, G.L. Tyler (USA); G.E. Hunt (Chairman), J.E. Guest (UK); M.Ya. Marov (USSR); J.E. Blamont, A. Brahic (France)

A. VOYAGER ENCOUNTER WITH SATURN
Chairman: G.E. Hunt

| | | |
|---|---|---|
| 1. | Magnetosphere of Saturn | F. Scarf |
| 2. | Satellites of Saturn | D. Morrison |
| 3. | Rings of Saturn | J. Cuzzi |
| 4. | Structure and Composition of the Atmosphere of Saturn | T. Owen |
| 5. | Meteorology of Saturn | R. Beebe |
| 6. | Titan | T. Owen |

B. PIONEER VENUS AND VENERA STUDIES OF VENUS

Chairman: B.A. Smith

| | | |
|---|---|---|
| 1430-1500 | Surface of Venus | H. Masursky |
| 1500-1530 | Chemistry and Clouds of Venus | R. Prinn |
| 1530-1600 | Atmospheric Composition and Evolution of Venus Atmosphere | T. Donahue |
| 1630-1700 | Interaction of Venus and Solar Wind | L. Brace |
| 1700-1720 | Ground Based Radar Mapping | D. Campbell |
| 1720-1740 | Pioneer Venus Radar Mapping | G. Pettengill |
| 1740-1800 | The Venus Nature: A Discussion on New Findings | M. Marov |

24 August 1982

WORKING GROUP ON CARTOGRAPHIC COORDINATES

Presiding: M.E. Davies, Chairman

A meeting of the Working Group on Cartographic Coordinates and Rotational Elements of the Planets and Satellites was held with members of both Commissions 4 and 16 in attendance. The report of the Working Group is included in this report of Commission 16.

## REPORT OF THE IAU WORKING GROUP ON CARTOGRAPHIC COORDINATES AND ROTATIONAL ELEMENTS OF THE PLANETS AND SATELLITES: 1982

CHAIRMAN: M.E. Davies
MEMBERS: V.K. Abalakin, J.H. Lieske, P.K. Seidelmann, A.T. Sinclair, A.M. Sinzi, B.A. Smith
CONSULTANT: Y.S. Tjuflin

### 1. Introduction

The IAU Working Group on Cartographic Coordinates and rotational Elements of the Planets and Satellites was established as a consequence of resolutions adopted by Commisssions 4 and 16 at the IAU General Assembly at Grenoble in 1976. The first report of the Working Group was presented to the General Assembly at Montreal in 1979 and published in the *Trans. IAU*, pp.72-79, 1980. The report with appendices was published in *Celestial Mechanics*, 22, pp. 205-230, 1980. The guiding principles and conventions that were adopted by the Group and the rationale for their acceptance are presented in that report and its appendixes and will not be reviewed here.

This report will incorporate changes in the tables giving the directions of the north poles of rotation and the prime meridians of the planets and satellites expressed in standard equatorial coordinates of 1950.0. New tables give these same parameters with respect to the standard equator and equinox of J2000. The tables giving the sizes and shapes of the planets and satellites have also been revised.

### 2. Definition of Rotational Elements

The north pole is that pole of rotations which lies on the north side of the invariable plane of the solar system. At this time, two standard coordinate systems are in common use; they are designated B1950 and J2000. B1950 is defined with respect to the fundamental FK4 catalog and J2000 is defined with respect to the new FK5 catalog. B1950 has the standard epoch (J1950) of 1950 January 1.0, ET(JED 2433282.5) and J2000 has the standard epoch (J2000) of 2000 January 1.5 (JD 2451545.0), TDB. The variable quantities are expressed in units of days (86400 SI seconds) or Julian centuries of 36525 days.

The direction of the north pole is specified by the value of its right ascension $\alpha_o$ and declination $\delta_o$, whereas the orientation of the prime meridian is specified by the angle $W$ that is measured *along* the planet's equator in the positive sense with respect to the planet's north pole (i.e., in an easterly direction on the planet's surface) *from* the ascending node of the planet's equator on the standard equator *to* the point where the prime meridian crosses the planet's equator. Because the prime meridian is assumed to rotate uniformly with the planet, $W$ accordingly varies linearly with time. In addition, $\alpha_o$, $\delta_o$, and $W$ may vary with time due to a precession of the axis of rotation of the planet (or satellite). If $W$ increases with time, the planet has a *direct* (or prograde) rotation relative to the invariable plane; if $W$ decreases with time, the rotation is said to be *retrograde*.

In the absence of other information, the axis of rotation is assumed to be normal to the mean orbital plane; Mercury and most of the satellites are in this category. For many of the satellites, it is assumed that the rotation rate is equal to the mean orbital period.

The angle $W$ specifies the ephemeris position of the prime meridian, and for the planets or satellites without any accurately observable fixed surface features the adopted expression for $W$ defines the prime meridian and is not subject to correction. Where possible, however, the cartographic positions of the prime meridian is defined by a suitable observable feature and so the constants in the expression $W = W_o + \dot{W}d$, where d is the interval in days from the standard epoch, are chosen so that the ephemeris position follows the motion of the cartographic position as closely as possible; in these cases the expression for $W$ may require emendation in the future.

Recommended values of the constants in the expression for $\alpha_o$, $\delta_o$, and $W$ are given for the planets and satellites in Tables 1 and 2 for the standard equatorial coordinates with equinox B1950 at epoch J1950 and for the planets and satellites in Tables 3 and 4 for the standard equotorial coordinates with equinox J2000 at epoch J2000. In general these expressions should be accurate to one-tenth of a degree; however, two decimal places are given to assure consistency when changing coordinate systems. Zeros are added to rate values ($\dot{W}$) for computational consistency and are not an indication of significant accuracy. Three decimal places are given in the expressions for the Moon and Mars, reflecting the greater confidence in their accuracy. Expressions for the Sun and Earth, are given to a similar precision as those of the other bodies of the solar system for comparative purposes only.

3. Definition of Cartographic Coordinates Systems

Both planetocentric and planetographic systems of coordinates are used in the study of the planets and satellites. Both systems are based on the same fundamental reference axis but differ, as explained below, in the definitions of latitude and longitude. Planetocentric coordinates are used for general purposes and are based on a right-handed system of axes, whereas planetographic coordinates are used for cartographic purposes and depend on the adoption of additional parameters to define a reference surface, usually a spheroid, that approximates an equipotential surface of the planet.

For both systems, the fundamental reference $z$-axis is the mean axis of rotation and the planetary equator is the plane that is normal to this axis and passes through the center of mass of the planet. The $x$-axis is defined by the intersection of the equatorial plane with the plane of the prime meridian, whose position is defined in an arbitrary manner. The $y$-axis of planetocentric rectangular coordinates is defined so as to form a right-handed system.

*Latitude* is measured north and south of the equator; north latitudes are designed as positive. The planeto*centric* latitude ($\phi$) of a point is the angle between the equatorial plane and the line connecting the point to the center of mass. The planeto*graphic* latitude ($\phi'$) of a point on the reference surface is the angle between the equatorial plane and the normal to the reference surface at the point.

*Longitude* is measured around the equatorial plane from the prime meridian from $0^\circ$ to $360^\circ$. Planeto*centric* longitudes ($\lambda$) are measured positively to the east, whereas planeto*graphic* longitudes ($\lambda'$) are measured in the direction opposite to the rotation, i.e., positively to the west in the case of direct rotation. Planetocentric longitudes are measured from the ephemeris position of the prime meridian as defined by the rotational elements, whereas planetographic longitudes are measured from the cartographic position to the prime meridian as defined by the adopted longitude of some clearly observable surface feature. These two positions may normally be assumed to coincide, but it is conceivable that errors in the rotational elements may cause the cartographic position to drift away from the ephemeris position by a small amount $\Delta W$, where $\Delta W$ is measured positively to the east of the ephemeris position.

Planetcentric *radius* $(R)$ is measured from the center of mass to the point concerned. In the planetographic system the position of a point $(P)$ not on the reference surface is specified by the planetographic longitude and latitude of the point $(P')$ on the reference surface at which the normal passes through $P$ and by the *height* $(h)$ of $P$ above $P'$.

The reference surfaces for most of the planets are spheroids for which the radius of the equator $(A)$ is larger than the polar semiaxis $(C)$. For some planets and most satellites the reference surface is a sphere $(A = C)$, and the planetocentric and planetographic latitutes are then numerically the same. The polar axis of each reference surface is assumed to be the mean axis of rotation as defined by the adopted rotational elements since the accuracy of measurement cannot, at present, observe a motion of the axis of rotation with respect to the axis of figure.

The recommended values of the parameters for the reference surfaces for planets and satellites are given in Table 5. Radii for irregular-shaped satellites are given in Table 6.

It should be noted that east longitude on the Sun, Earth, and Moon is commonly considered to be in the positive direction and that longitudes are usually measured $0^{\circ}$ to $180^{\circ}$ east and west.

Table 1. Recommended Values for the Direction of the North Pole of Rotation and the Prime Meridian of the Sun and Planets (1982).

$\alpha_0$, $\delta_0$ are standard equatorial coordinates with equinox B1950 at epoch J1950. Approximate coordinates of the north pole of the invariable plane are $\alpha_0 = 272^\circ.40$, $\delta_0 = +66^\circ.99$.
T = interval in Julian ephemeris centuries (of 36525 days) from the standard epoch.
d = interval in ephemeris days from the standard epoch.
The standard epoch is 1950 January 1.0 ET, i.e., JED 2433282.5

| | | | | | | |
|---|---|---|---|---|---|---|
| Sun | $\alpha_0$ | = | 285°.90 | | | |
| | $\delta_0$ | = | 63.90 | | | |
| | W | = | 240.90 | + | 14°.1844000 d | |
| Mercury | $\alpha_0$ | = | 280.90 | - | 0.033 T | |
| | $\delta_0$ | = | 61.40 | - | 0.005 T | |
| | W | = | 184.74 | + | 6.1385025 d | (a) |
| Venus | $\alpha_0$ | = | 272.80 | | | |
| | $\delta_0$ | = | 67.20 | | | |
| | W | = | 213.63 | - | 1.4814205 d | |
| Earth | $\alpha_0$ | = | 0.00 | - | 0.640 T | |
| | $\delta_0$ | = | 90.00 | - | 0.557 T | |
| | W | | 99.87 | + | 360.9856123 d | (b) |
| Mars | $\alpha_0$ | = | 317.342 | - | 0.108 T | |
| | $\delta_0$ | = | 52.711 | - | 0.061 T | |
| | W | = | 11.504 | + | 350.8919830 d | (c) |
| Jupiter | $\alpha_0$ | = | 268.00 | - | 0.008 T | |
| | $\delta_0$ | = | 64.50 | + | 0.003 T | |
| | W | = | 80.60 | + | 870.5360000 d | (d) |
| Saturn | $\alpha_0$ | = | 38.50 | - | 0.034 T | |
| | $\delta_0$ | = | 83.31 | - | 0.004 T | |
| | W | = | 76.81 | + | 810.7939024 d | (d) |
| Uranus | $\alpha_0$ | = | 256.72 | | | |
| | $\delta_0$ | = | -15.04 | | | |
| | W | = | 360.00 | - | 554.9130000 d | |
| Neptune | $\alpha_0$ | = | 294.91 | | | |
| | $\delta_0$ | = | 40.53 | | | |
| | W | = | 360.00 | + | 468.7500000 d | |
| Pluto | $\alpha_0$ | = | 311.00 | | | |
| | $\delta_0$ | = | 4.00 | | | |
| | W | = | 360.00 | - | 56.3640000 d | |

Note (a) The 20° meridian is defined by the crater Hun Kal.
(b) The 0° meridian is defined by the transit circle at Greenwich, England.
(c) The 0° meridian is defined by the crater Airy-0.
(d) The equations for W for Jupiter and Saturn refer to the rotation of their magnetic fields (System III). On Jupiter, System I ($W_I = 17^\circ.7 + 877^\circ.900$ d) refers to the mean atmospheric equatorial rotation; System II ($W_{II} = 16^\circ.8 + 870^\circ.270$ d) refers to the mean

atmospheric rotation north of the south component of the north equatorial belt, and south of the north component of the south equatorial belt.

Table 2. Recommended Values for the Direction of the North Pole of Rotation and the Prime Meridian of the Satellites (1982).

$\alpha_0$, $\delta_0$, T, and d have the same meanings as Table 1 (epoch 1950 January 1.0 ET, i.e., JED 2433282.5)

**Earth — Moon**

$$\begin{aligned}
\alpha_0 &= 270^{\circ}.000 - 3^{\circ}.878 \sin E1 - 0^{\circ}.120 \sin E2 \\
&\quad + 0.070 \sin E3 - 0.017 \sin E4 \\
\delta_0 &= 66.534 + 1.543 \cos E1 + 0.024 \cos E2 \\
&\quad - 0.028 \cos E3 + 0.007 \cos E4 \\
W &= 244.375 + 13.1763581\, d + 3.558 \sin E1 \\
&\quad + 0.121 \sin E2 - 0.064 \sin E3 \\
&\quad + 0.016 \sin E4 + 0.025 \sin E5
\end{aligned}$$

where

$$\begin{aligned}
E1 &= 12^{\circ}.112 - 0^{\circ}.052992\, d, \quad E2 = 24^{\circ}.224 - 0^{\circ}.105984\, d \\
E3 &= 227.645 - 13.012000\, d, \quad E4 = 261.105 + 13.340716\, d \\
E5 &= 358.000 - 0.985600\, d,
\end{aligned}$$

**Mars — Phobos**

$$\begin{aligned}
\alpha_0 &= 317^{\circ}.31 - 0^{\circ}.108\, T + 1^{\circ}.79 \sin M_1 \\
\delta_0 &= 52.70 - 0.061\, T - 1.08 \cos M_1 \\
W &= 270.23 + 1128.8444790\, d + 0.66 \times 10^{-8} d^2 \\
&\quad - 1.42 \sin M_1 - 2.31 \sin M_2
\end{aligned}$$

**Deimos**

$$\begin{aligned}
\alpha_0 &= 316.29 - 0.108\, T + 2.98 \sin M_1 \\
\delta_0 &= 53.33 - 0.061\, T - 1.78 \cos M_1 \\
W &= 69.97 + 285.1619030\, d - 0.4 \times 10^{-9} d^2 - 2.66 \sin M_1
\end{aligned}$$

where $M_1 = 206^{\circ}.34 - 0^{\circ}.435764\, d$ $M_2$ $88^{\circ}.80 + 1128^{\circ}.409670\, d + 0.66 \times 10^{-8} d^2$

**Jupiter: Amalthea**

$$\begin{aligned}
\alpha_0 &= 268^{\circ}.00 - 0^{\circ}.008\, T \\
\delta_0 &= 64.50 + 0.003\, T \\
W &= 50.20 + 722.6303746\, d
\end{aligned}$$

**Io**

$$\begin{aligned}
\alpha_0 &= 268^{\circ}.00 - 0^{\circ}.008\, T + 0^{\circ}.094 \sin J1 + 0^{\circ}.024 \sin J2 \\
\delta_0 &= 64.50 + 0.003\, T + 0.040 \cos J1 + 0.011 \cos J2 \\
W &= 262.72 + 203.4889538\, d - 0.085 \sin J1 - 0.022 \sin J2
\end{aligned}$$

**Europa**

$$\begin{aligned}
\alpha_0 &= 268.03 - 0.008\, T + 1.086 \sin J2 + 0.060 \sin J3 \\
&\quad + 0.015 \sin J4 + 0.009 \sin J5 \\
\delta_0 &= 64.52 + 0.003\, T + 0.468 \cos J2 + 0.026 \cos J3 \\
&\quad + 0.007 \cos J4 + 0.002 \cos J5 \\
W &= 158.44 + 101.3747235\, d - 0.980 \sin J2 - 0.054 \sin J3 \quad \text{(a)} \\
&\quad - 0.014 \sin J4 - 0.008 \sin J5
\end{aligned}$$

**Ganymede**

$$\begin{aligned}
\alpha_0 &= 268.15 - 0.008\, T - 0.037 \sin J2 + 0.431 \sin J3 \\
&\quad + 0.091 \sin J4 \\
\delta_0 &= 64.57 + 0.003\, T - 0.016 \cos J2 + 0.186 \cos J3 \\
&\quad + 0.039 \cos J4 \\
W &= 196.82 + 50.3176081\, d + 0.033 \sin J2 - 0.389 \sin J3 \quad \text{(b)} \\
&\quad - 0.082 \sin J4
\end{aligned}$$

Callisto

$$\begin{aligned}\alpha_o &= 268.68 - 0.008\ T - 0.068 \sin J3 + 0.590 \sin J4 + 0.010 \sin J6\\ \delta_o &= 64.83 + 0.003\ T - 0.029 \cos J3 + 0.254 \cos J4 - 0.004 \cos J6\\ W &= 157.55 + 21.5710715\ d + 0.061 \sin J3 - 0.533 \sin J4 - 0.009 \sin J6 \qquad \text{(c)}\end{aligned}$$

where $J1 = 19^\circ\!.20 + 4850^\circ\!.7\ T$, $J2 = 120^\circ\!.80 + 1191.3\ T$, $J3 = 349^\circ\!.50 + 262^\circ\!.1\ T$, $J4 = 198.30 + 64.3\ T$, $J5 = 241.60 + 2382.6\ T$, $J6 = 317.70 + 6070.0\ T$

Saturn: Mimas

$$\begin{aligned}\alpha_o &= 38^\circ\!.50 - 0^\circ\!.034\ T + 13^\circ\!.13 \sin S1\\ \delta_o &= 83.31 - 0.004\ T - 1.53 \cos S1\\ W &= 246.80 + 381.9945550\ d - 13.04 \sin S1 - 43.41 \sin S6 \qquad \text{(d)}\end{aligned}$$

Enceladus

$$\begin{aligned}\alpha_o &= 38.50 - 0.034\ T\\ \delta_o &= 83.31 - 0.004\ T\\ W &= 301.41 + 262.7318996\ d \qquad \text{(e)}\end{aligned}$$

Tethys

$$\begin{aligned}\alpha_o &= 38.50 - 0.034\ T + 9.36 \sin S2\\ \delta_o &= 83.31 - 0.004\ T - 1.09 \cos S2\\ W &= 31.76 + 190.6979085\ d - 9.29 \sin S2 + 2.16 \sin S6 \qquad \text{(f)}\end{aligned}$$

Dione

$$\begin{aligned}\alpha_o &= 38.50 - 0.034\ T\\ \delta_o &= 83.31 - 0.004\ T\\ W &= 121.54 + 131.5349316\ d \qquad \text{(g)}\end{aligned}$$

Rhea

$$\begin{aligned}\alpha_o &= 38.22 - 0.034\ T + 3.00 \sin S3\\ \delta_o &= 83.34 - 0.004\ T - 0.35 \cos S3\\ W &= 14.52 + 79.6900478\ d - 2.98 \sin S3 \qquad \text{(h)}\end{aligned}$$

Titan

$$\begin{aligned}\alpha_o &= 34.30 - 0.034\ T + 2.58 \sin S4\\ \delta_o &= 83.71 - 0.004\ T - 0.30 \cos S4\\ W &= 79.10 + 22.5769768\ d - 2.56 \sin S4\end{aligned}$$

Hyperion

$$\begin{aligned}\alpha_o &= 33.42 - 0.034\ T + 4.89 \sin S5 + 2.75 \sin S4\\ \delta_o &= 83.80 - 0.004\ T - 0.57 \cos S5 - 0.32 \cos S4\\ W &= 344.75 + 16.9199514\ d - 4.86 \sin S5 - 2.73 \sin S4 + 9.09 \sin S7 \qquad \text{(i)}\end{aligned}$$

Iapetus

$$\begin{aligned}\alpha_o &= 289^\circ\!.26\\ \delta_o &= 78.73\\ W &= 275.36 + 4.5379571\ d \qquad \text{(j)}\end{aligned}$$

where $S1 = 68^\circ\!.60 - 36505^\circ\!.5\ T$, $S2 = 314^\circ\!.50 - 7225^\circ\!.9\ T$, $S3 = 134^\circ\!.90 - 1016^\circ\!.3\ T$, $S4 = 57.40 - 52.1\ T$, $S5 = 22.60 - 239.2\ T$, $S6 = 64.90 + 506.2\ T$, $S7 = 277.40 + 20528.5\ T$

Uranus: Miranda

$$\begin{aligned}\alpha_o &= 256^\circ\!.72\\ \delta_o &= -15.04\\ W &= 59.20 - 254^\circ\!.5968883\ d\end{aligned}$$

Ariel

$$\begin{aligned}\alpha_o &= 256.72\\ \delta_o &= -15.04\\ W &= 47.30 - 142.8356047\ d\end{aligned}$$

Umbriel

$$\begin{aligned}\alpha_o &= 256.72\\ \delta_o &= -15.04\\ W &= 146.40 - 86.8688136\ d\end{aligned}$$

| | | | | |
|---|---|---|---|---|
| | Titania | $\alpha_0$ | = | $256.72$ |
| | | $\delta_0$ | = | $-15.04$ |
| | | $W_0$ | = | $202.00 - 41.3513623\ d$ |
| | Oberon | $\alpha_0$ | = | $256.72$ |
| | | $\delta_0$ | = | $-15.04$ |
| | | $W_0$ | = | $3.20 - 26.7394375\ d$ |
| Neptune: | Triton | $\alpha_0$ | = | $294.89 - 20.09 \sin N$ |
| | | $\delta_0$ | = | $36.93 + 15.26 \cos N$ |
| | | $W_0$ | = | $132.30 - 61.2575147\ d + 10^\circ.52 \sin N$ |

whwere $N = 158^\circ.34 + 62^\circ.0\ T$

| | | | | |
|---|---|---|---|---|
| Pluto: | Charon | $\alpha_0$ | = | $311^\circ.00$ |
| | | $\delta_0$ | = | $4.00$ |
| | | $W_0$ | = | $133.00 - 56.3640000\ d$ |

Note (a) The 180° meridian is defined by the crater Cilix.
(b) The 128° meridian is defined by the crater Anat.
(c) The 326° meridian is defined by the crater Saga.
(d) The 162° meridian is defined by the crater Palomides.
(e) The 5° meridian is defined by the crater Salih.
(f) The 299° meridian is defined by the crater Arete.
(g) The 63° meridian is defined by the crater Palinurus.
(h) The 340° meridian is defined by the crater Tore.
(i) The 196° meridian is defined by the crater Bahloo.
(j) The 276° meridian is defined by the crater Almeric.

Table 3. Recommended Values for the Direction of the North Pole of Rotation and the Prime Meridian of the Sun and Planets (1982).

$\alpha_0$, $\delta_0$ are standard equatorial coordinates with equinox J2000 at epoch J2000. Approximate coordinates of the north pole of the invariable plane are $\alpha_0 = 272^\circ.38$, $\delta_0 = +66^\circ.99$.

T = interval in Julian centuries (of 36525 days) from the standard epoch.

d = interval in days from the standard epoch.

The standard epoch is 2000 January 1.5, i.e., JD 2451545.0 TDB.

| | | | | |
|---|---|---|---|---|
| Sun | $\alpha_0$ | = | $285^\circ.96$ | |
| | $\delta_0$ | = | $63.96$ | |
| | $W_0$ | = | $84.11 + 14^\circ.1844000\ d$ | |
| Mercury | $\alpha_0$ | = | $281.02 - 0.033\ T$ | |
| | $\delta_0$ | = | $61.45 - 0.005\ T$ | |
| | $W_0$ | = | $329.71 + 6.1385025\ d$ | (a) |
| Venus | $\alpha_0$ | = | $272.78$ | |
| | $\delta_0$ | = | $61.21$ | |
| | $W_0$ | = | $159.91 - 1.4814205\ d$ | |
| Earth | $\alpha_0$ | = | $0.00 - 0.641\ T$ | |
| | $\delta_0$ | = | $90.00 - 0.557\ T$ | |
| | $W_0$ | = | $100.21 + 360.9856123\ d$ | (b) |

| | | | | | |
|---|---|---|---|---|---|
| Mars | $\alpha_0$ = | 317.681 | − | 0.108 T | |
| | $\delta_0$ = | 52.886 | − | 0.061 T | |
| | $W$ = | 176.655 | + | 350.8919830 d | (c) |
| Jupiter | $\alpha_0$ = | 268.05 | − | 0.009 T | |
| | $\delta_0$ = | 64.49 | + | 0.003 T | |
| | $W$ = | 284.95 | + | 870.5360000 d | (d) |
| Saturn | $\alpha_0$ = | 40.66 | − | 0.036 T | |
| | $\delta_0$ = | 83.52 | − | 0.004 T | |
| | $W$ = | 38.90 | + | 810.7939024 d | (d) |
| Uranus | $\alpha_0$ = | 257.43 | | | |
| | $\delta_0$ = | -15.10 | | | |
| | $W$ = | 261.62 | − | 554.9130000 d | |
| Neptune | $\alpha_0$ = | 295.33 | | | |
| | $\delta_0$ = | 40.65 | | | |
| | $W$ = | 107.21 | + | 468.7500000 d | |
| Pluto | $\alpha_0$ = | 311.63 | | | |
| | $\delta_0$ = | 4.18 | | | |
| | $W$ = | 252.66 | − | 56.3640000 d | |

Note (a) The 20° meridian is defined by the crater Hun Kal.
(b) The 0° meridian is defined by the transit circle at Greenwich, England.
(c) The 0° meridian is defined by the crater Airy-0.
(d) The equations for W for Jupiter and Saturn refer to the rotation of their magnetic fields (System III). On Jupiter, System I ($W_I$ = 67°1 + 877°900 d) refers to the mean atmospheric equatorial rotation; System II ($W_{II}$ = 43°3 + 870°270 d) refers to the mean atmospheric rotation north of the south component of the north equatorial belt, and south of the north component of the south equatorial belt.

Table 4. Recommendation Values for the Direction of the North Pole of Rotation and the Prime Meridian of the Satellites (1982).

$\alpha_0$, $\delta_0$, T, and d have the same meanings as in Table 3 (epoch 2000 January 1.5, i.e., JD 2451545.0 TDB)

Earth: Moon

$$\alpha_0 = 270°000 - 3°878 \sin E1 - 0°120 \sin E2 + 0.070 \sin E3 - 0.017 \sin E4$$

$$\delta_0 = 66.534 + 1.543 \cos E1 + 0.024 \cos E2 - 0.028 \cos E3 + 0.007 \cos E4$$

$$W = 38.314 + 13.1763581\,d + 3.558 \sin E1 + 0.121 \sin E2 - 0.064 \sin E3 + 0.016 \sin E4 + 0.025 \sin E5$$

where E1 = 125°045 − 0°052992 d, E2 = 249°390 − 0°105984 d,
E3 = 196.694 − 13.012000 d, E4 = 176.630 + 13.340716 d,
E5 = 358.219 − 0.985600 d

Table 4. Recommended Values for the Direction of the North Pole of Rotation

| | | | |
|---|---|---|---|
| Mars: | Phobos | $\alpha_o = 317\overset{\circ}{.}65 - 0.108\ T + 1.80 \sin M_1$ | |
| | | $\delta_o = 52.87 - 0.061\ T - 1.08 \cos M_1$ | |
| | | $= 32.84 + 1128.8444790\ d + 0.66 \times 10^{-8} d^2 - 1.43 \sin M_1 - 2.32 \sin M_2$ | |
| | Deimos | $= 316.62 - 0.108\ T + 3.00 \sin M_1$ | |
| | | $53.50 - 0.061\ T - 1.78 \cos M_1$ | |
| | | $= 79.55 + 285.1619030\ d - 0.4 \times 10^{-9} d^2 - 2.68 \sin M_1$ | |

where $M_1 = 169\overset{\circ}{.}51 - 0\overset{\circ}{.}435764\ d$ $M_2 = 190\overset{\circ}{.}71 + 1128\overset{\circ}{.}409670\ d + 0.66 \times 10^{-8}\ d^2$

| | | | |
|---|---|---|---|
| Jupiter: | Amalthea | $= 268\overset{\circ}{.}05 - 0\overset{\circ}{.}009\ T$ | |
| | | $= 64.49 + 0.003\ T$ | |
| | | $= 208.06 + 722\overset{\circ}{.}6303746\ d$ | |
| | Io | $= 268\overset{\circ}{.}05 - 0\overset{\circ}{.}009\ T + 0\overset{\circ}{.}094 \sin J1 + 0\overset{\circ}{.}024 \sin J2$ | |
| | | $= 64.50 + 0.003\ T + 0.040 \cos J1 + 0.011 \cos J2$ | |
| | | $= 200.39 + 203.4889538\ d - 0.085 \sin J1 - 0.022 \sin J2$ | |
| | Europa | $= 268.08 - 0.009\ T + 1.086 \sin J2 + 0.060 \sin J3 + 0.015 \sin J4 + 0.009 \sin J5$ | |
| | | $= 64.51 + 0.003\ T + 0.468 \cos J2 + 0.026 \cos J3 + 0.007 \cos J4 + 0.002 \cos J5$ | |
| | | $= 34.97 + 101.3747235\ d - 0.980 \sin J2 - 0.054 \sin J3 - 0.014 \sin J4 - 0.008 \sin J5$ | (a) |
| | Ganymede | $= 268.20 - 0.009\ T - 0.037 \sin J2 + 0.431 \sin J3 + 0.091 \sin J4$ | |
| | | $= 64.57 + 0.003\ T - 0.016 \cos J2 + 0.186 \cos J3 + 0.039 \cos J4$ | |
| | | $= 42.79 + 50.3176081\ d + 0.033 \sin J2 - 0.389 \sin J3 - 0.082 \sin J4$ | (b) |
| | Callisto | $= 268.72 - 0.009\ T - 0.068 \sin J3 + 0.590 \sin J4 + 0.010 \sin J6$ | |
| | | $= 64.83 + 0.003\ T - 0.029 \cos J3 + 0.254 \cos J4 - 0.004 \cos J6$ | |
| | | $= 259.90 + 21.5710715\ d + 0.061 \sin J3 - 0.533 \sin J4 - 0.009 \sin J6$ | (c) |

where $J1 = 283\overset{\circ}{.}90 + 4850\overset{\circ}{.}7\ T$, $J2 = 355\overset{\circ}{.}80 + 1191\overset{\circ}{.}3\ T$, $J3 = 119\overset{\circ}{.}90 + 262\overset{\circ}{.}1\ T$,
$J4 = 229\overset{\circ}{.}80 + 64.3\ T$, $J5 = 352\overset{\circ}{.}25 + 2382.6\ T$, $J6 = 113\overset{\circ}{.}35 + 6070.0\ T$

| | | | |
|---|---|---|---|
| Saturn: | Mimas | $= 40\overset{\circ}{.}66 - 0\overset{\circ}{.}036\ T + 13\overset{\circ}{.}56 \sin S1$ | |
| | | $= 83.52 - 0.004\ T - 1.53 \cos S1$ | |
| | | $= 340.81 + 381.9945550\ d - 13.48 \sin S1 - 44.85 \sin S6$ | (d) |
| | Enceladus | $= 40.66 - 0.036\ T$ | |
| | | $83.52 - 0.004\ T$ | |
| | | $= 1.18 + 262.7318996\ d$ | (e) |
| | Tethys | $= 40.66 - 0.036\ T + 9.66 \sin S2$ | |
| | | $= 83.52 - 0.004\ T - 1.09 \cos S2$ | |
| | | $= 10.77 + 190.6979085\ d - 9.60 \sin S2 + 2.23 \sin S6$ | (f) |
| | Dione | $= 40.66 - 0.036\ T$ | |
| | | $= 83.52 - 0.004\ T$ | |
| | | $= 356.68 + 131.5349316\ d$ | (g) |

Rhea $\alpha_0 = 40.38 - 0.036\,T + 3.10 \sin S3$
$\delta_0 = 83.55 - 0.004\,T - 0.35 \cos S3$
$W^{o} = 232.47 + 79.6900478\,d - 3.08 \sin S3$ (h)

Titan $\alpha_0 = 36.41 - 0.036\,T + 2.66 \sin S4$
$\delta_0 = 83.94 - 0.004\,T - 0.30 \cos S4$
$W^{o} = 189.64 + 22.5769768\,d - 2.64 \sin S4$

Hyperion $\alpha_0 = 35.52 - 0.036\,T + 5.05 \sin S5 + 2.84 \sin S4$
$\delta_0 = 84.03 - 0.004\,T - 0.57 \cos S5 - 0.32 \cos S4$
$W^{o} = 103.88 + 16.9199514\,d - 5.02 \sin S5 - 2.82 \sin S4$
$+ 9.39 \sin S7$ (i)

Iapetus $\alpha_0 = 288^{\circ}.57$
$\delta_0$ 78.82
$W^{o}$ 351.15 + 4.5379571 d (j)

where $S1 = 177^{\circ}.40 - 3650^{\circ}.5\,T$, $S2 = 300^{\circ}.00 - 7225^{\circ}.9\,T$, $S3 = 345^{\circ}.20 - 1016^{\circ}.3\,T$,
$S4 = 29^{\circ}.80 - 52.1\,T$, $S5 = 261^{\circ}.45 - 239^{\circ}.2\,T$, $S6 = 316^{\circ}.45 + 506^{\circ}.2\,T$,
$S7 = 100.10 + 20528.5\,T$

Uranus: Miranda $\alpha_0 = 257^{\circ}.43$
$\delta_0 = -15.10$
$W^{o} = 243.81 - 254^{\circ}.5968883\,d$

Ariel $\alpha_0 = 257.43$
$\delta_0 = -15.10$
$W^{o} = 72.35 - 142.8356047\,d$

Umbriel $\alpha_0 = 257.43$
$\delta_0 = -15.10$
$W^{o} = 224.97 - 86.8688136\,d$

Titania $\alpha_0 = 257.43$
$\delta_0 = -15.10$
$W^{o} = 303.03 - 41.3513623\,d$

Oberon $\alpha_0 = 257.43$
$\delta_0 = -15.10$
$W^{o} = 194.50 - 26.7394375\,d$

Neptune: Triton $\alpha_0 = 295.34 - 20.12 \sin N$
$\delta_0 = 37.05 + 15.26 \cos N$
$W^{o} = 297.25 - 61.2575147\,d + 10.56 \sin N$

where $N = 189^{\circ}.66 + 62^{\circ}.0\,T$

Pluto: Charon $\alpha_0 = 311.63$
$\delta_0 = 4.18$
$W^{o} = 25.66 - 56.3640000\,d$

---

Note (a) The 182° meridian is defined by the crater Cilix.
(b) The 128° meridian is defined by the crater Anat.
(c) The 326° meridian is defined by the crater Saga.
(d) The 162° meridian is defined by the crater Palomides.
(e) The 5° meridian is defined by the crater Salih.
(f) The 299° meridian is defined by the crater Arete.
(g) The 63° meridian is defined by the crater Palinurus.
(h) The 340° meridian is defined by the crater Tore.
(i) The 196° meridian is defined by the crater Bahloo

(j) The 276° meridian is defined by the crater Almeric.

Table 5. Recommended Reference Spheroids for Mapping the Planets and Major Satellites (1982).

| Planet | Satellite | Equatorial Radius (km) | Flattening |
|---|---|---|---|
| Mercury | | 2439 | 0 |
| Venus | | 6051 | 0 |
| Earth | | 6378.140 | 0.00335281 |
| | Moon | 1738 | 0 |
| Mars | | 3393.4 | 0.0051865 |
| Jupiter | | 71398 | 0.0648088 |
| | Io | 1815 | 0 |
| | Europa | 1569 | 0 |
| | Ganymede | 2631 | 0 |
| | Callisto | 2400 | 0 |
| Saturn | | 60000 | 0.1076209 |
| | Mimas | 198 | 0.019 |
| | Enceladus | 253 | 0.020 |
| | Tethys | 525 | 0 |
| | Dione | 560 | 0 |
| | Rhea | 765 | 0 |
| | Titan | 2575 | 0 |
| | Iapetus | 725 | 0 |
| | Phoebe | 110 | 0 |
| Uranus | | 25400 | 0.030 |
| | Ariel | 400 | 0 |
| | Umbriel | 275 | 0 |
| | Titania | 500 | 0 |
| | Oberon | 450 | 0 |
| | Miranda | 150 | 0 |
| Neptune | | 24300 | 0.0259 |
| | Triton | 1600 | 0 |
| Pluto | | 1500 | 0 |
| | Charon | 600 | 0 |

Note: The equatorial radii for Mercury, Venus, Moor, Mars, Io, Europa, Ganymede, Callisto, Mimas, Enceladus, Tethys, Dione, Rhea, and Iapetus are used in current mapping programs, and those for Jupiter and Saturn are used in sequencing and analyzing data from current flight missions. The values for Mars and Pluto differ from those recommended by the IAU in 1976 (*Trans. IAU*, 16B, p. 60). The reference spheroid for Mars (3393.4 km radius) has been used in all mapping programs since 1973, although the IAU 1976 radius (3397.2 km) is probably a better value. In 1976 Pluto's satellite, Charon, had not been discovered.

Table 6. Recommended Reference Shapes for Mapping Irregular Satellites (1982)

| Planet | Satellite | Equatorial Radius, A (km) | Equatorial Radius, B (km) | Polar Radius, C (km) |
|---|---|---|---|---|
| Mars | Phobos | 13.5 | 10.7 | 9.6 |
| | Deimos | 7.5 | 6.0 | 5.5 |

| | | | | |
|---|---|---|---|---|
| Jupiter | Amalthea | 140 | 105 | 80 |
| Saturn | Hyperion | 200 | 125 | 110 |
| | Janus | 110 | 95 | 80 |
| | Epimetheus | 70 | 57 | 50 |

## COMMISSION 19: ROTATION OF THE EARTH (ROTATION DE LA TERRE)

Report of Meetings 18, 19, 23 and 24 August 1982

President: P. Paquet  Secretary: W.J. Klepczynski

### 18 August 1982 (Morning)

After the opening of the meeting, the President reported on some of the highlights of the past three years. These included: the availability of the observations made within the People's Republic of China (PRC); the successful completion of the MERIT Short Campaign; and, the establishment of the IAU/IAG Joint Working Group on the establishment and maintenance of a Conventional Terrestrial Reference System (COTES). He emphasized the need to still continue observations with the classical techniques, and noted the improvements which were being incorporated into the classical instrumentation. He also looks forward to the completion of the Hipparcos catalogue and the consequent improvement that should result in some optical determinations of Earth Rotation Parameters (ERP).

B. Guinot summarized the parts of the Annual Report of the Bureau International de l'Heure which pertained to Commission 19. He stressed the involvement of the BIH in the Merit Short Campaign. The BIH acted as the coordinator of the campaign and furnished weekly summaries of ERP's derived through analyses of observations contributed by the new techniques. He also noted the steady improvement in the ERP's published by the BIH over the years. In Polar Motion (PM), the pair variance improved from 0".015 in 1967 to 0".006 in 1982. The most recent values reflect the inclusion of improved Doppler data, observations from the PRC, and Connected Element Radio Interferometry. In Universal Time (UT), the formal standard deviation (not pair variance) has improved from 1.3 ms in 1967 to about 0.6 ms in 1982. He pointed out that a weaker smoothing of UT1 was introduced in Circular D on 1 January 1982. The new smoothing interval is 30 days as opposed to 90 days. The weaker smoothing is expected to improve the comparison of observations with BIH data by not artificially masking the short period effects. He also stressed the importance of the role that the G.E. Mark III Information System played during the MERIT Short Campaign in allowing fast, efficient international exchange of data.

K. Yokoyama, the new Director of the IPMS, reported on its activity during the last three years. He noted the dissolution of the ILS chain in June 1982 with the termination of observations at Gaithersburg and Ukiah by the National Geodetic Survey. During the MERIT Short Campaign, the IPMS acted as an analysis center for the classical techniques. He reported that the IPMS analysis noted the same short period variations in UT as other new techniques when compared with the strongly smoothed BIH data. He also noted the improvement in precision and accuracy of ERP's derived from optical astrometry due to the inclusion of data from the PRC. The completion of the re-reduction of past ILS observations from 1899 to 1978 was mentioned.

Ye Shu-hua reported on activities within the PRC. There are now six observatories contributing to Earth rotation studies and observations: Shanghai Observatory (responsible for the issuing of UT1); Purple Mountain Observatory (in Nanjing); Beijing Observatory and its Tienjin Latitude Station (responsible for Polar Motion); Wuchan Time Observatory (affiliated with the Institute of Geodesy and Geophysics); Shaanxi Observatory (responsible for broadcasting HF and LF time signals); and, Yunan observatory (in Kunming). Presently, the internal consistency for UT, determined within the PRC, is about 1 ms and for PM about 0".01. While three months predictions for UT1 and PM are good to ±5 ms and ±0".01, respectively. Local atomic time scales are maintained at Shanghai, Shaanxi and Beijing Observatories. Photoelectric astrolabes are now in operation at Beijing and Shanghai Observatories.

S. Debarbat reported on her comparison of Paris Observatory Astrolable observations with those of the Green Bank Connected Element Radio interferometer. She found that over the two and a half years of comparison, the external precision of the two techniques remained constant regardless of whether one uses a short or long span of dates. The one day averaged values for the Paris astrolabe have an external precision of 4.5 ms in UT and 0".059 in PM, while the three day averaged values of the Green Bank Interferometer have an external precision of 2.0 ms in UT and 0".034 in PM.

R.O. Vincente noted the success of the MERIT Short Campaign and its effective communications. He cautioned that extreme care must be exercised in analyses based on short intervals of data such as the three months of the Short Campaign. Furthermore, he recommended that investigators use continuous data points instead of averaged or smoothed values. He showed by several examples that significantly different values of the Chandler frequency could be obtained as a function of the time span of the data set used in the analysis.

C. Oesterwinter, reporting for R.J. Anderle and C.A. Malyevac, noted that PM, as determined from the "OCSAR" type Navy Navigation Satellites (TRANSIT) had deteriorated in the last two years due to increased drag effects arising from the high levels of solar activity. The use of rubidum frequency standards in portable tracking equipment should improve the overall performance of the observing network. A new procedure was introduced to derive UT based on discrepancies of the predicted Right Ascension of the NAVSAT satellites rather than on the basis of the prediction errors in the longitude of the node. While the revised procedure produces useful results, they are still not conclusive.

V. Dehant presented the results of a study, co-authored by P. Paquet, which analyzed the variations in the height of the Brussels station as deduced from Doppler observations during the period 1972-1980. The variations, which are related to unmodelled perturbations of the troposphere and ionosphere, can be interpreted as an estimation of the stability of the Doppler system. The main variations are of medium and long period: an annual component, one of about the 11.75 year period of the solar activity and one of four months (121-128 days).

J. Fanselow reported on JPL's effort to develop a near real time UT and PM service in support of the Deep Space Network. The nominal operational capability is for weekly measurements of ERP's available within one week after observations with an accuracy of 10-20 cms.

B.E. Schutz reported on the work that he and his colleagues, B.D. Tapley and R.J. Eanes, have done at the University of Texas at Austin on deriving ERP's from satellite laser ranging. Presently, the satellite force model LPM81.12 (page D-13, BIH Annual Report 1981) is being improved and a new model is expected to go into use after January 1983. Because UT1 is correlated with the satellite's longitude of the ascending node, UT1 changes can only be identified if an adequate model for the nodes exists. By basing a solution for the node on one year of data, one can convert node residuals into UT1 changes. Comparison of these UT1 residuals with weakly smoothed UT1 of BIH show good correlation.

Ye Shu-hua reported on the work of Gu Zhen-nian, Wang Shu-he and Zjeng Da-wei from Shanghai Observatory on analyzing the differences in PM deduced from several series of classical optical observations and that deduced from Doppler (DMA) and Satellite laser ranging (IASOM).

## 18 August 1982 (Afternoon)

W. Markowitz noted that it is almost impossible to determine exactly the periods of the annual and Chandler terms of PM. He bases this conclusion on the assumption that the annual and Chandler terms have random variations which preclude their separation and the fact that there are random observational errors further complicates the problem. By using short spans of data, which cover the beat period between the two periods (6-7 years) one can get reasonably good resolution of the two terms.

L.V. Morrison and F.R. Stephenson reported on their computation of the secular and decade fluctuations in the Earth's rotation between 700 BC and AD 1978 from timings of lunar eclipses and occultations.

T. Van Flandern reported on the work which he and his colleague M.R. Lukac, did in computing new values of $\Delta T$ back to 1820 utilizing observations of 150,000 lunar occultations. Their analysis indicates that the Earth's rate of rotation changed by three parts in $10^{10}$ (1 s/cy) during an interval of less than a year in early 1896.

M. Feissel showed that a very strong correlation exists between variations in the length of the day and in the angular momentum of the atmosphere. The variations seem to have a period between 50 and 55 days.

B. Guinot discussed the Chandlerian nutation from 1900 to 1980 following the lines of a study published in 1972. Based on the assumption that the annual term of the annual variation remains stable, he showed that a change in the amplitude of the Chandlerian motion occurred in 1980.

C. Kakuta conjectured that thermal energy dissipation under the crust, caused by earthquakes, might excite variations in the Earth's rotational motion.

S.K. Runcorn cited the importance of astronomical data on fluctuations in the rotation rate of the Earth and on the Chandler wobble because only the astronomical data, along with geomagnetic secular changes, gives evidence of the dynamical processes in the Earth's interior. The westward drift of the magnetic pole reflects the westward drift of the inner core. Furthermore, no quantitively satisfactory explanation has been offered for the irregular changes in the length of the day other than that they represent transfer of angular momemtum between the core and the mantle. It seems that only the outer part of the core interchanges its angular momentum with the mantle over periods of 10-100 years. In addition, the time scale of the irregular changes in the length of day (as presented by Van Flandern) is of importance to the geomagnetician. For this is related to the time taken for a magnetohydrodynamic disturbance to travel over a distance in the core comparable to the length of the non-dipole field. It is also related to the free decay time-constant of the semi-conducting mantle.

C.F. Yoder discussed the free rotational librations of the Earth's inner core. There are six possible free motions of the Earth's inner core. Three involve displacements of the inner core center of mass with respect to the mantle. The remaining three free modes are rotations of the inner core figure with respect to the mantle figure: a Chandler-like wobble, a free nutation of the inner core spin axis and a free libration in longitude. The free wobble exists even if no external torques act on the Earth's inner core while the remaining free motions require an external restoring torque. M. Slade and C. Yoder found that the gravitational interaction between the figures of the core and the mantle will drive the free nutation and longitude librations and reduce the period of the inner core wobble to about eight years for a density jump of about 0.6 $gm/cm^3$ at the inner core-fluid core boundary. The equivalent free nutation period of the inner core spin axis is about 12 years.

## 19 August 1982
## Joint Meeting with Commission 31 - Project MERIT
## (Session I - Chairman, P. Paquet)

Progress reports by the Working Groups on the rotation of the Earth (MERIT) and on the new COnventional TErrestrial reference System (COTES) were presented.

The first session consisted of three introductory reviews as follows:

| | |
|---|---|
| G.A. Wilkins: | The objectives and organization of Project MERIT |
| I.I. Mueller: | The objectives and proposals of the Working Group on the conventional terrestrial reference system (COTES) |
| M. Feissel: | A review of the MERIT Short Campaign and of the plans for the MERIT Main Campaign (1983/1984). |

The second session consisted of six reports on the activities and plans for each of the observing techniques and then a consideration of a resolution for submission to the General Assembly. The reports were as follows:

| | |
|---|---|
| K. Yokoyama: | Optical astrometry |
| C. Oesterwinter: | Doppler tracking of satellites |
| R. Schultz: | Satellite laser ranging |
| O. Calame: | Lunar laser ranging |
| W.J. Klepczynski: | Connected element radio interferometry |
| D.S. Robertson: | Very long baseline radio interferometry. |

The draft resolution was amended in minor respects and then adopted without objection. It was subsequently endorsed by the General Assembly.

G. Wilkins, the chairman of the IAU/IUGG Joint Working Group on the Rotation of the Earth explained that Project MERIT is a programme of international collaboration to Monitor Earth Rotation and to Intercompare the Techniques of observation and analysis. A Short Campaign was held in August - October 1980 and the results have been reported in detail. The Main Campaign will take place from 1 September 1983 to 31 October 1984. Current information about the project is distributed in the MERIT Newsletter which is available from the Royal Greenwich Observatory.

I.I. Mueller drew attention to the resolution of IAU Colloquium Number 56 that led to the setting up of the Joint Working Group of which he is chairman; the group is to prepare a proposal for the establishment and maintenance of a COnventional TErrestrial reference System. For the new system it will be necessary to adopt precise coördinates for a set of well-distributed observatories and to relate the new system to the celestial and inertial systems that are implicit in the various techniques for precise positioning. In particular it will be necessary to determine earth-rotation parameters (ERPs) to a precision of, say, 5 cm (0".002) with a time resolution of 12 hours (6 hours if possible). Extra observations will be made during the MERIT Main Campaign to look for short-period variations and to obtain data at selected sites by at least two techniques. Several mobile SLR and VLBI systems will be used. The proposals were endorsed by the International Association of Geodesy at its General Meeting in Tokyo in May.

M. Feissel reviewed the MERIT Short Campaign and identified areas which should be improved for the Main Campaign. The purpose of the Short Campaign was to test the different observing techniques and establish a reliable, efficient communication system. Quick-Look estimates for ERP's were usually available within one week; most results were available within two weeks. Communication of data was done either through the G.E. Mark III Information System, a worldwide computing system, or through TELEX messages. It seemed that the general organization of the Short Campaign was satisfactory. In regards to improvements, it is recommended that there be fewer Operational Centers and that the coordinators establish better links with the Operational Centers of their respective techniques. Analysis of the data showed that the period covered by the Short Campaign included a reasonably significant perturbation in the path of the pole. Some analyses showed the existence of systematic differences between techniques and also pointed out that different Analysis Centers, using the same data, showed systematic differences in ERP's amounting to about 2 ms. This pointed out the need to improve homogeneity in the reduction procedures and the importance of the work of the Standards Committee.

In closing the first session the President of Commission 19 congratulated the Working Groups on the progress that has been made since the last General Assembly.

Please refer to Report of Commission 31 for minutes of Session II.

## 23 August 1982

This session was primarily concerned with new instrumentation. The first three reports dealt with recent improvements to the astrolabe. Two reports on the forthcoming Hipparcos satellite were presented. The session also included reports on Connected Element Radio Interferometry and the Canadian Polar Motion stations.

G. Billaud reported on the photoelectric astrolabe designed and built at CERGA. The design goal was to overcome the three major imperfections of the classical Danjon astrolabe: 1) the temperature induced variation of the angles of the glass prism; 2) the chromatism and astigmatism of the Walloston prism images; and, 3) the personal equation of the observer. The instrument has a photomultiplier to replace the observer, a Maksutov-Cassegrain reflector instead of a refractor, and a modified Hog's slit micrometer. Preliminary results obtained from observations made from 9 February - 19 March 1982 indicate an accuracy of 3.8 ms in UT0-UTC, i.e., standard deviation of a comparison with the BIH.

Z.-y Li of Shanghai Observatory discussed the development of the photoelectric astrolabe within the PRC. A prototype Type I Photoelectric Astrolabe was constructed in 1972 by the Nanjing Astronomical Instrument Factory. It was installed at Shaanxi Observatory. Three Type II instruments were subsequently developed in 1974 and installed at Beijing, Shanghai and Yunan. The Type II instrument is divided into two parts. One is in the dome and consists of the photomultiplier box and telescope; the other part is located in a room under the dome and consists of the telescope controls, amplifiers and power supplies. The main telescope is sealed and exhausted to a low vacuum to eliminate refraction inside the tube and contamination of the mercury. In regard to accuracy, the rms deviation for time is 2.5 ms when compared to the smoothed Chinese UT1 system and in latitude it is 0".02 when compared to a Vondrak smoothed curve, fit through the observations.

C. Kakuta described the photoelectric astrolabe of Mizusawa. It is a new Tsubokawa automatic astrolabe of a pendulum type with an electronic transit detector. The instrument is based on the same principle as the original one at Kanozan Geodetic Observatory. Several improvements have been made and a real time digital filter is designed to eliminate motions of the stars. Presently, the instrument is being evaluated and the observational results will be compared with those derived from the Danjon astrolabe.

J. Popelar reported on three years (1979-1981) of PZT and Tranet satellite Doppler observations at two collocated stations near Ottawa and Calgary. A comparison of the variation in station coordinates and relative precision of the two techniques was made. In regard to station longitude, the Calgary PZT shows long term variations in the 3-4 m range while for the Ottawa PZT they are below the 2 m level. The two Tranet stations show long term variations in the horizontal station coordinates at the 2 m level with no apparent correlation between the stations. However, the height variations exceed 2 m and are clearly correlated. Residual variations of the 5 day averages from smoothed long term curves yield standard deviations around the 1 m level for both techniques. Popelar pointed out that careful hardware calibrations and careful modelling are necessary for best Tranet results.

W.J. Klepczynski described improvements to the Green Bank 35 km interferometer system. A new element, located orthogonal to and 32 km distant from the present baseline, will be in operation in the fall of 1983. This will allow the interferometer to resolve all three components of the Earth's rotation. A new source position catalogue was recently introduced. Using this new catalogue to re-reduce past data obtained during the best observing weather, the internal mean errors for UT0-UTC improved to 0.4 ms. In the near future, a new atmospheric model will be used in the reduction of observations. This should help in reducing the annual term exhibited in comparisons with the BIH. Water vapor radiometers, which will measure the differential water vapor seen along the line-of-sight between two antennas are being constructed and should be in operation within 4-6 months. They will be useful in reducing the day-to-day scatter in UT and PM as determined by the interferometer.

J. Kovalevsky reported on the status of the Hipparcos satellite. Launch is scheduled for 1986. It is hoped to have the first catalogue, based on Hipparcos observations, completed by 1991. A preliminary source list is scheduled to be completed next year, with final source catalogue to be completed by 1985. There still are a few small problem areas. One is the existence of possible systematic errors due to chromatic effects caused by non-alignment of optical parts; the other is the introduction of spurious signals in the detector (glare) caused by the presence of a stray bright star in the field of view. By choosing an attitude control system controlled by jets of gas, a possible source of error, called jitter, has been eliminated.

M. Feissel discussed the list of stars selected for the Hipparcos program from those supplied by members of Commission 19. Initially, 10,000 stars were proposed, but the list was cut to 5,508 stars. The majority of the stars are between $10^{\circ} \leq \delta \leq 60^{\circ}$ and 6-7 magnitude. The stars in this program are to be used primarily for geodetic purposes.

24 August 1982 (Scientific)

P. Hill discussed the plans of the Groupe de Rechereche de Geodesie Spatiale (GRGS) for VLBI. Primary aim of the project is for global geodynamics and astrometry. Initial plans call for participation in the ERIC (EISCAT Radio Interferometric Campaign) Project and include the acquisition of two Mark II terminals and a compact French hydrogen maser. Future plans call for the development of a large fixed antenna and a 15 m transportable antenna with Mark III terminals.

H. Enslin reported on the work of C. de Vegt to construct a photographic PZT reference star catalogue which will connect the PZT star catalogues of Calgary, Herstmonceux, Hamburg, Ondrejov and Potsdam. Observations with a 23 cm astrograph, along a $6^{\circ}$ belt centered at $52^{\circ}.5$ N, have begun. It is planned to adjust the declination system to an extragalactic reference frame through the optical counterparts of 45 compact extragalactic radio sources.

B. Kolaczek reviewed the secular and irregular variations of the mean latitudes of IPMS/BIH stations. She found that the variations of mean latitudes showed some regional similarities. She also computed the local annual term of latitude variations for the period 1962-1980 for the variations in latitude and found that they are very changeable for some stations and more stable for about 25 stations.

Ya. Yatskiv reported on the work of L.D. Kostina and V.I. Sakharov, on the long period variations in the Chandler wobble. For the period 1900-1975, they have fit the Chandler wobble by 2 two parameters ($\alpha, \beta$) exponential functions. For the two exponential functions, they found that the growth coefficientc ($\alpha$'s) were different, but that the damping coefficients ($\beta$'s) were similar. This leads them to conclude that although there may be different excitation mechanisms, the Earth's damping mechanism has remained stable.

L. Randic has been comparing the latitude differences between pairs of stations in the northern and southern hemispheres. For the pairs, Mizusawa - Mt. Stromlo and Washington - Santiago, he finds that the latitude differences are not constant and appear to exhibit an annual term. He discussed some possible causes for this, among them are meterological effects and geophysical processes.

W. Markowitz noted the differences between the astronomical and geodetic systems of reckoning PM. In the astronomical system, the y axis is located at 90° W longitude with longitudes increasing to the West. In the geodetic system, the y axis is located at 90° E with longitudes increasing to the East. He noted that it seems harmful to maintain both systems and recommended that astronomers adopt the geodetic system for PM.

P. Paquet summarized a study prepared by C. Dramba which attempts to develop an analytical representation for PM which includes the effects of an elastic Earth.

## 24 August 1982 (Business)

The President, P. Paquet, reported on the election of officers. Ya. Yatskiv was elected President and W.J. Klepczynski, Vice-President. In addition to Yatskiv and Klepczynski, the new organizing committee will be composed of F. Barlier (82), H. Enslin (80), C. Kakuta (80), B. Kolaczek (80), I. Mueller (82), V. Naumov (80), P. Paquet (past President), E. Proverbio (80), M. Rochester (82), Shu-hua Ye (82), and K. Yokoyama (IPMS).

Twenty-three new members were approved into the Commission. Two members resigned during the last triennum, J.L. Locke and J.D. Mulholland and two members died, E. Buchar and J.O. Fleckenstein. Three new consultants were added to the commission: A. Stolz (Australia), N. Sidorenkov (USSR), Sh-y. Zhu (PRC). B. Kolaczek was appointed as Commission 19's representative to the Directing Board of the IPMS and H. Enslin's appointment as Commission 19's representative to FAGS was continued.

### Resolution Adopted at the Joint Session of Commissions 19 and 31 on Project MERIT 18 August 1982

For details, refer to report of Commission 31.

### Resolution Adopted at the Joint Session of Commission 19 and 31 on Regularized UT1, 19 August 1982

For details, refer to the report of Commission 31.

### Resolution Adopted at the Joint Session of Commissions 4, 19 and 31 on Nutation and the Definition of GMST, 20 August 1982

For details, refer to the report of Commission 4.

### Resolution Adopted at the Joint Session of Commissions 4, 19 and 31 on the System of Astronomical Constants and Ephemerides, 20 August 1982

For details, refer to the report of Commission 4.

## REPORT OF JOINT MEETING OF COMMISSIONS 19 (ROTATION OF THE EARTH) and 31 (TIME) HELD ON 1982 AUGUST 19.

CHAIRMAN : 1st session (09.30-11.00) P. Pâquet
2nd session (11.30-13.15) S. Iijima

Progress reports by the Working Groups on the rotation of the Earth (MERIT) and on the new conventional terrestrial reference system (COTES).

The first session consisted of three introductory reviews as follows :

G.A. Wilkins : the objectives and organisation of Project MERIT
I.I. Mueller : the objectives and proposals of the Working Group on the conventional terrestrial reference system (COTES)
M. Feissel : a review of the MERIT Short Campaign and of the plans for the MERIT Main Campaign (1983/4).

The second session consisted of six reports on the activities and plans for each of the observing techniques and then a consideration of a resolution for submission to the General Assembly. The reports were as follows :

K. Yokoyama : optical astrometry
C. Oesterwinter : Doppler tracking of satellites
R. Schutz : satellite laser ranging
O. Calame : lunar laser ranging
W.J. Klepczynski: corrected-elements radio interferometry
D.S. Robertson : very-long baseline radio interferometry.

The draft resolution was amended in minor respects and then adopted without objection. It was subsequently endorsed by the General Assembly and is given as resolution No. R5 on page 39.

Wilkins , the chairman of the IAU/IUGG Joint Working Group on the Rotation of the Earth explained that Project MERIT is a programme of international collaboration to monitor earth rotation and to intercompare the techniques of observation and analysis. A Short Campaign was held in 1980 August to October and the results have been reported in detail (Wilkins and Feissel, 1982). The Main Campaign will take place from 1983 September 1 to 1984 October 30. Current information about the project is distributed in the MERIT Newsletter which is available from the Royal Greenwich Observatory.

Mueller drew attention to the resolution of IAU Colloquium No. 56 (Gaposchkin and Kolaczek, 1981) that led to the setting up of the Joint Working Group of which he is chairman ; the group is to prepare a proposal for the establishment and maintenance of a conventional terrestrial reference system. For the new system it will be necessary to adopt precise coordinates for a set of well-distributed observatories and to relate the new system to the celestial and inertial systems which are implicit in the various techniques for precise positioning. In particular it will be necessary to determine earth-rotation parameters (ERPs) to a precision of, say, 5cm (0".002) with a time resolution of 12 hours (6 hours if possible). Extra observations will be made during the MERIT Main Campaign to look for short-period variations and to obtain data at selected sites by at least two techniques. Several mobile SLR and VLBI system will be used. The proposals (COTES, 1982) were endorsed by the International Association of Geodesy at its General Meeting in Tokyo in May.

Feissel summarised the principal conclusions that had been drawn from the MERIT Short Campaign, which had stimulated faster development of the new techniques and had shown where improvements are required in the operational procedures. In order to bring the reference systems into better agreement it will be necessary to adopt a standard set of constants and models and to go on looking for sources of systematic errors in the techniques. Special efforts are also needed to eliminate the "weekend effect", to improve the links between the operational and coordinating centres, and to disseminate the information more widely, as well as to look for short-period variations and to make colocated observations.

In closing the first session the President of Commission 19 congratulated the Working Groups on the progress that has been made since the last General Assembly.

Yokoyama pointed out the Short Campaign had led to an improvement in the results from optical astrometry and that a further improvement could be expected during the Main Campaign as a result of the use of more accurate star catalogues for the astrolabes and PZTs and better corrections for nutation and tidal effects. He hopes to determine short-period variations from the observations at the better stations.

Oesterwinter reported that data from the Doppler tracking of satellites had been analysed in an attempt to detect plate motions ; the values found for the Australian and Pacific plates appeared to be significant and in general agreement with geological studies. The lack of proper information about local changes in the positions of the antennae had hindered the analyses. In reply to a question, Guinot stated that a new MEDOC II system is being developed ; it will include four new French stations and should come into operation during the Main Campaign.

Schutz reviewed the current status of the Satellite laser ranging stations that supplied data to the analysis centres at Greenbelt and Austin in the USA. For many stations the precision is about 10cm and for one or two it is 2-3cm. There is a very noticeable weekend effect in the frequency of observations ; a better balance between day and night observations will also be required if reliable determinations of short-period variations are to be made. More effort must be made to reduce both systematic and random errors. Guinot commented that the monthly reports from the University of Texas are very useful.

Calame said that the lunar laser ranging contribution to MERIT was carried out in the framework of the EROLD campaign, but was limited to the determination of UT since data were only available from the McDonald Observatory. She expressed the hope that LLR data would also be available from Grasse, Wettzell and Orroral during the Main Campaign, and she presented solutions for the annual and semi-annual terms in UT from McDonald data for the period 1971-1981.

Klepczynski spoke mainly about the connected-element radio interferometer at Green Bank since this was the only CERI system in regular use for the determination of earth-rotation parameters. Some results had been obtained from Cambridge, England, during the Short Campaign and he hoped that some time would be available on the Very Large Array at Socorro during the Main Campaign. The Green Bank system is being extended to allow the determination of all three parameters, and a water-vapour radiometer is being installed in order to reduce the errors due to the troposphere. In response to a question he said that it might be possible to obtain a solution with a precision of 0.5 ms at an interval of 8 hours.

Robertson showed results from VLBI and claimed that UT could be obtained with a precision of the order of 0.1 ms from 24 hours of observations in spite of bad weather and bad ionospheric conditions. The POLARIS network in the USA requires one more element in order to obtain all three parameters ; observations will be made every five days. In response to a question he said that this network when used with another element at Wettzell, Germany, should give results with precisions of 0.1 ms and 5 cm from 24 hours of observation.

In the ensuing general discussion Wilkins stated that the third and final draft of the MERIT recommendations on standards should be available during the autumn and that recommendations on the monumentation of sites could be prepared since this will be very important in the establishment of the new conventional terrestrial reference frame.

The meeting then adopted a resolution that noted the success of the MERIT Short Campaign, recognised the potential value of the MERIT Main Campaign, and endorsed the two resolutions concerning the COTES proposals for extra observations that were adopted recently by IAG. (See Resolution No. R5 on page 40.) The President of Commission 31 closed the meeting by thanking all those who had presented reports and contributed to the discussions.

References

COTES, 1982. Reference Frame Requirements and the MERIT Campaign : proposal for extra observations. Special issue of CSTG Bulletin.

Gaposchkin, E.M. and B. Kolaczek (eds), 1981. Reference Coordinate Systems for Earth Dynamics. D. Reidel Publishing Co., Dordrecht, Holland.

Wilkins, G.A. and M. Feissel (eds), 1982. Project MERIT : Report on the Short Campaign and Grasse Workshop with observations and results on earth-rotation during 1980 August-October. Royal Greenwich Observatory, Herstmonceux, Sussex, UK.

# COMMISSION 20: POSITIONS AND MOTIONS OF MINOR PLANETS, COMETS, AND SATELLITES (POSITIONS ET MOUVEMENTS DES PETITES PLANETES, DES COMETES ET DES SATELLITES)

## Report of Meetings, 18, 20, and 25 August 1982

PRESIDENT: E Roemer (Acting) SECRETARY: C J van Houten

### 18 August 1982

ADMINISTRATION I

The meeting was opened by the Vice President, who noted that because of unusual circumstances, the President of the Commission, G Sitarski, had been unable to participate in the planning for meetings of the Commission at the General Assembly, and was now unable to be present. Those in attendance were invited to join in extending greetings to him by signing a card. C J van Houten was then confirmed as secretary, and it was agreed that no interpreter was needed. All present stood in silent respect as the names of three members or former members of the Commission who had died since the previous General Assembly were read: P Herget, H Hirose, J G Porter.

Proposed new officers were E Roemer as President and C J van Houten as Vice President. Working Group chairmen proposed as continuing for the 1982-85 triennium were Ľ Kresák, comets; Y Kozai, satellites; and G E Taylor, prediction of occultations. Proposed members of the new Organizing Committee were Yu V Batrakov, M P Candy, Y Kozai, Ľ Kresák, L K Kristensen, B G Marsden, H Scholl, G E Taylor, D K Yeomans and Y Z Zhang. The names of a number of individuals who had been nominated as new members or consultants of the Commission were then announced; in some instances, eligibility would depend on confirmation of membership in the IAU. Additional nominations could still be considered, with the list of new members and consultants to be finalized at a later meeting of the Commission. The chairman then announced that a subvention of 10 000 SwFr for the Minor Planet Center was included in the proposed IAU budget for the triennium; approval was hoped for.

The chairman then reported on negotiations in progress with the Working Group on Planetary System Nomenclature (WGPSN) with the aim of clarifying the allocation of responsibilities for assignment of designations and names to bodies in the solar system. Responsibility for designations and names of satellites had grown into a matter of considerable concern to some members of Commission 20. At the request of the chairman, Roemer, Kozai and K Aksnes were confirmed as representatives of the Commission on a committee that would try to develop a plan responsive to concerns of WGPSN, Commission 20, and Commission 6.

Several recommendations were in process of refinement and would be presented for Commission consideration at a later administrative session. Topics addressed included a proposal to give a limited review to all proposed new minor planet names, recommendations by the ad hoc committees on Minor Planet Programs and on Magnitude Ephemerides of Minor Planets, and support for further efforts to recover lost numbered minor planets. On this last topic, it was noted that with the reobservation of (1009) Sirene by J B Gibson at Palomar, in close agreement with the prediction by L K Kristensen, only six numbered planets were still lost.

SCIENTIFIC SESSION: Comets

L Kresák, chairman of the Working Group on Comets, opened the session with a brief overview of progress since completion of the triennial report of the Commission. Among the most noteworthy events were the recoveries of two long-lost comets, P/du Toit-Hartley (1945 II = 1982b, 1982c) and P/Peters-Hartley (1846 VI = 1982h), both with the UK Schmidt telescope, Siding Spring.

B G Marsden then reported on the paucity of astrometric observations of comets, noting that during the past few years there has been a dramatic reduction in the number of astrometric observations made of comets. The problem applies not just to faint comets but to bright ones as well. It exists in both hemispheres, but, if anything, the Southern Hemisphere seems to be in slightly better shape than the Northern. The reduction in number of observations seems to coincide with the transfer of publication of observations from the IAU Circulars to the MPCs. Early observations of new discoveries and of recoveries are still published on the IAU Circulars, but with the publication of both series of circulars at the same location, it is clearly more efficient to include the bulk of the observations of comets along with those of minor planets. This facilitates the collection of data in machine-readable form. Files of comet observations were, in fact, included in the version of the magnetic tape of minor planet observations issued in March 1982. The MPCs do not enjoy the prominence given to the IAU Circulars, with their appearance on bulletin boards at observatories around the world, but they do have a moderately large circulation, are still published rather rapidly, and--as far as observations are concerned--have been declared as a 'refereed journal'. Nowadays, dozens of astronomers in other disciplines--radio astronomers, those working with the International Ultraviolet Explorer, even those attempting to bounce radar pulses off comets--are plaguing the Minor Planet Center/Central Telegram Bureau for improved orbit/ephemeris data on comets--data that cannot be supplied if astrometric observations are not made and reported quickly.

Marsden next called attention to nomenclature problems that attend the discovery of comets found long after the actual time of observation. The recent announcement of the discovery of two sungrazing comets in the SOLWIND satellite/coronagraph data of 1981 has created a minor nomenclature problem. Since its introduction in 1870, the system of provisional letter designations has never been applied in retrospect. If orbits can be determined for comets reported belatedly, those comets have been incorporated in the definitive Roman numeral designation system. But the 1981 Roman numerals will not be assigned until 1983. The letter-designation system could obviously be extended, but there is some question as to whether the year of discovery or of the original observation should be used. Further, if designations are to be given in retrospect, it will be difficult to control the system. Provisional designations could clearly be given to the comets recently identified on the Palomar Sky Survey, but what is to be done about the unconfirmed comet reports that occasionally appear in the IAU Circulars and other literature? It was proposed that a special committee of Commission 20 study the matter, prepare and publish a list of provisional designations, and continue to monitor the situation in the future.

Z Sekanina then gave a brief scientific report concerning his recent studies of Comet Howard-Koomen-Michels (1979 XI) and P/du Toit-Hartley. A satisfactory orbital solution was presented for comet 1979 XI, showing that it did indeed fall into the sun. The surviving dust tail was studied as a function of the particle production vs. heliocentric distance, the span of accelerations by solar radiation pressure, and particle-sublimation effects. Extremely refractory material was reported to exist in the submicron range of particle sizes. The motions and brightness variations of the two nuclei of P/du Toit-Hartley (1982b,c) were investigated on the basis of Sekanina's model of split comets. The fragments, $1 \times 10^6$ km apart at the time of discovery, were found to have separated shortly before perihelion passage during the comet's (unobserved) 1976/77 return to the sun, with a relative nongravitational acceleration near $3 \times 10^{-5}$ of the solar attraction. The anomalous brightness behavior of the companion nucleus at the time of discovery may have been a result of the object's highly oblate shape and morphologically heterogeneous surface.

D K Yeomans invited the participation of astrometric observers in the astrometry network of the International Halley Watch (IHW). Five spacecraft, none of which has any internal guidance capability, will require information from groundbased astrometric observations for midcourse corrections. Fast astrometry will be required at certain critical times in the missions. Several meetings were being held at the General Assembly to acquaint potential collaborators with the goals of IHW. D W Dunham then called attention to the need for early groundbased astrometry for guidance also of the ISEE 3 spacecraft, which is to be redirected for a rendezvous in September 1985 with P/Giacobini-Zinner.

SCIENTIFIC SESSION: Satellites

Y Kozai, chairman of the Working Group on Satellites, opened the session by distributing copies of a progress report, which was an extension of that published in Trans. IAU XVIIIA, material presented to Commission 7 and notes about very recent work having been added. Many faint satellites as well as faint rings for Jupiter and Saturn were discovered through both ground-based and spacecraft observations in the past three years. Most of the newly discovered satellites have very interesting dynamical characteristics: there are two co-orbital pairs, two seem to shepherd the F ring of Saturn, and three are moving near the equilateral triangular points of Dione and Tethys. Many astrometric observations of satellites with respect to the centers of planets or with respect to other satellites were made and published. Also, mutual phenomena which took place in 1979 and 1980 for Jupiter and Saturn were observed and analyzed to determine accurate orbital elements. Motions of the Galilean and other satellites still attract the interest of theoreticians, and several new theories which well represent observations were developed and published. A number of papers addressed the question of the dynamical evolution of satellites, particularly how so many commensurable relations of the mean motions might have arisen through secular changes in orbital semimajor axes.

S Ferraz-Mello then reported on the current status of studies of the motion of the Galilean satellites. The minimal standard residuals in the coordinates of the Galilean satellites that have been obtained from photographic observations correspond to a standard error for a single exposure of 3$\mu$m on the plates. To obtain the necessary precision (100-200 km), multiple exposures (4 or more) and long-focus telescopes (6 m or more) are needed. Mutual events may also be observed with such a precision, and hope exists for good results in 1985, when many events will be observable. The best current ephemerides are those obtained with the Sampson-Lieske theory and Lieske's elements E-2. Progress is being made toward development of new theories. Substantial improvement, however, remains strongly tied to the existence of good observations and to the continuity of the present observational efforts.

P K Seidelmann, with the collaboration of R S Harrington, assessed the current understanding of the dynamics of the recently discovered small satellites of Jupiter and Saturn. Some of the satellites had been discovered during the Voyager 1 and 2 encounters, while the 1980 ring plane crossing of Saturn provided both additional Earth-based discoveries and clarification of observations made at the ring-plane passage in 1966. Only isolated observations have been made of some satellites, with the consequence that accurate orbits cannot be determined. A number of satellites are in librational motion, sharing orbits with large and small satellites. Additional observations are required for determination of the librational periods. The dynamics of the small satellites and of ring particles present some interesting and challenging problems in celestial mechanics.

It was noted in discussion that several of the satellites discovered during the Voyager encounters have been observed, or should be observable, from the Earth. The Space Telescope Wide-Field/Planetary Camera with a CCD detector has been used successfully by at least two groups for observations of faint satellites, and Ch. Veillet has recorded several of the small Saturnian satellites photographically with the Danish-ESO 1.5-m reflector at La Silla.

Marsden then called attention to a problem concerning publication of astrometric observations of satellites, in that from time to time the Minor Planet Center/Central Telegram Bureau receives long lists of astrometric observations of satellites. Reports of new discoveries of satellites are published rapidly in the IAU Circulars as a matter of course. Some of the other observations have been published there also, but it is clearly inappropriate to continue this practice or to utilize the MPCs. It is proposed that the Working Group on Satellites examine the situation with respect to the publication of observations of satellites, specifically with regard to the possibility of issuing a companion publication to the MPCs. Such a publication could facilitate the collection of observations in machine-readable form. It could also serve as a forum for publishing orbital data on satellites and could play a useful role in making the decision as to if and when a new discovery should be given a permanent designation.

On behalf of J D Mulholland, Roemer called attention to problems posed by collections of astrometric plates of satellites for which funds for measurement and reduction were lacking. In principle, unmeasured plates from Mulholland's collection are available for loan to responsible individuals with reasonable ideas for their use. It is anticipated that a 'catalog' will eventually be compiled so that potentially interested individuals might be better aware of what is available.

## 20 August 1982

ADMINISTRATION II

Roemer presided over a special administrative session convened to consider recommendations concerning nomenclature developed by the special committee composed of representatives of Commissions 6 and 20 and of WGPSN. A scheme had been worked out to clarify the allocation of responsibilities for the several steps in the assignment of temporary and permanent designations, and, eventually, of names to newly discovered bodies in the solar system, including satellites, minor planets, comets, and rings. The long-standing responsibilities of Commission 20 with respect to designations and names for minor planets and comets were affirmed in the scheme. In addition, Commission 20 would be responsible for decisions concerning assignment of permanent designations to satellites and for indicating thereby when assignment of names by WGPSN was appropriate. The proposed scheme was approved without dissent.

[Following ratification by members of Commissions 6 and 20 and of WGPSN, the scheme was submitted to and received the approval of the IAU Executive Committee. Details of the scheme are included in IAU Information Bulletin No. 49, January 1983, and in the Astronomers' Handbook, Section 4 of this volume.]

Roemer then reported on recommendations concerning names for recently discovered satellites of Jupiter and Saturn that had been developed by WGPSN. Following considerable discussion, the Commission voted unanimously to recommend permanent designations and to endorse proposed names as follows:

| | | | | | |
|---|---|---|---|---|---|
| J XIV* | 1979 J2 | Thebe | S X | 1980 S1 | Janus |
| | | | S XI | 1980 S3 | Epimetheus |
| | | | S XII | 1980 S6 | ("Dione B") |
| | | | S XIII | 1980 S13 | Telesto |
| | | | S XIV | 1980 S25 | Calypso |

(*Note that this Roman numeral designation differs from that recommended by WGPSN, Trans. IAU XVIIIA, 669, 1982)

The Commission also voted, with one dissent, to support the recommendation of those most knowledgeable about the dynamical studies, that assignment of permanent designations and names for the following satellites was not yet timely: 1979 J1, 1979 J3, and 1980 S28.

SCIENTIFIC SESSION: Minor Planets

B G Marsden, Director of the Minor Planet Center, opened the session with a report on the status of work on minor planets:

(1) Observations. In contrast to the situation with comets, there is currently no shortage of accurate astrometric observations of minor planets. The August 1982 monthly batch of MPCs contains almost 4500 observations, and the accompanying 223 preliminary orbits amply document the value of several important programs. One can, however, question the usefulness of observations of very faint objects when no effort is made to follow objects up a month later; the limiting magnitude of the program at the Crimean Astrophysical Observatory, B = 18.0, might be taken as the general limiting magnitude for useful observing programs, unless they are conducted on the scale of the Palomar-Leiden survey and some of the recent Palomar-UK Schmidt surveys by E Helin and S J Bus.

(2) Identifications. There is also no lack of activity in finding and establishing identifications. These have led to the numbering of no fewer than 300 new minor planets during the past twelve months. L D Schmadel has recently found more than 2000 previously unsuspected identifications with numbered minor planets. Several observatories have extensive files of old plates; measurement of positions of unnumbered minor planets for which only approximate positions have been available would facilitate the recognition of further identifications.

(3) Lost planets. The above-mentioned activity has reduced the number of lost numbered minor planets to six: 473, 719, 724, 878, 1026 and 1179.

(4) Catalogues. The Minor Planet Center has recently published updates for the 1961 catalogue of orbits of unnumbered minor planets and the 1963 edition of the 'Identifizierungsnachweis'. The new orbit catalogue contains 2471 entries, the listing of discoveries, 29 157 provisional designations, of which 8550 are identified with numbered objects.

(5) Palomar-Leiden and McDonald surveys. The second (and final) phase of the P-L survey is now essentially complete. Of the 516 originally unidentified objects in the McDonald survey, 239 (46%) are now numbered or have been identified with numbered objects; of the 27 such objects brighter than B = 15, 18 are numbered.

(6) Verification. H Scholl has recently derived osculating elements for some 2300 minor planets at 200-day intervals from 1882 to 2017. One of the uses of this calculation is the checking of all the observations. Residuals have been calculated for the 206 329 observations of numbered minor planets in the Minor Planet Center's machine-readable file, and cases larger than 100" have been isolated.

(7) Ephemerides. Since many users now calculate their own ephemerides and procedures for setting telescopes directly from osculating elements, some thought might be given to abbreviating, possibly even eliminating, ephemerides, particularly the quadrature-to-quadrature ones given in detail, only three to a page, in the MPCs. Possibly the information on distances and magnitudes could be shortened, so that six ephemerides could appear on one page. It should be noted that ephemerides have two uses: one is to enable observations to be made of a specific object, and the other is to identify in retrospect whatever minor planet trails may appear on an exposure. Because of the second use, it would be rather unsatisfactory to represent ephemerides using Chebychev polynomials; in any case, the calculation of an unperturbed ephemeris from readily available osculating elements nowadays involves scarcely more effort than interpolating in a published ephemeris.

A straw vote showed a nearly even split between those who preferred that publication of ephemerides in the present form be continued in the MPCs and EMP and those who were prepared to compute their own ephemerides from osculating elements. There was little support for the use of Chebychev polynomials. A suggestion that approximate values of the elements a, e and i be included with the ephemerides in the MPCs was warmly received.

J Kovalevsky then presented the report of the ad hoc Committee on Program Definition. After some discussion, a study session was scheduled; a refined version of the recommendations of the committee was adopted by the Commission and is included in the report of the final administrative session.

E Bowell reported on the findings of the ad hoc Committee on Magnitude Ephemerides of Minor Planets. Lively discussion prior to the Patras meeting, together with a work session in Patras, made it clear that it was premature to make a recommendation to the Commission for adoption in 1982. Consequently, the committee requested that a decision be deferred until 1985. There was agreement that the calculation of minor planet magnitudes should be made by means of an easily applied two-parameter formula that includes the opposition effect as well as a non-linear drop-off in brightness at large phase angles, and that applies to both minor planets and satellites. In order to increase the range of expertise within the ad hoc committee, co-option of several additional members was recommended. Pending adoption of a new system, observers were urged to give all results in the current IAU system in addition to whatever alternative system they might choose to use.

C J van Houten reported on the completion of the Palomar-Leiden survey. The original aim of this survey was to provide statistical data on the orbits of minor planets of magnitude between 16 and 20 in opposition; these results have been published. During the first Tucson colloquium on minor planets, it appeared, however, that the interest in the orbits themselves was considerable. It was decided, therefore, to improve these orbits as much as possible, by improving the original positions and searching for additional positions on the PLS plates. In addition, the plates containing the photometric calibration field were also blinked and yielded 169 orbits. The present number of PLS orbits is close to 2400, 144 new first-class orbits having been added. The program is near completion.

I van Houten-Groeneveld provided information on the progress of the 1977 Trojan campaign. A strip centered on and perpendicular to the ecliptic had been photographed by T Gehrels with the 1.2-m Schmidt telescope of the Palomar Observatory. The center of the field was near the following libration point of the Sun-Jupiter system, at a distance on the sky of 65° from Jupiter. The plates of the two central epochs have been blinked, yielding discovery of 36 Trojan and Hilda objects. The number of other minor planets found totals about 1800. The available plates allow an extension of the observed arc to 19 days; orbital elements will be determined.

H Scholl then described the opportunity for accurate astrometric observations of minor planets offered by the satellite HIPPARCOS. Objects brighter than B = 13 can be observed close to quadratures during the interval 1987-1990. The accuracy of minor planet positions determined by HIPPARCOS is expected to be in the range 0\".01-0\".1. The minor planets 1, 2, 3, 4, 6, 7, 8, 9, 10, 11, 14, 15, 16, 18, 20 and 29 can be observed at each quadrature. In addition, the minor planets 5, 13, 19, 22, 23, 28, 31, 39, 40, 42, 44, 51, 63, 88, 89, 192, 216, 324, 349, 354, 471, 511, 532 and 704 can be observed at least at two quadratures. Ephemerides with an accuracy of 1" will be needed for those minor planets accepted by ESA for observation by HIPPARCOS. Observers are therefore requested to observe the listed minor planets as often as possible. The purposes of the accurate observations of minor planets include (1) improvement of orbits, (2) determination of the equator and equinox in the HIPPARCOS star catalogue, (3) detection of a possible rotation in the HIPPARCOS star system, and (4) determination of the masses of (20) Massalia and (44) Nysa. In addition, there is the possibility of obtaining accurate orbits of interesting minor planets indirectly, in that HIPPARCOS may observe a sufficiently large number of suitable reference stars. Reference stars for the minor planets 348, 486, 526, 1023 and 1825 should be observed by HIPPARCOS. With accurate orbits for these minor planets, masses for the minor planets 1, 4, 65, 52, and 10, respectively, should be derivable.

D L Matson reported on the opportunity for observations of minor planets presented by the Infrared Astronomy Satellite (IRAS). The mission of the satellite, scheduled for launch in late 1982 or early 1983, is to map the celestial sphere at wavelengths of 10, 20, 50 and 100 $\mu$m. In the course of the survey, IRAS is expected to make ~180 000 observations of thermal emission from some 30 000 minor planets. Positional accuracy, however, will be between ~0.5 and ~1 arcminute, depending upon how many of the satellite's sensors detect each asteroid during a given observation. In addition to the many main-belt asteroids expected to be observed, near-Earth, fast-moving objects will also be discovered. It is urgent that ground-based astrometric observers obtain follow-up observations promptly, so that orbits for these objects can be determined.

After some brief remarks about problems that had arisen concerning names proposed for minor planets, Marsden recommended that the Commission adopt a policy calling for review of all names proposed for minor planets, even when names are proposed by discoverers. The review could be done as indicated in the 1979 Commission 20 resolution, except that in the case of a name proposed by the discoverer, the waiting period for naming a newly numbered object could be reduced. He also recommended that the length of names be limited to sixteen characters, including spaces and hyphens. To save possible embarrassment, all who propose names for minor planets were urged to be cautious about situations that could lead to premature publicity.

<u>25 August 1982</u>

SCIENTIFIC SESSION: Prediction of Occultations

The chairman of the Working Group on Occultations, G E Taylor, noted in his opening remarks that despite a considerable effort from both professional and amateur observers the occultation campaigns had only limited success during the preceding three years. Although occultations by 13 minor planets had been observed since 1979, in many cases only one chord had been observed, or chords on only one half of the disk of a minor planet. Lack of adequate fast astrometry in the Southern Hemisphere was a factor. Participation of more observers able to provide precise relative positions of minor planet and star as soon as they are in the same astrometric field could increase the success rate, as could the participation of more observers with transportable equipment for photoelectric timing. With information about the larger satellites of Jupiter and Saturn available from the Voyager encounters, the value of predictions of future occultations by these bodies seems limited.

R L Millis made a number of suggestions concerning improvement of occultation timings. Since photoelectric observations are much to be preferred over visual ones, and results from permanent observatories do not give enough information, the participation of many properly equipped mobile observers is essential. Inexpensive systems that give timings to $0^s.01$ precision recorded on magnetic tape and usable on Celestron telescopes of 8- to 14-inch aperture are being developed at the Lowell Observatory. Inquiries concerning these systems are invited. For the foreseeable future, however, visual observers will far outnumber those prepared to make photoelectric observations, and their participation continues to be useful.

L Wasserman considered how the limited manpower available for observation of occultations could best be used. Some selection criteria must be applied to predicted events by the observers, if not by the predictors. Top priority should be given to occultations by minor planets of large linear or angular size and to events chosen on the basis of expected duration, relative brightness of star and minor planet, or small amplitude of the lightcurve as an index of regular shape. Searches should be made for occultations by particularly important minor planets that involve stars fainter than those found in typical astrometric catalogs. And more help is needed in last minute astrometric refinement of predicted occultation ground tracks.

E Bowell then reported on occultations of stars by comets. Rather than attempt to observe an occultation by a cometary nucleus, for which the chances of success are slight, it should be of great interest to determine the distribution of material in the coma, in a region close to the nucleus, by measuring directly the optical depth in the continuum. Some information on the single-particle scattering function might also be inferred from such measurements. Extended, dusty comets are good candidates for such observations. The astrometric accuracy required is not as great as that for occultations of stars by minor planets because most cometary comae are much larger and therefore produce a wider ground track. A plea was made for early recovery of comets so that an ephemeris of 1" accuracy could be derived and used to make timely predictions of occultations.

L K Kristensen requested as many observations as possible of occultations of stars by (51) Nemausa. The purpose is to determine the rotation in the proper motion systems of the future fundamental catalogs that will be based on space astrometry. Occultations provide precise observations of planet positions for a period of time longer than the mission length of the HIPPARCOS satellite if the occulted stars have been on the observing program of the satellite. Occultations by (51) Nemausa until the year 2017 have been found by G E Taylor. A current campaign of lightcurve observations will, ultimately, give rotation and flattening of (51) Nemausa, which will increase the accuracy of the derived star positions.

ADMINISTRATION III

Roemer then presided over the final session of the Commission, at which a number of administrative decisions were finalized. Officers and members of the Organizing Committee were confirmed in accordance with nominations announced at the first session of the Commission. New members admitted by vote of the Commission include M F A'Hearn, N A Belyaev, D Benest, R Bien, F Börngen, R L Branham, A Carusi, E Helin, P D Hemenway, J Hers, A S Khatsiashvili, C-I Lagerkvist, N R Lomb, A M Nobili, P R Weissman, and I P Williams. (J-X Zhang had been co-opted to membership during the triennium, upon full adherence of China to the IAU.) Consultants for 1982-85 were confirmed as follows: C M Bardwell, K I Churyumov, R W Farquhar, V A Izvekov, Z M Pereyra, C F Peters, V Protitch-Benishek and H J Reitsema.

Composition of Working Groups was confirmed as follows:

Orbits and Ephemerides of Comets: J C Bennett, M P Candy, A C Gilmore, E I Kazimirchak-Polonskaya, Ľ Kresák (chm), B G Marsden, E Roemer, G Sitarski, Ko Tomita, R M West, P Wild and D K Yeomans.

Satellites: K Aksnes, Yu V Batrakov, A Bec-Borsenberger, O Calame, C Cristescu, S Ferraz-Mello, R Greenberg, P A Ianna, Y Kozai (chm), J H Lieske, R L Millis, B Morando, J D Mulholland, D Pascu, M Rapaport, E Roemer, J L Sagnier, P K Seidelmann, V A Shor, A T Sinclair, D T Vu; J E Arlot (consultant).

Prediction of Occultations: M P Candy, D W Dunham, J L Elliot, A C Gilmore, J Hers, A R Klemola, Y Kozai, B Morando, V A Shor, G E Taylor (chm), L H Wasserman, R M West; H J Reitsema (consultant).

The membership of the ad hoc Committee on Magnitude Ephemerides of Minor Planets (and satellites), augmented to include representation from Commissions 15 and 16, was approved as follows: Yu V Batrakov, E Bowell (chm), C Chapman, T Gehrels, A Harris, B Hapke, Ľ Kresák, B G Marsden, D Morrison, E Roemer, L D Schmadel and J Veverka. The committee will examine various formulae that meet the several criteria specified in the interim report and recommend a formula for the prediction of magnitudes to be considered for adoption by the IAU in 1985.

Establishment of a Satellite Nomenclature Liaison Committee (SNLC) was then formally approved. This committee will work with WGPSN on matters of satellite nomenclature. The membership is to include the President and Vice President of the

Commission as well as the Chairman and two other members of the Working Group on Satellites. The members of the committee for 1982-85 are E Roemer, C J van Houten, Y Kozai, K Aksnes and P K Seidelmann. Aksnes will serve as chairman and occupy the voting seat on WGPSN accorded to a representative of Commission 20; Seidelmann was designated as his alternate. This committee also was asked to consider the style of the permanent designations for satellites, some preference having been expressed in WGPSN for Arabic rather than Roman numerals. Following some discussion of the merits of the alternatives, a straw vote of Commission 20 members showed a strong preference for Roman numerals over Arabic.

Formation of a Comet Nomenclature Committee (CNC) was also approved, with membership to include the President of the Commission (Roemer), the Chairman of the Working Group on Comets (Kresák), and the Director of the Central Bureau for Astronomical Telegrams (Marsden, chm). The committee is to study such matters as designations for comets discovered on old plates and for the comets found recently in the data from the SOLWIND satellite coronagraph, as well as the problem that would arise should more than 26 comet discoveries/recoveries occur within one year. The committee was also asked to prepare and publish a list of comets that received only provisional designations, including those originally identified as minor planets. A straw vote of Commission members concerning designations for comets discovered on old plates showed a strong preference for the year to be the one in which the plate was taken, rather than the one in which the discovery occurred.

The matter of the policy for review of minor planet names was addressed next, approval being given to the following recommendation:

> "All names proposed for minor planets will be reviewed for suitability, even when names are proposed by discoverers. The review will be done as indicated in the 1979 Commission 20 resolution, except that in the case of a name proposed by the discoverer the six-month waiting period for a newly numbered object can be reduced to two months. Names shall be limited to a maximum length of sixteen characters, including spaces and hyphens." [The review committee is composed of the President and Vice President of the Commission and the Director of the Minor Planet Center.]

Attention was then given to the report of the ad hoc Committee on Minor Planet Program Definition [E Bowell, J Kovalevsky(chm), L K Kristensen, B G Marsden, J Schubart, V A Shor, G E Taylor, C J van Houten, J G Williams] . The revised report was read in its entirely so that all members of the Commission could be aware of recent changes. Following further discussion, the report and recommendations of the committee were adopted by the Commission as follows:

### Programs of Observations of Minor Planets

> "The ad hoc committee on programs of observations of minor planets set up by Commission 20 wishes, first, to stress the importance of minor planet observations in understanding the structure, the formation, and the evolution of the solar system. An improvement of the general statistical description of the minor planet belt, of its families, and of its marginal membership (objects with orbits of high inclination or high eccentricity, Mars and Earth crossers, Trojans, Chiron or Hidalgo type planets, inactive cometary nuclei, etc.) provides clues to the dynamical evolution of the solar system and possibly to its formation. It is also a key to the understanding of the relation between various populations of the solar system (small bodies, comets, satellites and cratering processes).
>
> "Accurate observations of particular minor planets have a major impact

on the knowledge of the physics of these very poorly known objects (masses, possible existence of satellites, planning of eventual space-craft missions). Celestial mechanics greatly benefits from the refined analysis of the motion of some selected minor planets and from accurate statistical descriptions of minor planet families.

"Finally, some selected minor planets are useful for determining the inertial celestial reference frame; these bodies have to be observed very accurately.

"The committee requests that Commission 20 maintain and periodically update lists of minor planets or classes of objects to be observed or searched for with given accuracy. A first approximation to these is given in the appendix.

"As a general remark, the ad hoc committee wishes to propose the following guidelines to observers:

1. Imprecise 'quick look' observations are not useful except to point out in real time the position of a newly discovered unusual object. All published observations made in the course of a given program should be reduced. Quick look observations of common objects are discouraged.
2. Most numbered planets have good ephemerides that are sufficient for many years in advance. Those with unreliable orbits should be identified in 'Efemeridy Malykh Planet'. Two or three observations of each planet every five years is all that is needed to maintain well determined orbits.
3. Whenever maximum accuracy is needed, the best available star catalog should be used. The name of the catalog should be given so that in further studies corrections for systematic catalog errors can be introduced. Other factors that limit precision should be noted. All observations should be carried out, reduced, and published in accordance with the instructions specified for each program.
4. Whenever an observation is published, observers are requested not to give more than one decimal figure beyond the accuracy of the observation.

"Finally, the committee is quite aware of the fact that it has not been able to collect information on all existing programs, so that the omission of any program is not to be considered as a sign of disapproval. On the contrary, it would like to encourage well planned new programs and requests that Commission 20 collect information to improve the lists."

Appendix: Lists and classes of minor planets recommended for observation

1. Objects currently under study for dynamical purposes and for which the most precise positions are needed. Random errors should be better than 0."5, but systematic errors due to instrumental or atmospheris dispersion should be much less. The observations should be referred to a well defined system (e.g., AGK3) and plate constants or dependences must be stated. Long and uniform series of at least 25 observations made by several observatories are necessary at every opposition and should extend as far as possible toward quadrature unless otherwise specified in the program instructions. The following programs belong to this class:

- The 20 selected Leningrad objects.

- Branham's objects.
- (51) Nemausa.
- Objects making close approaches to larger planets, whose masses might thereby be determined, including (197), (348), (486), (526), (1023), (1825).
- Objects to be observed in connection with HIPPARCOS (See Scholl's presentation, above.)

2. Objects currently eligible for occultation predictions. Four or five observations per opposition, well distributed in time and extending through quadrature, are requested with an accuracy of 1".

(Consultations are in progress about the list of objects.)

3. Numbered objects of general dynamical interest, for which a few observations to an accuracy of ~ 2", or better if possible, are required. Three or four such observations per opposition are sufficient. Examples are:

- Trojans.
- Mars crossers.
- Minor planets beyond the Hecuba gap.
- The 3/1, 2/1 and (N + 1)/N librators.
- High inclination ($i > 30°$) and/or high eccentricity ($e > 0.4$) objects.
- Hungarias and Phocaeas.

4. Numbered objects that need an improvement in their ephemerides. Four or five observations per opposition with an accuracy of 2" are necessary.

(Consultations are in progress about compilation and publication of a list of such objects.)

5. Unnumbered minor planets with reasonably reliable ephemerides. Four or five observations per opposition with an accuracy of 2" are necessary until they are numbered. Such objects should be identified and ephemerides provided in the MPCs.

6. Search for new asteroids. For each type of body, the aim should be for statistically significant sampling, in particular, for completeness to a given magnitude or in a given region. Any discovered object should be followed up by several observations with an accuracy of 2" during each of the two succeeding months, so as to provide a basis for good ephemerides for the next oppositions. If only single observations are obtained, the daily motion should be given to facilitate identification. The magnitude also is needed as an identifying characteristic in all cases.

- Fast moving objects
- Objects with a large component of motion in declination at opposition.
- Objects with direct motion at opposition.
- Unidentified objects with typical motion at opposition and $m < 18$.

A committee was to be appointed by the President of the Commission to supervise completion and updating of the lists of minor planets in need of observations. [Following consultations, appointments have been made as follows: President of the Commission (Roemer, chm), Director of the Minor Planet Center (Marsden), ITA representative (V A Shor), L K Kristensen, E Bowell.]

The Commission then approved the following recommendation in support of further efforts to recover lost numbered minor planets:

"Commission 20 encourages the accurate measurement of positions of minor planets from the extensive plate collections such as those at the Budapest, Goethe Link, Johannesburg, Lowell, Simeis and Turku observatories. Specifically, there is a need for measurement of plates of unnumbered minor planets for which only approximate positions were previously available. Further, in order to support the current efforts to recover the few remaining lost minor planets or to make a linkage with a possible accidental rediscovery, Commission 20 encourages the early reexamination and remeasurement of all existing plates that contain or may contain images of those lost planets and the prompt publication of the results in the Minor Planet Circulars. Commission 20 notes with appreciation that Richard M West, Garching, has volunteered to process plates of the lost numbered planets should the original observatories be unable to do so."

Noting that a large number of IAU symposia and colloquia were already under consideration for the 1982-85 triennium, Roemer invited those planning meetings of potention interest to members of the Commission, or with proposals for meetings farther in the future, to inform her.

F K Edmondson then proposed a vote of thanks to the acting president for her excellent organization in the difficult situation occasioned by the absence of the commission president. This was adopted by acclamation, and the final session of Commission 20 at the XVIIIth General Assembly of the IAU was then adjourned.

The following corrections to the report of the Commission published in Trans. IAU XVIIIA (1982) have been noted:

p. 205 The orbital periods and semimajor axes of the three Jovian satellites should read $7^h\ 08^m\!.0$, $1.79R_J$; $16^h\ 11^m\!.3$, $3.08\ R_J$; and $7^h\ 04^m\!.5$, $1.78R_J$ for 1979 J1, 1979 J2 and 1979 J3 respectively.

p. 210 (top) for Hygeia read Hygiea

## COMMISSION 21 : LIGHT OF THE NIGHT SKY (LUMIERE DU CIEL NOCTURNE)

Reprot of Meetings, 19 and 24 August 1982

PRESIDENT : H. Tanabe SECRETARY : Ph. Lamy

19 August 1982

ADMINISTRATIVE SESSION

### 1. Welcome Address

The president expressed his pleasure to see colleagues of Commission 21 in Greece, which is after all the cradle of astronomy. He emphasized that the past triennium was marked by active international cooperation and hoped that such cooperative work will further develop in the coming triennium.

The president invited the participants to stand in silence in memory of two members deceased in the past three years : Kazimierz Kordylewski and Donald William Schuerman, after recalling their contributions to our research field.

Besides the formal meetings of the commission and Joint Discussion V, in which Commission 21 participates, information was given on a related informal meeting on the Halley Optical Probe Experiment (HOPE) for the Giotto Mission to Halley's Comet organized by A.C. Levasseur-Regourd. The president requested to all speakers at the meetings to speak slowly to permit better understanding by the participants from various countries.

### 2. Commission Membership

The names of officers for the triennium 1982-1985, nominated by the IAU Executive Committee on proposal by the current officers are given :

President : R.H. Giese
Vice-President : K. Mattila

The president summarized the interests of the new president in zodiacal light and space observations, and the new vice-president in diffuse galactic and extragalactic light.

The list of the new organizing committee was accepted by the commission members attending the session, and was transmitted for approval by the Executive Committee.

Organizing Committee : R. Dumont, Yu.I. Galperin, M.S. Hanner, A.C. Levasseur-Regourd, F. Sánchez-Martinez, H. Tanabe, J.L. Weinberg

The following existing IAU members were co-opted for Commission 21 membership : G.S. Ivanov-kholodny, C.F. Lillie, V.B. Nikonov, L.E. Pasinetti, C. Sánchez-Magro. Some of the names proposed by their National Committee were unknown to the members attending the session. After a discussion on this question, they were accepted.

Admissions of the following applicants to Commission 21 membership were accepted under the condition that they are admitted into the Union : P. Alvarez, W. Hofman, T. Nishimura, H.W. Ripken, M. Schüssler, R.H. Zerull.

Three names were deleted from Commission 21 membership : K. Kordylewski†, Th. Schmidt, D.W. Schuerman†.

A new list of 27 consultants was approved for the next triennium.

3. Reports on Astronomy 1982

The president thanked the members for sending the information on their activities and acknowledged the help provided by R. Dumont, R.H. Giese and K. Mattila in preparation of the report. The 8-page limitation did not allow the president to include all the information he would have liked and many of airglow studies had to be excluded.

F.E. Roach pointed out that airglow is now part of the ionospheric sciences for which Commission 21 is not concerned. In a general discussion, it was recognized that the domain of competence of Commission 21 is evolving. An opinion was expressed that the way of the commission may have to be reconsidered in the coming decade.

At a suggestion of K. Mattila, president-elect R.H. Giese expressed that he will nominate, in advance, the persons to share writing the next triennial report to maintain its high quality.

4. The Newsletter of the Commission

The president issued three newsletters; Nos. 4, 5 and 6, during this triennium. Newsletter No. 4 was devoted to the report of the commission meetings at Montreal General Assembly, No. 5 was a compilation of laboratory reports from each member, and No. 6 included the triennial commission report and its full references.

Warm approval and satisfaction were expressed by the members attending the session for the excellent work performed by the president.

It was agreed that one newsletter per year is adequate.

5. Relationship with other Commissions

The president recalled the participation of Commission 21 together with Commission 22 to the IAU Symposium No. 90 "Solid Particles in the Solar System" held in Ottawa in 1979.

In 1980, the president responded questions on airglow problems asked by Commission 50 (Protection of Existing and Potential Observatory Sites).

Commission 21 participates in Joint Discussion V "Origin and Evolution of Interplanetary Objects" to be held during this General Assembly, together with Commissions 15, 20 and 22.

6. Other Topics

H.W. Ripken pointed out the lack of information on the detailed programs of the scientific meetings of all commissions. After a discussion, it was agreed that the incoming president will write a letter to the General Secretary on this problem.

The administrative session came to an end with the president expressing his thanks to his vice-president, the organizing committee, the members and consultants of Commission 21 for their kind cooperation during this triennium.

The incoming president R.H. Giese, on behalf of all members, thanked very warmly the retiring president H. Tanabe.

SCIENTIFIC SESSION

K. Mattila : *Recent research on the galactic and extragalactic components of the light of the night sky (invited review paper).* — An abstract of the paper has been published in Reports on Astronomy 1982 (Commission 21).

R.H. Giese : *The space observation of the zodiacal light (invited review paper).* — Available are more than 30 observations from 1 AU (rockets, satellites, manned spacecraft) but only few (Pioneer 10/11, Helios 1/2) from spaceprobes near the ecliptic. These, together with ground-based observations, yield as a (good but only) first approximation an ellipsoidal (or fan like) distribution of dust number density n(r) with solar distance and a power law $n \sim r^{-1.3}$ in ecliptic. The volume scattering function of the zodiacal light particles (10 to 150 μm) is compatible with recent laser scattering measurements of irregular, fluffy and meteoritic particles. To gain more insight (bimodal dust population, changes of physical properties etc.) an out-of-ecliptic photopolarimeter was 70% completed but cannot be flown due to cancellation of the NASA spacecraft for International Solar Polar Mission Space observations in the near future should concentrate to UV ($\lambda < 3000$ Å) surveys from 1 AU and, even more important, to investigations of the relatively unobserved regions close ($\lesssim 20\ R_\odot$) to the sun, where strong interactions of dust with solar radiation and solar wind take place.

S. Koutchmy, Ph. Lamy, G. Nikolsky : *Preliminary results from the PCN experiment aboard Saliout 7.* — The "Photographie du Ciel Nocturne" experiment consists in a modified Nikon camera equipped with a 58 mm objective at F/1.2. Color photographs taken with high speed Kodak Ektachrome film (400 ASA) were obtained in June 1982. They include : (1) the E-layer, stratified with the usual oxygen (green line) layer above an orange layer probably composed of sodium, observed for the first time; (2) noctilucent clouds photographed from space for the first time; (3) the zodiacal light between Mercury and Venus. A preliminary analysis of this latter photograph confirms that the plane of symmetry is close to the orbital plane of Venus and reveals a flatter zodiacal cloud than presently expected.

R. Robley : *Does the sun disturb the zodiacal cloud ?* — About the position of the plane of symmetry of the zodiacal cloud the discrepancy obtained by ground-based observations has not been cancelled by the space observations : the results from Helios 1 and 2 are different from these of the D2A satellite. Examining the Gegenschein observations made by Tanabe (1965), it appears that the Gegenschein centre relatively to the ecliptic plane has not the same position in 1957 (solar activity maximal) as in 1962. It is proposed to explain this effect by the transverse, relatively to the ecliptic, component of the solar wind ejected through the polar coronal hole; at the opposite of the protons ejected radially (Misconi, 1976) the transverse compoment might, under certain circumtances, disturb the orbital plane of the dust grains.

24 August 1982

SCIENTIFIC SESSION

H.W. Ripken : *Interaction processes between the solar wind and zodiacal dust particles, and relevant observational consequences.* — Solar wind ions impinge on zodiacal dust grain surfaces, penetrate, neutralize, and desorb from saturated surfaces into free space. Before ionization occurs the neutral particles resonantly scatters solar radiation. To model this radiation and, by comparisons with observations, gain information on dust-desorbed neutrals and dust grains, the neutral particle density and the velocity distribution function have to be computed. Methods and results are presented. Resonance radiation calculations for H and He yield integral and spectral EUV intensities and provide information on the integral dust cross section, spatial and temporal density gradients, the velocity distribution function, and particle erosion rates.

J. de Buitrago, F. Sánchez-Martinez : *Dynamics of zodiacal dust and new inversion techniques.* — The zodiacal cloud, is considered as a system essentially ruled by gravitation and radiation forces acting as a perturbation. Within this context, it is found that spatial distribution of the kind $n(r, \beta_\odot) \sim r^{-\nu} \cdot f(\beta_\odot)$, $(\nu > 1)$, are adequate and consistent. In this picture, the angular distribution $f(\beta_\odot)$, is not determined by the dynamics, remaining mostly dependent upon the injection mechanism. If we accept $\nu = 1$ as an approximation, it can be shown that for observations in the helioecliptic plane, the Brightness Integral can be formally transformed into a second order integral equation, with $f(\beta_\odot)$ as the incognite function, which can be solved numerically. Figures of $f(\beta_\odot)$ and the obtained zodiacal cloud profile are presented.

Ph. Lamy, J.M. Perrin : *A new model of light scattering by large irregular grains.* — This new model combines the eikonal model for the diffraction part and geometrical optics for the scattering at large angles (as developed by M. Wolff for planetary regolith). Progress has been made by obtaining the exact expression of the eikonal function. The experimental results obtained by K. Weiss by laser scattering of irregular grains are well reproduced both in intensity and polarization, in particular the broadening of the diffraction lobe.

Ph. Lamy : *On the bulk density of interplanetary dust grains.* — The sources of all possible information on the density of interplanetary grains are reviewed. It is shown that the sources leading to small values (i.e. $< 1$ g $cm^{-3}$) are either indirect or unreliable (meteors, space experiments, early dust collections). On the contrary, reliable sources, such as the extraterrestrial grains collected by Brownlee and the analysis of the lunar microcraters, indicate "normal" densities, typically 2 g $cm^{-3}$, definitely larger than 1 g $cm^{-3}$.

R. Dumont : *Recent progress of the inversion technique in zodiacal light interpretation (invited review paper - read by A.C. Levasseur-Regourd).* — An increasing amount of works have been made during this triennium towards an inversion of the brightness integral or towards some localization of the contributions along the line-of-sight. A main distinction is between the "local inversion" (where the instrument is located) and the "extended inversion". Practical results of recent works are the shape of the phase function (moderate forward enhancement) and the increase with heliocentric distance of the local polarization degree at 90° scattering angle.

A.C. Levasseur-Regourd : *P.I.R.A.M.I.G./Saliout 7 camera - Photographic high sensitivity camera for visible and near Infra-Red light, studying Atmosphere, Interplanetary Medium and Galaxies.* — P.I.R.A.M.I.G. has been developed at Laboratoire d'Astronomie Spatiale (LAS) and is used by a team of scientists of Service d'Aéronomie LAS and Institut d'Astrophysique in France and of USSR. It is devoted to the detection of large faint sources and discriminates between atmospheric and interplanetary/galactic sources through orientation of the orbital station. During the French "Spationaut" mission in June 1982, more than 300 pictures (including sensitometry and duration) were taken. Preliminary results show the spatial extension of OH waves, the vertical structure of emissive layers, and absorbing regions close to galactic center. Results about zodiacal light polarization in infra-red, symmetry surface, Gegenschein and libration points may be expected.

P.V. Kulkarni : *Low latitude airglow (invited review paper)* — With the new interest in the F region of the atmosphere where detection of depletions and irregularities has become important, airglow in the F region is gaining importance. Column emission rate of 1356 Å or 7774 Å of oxygen gives the peak electron density while addition of 6300 Å measurement gives the height of maximum electron density of the F region. With high spatial resolution ($\sim$ 0°1) at an appropriate site (±18°±5° magnetic latitude) it is possible to map the "bubbles" and "plumes" in the tropical

F region. It was shown that by measuring a line and a branch emission of an OH band, temperature of the mesospheric region can be monitored and the periodicities found could be connected to the gravity waves in the atmosphere.

P. Alvarez, F. Sánchez-Martinez : *Spectrum of the atmospheric continuum.* — During 1978 and 1979 the spectrum (4000 Å - 8500 Å) of the light of the night sky in the celestial north pole direction has been monitored at Izaña (Tenerife). These measurements show a well correlated (0.75) atmospheric continuum emission between 4000 Å and 7000 Å, but not correlated with the same emission at 8200 Å. This fact has enabled us to determine the profile of this emission between 4000 Å and 7000 Å, making no assumptions on the extraterrestrial components. It shows a broad maximum at 5500 Å, bluer than the obtained in 1972 by Sternberg and Ingham. The zenithal mean absolute emission of the atmospheric continuum obtained at 5500 Å is 0.25 R/Å and at 8200 Å is about 0.35 R/Å.

A.S. Asaad : *Recent observation of zodiacal light in Egypt.* — Results of zodiacal light observation at Abu-Simbel since 1975 show an increase of brightness with decreasing solar activity in different colours. Colour indices B - V, B - R increase with sun's elongation during quiet solar activity while colour index B - V decreases during high activity and no variation is noted in between. Lunar variation of zodiacal light is noted, showing a peak at new moon. Such variations are not yet observed from outer space. If however sufficiently accurate and long period observations from outer space are obtained, then the variations noted by many observers may be attributed to the Earth's atmosphere. Lunar variations can be explained on this basis.

H. Tanabe, A. Takechi, A. Miyashita : *An interpretation of the false zodiacal light.* — The false zodiacal light has been observed by several zodiacal light observers since the beginning of the last century, but its full explanation is not found yet. From our photographic and spectroscopic observations made in 1978 and from descriptions of previous observers, we have reached the conclusion that the false zodiacal light may be a real zodiacal band seen wider than usual due to the combination of the luminance difference threshold of the human eye and a special distribution of the airglow (5577 Å) in the sky : namely, that the airglow be faint in the region of faint astronomical light.

F.E. Roach : *Overview of a half century of the light of the night sky.* — No abstract available.

## COMMISSION 22 - METEORS AND INTERPLANETARY DUST (METEORES ET LA POISSIERE INTERPLANETAIRE)

### Report of Meetings, 18 and 24 August 1982.

PRESIDENT: W.G. Elford

SECRETARIES: B.A. McIntosh
B.A. Lindblad

### 18 August 1982

The President welcomed 14 members and 2 consultants. Apologies were received from L. Kresak, B.A. Lindblad and Z. Sekanina.

I. OFFICERS AND MEMBERSHIP

The following nominations for Officers of the Commission were endorsed : President, O.I. Belkovich; Vice President, D.E. Brownlee; Organizing Committee: P.B. Babadzhanov, W.J. Baggaley, W.G. Elford, H. Fechtig, M.S. Hanner, J.A.M. McDonnell, J. Stohl, K. Tomita.

The President announced the names of six new members: A.T. Blackwell, E.Grun, M. Hajdukova, P. Lamy, H. Miles, H.W. Ripken. Four other nominations were held over until the next business meeting pending confirmation of membership. Consultants appointed for the next term are: G. Cevolani, K.B. Hindley, F. Horz, J.W. Mason, T. Nasagawa, G. Schwehm, J. Trulsen, D.K. Yeomans.

The President reported the deaths of three members of the Commission since the last meeting: V. Guth, H. Hirose, D.W. Schuerman.

II REPORT OF THE COMMISSION

The published report to be considered at the next business meeting. An extended report was tabled.

III PROJECTS AND WORKING GROUPS.

1. International Halley Watch

Dr I. Halliday reported on the work of the Steering Committee for the International Halley Watch. While there is no separate discipline for the study of meteors associated with Comet Halley it was agreed that Commission 22 should be responsible for encouraging regular observation of the Orionid and $\eta$-Aquarid showers during the next 5-10 years.

It was pointed out that although there may be no dramatic change in the observed fluxes, studies of the structure of these streams during the Halley apparition are very important in complementing the comet observations. Dr Millman reported that he is preparing a paper summarizing the present state of knowledge concerning meteor showers associated with comet Halley, and he agreed to distribute reprints to members of Commission 22.

It was noted that Dr Z. Sekanina of J.P.L. (USA) is the archivist for meteor data relating to the Halley apparition. A resolution on IHW was deferred until the next meeting.

2. Project GLOBMET

The President outlined the background to GLOBMET (Global Meteor Observation System) which had been proposed by the Soviet Geophysical Committee and supported by resolutions at the IAMAP meeting in Hamburg (August, 1981) and the IAGA meeting at Edinburgh (August, 1981). A committee had been formed within SCOSTEP with representatives from IAGA, IAMAP, IAU and URSI to produce a GLOBMET planning document. A draft document was now before Commission 22 for comment, with the aim of having a document in final form to present to the IUGG Assembly in August 1983.

The main objective of the project is to promote a network of radio meteor observatories to study: (1) the atmospheric circulation in the meteor region, (2) the influx and distribution of meteors in the neighbourhood of the earth, (3) models of meteor/atmosphere interaction.

Several speakers pointed out that the data sought by workers concerned with meteor wind studies were not always suitable for astronomical studies. It was agreed that a working group, (C.S.L. Keay, O.I. Belkovich, B.A. McIntosh and J. Stohl) examine the proposed astronomical objectives and bring a recommendation to the next meeting.

3. European and Canadian Fireball Networks

The meeting noted with concern the decision to close down two stations of the European network that are operated by the University Observatory of the Viennese Institute, and also the doubt about confirmed financial support for the MORP network (Canada). It was agreed that a working group (Z. Ceplecha, I. Halliday, C.S.L. Keay) prepare resolutions of support for these networks for presentation to the next meeting.

24 August 1982

The meeting was attended by 20 members, 2 consultants and 3 observers. Apologies were received from R.H. Giese, P. Lamy and Z. Sekanina.

I MEMBERSHIP

In addition to the six new members noted at the previous meeting, the following were recorded as new members: I. Hasegawa, R.L. Newburn, E.F. Tedesco, A.K. Terenteva. The name of I.P. Williams was added to the list of consultants. The total membership was now 87 plus 10 consultants.

II REPORT OF THE COMMISSION

The President tabled the draft report that is to appear in the Transactions, and expressed his thanks to those who had prepared the various sections of the report. Due to space constraints in the Transactions the printed report will be a significant abbreviation of the material submitted. The full report will be posted to all members. The draft report was approved without discussion.

III RESOLUTIONS

Three working groups presented the following resolutions which were adopted:

1. Resolutions concerning the European and Canadian Fireball Networks

Noting the importance of the recent contributions from the Canadian and European Fireball Networks, Commission 22

(a) recommends that continued support be provided for these observational programs in order that a more extensive body of reliable data may be gathered for critical analysis. Although the recovery of meteorites from these two

major camera networks is likely to remain limited to rare events, progress in the interpretation of fireball data is now leading to significant studies of both the astronomical properties of meteorites and the physics of their atmospheric interaction.
(b) recommends that the University Observatory of Vienna should give its fullest support to the continuation of operation of the two Austrian stations of the European Fireball Network because of their key position in the Network and their essential role in the location and recovery of meteorites which fall on Austrian soil.

2. Resolution concerning International Halley Watch

Noting the major effort which will be made to observe Comet Halley, organized by the International Halley Watch, Commission 22
recommends that meteor astronomers pay particular attention to meteor showers associated with the comet, the Orionid and Eta-Aquarid showers, since a better understanding of the meteroids from the comet can contribute to an increased knowledge of the comet itself.

3. Resolutions concerning project GLOBMET

Noting the request from SCOSTEP for comments on the GLOBMET document, and recognizing that data from GLOBMET can be of astronomical importance, Commission 22
recommends that,
(1) the GLOBMET document emphasize that the data requirements for astronomical purposes may differ from those for geophysical purposes,
(2) the Manual on Meteor Radar Observations include a section on Parameters of Astronomical Importance,
(3) a list of the parameters of astronomical importance prepared by the Working Party be sent to the convenors of the GLOBMET project.

IV. PROJECTS

Meteor Data Centre, Lund, Sweden

The President reported that a Meteor Data Centre had been set up by Dr B.A. Lindblad at the Lund Observatory in Sweden, and that the Executive Committee had allocated SwFr 1800 per annum for the three calendar years 1983-85 towards the cost of maintaining the Centre.

Dr Lindblad stated that the purpose of the Meteor Data Centre is to collect and store data obtained by photographic and radar methods on atmospheric trajectories and heliocentric orbits of meteors and fireballs. Limited amounts of data relating to other meteor programs such as radar-visual studies, meteor radar rates etc., are also available. Computer programs for stream searching and listing will also be available.

Orbital and trajectory data have been obtained mainly from published sources. There is also a limited amount of unpublished data. Data can be supplied on IBM cards, tape or as a computer listing. Duplicate copies of the photographic meteor data are also available on request from Dr Z. Ceplecha, Ondrejov Observatory, Czechoslovakia.

V GENERAL BUSINESS

1. Circulation of Agenda of Business and Scientific Meetings

Dr Ripken pointed out that the Agenda of the Business and Scientific meetings had only become available a day or two before each meeting. He suggested that such material should be supplied to the Secretariat prior to the General Assembly so that the information could be made available to all participants, possibly by inclusion in the registration material. It was agreed that the President should consult with the General Secretary on this matter.

2. Appreciation

The President thanked those present for their participation. The incoming President expressed the members' thanks to the retiring President for his work on behalf of the Commission during the past three years.

VI. SCIENTIFIC PRESENTATIONS

Eleven short contributions were presented during this session.

| | |
|---|---|
| I.P. Williams : | The evolution of the Quadrantid and Geminid meteor streams. |
| O.I. Belkovich, Sulejmanov and Tokhtasjev : | Structure of the Geminid and Quadrantid meteor streams. |
| K.B. Hindley : | Recent studies of the Quadrantid and Taurid meteor streams. |
| H. Miyashita and H. Tanabe : | Recent activity of meteor showers observed at the Tokyo Astronomical Observatory. |
| A. Hajduk and G. Cevolani : | Zones of increased particle flux density around the orbit of Comet Halley. |
| J.W. Mason : | Recent studies of the Perseid meteor streams. |
| R.L. Hawkes, J. Jones and Z. Ceplecha : | The height and orbital distributions of TV meteors. |
| P. Pecina and Z. Ceplecha : | New aspects in single body meteor physics. |
| B.A. McIntosh : | Infrasound from meteors : a negative result. |
| C.S.L. Keay : | A microprocessor radar meteor echo analyser. |
| J. Stohl : | On the distribution of sporadic meteors. |

# COMMISSION 24: PHOTOGRAPHIC ASTROMETRY (ASTROMETRIE PHOTOGRAPHIQUE)

## Report of Meetings on 18, 19 and 23 August 1982.

PRESIDENT: H. Eichhorn SECRETARY: J. L. Russell

Most of the business aspects of the Commission were conducted during the business meeting on 18 August. President Eichhorn welcomed the members of the commission and asked them to stand in memory of those colleagues who have died since the last meeting: Johannes Haas, Paul Herget and Jan Schilt, as well as of Valfrids Oswaldo who was not a member of our Commission at the time of his death.

The President then announced the appointment of Jane L. Russell as Secretary of the Commission. As a result of the elections and carry over, the following will be the Officers and the Organizing Committee for the trienuium 1982-1985:

President: Wilhelm Gliese
Vice President: Arthur R. Upgren
Organizing Committee: T. E. Corbin, H. Eichhorn. G. D. Gatewood, I. I. Kanaev, T. E. Lutz, J. D. Stock, W. F. van Altena.

The following were unanimously approved as new members of the Commission:

A. N. Argue, R. Branham, Liang-tia G. Chiu, J. Fanselow, F. Firneis, A. Fresneau, P. D. Hemenway, J. Hughes, W. H. Jefferys, K. Johnston, L. Stange, J. W. Stein, R. C. Stone, C. Turon, H. G. Walter, G. Westerhout.

The Commission approved the following report of the Working Group on Parallax Standards:

### REPORT OF THE WORKING GROUP ON PARALLAX STANDARDS TO I.A.U. COMMISSION 24

A. R. Upgren, Chairman

The Working Group on Parallax Standards was established at the meeting of Commission 24 during the XVII General Assembly of the International Astronomical Union in Montreal in August 1979. Its purpose is to investigate and propose methods of evaluating and reducing systematic parallax errors arising between observatories. Its concentration has been on the establishment of a number of star fields which would serve as standards observed by all parallax programs in the usual manner in which they obtain parallax observations.

Two closely related objectives emerged in the course of the group's deliberations. The first of these is the necessity for continued monitoring and calibration of each telescope and the second is the evaluation of parallax differences between observatories. Fields centered on open clusters are best suited for the first objective since many more stars are available for measure than the terms necessary for modeling each of the telescopes.

The Working Group recommends that the first of these objectives be met by regular evening and morning parallax observations of three fields each centered on an open cluster. The three fields are roughly equally separated in right ascension and are accessible to observatories at all latitudes. The fields are:

| Object | R.A. 1980 Dec. | | Object at Center |
|---|---|---|---|
| Pleiades | $3^h45^m.1$ | +24°08' | BD +23°520 |
| Praesepe | 8 39.2 | +19 45 | BD +23°2166 |

In selecting a larger list of individual stars and star fields, the Working Group sought to optimize the following features: widespread and uncorrelated distributions in magnitude, color, right ascension and declination. To a lesser extent, the absence of detectable orbital motion in the target star or stars, availability of past parallax observations, minimal disruption of presently active parallax programs and astrophysical interest of the target star or stars were also taken into consideration. A preliminary examination of each field has been made in order to be sure that an adequate distribution of suitable reference stars is available for measure.

In order to accomodate all programs, the list of stars has been divided into three sections. The first section contains those stars which it is hoped will be observed on all programs and includes mostly stars already being observed by a majority of them. The second section is a supplementry list intended to increse the number of standards for the programs which can accomodate a larger number. Both sections are limited to target stars between apparent magnitudes 7 and 14 because this is the magnitude interval accessible to most active parallax telescopes with a need for magnitude reduction devices which is small compared to that for stars covering a wide range in apparent magnitude. The third section includes brighter stars, (magnitude 6 and brighter) most of which are among the brightest and closest to their spectral type, along with a smaller number of very bright but distant stars whose true parallax is known in advance to be near zero. This list is proposed with the awareness that ground based telescopes and future space satellites with astrometric capability may both be helpful in establishing a zero point for the system, free of systematic effects at all magnitudes. It also focuses the continued need for parallax programs to respond to problems of stellar astrophysics insofar as they are able to do so, and in the course of calibration, to provide the astronomical community with at least a few highly precise parallaxes of many kinds of stars. Four of the brightest and nearest stars do not appear on this list. They are Sirius, Procyon, Altair and Alpha Centauri. They have parallaxes presently known with one or two percent error and since all four are binaries with considerable orbital motion, the possible uncertainty in their orbital elements makes further improvement in their parallaxes doubtful and not worth the effort involved in their redetermination.

The stars are listed in three tables numbered as indicated above. Each table lists one or more identifications of the central or target star. For all stars, an identification chart is available. Positions for the central star or stars are given for both 1950 and 2000 and B and V magnitudes are both given for ease of use by those with either photographic or visual refractors.

The Working Group recommends a periodic review of the observations of each participating observatory or satellite program, including measures with an automatic machine as necessary. An initial evaluation of the observations, including a critical examination of the available reference star frames, should be made after about three to five years. This plan should be modified as needed after an evaluation is made at that time.

We urge all observatories active in parallax observations to extend their programs to include the three cluster fields, the primary standards listed in Table I and as many of the secondary and bright star fields as possible.

Table I

Primary List of Parallax Standard Fields

| Field | Principal Star | 1950 R.A. | 1950 Dec. | 2000 R.A. | 2000 Dec. | V | B | B-V |
|---|---|---|---|---|---|---|---|---|
| 1. | HD 225213 | $0^h02^m.5$ | $-37°36'$ | $0^h05^m.2$ | $-37°19'$ | 8.63 | 10.08 | 1.45 |
| 2. | van Maanen 2 = Gl 35 | 0 46.5 | +5 09 | 0 49.2 | + 5 25 | 12.37 | 12.93 | 0.56 |
| 3. | BD +14°699 = vA 459 | 4 25.2 | +13 46 | 4 28.1 | +13 52 | 9.52 | 10.45 | 0.93 |
| | BD +13°685 = vA 472 | 4 25.3 | +14 10 | 4 28.2 | +14 16 | 9.03 | 9.87 | 0.84 |
| 4. | HD 33793 = Kapteyn's Star | 5 09.7 | -45 00 | 5 11.4 | -44 56 | 8.81 | 10.37 | 1.56 |
| 5. | BD -3°1123 | 5 28.5 | - 3 43 | 5 31.2 | - 3 40 | 7.97 | 9.44 | 1.47 |
| 6. | G87-7 = Gl 246 | 6 44.2 | +37 35 | 6 47.4 | +37 31 | 12.10 | 12.02 | -0.08 |
| 7. | BD -20°2665 | 8 42.5 | -21 16 | 8 45.0 | -21 27 | 10.16 | 11.38 | 1.22 |
| 8. | Ross 434 = Gl 366 | 9 41.7 | +76 17 | 9 47.1 | +76 03 | 10.63 | 12.06 | 1.43 |
| 9. | HD 95735 = Lal 21185 | 11 00.6 | +36 18 | 11 03.5 | +36 01 | 7.50 | 9.00 | 1.50 |
| 10. | Ross 128 = Gl 447 | 11 45.1 | + 1.06 | 11 47.4 | + 0 49 | 11.10 | 12.86 | 1.76 |
| 11. | BPM 36430 | 11 53.7 | -48 24 | 11 56.4 | -48 40 | 12.85 | 12.65 | -0.20 |
| 12. | BPM +2°2538 = Ross 456 | 12 23.0 | + 1 34 | 12 25.3 | + 1 17 | 9.61 | 10.06 | 0.45 |
| 13. | Ross 476 = Wolf 485B | 13 27.5 | - 8 27 | 13 30.2 | - 8 42 | 14.34 | 15.97 | 1.63 |
| | BD - 7°3632 = Wolf 485A | 13 27.7 | - 8 19 | 13 30.5 | - 8 34 | 12.31 | 12.38 | 0.07 |
| 14. | Proxima Cen = Gl 551 | 14 26.3 | -62 28 | 14 30.2 | -62 41 | 11.05 | 13.02 | 1.97 |
| 15. | G202-48 = Gl 625 | 16 24.2 | +54 25 | 16 25.2 | +54.18 | 10.14 | 11.75 | 1.61 |
| 16. | G240-72 | 17 48.9 | +70 53 | 17 52.0 | +70 52 | 14.15 | 14.55 | 0.40 |
| 17. | Barnard's Star = Gl 699 | 17 55.4 | + 4 33 | 17 58.1 | + 4 32 | 9.54 | 11.28 | 1.74 |
| 18. | Wolf 940 | 21 44.1 | - 0 24 | 21 46.4 | - 0.10 | 12.67 | 14.29 | 1.62 |
| 19. | Ross 248 = Gl 905 | 23 39.4 | +43 55 | 23 42.1 | +44 11 | 12.29 | 14.21 | 1.92 |
| 20. | G130-6 = Gl 905.2B | 23 41.3 | +32 16 | 23 44.0 | +32 32 | 12.90 | 13.05 | 0.15 |
| | G130-5 = Gl 905 2A | 23 41.4 | +32 19 | 23 44.1 | +32 35 | 11.73 | 13.28 | 1.55 |

Table II

Secondary List of Parallax Standard Fields

| Field | Principal Star | 1950 R.A. | 1950 Dec. | 2000 R.A. | 2000 Dec. | V | B | B-V |
|---|---|---|---|---|---|---|---|---|
| 1. | CoD -31°662 | $1^h30^m.0$ | -30°56' | $1^h32^m.2$ | -30°40' | 10.78 | 11.42 | 0.64 |
| 2. | Wolf 110 = G3-36 | 2 00.1 | + 5 28 | 2 02.5 | + 5 42 | 12.33 | 13.77 | 1.44 |
| 3. | BD +33°529 = Wolf 1323 | 2 49.0 | +34 12 | 2 52.0 | +34 24 | 9.55 | 10.89 | 1.34 |
| 4. | Ross 578 = G160-5 | 3 35.8 | -11 37 | 3 38.4 | -11 27 | 13.01 | 14.69 | 1.68 |
| 5. | LHS 1565 | 3 34.3 | -44 40 | 3 36.1 | -44 30 | 13.03 | 14.93 | 1.90 |
| 6. | Stein 2051A = Gl 169.1A | 4 26.7 | +58 53 | 4 31.3 | +58 59 | 11.09 | 12.73 | 1.64 |
| | Stein 2051B = Gl 169.1B | 4 26.7 | +58 53 | 4 31.3 | +58 59 | 12.44 | 12.75 | 0.31 |
| 7. | Ross 41 = Gl 203 | 5 25.3 | + 9 37 | 5 28.1 | + 9 39 | 12.48 | 14.12 | 1.64 |
| 8. | Ross 986 = Gl 268 | 7 06.6 | +38 38 | 7 10.2 | +38 33 | 11.48 | 13.19 | 1.71 |
| 9. | Wolf 1421 = Gl 289 | 7 45.2 | +20 30 | 7 48.2 | +20 22 | 11.47 | 12.99 | 1.52 |
| 10. | BD +31°1684 | 7 50.4 | +30 46 | 7 53.4 | +30 38 | 8.30 | 8.92 | 0.62 |
| 11. | G51-15 | 8 26.9 | +26 57 | 8 30.3 | +26 47 | 14.81 | 16.87 | 2.06 |
| 12. | BD +42°1922 | 8 41.9 | +41 52 | 8 45.5 | +41 41 | 8.60 | 9.54 | 0.94 |
| 13. | Ross 85 = Gl 361 | 9 38.5 | +13 26 | 9 41.3 | +13 12 | 10.40 | 11.90 | 1.50 |
| 14. | G163-27 | 10 55.1 | - 7 15 | 10 57.4 | - 7 31 | 14.33 | 14.65 | 0.32 |
| 15. | BD -17°3526 | 11 57.6 | -18 07 | 12 00.3 | -18 23 | 10.78 | 11.95 | 1.17 |
| 16. | HD 134439 | 15 07.2 | -16 13 | 15 10.0 | -16 24 | 9.02 | 9.81 | 0.79 |
| | HD 134440 | 15 07.2 | -16 13 | 15 10.0 | -16 24 | 9.43 | 10.28 | 0.85 |
| 17. | Ross 1057 = G257-20 | 15 42.2 | +76 10 | 15 41.1 | +76 03 | 12.22 | 13.87 | 1.65 |
| 18. | Ross 808 = G180-23 | 15 59.6 | +36 57 | 16 01.5 | +36 48 | 14.36 | 14.53 | 0.17 |
| 19. | CoD =38°10980 | 16 20.2 | -39 07 | 16 23.4 | -39 14 | 11.00 | 11.14 | 0.14 |
| 20. | Wolf 1346 = Gl 794 | 20 32.2 | +24 54 | 20 34.3 | +25 04 | 11.53 | 11.47 | -0.06 |
| 21. | Ross 769 = Gl 817 | 21 02.2 | -17 08 | 21 05.1 | -16 56 | 11.44 | 12.88 | 1.44 |
| 22. | Wolf 1106 = G231-27 | 21 06.8 | +59 34 | 21 08.3 | +59 46 | 13.30 | 14.85 | 1.55 |
| 23. | BD +8°4638 | 21 14.1 | + 9 11 | 21 16.4 | + 9 23 | 7.95 | 8.97 | 1.02 |
| 24. | CoD -49°13515 | 21 30.2 | -49 13 | 21 33.4 | -48 59 | 8.70 | 10.16 | 1.46 |
| 25. | G261-45 = Gl 836.2 | 21 36.7 | +82 49 | 21 34.3 | +83 02 | 13.02 | 13.00 | -0.02 |
| 26. | BD +42°4471 | 22 38.3 | +42 45 | 22 40.4 | +43 00 | 9.81 | 10.92 | 1.11 |

Table III

Parallax Standards - Bright Stars

| Field | Principal Star | 1950 R.A. | 1950 Dec. | 2000 R.A. | 2000 Dec. | V | B | B-V |
|---|---|---|---|---|---|---|---|---|
| 1. | β Cas | $0^h06^m.5$ | +58°52' | $0^h09^m.2$ | +59°09' | 2.25 | 2.60 | 0.35 |
| 2. | α Eri | 1 35.8 | -57 30 | 1 37.7 | -57 15 | 0.47 | 0.28 | -0.19 |
| 3. | α Per | 3 20.8 | +49 40 | 3 24.3 | +49 51 | 1.79 | 2.27 | 0.48 |
| 4. | ε Eri | 3 30.5 | - 9 38 | 3 32.9 | - 9 28 | 3.73 | 4.62 | 0.89 |
| 5. | α Tau | 4 33.0 | +16 24 | 4 35.9 | +16 30 | 0.86 | 2.39 | 1.53 |
| 6. | $\pi^3$ Ori | 4 47.1 | + 6 52 | 4 49.8 | + 6 57 | 3.19 | 3.64 | 0.45 |
| 7. | β Ori | 5 12.1 | - 8 16 | 5 14.5 | - 8 12 | 0.08 | 0.05 | -0.03 |
| 8. | α Ori | 5 52.5 | + 7 24 | 5 55.2 | + 7 24 | 0.80v | 2.66v | 1.86 |
| 9. | α Car | 6 22.8 | -52 40 | 6 24.0 | -52 41 | -0.73 | -0.57 | 0.16 |
| 10. | δ CMa | 7 06.4 | -26 19 | 7 08.4 | -26 24 | 1.84 | 2.52 | 0.68 |
| 11. | β Gem | 7 42.3 | +28 08 | 7 45.3 | +28 01 | 1.15 | 2.15 | 1.00 |
| 12. | α Leo | 10 05.7 | +12 12 | 10 08.4 | +11 58 | 1.36 | 1.25 | -0.11 |
| 13. | HD 103095 | 11 50.1 | +38 05 | 11 53.0 | +37 43 | 6.45 | 7.20 | 0.75 |
| 14. | γ UMa | 11 51.2 | +53 58 | 11 53.8 | +53 42 | 2.44 | 2.44 | 0.00 |
| 15. | δ UMa | 12 13.0 | +57 18 | 12 15.4 | +57 02 | 3.31 | 3.39 | 0.08 |
| 16. | β CVn | 12 31.4 | +41 38 | 12 33.8 | +41 21 | 4.27 | 4.86 | 0.59 |
| 17. | β Com | 13 09.5 | +28 08 | 13 11.9 | +27 53 | 4.26 | 4.84 | 0.58 |
| 18. | η Boo | 13 52.3 | +18 39 | 13 54.7 | +18 24 | 2.69 | 3.27 | 0.58 |
| 19. | α Boo | 14 13.4 | +19 26 | 14 15.7 | +19 11 | -0.06 | 1.17 | 1.23 |
| 20. | α Lyr | 18 35.2 | +38 44 | 13 36.9 | +38 47 | 0.04 | 0.04 | 0.00 |
| 21. | α Cyg | 20 39.7 | +45 06 | 20 41.4 | +45 16 | 1.26 | 1.35 | 0.09 |
| 22. | 61 Cyg A | 21 04.7 | +38 30 | 21 06.9 | +38 44 | 5.19 | 6.38 | 1.19 |
| | 61 Byg B | 21 04.7 | +38 30 | 21 06.9 | +38 44 | 6.02 | 7.40 | 1.38 |
| 23. | ε Ind | 21 59.5 | -57 00 | 22 03.4 | -56 47 | 4.67 | 5.73 | 1.06 |
| 24. | α Gru | 22 05.1 | -47 12 | 22 08.2 | -46 58 | 1.73 | 1.60 | -0.13 |
| 25. | HD 216803 | 22 53.6 | -31 50 | 22 56.4 | -31 34 | 6.49 | 7.59 | 1.10 |
| 26. | α PsA A | 22 54.9 | -29 53 | 22 57.6 | -29 37 | 1.16 | 1.25 | 0.09 |

Remarks

1. Nearest $\delta$ Scuti-type variable star.
2. Nearest evolved B star.
3. Brightest member of the nearby Per cluster.
4. Nearby dK star.
5. Nearest late gK star.
6. Nearest unevolved single dF star.
7. Distant blue supergiant; true parallax is < 0".002.
8. Distant red supergiant; true parallax is < 0".002.
9. Nearest supergiant.
10. Distant supergiant; true parallax is < 0".002.
11. Nearest normal giant.
12. Nearest B main sequence star.
13. Groombridge 1830; neaby sdG star.
14. Nearby A0V standard star; member of UMa cluster.
15. Nearby A3V standard star; member of UMa cluster.
16. Weak-line G0V star, otherwise nearly identical to No. 17.
17. Strong-line G0V star, otherwise nearly identical to No. 16.
18. Nearby old subgiant star.
19. Nearest high-velocity giant star.
20. Nearby A0V flux-calibration standard star.
21. Distant supergiant; true parallax is < 0".002.
22. Nearby pair of high-velocity dK stars.
    One of the pair may show orbital motion.
23. Nearby single dK star.
24. Nearby B main sequence star.
25. $\alpha$ PsA B; proper motion dK companion to No. 26.
26. Nearest normal dA star.-

The Commission expressed its thanks to the Working Group.

The Commission also approved the report of its working group on the identification of radio/optical astrometric sources and thanked them for their work. The following is a summary of their report:

SUMMARY OF REPORT OF I.A.U. COMMISSION 24
WORKING GROUP ON THE IDENTIFICATION
OF RADIO/OPTICAL ASTROMETRIC SOURCES

A. N. Argue, Chairman

The objective is to select suitable sources to act as benchmark in establishing a coordinate reference frame based on extragalactic objects. The system must be accessible to both radio and optical astrometry, therefore objects must be suitable for both methods. Approximately 100 objects uniformly spread over the sky are required.

With only a very small number of exceptions, extragalactic sources are too faint for HIPPARCOS, so that bootstrapping by ground based telescopes or, in particular, by Space Telescope will be necessary. Coordinate with Space Telescope requires

(i) Selection of 100 sources

(ii) Selection of HIPPARCHOS bootstrap stars

(i) A data-base is being compiled at the Department of Astronomy, U. of Texas at Austin, containing 600 sources with radio and optical data. The data are being continually augmented and updated, particularly in the radio domains, but the liklihood is that most of the objects finally selected already feature in this data-base.

(ii) the most likely one-third of those sources in (i) have been tenatively selected, and suitable bootstrap stars identified as HIPPARCOS candidates. These candidates must conform to certain well defined criteria as to cleanliness etc. to be acceptable to HIPPARCOS and Space Telescope. Following careful examination on Sky Survey prints they are being tested for absence of duplicity within the range 0.3 to 15 arc sec at Hamburg, and for absence of structure at the resolution limit of Space Telecope by speckle interferometry by groups at Lowell Observatory and Imperial College, London.

The Commission endorsed the proposal to hold an IAU Symposium entitled "Astrometric Techniques" from 9-12 January in Merida, Venezuela. (The organization of this Symposium was subsequently approved by the Executive Committee of the IAU.) The Commission voted to continue its joint Working Group on the Stellar Reference Frame (H. Eichhorn, chairman) with the same composition as in the triennium just passed.

The president appointed Messrs. Murray, Lutz and Stock to function as a Resolution Committee, charging it especially with solicitation for suggestions how this Commission could help avert the threatened closing of the Sydney (Australia) Observatory. During the meeting of August 19, D. King from the Sydney Observatory reported that the Observatory had, in fact, already been closed, but that the $12^{o}$ of the Sydney photographic catalogue which had already been measured would be reduced by April 1983, and suggested that Commission support best be directed toward urging the government of New South Wales to keep on the payroll those astronomers who were formerly on the staff of the Sydney Observatory, so that they could continue and finish the ongoing work. Such a resolution passed unanimously. The incoming president (Gliese) was asked to convey the best wishes of the Commission to the Lembang Observatory in Java on the occasion of its anniversary.

In the meeting on 19 August, scientific papers were presented by W. J. Luyten on Catalogues of Large Proper Motion, by T. Corbin on the reference stars for the Halley Watch:

The USNO is committed to provide improved positions for the Astrometry Network of the International Halley Watch covering the path from $21^h\ 00^m$, $-10\overset{\circ}{.}7$ to $18^h\ 49^m$, $-35\overset{\circ}{.}0$. These are the predicted positions of the comet on 10 February 1986 and 30 March 1986, respectively. It is intended that the reference stars will cover an area extending $2^{\circ}$ on either side of the comet's path with an average density of about 5 per square degree. The stars have been chosen from the SAO and are all in the magnitude range 6.0 m 8.9 except for a few fainter stars that were needed to fill gaps in the distribution. All known double stars that meet the conditions $m_2 - m_1 < 4.0$, and $2\overset{''}{.}0 \leqq d = [54'' - 2(m_1 + m_2)]$ have also been eliminated. Here $m_1$ and $m_2$ are the magnitudes of the two components and d is their separation.

This was followed by a paper by T. Corbin and G. Douglass on the astrographic catalogue: The U.S. Naval Observatory is currently contributing to the effort to make all of the AC available in machine readable form. Strasbourg has completed the region $-2^{\circ}$ to $+32^{\circ}$ and has kindly provided us with a copy of the data. We have finished $-4^{\circ}$ to $-12^{\circ}$ and $-41^{\circ}$ to $-52^{\circ}$. Our efforts will now concentrate on the $-12^{\circ}$ through $-40^{\circ}$ region. The data will be used at USNO to provide finding positions for the measurement of the 8" Dual Astrograph plates. Finally, the AC positions will be reduced to the system of the FK4 through the use of a reference system based on the IRS (International Reference Stars) positions and proper motions.

J. Stock gave a paper on plate overlap reduction by using spherical projection; Zhao one on Astronomy at the Shanghai Observatory, and P. Hemenway delivered a paper on HIPPARCOS.

During the Scientific Meeting of 23 August, papers were presented by A. Reiz, A. Nobili and C. Dahn, the latter on parallax measurements of faint stars using CCD cameras.

This is an abstract of Nobili's paper: Geophysical use of synchronous Earth satellites astrometry. Geosynchronous satellites currently used for telecommunication purposes can be easily tracked with moderate size large field telescopes measuring the satellites angular coordinates with respect to some catalogue field stars and the accuracy is between 1 and 2 arcsec. These data can be analyzed to measure the ellipticity of the Earth's equator. A recent paper (Catalano, McCrosky, Milani and Nobili, 1982, Journal of Geophysical Research, in press) shows that the organization of an international optical tracking campaign with the accuracy mentioned above would allow an improvement in the knowledge of the long wavelength geoid by at least one order of magnitude, that is very interesting for many geophysical purposes.

The following paper was transmitted with a request for publication:

## A DETERMINATION OF STELLAR PROPER MOTIONS WITH RESPECT TO GALAXIES AT PULKOVO

N. M. Bronnikova, A. N. Deutsch, N. V. Fatchikhin
O. N. Orlova and N. A. Shakht
Pulkovo Observatory, U.S.S.R.

Absolute proper motions of stars determined with respect to galaxies can be used both for determination of the solar motion parameters, galactic rotation constants and for a calculation of the correction to the precession constant.

At present four programs for the measurement of stellar proper motion with respect to galaxies are in progress. The LICK [1] and Yale-Columbia [2] programs use 51-cm aperture astrographs, Lohrmann program [3] uses the 2-m Reflecting Telescope. The program of the Pulkovo observatory [4] uses various Normal Astrographs and other more long-focus instruments at a few observatories in the northern and southern hemispheres.

The Pulkovo zone envolves 157 areas within $-5^{\circ}$ to $+90^{\circ}$ in declination [5]. Besides areas with galaxies the Pulkovo program foresees separate photographs of the fundamental stars of the catalogue of faint stars (FKSZ. The list of 240 FKSZ was published by A. N. Deutsch [6].

A number of papers have been published dealing with the proper motions of stars related to galaxies, obtained at Pulkovo by N. V. Fatschichin [7], at Taschkent by A. G. Rachimiv [8], at Golosseevo by S. P. Rybka [9] and N. V. Harchenko [10].

At present a new catalogue of absolute proper motion of stars from observation with the Pulkovo Normal Astrograph is nearing completion. We have got 95% of second-epoch plates with galaxies and 73% of second-epoch plates with FKSZ. The epoch differences are more than 20 years. We hope to obtain all second-epoch plates in 1983 and then we shall have 3 plate pairs for each of 157 areas with galaxies and 2 pairs for 24 areas with FKSZ. All plates with FKSZ and one of the three pairs with galaxies is taken with diffraction gratting.

270 plate pairs for 135 areas with galaxies have been measured. For 80000 stars up to the 16 magnitudes have been determined relative proper motion. The reductions have been calculated for a conversion of relative proper motions to absolute ones using galaxies, stars of the AGK3 catalogue on 135 areas. Besides we applied a statistical method using Binnendyk's secular parallaxes.

Using the reduction on 20 areas distributed on the hemisphere uniformly we have calculated the secular parallaxes for stars $13\overset{m}{.}7$ and $15\overset{m}{.}7$. The meaning parallaxes are given next table.

| m/b | $27^{\circ}$ | $61^{\circ}$ |
|---|---|---|
| $13\overset{m}{.}8$ | $+0''.0089$ | $+0''.0109$ |
| | $\pm$ 21 | $\pm$ 18 |
| $15\overset{m}{.}7$ | $+0''.0059$ | $+0''.0098$ |
| | $\pm$ 22 | $\pm$ 20 |

where b is a galactic latitude, m is a magnitude of stars.

The average number of well measured galaxies in an area is 4, the stars AGK3 is 20. According to preliminary data [11.12] the probable error of a star of $14^{m}$ - $15^{m}$ as determined from one pair equals $\pm 0''.006$. The error of the measurement of galaxies is about twice as large as that of stars of $14^{m}$ - $15^{m}$. The probable error of the reduction determined using 3 pairs is $\pm 0''.003$ [13].

References


Deutsch, A.N. 1954. Trans. IAU, 8, p. 789.

Deutsch, A.N. 1980. Isvestia Glavn. Astr. Obs. 198, p. 124.

Deutsch, A.N., Lavdovsky, V.V., Fatschichin N.V. 1955. Isvestia Glavn. Astr. Obs., 154, p. 14.

Deutsch, A.N., Lavdovsky, V.V. and Orlova, O.N. 1975. In book "Sovremennye problemy Astrometrii". p. 214, Moskwa.

Deutsch, A.N., Lavdovsky, V.V. and Orlova, O.N. 1978. In book "Novye idei v Astrometrii". p. 72, Leningrad.

Fatschikhin, N.V. 1974. Trudy Glavn. Astr. Obs., 81, p. 2.

Fatchikhin, N.V. 1982. Izvestia Glavn. Astr. Obs., 200, 107. Pulkovo, 1982.

Harchenko, N.V. 1980. In book "Astrometria i Astrophizika". 35, p. 3, Kiev.

Harchenko, N.V. 1980. In book "Astrometria i Astrophizika". 42, p. 70, Kiev.

Rachimov, A.G. 1978. In book "Kinematisheski i dinamisheskie harakteristiki otdelnych zvezdnych system". 109, 59, Tashkent.

Rybka, S.P. 1978. In book "Astrometria i Astrophizika", 35, p. 3, Kiev.

Sanding, H.-U. 1976. Mitteil. Lohrman Obs., N33, p. 943.

Wesselink, A.J. 1974. "New Problem in Astrometry". Ed. W. Gliese, p. 201.

Wright, W.H. 1950. Proc. Amer. Phil. Soc., 94, p. 1.

COMMISSION 25: STELLAR PHOTOMETRY AND POLARIMETRY
(PHOTOMETRIE ET POLARIMETRIE STELLAIRES)

Reports of Meetings 18 and 21 August 1982

PRESIDENT : J.A. Graham
VICE-PRESIDENT: J. Tinbergen

SECRETARIES: J.A. Graham
M.F. McCarthy

At the Patras General Assembly, activities of Commission 25 consisted of 5 sessions. 3 sessions were joint meetings with other Commissions and 2 were held by Commission 25 alone. Reports of two of the joint meetings, "Systematic Effects in Abundance Determinations for Metal-Poor Stars" and "Stellar Absolute Energy Distributions" will be given elsewhere. Here, we report on the joint meeting with Commission 37, "Photometry in Crowded Fields" and the two Commission 25 meetings "New Standard Stars Results" and "Commission 25 Business".

## PHOTOMETRY IN CROWDED FIELDS

This joint meeting was with Commission 37 (Stars Clusters and Associations). It was organized by Dr. Alan Penny (UK) who also presided except during the period of his own presentation when the undersigned took the chair. The following talks were given.

1) Photometry of Magellanic Cloud Clusters -- M.F. Walker (USA)

Dr. Walker outlined the program undertaken in collaboration with A. Blecha and J. Anderson to study Magellanic Cloud clusters with the Danish 1.5m telescope on La Silla. A McMullen electronographic camera is used to obtain electrographs which are subsequently analyzed with either a Joyce Lobel measuring machine or the Geneva two-dimensional scanning system. Results for the cluster Hodge 11 were presented and a discussion given of methods to deal with the problem of field crowding.

2) The KPNO Rich Field Program -- J.R. Mould (USA)

The Kitt Peak Rich Field program has become widely used in deriving magnitudes and colours from digital images. Dr. Mould gave an account of its operation from the user's viewpoint. All images in the field are replicas of some point-spread function which is determined as a first step towards subsequent analysis. The program accounts for elongation of images if present.

3) RR Lyrae Stars in the Central Bulge of Our Galaxy -- B. Helt (Denmark)

Dr. Helt described a search for RR Lyrae variable stars in the central regions of our Galaxy. The Danish 1.5m telescope and 40mm electronic camera was used to observe 3 fields, each 10' diameter. The search was carried out both by blinking plates and by forming positive/negative composite images. Many new variables have been discovered in the Sgr I field. Dr. Helt emphasized that it is essential to always check the linearity of the electronic camera by observing, for example, standard sequences during each observing run.

4) Gaussians and Lorentzians to Fit a Multiple Stellar Image -- A. Penny (UK)

This talk described the use of single 2D Gaussian profiles and multiple 2D Franz-Lorentz profiles by a suite of programs on the Royal Greenwich Observatory Starlink node. The single Gaussian was surprisingly resistant to the presence of a nearby star. The multiple Lorentz profiles fitted actual star images well.

5) Crowded Field Electronography -- P. Linde (Sweden)

In this paper, the techniques used at Lund Observatory for dealing with photometry in crowded fields were discussed. Methods for fitting and checking the point-spread function were described and reference was made to the linearity and the relation between photoelectric and electronographic magnitudes.

J.A. Graham
18 August 1982

NEW STANDARD STAR RESULTS

The scientific session of Commission 25 on August 18 was concerned with "New Standard Star Results". With the many new area detectors available, a strong need has arisen for faint photometric and polarimetric standard stars which can serve as calibration standards. Preferably such stars should be distributed over a wide range of magnitude, be spread over the sky sufficiently so that some are accessible at any time,and be well tied into established photometric systems.

A new Catalogue of Photometric Sequences is being prepared by N. Argue (UK), E. Miller (USA) and W. Warren Jr. (USA). N. Argue was prevented by unforseen circumstances from presenting a report on this project but later communicated results about progress. The work is actually the third supplement to two earlier catalogues. It is complete now to 1982.0 and will contain 1271 new sequences, mostly UBV(RI) to limiting magnitude 23. The magnetic tape will soon be available from Wayne Warren at the World Data Center, NASA-Goddard. Later on, copies of the tape will be available from the Centre de Données Stellaires, Strasbourg and a printed version will be circulated.

Recent work concerning UBVRI Photoelectric Sequences Around the Celestial Equator was described by A. Landolt (USA). Observations have been made of 223 stars on the UBVRI systems of Johnson and Cousins. The stars average 20.7 measures each on 12.2 different nights and fall in the magnitude range $7 < V < 12.5$.

Because of illness, D. Kilkenny (S. Africa) was unable to present his paper "Standard Star Work at the South African Astronomical Observatory". In his place, I. Glass (S. Africa) gave a brief summary of programs in progress at SAAO and presented some observations of standard stars in the infrared JHKL bands.

Finally, M. Breger (USA) gave an account of work done on Highly Precise Standards for Polarization Measurements. The widely used polarized and unpolarized standards by Serkowski have been extensively reobserved to extend their accuracy to a few parts in $10^4$ in $p(\lambda)$. Among polarized standards, 3 early-type supergiants (55 Cyg, 9 Gem, HD 183143) show definite polarimetric variability. The unpolarized standard HD 202573 is slightly polarized with $p(V) = 0.054 \pm 0.010\%$. These stars should not be used as standards. Special attention was given to the rotation of position angle with wavelength in this work.

J.A. Graham
18 August, 1982

COMMISSION 25 BUSINESS MEETING

The President (J.A. Graham) called the meeting to order at 9.05 a.m. and appointed the undersigned to serve as Recording Secretary for the session. About 25 members were present. President Graham announced the following Agenda which was approved unanimously.

1) Opening Remarks. Commemoration of Departed Colleagues.
2) Report of Commission in Trans. IAU XVIIIA.
3) New Members of the Commission.
4) New Commission Officials.
5) Resolutions.
6) Scientific Results.

The President commemorated our departed President of the Union, Dr. M.K. Vainu Bappu and recalled his early work in Hγ photometry. Though he was a never a member of Commission 25, Dr. Bappu had an initial and continuing interest in our field and we will miss his presence and direction. The President also commemorated the departure from us of Dr. Harold L. Johnson, whose contributions to photometry were so many and so important for our science, Dr. Graham characterized Johnson's spirit as one which, with excellence as his norm, was alert and active. He was always a person who looked for and found new fields to conquer. Next President Graham eulogized our departed colleague K. Serkowski whom he called "one of our polarimetric pioneers". Then President Graham invited all members to stand for a moment of silence to honor the memory of these departed colleagues.

The President's Triennial Report was next considered. Pres. Graham thanked all members who contributed and acknowledged especially the assistance of Drs. Bessell, Chugajnov, Hauck, Hyland and Tinbergen in its preparation. M. McCarthy moved that the report be accepted; this was seconded by T. Gehrels and approved.

President Graham next read a list of new officers proposed to serve the Commission during the next three years; this read as follows:

President : J. Tinbergen
Vice-President: F. Rufener

Scientific Organizing Committee: M. Bessell, P. Chugajnov, J. Dachs, P. Edwards, J. Graham, B. Hauck, D. Kilkenny, A. Landolt, I. McLean, V. Piirola, N. Shakhovskoj.

M. Bessell pointed out that there was no person on the SOC whose work was chiefly infrared photometry. I. Glass (S. Africa) was suggested and added to the above list. Approval of the nomination list was moved formally by A. Landolt, seconded by M. McCarthy and carried. The President then thanked the retiring members of the SOC for their years of service.

The President suggested that recommendations from members could be placed in this account of the proceedings rather than be submitted to the formalities of the Resolutions Committee, unless there was some pressing item of importance. There were no resolutions to be submitted formally on this occasion.

Vice-President Tinbergen, President-Elect of Commission 25 asked members to send him notes for the periodic progress reports which he plans to circulate to members of the Commission. The new address of our incoming President is as follows:

Dr. J. Tinbergen
Kapteyn Sterrenwacht Werkgroep
Mensingheweg 20
9301 KA - Roden
THE NETHERLANDS

The final business matter was the reading of the list of new members. These are L. Baldinelli, B. Carney, D. Hall, M. Das, J. Denoyelle, J.C. Forte, I.S. Glass, A. Grauer, M. Grenon, H. Hensberge, R. Hilditch, D. Kilkenny, Z. Kviz, P. Laskarides, R. Lenzen, V. Oskanyan, A. Penny, R. Pfeiffer, A.G.D. Philip, D. Sullivan, H. Trodahl. Consultants for the coming triennium are J-Y. Daniel and E. Miller.

A rousing round of applause was offered to thank John Graham for his splendid service of Commission 25. Congratulations and promises of support were also offered to our new President Jaap Tinbergen as he begins his work for us all.

The final portion (40 mins) of the session was devoted to scientific presentations. The undersigned wishes to point out that our meetings were attended by many guest members of the IAU and Invited Participants thus demonstrating the appeal and importance of the twofold branches of our science: stellar photometry and polarimetry.

Dr. E.F. Milone (Canada) described the new photometer in use with the 16" telescope at Calgary. The photometer incorporates a rapid alternate detecting system. It uses a chopping secondary mirror whose position angle can be varied 360 degrees in the plane of the sky. Four-channel pulse counting electronics are synchronized to four positions of the secondary mirror permitting effectively simultaneous measurement of a comparison star, variable star and two sky backgrounds using a single photomultiplier, filter and aperture. The system adequately compensates for the effects of transparency, sky brightness and detector sensitivity variations and has been successfully used with a cloud cover as much as 3 magnitudes.

Next, Dr. P. Edwards (New Zealand) spoke on Atmospheric Extinction Measurements. He described cooperative work undertaken in collaboration with the meteorologists to determine extinction values in UBV,uvby and narrow band regions. Dr. Edwards outlined the procedures followed and also mentioned the complications caused this year by the eruption of 2 volcanoes (one in Mexico and one in Java). Very large extinction values were noted in Texas and Hawaii by some commission members. It was recommended that, before blithely applying mean extinction coefficients, all colleagues should be careful to test the quality of the night with adequate and necessary measurement of standard stars. It was noted that a working group had been formed within Commission 50 (Protection of Existing and Potential Observing Sites) to determine extinction effects due to volcanoes. There was an active discussion of the importance of archival storage and possible publication of extinction data.

The scientific portion of the session concluded with a lively talk by Dr. Tom Gehrels (USA) about the program of the Spacewatch Camera. This presently is the 0.9m Newtonian telescope of the Steward Observatory on Kitt Peak. The plan is to replace it with a specially designed 1.8m prime focus reflector. It will use a CCD detector in the continuous scanning mode primarily for the discovery of near earth asteroids but also for stellar and galactic observations. During the light part of the moon, the telescope is to be used with a radial velocity spectrometer for the search of planets of other stars. This program was initiated by Krzysztof Serkowski and after his death continued by Robert McMillan.

This concludes the Secretary's Report of the Commission 25 Business Meeting.

Martin F. McCarthy, S.J.
21 August 1982

## COMMISSION 26: DOUBLE AND MULTIPLE STARS (ETOILES DOUBLES ET MULTIPLES)

### Report of Meetings, 18 and 23 August 1982

PRESIDENT: O. G. Franz SECRETARY: J. Dommanget

#### 18 August 1982

The President opened the meeting and welcomed those in attendance. J. Dommanget was appointed to serve as Secretary. All present stood briefly in silence to honor the memory of two distinguished colleagues, P. M. Djurkovic and U. Güntzel-Lingner, lost by death since the 1979 meeting.

1. Commission officers and membership

By written nomination and secret balloting, the Commission endorsed the following officers for 1982-1985:

President: M. G. Fracastoro
Vice President: K. D. Rakos
Organizing Committee: J. Dommanget, W. D. Heintz, and H. A. McAlister

Continuing for a second term on the Organizing Committee will be A. Poveda and C. D. Scarfe. Franz will serve one term as an ex-officio member.

The President announced the resignation from the Commission of A. S. da Silva and M. C. Johnson. He then presented, and the Commission endorsed, the following list of proposed new members: E. Cabrita, R. B. Culver, A. N. Dadaev, F. C. Fekel, O. E. Ferrer, C. A. Hernandez, G. Ishida, M. I. Kumsiashvili, N. L. Magalashvili, R. Pannunzio, G. N. Salukvadze, A. M. Shulberg, M. Tapia, A. Valbousquet, and K. A. van der Hucht.

2. Double star work with the 26-inch Johannesburg refractor

At the invitation of the President, J. Hers attended the meeting and commented briefly on the status of the Johannesburg refractor. He expressed his willingness to help explore conditions under which some use of this intrument for visual double star work might be resumed. Among these conditions would have to be expressions of real interest by double star observers from abroad to come to Johannesburg for short or longer observing visits if given some financial help. Statements of interest and specific information on necessary assistance (e.g., travel expenses, living expenses, accommodations, etc.) should be communicated directly to J. Hers, Post Office Box 48, Sedgefield, South Africa 6573.

3. Name of Commission 26

A proposal by Dommanget to change the name of the Commission from "Double Stars (Etoiles Doubles)" to "Double and Multiple Stars (Etoiles Doubles et Multiples)" was adopted unanimously and forwarded to the Secretary General for formal action by the General Assembly during its second session at Patras.

4. Proceedings of IAU Colloquium No. 62

R. Harrington reported that some authors of contributed papers have shortened their manuscripts to reduce printing costs, while others have agreed to pay charges for excess pages. Publication of the proceedings still depends on the availability of the necessary funding.

5. Nomenclature of components of multiple systems

At the request of the President, Heintz submitted a statement concerning the designation of the components of multiple stars. Following several comments by Commission members, the President recommended that further discussion be postponed until the second session of the Commission to allow all interested members time to review the problem and to analyze the statement of Heintz.

6. Supplementary Rules (By-laws)

The President recalled that during its meeting of 19 May 1981 at Flagstaff, Arizona, the Organizing Committee discussed and endorsed new Supplementary Rules (By-laws) to define and clarify the nominating and balloting procedures for Commission offices. These proposed by-laws, intended not to replace or alter, but to supplement the existing Internal Working Rules, were subsequently submitted to all Commission members for their review and written comments. While those who responded expressed unanimous support for the intent of these by-laws, some difficulties and certain inconsistencies with existing working rules were noted, indicating the need for further deliberations before a final version of the by-laws could be prepared.

The President's remarks were followed by extensive and often spirited discussion, particularly concerning voting by proxy and the problems which balloting by mail may pose if unforeseen and unpredictable situations should arise. The Commission then decided, by unanimous vote of those present and voting, to adopt the new by-laws "in principle." The retiring president was directed to form an *ad hoc* committee for the purpose of revising the proposed Supplementary Rules, perhaps combining them with existing Internal Working Rules, and submitting a revised text to the Commission for ratification as soon as possible after the Patras General Assembly.

## 23 August 1982

I. SCIENTIFIC PRESENTATIONS

1. A. H. Batten: Recent Results on Σ2367 and 70 Ophiuchi
2. C. E. Worley: Status Report on the Double Star Index, Observation, and Orbit Catalogs
3. J. Dommanget: A New Catalog of Ephemerides of Radial Velocities for Orbital Visual Binaries
   : The Observation of Radial Velocities of Visual Binaries
4. C. D. Scarfe: Zeta Herculis - The Third Dimension

II. ADMINISTRATIVE MATTERS

1. Co-sponsorship of symposia and colloquia

The Commission voted to offer co-sponsorship for the following proposed meetings:

a) Symposium on "Calibration in Astronomy"; tentative plans being coordinated by A. H. Batten.

b) Colloquium on "MK Spectral Classification: Criteria and Applications"; R. F. Garrison, Chairman, Colloquium Organizing Committee.

c) Colloquium on selected topics concerning the "Masses and Dynamics of Double and Multiple Stars," to be held at Lembang in June 1983; B. Hidayat, Chairman, Local Organizing Committee.

2. Next IAU Double Star Colloquium

A tentative proposal to hold a "Double Star Colloquium" in 1984 in Italy was presented by Fracastoro and accepted by the Commission. Precise date, location, and specific topic will be determined by a colloquium organizing committee at a later date.

3. Nomenclature of components of multiple stars

In response to the statement presented by Heintz during the session of 18 August 1982, separate proposals on the designation of components of multiple systems were presented by Couteau and Valbousquet. Heintz reiterated that the purpose of his statement was not to make a specific proposal, but to call attention to this problem and the need for its consideration and solution. Since it was clear that such a solution could not be found during the remainder of the Commission meeting, the President urged those most interested in and affected by this problem to confer directly and to develop a mutually acceptable nomenclature.

# COMMISSION 27: VARIABLE STARS (ETOILES VARIABLES)

## Report of Meetings: August 19 and 23, 1982

PRESIDENT: J.D. Fernie SECRETARY: Nancy R. Evans

### BUSINESS AND SCIENTIFIC MEETING, August 19 1982

1. Nominations.

The commission agreed to nominate the following Organizing Committee for 1982-85:

President: N. Baker
Vice-President: B. Szeidl
Members: M. Breger, A.N. Cox, R.E. Gershberg, M. Jerzykiewicz, L.N. Mavridis, L.N. Mirzoyan, J.R. Percy, A.M. van Genderen, B. Warner, J.D. Fernie (ex-officio).

2. New Members.

It was agreed that the executive be authorized to review the list of proposed new commission members and decide on their admission.

3. General Catalogue of Variable Stars.

The President read a report by Dr. Samus' concerning the continuing work on the GCVS. The fourth edition of the catalogue is well underway, and will include a listing of extragalactic variables and one of cross-listings. The new Catalogue of Suspected Variables, containing 14811 stars, has been published, and a copy was shown at the meeting. The Catalogue is also available on magnetic tape through the Strasbourg Centre, as will be the 4th edition of the GCVS. The total number of designated variable stars now amounts to 28254.

The Commission is much indebted to the Sternberg Astronomical Institute for its work on these catalogues.

4. Information Bulletin on Variables Stars.

A report by the editor of IBVS, Dr. Szeidl, was read to the meeting. Dr. Szeidl noted that as of August 6, 1982, 2184 issues of the Bulletin had been published, including more than 500 since the previous General Assembly. With only Dr. Szeidl, a secretary, and Dr. Szabados available to produce the Bulletin, the task is a heavy one, and Dr. Szeidl emphasized that material submitted to the Bulletin must be in camera-ready form, particularly if the turn-around time of two to four weeks is to be maintained.

Dr. Szeidl proposed that the Bulletin continue to publish material on eclipsing binaries, a proposal wholeheartedly supported by the meeting. It was moved and unanimously agreed that Dr. Szeidl be thanked for his production of the Bulletin and asked to continue the work.

5. Archives of Unpublished Data.

Dr. Breger reminded the meeting of the existence of the archives for unpublished photoelectric data, and the manner in which the system works. Summaries of existing data files are published in the PASP every eighteen months. A third data bank, that at the Strasbourg Centre, has been added to the system.

6. Hipparchos Satellite Project.

Dr. C. Jaschek pointed out to the meeting that the Tycho portion of the Hipparchos satellite is expected to obtain photoelectric magnitudes for many thousands of the brightest stars over a three-year period, with typically fifty observations per star. This could be of immense benefit to variable star observers. Technical details are available from Dr. E. Hoeg of the Copenhagen Observatory.

7. Cepheid Colloquium.

The President announced that it was hoped to celebrate the 200th anniversary of the discovery of cepheid variables by holding an IAU Colloquium on cepheids at the University of Toronto/David Dunlap Observatory probably in August 1984.

8. Scientific Session.

The meeting closed with three short review papers:

A. Maeder: Hubble-Sandage Variables
J. Percy : Early-type Variables
A. Cox : Double-mode RR Lyrae Stars

MEETING OF FLARE-STAR WORKING GROUP, August 23, 1982

The meeting was arranged by Dr. Mavridis and was held at the Stephanion Observatory. Sixteen persons attended. The following papers were presented:

V.A. Ambartsumian: The Main Aims of the Program of Observations of Flare Stars in Open Clusters at the Burakan Observatory.
L.N. Mirzoyan : Some Results of Three-Colour Photographic Observations of Flare Stars in Open Clusters.
P.V. Kulkarni : Infrared Flare Activity in the Rapid Burster 1730-333.
C.A.O. Torres : Evolutionary Status of the Pleiades from the K-type BY Dra Stars.
B.R.P. Pettersen : Time Scales of Variations in Continuum and Hydrogen Emission Lines During Flares.
L.N. Mavridis, G. Asteriadis, and F. Mahmoud: Long-Term Variations of the Flare Activity and Quiet-state Luminosity of the Flare Stars EV Lac and BY Dra.

SCIENTIFIC MEETING, August 23, 1982

An informal scientific session produced the following speakers and papers:

D. Kilkenny: The RCB Stars.
P.W. Hill : Extreme Helium Stars.
N.R. Evans : Recent Observations of SU Cas.
W. Gieren : Continued Discussion of SU Cas.
C.T. Bolton: V986 Oph.
J.A. Mattei: The AAVSO Data Bank and Publications.
J. Rountree: The BW Vul Observational Campaign.
S. Barcza : Problems in Calibrating RR Lyrae UBV Colours in Te and log g.
G. Bakos : Light Variations in 57 Peg.
M. Frolov : The Classification of $\delta$ Scuti Stars.
J. Cuypers : The $\beta$ Cephei Star $\beta$ Cru.

# COMMISSION 28: GALAXIES (GALAXIES)

## Report of Meetings, 18, 19, 20, 23, 24, and 25 August 1982

PRESIDENT: B. E. Westerlund SECRETARY: N. Bergvall

Commission 28 was involved in twelve sessions, three of them all-day joint discussions:

Evolution in Old Stellar Populations in Galaxies (Commissions 25, 28, 29, 33, 34, 35, 37, 45, 47), August 19.

The Extragalactic Distance Scale and the Hubble Constant (25, 28, 40, 45, 47, 48), August 23.

Active Galactic Nuclei (28, 40, 48), August 18.

Details of these joint discussions will appear in Highlights of Astronomy, volume 6. Of the remaining sessions, three were joint meetings: Dark Haloes of Galaxies (with Commission 33), Beams and Jets in Extragalactic Sources (with Commission 40), and The Local Supercluster and its Environment (with Commission 47). The remaining sessions were Commission 28 alone. Programs for these meetings and abstracts of some of the papers presented follow the minutes of the business meeting below. Other abstracts or summaries may be found in the Highlights of Astronomy.

### 20 August 1982

BUSINESS MEETING

1. NEW OFFICERS AND MEMBERS

The Commission unamimously approved the election of the new President, V. C. Rubin, and Vice-President, P. C. van der Kruit. The President proposed, and the Commission unanimously accepted, the list of names proposed for the new Organizing Committee.

The list of IAU members who wish to join Commission 28 was read and adopted. A second list of proposed new members to the Union, wishing to join Commission 28, was read and adopted, after a short discussion.

II. REPORT OF THE COMMISSION

The President asked for any comments regarding the report of the Commission, which had been distributed to the Commission members present in Patras before the meeting. The report was adopted without discussion.

III. THE WORKING GROUPS

A large amount of administration work goes into the Working Groups and compared to this some members considered the scientific benefit to be too meager. As arguments against this, the importance of the Working Groups for initializing cooperation between different fields and for circulating abstracts was stressed. After some discussion the Commission, recognizing the importance of circulating information, recommended that the Working Groups should continue to exist.

IV. RESOLUTIONS AND RECOMMENDATIONS

D. Lynden-Bell pointed out the fact that, although precession influences a measured orientation of a galaxy, very few observers quote the epoch they have used, when presenting their results. The problem was discussed, and it was recommended that the epoch should be quoted whenever a position angle is published.

V. Rubin drew attention to the ambiguity existing in connection with the definition of the inclination angle of a galaxy. After a short discussion the following internal resolution was adopted:
"Define the inclination i of a galaxy such that i = 0 when the angle between the line-of-sight to the galaxy and the rotation axis of the galaxy equals $0^{\circ}$ (i.e. face on), and i = $90^{\circ}$ when the angle between the line-of-sight and the rotation axis of the galaxy equals $90^{\circ}$ (i.e. edge-on)".

V. MISCELLANEOUS

The problem of defining coordinates for objects within a galaxy was discussed. A group, consisting of H. R. Dickel, P. W. Hodge, D. Lynden-Bell and S. M. Simkin was formed in order to treat the problem in more detail.

The question of whether supernovae should be included in Commission 28 was discussed. It was suggested that a Working Group on supernovae should be established. The Organizing Committee was asked to find a proper solution to the problem.

T. Kinman had asked the Commission to urge all observers to publish their redshifts as soon as they are available. It was recognized that systematic collection of redshift data is in progress at a few places. The Commission recommended the incoming President to contact H. Rood in order to encourage him to continue his collection of redshifts and to publish them at regular intervals.

K. Freeman reported from the Bescançon symposium "Internal Kinematics and Dynamics of Galaxies".

PROGRAMS OF SCIENTIFIC SESSIONS

NEW RESULTS IN EXTRAGALACTIC RESEARCH August 20

S. Gottesman: Neutral Hydrogen in the Dwarf Elliptical Galaxies NGC 185 and NGC 205
F. Schweizer: Merging Galaxies and the Formation of Ellipticals
H. van Woerden: Distribution and Motion of HI in Lenticular Galaxies
H. van Woerden and R. Warmels: Comparison of HI Diameters in Virgo Cluster and Field Galaxies
L. Dressel: NGC 315: High- Velocity Clouds in an Active Elliptical Galaxy?
J. Hesser: Mass and Distance of NGC 5128
K. C. Freeman: Dynamics of Galactic Bulges
B. F. Burke: Gravitational Lenses and Dark Matter in the Universe
J. Hutchings: QSO- Galaxy Interactions and QSO Morphology
D. M. Worrall: The Broad- Band X- Ray Spectrum of a QSO Sample and Multifrequency Observations of BL Lac Objects
H. R. Miller: A Recent Photometric Study of Seyfert Galaxies
G. A. Gurzadyan and V. G. Gurzadyan: On the Dynamical Structure of QSOs Derived from Emission Line Data
A. Pedlar: A Hα Filament Associated with Radio Emission in NGC 1275
E. Valtaoja: A 15.7 Min Periodicity in OJ 287?
M. G. Pastoriza: Metallic Abundances of E and SO Galaxies and Globular Clusters of the LMC and SMC
Ch. J. Peterson: The Motion of the Local Group

DARK HALOES OF GALAXIES August 20 Joint Meeting with Commission 33

See Commission 33 for presented papers

THE LOCAL SUPERCLUSTER AND ITS ENVIRONMENT August 24 Joint Meeting with Commission 47

B. Tully: Morphology of the Local Supercluster
J. Huchra: The Dynamics of the Local Supercluster
E. Salpeter: Theoretical Models for Virgocentric Infall
G. Abell: Properties of Superclusters
J. Einasto, H. G. Corwin, J. Huchra, R. H. Miller, and M. Tarenghi: Local Cells of the Universe; Voids of Diameter 200 Mpc (presented by M. Tarenghi)
J. Zeldovich: Remarks on the Structure of the Universe

GIANT HII COMPLEXES OUTSIDE OUR GALAXY August 24

1. Observational results and their primary interpretations in the following domains:
   Radio: D. S. Heeschen
   X-rays: L. van Speybroeck
   Visible: M. Rosa
   UV: P. Benvenuti
   IR: C. M. Telesco
2. Confrontation with related investigations in our Galaxy and general interpretations: F. P. Israel

BEAMS AND JETS IN EXTRAGALACTIC SOURCES August 25 Joint Meeting with Commission 40

See Commission 40 for presented papers

ABSTRACTS OF SCIENTIFIC PAPERS:

S. T. Gottesman: Observations have been made in collaboration with D. W. Johnson of the atomic hydrogen content of the dwarf, peçuliar elliptical galaxies NGC 185 and NGC 205. An angular resolution of about 1!1 was obtained with the VLA with a velocity resolution of approximately 6 km/sec.

The gas in NGC 185 does not appear to be rotating. Rather, a single not highly resolved feature was observed with a mass of $1.8 \times 10^5$ $M_{\odot}$.

In the case of NGC 205 an extended feature with structure is found with a mass of $4.1 \times 10^5$ $M_{\odot}$. The gas appears to be in the form of a rotating disk with a maximum observed velocity of 29 km/sec at a radial distance of ~1!5.

In each system, the centroid of the gas is significantly offset from the center of the galaxy. Also, the gas does not coincide with the well known dust clouds observed in NGC 185. In the case of NGC 205 the gas appears to be more closely associated with the dark matter. It is difficult to understand how .01% of the total mass of these galaxies has failed to coalesce at the bottom of the galaxien potential well. Perhaps the explanation is associated with the bursts of star formation that are occurring in these peculiar elliptical systems.

L. L. Dressel: We report the detection of HI in absorption against the nuclear continuum source in the giant elliptical galaxy NGC 315. This is the first reported observation of individually resolved high-velocity clouds in or near an elliptical galaxy. The absorption line is highly blue-shifted in the reference frame of the galaxy, with a projected velocity toward the galaxy of 490 km $s^{-1}$. It is very strong (244±4 mJy) and consists of two components with very narrow velocity widths of 2.5 km $s^{-1}$ (FWHM) each. The peak optical depths are 0.9 and 0.2. If the linewidths are due solely to thermal broadening, then the kinetic temperature must be 150 K. The neutral gas column densities implied are $2 - 5 \times 10^{20}$ hydrogen atoms $cm^{-2}$. These absorbing clouds are thus very similar to the diffuse HI clouds observed in absorption in the Galaxy. We did not detect OH in absorption in these clouds to a limiting peak optical depth of 0.04, but this result is consistent with either a galactic or extragalactic origin of the clouds. They may have condensed out of a large unobserved galactic halo of gas, or they may reside in or have been captured from the intracluster medium.

The investigation has been carried out in collaboration with T. M. Bania and M. M. Davis.

J. E. Hesser: Spectroscopic observations with the CTIO 4-m telescope and SIT vidicon have confirmed 17 of 18 visually selected candidates as globular clusters. The cluster system, as presently known, extends to a projected radius of ~ 36(D/5 Mpc) kpc, where D is the distance to NGC5128 in Mpc. Evidence has been found for a radial B-V color gradient, with the redder, presumably more metal enriched, clusters lying nearer the nucleus. The heliocentric radial velocities range between ~ 300 - 860 km $s^{-1}$. Star counts on UK Schmidt plates suggest a total cluster population of ~ 600. Observations in May, 1982, of 13 objects within the color range of the confirmed clusters but chosen without regard to possible spatial extent reveal two more clusters. This supports the prediction of the star counts concerning the presence of many more clusters than have been visually identfied by their slightly non-stellar appearance. From the velocities we estimate a mass of $\sim 10^{12}$ $M_\odot$. Scaling the combined Galactic and M31 globular cluster luminosity function to N ~ 600, E(B-V) = 0.10 mag allows a comparison with the NGC 5128 cumulative luminosity function for confirmed clusters; this comparison yields a distance of ~ 3 Mpc.

The investigation has been carried out in collaboration with H.C. Harris, S. v.d. Bergh, and G. L. H. Harris.

Diana M. Worrall: Einstein Observatory IPC X-ray measurements of QSOs have been used by several authors to estimate the contribution of QSOs to the diffuse X-ray background (XRB). However, the shortage of information on quasar spectra is a major difficulty since one compares IPC fluxes essentially measured at 1 keV with an XRB known above 3 keV. X-ray spectra above 2 keV are known with good statistics for the two brightest QSOs, and with less precision for a sample of 14 others. A soft spectral excess, such as is observed in BL Lac-type objects, would result in an overestimate of the QSO contribution to the XRB. Frank Marshall (Goddard Space Flight Center) and I have compared the average ~ 7 keV flux for an optically-selected sample of 25 QSOs detected, just above threshold, with the HEAO-1 A2 experiment, with their average ~ 1 keV flux measured with the IPC. Our preliminary analysis gives a 90 % confidence range for $\alpha(f_\nu \sim \nu^{-\alpha})$ of 0.62 (+0.36, -0.24) for $N_H \leqslant 3\ 10^{20}$ atoms $cm^{-2}$. Higher column densities for OSOs are not anticipated, but if $N_H$ were ~ $10^{22}$ atoms $cm^{-2}$, then $\alpha \geqslant 0.75$. Although, we find no evidence for the soft energy excesses typically seen in BL Lac-type objects, and our spectrum is similar to those exhibited by Seyfert galaxies, the spectrum is soft enough that, if these objects are typical of the higher redshift more radio-quiet QSOs, then QSOs may perhaps be excluded as being the dominant origin of the diffuse X-ray background. (See De Zotti et. al., 1982, Ap. J., 253, 47).

Fred Bruhweiler (Computer Sciences Corporation) and I have observed the BL Lac-type objects OJ 287, OQ 530, 3C 371 and ON 325 with the IUE satellite. This has been done in conjunction with collaborators who have made nearly-simultaneous observations at visual, infrared and radio wavelengths. We have applied synchrotron self-Compton emission models to the data. Two relevant parameters for each source, the self-absorption frequency of the synchrotron source and the source size at this frequency (or source variability time scale), are not known. The smallest synchrotron components, which are the ones producing the most Compton radiation, may be self-absorbed at radio frequencies above those at which the sources are mapped. The use of reported radio source sizes may thus lead to erroneous conclusions. We find, however, that our higher energy spectral measurements provide sufficient evidence alone to show that either relativistic beaming is present in these sources, or that a more complex model involving an electron reacceleration mechanism is required.

G.A. Gurzadyan and V.G. Gurzadyan: Two facts concerning the behaviour of resonance ultraviolet doublet of ionized magnesium 2800 MgII in emission in the spectra of quasars (QSOs) and active galactic nuclei attract one's attention. The first one is attributed to the ratio of observed flux in MgII doublet to the flux in hydrogen $L_\alpha$-line: $F(MgII/F(L_\alpha)) \equiv Q$. For usual stars these lines are generated in the stellar chromospheres and always Q > 1, varying in a band 2 - 20; for instance, Q(Sun) = 4.5. However for QSOs, as a rule, Q < 1; for investigated QSOs Q $\simeq$ 0.2 The second fact concerns the width of MgII and Hβ emission lines. For QSOs their width corresponds to very high velocities $\sim$ 5000 km $s^{-1}$. Here the following problem is examined: is it possible on the basis of these two extraordinary facts peculiar only to QSOs and Seyfert galaxies to decipher their internal structure and to throw light on the nature of emission regions? We point out that the observational data can be explained by a presence of star-mass ionized cloud system with Maxwellian velocity distribution. A model of generation of these clouds is proposed.

A. Pedlar: NGC 1275 (Perseus A, 3C84) is surrounded by a series of Hα filaments which Cowie et al. (MNRAS 191, 399) believe to be due to cooling accretion flows from hot ($10^7$K) X-ray emitting gas in the Perseus cluster. We have mapped the λ73 cm radio continuum emission from NGC 1275 using MERLIN with a resolution of 1 arcsec. The resulting structure shows a 5 kpc radio jet emanating from the nucleus in PA ~ 160°. This aligns approximately with the milliarcsecond jet in the nucleus (PA ~ 170°). Of particular interest is a filament in PA ~ 170° which aligns precisely with the VLBI jet, and is also unique in that it shows by far the largest velocity gradient in any of the filaments and extends ~1 arc min from the nucleus. We attribute this to the action of a beam which originates in the nucleus and consider the ionised gas to have been entrained into the beam and transported to the outer part of the galaxy before it cools to ~$10^4$ K and emits Hα lines. We therefore feel that this particular filament is not caused by a cooling accretion flow, but is associated with ejection from the nucleus. Consequently optical comparisons between this filament and others in the halo would be useful to confirm our hypothesis.

A full account of this project is about to be submitted to MNRAS. The work was done in collaboration with R. Booler and R. Davies.

E. Valtaoja: A. 15.7 minute periodic flux variation was noticed in OJ 287 in April 1981 at 37 GHz. The source has been subsequently monitored for 142 hours att 22 GHz and 37 GHz as well as in optical photometry. The 15.7 minute period was discovered at two simultaneous photometric observations in February 1982, and the same period was also seen at 22 GHz at the same time. The optical variations were observed in phase with each other at both observatories, but the radio variations were out of phase by about one third of the phase. The peak-to-peak amplitude of the variation is about 5 % of the total intensity, and the period is 943 ± 10 seconds. The results support a spinar model for OJ 287.
The investigation has been carried out in collaboration with H. Lehto, P. Terrikorpi, S. Haarala, T. Korhonen, M. Valtonen, H. Teräsranta, E. Salonen, S. Urpo, M. Tiuri and V. Piirola.

G. Pastoriza: Metallicity of 8 E and SO galaxies as well as that of red globulars of the LMC and SMC were obtained by means of DDO integrated photometry calibrated with galactic globular clusters (Bica and Pastoriza, 1982). A correction was obtained in order to reduce the colors of the galaxies to zero redshift. The relation metallicity vs $M_V$ for the galaxies is analyzed (adding to our sample the observations of McClure and Van den Bergh, 1968 and Faber, 1973). For the Magellanic Clouds we found metallicity ranging from intermediate to poor.
The investigations has been carried out in cooperation with H. A. Dottori and E. L. D. Bica.

Ch. J. Peterson: The Rubin et al. (1976, Astron. J., 81, 687, 691) study of a distant sample of Sc I galaxies produced a solution for the motion of the Local Group that was (and still is) at variance both with several other determinations which have used other selections of galaxies and with the motion deduced from the anisotropy of the cosmic background radiation. The weakest part of their study was reliance (in part) on photographic magnitudes (which they corrected) from the Zwicky et al. (1961-1968) Catalogue of Galaxies and Clusters of Galaxies. We now have Kitt Peak photoelectric photometry of their galaxies and reanalysis for the motion by the method of Jackson (1982, Ph. D. thesis, University of California - Santa Cruz) shows that use of the more reliable photoelectrically derived magnitudes does not remove the differences between the motion vector from the Rubin et al. galaxies (V= 671 $\pm$ 195 km $s^{-1}$ toward l = 200$^{\circ}$, b = - 12$^{\circ}$) and the motion derived from closer samples of galaxies (e.g. de Vaucouleurs et al. 1982, Astrophys. J., 248, 395, 408) and that derived from the cosmic background radiation (Boughn et al. 1981, Astrophys. J. (Letters), 243, L 113). The new observational data remove any last doubt that the Rubin et al. motion is an artifact of systematic errors in the Zwicky magnitudes. Jackson's work further eliminates as sources of error any bias in the Rubin et al. selection of galaxies in terms of either apparent or absolute magnitudes, use of a simple cosecant law for the galactic absorption, or the methodology of analysis. We suggest that the differences in derived motions may merely reflect the complex flow patterns and inhomogeneities of the galaxy distribution in the local region of the present day universe.

This work has been supported by the U.S. National Science Foundation.

## Report of Working Groups

I. THE MAGELLANIC CLOUDS

The Working Group met on 25 August. Dr M. W. Feast was re-elected Chairman.
The following papers were read:

D. Lynden-Bell: Which Way are MC Going?
R. J. Cohen and R. Davies: The Magellanic Stream
M. T. Brück: The SMC Halo
T. Lloyd Evans: C and MS stars in the MC
M. F. Walker: Santa Cruz MC Programmes
M. S. Bessell: Red Variables in MC
J. Isserstedt: Star Formation in MC (film)
Ch. Fehrenbach: MC Catalogues
A. F. J. Moffat: The Core of 30 Doradus
A. F. J. Moffat: WR stars in MC
M. Kontizas: Kron 3 and NGC 152
M. Kontizas: Dynamics of SMC Clusters
N. R. Walborn: Supergiant Planetaries in MC
K. Nandy: U-V extinction in SMC
P. Dubois: The Thickness of the SMC
B. Y. Mills: MC SN Remnants
D. Crampton and J. B. Hutchings: MC XR Binaries
C. Chiosi: Star Formation in MC

II. GALAXY PHOTOMETRY AND SPECTROPHOTOMETRY

The Group convened in the afternoon of August 24 jointly with the W. G. on Internal Motions. The Chairman of the W. G. acted as chairman of the session, which started with the business. It was unanimously agreed that the W. G. should continue to exist. It was also decided to elect the W. G. Chairman for the next triennium by consulting all members by mail. A drafted resolution aiming to simplify the notation of units commonly used in surface photometry work was presented by de Vaucouleurs and Capaccioli. They also proposed to constitute a task force or consortium for a careful photometric mapping of the Virgo Cluster galaxies.

The following papers were presented:

S. Okamura: Recent Work on Galaxy Photometry in Japan
J.-L. Nieto: Recent Results from High-Resolution Observations of Central Regions of Galaxies
Ch. Peterson: Systematic Errors in the Magnitudes of Zwicky Catalogue Galaxies
H. Lorenz: Image Processing by Means of A Digital Filter Technique
M. Capaccioli: Some Properties of Early-Type Galaxies Inferred from Photometric Mapping

ABSTRACTS OF PAPERS:

S. Okamura: The present status of galaxy photometry in Japan was briefly reviewed. The major programs carried out recently or in progress now at several institutions were referred to, and the results of some recent works were reported. Preliminary results were presented of the surface photometry program by the Tokyo group in which they measured the two-dimensional luminosity distribution of some 260 galaxies of various morphological types in Virgo and Ursa Major regions in the V-band on the basis of the photographic plates taken with the 105-cm Schmidt at Kiso Branch of the Tokyo Observatory. They could support the consortium proposed at the meeting by Capaccioli and de Vaucouleurs by extending this project.

J. L. Nieto: Two sets of observations are presented which were made at the prime focus of the CFH telescope. The resolution of these data is very good, which confirms the high quality of the images obtainable at this site.
1 - NGC 6251 exhibits a very narrow dust lane almost perpendicular to the major axis of the optical image. The dust lane is parallel to the radio jet. This suggests a random orientation of the nuclei of galaxies with respect to their main bodies, represented for instance by the rotation axis of the dust and the gas.
2 - NGC 3379 was observed with resolutions between 0.5 and 3 arc-seconds. Adding all the relevant data published in the literature to these new data allowed the building of the high resolution profile of this galaxy. Some evidences for the existence of two populations as well as a peak of light in the center of this galaxy are presented.

Ch. Peterson: From a comparison of photoelectrically derived total magnitudes of a sample of 119 Sc galaxies with the photographic magnitudes of the Zwicky et al. Catalogue, we find there exists in the photographic data both a scale error and a correction based on the average surface brightness m':

$$B(pe) = A+(1.079\pm0.008)m(pg,Zw)-(0.464\pm 0.003)m'(pg,Zw)$$

These coefficients are essentially the same as found by de Vaucouleurs and Pence (1979) from a study of Virgo region galaxies. The zero point A is essentially the same for the galaxies of the catalogue volumes I-V. Volume VI differs by 0.3 magnitudes. Prior investigators, on the other hand, have suggested that Volume I systematically differs from the remaining Volumes. It is suggested that this apparent conflict is due to a need to correct Zwicky data also for galaxy type.

H. Lorenz: On the basis of the HAAR transformation a processing of astronomical images is described which separates the noise optimally and reduces the redundancy of the data up to 100. The system fits its own bandwidth to the incoming signal and steeres a two-dimensional KALMAM-FILTER: The objects in the field are picked up in the domain of the HAAR-tranform. Astronomically relevant parameters are derived from the expansion of the intensity distribution into potential moments.
A measuring system is planned in cooperation with VEB Carl-Zeiss Jena. This system will include software packages for different applications and it will ensure mechanical accuracy and stability (air bearings, laser metering).

M. Capaccioli: A brief account is given of the work done at Padova-Asiago in the field of galaxy surface photometry, with emphasis to the study of the properties of early-type galaxies.
1 - A photometric and geometrical investigation of ten galaxies in Virgo is presented which, together with previously published studies, stresses the need for further, more accurate work.
2 - From the analysis of the available luminosity profiles for E galaxies we find a systematic behaviour of the residuals with respect to the $r^{1/4}$ law.
3 - We have re-discussed the luminosity law which represents the bulges of SO's, finding it steeper than before thought.

III. INTERNAL MOTIONS IN GALAXIES

A short business session was held at the joint meeting of the Working Group on Internal Motions in Galaxies with the Working Group on Galaxy Photometry and Spectrophotometry. It was agreed that the Working Group on Internal Motions in Galaxies served a very useful purpose in circulating preprint information. In the absence of alternative nominations S. T. Gottesman agreed to serve for another 3 years as chairman of that Working Group. However, he noted that he wished this to be his last term in that position.

A total of 9 papers concerned with internal motions in galaxies were read before the two Working Groups. S. Simkin reported on the alignment of the axes of radio sources with the rotation axes of galaxies. She finds that for high luminosity sources the axes are closely aligned. S. Simkin also read a paper by Pronik reporting variability, over a period of 10 years, in the Hβ flux and forbidden line strengths observed in the galaxies NGC 1275 and NGC 3227. Pronik and Merculova, as reported by Simkin, find that over a shorter period of 4 days the Hβ flux of NGC 1275 may vary by a factor of 2. Terzian reported on the Arecibo program to measure the HI properties of binary and multiple galaxies. A tendency has been found for the total mass to light ratios to be similar within pairs. Brinks showed his film on the kinematics and structure of the atomic hydrogen in M 31. He noted several features of interest including the non-circular motions of the inner spiral structure. U. J. Schwarz reported on a Westerbork study of HI in the peculiar system NGC 3718/3729. Many of the properties can be explained by a warped edge-on ring. J. M. van der Hulst presented new VLA observations of the barred spirals NGC 1097 and NGC 1365. For the latter he finds no gas in the central region, whereas NGC 1097 reveals HI in the bar zone. However, its structure is complicated by a possible interaction with a nearby companion. S.T. Gottesman reported on the VLA observations of the HI in the barred systems NGC 3992 and NGC 4731. The barred central region of NGC 3992 showed little or no HI while the bar of NGC 4731 appears quite gas rich. The final paper was read by G. L. Pilbratt on CO observations of M51 obtained with the Onsala telescope. These data show little contrast between the arm and inter arm regions. However, further observations are in progress.

## IV. SPACE SCHMIDT SURVEYS

This Working Group met on the afternoon of August 25. The session began with a brief business meeting, during which it was decided that the current chairman, K. Henize, and vice-chairman, F. Bertola, would continue in their offices for the next 3 years.

The scientific session was initiated by H.J Smith, who reviewed the instrument concepts and scientific importance of the Space Schmidt program and reviewed the status of the multi-national proposal for a Space Schmidt Telescope (SST) which is coordinated by J. Wray at the University of Texas.

R. Barbon reviewed the extensive studies taking place at Asiago/Padova, in cooperation with Aeritalia Corporation, concerning the design of the main structure of the SST. (See abstract).

H. Heckathorn described progress at the Naval Research Laboratory (NRL) on the development of large-format high-resolution UV electrographic detectors for the SST (See abstract).

In a separate paper, Heckathorn described follow-up studies of very hot, faint stars discovered by the NRL S201 UV telescope carried on the Apollo 16 and Skylab missions.

In the first of three papers on telescopes and/or techniques related to the objectives of the SST, T. Gull reviewed the characteristics and scientific objectives of the UV telescopes to be carried on the NASA OSS-3 payload.

H. Lorenz described a method for producing spectra of very faint objects using a multi-slit field spectrograph with the 6-meter telescope. (See abstract).

K. Kodaira described the 60-cm f/4 UV reflector, VUVT, which is under study for flight on the Japanese Ultraviolet-Telescope Satellite (UVSAT).

## ABSTRACTS OF PAPERS:

R. Barbon: Most of the efforts in Italy have been so far devoted to a feasibility study of the telescope main frame made by AERITALIA, a well known space contractor. Although the study resulted positive toward the fabrication and operation of the instrument, the total cost estimate for the hardware to be built in Italy came out much higher than planned. A new approach, which makes use of graphite-epoxy, looks more promising since this material, with an extremely low coefficient of thermal expansion, leads to a structure with lower weight and passive thermal control only, thus reducing the cost considerably. AERITALIA has been asked to provide an engineering verification of the new design and a cost estimate. In the mean time the Italian groups involved in this project have pointed out the areas of interest for the research programs to be accomplished during the first flights as follows. 1. Spiral structure in nearby galaxies: theories vs. observations. 2. Progenitors of type I SNe. 3. Young stellar population in elliptical galaxies. Morphology of diffuse nebulae.

H. M. Heckathorn and G. R. Carruthers: The Naval Research Laboratory has a program to develop electrographic detectors for use in a variety of space astronomy and other space science applications. Of particular interest to this Working Group is our continuing development of UV-sensitive electrographic detectors having large-diameter, semitransparent photocathodes which are designed for use at the focal surface of a large auxiliary optical system such as the folded all-reflecting Schmidt. Our presentation will describe technical developments which have taken place since the publication by NASA of the "Spacelab Wide Angle Telescope Working Group Report" in 1979. Four specific developments will be discussed in detail.

(1) Successful fabrication of two large diameter (177-mm), thin (1-mm), curved (2000-mm radius), and UV-transmitting (calcium fluoride) faceplates.

(2) Successful design of a permanent magnet focussing assembly of the size and field uniformity required by the Space Schmidt detector.

(3) Successful modification of our prototype large-format (70 % scale) detector to allow for (1) continuous vacuum pumping, with a demountable but sealed tube structure, and (2) the imposition of a thin, electron-permeable barrier membrane (located between the photocathode and the recording emulsion) which is both optically opaque and impervious to the products of film outgassing.

(4) Continued development of medium- and large-format electrographic detectors sensitive in the mid-UV spectral region (1700-3200 Å) which make use of "solar-blind" cesium telluride photocathodes coupled with UV-grade fused silica faceplates.

At this point, we feel that technical feasibility has been demonstrated for all aspects of the far-UV large-format electrographic detector for the Space Schmidt save one: verifying that the thin faceplate can survive the vibration and acoustic environment of a Shuttle launch. We intend to conduct vibration tests on one of the two calcium fluoride faceplates, in an evacuated housing and mounted on a prototype image tube structure as planned for the Space Schmidt. The other faceplate will be used (with a cesium iodide semitransparent photocathode) in a laboratory full-scale prototype detector, when and if funds for the construction of this detector (and procurement of the PFMA) become available.

V. Afanasiev, I. Karachentsev, S. Dodonov, I. Terebish, H. Lorenz: A "multi slit method" is described which improves by means of night sky masking the limiting magnitude. The spectrograph is based on focal reducer optical arrangements. A limiting magnitude of 22.5 mag in B has been achieved for spectra with a reciprocal linear dispersion of 650 Å/mm. The preliminary results indicate a substantional flattening of the log M/log S diagram for quasars at magnitudes of $21^m$ - $22^m$.

## COMMISSION 29: STELLAR SPECTRA (SPECTRES STELLAIRES)

Report of the Business and Scientific Meetings, 18-25 August 1982

PRESIDENT: W. Bonsack SECRETARY: C. Jaschek

### Business Meeting

The triennial business meeting of the Commission was opened by the President, who first gave a report on the activities of the Commission since the last meeting. The principal activities consisted of sponsorship of symposia and colloquia, of working groups on specialized topics, and of the International Register of Stellar Spectroscopists. The Commission participated in six IAU-sponsored meetings in 1979-82, and the Organizing Committee has approved sponsorship of two more in the near future. The IRSS is a list of stellar spectroscopists giving names, addresses, and research specialties, without regard to IAU membership. It is edited by W. Bonsack and A. Slettebak, and is co-sponsored by Commission 45. A current computer file of the contents of the IRSS is maintained at the University of Hawaii. Two editions have been printed and distributed without charge to all persons listed and to 150 institutions; the current file contains 260 names.

The President also reported that the members of the Commission currently number 185, with approximately 15 new members to be added. Four members of the Commission have died since 1979, including the President of the IAU, Prof. Bappu. There have been no resignations.

The future of the IRSS was discussed, and it was voted to continue the sponsorship by the Commission; Bonsack and Slettebak are willing to continue as editors.

Reports on the Working Groups sponsored by the Commission were given by the Chairs of the WG's.

WG on Be stars: The report was presented in written form by M. Jaschek. The WG organized an IAU Symposium on Be Stars (no. 98, Munich, FRG), and publishes a newsletter. A scientific meeting was planned for the General Assembly. The WG is cosponsored by Comm. 45.

WG on Ap stars: The report was presented in oral and written form by C. R. Cowley. The WG participated in organizing the 23rd Liege International Colloquium, and has joined with an informal European group in publishing a newsletter. A paper giving recommendations for abundance determinations in Ap stars is being prepared by Adelman and Cowley. A scientific meeting has been planned for the General Assembly. The WG is cosponsored by Comm. 45.

WG on Standard Stars: This WG was recommended at the 1979 business meeting, and was organized by L. Pasinetti, with the collaboration of a provisional committee. The WG will deal with the problem of identifying "normal" standard stars, and providing standard reference data, for various problems in stellar astronomy. A scientific meeting was organized for the General Assembly, at which the need for standard stars and data in various fields were summarized. The WG is cosponsored by Commissions 30 and 45.

The president asked the Commission if it wished to formally sponsor the WG on

Standard Stars. By a show of hands, the Commission voted to sponsor the WG.

C. Jaschek proposed that the Commission approve the proposal of A. Batten to co-sponsor a Symposium on Calibration in Astronomy. Garrison and Underhill suggested that the topic is too broad, and should be restricted severely. It was agreed to refer the question to the WG on Standard Stars for recommendation (Note: after a further urgent request from Dr. Batten for a decision by the Commission, the O.C. declined to sponsor the meeting, on the grounds stated above).

The election of officers and the Organizing Committee of the Commission was carried out. Bonsack presented the selection of the O.C. for the new officers: President, Dr. Jun Jugaku; Vice President, Dr. Giusa Cayrel-de-Strobel. The Commission confirmed this selection. A slate of 14 candidates for the O.C. was presented, which included six from the previous O.C. who were willing to continue, and eight additional who had been selected to give geographic balance, and who had indicated willingness to serve if elected. The president called for further nominations, but none were offered. Voting was by paper ballot, each member present voting for up to eight candidates for the eight offices; the eight with the most votes were declared elected. Those elected were Y. Andrillat, P. Conti, C. Cowley, C. Jaschek, H. Levato, M. Smith, R. Viotti, and S. Wolff.

The President next reported on Colloquia and Symposia which had been proposed for sponsorship or co-sponsorship. The titles of the meetings with their proposers are "Observational Tests of Stellar Evolution Theory" (A. Maeder); "Origins of Non-radiative Heating/Momentum in Hot Stars" (A. Underhill); "MK Spectral Classification: Criteria and Applications" (R.F. Garrison); "Structure and Evolution of the Magellanic Clouds" (S. v.d. Bergh, K.S. de Boer); and "Late-Type Chemically-Peculiar Stars" (P.C. Keenan); "Luminous Stars, Associations, and Galaxies" (P.S. Conti). The first two of these meetings had already been approved for sponsorship by the OC. Drs. Garrison and Underhill described their proposed meetings. At the request of Dr. Keenan, his proposed meeting was discussed, including the invitation conveyed by Mrs. Harmer to hold it at the Royal Greenwich Observatory. It was recommended that the meeting topic should be broad enough to include all chemically peculiar stars of late type, which would attract more participants than could be accommodated by RGO. C. Jaschek proposed that the meeting be held at Strasbourg. Bonsack agreed to convey these comments to Keenan and Harmer.

## Scientific Meetings

SYSTEMATIC EFFECTS IN ABUNDANCE DETERMINATIONS FOR METAL-POOR STARS (Joint meeting of Commissions 25, 29, 37, 45). The meeting was intended to explore the technical problems which have led to significant discrepancies in the abundances obtained for the more metal-rich globular clusters by different techniques. Invited papers were presented by M.S. Bessell on "Low-resolution and Photometric Abundance Studies"; by R.C. Peterson on "High-resolution Abundance Studies", and by B. Gustafsson on "Model Atmospheres and Spectrum Synthesis". Short papers were contributed by A. Underhill, M. Spite, M. Grenon, R.P. Kraft, and R. Rutten. A summary was given by G. Cayrel. The meeting was organized by a committee chaired by W. Bonsack; the invited papers will be published in the Publications of the Astronomical Society of the Pacific.

SCIENTIFIC MEETING OF THE WORKING GROUP ON Ap STARS (Joint meeting of Commissions 29 and 45). The scientific meeting was orgainzed and chaired by C. Cowley, and consisted of short review papers on Digital Data Processing, Metal-Weak A and B Stars, High-Velocity A and B Stars, CP Stars in Clusters, Turbulence and Diffusion Theory, Microturbulence and Abundances, Magnetic Fields, Abundance Patches, and Rapid Var-iations. The papers emphasized the progress in the field since the Liege Symposium in 1981. In addition to the scientific program, a new Organizing

Committee was nominated and elected by those present. C. Cowley was re-elected chairman.

MEETING OF THE WORKING GROUP ON STANDARD STARS (Joint Meeting of Commissions 29, 30, and 45). The program began with a business session, in order to discuss the program of the WG, and to elect an Organizing Committee. L. Pasinetti chaired the meeting, and gave an introduction listing possible items for the program, including (1) revision of existing lists of standard stars, (2) study of standard stars with modern techniques and criteria, (3) the spectral type of the sun and solar analogues, and (4) distribution of standard stars around the sky. The invited papers reviewed the current situation in some of the above respects: standard velocity stars (A.H. Batten), standard stars for polarization (M. Breger), identifications and equivalent width measurements from photographic plates (C. Morossi), and UV catalogues and standards (K.A. van der Hucht). A paper on the mid-UV spectrum of Procyon was contributed by F. Castelli, et al. The Organizing Committee elected for 1982-85 is A.H. Batten, M. Breger, R. Garrison, M. Gerbaldi, I.N. Glushneva, M. Hack, C. Jaschek, L.E. Pasinetti (chair), F. Spite, and K.A. van der Hucht.

MEETING OF THE WORKING GROUP ON Be STARS (Joint Meeting of Commissions 29 and 45). Proceedings will be reported by Commission 45.

ATOMIC AND MOLECULAR DATA IN STELLAR SPECTRA (Joint Meeting of Commissions 14, 29, and 36). Proceedings will be reported by Commission 14.

## Joint Discussions

Commission 29 participated in two Joint Discussions held at the General Assembly: JD II, "Evolution in Old Stellar Populations in Galaxies," and JD VII, "Mass Loss Phenomena".

## COMMISSION 30: VITESSES RADIALES (RADIAL VELOCITIES)

### Compte Rendu de séances, 21 Août 1982

PRESIDENT: M. Duflot SECRETAIRE: J. Andersen

### SEANCE ADMINISTRATIVE

#### 1. Bureau 1982 - 1985 de la Commission

Les propositions suivantes du Comité d'Organisation sont approuvées à l'unanimité des membres présents :

Président : A.G. Davis Philip
Vice Président : J. Andersen
Membres du Comité d'Organisation : M. Duflot, D.P. Hube, E. Maurice, M. Mayor, J. Sahade

#### 2. Motions de soutien

La Commission a décidé de participer avec la Commission 29 à la création du Working Group sur les étoiles standards proposé par la Commission 45, et de nommer A.H. Batten comme représentant.

D'autre part la Commission a décidé de soutenir la demande de A.H. Batten pour un Symposium sur la calibration en astronomie qui se tiendrait en 1984 à Como en Italie.

Les deux motions suivantes ont donc été approuvées

- Commission 30 cosponsors the Working Group on standard stars created by Commission 45.
- Commission 30 supports the proposal that a Symposium on "Calibration in Astronomy" be held in 1984.

#### 3. Futur Symposium sur les Vitesses Radiales

A.G.D. Philip a proposé un Symposium sur les Vitesses Radiales qui pourrait avoir lieu en automne 1984 (Albany). La Commission donne son accord pour que A.G.D. Philip continue à préparer ce symposium et fasse la demande officielle auprès du Comité Exécutif de l'UAI.

#### 4. Etoiles de vitesse radiale standard

Après un rapport présenté par A.H. Batten, la Commission a approuvé les décisions suivantes :

(a) The star HD 184 467 is now known to be a 2 sp binary (P $\sim$ 480 d ).
- The star HD 35410 is probably a 1 sp binary.
- The star HD 14969 is a binary.

These stars should not be used as standards and deleted from the list of standard stars.

(b) The following stars should be considered as secondary standards, suitable for use only at low dispersion :

| Bright Stars | Faint Stars |
|---|---|
| HD 36673 | HD 3765 |
| 84441 | 9138 |
| 102870 | 26162 |
| 186791 | 29587 |
| 206778 | 51250 |
| | 65583 |
| | 92588 |
| | 103095 |
| | 115521 |
| | 126053 |
| | 144579 |
| | 145001 |
| | 154417 |
| | 187691 |
| | 223311 |

(Some of them may be variable and some may have velocities different from the adopted values).

Rappelons qu'en 1958 (Transactions UAI 10), la sous-commission 30a "recommended that the stars α Per, HD 20902, and α Car, HD 45348, should be dropped from the list of standard velocity stars".

5. Working Group sur les vitesses radiales moyennes

C. Jaschek a souligné l'importance de la demande faite par la communauté astronomique d'un catalogue de vitesses radiales moyennes.

Mme M. Barbier, présidente du WG, a communiqué les conclusions du groupe qui s'est attaché à trouver une solution qui permette d'obtenir les VR moyennes au moyen d'un ordinateur. Cette solution est discutée.

Afin de pouvoir prendre une décision définitive, les deux premières heures du catalogue seront réalisées, le plus rapidement possible, sur les bases proposées, et soumises au Comité d'Organisation de la Commission. Mme M. Barbier demande le maintien du WG actuel (M. Barbier, A.H. Batten, M. Duflot, D. Evans, R. Griffin) et J. Andersen accepte de s'y joindre.

## SEANCE SCIENTIFIQUE

During the scientific session, the following papers were given :

J. Dommanget : Informations et problèmes relatifs à l'observation des vitesses radiales des étoiles doubles visuelles.

The necessity was stressed to obtain radial velocity observations in order to define orientations of the orbital planes and any possible systematic in these.

Ch. Fehrenbach : Le prisme objectif de 60 cm. Résultats actuels.

La mesure des clichés obtenus au prisme objectif à champ normal de 60 cm est faite par enregistrement digitalisé de pointés photométriques et par corrélation. La méthode a été améliorée par la mise en mémoire dans l'ordinateur de 6 spectres étalons. La précision des mesures est de 4 km $s^{-1}$ pour une moyenne de trois clichés et ceci pratiquement quel que soit le type spectral, alors que la dispersion n'est que de 200 A $mm^{-1}$. On obtient sur plaque IIIaJ des étoiles de magnitude 12 avec deux poses de 40 m (méthode de retournement).

W.I. Beavers : Radial velocity program and future plans at the Erwin Fick Observatory.

Exploiting the advantages of a dedicated coudé instrument, experiments are made with special masks for various types of objects. Observations are concentrated on bright stars, including checks of long term stability of standard and other stars.

M. Mayor : Reports on CORAVEL measurements.

A brief report was given on the capabilities of CORAVEL to supply metal abundances and rotational velocities as a by-product, and of the large-scale programmes undertaken, especially those with CORAVEL II in the southern hemisphere (LMC + SMC supergiants and cepheids, ω Cen , 47 tuc , Pop.II stars).

D.W. Latham : The digital stellar speedometer at the F.L. Whipple Observatory.

Recording one order of a cross-dispersed echelle spectrum with a photon counting Reticon detector in one of several similar instruments designed for determination of galaxy redshifts, stellar radial velocities are determined with good precision by software correlation with a library of standard spectra. Limiting magnitude is about $17^{m}.5$ with the MMT.

J. Andersen : Radial Velocities of Bright Stars ; Future Instrument Developments.

Through a collaboration on coudé and CORAVEL observations from ESO, La Silla, only 9 observations now remain to complete all HR stars, and observations are started on some 4000 F stars of known age and metal abundance. The gross inadequacy of the existing set of standard stars to check observations of B, A, and F stars of different rotations was stressed.

Combining existing optical, electronic, and software advances (Mayor,Latham) it should be possible to build instruments allowing either rapid, on-line results on bright stars with a small telescope, or reaching $B = 18^{m} - 19^{m}$ on a 4 m telescope (old SMC/LMC and dwarf galaxy giants, globular clusters, ....).

Ch. Fehrenbach : Les vitesses radiales dans le projet HIPPARCOS.

Radial velocities of the northern early-type HIPPARCOS stars will be determined using the objective prism-technique (see above) in fields of sufficient star density, and coudé observations otherwise.

F. Fekel : Standard Stars Observed at the McDonald Observatory.

Results for some bright A-F stars found to be constant over several years of use as standards in spectroscopic binary programmes were given.

D. Dravins : Astrometric Determination of Absolute Heliocentric Radial Velocities in Moving Clusters.

It is proposed to use HIPPARCOS determinations of proper motions and convergent points for the Hyades and Ursa Major to compute absolute radial velocities of individual stars to about 400 m/s. Comparison with the direct measurements will yield information on other sources of systematic line shifts (convection, relativistic redshift, ....).

# COMMISSION 31: TIME
# (L'HEURE)

Report of Meetings, 19, 20, 23 August 1982

PRESIDENT: S. Iijima SECRETARY: D.D. McCARTHY

## General

Prior to Commission meetings, a short session of the Organizing Committee was held on the afternoon of 17 August 1982 in which items and proposals to be submitted for consideration by the Commission were reviewed.

Twenty-five members of Commission 31, four future members, three consultants, and four invited participants attended Commission 31 meetings.

## 19 August 1982

In a joint meeting of Commissions 19 and 31 the progress of Project MERIT was reviewed. Talks by G. Wilkins I.I. Mueller, and M. Feissel are summarized in the report of IAU Commission 19.

These discussions were followed by presentations by each of the principal coordinators of the project.

K. Yokoyama pointed out the Short Campaign had led to an improvement in the results from optical astrometry and that a further improvement could be expected during the Main Campaign as a result of the use of more accurate star catalogs for the astrolabes and PZTs and better corrections for nutation and tidal effects. He hopes to determine short-period variations from the observations at the better stations.

C. Oesterwinter reported that data from the Doppler tracking of satellites had been analyzed in an attempt to detect plate motions; the values found for the Australian and Pacific plates appeared to be significant and in general agreement with geological studies. The lack of proper information about local changes in the positions of the antennae had hindered the analyses. In reply to a question B. Guinot stated that a new MEDOC II system is being developed; it will include four new French stations and should come into operation during the Main Campaign.

B. Schultz reviewed the current status of the satellite laser ranging stations that supplied data to the analysis centers at Greenbelt and Austin in the USA. For many stations the precision is about $\pm 10$ cm and for one or two it is $\pm 2$-3 cm. There is a very noticeable weekend effect in the frequency of observations; a better balance between day and night observations will also be required if reliable determinations of short-period variations are to be made. More effort must be made to reduce both systematic and random errors. B. Guinot commented that the monthly reports from the University of Texas are very useful.

O. Calame said that lunar laser ranging contribution to MERIT was carried out in the framework of the EROLD campaign, but was limited to the determination of UT since data were only available from the McDonald Observatory. She expressed the hope that LLR data would also be available from Grasse, Wettzell and Orroral during the Main Campaign, and she presented solutions for the annual and semi-annual terms in UT from McDonald data for the period 1971-1981.

W.J. Klepczynski spoke mainly about the connected-element radio interferometer at Green Bank since this was the only CERI system in regular use for the determination of Earth-rotation parameters. Some results had been obtained from Cambridge, England, during the Short Campaign and he hoped that some time would be available on the Very Large Array at Socorro during the Main Campaign. The Green Bank system is being extended to allow the determination of all three parameters, and a water-vapor radiometer is being installed in order to reduce the errors due to the troposphere. In response to a question he said that it might be possible to obtain a solution with a precision of ±0.5 ms at an interval of 12 hours.

D.S. Robertson showed results from VLBI and claimed that UT could be obtained with a precision of the order of ±0.1 ms from 24 hours of observation in spite of bad weather and bad ionospheric conditions. The POLARIS network in the USA requires one more element in order to obtain all three parameters; observations will be made every five days. In response to a question he said that this network when used with another element at Wettzell, Germany, should give results with precisions of ±0.1 ms and ±5 cm from 24 hours of observation.

In the ensuing general discussion G. Wilkins stated that the third and final draft of the MERIT recommendations on standards should be available during the autumn and that recommendations on the monumentation of sites would be prepared since this will be very important in the establishment of the new conventional terrestrial reference frame.

The 47 voting participants then unanimously endorsed the following two resolutions.

The International Astronomical Union
taking note of the success of the MERIT short campaign, and
recognizing that the results to be obtained during the MERIT main campaign will be of short and long-term benefit for the studies of the planet Earth
endorses the following two resolutions adopted by the International Association of Geodesy (IAG), at its General Meeting at Tokyo (May 1982):

l'Union Astronomique Internationale
prenant note du succes de la campagne courte du programme MERIT et
reconnaissant que les résultats à obtenir au cours de la campagne principale de MERIT apporteront des contributions à long et à court termes pour les etudes de la planète Terre
soucrit aux deux résolutions suivantes adoptées par l'Association Internationale de Géodésie (AIG) lors de sa réunion générale de Tokyo (Mai 1982):

AIG Resolution 1

The International Association of Geodesy
noting that the results to be obtained during the MERIT main campaign will be of long-term benefit to geodesy and its applications
strongly endorses the proposals of the COTES and MERIT Working Groups that during the campaign (a) very long baseline radio interferometric and satellite and lunar laser ranging systems be used for co-located observations

of high precision at the recommended sites, and (b) observations be made intensively for a limited period to detect any short-period variations in the derived Earth-rotation parameters, and

urges that the appropriate resources and facilities be made available for these activities by the countries involved.

AIG Resolution 2

The International Association of Geodesy

considers that it is important that the new terrestrial reference frame to be derived from high-precision observations during the MERIT Main Campaign should be extended and related to existing services as accurately and quickly as possible, and

urges that coordinated precise positioning observations be made during the campaign by satellite radio-tracking systems at the Very Long Baseline Interferometric, Lunar and Satellite Laser Ranging sites as well as at a larger number of well distributed sites around the world.

It recommends (a) that the coordinates of the stations of the International Latitude Service be determined in the Reference System (NWL 10F) adopted for the analysis of observations by the Doppler method of artificial satellite of TRANSIT or NOVA type, and (b) that special support be given to BIH acting as the coordinating center for the MERIT international campaign.

In a meeting devoted to regularized UT1: B. Guinot described the zonal tidal effects on the rotation angle of the Earth. Up to ±0.8 ms error may appear in UT1 because of failure to account for this effect. D.D. McCarthy then reported on the need for clarification of the use of UT1*. Comparison of U.S. Naval Observatory UT1 and UT1 derived from BIH UT1* shows that USNO UT1 is in better agreement with UT1 derived from the regularized UT1.

S. Yatskiv showed that regularized UT1 could be used practically in an ephemeris terrestrial system used as an reference system intermediate between the Conventional Inertial System and the Conventional Terrestrial System.

D.S. Robertson reported that he found no improvement when he compared VLBI results with UT1 derived from UT1*.

A discussion then followed which concerned some of the details of a proposed resolution regarding the zonal tidal model and the need for more careful clarification. Action on the proposal was postponed to a later meeting.

ADMINISTRATIVE SESSION

In the first business meeting of Commission 31, S. Iijima presented the President's Report. Members stood in silence to honor the memory of N. Matsunami and A. V. Shiryaev who passed away since the last meeting in Montreal. This was followed by a report delivered by B. Guinot on the Bureau International de l'Heure. He stated that there were no major changes in the formulation of TAI, and that clocks and frequency standards from N. America, Europe and N. Africa continued to be intercompared using Loran-C. The ALGOS algorithm produces EAL from which TAI is formed using primary cesium standards. Steps in frequency of $\pm 0.2\times10^{-13}$ are made with about two steering corrections per year.

An annual effect in the difference between TAI and the primary standards was noted. The amplitude is about 1 microsecond and it is not clear what the cause is although Guinot feels that it is probably due to commercial clocks. Work is now in process on a new version of ALGOS.

The maximum error in TAI for averaging times greater than two months is $\pm 1 \times 10^{-13}$. The BIH expects to cooperate in Project LASSO. Guinot noted that the relationship with the International Bureau of Weights and Measures provides the services of one physicist and remains very satisfactory.

W. Markowitz stated that his report on the meeting of the Consultative Committee for the Definition of the Second (CCDS) held in Paris in September 1980, had already been published and is available in the IAU Information Bulletin of January 1981.

J. Pilkington delivered the report of H.M. Smith on the International Radio Consultative Committee (CCIR) Study Group 7. He mentioned that CCIR reports have not been so widely known, but contain much useful information for investigators in the field. The quality of the work of BIH was also recognized by the CCIR and this was reported to the Commission.

S. Aoki concluded the session by presenting a discussion on the aspects of legal time. This included a compilation of the reports from Commission members regarding existing legislation in their respective countries concerned with time. His report will be published elsewhere, e.g., in the Time and Latitude Bulletins of the Tokyo Astronomical Observatory.

<u>20 August 1982</u>

A joint meeting with Commissions 4,7,8 and 19 was held for the discussion of resolutions related to the adoption of new astronomical coordinate systems, constants and theories in 1984. (See the report of Commission 4.)

In an extra meeting of Commissions 19 and 31 regarding Regularized UT1 which followed, I.I. Mueller briefly reviewed the discussion of the previous day. B. Guinot presented a new resolution regarding the use of a zonal tide model. The following resolution was passed with 21 in favor and 1 abstention:

International Astronomical Union Commissions 19 and 31
<u>considering</u> the need to clarify the role of the zonal tides in the processing and publication of Universal Time
<u>recommend</u> (1) that, when the effect of the short period zonal tides, i.e., with (periods less than 35 days), is computed for the above purposes, the attached expression based on Yoder, Williams and Parkes study (<u>J. Geophys Res</u>., <u>86</u>, 881 1981) be used, and (2) that the letter R be added to the notation of the relevant quantities to indicate that the correction for the short period zonal terms has been made (for instance UT1R).

l'Union Astronomique Internationale Commissions 19 et 31
<u>Considerant</u> le besoin de représenter clairement le rôle des marées zonales dans le calcul et la publication de temps universel,
<u>recommandent</u> que lorsque l'effet des marées zonales à court terme (période inferieure à 35 jours) est calculé dans les buts précités, on utilise l'expression ci-jointe, tirée des travaux de Yoder, Williams et Parke (<u>J. Geophys Res</u>., <u>86</u>, 881 (1981), et que le suffixe R soit ajouté à la notation des quantities concernées pour signifier que la correction des effets à court terme a été effectuée (exemple: UT1R).

TERMS DUE TO ZONAL TIDES, WITH PERIODS UP TO 35 DAYS

Based on Yoder, Williams and Parke: 1981, J. Geophys Res., Vol 86, 881 with $\kappa/C = 0.94$

UT1R, $\Delta$R, $\omega$R represents the regularized forms of UT1, of the duration of the day $\Delta$, and of the angular velocity of the Earth, $\omega$.

The units are $10^{-4}$ s for UT, $10^{-5}$ s for $\Delta$, and $10^{-14}$ rad/s for $\omega$.

| | ARGUMENT | | | | | PERIOD | UT1-UT1R | $\Delta - \Delta$R | $\omega - \omega$R |
|---|---|---|---|---|---|---|---|---|---|
| N | l | l' | F | D | $\Omega$ | Days | Coefficient of Sin (Argument) | Coefficient of Cos (Argument) | |
| 1 | 1 | 0 | 2 | 2 | 2 | 5.64 | -0.02 | 0.3 | - 0.2 |
| 2 | 2 | 0 | 2 | 0 | 1 | 6.85 | -0.04 | 0.4 | - 0.3 |
| 3 | 2 | 0 | 2 | 0 | 2 | 6.86 | -0.10 | 0.9 | - 0.8 |
| 4 | 0 | 0 | 2 | 2 | 1 | 7.09 | -0.05 | 0.4 | - 0.4 |
| 5 | 0 | 0 | 2 | 2 | 2 | 7.10 | -0.12 | 1.1 | - 0.9 |
| 6 | 1 | 0 | 2 | 0 | 0 | 9.11 | -0.04 | 0.3 | - 0.2 |
| 7 | 1 | 0 | 2 | 0 | 1 | 9.12 | -0.41 | 2.8 | - 2.4 |
| 8 | 1 | 0 | 2 | 0 | 2 | 9.13 | -0.99 | 6.8 | - 5.8 |
| 9 | 3 | 0 | 0 | 0 | 0 | 9.18 | -0.02 | 0.1 | - 0.1 |
| 10 | -1 | 0 | 2 | 2 | 1 | 9.54 | -0.08 | 0.5 | - 0.5 |
| 11 | -1 | 0 | 2 | 2 | 2 | 9.56 | -0.20 | 1.3 | - 1.1 |
| 12 | 1 | 0 | 0 | 2 | 0 | 9.61 | -0.08 | 0.5 | - 0.4 |
| 13 | 2 | 0 | 2 | -2 | 2 | 12.81 | 0.02 | -0.1 | 0.1 |
| 14 | 0 | 1 | 2 | 0 | 2 | 13.17 | 0.03 | -0.1 | 0.1 |
| 15 | 0 | 0 | 2 | 0 | 0 | 13.61 | -0.30 | 1.4 | -1.2 |
| 16 | 0 | 0 | 2 | 0 | 1 | 13.63 | -3.21 | 14.8 | -12.5 |
| 17 | 0 | 0 | 2 | 0 | 2 | 13.66 | -7.76 | 35.7 | -30.1 |
| 18 | 2 | 0 | 0 | 0 | -1 | 13.75 | 0.02 | -0.1 | 0.1 |
| 19 | 2 | 0 | 0 | 0 | 0 | 13.78 | -0.34 | 1.5 | -1.3 |
| 20 | 2 | 0 | 0 | 0 | 1 | 13.81 | 0.02 | -0.1 | 0.1 |
| 21 | 0 | -1 | 2 | 0 | 2 | 14.19 | -0.02 | 0.1 | -0.1 |
| 22 | 0 | 0 | 0 | 2 | -1 | 14.73 | 0.05 | -0.2 | 0.2 |
| 23 | 0 | 0 | 0 | 2 | 0 | 14.77 | -0.73 | 3.1 | -2.0 |
| 24 | 0 | 0 | 0 | 2 | 1 | 14.80 | -0.05 | 0.2 | -0.2 |
| 25 | 0 | -1 | 0 | 2 | 0 | 15.39 | -0.05 | 0.2 | -0.2 |
| 26 | 1 | 0 | 2 | -2 | 1 | 23.86 | 0.05 | -0.1 | 0.1 |
| 27 | 1 | 0 | 2 | -2 | 2 | 23.94 | 0.10 | -0.3 | 0.2 |
| 28 | 1 | 1 | 0 | 0 | 0 | 25.62 | 0.04 | -0.1 | 0.1 |
| 29 | -1 | 0 | 2 | 0 | 0 | 26.88 | 0.05 | -0.1 | 0.1 |
| 30 | -1 | 0 | 2 | 0 | 1 | 26.98 | 0.18 | -0.4 | 0.3 |
| 31 | -1 | 0 | 2 | 0 | 2 | 27.09 | 0.44 | -1.0 | 0.9 |
| 32 | 1 | 0 | 0 | 0 | -1 | 27.44 | 0.53 | -1.2 | 1.0 |
| 33 | 1 | 0 | 0 | 0 | 0 | 27.56 | -8.26 | 18.8 | -15.9 |
| 34 | 1 | 0 | 0 | 0 | 1 | 27.67 | 0.54 | -1.2 | 1.0 |
| 35 | 0 | 0 | 0 | 1 | 0 | 29.53 | 0.05 | -0.1 | 0.1 |
| 36 | 1 | -1 | 0 | 0 | 0 | 29.80 | -0.06 | 0.1 | -0.1 |
| 37 | -1 | 0 | 0 | 2 | -1 | 31.66 | 0.12 | -0.2 | 0.2 |
| 38 | -1 | 0 | 0 | 2 | 0 | 31.81 | -1.82 | 3.6 | -3.0 |
| 39 | -1 | 0 | 0 | 2 | 1 | 31.96 | 0.13 | -0.3 | 0.2 |
| 40 | 1 | 0 | -2 | 2 | -1 | 32.61 | 0.02 | -0.0 | 0.0 |
| 41 | -1 | -1 | 0 | 2 | 0 | 34.85 | -0.09 | 0.2 | -0.1 |

l = 134°.96 + 13°.064993(MJD-51544.5) Mean Anomaly of the Moon
l' = 357°.53 + 0°.985600(MJD-51544.5) Mean Anomaly of the Sun

$F = 93\overset{\circ}{.}27 + 13\overset{\circ}{.}229350(MJD-51544.5)$   $L - \Omega$; L: Mean Longitude of the Moon
$D = 297\overset{\circ}{.}85 + 12\overset{\circ}{.}190749(MJD-51544.5)$   Mean Elongation of the Moon from the Sun
$\Omega = 125\overset{\circ}{.}04 - 0\overset{\circ}{.}052954(MJD-51544.5)$   Mean Longitude of the Ascending Node of the Moon

A short discussion followed concerning a possible sign change in the pole coordinates to reflect the adoption of a positive East longitude convention. No action was taken.

Joint meetings were also later held with Commissions 4 and 19 which were concerned with astronomical constants and ephemerides. (See report of Commission 4.)

23 August 1982

In a scientific session H.F. Fliegel reported on Time Transfer using the Global Positioning System (GPS). This system of navigational satellites is designated by the United States Department of Defense as the successor to the TRANSIT satellite system. GPS was originally proposed as a constellation of 24 satellites, which would have provided a continuous worldwide navigational capability. It is now to be implemented as an 18-satellite system, of which 4 satellites will almost always be more than $5^\circ$ above the horizon. Four of the proposed 18 satellites are now operating, and are designated Navstars 3, 4, 5 and 6. They transmit two codes: a P (precision) code, which is intended to be available in the final system only by authorized users; and a C/A (clear access) code, which is, and which is expected to remain, available to all users.

Fliegel reviewed the time transfer experiments made since the last IAU General Assembly using the C/A code of the GPS system, especially between the U.S. National Bureau of Standards and the U.S. Naval Observatory.

B.E.H. Serene presented a paper entitled "The LASSO Experiment Readiness Prior to SIRIO-2 Launch". The launch is scheduled for 10 September 1982. In the paper, Serene described the laser pulse system. It is expected that $\pm 1$ ns accuracy can be achieved in time transfers among Austria, France, East and West Germany, Italy, Netherlands, Spain, USA, and ESA. The first activity will be operated from CERGA in order to use the laser ranging facilities there. This will consist of getting the satellite into position over longitude $25^\circ$ W, allowing the crystal to warm up and conducting experiments with signal detections and threshold sensitivity.

S. Leschiutta then reported on LASSO follow-on activities. Scientific needs which will be addressed include navigation, precise time transfer, VLBI, and the metrological system.

The error sources in LASSO are expected to be due to the width and shape of the pulse, the stability of the amplitude, and optical and electrical delays.

Other effects on the the satellite orbit such as attitude control, and possible weather dependence may be a source of concern. All of this activity will require coordination of synchronous activity of many groups. The advantages include the facts that there is no ionosphere effect and no problems with military usage. Accuracy of the time transfer is hoped to be about $\pm 1$ ns. Further activity may be concerned with the development of microwave and optical systems. Leschiutta then went on to describe the planning of a second generation SIRIO, the POP satellite (POPSAT) for positioning, polar motion, Earth rotation and the Precise Range and Range Rate Satellite (PRARE).

S. Starker presented a paper entitled "The Use of Spread Spectrum Techniques for Time Transfer". He pointed out that pulse techniques require a strong pulse and a high signal-to-noise ratio, while spread spectrum techniques require relatively low power and low signal-to-noise. He described an experiment which used the OTS-2 satellite which, despite a problem with multipath, gave clock difference accuracy of $\pm 20$ ns.

J.A. Galt reported on Satellite Time Comparisons by CW Techniques. He discussed the use of the Symphonie Satellite in which precisions of $1 \times 10^{-14}$ were obtained routinely between Canada and PTB. This method is no longer used since satellite ground stations are not available. Galt also mentioned the ANIK A1 satellite which is experimental but has achieved $\pm 1$ ns precisions over very small distance.

C.O. Alley presented a paper entitled "Sub-ns Time Comparisons Using 100 ps Laser Pulses" in which he described a joint U.S. Naval Observatory - University of Maryland experiment in time transfer. A standard deviation of 100 comparisons of $\pm 200$-400 ps was achieved and the accuracy is estimated to be 1-2 ns over a 26-km separation. Alley also described the Shuttle Time and Frequency Transfer (STIFT) experiment which has a goal of $\pm 1$ ns in time and $\pm 1 \times 10^{-15}$ in frequency transfer.

A paper entitled "Time Transfer by VLBI" was presented by J. Fanselow in which he listed the delays encountered in VLBI time transfer. These include geometric, media (ionospheric, tropospheric,....) antenna, instrument, as well as the clock difference. The Deep Space Network of the Jet Propulsion Laboratory has a water vapor radiometer capable of $\pm 3$ ns. Within two years it is expected that this will be reduced to a precision better than $\pm 1$ ns and that time transfer accurate to hundreds of picoseconds may be possible.

Ya. S. Yatskiv presented a paper entitled "Pulsar Time Scale" by V.G. Il'in, Ya. P. Ilyasov, A.D. Kuzmin, S.B. Pushkin, G.N. Palij, T.Y. Shabanova, Yu. P. Shitov. The authors propose to use pulsars for possible time transfer at an accuracy level of about 200 microseconds. The pulsars are PSR 0834 + 06 with a period of $1^s.73$ and a stability of $6.8 \times 10^{-15}$; PSR 0950 + 08 with a period of $0^s.25$; and PSR 1919 + 21 with a period of $1^s.33$. Discussion of the paper pointed out the existence of random noise in the rates of pulsar timings and the requirement for large instruments and possible secular rates in pulsar timing.

S.-h. Ye presented a paper entitled "Time Comparison Equipment via Laser Pulses" by F.-m. Yang, Q.-x. Zhuang, J.-y. Su, D.-t. Tan, Z.-y. Li, and J.-x. Cai. She reported on time transfer experiments over a 25.2 km distance which achieved an external precision of $\pm 1.5$ ns. This was done using Rubidium clocks and it is expected that results would be improved with improved clocks. She feels that an internal precision of $\pm 0.5$ ns is possible and is eager to participate in experiments with other laboratories.

In a joint session with Commission 4 the following resolution on UT1 as the argument of almanacs was discussed and passed with 27 in favor and one abstention:

International Astronomical Union Commissions 4 and 31

Noting that the present method of keeping UTC within $0^s.9$ of UT1 by means of leap seconds both provides the second of SI and meets the needs for safe celestial navigation,

Recommend that the Air and Nautical Almanacs continue to be published with UT1 as the argument.

Explanatory Note This resolution concerns only the continued use of UT1 as the argument of navigational almanacs. It does not deal with the use of less precisely defined terms, such as GMT and UT, which are used as tabular headings in the sense of UT1. Recommendations concerning their use were made in Resolution Number 1, adopted by Commissions 4 and 31 in 1976 at Grenoble.

l'Union Astronomique Internationale Commissions 4 et 31

notant que la methode actuelle qui maintient le UTC a moins de 0.9s du UT1 au moyen de secondes intercalaires procure la seconde du S1 et remplit les besoins pour une navigation astronomique sure,

recommandent que les éphémérides aeronautiques et nautiques, continuent à être publiées en utilisant UT1 comme argument.

Note d'explication Cette résolution ne se rapporte qu'á la poursuite de l'usage de UT1 comme argument des éphémérides de navigation. Elle ne traite pas de l'usage de termes moins précisement definis, tels que GMT, UT qui sont utilisés dans les tables avec les sens de UT1. Des recommandations sour l'emploi de ces termes apparaissent dans la Résolution no 1, adoptée par les commissions 4 et 31, en 1976, à Grenoble.

A resolution recognizing the CCDS Declaration on TAI was discussed and tabled.

Following those items S. Aoki presented a paper entitled "Operational Procedure of TAI". He discussed precise definitions of the geocentric frame and the geoid which are key concepts in the relationships between TAI, TDT and TDB. A proper reference frame of a non-rotating massless particle, whose trajectory coincides with the geocenter, is introduced as a representative geocentric frame. By this frame, TDT is defined through a theory of dynamics from TDB without considering any geophysical effect such as the geopotential, the Earth rotation, and the polar motion.

On the other hand, an atomic time can be connected to TDT by taking into account all geophysical effects within this frame. From our present knowledge of geophysical phenomena, it is possible to identify TAI with TDT within the accuracy of the order of $10^{-16}$. He noted, however, that TAI is in principle a "coordinated" time scale defined by certain operational procedure which may be changed according to our knowledge about the Earth.

W. Markowitz then described the relationship between Ephemeris Time, TDT and the present extrapolation of Ephemeris Time obtained by using TAI. He reported that it is not yet possible to detect a linear drift in the number of periods of Cesium-133 radiation in one Ephemeris Time second.

A paper prepared by the Hydrographic Department of Japan showing lunar occultation observations of ET-TAI was presented by A. Yamazaki. On 1 January 1981, the Hydrographic Department of Japan (JHD) took over the services of the international center for the receipt and processing of timings of Lunar occultations from H.M. Nautical Almanac Office.

Since then JHD:

1. set up "International Lunar Occultation Center" in the Astronomical Division;

2. accepted 17,866 timing data in total (11,362 in 1981 and 6,504 in 1982) as of July 31, 1982;

3. checked and corrected these data and converted some of them into machine-readable forms;

4. computed the predictions of total and grazing occultations and sent to some institutions and groups;

5. edited a pamphlet "Guide to Lunar Occultation Observations" and distributed to all the observers.

The processing of the data is now in progress and results will be obtained soon.

ADMINISTRATIVE SESSION

The officers and Commission representatives were selected as follows:

President: G. Hemmleb

Vice President: D.D. McCarthy

Organizing Committee: N.S. Blinov, C.C. Costain, H.F. Fliegel, B. Guinot, S. Iijima, Y. Miao, A. Orte, P. Paquet, E. Proverbio, J. Pilkington

Representative to CCDS: W. Markowitz (J. Benavente, alternate)

Representative to CCIR Study Group 7: H.M. Smith (J. Pilkington, alternate)

BIH Directing Board: P. Paquet, S. Iijima

New Commission Members: D.W. Allan, S. Aoki, K. Dorenwendt, M.-k. Fujimoto, C. Kakuta, A.M. Lozinskij, D. Matsakis, D.S. Robertson, J.McA. Steele.

Consultants: C.O. Alley, J.A. Barnes, P. Giacomo, R. Kaarls, Y. Koga, R. Lake, S. Leschiutta, B.S. Mathur, P. Mourilhe Silva, J. Rutman, Y. Saburi, S. Starker, J. Terrien.

The commission expressed its thanks to H.M. Smith for his past service on the BIH Directing Board.

# COMMISSION 33: STRUCTURE AND DYNAMICS OF THE GALACTIC SYSTEM (STRUCTURE ET DYNAMIQUE DU SYSTEME GALACTIQUE)

Report of Meetings on 18, 20, 21, 23, 24 and 25 August 1982

ACTING PRESIDENT: R. Wielen SECRETARY: W.B. Burton

## 18 August 1982

### Business Meeting of Commission 33

Due to illness, the President of the Commission, G.G. Kuzmin, was unable to attend the General Assembly. Therefore, the Vice-President, R. Wielen, was appointed as Acting President.

The Commission unanimously approved the election of the new President, R. Wielen, and Vice-President, W.B. Burton. The Commission unanimously accepted the list of names proposed for the new Organizing Committee (continuing until 1985: W. Iwanowska, E.K. Kharadze, G.G. Kuzmin as outgoing president, D. Lynden-Bell; for the period 1982-1988: L. Blitz, G. Lyngå, M. Mayor, M. Miyamoto). The list of IAU members who wish to join Commission 33 was read and adopted: G. Asteriadis, L. Balazs, B. Baud, L. Blitz, H.V. Cane, J.Y. Caswell, N.F. Comins, J.K. Denoyelle, G.F. Gilmore, K. Hamajima, S. Inagaki, M. Iye, R.N. Manchester, T. Papayannopoulos, B.J. Robinson, H. Schwerdtfeger, J.A. Sellwood, A. Strobel, C. Terzides, K.O. Thielheim, W. Tobin, S.S.O. Wramdemark, Y. Yoshii.

Recent and future meetings were discussed. W.B. Burton reported on the IAU Symposium No. 100 "Internal Kinematics and Dynamics of Galaxies" in Besancon and on the Leiden Southern Workshop; L. Blitz summarized the Vancouver Workshop on "The Milky Way. Structure, Kinematics and Dynamics" (Editor of the proceedings: W.L.H. Shuter; to be published in the Reidel Astrophysics and Space Science Library). R. Wielen reported that Commission 33 is sponsoring an IAU Symposium on "The Milky Way as a Galaxy" in Groningen, June 1983, and is co-sponsoring an IAU Colloquium on "Nearby Stars and the Stellar Luminosity Function" in Middletown, June 1983. Commission 33 also supported a proposal for a Colloquium on "Statistical Methods in Astronomy" in Strasbourg. R. Wielen described the program for the scientific sessions of Commission 33 at the Patras General Assembly.

After a lengthy discussion, Commission 33 decided to establish a Working Group on Galactic Constants. The purpose of the working group is (1) to consider whether any changes in the system of galactic constants $R_0$, A, B, $\Theta_0$ should be recommended and (2) to produce a critical review about the available data on these galactic constants and on other quantities (e.g. the solar motion). F. Kerr will be the chairman of the working group, D. Lynden-Bell the vice-chairman.

### Observing Program for the Astrometry Satellite HIPPARCOS

Chairman: R. Wielen. The following papers were presented:

R. Wielen: Introduction: Possible contributions of IAU Commission 33 to the HIPPARCOS observing program
C.A. Murray: Scientific organisation of the HIPPARCOS project

C. Turon: Compilation of the Input Catalogue for HIPPARCOS
A.R. Upgren: Trigonometric parallaxes from HIPPARCOS
R. Wielen: Proper motions from HIPPARCOS
J.P. Garrigue, J. Guibert, M.O. Mennessier, N.Q. Rieu: HIPPARCOS and the variable stars
E. Høg: TYCHO, an astrometric and photometric survey

20 August 1982

Joint Meeting of Commissions 33 and 28. Chairman: D. Lynden-Bell. The following papers were presented:

J.P. Ostriker: Review of dark haloes
H. van Woerden: Masses of galaxies from 21 cm curves and binaries
A.J. Kalnajs: Interpretation of velocity curves
J.E. Baldwin: Velocity curve of M31
A. Pedlar: Mass-to-light ratios in small groups of galaxies
J. Hunter: Velocity curve for a galaxy with satellites
E.E. Salpeter: Evidence on dark haloes and M/L in clusters of galaxies
D.N. Schramm: Cosmological constraints on the constitution of dark haloes
R.B. Tully: Interpretation of Virgo Cluster dynamics

21 August 1982

Selection of Objects for Observations by HIPPARCOS and Space Telescope for the Establishment of an Extragalactic Reference Frame

Joint Meeting of Commissions 24, 7, 8, 33, 40 and 44. Chairman: A.N. Argue. See report of Commission 24.

23 August 1982

The Large-Scale Distribution of Gas in the Galaxy

Joint Meeting of Commissions 33 and 34. Chairman: W.B. Burton. The following papers were presented:

L. Blitz: Mass and distribution of hydrogen, especially in the outer parts of the Galaxy
I. Kazès: Distribution and physical state of HI seen in absorption
D.B. Sanders: Distribution of CO in our Galaxy and in nearby spirals
B. Robinson: Distribution of CO from southern-hemisphere surveys
R.J. Cohen: Distribution of molecules other than CO and $H_2O$
W.L.H. Shuter: The all-plane variation of sin$\ell$ with v
B.E. Turner: Mapping of the 1720 MHz OH line
H. Bloemen: $\gamma$-rays and the galactic gas distribution
C. Heiles: Supershells as a characteristic of the galactic gas morphology
R. Giovanelli: High-velocity clouds
I. Mirabel: High-velocity clouds near the direction of the galactic center
A. Winnberg: The total mass in the inner Galaxy
I. Kazès: HI absorption lines from molecular clouds

## 24 August 1982

## Recent Results on Galactic Radio Sources and mm Waves

Joint Meeting of Commissions 40, 33 and 34. Chairman: G. Swarup. See report of Commission 40.

## 25 August 1982

## Dynamics

Chairman: R. Wielen. The following papers were presented:

G. Contopoulos: Ordered and ergodic motions in galaxies
E. Athanassoula: Periodic orbits and ergodicity in barred galaxy models
S. Ninković: Contribution to the study of the structure and dynamics of the galactic halo

A full account of the Joint Discussion II on the "Evolution of old stellar populations", co-sponsored by Commission 33, is given in the "Highlights of Astronomy".

## COMMISSION 34: INTERSTELLAR MATTER (MATIERE INTERSTELLAIRE)

Report of Meetings, 20 - 25 August 1982

PRESIDENT: V. Radhakrishnan
VICE-PRESIDENT: M. Peimbert SECRETARY: D.C.V. Mallik

### Business session, 20 August

The President and members honoured the memory of K. Serkowski, a member of the Commission, who died on October 7, 1982, and of B.M. Tinsley who died on 23 March 1981.

The President then gave a short report on the Commission's activities during the triennium. The Commission 34 Report for 1982 had been prepared in consultation with the members of the Organising Committee. The President thanked all the contributors for their help, and mentioned that the Report had been commented upon very favourably by the General Secretary of the Union.

#### Membership

According to the list passed on to the President by the IAU Secretariat the membership of the Commission stood at 478, registering a healthy growth since the last General Assembly. Four more names were added during the Business Session bringing the total to 482. The President noted that although this made the Commission one of the largest in the Union the number was not unmanageable in any sense and no measure was needed to curb it.

#### Organising Committee

Next, the slate of Officers and Organising Committee for the next triennium was approved with one change necessitated by B.D. Donn's preference to step down from the O.C. Following a proposal by M. Peimbert, his place was taken by J.S. Mathis. The Committee as proposed for 1982-85 now consists of M. Peimbert President; J. Lequeux, Vice-President; S. D'Odorico; B.G. Elmegreen; E.B. Kostyakova; J.S. Mathis; U. Mebold; V. Radhakrishnan; P.A. Shaver; Y. Terzian, P.G. Wannier.

#### Change of Name of Commission

The President proposed that the name of Commission 34 should be changed to simply "Interstellar Matter" from "Interstellar Matter and Planetary Nebulae" since planetary nebulae are only one of various types of objects the Commission deals with and as such there was no need for their mention in the Commision title. In the discussion that followed, L.H. Aller, perhaps the oldest member of the Commission present in the Business Meeting, noted that the proposed new name of the Commission was really not new since the Commission was called by this name in the early forties when he became a member. Peimbert noted that planetary nebulae were different in the sense that their study includes the study of the central stars also. He felt that a Working Group on Planetary Nebulae may be formed to coordinate the work in this special area. By agreement the decision on the formation of the WG was postponed to the next IAU General Assembly. The

President allayed the fears of some of the members by stating that the change in the name did not imply any change in the activities of the Commission, and that planetary nebulae would continue to be one of the important areas of interest of the Commission. The members agreed with the proposed change of name.

Symposia/Colloquia

The Vice-President then listed the following meetings under consideration which may be of interest to members of the Commission:

1. "Milky Way as a Galaxy", Groningen, The Netherlands, May-June 1983.
2. "Magellanic Clouds", Tubingen, FRG, 1983.
3. "Nearby Interstellar Medium", U.S.A., 1983-84. The venue is yet to be decided.
4. "Star Forming Regions", Tokyo, Japan, 1985.
5. "Planetary Nebulae", Mexico, D.F., Mexico or Santa Cruz, U.S.A., 1987.

Other Matters

The President then called on Helene Dickel, coordinator of the Working Group on Nomenclature, to give a report of their activity. Dickel mentioned the efforts of the WG towards evolving a standardised nomenclature in areas of interest to Commission 34, and reported in particular on the publication of two compendiums: (1) The First Dictionary of Nomenclature of Celestial Objects by A. Fernandez, M.C. Lortet and F. Spite, and (2) A Compilation of Current Designation Practices in Astronomy by J. Mead. In view of the good work done by the Working Group it was decided that the Group should continue. In response to requests for a Working Group on Calibration of HI Profiles, U. Mebold, a member of the OC, was named as the coordinator to form such a group.

Scientific Sessions

The President spoke briefly on the planning of the scientific sessions during the current General Assembly. The programme, he said, was decided after extensive correspondence with members of the OC and Presidents of other Commissions. Commission 34 is involved in Joint Discussion II on Evolution of Old Stellar Populations in Galaxies and five Joint Meetings with various commissions. In addition, a session of contributed papers on Important Recent Results and a session on Planetary Nebulae have also been organised. The members expressed satisfaction at the general planning of these sessions. However, some members felt that a single session on contributed papers was highly inadequate and at least two such sessions should be planned in future. It was suggested that more papers could be accommodated by having poster sessions in addition to oral presentations, although opinion was divided on the suitability of such a session during an IAU General Assembly.

Brief reports on the scientific sessions are given below.

## JOINT COMMISSION MEETINGS

### The Violent Interstellar Medium

Commissions 34, 40, and 48

Chairman: R. McCray 20 August 1982

Following the brief introductory remarks by the Chairman, de Boer reviewed the data on the high velocity gas in the galactic disk and halo. Giving a brief historical perspective on the pioneering work in the optical region by Adams, Spitzer, Munch and others, he went on to describe in some detail the wealth of information obtained in UV by Copernicus and IUE. This included the observations of high velocity motions in galactic H II regions. Although there is reasonable coverage of the velocity data in many parts of the sky, a consistent theoretical model describing the motions of the gas both in the disk and the halo is yet to emerge. de Boer mentioned in this connection the theoretical work of Shapiro and Field and of Bregman on the Galactic Fountain. Very recently, observations by de Boer and Savage of the interstellar lines in the spectrum of Barnard 29 in M13 have suggested that the halo gas does not corotate with the disk as was believed earlier. The data suggest a decrease in the rotational velocity with increasing distance from the plane.

The high velocity motions and complex structure of the Carina region were discussed by Walborn who showed spectacular [S II] interference filter photographs of the region. The data on both the optical and UV interstellar lines of the region indicate high velocities in the range -200 $kms^{-1}$ to +150 $kms^{-1}$. The Carina region is also a source of soft X-rays as observed by the Einstein Observatory. According to Walborn, the sources of the kinetic energy in the region are derived from the energetic wind of the Of and WR stars found in the region. The question as to whether the region contains an unrecognised SNR has not yet been settled.

Ever since the discovery of $H_2O$ maser sources in Orion and W49 more than a decade ago, observations in infrared and millimeter waves have clearly shown the existence of high velocity mass outflows in molecular clouds. More recent indications have come from the data on high rotational lines of CO seen in the submillimeter regions of the spectrum and from quadrupole lines in the vibration-rotation spectrum of $H_2$. Reviewing the violent activity in dense molecular clouds, Genzel emphasized that the high excitation observed in molecular lines in these regions indicates the presence of hydromagnetic shocks. Further, a bipolar structure of the molecular outflows is seen to be an extremely common occurrence in these clouds as also in compact IR sources, very compact HII regions and in the neighbourhood of T Tauri stars and Herbig-Haro objects. Genzel described in detail the kinematic picture of the Kleinmann-Low region in Orion that has emerged from these observations. It seems unlikely that the outflows are radiatively driven in this region since the momentum in the flow is one to two orders of magnitude higher than the momentum in the radiation field. However, the radiative energy available is more than an order of magnitude higher than the kinetic energy in the outflows and a satisfactory mechanism is yet to be invoked for an efficient conversion of this energy.

The observational status of the soft X-ray background was discussed in the talks by Clark and Kraushaar. Clark reviewed the SAS 3 results and stressed that the observed C-band intensity can be fitted by contributions from two components -a local one from a hot bubble in which we are immersed, and a remote or extended one due to the widely distributed galactic corona. The extended component is absorbed in transmission as is evident from an anticorrelation of the C-band intensity with the column density of the neutral gas. The effective absorption cross-section of the gas is, however, less than the photoelectric absorption, and this indicates a high degree of 'clumping' in the gas. Kraushaar's review included the higher energy M-band data. The existence of emission lines in this band has conclusively proven the thermal nature of the X-rays. Although the C-band and B-band intensities anticorrelate with neutral hydrogen column density, the M-band intensity, according to Kraushaar, does not show such an anticorrelation. Moreover, the origin of the M-band intensity is unknown since it requires substantial amounts of gas at $10^{6.4}$ K rather than at $10^{5.7}$ K as predicted by the McKee-Ostriker theory of a supernova-explosion-dominated ISM.

Kraushaar conjectured that a part of the intensity might be extragalactic in origin, while the rest could be contributed by local M dwarfs, although a fit to the data would then imply a local density of M dwarfs much higher than the values quoted in stellar surveys.

Cash reviewed the observations of X-ray supershells in the Cygnus, Orion/Eridanus and Carina inner regions. These shells are invariably associated with Giant Molecular Clouds and OB associations with their outer boundaries usually delineated by optical Hα filaments. The absence of these shells in the Gum nebula and Carina outer regions impose stringent requirements on the thermal properties of the X-ray emitting gas. The optical supershells were reviewed by Meaburn who also showed beautiful Hα photographs of regions in the LMC and SMC containingthese structures. Meaburn emphasised that these are super-supershells in the sense that their physical dimensions are $\sim$ a kiloparsec, much larger than the X-ray supershells that Cash described. As a possible mechanism for producing these, Meaburn referred to the theory of stochastic self-propagating star-formation given by Gerola and Seiden. He concluded his review with a brief summary of recent speckle interferometric observations of R136 at the centre of 30 Doradus.

The concluding review of the session was given by Ostriker on the three-phase model of the interstellar medium. He spoke briefly on the original version of the model by McKee and himself and described recent improvements of it due to Cowie, McKee and himself. He discussed the theoretical and observational virtues of the model and also its problems.

The papers presented at the above session are expected to be published in the Highlights of Astronomy Vol. VI.

## Spectroscopic Investigation of Grains

Commissions 15 and 34

Chairman: B.D. Donn 21 August 1982

The session opened with a review by Knacke on interstellar silicates. The existence of silicates in both interstellar and circumstellar dust seems to be well established through the identification of infrared bands at 9.8 μm and 18 μm. The silicates are present in both diffuse and molecular clouds and their composition appears to be quite uniform since good fits to the observed spectra can be achieved by varying only the temperature and column density of dust in these sources. Polarization measurements show the silicate particles to be aligned and give quantitative limits on the band strengths. Knacke described the laboratory work on Brownlee particles whose spectra show remarkable similarities with the interstellar spectra, and condensation experiments that suggest silicates with spectra resembling the interstellar can condense from the gas. According to him, modelling shows that the visual and UV extinction and polarization data can be well fitted by silicates and graphite-like particles.

Kratschmer reported on the laboratory measurements on ice and ammoniated ice mixtures in the 4-18 μm range. Application of the data to the observed interstellar extinction characteristics suggests that for the BN object the 3.1 μm feature together with its long wavelength wing may be described by spherical particles composed of 80% $H_2O$ and 20% $NH_3$ with radii of 0.3-0.4 μm. Referring to the absence of the 12 μm absorption feature of ice in some IR sources, he said that this might be due to a deviation of the IR continuum beyond 10 μm from a true blackbody, since the laboratory work did not indicate appreciable decrease in the strength of this feature due to the presence of admixtures.

Donn described the experimental research on grains carried out at Goddard, University of Alabama and York University. The program at Goddard is primarily a study of the condensation of refractory particles from the vapour and the determination of their various properties. A secondary phase of the program is a study of ice mixtures. These experiments have helped elucidate the nature of the particles in the interstellar medium that produce the 10 μm and 20 μm absorption bands and also the 3.4 μm feature. He mentioned the work of Wdowiak on laboratory condensates. They showed absorptions at 4500 Å, 6273 Å, 6282 Å and 5780 Å, and on warmup, emission features are seen in the range 5000-7500 Å. Finally Donn referred to the laboratory work by Duley on MgO grains of size <20 nm which suggests the presence of an absorption band at 220 nm similar to the interstellar UV absorption band. Mentioning the problems with graphite, Donn noted that further work on MgO would be desirable.

Pirronello described experiments on irradiation of ice mixtures with 1.5 MeV helium ions to study the interaction of energetic particles with frozen gases. The last paper of the session was given by d'Hendecourt on expulsion of molecules from grains in dense clouds. Remarking on the existence of appreciable amounts of CO, $H_2O$, and $NH_3$ in these clouds, he suggested that grain-grain collisions at relative velocities above 40 $ms^{-1}$ could produce molecular ejection from grains in clouds.

## Large-Scale Distribution of Gas

Commissions 33 and 34

Chairman: W.B. Burton 23 August 1982

The mass and distribution of atomic hydrogen, distribution of CO from northern and southern surveys, and distribution of molecules other than CO and $H_2$ were discussed in reviews by Blitz, Sanders, Robinson and Cohen respectively. Heiles spoke on the HI supershells as a characteristic of the Galactic gas morphology. Kazes described work on HI observed in absorption. High-velocity clouds were discussed by Giovanelli. A review on Gamma rays and the Galactic gas distribution was presented by Bloemen, and Young gave the final review on a comparison of the Galactic gas distribution with that in nearby spirals. A detailed report of this meeting will be found in the Commission 33 report in the Proceedings.

## Recent Results in Galactic Radio Astronomy

Commissions 34 and 40

Chairman: V. Radhakrishnan 24 August 1982

Eighteen contributed papers were presented on topics which included Pulsars, Supernova Remnants, H II regions and Spectral line observations of interstellar molecules. Among the more noteworthy results were the following.

Lyne reported on proper motion measurements on pulsars and an interesting correlation that has been found between their observed velocities and derived magnetic fields. Mills presented observations of SNR in the LMC and SMC obtained with the recently commissioned Molonglo Synthesis Telescope. These observations clearly suggest that the majority of SNRs have not decelerated, contrary to the generally held view. Very beautiful observations of the Galactic Centre, obtained with the VLA, showing evidence of a spiral-like morphology for Sgr A - West and a shell-structure for Sgr A - East were presented by Ekers. Jennison described the results obtained with the Mark I - Kent Low Frequency system which showed a jet-like structural anomaly in the SNR Cas A.

A complete list of all the papers will appear in the Comm. 40 reports in the Proceedings.

## The Interstellar Medium from Space

Commissions 34 and 44

Chairman: W. Grewing 25 August 1982

The existence of a million-degree component of the interstellar gas is now well established. Garmire discussed the soft X-ray sky maps that have been obtained with the proportional counters aboard the HEAO 1 spacecraft, and argued that some of the extended features seen in these maps can be accounted for if the solar system is surrounded by million-degree gas extending out to a distance of about 100 pc. Within this volume must be a number of cool, evaporating clouds. At its edge one finds an expanding shock.

The same local region in space was then discussed by Bruhweiler who reviewed the results from recent UV experiments. It had been known already that the H I column density is very low along a few lines of sight. A more systematic study has now shown that low H I column densities are found in the entire hemisphere between l = 90 and l = 270 degrees to distances of many tens of parsecs. Conversely, looking towards the galactic center, the line of sight seems to pass through a local cloud before penetrating again through a region with little H I. Bruhweiler went on to discuss the implications that the very low H I densities have for the EUV radiation field in interstellar space.

The state of the interstellar gas at the outskirts of our galaxy was considered by Blades who reviewed the recent observations of the galactic halo, and haloes around external galaxies. In particular, the IUE observations of distant background targets provide clear evidence for gas well outside the plane of our galaxy. In addition to the predicted highly ionised species, one also finds evidence for neutral or only singly ionised constituents, indicating the presence of cool clouds. Indeed, in the very high resolution optical observations, many of the absorption components are seen to break up into several sub-components that are extremely narrow.

The contributions that infrared observations can make to extend our understanding of the interstellar medium were highlighted by Harwit, who chose the Orion complex as an example. The importance of various fine-structure transitions was stressed, and their observed strengths were compared to model predictions for H II regions. Turning to very recent results, Harwit reported the detection of the diffuse [C II] 156 μm emission from the galactic plane that was measured in a very beautiful experiment in which the moon was used as a reference source while it moved across the plane.

Ever since the interstellar reddening law was known from UV wavelengths through the optical to near and medium IR wavelengths, the question has been with us as to what materials cause the extinction. As Mathis pointed out in the concluding talk of the session, a vast amount of data has been compiled over the past years, mostly from UV observations. He stressed that the features in the extinction curve are still a puzzle. While they can be attributed to a number of materials, all those materials would also produce additional features which are not observed. Mathis went on to discuss what can be concluded from the observed polarisation properties, from the peculiar extinction observed e.g., in dark clouds, and from the correlation with the atomic and molecular hydrogen densities. Some systematic effects are observed in the properties of circumstellar dust depending on the O/C ratio in the associated stars. Mathis concluded his review by stressing the need for further IR and FIR observations to be combined with observations in the UV.

## COMMISSION 34 MEETINGS

### Important Recent Results

Chairman: Y. Terzian 20 August 1982

The following contributed papers were presented:

Stellar Winds and Their Impact on the Interstellar Medium - Howard A. Smith.

Multi-colour Polarimetric and Photometric Studies of the Rho Ophiuchi Dark Cloud - Frederick J. Vrba, G.V. Coyne, S. Tapia.

The Ratio of Deuterium to Hydrogen in Interstellar Space - Claudine Laurent, Alfred Vidal-Madjar, Cecile Gry, Paul Bruston, Roger Ferlet, and Donald York.

Asymmetric Broad $HCO^+$ Line Wings in Cores of Molecular Clouds - Aa. Sandqvist, A. Wooten, R.B. Loren, P. Friberg, and A. Hjalmarson.

The Diffuse Interstellar Lines at 5780 and 5797 Å - D. McNally.

Infrared and Radio Observations of Broad-Winged CO Sources - Howard A. Smith, Harley A. Thronson, and Charles J. Lada.

Warm Neutral Halos Around Molecular Clouds - P.G. Wannier, and M.R. Morris.

Detailed Models of Non-Spherical H II Regions - Robert Rubin.

Anomalous [SII] 6716/Hα Intensity Ratios in the Faint Galactic Background. - R.J. Reynolds.

Models of Molecular Clouds: Multi-transition Studies of CS and $H_2CO$. -Neal J. Evans II, Lee G. Mundy, Ronald L. Snell, and Paul F. Goldsmith.

### Planetary Nebulae

Chairman: M. Peimbert 21 August 1982

Terzian reviewed the observations of nebular shells. He noted that there had been major progress in observational techniques in the last few years. UV spectroscopy of the nebulae by IUE on the one hand, and high-resolution ($\gtrsim$ 0\".5) radiomaps obtained by the VLA on the other have enriched enormously our knowledge of the nebular shells. Recent observations have also revealed morphological details showing many of the nebulae to have more than one shell and sometimes halos. In her review of the central stars, Acker said that spectra of more than a hundred nuclei are now available. They show a wide range of characteristics with WR types perhaps dominating in number. Several of the visible nuclei are of late spectral types which, according to her, indicates the likely presence of hot companions. The binary nature of at least 16 planetary nuclei has been confirmed and 17 more are suspected to be binaries. She discussed the physical parameters of these stars and emphasised how greatly UV observations have helped in this respect.

Peach described briefly the important atomic processes in nebular shells and discussed the advances made in the calculations of atomic data in the last few years. The work of Mendoza on transition probabilities, electron excitation rates and photoionization cross-sections, of Nussbaumer and Storey on recombination processes, and of McCarroll and associates and Dalgarno and associates on charge exchange was mentioned in this connection. Aller spoke on the chemical abundances in planetaries and said that recent work had shown these to be sufficiently different from one nebular type to another so that an average chemical composition was no longer a meaningful characterisation Aller emphasised the advantages gained in recent times by the use of UV and infrared observations to obtain abundances and discussed the improved methods developed to do so.

Serrano described the planetary nebulae in a galactic context. Mentioning the difficulties in the determination of distances for the galactic planetaries, he quoted results from the surveys of such objects in Local Group galaxies to show that the total number in the Galaxy should be close to 10,000. This, he said, favoured the calibration of distances by Cudworth over the one used by Cahn and Kaler. Serrano discussed the recent work on the determination of scale heights, space densities and birthrates of these nebulae in the solar neighbourhood and pointed out that the low value of the scale height obtained in general was a puzzle. He also discussed the significance of the carbon and nitrogen abundances of planetaries from the point of view of chemical evolution of galaxies.

The session ended with a contributed paper by Walborn on a suspected extended planetary nebula in LMC.

## COMMISSION 35: STELLAR CONSTITUTION (CONSTITUTION DES ÉTOILES)

### Report of Meetings 20 and 23 August 1982

PRESIDENT: R. J. Tayler VICE-PRESIDENT: A. N. Cox

### 23 August 1982

BUSINESS MEETING

At the start of the meeting members stood in silence for a few minutes in memory of B. M. Tinsley, who had died since the Montreal meeting.

The President reported that the vote which he had organised earlier in the year had produced the following list of nominations for membership of the Organising Committee:

PRESIDENT A. N. Cox (Los Alamos Scientific Laboratory, New Mexico)

VICE-PRESIDENT D. Sugimoto (University of Tokyo)

ORGANISING COMMITTEE P. Bodenheimer, C. Chiosi, D. J. Faulkner, A. Maeder, J-L. Tassoul, R. J. Tayler, A. V. Tutukov, V. Weidemann, J-P. Zahn.

The composition of the Organising Committee was agreed. It was explained that the rule adopted at the Montreal meeting about the rotation of membership of the Organising Committee meant that D. J. Faulkner, R. J. Tayler, A. V. Tutukov and J-P. Zahn would leave the Organising Committee in three years' time. The ballot of members of the Commission had attracted not much short of a 50 per cent vote; this was quite a high vote as there are a number of senior members of the Commission who are no longer active in the field of stellar constitution.

The membership of the Commission was revised. Five new names had been added by the Executive Committee since the Montreal meeting; four of these were members from China (Nanjing) who were added when the arrangements for Chinese adherence to the IAU were settled and the fifth was an existing member who had intended to join at Montreal. The President had inspected the curricula vitae of all the new members who were now proposed by National Committees and he was satisfied that they were all active in the Commission's field. In addition some existing members of the IAU wished to join the Commission. The election of all of the new members was agreed. No member of the Commission, other than B. M. Tinsley, had been lost by death or resignation in the past three years. After the election of the new members, the Commission membership is 247.

The President announced that the Draft Report of the Commission had once again consisted of short reviews on specific topics with contributions by V. Castellani, J. P. Cox, A. Maeder, J. J. Monaghan, A. V. Tutukov and J. C. Wheeler. The small number of free offprints which he had received had been distributed to members of the outgoing and incoming Organising Committees and to the authors of contributions.

There was a discussion about the number of meetings of the Commission at Patras with some members feeling that there should have been more meetings. The President explained that there were two reasons why he had not tried to organise further meetings. The first was that Commission 35 was involved in three Joint Discussions which were occupying three full days. The second was that he had received no suggestions for topics for Commission meetings following his circular letter earlier in the year. Dr Cox said that he would welcome the views of commission members on the number of meetings to be held at the next General Assembly and on topics for the meetings.

## Scientific Meetings

MIXING IN STELLAR INTERIORS 20 AUGUST 1982

This meeting was organised by A. Maeder and had the following programme:

J-P. Zahn Mixing processes induced by stellar rotation
A. Renzini Surface chemical composition and mixing
A. Maeder Cluster observations and models with mixing

A fourth paper by G. Shaviv could not be given because he was unable to be at Patras but a short communication by I. W. Roxburgh on a new instability in stellar interiors was added to the programme.

Zahn said that mixing in stellar interiors could be produced either by advection by large scale circulation or by instability leading to turbulence. The main cause of such instability was differential rotation, which could arise even in a star which was initially uniformly rotating as a result of contraction or expansion, of mass or angular momentum loss or of meridional circulation. He discussed four types of instability; shear flow, dynamical (Rayleigh), baroclinic and double diffusive. In particular in discussing baroclinic instability he stressed that boundary conditions are very important so that local stability criteria, which are relatively easy to obtain, are not very useful. Roxburgh also discussed baroclinic instability in the case of the Sun. He said that, if diffusion was present, there could be an overstability which might occur even if the well known Goldreich, Schubert, Fricke instability were absent. He suggested that the present solar angular velocity profile might be such as to give marginal instability, with the solar centre rotating much more rapidly than the surface, and that this might ease or solve the solar neutrino problem.

Renzini discussed the comparison between surface element abundances in red giants and the predictions of stellar evolution theory, in an attempt to decide whether mixing processes other than convection are needed. He showed that there is no compelling evidence for additional mixing in the first dredge-up stage in giants with masses $< 3M_{\odot}$. There are problems in Cepheids and GF supergiants, but these cannot simply be solved with an additional mixing process as the observations do not seem to conserve the total number of CNO nuclei which should not be changed in H burning; some abundances are probably in error. There is apparently a need for a third dredge-up in low mass stars but this is not predicted by standard models.

Maeder explained that along the main sequence from 1 $M_{\odot}$ to 60 $M_{\odot}$ the observed main sequence band is wider than expected from standard theory. The effect increases with mass and it is not due to variation in chemical composition or to atmospheric effects. He discussed several detailed comparisons of theory and observation. In NGC 752 no hook is visible in the evolutionary tracks and in 34 clusters in the age range from the Hyades to the Pleiades there are ten times more stars about the main sequence band than expected. He discussed diffusion due to overshooting from convection zones and argued that this was necessary to explain the observations.

SHORT COMMUNICATIONS 23 AUGUST 1982

Following the Business Meeting a Scientific Meeting was devoted to any topics of interest to members of the commission with the following programme:

| | |
|---|---|
| K. A. van Riper | Stellar collapse |
| J. J. Cowan | r-process production in supernovae |
| A. N. Cox | Theoretical studies of Mira variables |
| Q-Y. Qu | Abnormal neutron stars and quark abnormal neutron stars |
| C. M. Sharp | Molecular opacities |
| J. L. and M. Tassoul | Meridional circulation in early type stars |
| G. Ruben | Meridional mixing in supernovae |
| D. Sugimoto | Mass loss |

van Riper discussed the pre-supernovae collapse of a massive star. The calculations of different groups disagree about whether or not an explosion occurs, which is not surprising as the gravitational energy released in a neutron star collapse is ten times the observed energy of a supernova. He discussed the role of nuclear disintegration, neutronization and neutrino trapping and explained that a core bounce should occur when it is slightly above nuclear density. Whether or not the envelope is ejected depends critically on the entropy distribution determined by earlier stellar evolution and on whether or not there is a phase transition just below nuclear density.

Cowan explained that although it is generally accepted that the r-process occurs in supernovae, its precise site is uncertain. There is a possibility that it can occur in the helium zone which is shocked as a result of the explosion. Provided that there is enough $^{13}C$ in the helium zone, neutrons provided by $(\alpha,n)$ reactions can produce the first two r-process peaks; the failure to produce the third peak may be due to errors in $\beta$-decay rates. An essential feature is that this is a non-equilibrium r-process. Provided a relatively small fraction of the helium zone has the correct conditions for the r-process, observed abundances can be explained.

Cox discussed some problems in the interpretation of the properties of Mira variables in 47 Tuc, in particular whether they pulsate in the fundamental mode or in the first overtone and suggested that a good treatment of time-dependent convection would be needed to solve the problem. Qu described an abnormal state of nuclear matter which could arise above nuclear density as a result of coupling between nucleons and mesons. In such circumstances the effective mass of the nucleon becomes very small and abnormal neutron stars may be possible with higher maximum masses than those of normal neutron stars. Both abnormal neutrons and abnormal protons can be important and models are also possible with a quark interior and an abnormal neutron exterior. He suggested that the change of state from a normal to an abnormal neutron star could produce a supernova explosion.

Sharp described his calculations of molecular opacities which might be relevant in the atmospheres of late type stars. He had considered a solar composition and had included 22 elements and 60 isotopes with 36 diatomic molecules and 202 isotopic variants and 9 triatomic molecules and 119 isotopic variants. J-L.Tassoul discussed the problem of finding a rotation law and a circulation pattern so that radiative equilibrium prevails at every point in a star. He described how a baroclinic structure combined with a turbulent dissipative flow can lead to acceptable solutions.

Ruben considered the hydrodynamics of a rotating supernova. He explained that even though the rotation energy is very much less than the collapse energy, there are very complicated motions. When the shock wave reaches the surface of the star, there is a strong meridional movement with a magnitude which is 10 - 20% of the radial velocity. Sugimoto said that, when considering problems in which substantial mass loss from a star is supposed to occur, it is necessary to take account of the energy involved in getting matter from the central regions to the surface so that mass loss can occur. He believed that this was very important in mass loss from neutron stars in X-ray burster models.

In addition to the Commission meetings, Commission 35 was concerned in three Joint Discussions on Solar Luminosity Variations, Evolution in Old Stellar Populations in Galaxies and Mass Loss Phenomena.

## COMMISSION 36: THE THEORY OF STELLAR ATMOSPHERES (LA THEORIE DES ATMOSPHERES STELLAIRES)

### Report of the Meetings on August 23rd and 25th

CHAIRMAN: G. Traving SECRETARY: K. Kodaira

### Bussiness Meeting August 25th, 9:000 m

I. COMMISSION ACTIVITIES

The President outlined the activities of the Commission in the period following the 17th General Assembly.

II. MEMBERSHIP

a) DECEASED: Heidman, N.; Myerscough, V.P.

b) VOLUNTARY RESIGNATIONS: Fischel, D.; Thompson, R.I; Unsöld, A.

c) The president reported the names of those members, who failed to confirm their membership.

d) The membership of the following new members was approved: Crivellari, L.; Cannon, C.J.; Dupree, A.K.; Erikson, K.; Froeschlé, C.; Haisch, B.M.; Hartmann, L.W.; Heasley, J.N.; Hubeny, I.; Koester, D.; Mnatsakanian, M.; Nagirner, D.I.; Rachkovsky, D.N.; Rutten, R.J.; Simon, T.; Stencel, R.E.; Vardavas, I.M.; Weber, S.V.; Wilson, S.J.; Yanovitsky, E.G..

III. NOMINATION OF PRESIDENT AND VICEPRESIDENT

The nomination of B. Gustafsson as President and of K. Kodaira as Vice-President was approved unanimously.

IV. ORGANIZING COMMITTEE

The following names for the new Organizing Committee were approved: Gray, D.; Hearn, A.G.; Kalkofen, W.; Kudritzki, R.; Mihalas, D.; Sapar, A.; Seaton, M.; Traving, G.; Underhill, A.B..

V. FUTURE ACTIVITIES

Plans for future activities of the Commission were discussed, but no decisions taken. The business meeting adjourned at 10:05 am.

### SCIENTIFIC SESSIONS

1) JOINT SESSION (with Commissions 14 and 29): August 23rd, 8:30 am
The minutes of this session will be published elswhere.

2) MEETING: August 23rd, 3:15 pm
Radiative Transfer
Chairman: W. Kalkofen
Introduction to papers on transfer in spherical atmospheres: W. Kalkofen
Papers Presented:
(i) Construction of Models: R. Wehrse
(ii) Discrete Space Theory: A. Peraiah

Introduction to papers on fast methods of solving the transfer equation: W. Kalkofen
Papers Presented:
(i) Probabilistic Radiative Transfer: R. Canfield
(ii) Use of Approximate Lambda-Operators: G. Scharmer
(iii) An Integral Equation Method: P. Ulmschneider
The material presented at this session together with further papers on transfer in spherical media and on fast methods are to be published in a forthcoming book on "Methods in Radiative Transfer". The expected publication date is 1983.

3) <u>MEETING: August 25th, 3:15 pm</u>
The impact of new observational data on the theory of stellar atmospheres.
Chairman: G. Traving
Papers Presented:
(i) The Impact of UV-Observations on the Theory of Late-Type Stellar Atmospheres: D. Reimers (presented by L. Hartmann)
(ii) The Impact of X-Ray Observations: A.K. Dupree (presented by W. Kalkofen)
(iii) The Impact of UV-Observations on the Theory of Early-Type Stellar Atmospheres: A.B. Underhill
The papers presented in this session form part of the report of Commission 36 which has been published in the "Transactions IAU Vol. 18 A", pg. 479.

# COMMISSION 37: STAR CLUSTERS AND ASSOCIATIONS (AMAS STELLAIRES ET ASSOCIATIONS)

PRESIDENT: G. Lyngå SECRETARY: D.C. Heggie

In addition to the Commission sessions described below, the Commission also participated in the following:

1. Joint Discussion II: Evolution in old stellar populations in galaxies (chairman: G. Lyngå)

2. Joint Discussion VII: Mass-loss phenomena (chairman: A.G. Hearn)

3. Session 251: Photometry in crowded fields (chairman: A. Penny)

4. Session 291: Systematic effects in abundance determinations for metal-poor stars (chairman: W.K. Bonsack)

Reports of the Joint Discussions will be found in Highlights of Astronomy, reports of Session 251 will be found in this volume in the proceedings of Commission 25 and reports of Session 291 are being printed in the Publications of the Astronomical Society of the Pacific.

## 23 August 1982

### BUSINESS SESSION

1. The President summarised the changes in the membership of the Commission, and the following new organising committee was elected to serve for the three-year period 1982-85:

| | |
|---|---|
| K.C. Freeman | (President) |
| G.L.H. Harris | |
| D.C. Heggie | (Vice-president) |
| G. Lyngå | (Past President) |
| P.E. Nissen | |
| C.A. Pilachovski | |
| G.N. Salukvadze | |

2. A brief report was given of IAU Colloquium No. 68, "Astrophysical Parameters for Globular Clusters" (Schenectady, New York, U.S.A., October 8-10, 1981), and the President listed a number of other meetings which the Commission had also decided to support. It was decided to support the proposal by R.M. Humphreys for a meeting on "Luminous Stars, Associations and Galaxies", tentatively planned for autumn, 1984, in Europe.

3. The meeting discussed the following proposal for the numbering of stars in clusters, which was prepared by a Working Group consisting of A. Moffat (chairman), G. Lyngå, J.-C. Mermilliod, and S. van den Bergh.

Proposal:

(1) All star cluster investigators should always publish finding charts that unambiguously indicate the stars observed (preferably as in (2) below), if not given explicitly elsewhere. Even the measure of only a few stars justifies a chart. The chart should be labelled "N" at the top for north and "E" at the left for east with a line segment at the edge showing a simple number of arc units on the sky. Editors of appropriate journals should be urged to insist that authors comply.

(2) Numbering should be made in a systematic fashion. The preferred, most flexible way is to use XY coordinates in minutes of arc, centered on some arbitrary but precisely defined star or other position near the cluster centre and aligned so that X, Y are parallel to and go in the same sense as RA and Dec, respectively, for 1950.0. Once the origin has been chosen it must remain the same for all future work on a given cluster. An appropriate level of precision for the tabulated X, Y values should be adopted in order to avoid ambiguities. To facilitate easy use, the charts should indicate the X, Y axes with tic marks at intervals of one arc minute or better, depending on the degree of concentration of the cluster. The scale should nominally be 1' per cm (= 6" per mm) or a simple multiple thereof. Once one good chart is published in this way for a given cluster, subsequent papers on the same cluster can, if appropriate, refer to it by X, Y coordinates for identification of stars, without having to make a new chart. Cross-identifications to old numbering systems should be supplied.

(3) A photographic atlas containing charts of all existing Galactic open and globular star clusters should be made by some central body (e.g. CDS) as a basis for (2). The charts would be left blank except for the addition of X, Y axes and tic marks. J.-C. Mermilliod has indicated he might supervise the undertaking of this task. Plates for Northern clusters will be obtained with the CERGA Schmidt telescope by Dr Heudier.

In the discussion that followed, J.-C. Mermilliod exhibited an example of a suitable chart prepared from a Schmidt plate, though how the atlas of charts should be published, and what the limiting magnitude should be, had not been settled. The atlas would cover perhaps 200 or so of the most frequently studied Galactic open and globular clusters. Astronomers working to fainter magnitudes than the chart limit, could prepare their own charts and send copies to the CDS (Centre de Données Stellaires) at Strasbourg. A suitable model for such charts is the well-used catalogue containing UBV magnitudes of stars in clusters by Hoag et al. (1961, Publ. U.S. Naval Obs. 17, part 7).

The proposal of the Working Group was accepted.

4. B. Balázs reported on the status of the Catalogue of Star Clusters and Associations, and the problems caused by the exponential growth of literature in this field. By 1983 it was hoped that data published up to 1976 would be incorporated, and that the backlog would disappear by 1987. To facilitate this, Dr Balázs repeated his appeal at Grenoble in 1976, i.e. "recognising the importance of the Catalogue of Star Clusters and Associations to all astronomers, Commission 37 urges all authors to send copies of their papers and these objects to the Editors of the Catalogue, B. Balázs or J. Ruprecht".

G. Lyngå stated that the Lund-Strasbourg computer-based catalogue of open cluster data would be updated in 1983. J.C. Mermilliod reported the addition of new UBV data for new clusters to his catalogue (27.153.023), and A.G. Davis Philip stated that he and R.E. White were engaged on a new edition of an atlas of colour-magnitude diagrams of globular clusters.

SCIENTIFIC SESSION

This consisted of a poster session followed by a discussion supervised by the chairmen.

1. Open Clusters (chairman: P.E. Nissen)

The following poster papers were displayed:

R. Cayrel, G. Cayrel and B. Campbell: New Observational Study of Lithium depletion along the main sequence of the Hyades.

Observations made at the coudé focus of the Canada-France-Hawaii Telescope have allowed us to establish that the equivalent width of the lithium resonance doublet at 6707 Å decreases from 79 mÅ at spectral type G1V down to less than 2 mÅ at spectral type K1V. This represents a depletion by a factor one hundred over this range.

A.M. Jakobsen: Differential Ages in Open Clusters.

A relationship between uvby, Hβ Strömgren photometric observations and the ages and masses of B-type stars has been established. In some open clusters an age spread from 10 million to 100 million years between the early (high mass) and the late (low mass) B-type stars is indicated. NGC 457, 663, 1893, 1960, 2264, 6910, h+χ Per and Stock 2 are extensively observed.

M.L. Roth: Ages of Open Clusters (a Catalogue).

A.R. Upgren and E.W. Weis: Photometry and luminosity function of the Hyades.

D.A. VandenBerg and T.J. Bridges: Zero-age main sequences applied to the Pleiades, Praesepe, and Hyades.

Zhao Jun-Liang: Determination of membership in open clusters by means of statitistical decision theory.

2. Globular Clusters (chairman: J. Mould)

The following poster papers were displayed:

G. Alcaino: NGC 6256: Definitively a Galactic globular cluster.

G. Alcaino: The globular cluster NGC 6544.

G. Alcaino and W. Liller: Main sequence photometry for globular clusters.

A. Ardeberg, H. Lindgren and P.E. Nissen: Metal abundance and age of NGC 6397 from uvby-β photometry of turn-off stars.

A. Chieffi and F. D'Antona: CNO self-pollution in globular clusters: Evolutionary effects.

Making the working hypothesis that presently evolving stars in globular clusters have been polluted in CN rich matter in the first epoch of the life of the cluster, we compute the evolution of stratified stars, in which the external layers are CN enriched. These stars evolve very rapidly when the hydrogen burning shell crosses the chemical discontinuity left by convection.

J.E. Hesser, W.E. Harris and B. Atwood: Photometry of globular clusters with the CTIO SIT Vidicon.

A color-magnitude diagram reaching $M_V \sim +9$ mag. has been constructed for NGC 104 = 47 Tuc. It shows excellent agreement in position and slope with the theoretical tracks computed by D.A. VandenBerg for Y = 0.2, Z = 0.006 ([Fe/H] = -0.5), $\alpha$ = 1.65, and $15 \le \tau_9 \le 18$. A preliminary V magnitude luminosity function for NGC 2257 in the LMC confirms Stryker's identification of the location of its turnoff and supports her contention that it is as old as M92.

W. Liller and G. Alcaino: A search for globular clusters around the edge-on spiral galaxies of the Sculptor Group NGC 55 and NGC 253.

T. Lloyd-Evans: MS stars in ω Centauri.

S.J. Shawl and J.E. Hesser: Radial velocities from integrated light observations of globular cluster Hα absorption lines.

Radial velocities of $\sim 5$ km $s^{-1}$ precision determined with the CTIO single-étalon Fabry-Perot interferometer for ~ 85 galactic globular clusters are generally in good agreement with those in Webbink's compilation (Astrophys. J. Suppl. 45, 259, 1981). Results we reported at IAU Coll. 68 from 120 Å $mm^{-1}$ image-tube spectra for NGC 6171, 6528, 6624, 6652, 6681 and 6723 are confirmed by the higher accuracy F-P measurements.

25 August 1982

SCIENTIFIC SESSION: CLUSTER CORES (chairman: D.C. Heggie)

The following reviews were presented:

D. Lynden-Bell: Theory of core evolution.

Lynden-Bell showed how self-similar models elucidated core collapse and re-expansion. At the moment of final core collaps the density is of the form $\rho \propto r^{-\alpha}$ with $\alpha$ = 2.22. After core collapse, binary formation and evolution at the centre provides the energy for a re-expansion with the central regions forming a singular almost isothermal sphere with $\rho \propto r^{-2}$. This law bends over to the $r^{-2.22}$ halo at a transition radius $r_c$ which increases as

$$|t-t_o|^{\frac{2}{6-\alpha}}.$$

In this re-expansion the characteristic scales follow the same law as the collapse phase (Lynden-Bell and Eggleton MNRAS 1980, 191, 483):

$$v_c^2 \propto |t_o - t|^{(4-2\alpha)/(6-\alpha)} \; ; \; M_c \propto |t_o - t|^{(6-2\alpha)/(6-\alpha)} \; ;$$

$$E_c \propto |t_o - t|^{2(5-2\alpha)/(6-\alpha)} \; ; \; \rho_c \propto |t_o - t|^{-2\alpha/(6-\alpha)} \; .$$

In the above c refers to values at the edge of the core where

$$-\delta \log \rho / \delta \log r$$

changes from 2 towards 2.22.

D.H. Martins: Surface photometry of globular cluster cores.
Recent results of surface and stellar photometry of several galactic globular cluster cores were reviewed. Most attention was given to those clusters presumed to contain X-ray sources or which possess excess central surface brightness. In particular, discussion centered around NGC 6397, 6440, 6441, 6624, 6712, and 7078. Data analysis techniques were reviewed with emphasis placed on useful developments in the area of image processing.

Selected references:

Auriere, M. 1982. Astron. Astrophys. (in press).
Auriere, M., and Cordoni, J.-P. 1981. Astron. Astrophys. 100, 307.
Auriere, M., and Cordoni, J.-P. 1981. Astron. Astrophys. (Suppl. Ser.) 46 347.
Bahcall, N.A., Lasker, B.M., and Wamsteker, W. 1977. Astrophys. J. (Letters) 213, L105.
Da'Costa, G.S. 1979. Astron. J. 84, 505.
Grindlay, J.E. 1981. IAU Colloq. No. 68, Astrophysical Parameters for Globular Clusters, ed. A.G.D. Philip and D.S. Hayes.
Martins, D.H., Harvel, C.A., and Miller, D.H. 1980. Astron. J. 85, 521.
Martins, D.H. and Harvel, C.A. 1980. Astrophys. J. (Letters) 236, L77.
Martins, D.H., 1982. (in preparation).

The following poster papers were contributed:

A. Altamore, L. Angeletti, R. Capuzzo-Dolcetta and P. Giannone: UV observations of M5.

E. Bettwieser: Late stages of core collapse in globular clusters.

G.S. Bisnovatyi-Kogan, M.M. Romanova: Globular clusters and X-ray bursters.

D.C. Heggie: Post-collapse evolution of a gaseous cluster model.

S. Inagaki: The effects on binaries on the evolution of globular clusters.

H.E. Kandrup: Dynamical friction in a mean field approximation: a summary.

J. Katz: On the stability of two-component cores of globular clusters.

G.E. Kron, L.H. Wasserman and A.V. Hewitt: Abnormalities in the brightness profiles of the core regions of some globular clusters.

A.P. Lightman: What's happened in the cores of globular clusters?

N.M. Spasova: Globular cluster cores - structure and analysis.

## COMMISSION 38 : EXCHANGE OF ASTRONOMERS - (ECHANGE DES ASTRONOMES )
Committee of the Executive Committee

Report of Meetings, August 18, 19, 24.

PRESIDENT : J. DELHAYE SECRETARY : P.M. ROUTLY

### 1. First Meeting - August 18, 1982.

President J. Delhaye called the first meeting of Commission 38 to order at 11 00 hrs, Aug 18, 1982.

#### 1.a Recording Secretary

At President Delhaye's request, Dr. P.M. Routly accepted the assignment of recording the minutes of the Commission's meetings during the course of the XVIIIth General Assembly.

#### 1.b Members Present

The following members or observers, together with their countries of affiliation, were present at the August 18th meeting : H.M. Al-Naimy, IRAQ; T.M. Borchkhadze, USSR; J. Delhaye, France; M.K. Hamid, ...; E.A. Muller, SWITZERLAND; P.M. Routly, USA; C.R. Tolbert,USA; F.B. Wood, USA .

#### 1. c Agenda

Mr. Delhaye announced that there would be a Joint Meeting with Commission 46 on Aug 19 and a second meeting of Commission 38 on Aug 24. President Delhaye stated that the purpose of the Joint Meeting was to learn details of a new Visiting Lecturers Program which had been proposed to the Executive Committee by Commission 46 and had already been approved.

The following agenda to cover all meetings involving Commission 38 was agreed to by the attending members.

AGENDA

1. Report of the President
   1.1 - The activities of the Commission for the last three years.
   1.2 - Critical remarks which have been made.
   1.3 - Budget.
2. Open Problems
3. Joint Meeting with Commission 46, Aug 19 th
4. Election of Officers and Members of Commission 38 .

President Delhaye stated that he wished to discuss Items 1.1 and 1.3 immediately but defer Items 1.2, 2, and 4 to the secpnd meeting of Commission 38 on Aug 24.

#### 1.d President's Report - Agenda Items ≠1.1 and ≠1.3

President Delhaye distributed a report covering the years 1979-1982 which included a current listing of the members of Commission 38, a copy of the Guidelines defining the Exchange of Astronomers Program as adopted in 1980, Interim Reports ≠1 and ≠2 on the Commission's detailed operations dated June 1, 1980 and May 15, 1981, a final report to the Executive committee dated Nov 30, 1981, an addendum to the final report covering the period Dec 1, 1981 to July 15, 1982, and a brief summary of finances . Between the Montreal General Assembly and July 15, 1982, a total of 35 applications were received, 22 grants were awarded, 8 applications were denied, 3 were withdrawn, and 2 are still pending.

Total expenses to June 1st, 1982 amounted to approximately 61,319 SFR. Since the budget allocated to Commission 38 for the period 1979-82 was only 58,000 SFR, it was necessary for President Delhaye to approach the Executive Committee for supplementary funding. An additional sum of 10,000 SFR was authorized, raising the total budget to 68,000 SFR, permitting the operations of Commission 38 to balance.

While no formal action was needed or taken, the Committee Members in attendance endorsed President Delhaye's report and expressed the general feeling that the Exchange of Astronomers Program had been effectively and efficiently run during the past three years.

With regard to the future, President Delhaye reported that Commission 38's budget for 1982-85 has again been set at 58,000 SFR by the Executive Committee. Assuming the continued financial health of the IAU, it was understood that the Executive Committee would consider an augmentation to this budget should the demands on Commission 38 require it.

There being no further business to discuss, President Delhaye adjourned the meeting at 12.30 hours.

2. Joint Meeting with Commission 46 - August 19, 1982.

Interested members of Commission 38 met with Commission 46 on August 19, 1982 at 11.00 hours to discuss the Education of Astronomers from Developing Countries and Plans for the Visiting Lecturers Program. Results of these deliberations are described in paragraph 3.c below.

3. Second Meeting - August 24, 1982.

President Delhaye called the second and final meeting of Commission 38 to order at 15.00 hours on August 24, 1982.

3.a Members Present

The following members or observers, together with their countries of affiliation, were present at the Aug 24 meeting :T.M. Borchkhadze, USSR; A.A. Boyarchuk, USSR; J. Delhaye, FRANCE; S. Ferraz Mello, BRAZIL; H.F. Haupt, AUSTRIA; K.C. Leung, USA; P.M. Routly, USA; W. Shougnan ...; F.G. Smith, UK; C.R. Tolbert, USA; D. Wentzel, USA; F.B. Wood, USA.

3.b New Officers and Members of Commission 38 - Agenda Item #4

After careful examination and general discussion of the present composition of Commission 38, the following slate of officers and members for 1982-85 was proposed for ratification at the second session of the XVIII General Assembly :

| | | |
|---|---|---|
| President | F.B. WOOD | University of Florida, Department of Astronomy, Thomas W. Bryant Bldg Gainesville, Fla 32611. |
| Vice President | E.A. Muller | Observatoire de Genève, Sauverny, Suisse. |
| Organizing Committee | A.A. Boyarchuk; J. Delhaye; D.A. MacRae; P.M. Routly; J. Sahade; F.G. Smith; C.R. Tolbert; H.H. Voigt; S.H. Ye. | |
| Members | A.W. Al-Sabti; B.J. Bok; A. Florsch; G. Gordoli; H. Haupt; G. Keller; K.C. Leung; H. Li; M. Marik; S. Miyamoto; I.S. Nha; S.E. Okoye; A. Opolski; T.L. Page; A. Reiz; G. Ruben; E.V.P. Smith; G. Teleki; J.P. Wild. | |

3.c Joint Meeting with Commission 46 - Agenda #3

As a result of the Joint Meeting held on Aug 19, the following actions were formally endorsed by Commission 38 at its Aug 24 meeting.

1. Commissions 38 and 46 adopted the report of the joint Working Group on the Education of Astronomers from developing countries. The Working Group, having completed its charge, was formally terminated.

2. Commissions 38 and 46 resolved to establish a subcommittee to implement the recommendations of the Visiting Lecturers Program in the report of the Working Group and in Agenda item 13(d)2 of the 18th General Assembly of the IAU. Initialy, the membership of the committee will be J. Delhaye (38), S. Ferraz-Mello (46), D.C. Wentzel (46, Chairman), and F.B. Wood (38). The membership will be subsequently expanded to ensure active participation from Developing Countries in its work, with advice and consent from the Organizing Committees of Commissions 38 and 46.

It was agreed that the terms of reference for the subcommittee would be:

a. To seek individuals and their institutions willing to host a Visiting Lecturers Program.

b. To identify the local procedures to support the program; in particular, the local body with whom any future contrats should be made.

c. To outline a Visiting Lecturers Program agreeable to both host and subcommittee; at this time the subcommittee should identify several possible lecturers with experience in the field of the agreed programme, known experience of the developing world and preferably fluency in the local language. ( It is recognized that such an identification will not limit final selection. )

d. To submit, in light of c. above, a formal proposal to the IAU Executive Committee for a Visiting Lecturers Programme in a named Developing Country, to be based an a named host institution.

e. To submit at the time of d. above, the outline of a second programme, such that the General Secretary may seek support through a UNESCO contract.

f. To establish, at the time of d. above, an advisory panel with experience of local conditions, developing countries and the type of programme envisaged, in order to advise the Executive Committee on the final selection of lecturers and on the articulation and oversight of the programme.

It was further agreed that the countries to be initially contacted with little or no astronomy, would be : Egypt, Indonesia, Kenya, Korea, Nigeria, Paraguay, Peru, Portugal, Sri Lanka, Thailand.

3.d Criticisms and Problems - Agenda Items ≠ 1.2 and ≠ 2.

President Delhaye reported that a few members on the Executive Committee had raised several questions critical of Commission 38's operation.

a. Qualifications of Candidates. One of these questions had to do with assuring that candidates for Commission 38 grants were the best qualified available. It was suggested that Commission 38 have some way of intercomparing or evaluating candidates so that they might be ranked in some kind of order of merit or desirability.

The Commission members in attendance discussed this idea at some length and rejected it for the following reasons. First, it implies the existence of deadlines so that applications could be accumulated prior to the rating process. This contrasts to the present procedure where applications are acted upon as they are received. It was felt that the "deadline approach" would require too much lead time to be responsive to the schedules of prospective applicants. As to the qualifications of applicants, it was pointed out that present procedures already require the approval, support,and endorsement of the candidate by responsible officials at both the Host and Home Institutions. To require further approval was felt to be an unnecessary additional complication and one that cast doubt on the competence and veracity of the applicant's endorsers.

b. Per Diem and Dependents . Another question that was raised concerned per diem and dependents. It was suggested that the Exchange of Astronomers Program cover per diem costs as well as the costs of dependents in lieu of travel in some cases and in addition to travel in others. Commission Members agreed that such support would be ideal but was not feasible within the framework of the present budget.

In summary, the Commission decided to recommend thet the present guidelines be left as they are and that the Exchange of Astronomers Program continues to be operated without substantial change from current operational procedures.

3.c Thanks and Adjournment.

President Delhaye thanked the Commission and its officers for their work and support over the past three years. He also extended best wishes to the new officers and expressed the hope of seeing everyone again at the next General Assembly in India in 1985. There being no further business to discuss, President Delhaye adjourned the meeting at 16.45 hours.

P.M. ROUTLY, Secretary.

# COMMISSION 40: RADIO ASTRONOMY (RADIO ASTRONOMIE)

Report of Meetings, 18, 19, 20, 21, 23, 24 and 25 August 1982

PRESIDENT: G. Swarup SECRETARY: V.K. Kapahi

## Business Meetings

With the President, Govind Swarup, in the chair, business meetings of the Commission were held on 19 and 25 August. The following business was transacted.

### I. REPORT OF THE PRESIDENT

The activities of the commission during the preceding three years were briefly summarised by the President. Although the radio astronomy service seems to have got a reasonable deal, on the whole, at the World Administrative Radio Conference (WARC) in 1979, we must continue our efforts to obtain adequate protection for the radio astronomy bands from harmful transmissions by other services.

The President thanked all the contributors to the triennial report which had been published by the IAU and was available at Patras. As in the previous report astrophysical results obtained through the radio technique were covered by several other appropriate Commissions. The Commission 40 report deals mainly with basic measurements, radio surveys, and new developments in radio telescopes and instrumentation.

During the three years the Commission cosponsored IAU Symposia 95-*Pulsars;* 97-*Extragalactic Radio Sources;* 100-*Internal Kinematics and Dynamics of Galaxies;* 101-*Supernova Remnants and their X-ray Emission;* 104-*Early Evolution of the Universe and its Present Structure;* Colloquium 71-*Activity of Red Dwarf Stars.*

### II. LINES OF ASTROPHYSICAL IMPORTANCE

Brian Robinson summarised the activities of the working group on the protection of radio spectral lines. Based on comments from several astronomers, the group recommended that the lines at 36.392 GHz ($HC_3N$) and 150.74 GHz (nitric oxide) be deleted and several other lines be added to the 1979 list of astrophysically important lines (IAU Transactions Vol.XVII B). The new list included several lines above 275 GHz, a region of the spectrum pioneered by radio astronomers. Following the discussion that ensued, the WG met again on 24 August and the proposed changes to the 1979 list were again discussed in the 2nd business session. The revised new list of lines, approved by the Commission is shown in Table 1 in which the lines below 275 GHz are listed in order of increasing frequency while those above 275 GHz are ranked in order of their foreseen astrophysical significance. The revised membership of the WG is: A. Baudry, R. Booth, L. Doherty, M. Morimoto, V. Pankonin, V. Radhakrishnan, B. Robinson (Chairman), R. Schilizzi, V. Slysh, B. Turner, A. Webster and T. Wilson.

In view of the importance of obtaining official backing of IAU, a resolution (listed in the Proceedings of the XVIIIth General Assembly) concerning protection from harmful interference was unanimously approved by the members.

### III. WORKING GROUP ON VLBI

Ken Johnston proposed that a working group be formed to foster the interdisciplinary use of large steerable antennas for astrophysical, astrometric and geodetic research in Very Long Baseline Interferometry (VLBI). The proposal was well received by the members and a resolution recommending the formation of such a working group was approved. The resolution was later endorsed by the IAU General Assembly. The working group, whose present membership is K. Johnston (Chairman), S. Ananthakrishnan, D. Backer, E. Bajaja, R. Booth, N. Broten, M. Cohen, G. Grueff, D. Jauncey, P. Kaufmann, K. Kellermann, A. Kus, L. Matveyenko, M. Morimoto, G. Nicholson, I. Pauliny-Toth, R. Schilizzi, P. Wilkinson and S. Ye, will be concerned mainly with the collection and dissemination of information concerning the available facilities and instrumentation at different observatories, plans of experiments in astronomy and geodesy, and the need for symposia etc. with a view to encouraging international cooperation in the use of VLBI.

## IV. RADIO SOURCE NOMENCLATURE

A resolution recommending a system of naming radio sources by positions (the present IAU system) referred to epoch J 2000.0 and renaming of sources referred to epoch B 1950.0 was proposed by Bernie Burke. As no consensus could be reached at the meeting, it was agreed to form an ad-hoc working group that would discuss the issue in some detail and come up with a recommendation at the second business session. The working group, composed of B. Burke (Chairman), J. Baldwin, H. Dickel, K. Johnston, V. Kapahi, B. Mills, R. Strom and G. Westerhout met on 23rd August and concluded that it was not prepared to recommend any changes at this time. A reconvocation of the group might be appropriate at the next General Assembly of the IAU.

## V. PULSAR TIME SCALE

N. Kardashev on behalf of V.G. Il'in et al. proposed the creation of a new time scale, to be called Pulsar Time Scale, based on radio pulses from PSR 0834+16 as the main reference and 0950+08 and 1919+21 as other references. According to them, such a time scale would be more precise and stable than Universal Time and Ephemeris Time Scales. After some discussion it was agreed that an ad-hoc committee consisting of R. Manchester, D. Backer and A. Lyne should consider the merits of the proposal in detail and report their findings at the second meeting of the Commission. The gist of the Committee's recommendations was as follows. Any such proposal should be made jointly with Commission 31. Although a Pulsar Time Scale would be useful only for long time intervals it remained to be shown that the precision achieved would be comparable to that defined by TAI. More than 3 pulsars would be required to take care of possible large period variations in one or more pulsars. Accurate positions and proper motions will need to be known. There are difficulties in defining a standard pulse template; in general this template will vary with frequency. Dispersion Measures will also need to be accurately known. A moderately large antenna will be required to obtain the necessary sensitivity.

A decision on the proposal was therefore deferred.

## VI. IUCAF AND CCIR ACTIVITIES

Gart Westerhout summarized the activities of IUCAF which is now concerned with the actions to be taken in the post-WARC period. The impact of WARC-79 on Radio Astronomy has been well summarized in a paper published by V. Pankonin and R.M. Price (IEEE Trans. on Electromagnetic Compatibility, 1981, Vol. EMC-23, p.308). IUCAF will watch closely the preparations for the forthcoming WARC (1985) dealing with the efficient use of the geostationary orbit.

A brief report of the activities of CCIR Study Group-2 since WARC-79 was presented by Lorne Doherty. The Study Group met in Geneva in 1980 and 1981 and the Plenary Assembly was held in February 1982. A new report of CCIR, containing the work done at these meetings, will be published later this year (Recommendations and Reports of the CCIR, XV Plenary Assembly, Geneva 1982, Vol.II - Space Research and Radio Astronomy). There has been a major reorganization of texts dealing with radio astronomy, as a consequence of the results of WARC-79 and the addition of new technical material.

The Study Group-2 has formed a working party to prepare technical material for the 1985 Space WARC. Consideration of emissions from space, particularly out of band emissions, will be an important subject of discussion in the coming years. Other important areas of interest are preferred frequencies for spectral line observations, harmful interference criteria for synthesis and VLBI arrays, and various sharing problems with other services. Radio astronomers involved with CCIR work would welcome assistance from other colleagues in these areas.

VII. REPRESENTATION IN CCIR AND IUCAF

IAU has two representatives in CCIR and IUCAF. The commission chose J. Whiteoak as one of the new CCIR representatives in place of B. Robinson. L. Doherty will continue to be the other representative. For IUCAF, G. Swarup will replace G. Westerhout, while R. Schilizzi will continue to be the other representative.

VIII. TRANSMISSIONS AT 38 MHZ

A suggestion was made by Roger Jennison that since current observations at 38 MHz suffer severe interference from long distance propagation of transmissions, particularly from the USA and the USSR, a resolution should be passed recommending that every effort be exerted to remove all transmissions from the 38 MHz band. Several members were of the opinion, however, that any such resolution would have little effect at present, and it would be preferable to take up the matter first through the concerned national administrations. After considerable discussion the proposed resolution was 'reluctantly' withdrawn.

IX. INTERNATIONAL HALLEY WATCH

William Irwine made a brief presentation on the aims of the 'International Halley Watch' project and stressing the importance of coordinated radio observations of the comet in the coming years, with the most sensitive instruments. Those planning to make observations, both in the continuum and in molecular lines should get in touch with him at the Department of Physics and Astronomy, University of Massachusetts, Amherst, Massachusetts 01003.

X. NEW OFFICERS AND ORGANIZING COMMITTEE

The Commission unanimously approved the following slate suggested by the outgoing Organizing Committee: President: K. Kellermann, Vice President: J. Baldwin Organizing Committee: J. Ables, N. Broten, G. Dulk, B. Höglund, N. Kardashev, M. Morimoto, S. Okoye, G. Setti, R. Strom, G. Swarup, L. Weliachew, T. Wilson and S. Wang.

XI. NEW MEMBERS

Proposal to add 67 new members whose names had been put up on the notice boards earlier was approved by the Commission. The total membership of the Commission has now almost reached the 500 mark.

XII. MISCELLANEOUS

Paul Wild's announcement that the news of the Australian Government's decision to fund the proposed Australian Synthesis radio telescope had just reached Patras was greeted by members with applause and jubilation.

The second business meeting closed with Ken Kellermann, the new President, proposing a vote of thanks to all those responsible for organizing Commission activities in Patras and with a request to members to let him have their views regarding any changes they would like to see in the Commission's activities in future.

## Scientific Meetings

### I. JOINT DISCUSSIONS

Commission 40 cosponsored two Joint Discussions. *Active Galactic Nuclei,* with Commissions 28 and 48 on 18 August and *Extragalactic Distance Scale and the Hubble Constant* with Commissions 25, 28, 45, 47 and 48 on 23 August. The transactions of these meetings will be published by the IAU in 'Highlights of Astronomy'.

### II. JOINT MEETINGS

Commission 40 participated in as many as five Joint Meetings.
(i) *Early Years of Radio Astronomy* - with Commission 41 on 19 August.
(ii) *Solar Maximum Year* - with Commissions 10, 12 and 44 on 20 August.
(iii) *The Violent Interstellar Medium* - with Commissions 34 and 48 on 20 August.

The details of the above meetings may be found in the reports of other Commissions and/or in 'Highlights of Astronomy'.
(iv) A Joint Meeting with Commissions 33 and 34 entitled *Recent Results: Galactic and mm-wave* was held on 24 August. In the first session of this meeting, chaired by G. Swarup, the following papers were presented.

F.G. Smith: Pulsar Results from Jodrell Bank
V. Radhakrishnan: Recent Pulsar work at RRI, Bangalore
D.J. Smith: Electric Field in Pulsar Magnetospheres
R.N. Manchester: Pulsar Search in Supernova Remnants
T. Velusamy: (i) Pulsar Search at Ooty, (ii) Interstellar Scattering towards the Inner Galaxy
B.Y. Mills: Supernova Remnants in the Magellanic Clouds
R.C. Jennison: A Structural Anomaly in Cas A
R.D. Ekers: High Resolution Continuum and Recombination Line Radio Observations of the Galactic Centre.

The second session was chaired by V. Radhakrishnan and the following papers presented.

K.J. Johnston: Recent VLA Results on HII Regions
V. Hughes: VLA Observations of Cep A
F. Churchwell: The Kinetic Temperature Profiles of Sgr B2, Orion and DR21
I. Kazes: Possible Relation between HI Gas and Molecules
R.P. Norris: MERLIN OH Maps
C. Heiles: (i) Recent mm-wave Interferometric Results from Hat Creek, (ii) Magnetic Field Strengths in two Types of Region
H.R. Dickel: Observations of HCN in DR21 with the Hat Creek Interferometer
L.E.B.Johansson: Spectral Scan of Orion A and IRC+10216
B.J. Robinson: Southern Galactic Plane CO Survey
J. Guibert: $^{13}CO$ Observations toward the Second Galactic Quadrant
B. Irwine: Discovery of Interstellar $HOC^+$
N.J. Evans II: Models of Molecular Clouds: Multi-transition Studies of CS and $H_2CO$

(v) A Joint Meeting with Commission 28 entitled *Beams and Jets in Extragalactic Sources* was held on 25 August with R.D. Ekers in the chair. The following talks were given.

R.A. Laing: Large-Scale Radio Jets (review)
P.N. Wilkinson: Small-Scale Radio Jets (review)
L.B. Baath: VLBI monitoring of BL Lacs
G. Swarup: Jets in Quasars 3C9 and 3C280.1
H. Murdoch: Precessing Jets in 2300-189
J. van Gorkom: Jets in Edge-on Spirals?
G.K. Miley: Optical Studies of Jets (review)
W. van Breugel: Optical Observations of 3C293
K.J. Johnston: SS433-Recent Results (review)
P.A.G. Scheuer: Problems in Jet Physics (review)
R. Sanders: Reconfinement of Jets

The procedings of the above session will be published in 'Highlights of Astronomy'.

III. COMMISSION MEETINGS

i) *Recent Results: Extragalactic.* The talks given at this session (24 August), chaired by V.K. Kapahi, were:

B.F. Burke: Gravitational lenses: Probes of the dark matter
M.V. Gorenstein: High sensitivity VLBI maps of 0957+561 A, B
D. Walsh: Mass distribution of the gravitational lens 0957+561
R.J. Davis: MERLIN observations of 3C273
R.T. Schilizzi: Large scale radio structure of superluminal sources
K.J. Johnston: Recent results on BL Lac sources
J.E. Baldwin: Structure and spectra of giant radio galaxies
V.K. Kapahi: Radio and optical axes of elliptical galaxies
F.G. Smith: Is the Universe rotating?
D.C. Backer: VLBI at 89 GHz
S.H. Ye: The Shanghai-Effelsberg VLBI experiment
L. Weliachew: Neutral gas kinematics and OH masers in the central regions of M82
J.M. van der Hulst: Radio supernovae

ii) *Reports from Observatories.* In this session, whose chairman was K.I. Kellermann, the following talks were given.

J. Guibert: Technical developments in Paris/Meudon observatories
G.S. Downs: Radio astronomy with the NASA Deep Space Network
L.J. Malkamaki: An actively stabilized acousto-optical spectrometer
G. Westerhout: USNO-Green Bank astrometry project
M. Morimoto: The Nobeyama 45m radio telescope
A. Baudry: Programs with the Bordeaux mm-telescope
R.N. Manchester: Radio astronomy at CSIRO
M.R. Kundu: Recent results from Clark Lake
P. Kaufmann: Latest results from Itupetinga
R.T. Schilizzi: Report from the Netherlands Foundation for Radio Astronomy
N.W. Broten: The proposed Canadian Long Baseline Array
E.E. Baart: The Rhodes-Hartebeesthoek 2.3 GHz survey
P. Thomasson: First two years of MERLIN
R.D. Ekers: Status of the VLA and VLB array
R. Booth: Recent results from Onsala
B. Hooghout: The French/German mm-telescopes
A. Kus: New developments in Torun

TABLE 1. RADIO FREQUENCY LINES OF THE GREATEST ASTROPHYSICAL SIGNIFICANCE

| Line | | Frequency (GHz) | Line | | Frequency (GHz) |
|---|---|---|---|---|---|
| *I. Lines below 275 GHz* | | | | | |
| D | denterium | 0.3274 | $C_2H$ | ethynyl radical | 87.317 * |
| H | hydrogen | 1.4204 | HCN | hydrogen cyanide | 88.632 |
| OH | hydroxyl | 1.6122 | $HCO^+$ | formyl ion | 89.189 |
| OH | hydroxyl | 1.6654 | HNC | H. isocyanide | 90.664 * |
| OH | hydroxyl | 1.6674 | $N_2H^+$ | | 93.17 |
| OH | hydroxyl | 1.7205 | CS | C. monosulphide | 97.981 |
| CH | CH radical | 3.2638 | $^{12}C^{18}O$ | carbon monoxide | 109.782 |
| CH | CH radical | 3.3355 | $^{13}C^{16}O$ | carbon monoxide | 110.201 |
| CH | CH radical | 3.3492 | CO | carbon monoxide | 115.271 |
| $H_2CO$ | formaldehyde | 4.8297 | $H_2CO$ | formaldehyde | 140.840 |
| $H_2CO$ | formaldehyde | 14.488 * | CS | $(J = 3 \rightarrow 2)$ | 146.969 * |
| $H_2O$ | water | 22.235 | $H_2CO$ | formaldehyde | 150.498 |
| $NH_3$ | ammonia | 23.694 | $H_2O$ | water $(3_{13} \rightarrow 2_{20})$ | 183.310 * |
| $NH_3$ | ammonia | 23.723 | CS | $(J = 4 \rightarrow 3)$ | 195.962 * |
| $NH_3$ | ammonia | 23.870 | $^{12}C^{18}O$ | $(J = 2 \rightarrow 1)$ | 219.560 |
| SiO | silicon monoxide | 42.821 | $^{13}C^{16}O$ | $(J = 2 \rightarrow 1)$ | 220.399 |
| SiO | silicon monoxide | 43.122 | CO | $(J = 2 \rightarrow 1)$ | 230.538 |
| CS | C. monosulphide | 48.991 | CS | $(J = 5 \rightarrow 4)$ | 244.953 * |
| $DCO^+$ | formyl ion | 72.039 * | HCN | $(J = 3 \rightarrow 2)$ | 265.886 * |
| SiO | silicon monoxide | 86.243 | $HCO^+$ | $(J = 3 \rightarrow 2)$ | 267.557 * |
| $H^{13}CO$ | formyl ion | 86.754 * | HNC | $(J = 3 \rightarrow 2)$ | 271.981 * |
| *II. Lines above 275 GHz* * | | | | | |
| C | $(^3P_1 \rightarrow {}^3P_0)$ | 492.162 | HCN | $(J = 4 \rightarrow 3)$ | 354.484 |
| CO | $(J = 3 \rightarrow 2)$ | 345.796 | $HCO^+$ | $(J = 4 \rightarrow 3)$ | 356.734 |
| CO | $(J = 4 \rightarrow 3)$ | 461.041 | $N_2H^+$ | $(J = 3 \rightarrow 2)$ | 279.511 |
| CO | $(J = 6 \rightarrow 5)$ | 691.473 | $N_2H^+$ | $(J = 4 \rightarrow 3)$ | 372.672 |
| $^{13}C^{16}O$ | $(J = 3 \rightarrow 2)$ | 330.587 | $NH_3$ | $(1_0 \rightarrow 0_0)$ | 572.498 |
| $^{13}C^{16}O$ | $(J = 4 \rightarrow 3)$ | 440.765 | $H_2O$ | $(1_{10} \rightarrow 1_{21})$ | 556.936 |
| $^{12}C^{18}O$ | $(J = 3 \rightarrow 2)$ | 329.330 | $H_2O$ | $(4_{14} \rightarrow 3_{21})$ | 380.197 |
| $^{12}C^{18}O$ | $(J = 4 \rightarrow 3)$ | 439.088 | | | |

Note: Lines marked with an asterisk are additions to the 1979 list. Lines at 36.392 GHz ($HC_3N$) and 150.74 GHz (Nitric Oxide) have been deleted from the 1979 list.

## COMMISSION 41: HISTORY OF ASTRONOMY (HISTOIRE DE L'ASTRONOMIE)

Report of Meetings, 18, 19 and 20 August 1982

PRESIDENT: M.A. Hoskin. SECRETARY: O. Gingerich.

### 18 August 1982

### I. THE USE OF HISTORICAL RECORDS IN ASTRONOMICAL RESEARCH

Nearly fifty astronomers gathered to hear papers discussing areas of astronomical research in which evidence is drawn from historical records. A primary purpose of the session was to alert researchers to the problems involved in assessing the historical record -- its reliability, completeness, interpretation, etc. Speakers also hoped to interest historians who might lead them to historical data hitherto unrecognized.

F. Richard Stephenson spoke of supernovae and ancient eclipse records, and reported that *lunar* eclipses now seem to offer the strongest evidence for determining the slowing of Earth's rotation during the past 2500 years. R. Nadal discussed the first part of an investigation of the Hipparchus commentary on Eudoxus and Aratus, which should eventually shed light on the origins of the Ptolemaic star catalogue. Samuel Goldstein described his analysis of thirteen tables in the *Almagest*, which has led him to the controversial conclusion that these tables antedate Ptolemy's work by several centuries. Tao Kiang reported on the ongoing Chinese searches for astronomical phenomena in provincial histories. A paper read on behalf of S.R. Malin and D. Gubbins urged that historians keep on the look-out for geomagnetic data in pre-1850 records. Finally, John Eddy gave a general summary of astronomical data in a variety of early records, including some from American Indians.

### 19 August 1982

### II. JOINT MEETING WITH COMMISSION 40 ON THE EARLY HISTORY OF RADIO ASTRONOMY

This crowded session was arranged by a committee set up jointly with Commission 40 (radio astronomy), and commemorated the fiftieth anniversary of Karl Jansky's detection of cosmic radio signals. The speakers were: Robert Hanbury Brown, "Early radio astronomy at Jodrell Bank"; Peter Scheuer, "The development of aperture synthesis"; Bernard Mills, "The first radio source surveys and the log N - log S relationship"; William McCrea, "The influence of radio astronomy on cosmology"; Fred Haddock, "The development of radio astronomy in the United States"; F. Graham Smith, "Early radio source work at Cambridge"; and W.T. Sullivan, "The discovery of the 21-cm hydrogen line". Written versions of what were often very informal presentations are to be included in work to be published by Cambridge University Press.

### 20 August 1982

### III. BUSINESS SESSION

Members stood for a few moments to honour the memory of members recently deceased: Joseph Ashbrook, Arthur Beer, Maria-Luisa Righini-Bonelli, Willy Hartner, Henri Michel, B.E. Tumanjan, and F. Zagar. O. Pedersen (Denmark) was proposed as

President for 1982-85 and J.A. Eddy (USA) as Vice-President. H. Eelsalu (USSR), E.G. Forbes (UK, representing IUHPS), M.A. Hoskin (UK), J.P. Verdet (France), and K. Yabuti (Japan) were elected as members of the Organizing Committee. It was noted with pleasure that E. Poulle (France) and C. Wilson (USA), hitherto consulting members of Commission 41, were to be elected full members of IAU. R.L. Bishop (Canada), Z.-G. Chen (China), J.B. Carlson (USA), J. Darius (UK), P.S. Florides (Ireland), G.M. Idlis (USSR), C.J. Peterson (USA), N.A. Porter (Ireland), Z.-Z. Xi (China) and Q.-M. Zhao (China), together with E. Poulle and C. Wilson, were elected members of Commission 41.

As all consulting members lapse at each General Assembly, the following were elected consulting members for 1982-85: A. Aaboe, W.B. Ashworth, A.F. Aveni, E. Badolati, J.A. Bennett, R. Billard, P.G. Bulgakov, D.H. DeVorkin, A.I. Eremeeva, B.R. Goldstein, N.S. Hetherington, E.S. Kennedy, D.A. King, I.A. Klimishin, H.G. Körber, P. Kunitzsch, H. Labat, N.B. Lavrova, F.R. Maddison, Y. Maeyama, R. Mercier, N.I. Nevskaya, D.J. Price, B.A. Rosenfeld, G. Rosinska, G. Saliba, J. Samso, A.R. Serban, R.W. Smith, Z. Sokolovskaya, G.J. Toomer, G. L'E. Turner, A. Van Helden, J. Vernet Gines, B.L. van der Waerden, D.W. Waters, S.R. Weart, R.S. Westman, and D.T. Whiteside.

Part of the third session was devoted to a discussion, introduced by Olaf Pedersen on the basis of replies to a questionnaire, concerning tools needed by the historian of astronomy. There was support for proposals for a number of publications, amongst them: *The Alfonsine Tables* together with an explanatory appendix; the *Almagest* in the Latin translation of Gerard of Cremona; a facsimile edition of a medieval collection of astronomical manuscripts; a glossary of Latin astronomical terms; and algorithms to be used with computers. The Commission voted to have the Organizing Committee set up a working group to promote the publication of suitable facsimile editions, and the Committee was also instructed to work to establish a central catalogue of observations made before 1600.

Announcements were made concerning a number of projects. Hoskin reported that the *General history of astronomy*, which is being published by Cambridge University Press under the auspices of IAU and IUHPS, will now have each of the four volumes divided into either two or three parts, and that part 1 of vol. iv is now in the press. It is hoped that two parts will be published each year. Pedersen announced that the manuscripts, notes and correspondence of Ejnar Hertzsprung, now in the History of Science Department of Aarhus University, were available on microfiche, and Hoskin reported that plans were well advanced for making the papers of the third and fourth Earls of Rosse similarly available. It was pointed out that microfilms of the Herschel papers of the Royal Astronomical Society had been purchased by a number of institutions world-wide, and that historians and astronomers wishing to work on the papers no longer needed to incur the cost of travel to England. By charging for each print a little more than the actual cost, an institution could recover its initial outlay and make its archival holdings available world-wide at no expense to itself.

Other announcements included news of a catalogue of seventeenth-century books in European astronomical observatories being prepared by Giovanna Grassi; of a new critical edition of the star catalogue in the *Almagest* by Paul Kunitzsch; and of a forthcoming publication by R. Dvorak to replace the *Planetentafeln* of P.V. Neugebauer. Ephemerides of the Sun, Moon and planets from -1000 to -600 have been published for the Austrian Academy of Science by H. Hunger and R. Dvorak, though the accuracy at such a distant time interval must be regarded with considerable reserve. Among other topics discussed by the Commission was the possibility raised by Derek Howse of sponsoring at Greenwich in 1984 a symposium to celebrate the centenary of the International Meridian Conference.

## COMMISSION 42: CLOSE BINARY STARS (ÉTOILES DOUBLES SERRÉES)

### Report of Meeting, 18 August 1982

PRESIDENT: B. Warner SECRETARY: E. Budding

### Business Meeting I - August 18 1982

Minutes of the meeting held in Room B2 University Complex, Patras from 09.30 until 11.00.

1. The Chairman (President of the Commission - B. Warner) first listed the proposed agenda on the blackboard and requested comments from the floor. After none were given the proposed agenda was deemed acceptable and adopted. The Commission then elected E. Budding as Secretary for the meetings, whose responsibility was to note the minutes and act as correspondent for the General Assembly Journal "Astrocosmos".

2. The President asked for a moment of silence in memory of five Commission members: A. Beer, G. Chis, K. Serkowsky. M.B. Shapley and J.A.J. Whelan who had died since the last General Assembly.

The President then went on to consider the present membership, currently totalling 211 people, which he noted as a relatively large number. He had received no notifications of intended withdrawals, though some 41 new proposals had come to his attention - 7 by direct contact with himself and 34 through the IAU secretariat sent via National Committees. The President urged serious attention to the known activity of proposed Commission Members, mentioning that, for example, merely expressing an interest in the workings of the Commission should not be a basis for acceptance, particularly in view of the large current membership.

On the other hand, he knew of several names who he thought ought to become Commission Members and put some of these before the meeting, at the same time inviting further nominations from those present. The President pointed to the relative shortage of theoreticians among the membership and hoped that a better balance might be achieved. After some discussion of nominations it was announced that a list of those proposed would appear on the Commission's notice board, and the Organizing Committee would in due course make the selection. The President asked all those with information or comments on proposed new members to contact him, or other members of the Organizing Committee, during the week.

3. The President listed the existing Organizing Committee of the Commission (in no particular order) as follows: A.M. Cherepashchuk (USSR), A.H. Batten (Canada), B. Warner (South Africa), R.H. Koch (USA), Y. Kondo (USA), J. Smak (Poland), E.P.J. van den Heuvel (Holland), M. Kitamura (Japan). In a separate group were listed J.A.J. Whelan (GB - deceased), K. Gyldenkerne (Denmark), G. Larsson-Leander (Sweden), M.G. Fracastoro (Italy), S.D. Sinvhal (India) and L.B. Lucy (USA). The last four named had all indicated to the President, either directly or indirectly, a desire to leave the Organizing Committee.

The President then put the following names on the board G. Shaviv (Israel), K.D. Abyánkhar (India), M. Rodonó (Italy), J. Anderson (Denmark), K. Ch. Leung (USA); to stand in place of the members leaving the Committee. These were accepted by the meeting.

Finally the President proposed that the usual succession to his office be followed, i.e. Vice President A.H. Batten to become the next President. He also proposed that J. Smak become the succeeding Vice-President. These proposals were accepted nem con.

4. The President posed the question as to the real usefulness of the Commission Report. After thanking various Report Contributors he noted that a much more comprehensive record of the work on close binary systems could always be derived from the annual *Astronomy and Astrophysics Abstracts* where restrictions as to length or content were much less present. He also thought that the General Secretary may be able to offer some advice on the subject. T. Herczeg (supported by K. Leung) pointed out a possible usefulness of the Report to members of other Commissions of the Union. F.B. Wood commented on the possible historical value of the Report to future archivists. There was some discussion as to the nature or style of subsections which the President thought could have room for improvement. It was possible that certain lists of references could be more usefully diffused via a centre such as that at Strasbourg.

5. At the last General Assembly Dr. Larsson-Leander had announced his intention to give up the editorship of the "Bibliography and Program Notes" by this meeting. As it was generally agreed that these communications were very useful and should be maintained, a new editorship would therefore be required. The President praised the work of Dr. Larsson-Leander and his team of helpers for their unstinting efforts over so many years, and strong applause greeted the President's request for a vote of thanks. It was pointed out that the IAU does not provide any financial support for the "Program Notes" and finance might become a problem for a new editor. Dr. T. Herczeg was proposed as possible editor to replace Dr. Larsson-Leander, and he announced his willingness to accept the responsibility.

The main uses of the "Program Notes" were thought to lie in their ability to allow members to avoid duplication of effort and offer co-operation on known programmes of work. Dr. Bolton suggested that the notes could contain a questionnaire about current programmes, rather than rely on members writing letters. Yoji Kondo suggested a more basic questionnaire might be issued to determine in what areas the "Program Notes" were felt to be useful. The possibility of a nominal charge, both to help towards costs and to prevent unnecessary postage, was raised, but it was deemed technically difficult and at present unnecessary. The possibility of issue from Strasbourg was again raised in case of financial difficulties. Dr. Herczeg said that he would try to examine the various practical options.

Y. Kondo was invited to speak about setting up a co-ordinated program on Aur stars - to include as a special case the forthcoming eclipse of $\varepsilon$ Aur. He visualised that a great deal of observational material spanning a wide spectral range would be available after the end of the eclipse of $\varepsilon$ Aur in 1985. Also material would be available for $\zeta$ Aur, 31 and 32 Cyg, VV Cep and $\mu^1$ Sag. $\mu$ Gem was also mentioned as a possibly similar candidate. Members were asked to communicate directly with R. Stencel with a view to arranging a Colloquium in 1985. Asked about the requirement for further organizational support from the Commission, Kondo indicated his desire to spare collective effort. It was thought that he and Stencel could effectively deal with the proposal themselves.

The President then spoke on behalf of J. Mattei of the AAVSO to urge IAU support for the observational work of a large number of amateur observers. Some 150 cataclysmic variables have accumulated photometric data on them, and though some of the work prior to 1964 had been reduced and published there is a large backlog of observations requiring effective processing. A figure of some 4.5 million individual observations was mentioned, including a growing quantity of photoelectrically based data.

Support from the IAU would add prestige to the work; a fuller funding might then be obtained from other sources. The final output was visualised as comprising an Atlas of Light Curves, and magnetic tapes for the corresponding numerical information. A.H. Batten volunteered continued support for this work as incoming President.

The President summarized the discussion in terms of two points: (i) that there was unaninimous support by the Commission for the reduction and publication of these observations as soon as possible, and (ii) that the incoming President would make some private arrangements with Mrs. Mattei to this end. On the basis of (i), the Organizing Committee later submitted the following resolution to the Resolutions Committee of the IAU: "Considering the recent increased interest in cataclysmic variables, and the absence of long term light curves for correlation of observational and theoretical research, we note the data collected by the AAVSO from observers and groups worldwide, and we resolve that these already computerised observations would be a good source of reference and that they should be published". Signed by Janet A. Mattei, President, AAVSO and Brian Warner, President, Commission 42.

8. The President then read out a communication he had received from Dr. Katherine Kron concerning some old observations of the Algol type variables RR TrA, ST Car and X Pic, obtained in 1951-53 by J. Summerville at the request of A. Hog. The data consists of a large quantity of UBV photoelectric photometry, particularly of RR TrA. The problem was - what would become of these observations?

The President suggested that the material be sent to the variable star archives of the RAS, but appealed for any further comments. Chambliss remarked that, if it were possible, the observations could be better served by a more immediate analysis. The suggestion that an announcement in the *Inf. Bull on Variable Stars* about these data should be given, was supported by the President.

9. Under the heading of "any other business" the President stated that in view of the number of proposed contributions to the Scientific Meeting to be held on August 23 that the two sessions (3 and 4) be allowed to run continuously (i.e. without the allocated half-hour break between). About 16 titles had already been received and perhaps more might come. The President thought that the second Business Meeting could also be used for scientific presentations if the demand was sufficient.

In a final point at this Meeting, E. Budding commented on the forthcoming bicentenary of Goodricke's original noticing of the periodicity of Algol and wondered whether some special meeting might be arranged in honour of the occasion. The general feeling appeared to be that the suggestion had come too late for international organization, though no categoric decision was arrived at.

The Meeting was closed at $11^{h}00$.

Report on Scientific Sessions held on 23 and 24 August, 1982

PRESIDENT: Brian Warner SECRETARY: Edwin Budding

## Abstracts of Papers

### Optical Spectrophotometry of the HDE226868/Cygnus X-1 System

D.R. Gies and C.T. Bolton

We have carried out spectrophotometric measures on high signal-to-noise optical spectra obtained by orbital phase averaging lower signal-to-noise high dispersion spectra in ten phase bins. The equivalent widths of the absorption lines confirm the O9.7 Iab spectral classification of the primary. The strengths of all of the absorption lines vary by less than ten percent with orbital phase, except that the He II λ4542 line varies by approximately 25 percent with maximum strength coincident with superior conjunction of the supergiant.

Fourier transform analysis of several He I and Si III absorption lines yields $v\sin i = 109 \pm 2$ km s$^{-1}$ and $v_{macro} = 60 \pm 10$ km s$^{-1}$. If we assume synchronous rotation and a minimum distance of 1.9 kpc, then the light and velocity curves for the system can be combined to yield $30^{o} \leq i \leq 40^{o}$ and *minimum* masses of 24 and 10 solar masses for the supergiant and its companion.

X-ray heating of the hemisphere of the primary star that is facing the companion cannot account for the observed behaviour of the He II λ4686 emission profiles. However, we have been able to reproduce most of the significant features of the behaviour of these profiles with a model in which the stellar wind from the primary is strongly enhanced near the inner Lagrangian point. At the moment, there does not seem to be any need to invoke a disk around the companion to explain the He II emission line profile.

### The X-ray burster 1735-44 : a binary period?

J. Hutchings, D. Crampton and A. P. Cowley

SIT Vidicon spectra with 30 minute time resolution indicate a 2.86 hour cycle apparently periodic over two nights. Line intensities and velocities of ionized C and N vary systematically but in a complex way. Possible binary parameters and mechanisms for exciting the lines are discussed.

### The unique binary BE UMa

J. Hutchings, D. Crampton and A. P. Cowley

New radial velocity data refine the orbit and show a 0.15 phase error in earlier ephemerides. Analysis of the light curves considerably constrains model parameters, and suggests that the binary is in an unusual evolutionary state: a very hot subdwarf heating one side of an M giant companion.

### The radial-velocity orbit of HD77581 = Vela X-1 determined by cross correlation

E. J. Zuiderwijk

Radial velocities have been estimated for the early B-type super-giant HD77581 using 65 12Å/mm Coudé spectrograms obtained between 1973 and 1976 at the European

Southern Observatory in Chile (ref 1). The rectified intensity registrations of individual spectra were cross-correlated with a high signal-to-noise average of all spectra (the template), after sampling in wavelength intervals centred on the strongest spectral features at positions equidistant in log $\lambda$, and after proper normalisation. The difference in velocity between test spectrum and template follows directly from the position of the maximum of the cross-correlation function. The accuracy of the velocity estimate is between 2 and 3 km/s for most of the spectra. A more detailed description of the method and error determination will be given elsewhere.

Orbital elements of the X-ray source Vela X-1 (= 4U 0900-40) are well known:- $e = 0.089$ and $\omega = 168^o$ (ref 2) - and were used to transform the epoch of observation for each spectrum into true anomaly. The radial velocity variations as a function of true anomaly, however, do not follow the expected sine wave. Instead, during the transition of the X-ray source a disturbance of about 7 km/s magnitude appears to be superposed on the underlying sinusoid. The occurrence of this perturbation coincides with an increase in strength of the HeI triplet series (by about 30%), and with a distortion in the H$\beta$ profile. Therefore, observations made during X-ray transition were given a low weight in an orbital solution. Then $K_{opt} \sim 17$ km/s, as opposed to the 20.5 km/s found before (ref 1), and the mass of the X-ray source $M_X \sin^3 i \sim 1.3\ M_\odot$. For an inclination of 75° (ref 3) this yields: $M_X \sim 1.45\ M_\odot$.

References

Van Paradijs, J. *et al.*: 1977, Astron. Astrophys. Suppl. 30, p. 195.
Hayakawa, S.: 1982, Space Sci. Rev. 31, p. 283.
Avni, Y. and Bahcall, J.N. 1976, Astrophys. J. 197, p. 675.

## RY Sct: An International Collaborative Project

R. M. West

The $11^d$ period close binary system RY Sct is being observed during a concerted campaign. Photometric observations have been obtained in the USSR by Zakirov and Kumsiachvili and analysed by Cherepashchuk; polarimetric observations by Korol (USSR) and Landaberry (Brazil); high dispersion spectroscopy by Sahade (Argentina) and West (ESO); IUE spectra by West (ESO); 21-cm observations have been attempted by Bajaja (Argentina) and Kazès (France), and Belserene (USA) has analysed the historical light curve. Preliminary results indicate that this system, in which the primary star is of spectral-type B0, is surrounded by an expanding gas shell and the presence of gas streams and a rotating torus are indicated. From the evolutionary point of view, RY Sct is probably a younger, hotter and more massive analogue of the $\beta$ Lyrae system. The total mass is in excess of 30 solar masses. The programme continues and a full discussion is expected to become available within the next year or two.

## On the Age and Angular Momentum of W Ursae Majoris binaries

W. van Hamme

Homogeneous and accurate values of the parameters of W Ursae Majoris contact binaries are required to solve the problems of their structure and age. Photometric elements of many systems are available, but the difficulties of spectroscopic work, because of the rapid rotation of the components, makes reliable determination of absolute dimensions impossible. Nevertheless, the position of 47 W Ursae Majoris systems in a mass-radius diagram was constructed from the photometric elements and only one spectroscopic datum, i.e. the observed spectral type of the binary, and the spectral type-mass relation of Habets and Heintze (1981, Astron. Astrophys. Suppl. Ser. 46, 193). This position then was compared with the position of the theoretical Main Sequence band, constructed for 11 different chemical compositions from the evolutionary models of Mengel, Sweigart, Demarque and Gross (1979, Astrophys.

J. Suppl. 40, 733), and differential effects in age were studied. Angular momentum-mass and angular momentum-mass ratio diagrams also were constructed. The main conclusions are:

(i) the subdivision into A- and W-type systems does not purely correspond to a separation in age, e.g. some of the A-type systems (7 of the total number of 22 of our sample) share the same evolutionary state as the bulk of the W-type systems.

(ii) there is a clear separation in the A-type group between systems which can and which cannot be the evolutionary products of W-type systems.

### A Possible Source for W UMa Systems
E. Budding

A possible source for the W UMa binaries is described, in terms of a Case B type process of mass transfer between components, for which the original mass ratio ($m_2/m_1$) is quite small ($\lesssim 0.2$). The slow stage of the Case B process is concentrated on and discussed by means of a simplified analytical representation, which reduces, essentially, to a pair of simultaneous ordinary differential equations.

The progress of the resulting common envelope phase is studied by reference to a number of parameters, including some involved in braking mechanisms, which get rid of a certain proportion of orbital angular momentum.

Many observed features of W UMa systems seem able to be broadly accounted for in the scheme, which predicts that a large proportion ($\sim$ 10%) of seemingly single intermediate mass stars should actually be attended, not too distantly, by low mass companions.

### On the Light Curve Analysis of RS CVn Type Binaries
Osman Demircan

The present state of the light curve analysis of the RS CVn type binaries has been reviewed. Considerations on the wave-like distortion effect have been critically examined and under the spot model assumption an idea has been introduced to simplify the light curve analysis problem of RS CVn type binaries. The idea is that if spots cover only one part of the hemisphere (say $\Delta\lambda$ in longitude) of a component they ought to be out of sight in a certain time interval. In this interval of time ($180^o - \Delta\lambda$ in phase if $P_{rot} = P_{rev}$), the light changes should be free from spot effect and can be analysed by considering only the eclipse and proximity effects.

### DM Virginis - A Future RS Canum Venaticorum Star?
C. R. Chambliss and R. L. Walker

DM Virginis is a detached eclipsing binary system consisting of two almost identical F6-7 IV-V components. Its period is shown to be about 4.6694 days. Light curves in UBV are analysed by the Wood method. Consistent solutions are found, and these indicate that the eclipses are complete and nearly central. The primary is an occultation, and the duration of the complete phase of the eclipses is about 10 minutes. The absolute dimensions of the components are given, and these are shown to be very similar to those of the eclipsing binary CD Tauri and to the subgiant Procyon. DM Virginis is an important system, because its components are essentially free of the close proximity effects found in most eclipsing binary systems. Although this system cannot presently be regarded as an RS CVn system, it is seen as a precursor of this group of stars.

### New Results in the Copenhagen Binary Programmes
J. Andersen, B. Nordstrom and J. V. Clausen

The Copenhagen programme (the whole programme is carried out by a group of $\sim$8 people) of photometric and spectroscopic observations of double-lined eclipsing binaries aims at determining masses, radii, and luminosities of sufficient accuracy for critical comparison with stellar structure calculations. During the past couple of years, a large body of new observations has been collected. When these have been fully analysed the existing meagre body of high accuracy data should be at a level permitting a comprehensive discussion of the whole material, planned for 1983. However, some of the individual unpublished or preliminary results should be of immediate interest to other workers in the field, in particular:

1) Photoelectric photometric and radial velocity observations of TZ For are yielding the first accurate masses and radii for G giants;
2) EM Car will yield one of the very few existing reliable mass and radius determinations for O stars (about 20 $M_{\odot}$);
3) For the well-known bright system V Pup the interpretation of older radial velocity data as evidence for abundant circumstellar material is found to be incorrect;
4) Several of the newly studied systems have eccentric orbits, and the improved determination of $K_2$, the density concentration coefficient, contribute yet another parameter in the comparison with theories of stellar evolution;
5) Recent photoelectric radial velocity observations of two of the Serpentid group of rapid mass-exchange binaries studied in the satellite UV by Plavec *et al.* (SX Cas, RX Cas), provide reliable velocity curves for the secondaries which cast doubt on the accepted semi-detached nature of these systems - and thus on the mechanisms for the mass exchange itself.

## Spectroscopic Observations of the Eclipsing Triple System λ Tauri

Francis C. Fekel and Jocelyn Tomkin

Radial velocities from low noise Reticon observations obtained at McDonald Observatory of the University of Texas are used to determine the orbital elements of the short-period eclipsing components of this semi-detached system. Combining the minimum masses from the orbital solution with the inclination from the photometric solution, the B3 V star has a mass of 7.2 $M_{\odot}$ and the mid-A subgiant has a mass of 1.9 $M_{\odot}$.

The 33 day periodicity in the velocity residuals, discovered in previous investigations, is confirmed and is present in the secondary velocities as well as those of the primary. The presence of the 33 day periodicity in the velocities of *both* components means that it can be unambiguously ascribed to orbital motion around a third body. The K and f(m) for the 33 day orbit are 10.1 km $s^{-1}$ and 0.0034 $M_{\odot}$. From photometric considerations the orbits appear to be nearly coplanar. Therefore, the mass of the third star is 0.7 $M_{\odot}$ and is probably a K dwarf.

## The VW Cephei System

Albert P. Linnell

High speed photometry of VW Cephei discloses important features in the light curves and color curves (Linnell, 1982, *Ap. J. Suppl.*, in press). Light synthesis simulation leads to the following conclusions: (1) None of the proposals in the literature to explain W-type W Ursae Majoris light curves satisfactorily represents the VW Cep data. (2) The facing hemispheres of both components must have elevated temperatures, with the secondary slightly hotter. A tentative proposal is that these temperature excesses represent the observational signature of the energy exchange process.

## Period changes and the distortion of the light curve of the eclipsing variable CQ Cep with a WR component

J. M. Kreiner and J. Tremko

The variation of the period of the eclipsing variable CQ Cep has been found. The period has been shortening in the time interval from the beginning of the observation until J.D. 243 0000 by a value of 0.15 s $year^{-1}$. After J.D. 243 0000 the period became constant and has been undergoing small variations only. The period changes are ascribed to the non-stationary and non-isotropic mass outflow from the Wolf-Rayet component. The influence of the asymmetry of the light curve on the determination of the epochs of minima is discussed.

### The Theoretically expected X-ray Luminosity and the Binary Nature of WR Runaway Stars

D. Vanbeveren, W. van Rensbergen and C. de Loore

The X-ray luminosity expected in WR+ compact star binaries is calculated by combining the production of hard X-rays by accretion and absorption by a stellar wind, in order to decide if the column density in a WR stellar wind is large enough to absorb all X-radiation.

Stating that there are as many WR+ compact star binaries as WR+OB systems depends on the survival probability of the binary during the supernova explosion. This probability is critically investigated and a statistical study is presented based on massive close binary evolution in order to estimate the number of WR+ compact star binaries.

### UBV Light Curves of AR Lacertae during 1980-81 and 1981-82

I.-S. Nha, H. S. Park, Y. W. Kang, F. B. Wood, R. M. Williamon and D. B. Leviton

The UBV observations of the brightest RS CVn-type eclipsing binary star AR Lac were made at four observatories, two in Korea and two in the USA, in 1980-81 and 1981-82 seasons. As a result of the cooperation, two light curves in the yellow and in the blue are completed for each observing season.

The orbital period of AR Lac has decreased by $0\overset{d}{.}014$ since 1977. An analysis of our yellow light curves together with five other yellow curves available in the literature since 1975 shows that there seems no periodicity in the migration of the distortion waves. There is a gradual decrease of at least $0\overset{m}{.}1$ between 1976 and 1982 in the brightness of the cooler component if one assumes that the hotter component is constant.

### A Re-analysis of the Eclipsing Binary System VV Orionis

C. R. Chambliss

VV Orionis is an eclipsing binary consisting of B1 V and B4 V components with a period of 1.4854 days. The eclipses are complete and the primary is a transit. The observations of VV Ori in UBV and in uvby obtained by Chambliss in 1975-79 have been re-analysed with the use of an updated version of the WINK program of Wood. Several solutions were obtained using the normal points (ca. 80 per light curve). These establish fairly reliable values for the linear limb darkening coefficient of VV Ori A ($u_1$) and approximate values for the luminosity of the third component of this system. Attempts to determine non-linear coefficients of limb darkening for the primary component, however, proved to be unsuccessful. Solutions were also obtained using all observations (ca. 620 per light curve). Very close agreement was found between the values of the geometric elements determined from these solutions and those determined by use of the normal points only. The solutions based on all observations produced reliable values for $u_1$, typically about $0.30 \pm 0.04$. The results are in good agreement with theoretical limb darkening coefficients derived from model atmosphere calculations. The contribution of VV Ori C (orbital period = ca. 120 days) to the light of the system was also ascertained, and it was found that this could be best interpreted if this component has a spectral type of A3 V. Mean values for the geometric and photometric orbital elements of VV Ori are also derived. Since VV Ori A is one of the very few early-type stars for which reliable limb darkening coefficients can be empirically determined, it is viewed as being a system of considerable importance.

## On Selected Contact and Semidetached Systems

K. C. Leung, L. Z. Shen and D. S. Zhai

Photoelectric observations of AW Lac, BX Peg, BB Peg, and RZ Dra were obtained at Beijing Observatory. The data were analized with the Wilson and Devinney method at the University of Nebraska.

AW Lac : The spectral type of B2: for this system is rather uncertain. An extensive search for photometric solutions at many assumed values of mass ratio, q, were made. The most reliable photometric q (i.e. the sums of weighted squares of deviations, $\Sigma$ , at the minimum) were found between $0.9 < q \leq 1$. The degree of overcontact is large at about 70%. The period and the adopted spectral type combined would imply that this system is located close to the domain of Zero-age Contact systems. Unfortunately, we are unable to determine its absolute dimensions and compare them against the ZAMS mass-radius relation because of the lack of any radial velocity curves.

BX Peg and BB Peg : These systems have similar spectral types; G5 and F8, and periods; 0.2804 and 0.3615 days respectively. Their light curves are quite similar in shape, with the exception of the significant difference in the heights of the maxima for BB Peg. An extensive search for solutions at the assumed values of q's at 0.35, 0.45, 0.55, 0.7, 0.9, 1.2, and 2.8 were made. Coincidentally, both systems are having reliable q values at about 2.8. The degrees of overcontact are small at 12% and 10% respectively. They are both W-type systems. It is of interest to point out that while the two systems are similar in most aspects but there is such a large difference in the observed maxima for BB Peg.

RZ Dra : After extensive search for q values, we have improved significantly on the previously reported photometric q. The reliable q should be at 0.45 and the system is semidetached with its secondary filling the Roche lobe.

## The Period and Light Variations of WW Aurigae

J. Tremko and G. A. Bakos

Recent photoelectric observations by Kiyokawa and Kitamura (1975) have shown that the period of WW Aur remained constant for the last 30 years. On the other hand the light curve showed some variations of about 0.02 mag during and outside eclipses.

From our photoelectric observations we have derived three new epochs of primary and seven epochs of secondary minima. By means of an O-C diagram we have improved the period to a value P = 2.52501946 days from the old value, P = 2.52501922 days. We also found that the secondary eclipse is displaced from the mid-point of two consecutive primary minima by 0.0008 days indicating that an apsidal motion effect may be present.

Variations of the B-V index are obvious at primary minima. The colour index increases by 0.03 mag at mideclipse indicating that the more luminous and hotter component is being eclipsed. A blueing effect at secondary minima could not be observed since no two colour observations were available. Other light variations, possibly due to gas streams have been observed before the first and after the fourth contact. At present time these light changes are being investigated.

### Reference

Kiyokawa, M. and Kitamura, M.: 1975, Annal Tokyo Obs. Series 2, Vol. XV, No. 3.

## Non-Keplerian Effects in Triple Stars

S. Söderhjelm

This communication summarizes the consequences of a third companion (period $P_2$) to a close binary (period $P_1$). Tidal deformation of the close pair is taken into account, and both numerical and analytical methods are used. Apart from the well-known revolutions of the apsidal and nodal lines, a modulation of the binary eccentricity is important. The amplitude of this modulation grows with the relative inclination between the orbits, and above a "critical inclination" no circular orbits are possible. For point-mass triples, the critical inclination is around 40°, but it increases with the tidal distortion. When the binary apsidal motion alone becomes larger than the three-body contribution, the eccentricity stays nearly constant at all inclinations. Such facts must be included in any study of the orbital evolution of triple systems. The interaction between tidal friction and the three-body perturbations will proceed along qualitatively different tracks in different regions of the ($P_1$-$P_2$)-plane and for different relative inclinations. A full discussion is very complicated and remains to be done. As for the actual observation of three-body effects, the relevant time-scale is about $P_2{}^2/P_1$, which is often quite long. An important case is IU Aur, where the eclipse depths are secularly increasing due to a $400^y$ (approx.) nodal motion. In a few years, the inclination will pass through 90°, and photometric coverage is very desirable.

## Close Binary Systems before and after Mass Transfer: Origin of Spectroscopic Semidetached Binaries

J. P. de Grève and F. Vansina

For 26 semidetached binaries, the masses of both components before and after the Roche lobe overflow are determined, assuming a case B of mass transfer. When an amount of mass loss from the system of roughly 40% is adopted the initial mass ratios are most concentrated in the range 0.35 to 0.75. Only a few systems have initial mass ratios close to 1.

## SV Centauri - Change of the Light Curve since the Beginning of the Century

Z. Kviz

The comparision of the light curves of SV Centauri for the last 90 years indicates that the minima are getting deeper (especially the primary one) and probably wider. The primary minimum occasionally shows a flat bottom. Frequent monitoring of the depth and shape of the minima (mainly primary) is urgently needed.

## COMMISSION 44: ASTRONOMY FROM SPACE (L'ASTRONOMIE A PARTIR DE L'ESPACE)

### Report of Meetings

ACTING PRESIDENT (23-26 AUGUST): M. Oda

ACTING PRESIDENT (17-22 AUGUST): Y. Kondo SECRETARY: J. Rahe

In the absence of Commission 44 President R. J. van Duinen, Y. Kondo was designated Acting President until the arrival of Vice-President M. Oda.

### BUSINESS MEETING - 18 August 1982

The first item on the agenda was the election of new officers for the next triennial period and the co-option of new members. M. Oda (Japan), the incumbent Vice-President, was named new President, following the well-established protocol, and Y. Kondo (U.S.A.) was elected new Vice-President for the next three-year period. Newly elected members of the Organizing Committee are M. Grewing (F.R.G.), D. C. Morton (Australia), K. A. van der Hucht (The Netherlands) and A. J. Willis (U.K.); the remainder of the Organizing Committee for the next three years are A. A. Boyarchuk (U.S.S.R.), K. Fredga (Sweden), L. Peterson (U.S.A.), E. Peytremann (France), A. B. Underhill (U.S.A.), B. Valnicek (Czechoslovakia) and R. J. van Duinen (The Netherlands).

The Acting President reported with sorrow the tragic death of D. W. Schuerman in May 1982.

Commission 44 voted to co-sponsor the proposed IAU Colloquium "The Origin of Non-Radiative Heating/Momentum in Hot Stars", to be held in June 1984; the meeting was proposed by A. B. Underhill.

As no further business meetings were scheduled at the General Assembly, the following actions were taken by polling the members of the Organizing Committee:

Commission 44 agreed to endorse the following resolution. "Commission 44 strongly supports the use of the Space Shuttle science payload for observing Comet Halley during its 1985-86 apparition. Space Shuttle observations can be carried out when the Comet's position is near the sun and other means of observations are not feasible. Such measurements will be a valuable complement to ground-based observations and cometary missions. Multiple launches of the Shuttle cometary science payload will make a significant contribution to the world-wide efforts to study Comet Halley."

Commission 44 also agreed to co-sponsor the proposed IAU Colloquium "Local Interstellar Medium", to be held in June 1984 at the University of Wisconsin. The meeting, sponsored by Commission 34, was proposed by Y. Kondo.

SCIENCE MEETINGS

Commission 44 sponsored or co-sponsored a number of scientific meetings at the General Assembly, as listed below:

New Ideas About Stars Demanded by Observations from Space, 18 August, chaired by A. B. Underhill, sponsored by Commission 44.

Galactic Astrophysics & Gamma-Ray Astronomy, 19 August, G. Morfill, Commissions 44 & 48.

UV and X-Ray Observations of Interacting Binary Systems, 20 August, Y. Kondo, Commissions 42 & 44.

Solar Maximum Year, 20 August, E. Tandberg-Hanssen, Commissions 10, 12, 40 & 44.

Recent Solar Results, 21 August, V. Bumba, Commissions 10, 12 & 44.

WG on Objects for the Extragalactic Reference Frame, 21 August, Commissions 7, 8, 24, 33, 40 & 44.

Space Telescope, 23 August, F. Macchetto, Commission 44.

Future Space Program, 25 August, M. Oda, Commission 44.

The Interstellar Medium Observed from Space, 25 August, M. Grewing, Commissions 34 & 44.

Brief summaries of the meetings have been received from some of the chairmen and are reported in the following. The proceedings of the Joint Session "UV and X-ray Observations of Interacting Binary Systems" is to be published in the Highlights of Astronomy; the summary of this meeting will therefore not be included in the following report.

New Ideas About Stars Demanded by Observations from Space - Anne B. Underhill, Chairman.

The wide topic of New Ideas About Stars Demanded by Observations from Space was reviewed by A. G. Hearn who summarized results concerning the hot stars, by C. Jordan who summarized the results for cool stars, and by C. J. Crannell who summarized the results attained in the domain of high-energy astrophysics.

Hearn noted that the observations (1) that most, if not all, hot stars lose mass at a detectable rate, and (2) that many radiate X-rays at a detectable level causing us to modify our concept of what occurs in the outer atmospheres of hot stars. When one is determining the rate of mass loss, it is essential to take account of how the ionization balance changes outward. Radiatively driven instabilities in a wind may be important for generating the very hot gas which is required to radiate X-rays at the observed level.

Jordan reviewed the observations of cool stars under three headings: (1) surveys in the lines of Mg II and in the EUV range; (2) detailed studies of individual stars, and (3) results from X-ray surveys. She pointed out that many of the differences noted among the UV spectra from G and K stars appear to be associated with changes in coronal structure and in the presence or absence of stellar winds. However, there appears to be no sudden change in total energy input into the chromosphere-corona as one goes along the spectral sequence. A positive correlation between the rate of rotation and the degree of activity in the chromosphere-corona seems to exist for dwarf stars of types F, G, and K. The activity of stellar magnetic fields appears to be an important factor for controlling what is seen.

Crannell reviewed the instrumentation used to observe stars in the energy range from 10 keV to 1 GeV. A wide range of physical processes seems to be active including electron bremsstrahlung, nuclear de-excitation, and particle annhilation. Observations of stars in the high-energy range gives information chiefly about binary stars and neutron stars.

Galactic Astrophysics & Gamma-Ray Astronomy - G. Morfill, Chairman.

The aim of this meeting was to determine the role of galactic gamma-ray astronomy within the general concept of galactic astrophysics. The timing, at the end of the COS-B mission, was regarded as opportune, because it gives interested astrophysicists the possibility for interdisciplinary studies using the existing gamma-ray data base (e.g., comparison with infra-red, radio, x-ray, and other observations), as well as for theoretical studies. The next generation of gamma-ray detectors will probably not be in operation for another 5-10 years, and therefore it is hoped that the proceedings of this meeting can be used (in the intermediate time) as a basis for further studies, as a stimulation for more theoretical work and as an important contribution for defining the aims and operation of future gamma-ray missions. The interrelationship with other branches of astronomy, the astrophysical implications and the study of relevant physical processes using available measurements in the near-Earth environment were important results of the meeting.

The Results of the Solar Maximum Year - E. Tandberg-Hanssen, Chairman.

After an overview of the SMY by de Jager there followed reports on the FBS (Flare Buildup Study), the SERF (Study of Energy Release in Flares), and the STIP (Study of Traveling Interplanetary Perturbations) by Gaizauskas, Rust, and Dryer, respectively. A comprehensive report on the SMY activities in the USSR was given by Mustel, and Wang Shon Gian reported on solar observations in the People's Republic of China. After lunch, highlights from the Hinotori and SMM satellites were discussed by K. Tanaka, House and Leibacher. The session concluded with two comprehensive papers on Magnetic Field Studies and Radio Astronomical Studies during SMY by Hagyard and Pick, respectively.

The combined impression made by all these contributions was the wealth of information on solar activity gained during the SMY.

Space Telescope - F. Macchetto, Chairman.

The meeting, which was well attended, was entirely devoted to the Space Telescope project and the expected impact that it will have on astronomy.

Prof. Giacconi discussed the role of the Space Telescope Science Institute in leading the scientific, related activities of the Space Telescope. He described its tasks, responsibilities and management organization. He illustrated what an observer had to do to prepare for and to carry out an observing program and what he can expect in terms of support by the Science Institute. He described the selection procedures that the Science Institute will adopt and discussed the policies of data rights and archival research.

Dr. F. Macchetto illustrated the preparations that are being made in Europe to fully exploit the Space Telescope data and explained the role of the European coordinating facility as the European Center for Space Telescope Archival Data and Astronomical Software Exchange and Development.

Prof. C. R. O'Dell described the Space Telescope Observatory as a whole, illustrated the present status of development, and showed the excellent quality achieved in polishing the mirrors. He also assured everyone about NASA's commitment to launch the Space Telescope prior to the next IAU General Assembly.

The Principal Investigators for the scientific instruments described briefly their instruments and the expected scientific performance obtainable from them. The speakers were Prof. J. Westphal (Wide-Field Planetary Camera), Dr. F. Macchetto (Faint Object Camera), Dr. R. Harms (Faint Object Spectrograph), Dr. J. Brandt (High Resolution Spectrograph), Prof. R. Bless (High Speed Photometer), and Dr. R. L. Duncombe (for the Astrometry Team). The session was concluded by an enthusiastic talk by Prof. M. Longair who communicated the scientific excitement of this unique mission to the audience.

In support of this session the Space Telescope Science Institute published a volume which contains the text of each presentation and additional details on the Space Telescope and its complement of scientific instruments. This publication is available from the Science Institute.

The Interstellar Medium Observed from Space - M. Grewing, Chairman.

The existence of a million-degree-component of the interstellar gas is now well established. Dr. G. Garmire discussed the soft X-ray sky maps that have been obtained with the proportional counters aboard the HEAO I spacecraft, and argued that some of the extended features seen in these maps can be accounted for if the solar system is surrounded by a million-degree gas extending out to a distance of about 100 pc. Within this volume must be a number of cool, evaporating clouds. At its edge one finds an expanding shock.

The same local region in space was then discussed by Dr. F. Bruhweiler who reviewed the results from recent UV experiments. It had been known already that the H I column density is very low along a few lines of sight. A more systematic study has not shown that low H I column densities are found inthe entire hemisphere between $\ell$=90 and $\ell$=270 degrees to distances of many tens of parsecs. Conversely, looking towards the galactic center, the line of sight seems to pass through a local cloud before entering a region with little H I. Dr. Bruhweiler went on discussing the implications that the very low H I densities have on the EUV radiation field in interstellar space.

The state of the interstellar gas at the outskirts of our galaxy was considered by Dr. C. Blades who reviewed the recent observations of the galactic halo and halos around external galaxies. The IUE observations of distant background targets especially provide clear evidence for gas well outside the plane of our galaxy. In addition to the predicted highly ionized species one finds, however, also evidence for neutral or only singly ionized constituents, indicating the presence of cool clouds. Indeed, from very high resolution optical observations many of the absorption components are seen to break up into several subcomponents that are extremely narrow.

The contributions that infrared observations can make to extend our understanding of the interstellar medium were highlighted by Dr. M. Harwit, who chose the Orion complex as an example. The importance of various fine-structure transitions was stressed and the observed strengths were compared to model predictions for H II regions. Turning to very recent results Dr. Harwit reported the detection of the diffuse C II-156 micron emission from the galactic plane that was measured in a very beautiful experiment in which the moon was used as a reference source while it moved across the plane.

Ever since the interstellar reddening law became known from UV wavelengths throughout the optical to near and medium IR wavelengths, has the question been with us what materials are causing the extinction. As Dr. J. Mathis pointed out in the concluding talk of the session, a vast amount of data has been compiled over the past years, mostly from UV observations. He stressed that the features in the extinction curve are still a puzzle. While they can be attributed to a number of materials, all those materials would also produce additional features which are not observed. Dr. Mathis went on discussing what can be concluded from the observed polarization properties, from the peculiar extinction observed, e.g., in dark clouds, and from the correlation with the atomic and molecular hydrogen densities. Some systematic effects are observed in the properties of circumstellar dust depending on the O/C-ratio in the associated stars. Dr. Mathis concluded his review by stressing the need for further IR- and FIR-observations to be combined with observations taken at UV wavelengths.

Future Space Programmes - M. Oda, Chairman.

Future projects which had been approved or at least authorized in the scientific community were discussed. The Space Telescope and planetary missions were not included, since they were discussed in seperate meetings.

Dr. J. Rosendahl described near term and far term US programmes. They include the Space Telescope and GRO as approved projects, and AXAF whose early start is hoped for. The Explorer project, that is the moderate missions, includes the IR satellite, COBE, EUVE and XTE. The US participation in the ROSAT was decided. As for the Spacelab, several PI class experiments have been approved and infrared astronomy is considered as the next major area to pursue. As for the far term plans the Field's Committee Report was referred to. It was noted that a task force for the space platform or space station had been established.

Dr. Macchetto described the programmes of ESA. The ESA or ESA-related projects COS-B, ISEE, IUE, and GEOS-2 are under operation. Under development are EXOSAT, GIOTTO, Space Telescope, ISPM and HIPPARCOS. Several projects are under study including DISCO, KEPLER, POLO, ISO, MAGELLAN and X-80.

Dr. Grewing reported the progress of the German X-ray telescope, ROSAT.

Dr. Kardashev reported the USSR programmes for the 1980s. They are: 1) A gamma ray experiment in collaboration with France; 2) X-ray observations by the manned Salyut with Czechoslovakia; 3) A future plan for a Salyut X-ray telescope with Holland, France, ERG, and ESA; 4) An optical and X-ray telescope with France; 5) A future gamma burst experiments with Veneras and Prognoz stations; 6) A cryogenic submillimeter experiment with France at late 1980s; 7) The "Relict" experiment; 8) RACSAS (Radio Astronomical Cosmical system Aperture Syntheis); and 9) A radio interferometer with a distance of $10^6$ km.

Dr. Y. Tanaka reported the Japanese programmes in 1980s which follow the currently operating HAKUCHO and HINOTORI. They are approved X-ray observations ASTRO-B and ASTRO-C plus UVSAT as a planned project.

Dr. W. Cudlip reported the present status of IRAS.

A few reports were contributed including Dr. Samain's UV technology and Dr. Hudson's proposed super-high angular resolution techniques for X-ray observations.

COMMISSION 45: STELLAR CLASSIFICATION
(CLASSIFICATION STELLAIRE)

Report of Meetings, 18, 19, 20, 23, 24 and 25 August 1982

PRESIDENT: A. Slettebak SECRETARY: R. F. Wing

I. JOINT DISCUSSIONS

1. Evolution in Old Stellar Populations in Galaxies (19 August, with Commissions 25, 28, 29, 33, 34, 35, 37, and 47).

2. The Extragalactic Distance Scale and the Hubble Constant (23 August, with Commissions 25, 28, 40, 47, and 48).

3. Mass-Loss Phenomena (24 August, with Commissions 29, 35, 36, and 37).

II. JOINT MEETINGS

1. Systematic Effects in Abundance Determinations for Metal-Poor Stars (20 August, with Commissions 25, 29, and 37).

This meeting is outlined in the Report of Commission 29.

2. Stellar Absolute Energy Distributions (25 August, with Commission 25).

This meeting, which was organized by D. S. Hayes (Comm. 25) and R. F. Wing (Comm. 45), was chaired by Wing. The first session was devoted to absolute calibration techniques in the visual and infrared and to the energy distributions of the Sun and standard stars. The second session dealt with applications to specific photometric systems and comparisons of observed flux distributions to those calculated with model atmospheres. The following papers were presented:

H. Neckel, "The Solar Absolute Energy Distribution".

H. Tüg, G. W. Lockwood, and N. M. White, "A Direct Spectrophotometric Comparison, Sun-Vega" (read by Lockwood).

H. Moreno, "Secondary Spectrophotometric Standards for the Southern Hemisphere".

A. Gutierrez-Moreno, "Atmospheric Monochromatic Extinction at CTIO".

D. E. Blackwell, A. D. Petford, and M. J. Selby, "The Absolute Flux of Vega in the Infrared" (read by Petford).

D. S. Hayes, R. F. Wing, S. T. Ridgway, R. R. Joyce, and C. P. Rinsland, "Absolute Calibration of Infrared (1 - 4 μm) Monochromatic Fluxes" (read by Wing).

E. I. Terez, "Absolute Calibration of the Energy Distribution in the Vega Spectrum over the Region 3110-7500 Å; A. A. Arkharov and E. I. Terez, "The Measurement of the Absolute Spectral Energy Distribution of Vega in the Near-Infrared"; E. I. Terez, "On Stellar Spectrophotometric Standards" (read by Wing).

E. E. Mendoza, V., "The $\alpha$, $\Lambda$ Photometric System".

V. Straizys, "A Catalogue of Energy Distribution Curves Representing Different Spectral Types, Luminosities, and Metallicities".

A. G. D. Philip and D. S. Hayes, "Energy Distributions for Horizontal-Branch Stars" (read by Philip).

J. H. Goebel, "Stellar Infrared Spectrophotometry from the NASA Kuiper Airborne Observatory" (read by C. M. Telesco).

R. F. Wing, C. P. Rinsland, and R. A. Bell, "The Effective Temperature Scale for Late-Type Dwarfs" (read by Wing).

R. Wehrse, "On the Interpretation of Energy Distributions of Spherically Extended M (Super-)Giant Atmospheres".

A. D. Code, "Stars and Galaxies in the Ultraviolet".

## III. MEETINGS OF WORKING GROUPS

1. WG on Be Stars (23 August, with Commission 29).

M. Jaschek, Chairperson of the WG, organized and chaired this meeting. During the business portion of the meeting, the following matters were addressed:

(a) The desirability of holding a Symposium or Colloquium on the physical parameters of Be stars (R. Viotti). The consensus was that such a meeting would be premature at this time, in view of the recent IAU Symposium (No. 98) in April, 1981.

(b) Status report on the Be Newsletter (M. Jaschek). The discussion made it clear that the Newsletter is a great success and should be continued.

(c) New membership in the WG. The consensus is that membership should be open to all who are currently working actively on problems related to Be stars.

(d) Organizing Committee of the WG for 1982-85. An election held at the meeting resulted in the following seven individuals elected to the Organizing Committee:

A. Feinstein (Argentina)
P. Harmanec (Czechoslovakia)
L. Houziaux (Belgium)
A. M. Hubert-Delplace (France)
M. Jaschek (France)
A. Slettebak, Chairman (USA)
A. Underhill (USA)

Following the Business Meeting, seven short communications and two invited talks were presented:

J. R. Percy: "Short-Period Photometric Variations in Be Stars".

D. Briot et al.: "Intrinsic Parameters of Be Stars and some Line Continuum Correlations".

Y. Andrillat: "The Infrared Spectra of Be Stars".

G. Peters: "The Nature of the High-Velocity, Discrete Mass Loss in Be Stars".

N. M. Ashok, et al.: "Infrared Photometric Studies of Be Stars".

P. Granes and F. Vakili: "Comments on Current Work".

R. Polidan: "A Search for Distant Circumstellar Gas in Be Stars".

A. Underhill and J. C. Mermilliod: Invited talks on "The Evolutionary Status of Be Stars".

2. WG on Spectroscopic and Photometric Data (18 August).

The meeting was organized and chaired by C. Jaschek, Chairman of the WG. Following his report on the activities of the WG, there was discussion of future activities and the relationship of the WG to other groups.

The following members of the WG were proposed for 1982-85:

M. Barbier (France)
O. Dluzhnerskaya (USSR)
B. Hauck (Switzerland)
D. Hayes (USA)
N. Houk (USA)
C. Jaschek (France)
M. McCarthy (USA)
J. Mead (USA)
K. Nandy (UK)
L. Pasinetti (Italy)
D. Philip, Chairman (USA)
A. Slettebak (USA)
F. Spite (France)
V. Straizys (USSR)
S. Van't Veer (France)

3. WG on Ap Stars (18 August, with Commission 29).

The report of this meeting may be found in Commission 29 Report.

4. WG on Standard Stars (25 August, with Commissions 29 and 30).

The report of this meeting may be found in the Commission 29 Report.

IV. MEETING ON "OBJECTIVE-PRISM SURVEYS AND ATLASES" (20 August).

This meeting was organized and chaired by A. Slettebak, at the suggestion of R. West. The latter pointed out that ESO was considering the possibility of making an objective-prism spectral atlas of the southern sky, once the current ESO (R) atlas is finished, and suggested that a discussion of which survey would be most useful would be very desirable. One way of proceeding seemed to be to ask investigators at observatories with objective-prism facilities to describe their programs and future plans. The following speakers presented brief papers:

R. Cannon: "Current and Proposed Objective Prism Work with the UK Schmidt Telescope".

J. Graham: "Schmidt Programs and Prospects on Cerro Tololo".

N. Houk: "Progress Report on the Revised HD Classification Program".

A. Savage: "The Optimum Objective Prism Survey for the Identification of Faint Quasars".

M. Schmidt: "Programs and Plans for the Palomar Schmidt Telescope".

A. Slettebak: "Warner and Swasey Observatory Schmidt Programs".

B. Westerlund: "Plans for the Uppsala Southern Station Schmidt Telescope".

D. Crampton: "CFHT Blue Grens Quasar Survey".

H. Maehara: "Kiso Schmidt Telescope Programs".

## V. ADMINISTRATIVE MEETING (20 August).

### 1. Report of the President

Since the last General Assembly, Commission 45 co-sponsored the following meetings: IAU Symposium No. 98, "Be Stars", held in Munich, Germany from 6 to 10 April 1981, and IAU Colloquium No. 68, "Astrophysical Parameters for Globular Clusters", held in Schenectady, New York, USA, from 7 to 10 October, 1981.

Eight Circular Letters were sent from the President to the Commission membership during 1980-82, primarily for information purposes and to solicit input for our Commission's activities during the 18th General Assembly.

The Commission 45 Report for 1979-81 was published in the Transactions of the IAU, Vol. 18A. Section reports were written as follows: Classification using Slit Spectroscopy (R. F. Garrison), Objective-Prism Spectral Classification and Classification involving Automatic Methods (D. J. MacConnell), Classification using Multicolor Photometry (V. Straizys), Calibration of Spectral Types in terms of Physical Parameters of the Stars (P. C. Keenan), and Catalogues and Atlases (C. Jaschek).

### 2. Membership

We regret to report that the following members of Commission 45 died since the last General Assembly: M. K. V. Bappu, A. Beer, and H. L. Johnson. Resignations were submitted by K. Gyldenkerne and L. N. Mavridis. The following seven new members were proposed and accepted: D. C. Barry (USA), H. H. Guetter (USA), L. Huang (China, Nanjing), T. Lloyd-Evans (South Africa), S. Malaroda (Argentina), E. H. Olsen (Denmark), and H. H. Wu (China, Taipei). This brings the total Commission membership to 115.

### 3. Election of Officers and Scientific Organizing Committee for 1982-85

The new Officers and Scientific Organizing Committee were elected as follows: V. L. Straizys (USSR), President; R. F. Garrison (Canada), Vice-President; R. A. Bell (USA); M. Golay (Switzerland); A. Heck (Spain); M. Jaschek (France); D. J. MacConnell (USA); A. G. D. Philip (USA); and A. Slettebak (USA).

In some brief remarks, the new President, V. L. Straizys, suggested that the functions of Commission 45 might be extended to include the organization and coordination of research projects.

4. Reports of Working Groups of the Commission

Brief spoken or written reports of the activities of the Working Groups of Commission 45 were presented by the following Chairpersons:

C. Jaschek: WG on Spectroscopic and Photometric Data.

M. Jaschek: WG on Be Stars.

C. Cowley: WG on Ap Stars.

L. E. Pasinetti: WG on Standard Stars.

5. Other Reports

B. Hauck reported on his efforts to implement plate vault data retrieval.

C. Jaschek gave a brief report on the Workshop on UV Stellar Classification which was held at Vilspa, Spain in March 1982.

6. Proposed Future Symposia and Colloquia

Commission 45 has agreed to co-sponsor the following Symposia or Colloquia which have either been approved or are under consideration:

IAU Symposium No. 105: "Observational Tests of Stellar Evolution Theory"; Geneva, Switzerland; Sep. 1983.

IAU Colloquium: "Astronomy with Schmidt-Type Telescopes"; Asiago, Italy, Aug. 1983.

IAU Symposium: "Structure and Evolution of the Magellanic Clouds"; Tübingen, Germany F. R.; Sep. 1983.

7. International Register of Stellar Spectroscopists (IRSS)

This publication is edited by W. K. Bonsack and A. Slettebak and mailed to persons listed plus about 230 institutions. In view of support expressed at the meeting and also in the Administrative Meeting of Commission 29, it was decided to continue compiling and distributing this list.

# COMMISSION 46: TEACHING OF ASTRONOMY (ENSEIGNEMENT DE L'ASTRONOMIE)

## Report of Meetings held in Patras

PRESIDENT: D. G. Wentzel SECRETARIES: L. Houziaux, J. Pasachoff, R. Robbins, V. Ruzdjak

### Business Session, 18 August 1982

I. MEMBERSHIP 1982-1985
President: L. Houziaux
Vice Presidents: M. Rigutti, C. Iwaniszewska
Organizing Committee: J. Kleczek (Secretary of ISYA)
D. McNally (Editor of Newsletter, ICSU CTS representative)
D. G. Wentzel (Past-president of Commission, publisher of Newsletter)
E. V. Kononovich, M. Gerbaldi, R. R. Robbins (Astr. Education Materials)
S. E. Okoye, A. Sandqvist, S. Ferraz-Mello (Visiting Lecturer Program)

National Representatives

| | | | |
|---|---|---|---|
| Argentina | Dr. A. E. Ringuelet | Ireland | Dr. B. P. Byrne |
| Australia | Dr. A. W. Rodgers | Israel | Dr. R. Steinitz |
| Austria | Dr. H. F. Haupt | Italy | Prof. E. Proverbio |
| Belgium | Prof. L. Houziaux | Japan | Prof. N. Owaki |
| Brazil | Prof. S. Ferraz-Mello | Korea (Rep.) | Prof. Kyung Loh Yu |
| Bulgaria | Dr. N. S. Nicolov | Mexico | Dr. L. F. Rodriguez |
| Canada | Dr. J. R. Percy | Netherlands | Dr. J. A. van Paradijs |
| Chile | Dr. H. Moreno | New Zealand | Dr. N. A. Doughty |
| China | (Chinese Astronomical Society) | Nigeria | Dr. S. E. Okoye |
| | Prof. Feng Ke-jia | Norway | Dr. O. Elgaröy |
| | (Astronomy Union at Taipei) | Poland | Dr. C. Iwaniszewska |
| | Prof. Shen Chun-Shan | Portugal | Prof. J. Osorio |
| Columbia | Prof. E. Brieva | Rumania | Prof. N. Dinulescu |
| Czechoslovakia | Dr. J. Kleczek | Rep. of South Africa | Prof. A. H. Jarrett |
| Denmark | Dr. H. E. Jørgensen | Spain | Dr. Ma. A. Catala-Poch |
| Egypt | Dr. A. S. Asaad | Sweden | Dr. Aa. Sandqvist |
| Finland | Dr. K. Lumme | Switzerland | Prof. B. Hauck |
| France | Dr. L. Gouguenheim | Turkey | Dr. S. Hazer |
| Germany GDR | Dr. H. Zimmermann | United Kingdom | Dr. D. McNally |
| Germany GFR | Dr. H. Scheffler | U.S.A. | Dr. J. M. Pasachoff |
| Greece | Dr. L. Mavridis | U.S.S.R. | Dr. E. V. Kononovich |
| Hungary | Dr. M. Marik | Vatican City | Dr. J. Casanovas |
| India | Dr. S. M. Alladin | Venezuela | Dr. I. Ferrin |
| Indonesia | Dr. B. Hidayat | Yugoslavia | Dr. M. Karabin |

Regular Members. A. Acker, A. Z. Aiad, H. Andrillat, S. M. R. Ansari, I. Atanasijevic, V. Barocas, L. Bottinelli, W. Buscombe, J. M. Chamberlain, D. Clarke, S. Codina, E. A. Dibay, F. Dominko, D. L. DuPuy, F. Fuenmayor, H. Gurm, D. B. Hoff, S. Isobe, V. V. Ivanov, J. E. Kennedy, V. Kourganoff, R. C. Madison, J. C. D. Marsh, H. G. Miles, E. A. Müller, B. Onderlicka, W. Osborn, B. Peery, Qu, Qin Yue, M. Rego-Fernandez, L. J. Robinson, C. Roslund, A. E. Roy, A. R. Sadik, P. A. H. Seymour, J. L. Safko, T. Schmidt, H. L. Shipman, K. N. Taylor, K. P. Tritton, V. Vujnovic, W. Zealey, M. Zeilik. Members not heard from during 1979-82 were omitted from this list.

Consulting Members. M. L. Aguilar, C. E. Akujor, J. Bishop, M. K. Hemenway, N. T. Jiwaji, P. N. Okeke, L. I. Onuora, K. A. Portzevskii, A. R. Servan, J. Siroky, A. E. Troche-Boggino, R. H. Wilkinson.

Recommendation accepted: The preparation of the membership list before the IAU General Assemblies will be carried out by the first Vice-president of the Commission. Reason: the correspondence involved provides much useful information for his subsequent term as President of the Commission.

II. PUBLICATIONS

The National Reports and Astronomy Education Materials are incorporated in the Newsletters 11-13. The mailing list numbers about 220, with about 20% each to N. America, Western Europe, Eastern Europe, and roughly 10% each to Asia, Australia, Africa, South America. Additional copies are mailed by Houziaux and McNally to local teachers. Surface mail is used.

It was agreed to advertise the Newsletter in the UNESCO Bulletin and at this IAU General Assembly, but Wentzel stated that advertisement in international journals would yield a mailing list beyond the IAU budget and beyond the capability of the University of Maryland to publish it. The IAU budget for Commission 46 publications is Sfr. 5000 ≃ $2500 for 3 years.

Lengthy discussion centered on the most useful information in Astronomy Education Materials. How important is the attempt to make comments on each published work? Mavridis suggested adding the address of a contact person in each language who might provide more information. For instance, the French and Canadians publish separate lists.

Perkin Elmer's **Technical News** for September 1982 will have a discussion of astronomy and its value, with contributions by C. Townes, J. M. Pasachoff, W. Howard, J. P. Swings and E. M. Burbidge.

III. COOPERATION WITH INTERNATIONAL TEACHING ORGANISATIONS

Houziaux mentioned the newsletter of the ICSU Committee on Teaching of Science. Seven issues have appeared, including some news from the IAU. Since it is difficult to find technicians in many countries, a workshop and newsletter on technicians have been established. A CTS sub-committee has published six booklets (contact McNally) on learning strategies, including low-cost equipment. A future project may deal with computers and microcomputers.

An international conference on **Science, Technology, Education, and the Quality of Life** is set for 1985. We should participate to keep astronomy in view of the bodies that make international recommendations. Houziaux and McNally will work out a proposal. Wentzel has submitted an article to a related publication **Science, Technology and the Human Prospect.**

Houziaux has not heard from COSTAD since 1978 and did not even receive an answer to his query of their secretariat. Recommendation accepted: that we no longer appoint a delegate.

IV. COMMISSION PROJECTS

Houziaux will write to the organizers of the IAU Regional Conferences to point out the value of teaching sessions, in accordance with the Commission resolution at Montreal.

Project Contratype has led to the distribution of slide sets and posters to seven countries, supported by the V. M. Slipher Fund of the US National Academy of Sciences, and the circulation of 6 NASA movies to 14 countries. There were no new practical ideas on textbooks.

V. INTERNATIONAL SCHOOLS FOR YOUNG ASTRONOMERS

This is the most important project of the Commission. The most recent schools were 10-23 September 1979, at La Laguna, Tenerife, Spain (63 students including 12 from America, 8 professors), 17 September - 8 October 1980 at Hvar, Yugoslavia (35 students from Balcan and Mediterranean countries, 8 visiting faculty, stress on practical training, seminars by the participants), and 22 August - 9 September 1981 at Cairo, Egypt (28 students, 22 lecturers). The school in Kenya was cancelled for lack of applications, which might be related to the proximity in time and space to the Egypt ISYA, and because no local financial support could be obtained. The Commission asks ISYA Secretary Kleczek to specially inform all the National Representatives of countries neighboring to a planned ISYA. Wentzel reported preliminary plans for future ISYA. See below.

A list of the approved Rules and Guidelines for ISYA was circulated. It was stressed that each ISYA must be international in faculty and students.

Meetings of the Sub-committee for the ISYA, 19 and 23 August 1982.

The Sub-committee gave preliminary approval to the planning of two ISYA in 1983, after Drs. Hidayat and Fuenmayor provided preliminary budgets and topics. One ISYA is to be in May in Indonesia, with topics on solar physics and galactic structure, in conjunction with the June 1983 solar eclipse. One is to be in October in Venezuela, with some emphasis on space physics. The IAU budget for the ISYA is Sfr. 36000 ≈ $18000 for three years. The two ISYA preliminary budgets each include a $9000 contribution from the IAU, but this is not approved until final plans are presented.

Countries mentioned for a possible third ISYA in 1982-1985 are China, Portugal, India and Pakistan. While it seems attractive to involve China fully in IAU activities, the Sub-committee felt that an ISYA in Portugal would be more likely to come to fruition. Since funding must be sought for any third school, a decision should be made in early 1983.

Session on Visiting Lecturer Program, 19 August 1982

At a joint session with Commission 38, S. Ferraz-Mello presented the report of the Working Group on the Education of Astronomers from Developing Countries. The following proposal has been formulated and is the basis for a budget recommendation by the IAU Executive Committee for Sfr. 35000 ≈ $17000.

"The International Astronomical Union has decided on a special program to aid countries which wish to inititate or to make substantial improvements in astronomical activities.

The effectiveness of the program requires the collaboration of a University or similar institution from the country in question, committing itself to participate in a substantial manner and to continue astronomical activities after the completion of this program.

It is not required that the country concerned be a member of the IAU.

The application should be presented by the country to the Executive Committee of the Union and approved by this Committee.

Costs should be shared fairly, not necessarily equally: in principle, IAU provides the teachers and pays their travel expenses, and the country covers the living costs and the necessary equipment as much as possible.

A typical arrangement would consist of sending six to eight lecturers to one country in a period of three years, for stays of normally three months each. The three first courses will cover astronomy in general, at the advanced undergraduate level, the following courses can be adapted to the special interests of the host institution."

E. Müller stressed four points made by the Executive Committee: i) the scheme is not a way to provide paid jobs to young astronomers, ii) the host country must be not only interested but also committed to pursue education in astronomy after completion of the program, iii) the first project, funded by the IAU and the host country, must be a real success, so that funds may be available from UNESCO for later projects, iv) the teacher should be fluent in the language spoken by the students. While everyone agreed that a school, for instance, in Peru must have teachers fluent in Spanish, it was also clear that some schools could not possibly be held in the local language, for instance in Indonesia.

Iwaniszewska raised the example of Maria Luisa Aguilar, Peru, who wishes to be instructed how to train teachers. Akujor agreed to the need for teacher training. However, this seemed to be a minority view. Both also argued for seeking fellowships to encourage young people to become astronomy teachers, i.e. fellowships either locally or at established astronomical institutions, perhaps through exchange programs.

Other comments were to the effect that lectures should be presented at the appropriate level and there should be practical exercises (Ferraz-Mello), that a well-documented library be available before the lectures start (Moreno), that only experienced teachers be sent, who should keep their home salary while on leave of absence (Hidayat), and that follow-up funds after a project must be sought from UNESCO by the developing country itself (Müller) though perhaps helped by IAU approval (Ferraz-Mello).

In view of reported questions about the propriety of the IAU supporting development of astronomy where there is (almost) none, the Commission passed the following resolution.

Resolution: Whereas astronomy is an integral part of the physical sciences, both in education and in research, and whereas some countries now without astronomy may desire to introduce astronomy in their educational or scientific institutions, be it resolved that the IAU supports the introduction of astronomy in these countries and encourages activities by its Commissions toward this goal.

The following text was approved in principle at the joint Commission meeting, then worked out with the officers of Commission 46 and approved by a later meeting of Commission 38.

"Commissions 38 and 46 adopted the report of the joint Working Group on the Education of Astronomers from developing countries. The Working Group, having completed its charge, was formally terminated.

Commissions 38 and 46 resolved to establish a sub-committee to implement the recommendations of the Visiting Lecturers Program in the report of the Working Group and in Agenda item 13(d)2 of the 18th General Assembly of the IAU. Initially the membership of the committee is J. Delhaye (38), S. Ferraz-Mello (46), D. G. Wentzel (46, Chairman), and F. B. Wood (38). The membership will be subsequently expanded to ensure active participation from Developing Countries in its work, with advice and consent from the Organizing Committees of Commissions 38 and 46.

The terms of reference for the sub-committee will be: a) to seek individuals and their institutions willing to host a Visiting Lecturers Program; b) to identify the local procedures to support the program; in particular, the local body with whom any future contracts should be made; c) to outline a Visiting Lecturers Program agreeable to both host and sub-committee; at this time the sub-committee should identify several possible lecturers with experience in the field of the agreed program, known experience of the developing world and preferably fluency in the local language; (It is recognized that such an identification will not limit final selections.); d) to submit, in light of c) above, a formal proposal to the IAU Executive Committee for a visiting Lecturers Program in a named Developing Country, to be based on a named host institution; e) to submit at the time of d) above, the outline of a second program, such that the General Secretary may seek support through a UNESCO contract; f) and to establish, at the time of d) above, an advisory panel with experience of local conditions, developing countries and the type of program envisaged, in order to advise the Executive Committee on the final selection of lecturers and on the articulation and oversight of the program."

## Session on training of teachers, 20 August 1982

Iwaniszewska (Poland) summarized the 35 national reports appearing in Newsletters 11 and 13. Astronomy is a separate course at universities in only 5 countries. In the majority of countries training of teachers is performed at courses given irregularly by university departments (31%), summer courses (26%) and associations of teachers and astronomers (17%). Osorio (Portugal) commented that the situation has not improved very much, as only 5 countries have a complete course in astronomy.

Hauck (Switzerland) spoke about teacher-training in astronomy in France, Belgium and Switzerland and emphasized the importance of summer schools and 1-2 day courses performed in these countries. He also pointed out difficulties in some countries where there are many educational systems (Switzerland-26 educational systems). Wentzel commented that in the USA there are about 500 educational systems.

Gouguenheim spoke about the summer schools in France. The main goals are to teach the participants in basic astronomy and physics and to perform practical experimental work. She also demonstrated some of the equipment used in the practical work.

Owaki informed about the teachers training in Japan. He presented among others an example of a curriculum of Tokyo University for teachers of astronomy and the answers of school teachers to a questionaire regarding astronomy. He stressed that, after graduation, the contacts with teachers are kept in summer schools. Qu spoke about astronomy education in China. Since 1953 about 840 students have graduated at three Universities (in Nanking and Peking). However, it is difficult for graduated astronomers to find a job as there are no courses in astronomy in schools. Moreno stressed the potential of 3 large international telescopes to raise the interest for astronomy in Chile, and he informed about the courses for teachers of mathematics, physics and chemistry in astronomy. Also he pointed out the good interaction with visiting astronomers. Percy described some problems of teaching astronomy in Canada (two official languages, large territory) and presented his plan to publish a **"Teachers Guide to Astronomy"** which would contain material which the teachers can use directly (maps, tables etc.).

## Session on Computers, Television, Activities, 23 August 1982

Microcomputers in Astronomy Education, D. McNally (UK). These are now of sufficiently small cost to give all students access to computers and use them as a real tool, not merely for "computing for its own sake". BASIC is the most widespread language, available on all machines, with an increasing number of students arriving at universities familiar with it. Graphic capabilities are generally available and definitely worth having. In five specific areas computers are especially useful:

1) Teaching computing: don't let the computer center do this, because of point (2).
2) Providing sets of programs useful in astronomy: e.g. to compute precession, coordinates, radial velocities, curve fitting, least squares, etc. Eventually students should progress to a better language.
3) Demonstration programs: constellations, bodies moving under central forces, sky to various limiting magnitudes.
4) Difficult concepts: BB radiation, reddening, filters, statistics.
5) Control: linking computers, or computers plus instruments, for sophisticated functions e.g. pointing a telescope, guiding a microdensitometer in its measurements.

Programs obtained from outside sources are usually more trouble to adapt than writing your own from scratch. Incompatibilities in the software for various machines are still a major headache. Standards of documentation for interchange of programs between systems are not available.

Radbo (Sweden) has used a computer system for coupled equations, curve of growth abundance analysis for the solar atmosphere, galaxy synthesis. A disk of programs can be borrowed from Bjorn Sundelius (address c/o Mrs. Marie Radbo, Astronomy, Chalmers University, S-41296 Goeteborg).

Television. De Zeeuw (Netherlands) described **"Modern Astronomy"**, a 1981 TV course given in the Netherlands by Leiden Observatory and Teleac, the Dutch TV academy. This course is not part of an open university, because there is no provision for interaction. The program emphasises galaxies, but the 13 lessons of 30 minutes each (20 min of studio time then 10 min interview with an astronomer) are fairly comprehensive of an astronomy syllabus. Animation movies by Seth Szostak are included. 13 radio lessons were also produced, and a 450 page book by Nick de Kort, available for $28 US. It was a bigger success than expected, with 23000 books sold, and will be re-telecast in 1983 with minor changes, even though it was not really for the general public, but for those with some exposure already to astronomy, e.g. amateurs. In producing something like this, close interaction between the TV personnel and the astronomers responsible for content is essential, because some good material looks flat on TV and some bad ideas look good. Control of TV people is essential.

Discussion: Teleac is not very interested in translating this series to other languages. There was a British series on commercial TV that stimulated the distribution of 27,000 "starpaks", but the producers there now feel that Carl Sagan has done it all. In the US, it is probably impossible to get veto power over a TV producer. Address: TELEAC, Postbus 2414, 3500 GK Utrecht, The Netherlands.

Radbo: Sweden is an extended country with astronomy at only 4 universities, so an extensive set of correspondence courses have evolved in various fields (with the usual problem of having too little interaction). They added to this concept a series of videocassettes that can be mailed to the home of the student. The cassettes were longer and more detailed than the Netherlands programs, so a book was not as necessary (although a closely coordinated one was produced). A principal aim of the material was the continuing education of teachers. Machine incompatabilities were a big problem. The students for the course were first gathered at Gothenburg for planetarium demonstrations and introductions, then they got the 9 cassettes for 2 weeks at a time (plus a 150-page text). Contacts with the lecturers were maintained by regular correspondence and frequent phone calls, and the lecturers did some visiting. There were essay exams. Production costs were $60 K, and if the materials can be used for a 10 year period, the cost per course will be $7 K (versus $9 K per class for conventional techniques). They had 100 students the first time through. Dr. Curt Roslund did most of the manuscript. They have cassettes which can be borrowed for examination.

Pasachoff (USA) described the use of the Pioneer laserdisk system with 54,000 frames per disk. Quality of visuals is high and either a still or a movie sequence can be searched and projected in five seconds. The system can be interfaced with a computer. A series of 5 disks is being prepared by Center for Aerospace Education, Physics Dept., Drew University, Madison, NJ 07940, USA.
1) Voyager programs (final version price $320 for disk)
2) Apollo/Skylab/Shuttle
3) History of Spaceflight
4) General (all the slides from everyone's catalog)
5) Earth views (Landsat) and planetary geology
The videodisk player is about $600, the master disk is virtually indestructible. JPL has produced a Mariner disk (about 50,000 images) which is available now.

Wentzel mentioned that the American Institute of Physics produced and distributes 2-minute TV shows, 5 of them on astronomy. Robbins described StarDate, 2-minute radio programs produced at the U. of Texas, currently reaching about 2 million people per day, on a subscription basis to radio stations. Robbins also commented that 3D terminals now entering the market have tremendous education potential, although they are still quite expensive ($10,000). Nikolov (Bulgaria) reported on a program of 15-20 minute radio lectures that have been produced in his country over the last 4 years. Topics: Introductory material, solar system, stars, extragalactic.

Activities. Ruzdjak (Yugoslavia) described an astronomy activities program in his country, part of a "Science to the Young" movement. Astronomy is one of the competitions in both the 11-14 and 15-18 age groups, even though it is not normally in the school system. Practical work has been encouraged and has grown to 70% of the projects. Testing did not work out and has been abandoned.

Kleczek (Czechoslovakia) is writing a new edition of Minnaert's book "**Practical Work in Elementary Astronomy**". Wentzel asked that contributions be sent to Kleczek of new types of experiments, especially in the newer areas of astronomy. Suitable credit can be arranged.

Pasachoff reports some progress toward including astrophysics in the US Graduate Record Examination. Servan (Israel) described the modest growth of astronomoy education in his country. H. Couper (UK) reported on a one-day teach-in at Greenwich. Four astronomers talked to 50 students on what astronomy is really like.

Future sessions. Wentzel suggested that the next General Assembly include a session on the use of a planetarium. Should there be an IAU Symposium on teaching? McNally: such a symposium is useful if it concentrates on a particular topic, e.g. teaching of stellar structure. Wentzel: skeptical unless its form is a workshop allowing detailed personal exchanges of ideas. Houziaux: it is more efficient to have teaching sessions at regional IAU conferences.

## Session for local teachers, 16 August

This session, chaired by Prof. C. Antonakopoulos, University of Patras, was dedicated to the teaching of Astronomy at the high-school level. It was attended by representatives from USA, India, Portugal, Taiwan, Greece and several other countries.

The first speaker was Prof. L. Mavridis of the University of Thessaloniki who reviewed the educational programs of the Greek University and high school and explained the teaching of Astronomy at both of these levels.

The next speaker was Mrs. M. Laskaratou, the President of the Hellenic Astronomical Society, who discussed the efforts of the society to popularize Astronomy in Greece and to encourage the teaching of Astronomy in Greek highschools. She reported that the Greek shipowner A. Vergotis has promised to make a large donation for the establishment of an international research center for astronomy and seismology at his beautiful island of Cephalonia.

Prof. Papagiannis outlined the efforts of Boston University to stimulate astronomy education at the high-school level, in part by open houses and the Priscilla and Bart Bok annual awards for projects in Astronomy by high-school students. General discussion included the requests to teach some basic facts such as the cause of eclipses (Bhattacharya, India) and the parallel training of astronomers in related fields (Aller, USA).

# COMMISSION 47: COSMOLOGY (COSMOLOGIE)

## Report of Meetings, 19, 20, 21, 23 and 24 August 1982

PRESIDENT: G. Abell SECRETARIES: V. Trimble
I. Novikov

### 19 and 23 August 1982

### JOINT DISCUSSIONS

On 19 August Commission 47 cosponsored Joint Discussion II with Commissions 25, 28, 29, 33, 34, 35, 37, and 45 on the topic Evolution in Old Stellar Populations in Galaxies.

On 23 August Commission 47 cosponsored Joint Discussion III with Commissions 25, 28, 40, 45 and 48 on the topic The Extragalactic Distance Scale and the Hubble Constant.

The transactions of these Joint Discussions are found elsewhere in this volume. Joint Discussion II was under the principal sponsorship of Commission 37 and Joint Discussion III was under the principal sponsorship of Commission 28.

### 24 August 1982

### JOINT COMMISSION MEETINGS

On 24 August Commission 47 held a joint meeting, with Commission 28, on the topic The Local Supercluster and its Environment. The meeting, lasting for two sessions, was chaired by B.E. Westerlund and G.O. Abell; the program follows:

R.B. Tully: "Morphology of the Local Supercluster"
J.P. Huchra: "The Dynamics of the Local Supercluster"
E.E. Salpeter: "Dynamical Models of our Virgocentric Deviation From the Hubble Flow"
G.O. Abell: "General Superclustering"
J. Einasto, H.G. Corwin, Jr., J. Huchra, R.H. Miller and M. Tarenghi: "Local Cells of the Universe: Two Voids of Diameter 200 Mpc"
Y. Zeldovich: "Remarks on the Structure of the Universe"

Summaries of these papers are found in the report of Commission 28.

### 20 August 1982

### BUSINESS SESSION

The President announced, with sorrow, the passing of two members of the Commission since the last General Assembly:

J. Neyman
B.M. Tinsley

The President then summarized other activities of the Commission since the 17th Assembly. These included:

Principal sponsorship (with Commissions 28 and 48) of the I.A.U. Symposium No. 92, "Objects of High Redshift," held at the University of California, Los Angeles, California, USA, from 28 to 31 August 1979. The Symposium was attended by 146 astronomers from 15 countries. The proceedings have been published by Reidel.

Cosponsorship (with Commissions 28, 40, and 48) of I.A.U. Symposium No. 97, "Extragalactic Radio Sources," held at Albuquerque, New Mexico, from 3 to 7 August 1982. The Proceedings were published by Reidel.

Principal sponsorship (with Commissions 28 and 48) of I.A.U. Symposium No. 104, "Early Evolution of the Universe and its Present Structure," to be held at the Orthodox Academy, Kolymbari, Crete, from 30 August through 2 September 1982, immediately following the General Assembly at Patras. (The Symposium was attended by 191 astronomers from 22 countries; the Proceedings are being edited at this writing, and are expected to be published by Reidel in Spring 1983.)

Election of new officers and organizing committee members followed. Those elected, to serve from the end of the General Assembly until the end of the 19th Assembly in 1985, are:

| | |
|---|---|
| President: | J. Audouze |
| Vice President: | G. Setti |
| Organizing Committee: | G. Abell, L.P. Grishchuk, J. Gunn, S. Hayakawa, B. Jones, M. Rees, R. Syunyaev, and H. van der Lann |

The following new members were elected to the Commission: D. Basu, G. Chincarini, C.C. Dyer, J. Einasto, H.C. Ford, K. Kellerman, J. Lequeux, T. Lu, F. Melchiorri, W.T. Ni, J.C. Pecker, B.A. Peterson, Q.Y. Qu, H. Saar, V.F. Schwartzman, P.A. Shaver S.F. Shandarin, F.W. Stecker, S. Szalay, V. Trimble, E.L. Wright, and E.J. Zuiderẅyk.

H.P. Palmer has resigned.

<u>21 August 1982</u>

## SCIENTIFIC SESSIONS

In addition to the two joint discussions and joint commission meeting, the Commission held two scientific sessions, both on 21 August. The programs follow:

### Recent Observations of Cosmological Interest

Chair: G.O. Abell

G. Smoot: "Review of Recent Observations of the Large-Scale Anisotropy of the Cosmic Background Radiation"

P. Kronberg: "Absorption Lines, Faraday Rotation, and Magnetic Field Estimates for QSO Absorption-Line Clouds"

M.V. Gorenstein, I.I. Shapiro, N.L. Cohen, E.F. Falco, and A.E.E. Rogers: "Looking Through a Gravitational Lens: High-Sensitivity VLBI Maps of 0957+561"

M. Spite and F. Spite: "Lithium Abundance at an Early Age of the Galaxy"

Recent Observations of Cosmological Interest (Continued)

M. Longair: "Evidence for the Cosmological Evolution of the Stellar Components of Radio Galaxies"
D.A. Varshalovich: "High-Redshift Molecular Clouds and Absorption-Line Spectra of Quasars"
D. He: "QSOs in the Direction of the Virgo Cluster"

## The Early Universe

Chair: J. Audouze

Y. Zeldovich: "The Very Early Universe"
I. Novikov: "The Origin of the Primordial Perturbations in the Very Early Universe"
Q.Y. Qu: "Cosmological Studies in China"
D.N. Schramm: "Planetary-Mass Black Holes Formed During the Quark-Hadron Transition"
J. Audouze: "Some Aspects of Primordial Nucleosynthesis and Their Relation to Galactic Chemical Evolution"
J. Jones: "Population III Stars"
B. Jones: "On the Formation of Galaxies"

Full discussions of most of the material covered in the above papers will be published by the time, or shortly after the time, this volume is available; hence, summaries are not given here. On the other hand, a few very general comments on recent advances, especially to supplement or update the President's contribution to Reports on Astronomy, might be in order.

1. The Hubble Constant and Age of the Universe: Aaronson, Huchra and Mould have now observed a number of additional clusters in their ongoing study of the infrared Fisher-Tully relation and have revised downward somewhat their value for the global Hubble constant -- to about 85 $km \cdot s^{-1}$ $Mpc^{-1}$. Otherwise, the protagonists in the controversy over the distance scale and age of the universe have not budged from their positions a year ago. On the other hand, as was highlighted in Joint Discussion III, it is becoming increasingly difficult to reconcile what appear to be firm lower limits to the age of the universe from cosmochronology (either radioactive isotope ratios or globular cluster evolution studies) with a value of the Hubble constant of $H > 75$ $km \cdot s^{-1}$ $Mpc^{-1}$, and probably even with $H > 60$. The former point to an age near $16 \times 10^9$ yr, and "certainly" not less than 11 or $12 \times 10^9$ yr, while if $\underline{H}$ is near 100 $km \cdot s^{-1}$ $Mpc^{-1}$ Friedmann cosmologies with $\Lambda = 0$ demand ages less than $10^{10}$ yr, and probably ages for globular clusters less than $8 \times 10^9$ yr. Zeldovich reminded the audience at one session at Patras that even in Lemaître universes with $\Lambda > 0$, the "stillstand" in the $\underline{R(t)}$ curve would occur at redshifts of $z \approx 1/2$, inconsistent with observations. This is true, however, for only a density parameter, $\Omega$, near unity, while present observations do not rule out $\Omega = 0.04$, in which case a Lemaître universe with a stationary value of $\underline{R(t)}$ at a redshift of $z > 2$ might still be possible. There is still argument, however, over the possible role of selection effects in the determination of $\underline{H}$, and the calibration -- even of the infrared Fisher-Tully relation -- may not be as firm as one would hope. For example, very recent (October 1982) work by Sandage on the cepheids in M33 strongly suggests that the modulus of that galaxy should be increased by 0.7 mag. over Baade's value. There is not yet a consensus that the simple Friedmann big bang cosmologies are inconsistent with observations.

2. The Structure of the Universe: Considerable new evidence was presented, both at Patras and at the Crete symposium following the General Assembly, concerning the structures of superclusters and voids. On the theoretical side,

The Early Universe (Continued)

numerical simulations by Zeldovich and his associates have shown that "pancake"-like superclusters should eventually coalesce, giving the universe a cellular structure, with superclusters, perhaps irregular and filamentary, serving as interstices between the intrasupercluster voids. Observations by Einasto and his coworkers suggest just such a picture; moreover, Einasto is of the opinion that all superclusters are filamentary in structure. Tully's model of the Local Supercluster puts most of the galaxies in a roughly disk-shaped system, but with many in elongated (perhaps filamentary) clouds on either side. An interesting technique (percolation) for analyzing the surface distribution of galaxies provides a sensitive test of whether that distribution is filamentary or consists of some edge-on sheets of matter, or whether the galaxies clump into randomly shaped configurations. According to the Soviets, the application of the test, at least in some regions, favors the former. In any event, the observations appear to favor flat or filamentary superclusters, and are probably not consistent with the shapes to be expected from gravitational encounters between an initial random distribution of mass points.

3. The Microwave Background: New observations, both by the Berkeley and Princeton groups, do not confirm evidence for a quadrupole component to the microwave background radiation, although they do confirm what now appears to be the well-established dipole component. There are also new observations of the spectrum (Berkeley) that cast some question on the reality of the deviations from a black body previously reported, and which have generated so much discussion. There has been some success in interpreting the alleged spectral irregularities in terms of scattering by needle-shaped dust grains produced by "Population III" stars, but perhaps this kind of model will not be required.

4. The Early Universe: The past few years have seen a great flurry of activity in the theoretical study of the early universe. Quantum gravity, grand unified theories, and super-symmetry theories have been invoked in the hopes of solving a wide range of problems, including those of the baryon asymmetry, the monopole, and massive galactic halos. Grand unified theories give mass to the neutrino, and the super-symmetry theories provide massive photinos and massive gravitinos. Particularly interesting is the inflationary universe, first suggested by Guth in 1981. Here the very early universe begins with an exponential expansion similar to the steady-state cosmology. If correct, it could solve the horizon problem (how the universe could ever have been in communication to provide the observed large-scale homogeneity and isotropy) and the flatness problem (how, after expanding in volume by very many orders of magnitude, the universe can have a present density that is within a factor of a few of the critical closure density. It is beyond the expertise of this observer to evaluate these new ideas, but clearly they are bringing a great deal of excitement to theoretical cosmology and particle physics.

5. Editorial: The advances in our understanding of the cosmos over the past half-century have been astonishing, but perhaps a bit of caution and humility are in order. It is probably healthy to read a bit of the history of modern astronomy, and learn how quickly Herschel and his contemporaries were to draw firm conclusions about the nature of the nebulae. A theory does not have to be correct because it is simple or elegant, and the various protagonists in the controversy over the Hubble constant and age of the universe cannot all be correct. Is it certain that at the XXV General Assembly the Friedmann models of the universe will be regarded as firmly established as many seem to regard them today?

## COMMISSION 48: HIGH ENERGY ASTROPHYSICS (ASTROPHYSIQUE DE GRANDE ENERGIE)

President: F. Pacini

### Report of Activities during the XVIIIth General Assembly

Commission 48 was involved in the organization of the following Joint Discussions and specialized meetings:

1. Joint Discussion of Active Galactic Nuclei
2. Galactic Gamma-ray Astronomy
3. Violent Interstellar Medium
4. Impact of Space Research on Astrophysics and Extragalactic Systems, Cosmology

In addition, a Business Meeting was held on August 24, 1982. During the meeting, the proposed composition of the Organizing Committee was approved.

Further activities of Commission 48 around the XVIIIth IAU General Assembly involved the co-sponsorship of IAU Symposium 101: "Supernovae Remnants and their X-ray Emission" (Venice, Italy, August 30-September 2, 1982) and Symposium 104: "Early Evolution of the Universe and its Present Structure" (Crete, Greece, August 30-September 2, 1982).

## COMMISSION 49 : THE INTERPLANETARY PLASMA AND THE HELIOSPHERE (LE PLASMA INTERPLANETAIRE ET L'HELIOSPHERE)

Report of Meetings on 19 August, 1982

PRESIDENT : H.J. Fahr

Approximately 80 persons attended the meeting in the morning in which 4 review papers and 3 contributed papers were presented concerning the Inner and Outer Heliosphere. After a lively and extended discussion, it was decided to arrange a small workshop on 23 August in order to have a more detailed discussion about problems related to the 3D-structure of the solar wind.

In the afternoon, a meeting was held concerning the heliosphere interface, i.e. the region between supersonic solar wind plasma and unperturbed interstellar plasma. One review paper and three contributed papers were given, during which several aspects of the 3D-heliopause boundary were raised. It appears that the outstanding question is now whether or not there is a supersonic ISM-solar wind interaction. It was decided to further discuss this problem during a forthcoming workshop.

A business meeting with 20 people attending was held in the late afternoon. New members were approved unanimously and Professor S. Grzedzielski (Poland) was unanimously elected new Vice-president of Commission 49. The question of how to further exchange of ideas between Commissions was briefly discussed.

## COMMISSION 50 : PROTECTION OF EXISTING AND POTENTIAL OBSERVATORY SITES (PROTECTION DES SITES D'OBSERVATOIRE EXISTANTS ET POTENTIELS)

### Report of Meeting on Site Testing , 19 August 1982

PRESIDENT : F.G. Smith SECRETARY : A.J. Penny
The morning session was chaired by A. Hoag.

H. Wohl gave the first talk on "Solar site testing in the Canary Isles". He reviewed the history of the JOSO site testing effort since 1967 which began with the consideration of 40 sites, of which 15 were investigated in detail. In 1976 the search was concentrated on Izania on Tenerife and on Roque de Los Muchachos on La Palma. A comparison of the two sites used 40 cm telescopes looking at the sun for image motion and sharpness amongst other parameters. The Izania site was found to be superior. This was thought to be due to the Caldera on La Palma affecting the seeing more strongly than the topography of Tenerife does for the Izania site. Occasionally on Tenerife Fried's seeing parameter has been measured as large as 40cm which implies that better resolution would be obtained with telescopes larger than those used in the site testing. In a comment from the audience, the importance of such a large value or $r_o$ for active image compensating work was pointed out.

A. Penny then gave P. Murdin's report on "Protection and Investigation of the Roque de Los Muchachos site". Microthermal sensors at heights of 3m and 12m were used simultaneously for eight nights at two positions within the site. One was at Fuente Nova at the edge of the caldera where the original testing had been done and the other was 250m N and down hill from these, at a proposed site for the 4.2m telescope. When the wind is southerly, flowing over the rim of the caldera, turbulence occurs at both 3 and 12m heights at the rim, but only at the 3m height at the down hill site. The suitability of the proposed 4.2m site was thus confirmed. It is suggested that the JOSO comparison of the Roque and of Izania was affected by local conditions very close to the caldera rim. Further microthermal tests will be carried out to investigate local effects and to provide a comparison with telescope seeing measures.

M. Walker reported on "International Comparisons of Seeing". He reviewed the Polaris trail method which has new been used to test fifteen sites and has confirmed the pattern that isolated conical peaks in the five regions of the lower-middle latitude band have the best seeing. A recent investigation of the NW Cone (a site proposed for the California 10m telescope) on Mauna Kea and of a site off the Mauna Kea summit on the north shield was described. At the shield site microthermal activity was found and also the seeing was worse than on the NW Cone.

D. Crawford reviewed the Dark Sky Survey. A project has been funded to collect data from existing sites, to extend the Polaris trail method, and to measure sky brightnesses and night time water vapour levels. Seven night time water vapour meters have been built and distributed. A comparison between Mauna Kea and Mt Graham, Arizona, is being undertaken and further site testing is planned with the NTT 15m telescope in mind.

G. Lockwood spoke on "Experiences with long-term extinction measurements". Tests at Tucson and Flagstaff have found no effects from man-made pollution. Volcanic activity however has sometimes dramatic effects on aerosol extinction, and there was also evidence for climatic effects. In the south-western US, extinction is higher in summer than in winter. There appeared to be no change in the ozone extinction with the solar cycle.

J. Tremko reported on "Long-term variations of atmospheric transparency from measurements at Skalnate Pleso and Brno". From observations over 20 years, totalling 400 nights at Skalnate Pleso, long-term variations in the aerosol extinction were found, possibly correlating with the solar cycle. The Brno observations showed that in an urban environment the wavelength power law exponent of the aerosol extinction varies between 1.3 and 3.6.

A. Hoag reported on "Sky Brightness monitoring". Observations over the sky at a site and zenith measures at locations around the site can be used to estimate local and non-local contributions to the sky brightness. Work at Lowell and Kitt Peak was discussed and the effect of night sky emission lines and the solar cycle mentioned. Following this, M. Walker mentioned he had been monitoring the sky brightness at an isolated site in California and found a half magnitude increase in brightness from 1976 to 1980, which result matched the experience of the UK Schmidt in Australia.

J. Bhattacharyya reported on work in Bangalore by S. Venkatesh on the correlation between seeing as measured by a 1m telescope and microthermal activity. Cross-spectral analysis of the seeing with the microthermal activity and wind speed showed characteristic peaks at preferred frequencies with high coherence.

D. Carter described the New Zealand six site extinction monitoring program and C. Balnco described the Italian past and present site analysis program.

The afternoon session was chaired by F.G. Smith.

G. Alvarez reported on "Protection of the IAC Observatories in the Canary Isles". The aim is to make the Observatorio de Roque de Los Muchachos site an astronomical reserve. Already air routes above the site have been derouted and the installation of radar stations avoided. The implementation of Commission 50 recommendations is being discussed with the local government and the Spanish government is studying giving the IAC executive faculties concerning the protection of astronomical facilities. A sky brightness investigation, monitoring and control project has started. On Tenerife, the presence of artificial light is clear, but on La Palma there is no problem at present.

A. Asaad spoke on "Sky Brightness at Kottamia". Kottamia is suffering from light pollution caused by the growth of Cairo and looking to Cairo at a zenith distance of 45°, a sky brightness increase of 30 % is present. The projected growth of Cairo and the establishment of new towns will seriously affect the site. Options open to deal with this are to reduce the effect by light pollution control laws or by moving to a high site in Sinai.

D. Crawford spoke on "Tucson experience of light pollution ordinances". Two years ago stricter ordinances were adopted by the city and county and there has been good press coverage and good progress in light pollution control. The counties surrounding Mt Graham and the MMT are being lobbied and one man was working half-time on such lobbying. The time control of outdoor lighting was being pressed. A major argument which can be used is financial saving, and there is good cooperation with the professional lighting engineers. Much can be achieved by determined efforts.

P. Burke also spoke on the "Solar Power Systems". He described the characteristics of the proposed systems and the NAS technical report on it. The deleterious effects on optical, infrared and radio astronomy and on communication with deep space probes were described. There are international treaties preserving wavebands for radio astronomy but none for optical. Whilst NASA no longer have it as an active project, there is the longer term possibility that conditions might change to resurrect it. If this does happen it will probably be as an international large scale project. Astronomers will not have great political strength to oppose it, but can ensure that any project is thoroughly and properly analysed before being started.

P. Boyce also spoke on the Solar Power Systems. He reviewed the various ways by each its effect on astronomy could be greater than at first envisaged and pointed out that by the time any possible system is in operation the growth in ground based astronomy will mean there will be even more facilities affected. He reiterated the importance of educating governments and peoples of the value of astronomy.

C. Bolton said that the Canadian Astronomical Society and Canadian IAU National Committee wish to have a watching brief on the subject. He mentioned some of the technical problems not so far studied such as that of the rocket launches and of the scattering of light by the occasional volcanic dust layers.

F.G. Smith explained that this was the first time the IAU had discussed the problem and represented the start of IAU interest in the subject. The example of the work of Commission 50 on the protection of sites from ground based effects which had been effective by first studying the technical aspects and then drawing up recommendations pointed a way forward. The effects of all present and future space work should be investigated.

A draft resolution for submission to the General Assembly stating :
"Commission 50,

considering with deep concern the proposals for a Satellite Power System which would place very large structures into Earth orbit with the potential of destroying the conditions for a major part of astronomical research in many regions of the electromagnetic spectrum,

recommends national representatives to bring this concern to the notice of space agencies in their countries, and to ensure that the IAU is kept fully informed of proposed developments in space techniques which would involve dangers to astronomy."

was adopted unanimously. F.G. Smith said that the president of Commission 50 would be dealing with the administrative matters concerning this.

The meeting then turned to business matters. F.G. Smith announced his recommendations to the IAU of A. Hoag as president of this Commission and of S. van den Bergh as vice-president. He described the report of the Commission and on the Commission membership. A Working Group on the monitoring of extinction was set up with P. Edwards as chairman, who should be contacted by those wanting to join. A. Hoag mentioned the possibility of a Working Group for guidelines on site evaluation.

COMMISSION 51: SEARCH FOR EXTRATERRESTRIAL LIFE
(RECHERCHE DE LA VIE DANS L'UNIVERS)

PRESIDENT: M.D. PAPAGIANNIS

VICE PRESIDENTS: F.D. DRAKE and N.S. KARDASHEV

ORGANIZING COMMITTEE: R.D. BROWN, P. CONNES, G.D. GATEWOOD, L. GOLDBERG, J. JUGAKU, G. MARX, F. PACINI, M.J. REES, V.S. TROITSKY

## I. INTRODUCTION

The possibility that life might exist in many other places in the Universe has puzzled the minds of scientists and laymen alike for thousands of years. Nearly 2,400 years ago in his book On Nature the Greek philosopher Metrodorus of Chios (as best as I can translate it) was writing: "It goes against Nature, in a large field only one shaft of wheat to grow, and in an infinite Universe to have only one living world." At the end of the 17th century the famous Dutch physicist and astronomer Christian Huygens wrote the book Cosmotheoros in which he tried to show with scientific reasoning that the other planets must also harbour life. He also tried to reconcile science and theology arguing that uninhabited worlds are a great waste, which is uncharacteristic of God who has a purpose for everything. More recently, in the middle of the 19th century, the great German mathematician Karl Friedrich Gauss proposed to plant a colossal forest in the shape of an orthogonal triangle, to signal to astronomers of other worlds that the Earth is inhabited by intelligent beings who are well acquainted with the Pythagorean theorem.

For thousands of years all we could do about this intriguing subject was talk and philosophize. In our days, however, the situation has changed drastically. Now, our radio telescopes can receive radio messages transmitted by similar instruments thousands of light years away, and have already tried repeatedly to do that. Our spaceships are able to explore every part of our solar system, and have already landed on Mars to analyze its soil for signs of biological activity. The electronic and computer support of our optical and radio telescopes is improving at a fast pace, and in a few years, around 1985, we will have our first large telescope in space.

The first radio search was carried out in 1960 by Frank Drake, the Vice President of our Commission, who in his project OZMA looked for radio messages at the 21 cm line of hydrogen from two neighboring stars, epsilon Eridani and tau Ceti. Up to now, there have been about 35 radio and optical searches, which for the most part were individual efforts without any overall coordination or long term planning. They were carried out by a few brave astronomers who usually had to work without research grants and who often had to face the skepticism of even some of their own colleagues. In the 70's, however, the situation began to change with an accelerating pace.

## II. DEVELOPMENTS IN THE LAST 12 YEARS

The Byurakan conference in the Soviet Union in 1971 on CETI (Communication with Extra-Terrestrial Intelligence), which was co-sponsored by the USSR Academy of Sciences and the USA National Academy of Sciences, was an important landmark in the establishment of this new field. Credit must be given to the organizers of this meeting: the Soviet astronomers, V. Ambartsumian, N. Kardashev, I. Shklovsky and V. Troitsky, the American astronomers, F. Drake, P. Morisson and C. Sagan, who also edited the proceedings published by the MIT Press, and to R. Pesek of Czechoslovakia, who was also representing the International Astronautical Federation. Another important factor in gaining acceptance and recognition for this new field was the stubborn persistence of NASA in the search for extraterrestrial life which, in spite of repeated rejections by the U.S. Congress, was kept alive thanks to the dedicated leadership of a few NASA officials, such as G. Soffen and J. Billingham, and thanks to the tireless efforts of a few other scientists, such as B. Oliver.

A big step forward was also the Joint Session on "Strategies for the Search for Life in the Universe" which was held during the 1979 General Assembly of the IAU in Montreal. It was organized by an international committee chaired by M. Papagiannis, and was co-sponsored by three IAU Commissions: 16. Physics of Planets, 40. Radio Astronomy and 44. Space Astronomy. It represented the first time that the IAU was dealing with this subject. The whole-day session generated great interest and was attended by more than 300 astronomers. The following night, when in an open session chaired by Leo Goldberg, F. Drake and M. Papagiannis summarized the presentations and conclusions of the joint session for the general membership of the IAU, the large auditorium of the University of Montreal was packed with more than 1000 people. The success of these meetings in Montreal, showed the great interest that exists within the astronomical community for this new field and led to the formation of our new Commission at the next IAU General Assembly in 1982.

In the three year period between 1979 and 1982, several other events occurred which strengthened even further the case for the new Commission. The proceedings of the Montreal joint session, edited by M.D. Papagiannis, were published in 1980 by Reidel under the title <u>Strategies for the Search for Life in the Universe</u>. The MIT Press published in 1981 the volume <u>Life in the Universe</u>, which was edited by J. Billingham and contained the studies of several NASA workshops and the proceedings of a conference organized by NASA at the Ames Research Center in 1979. The book <u>Extraterrestrials - Where are They?</u>, edited by M. Hart and B. Zuckerman, was published by Pergamon Press in 1982 and contained the proceedings of a conference held in the fall of 1979 at the University of Maryland, USA. Another international conference on SETI, a follow-up of the Byurakan conference, was held in Tallinn, USSR in December of 1981. It was sponsored again by the USSR Academy of Sciences, which maintains an official SETI section, and was chaired by V.S. Troitsky. Its proceedings will be published soon.

A very significant development in this field was the publication by the U.S. Academy of Sciences in 1981 of the report: <u>Astronomy and Astrophysics in the 80's</u> which was prepared by a blue ribbon committee appointed by the Academy and chaired by G. Field. This report endorsed the establishment of an active program for the Search for Extraterrestrial Intelligence (SETI) and recommended to the Government the allocation of 20 million dollars for this program during the decade of the 80's.

All these developments came to a climax at Patras, Greece in August 1982 when the General Assembly approved the establishment of a new IAU Commission on this subject. A great pay-off to all these efforts came soon after. In October 1982, the U.S. Congress passed NASA's budget including, for the first time, a budget line for SETI. With the establishment of the IAU Commission and with government support for work in this field, the search for extraterrestrial life had finally come of age.

## III. THE ESTABLISHMENT OF THE NEW COMMISSION

Following the success of the Montreal meetings in 1979, several people wrote to P. Wayman and to M. Papagiannis inquiring if the IAU could become more active in this field. It is to the great credit of the Secretary General of the IAU that he took this subject under his wing, corresponded with many people, talked to other international unions, and finally brought it to the Executive Committee, which made the courageous decision that the IAU ought to assume the leadership in this field and approved the establishment of a new IAU Commission on this subject.

M. Papagiannis kept corresponding with P. Wayman throughout these three years, and at his request proposed the names for a 12-member advisory board, which were approved by the E.C. Through further consultations we settled on the name and on the objectives of the new Commission. On August 19, 1982, the advisory board met at Patras with P. Wayman who told us that the Executive Committee had approved the establishment of the new Commission and urged us to proceed with organizational matters.

The news about the new Commission was carried by the ASTROCOSMOS, the newspaper of the General Assembly, and a few announcements were posted in the halls of the University of Patras. The response was amazing as more than 100 IAU members asked immediately to join the new Commission. More members were added in the ensuing months through correspondence or personal contact. The Commission now has a membership of close to 200, about 180 IAU members and 20 consultants. The latter are distinguished scientists who share many common interests with the new Commission, but whose work is primarily in other disciplines such as Physics (E.M. Purcell, a Nobel laureate, USA), Life Sciences (G. Soffen, Director NASA, USA), Chemical Evolution and the Origin of Life (C. Ponnamperuma, USA), Astronautics, (R. Pesek, Czechoslovakia), Geophysics (B. Murray, USA), and others. The variety of fields represented simply confirms the fact that the search for extraterrestrial life is a subject of broad appeal that touches many disciplines.

It was decided that the name of the new Commission will be: IAU Commission 51 - Search for Extraterrestrial Life. Michael D. Papagiannis of Boston University was elected the first President of the new Commission, which was allowed to have two Vice Presidents, Nikolai S. Kardashev of the Center for Space Research of the USSR Academy of Sciences and Frank D. Drake of Cornell University. The other nine members of the initial advisory board became the Organizing Committee: Ronald D. Brown (Australia), Pierre Connes (France), George D. Gatewood (USA), Leo Goldberg (USA), Jun Jugaku (Japan), Gyorgy Marx (Hungary), Franco Pacini (Italy), Martin J. Rees (UK) and Vasevolod S. Troitsky (USSR). The IAU General Assembly at its second session on August 26, 1982 approved unanimously the establishment of the new Commission and the names of its officers.

## IV. OBJECTIVES AND FUTURE PLANS

The areas of interest of the new Commission include the following:

1. The search for planets in other solar systems.
2. The evolution of planets and their ability to sustain life.
3. The search for radio signals, intentional or unintentional, of extraterrestrial origin.
4. The search for, and study of, biologically relevant interstellar molecules.
5. The spectroscopic detection of biological activity from primitive forms of life.
6. The search for different manifestations of advanced civilizations.
7. The coordination and promotion of these activities at the international level, and the collaboration with other international societies (biological, astronautical, etc.) that have related interests.

In this last area we have already made substantial progress. M. Papagiannis was appointed the official representative of the IAU to the IAF (International Astronautical Federation) and as such attended its 33rd Congress in Paris at the end of September 1982. The IAF maintains an active SETI Committee, of which he was elected a member. In reciprocity, R. Pesek and J. Billingham, the Chairman and co-Chairman of this SETI Committee, were elected consultants of our new Commission, thus ensuring excellent communications between these two international bodies. Good contacts also have been established with ISSOL (International Society for the Study of the Origin of Life). Its Vice President, C. Ponnamperuma, has been elected consultant to out Commission and M. Papagiannis, who is already a member of ISSOL, will be proposed to become a member of its international advisory committee. We are continuing the dialogue with IUBS (International Union of Biological Sciences) which was initiated by P. Wayman, and we expect we will also be talking with COSPAR.

We are already working on several activities, including efforts:

1. To computerize our membership list and distribute it to all our members and consultants to promote closer contacts among all of them.
2. To publish a Newsletter about 3 times a year which will be mailed to all of our members and consultants. It is expected that they in turn will be submitting to the Newsletter brief news releases on their recent work which will speed up communications among the people working in this field. NATURE has also requested to receive copies of this Newsletter to communicate all new developments in this field to its wide readership.
3. To hold its first scientific meeting, hopefully an IAU Symposium, in the summer or fall of 1984. Our first choice for this event is Hungary which is centrally located and easily accessible from many parts of the world thus assuring a greater participation. At this first meeting, in addition to the normal scientific program, we also hope to have workshops which will try to plan and coordinate activities and research programs in our field around the world.

## V. RECENT RESEARCH WORK

According to a summary compiled in 1981 by J. Tarter of NASA-Ames, and some additional 1982 information from colleagues, in the 22 year history of radio searches from 1960 (project OZMA) to the summer of 1982, there have been at least 31 independent radio searches. These involved at least 12 different radio observatories in seven different countries (USA, USSR, Australia, Canada, France, W. Germany, Netherlands, and Japan). They have compounded about 5,000 hrs of observations, covering a variety of targets, frequencies, bandwidths, and resolutions. In the 60's we only had 4 radio searches, but in the 70's we had 21, and in the 80's we already had 6.

These six most recent ones were: A search along the galactic rotation axis by Lord and O'Dea at the 2.6mm line of CO with the millimeter radio telescope of the U. of Massachusetts. Two searches at the hydrogen line by Israel and Tarter (85 star fields) and by Shostak and Tarter (galactic center) using the Westerbork array in the Netherlands. The search at the hydrogen line by Tarter, Duquet, Clark and Lesyna of 210 solar-type stars with the Arecibo radio telescope. The observations of Biraud and Tarter at the OH line of 102 stars with the Nancay, France decimetric radio telescope. And finally, the ultra-high resolution observations by Horowitz of 250 solar-type stars at the hydrogen line and at twice the hydrogen line frequency with the Arecibo radio telescope.

Several other radio searches are also now being instigated. The Russians are building a network of 100 SETI dishes to obtain an all-sky coverage for broad band pulses. The Japanese are planning searches at the anti-maser line of formaldehyde at 4.83 GHz with the Nobeyama millimeter radio telescope, and Harvard University in the USA is converting its 26-m radio telescope to a SETI dedicated facility. This project is being supported by private funds from the Planetary Society and will soon be conducting "meridian transit" searches in the $-30^{\circ}$ to $+60^{\circ}$ declination range. The project is estimated to last 5 years with searches made at several "magic" frequencies (H, OH, etc.) using the "Suitcase SETI" spectrum analyzer developed by Paul Horowitz of Harvard and colleagues from Stanford University. This analyzer has a frequency resolution of 0.015 Hz over a bandwidth of about 1 kHz and is very effective in eliminating terrestrial interference. It presupposes, however, that the signals are transmitted exactly at the line frequency, having already been corrected by the transmitting civilization for all Doppler effects relatively to our Sun. When this equipment was used with the Arecibo radio telescope in May 1982, it achieved a sensitivity of $2 \times 10^{-27}$ W/m$^2$ in one minute, which means that it could have detected a radio signal of only 1kW power transmitted by an Arecibo-like antenna about 100 light years away.

Besides the radio searches, in the past 20 years we also had 3 optical searches.

A search by Wishnia with the Copernicus satellite for UV laser lines from 3 stars, a search by Schwartzman in the USSR for short optical pulses from peculiar objects, and an optical search by Freitas and Valdes in the United States for objects in the L4 and L5 regions of the Earth-moon system.

## VI. IMPORTANT TECHNOLOGICAL DEVELOPMENTS

Meanwhile great progress is being made in the development of a megachannel spectrum analyzer for our radio searches. We now have (NRAO, JPL, Harvard) several 65,500 channel analyzers, but a group at Stanford University, under the direction of Allen Peterson, is now building a revolutionary 8 million channel analyzer. The design, employing a novel approach to digital spectrum analyzers, has been completed and a first generation model with 72,000 channels is now ready for testing. This analyzer is being built in modular form which provides great flexibility and allows for continuous expansion. Its capacity will continue to grow in the coming years and is expected to reach its full strength of 8 million channels around 1987. It will have a bandwidth of 8 MHz and will be able to simultaneously analyze the input signals into several frequency resolutions ranging from 1 Hz to 1 kHz. A spectrum analyzer of this capacity and with these special features will accelerate immensely all radio searches that try to examine a large number of stars over a considerable frequency range, such as the "water hole" between the H and OH lines. It also would be ideal for radio searches for unintentional radio transmissions (such as our own TV and radar signals) from several nearby stars, which must span a wide frequency range with a high frequency resolution. The completion of this megachannel spectrum analyzer will be a major technological breakthrough for our radio searches.

Another area where great progress is being made, is the development of new instrumentation for the astrometric detection of planets in other solar systems. Up to now, the three observatories in the United States working on this problem (Sproul, Allegheny, and Naval), were making their astrometric observations using photographic methods. Two new techniques, however, are now being developed with great potential for the future. The one is an interferometric system that is being built by Shao of NRL and Staelin of MIT, and the other is a photoelectric system which is being developed by Gatewood and his colleagues at the Allegheny Observatory. A prototype of the interferometer was tested in 1980 tracking the position of Polaris, while a more advanced version capable of determining the angular distances between stars is now being tested at the Mt. Wilson Observatory. This technique, by measuring separately the positions of the red and blue fringes every few milliseconds, is capable of removing much of the effects of the atmospheric seeing and is expected ultimately to reach precisions of about 0.1 milliarcseconds (mas).

A prototype of the Multichannel Astrometric Photometer (MAP), as the Allegheny group calls their device, was also first tested in 1980, but a later version of the MAP is now ready to begin surveying activity at the Allegheny Observatory. The projected precision is 2-3 mas per hour, which surpasses the precision obtained with photographic methods over an entire year. The precision is now actually limited by the quality of the 30-inch objective lens of their refractor telescope, but a new lens figured for red light is now on order, largely thanks to the generosity of Bernard Oliver of the Hewlett-Packard Co., and is expected to become operational in 1984. This will further improve the precision of the system and will allow an intensive search to begin for Jupiter-like planets in any of the nearest 100 to 200 stars.

The accuracy of astrometric observations from the ground with the MAP is limited by atmospheric seeing to a precision of about 1 mas. The precision of the system, however, can be improved by about 3 orders of magnitude to about a microarcsecond when used above the atmosphere with a space telescope. A precision of 1 μas will finally allow the detection of Earth-like planets in other solar systems. The basic problem with space observations is the need for a good reference system.

Lockheed, however, which conducted a feasibility study of these new concepts, concluded that Earth-like planets can be detected with present technology using a double aperture, double MAP configuration, that would measure sufficient reference stars to obtain microarcsecond precision in less than 10 hours of space observations. It appears that we can expect great developments in the search for planets in other solar systems during the 80's, especially when the new techniques will be paired with space telescopes.

## VII. CONCLUSIONS

Astronomers for thousands of years have been trying to unravel the mysteries of the Universe, while biologists have been working on the mysteries of Life. In fact, however, life is an integral part and probably the finest manifestation of Nature, and therefore it is impossible to fully comprehend the Universe without understanding life's place in it.

At last we do have the science and technology to conduct serious searches for extraterrestrial life and we have already been doing so. The number and sophistication of these searches has been increasing steadily in recent years, and important technological developments anticipated for this decade promise to increase their effectiveness to much higher levels. We also have the necessary manpower, many dedicated astronomers and other scientists from around the world who are eager to undertake these tasks. The rapid swelling of the membership of our new Commission to nearly 200 attests to that. The time for this new field has come, and the decision of the IAU, the acknowledged spokesman for Astronomy around the world to establish this new Commission, has provided the necessary recognition and endorsement. We can expect, therefore, an exciting future for this new branch of Astronomy, the search for life (BIOS) in the Universe, which is rapidly becoming known as BIOASTRONOMY.

We must be aware, however, that the search for extraterrestrial life is likely to be an arduous task. The chances of finding extraterrestrial life, primitive or advanced, depend on how common they are and how easy it is to detect them. Primitive forms of life, judging from how fast they appeared on Earth, are probably relatively common in the Galaxy. The spectroscopic detection of their biological activities, however, in planets of other solar systems (one of the long-term objectives of our Commission) is a very difficult task that requires further technological developments. Galactic civilizations, on the other hand, are probably much easier to detect but they could be quite rare in the Galaxy, which would also make their discovery very difficult. It is even conceivable that an extensive search for extraterrestrial intelligence might ultimately lead us to the conclusion that We are the only technological civilization in the Galaxy. Even this, however, should not be taken as a failure of our efforts, because exciting as it might be to find other advanced civilizations, it is equally important to know that we are the only one in the entire Galaxy. The responsibility to preserve this unique civilization and the opportunity to infuse intelligent life into a barren Galaxy will be enormous challenges for the generations to come.

We stand at a historic cross-road. We have the technical capability , the skilled scientists, and above all the strong desire to move ahead with our search for extraterrestrial life. Let us proceed with this noble quest which will allow us to gain a far better understanding of the Universe in which we live.

ACKNOWLEDGEMENTS: On behalf of the entire membership of Commission 51, I wish to express my appreciation to Dr. Patrick Wayman, the General Secretary of the IAU during the period 1979-1982, for all his work and efforts to make this new Commission a reality. Also to all the members of the Executive Committee for their support in this effort and for the honor of electing me the first President of Commission 51. I am also grateful to the new Secretary General of the Union, Dr. Richard West, to our new President, Dr. R. Harbury Brown, and to the new E.C. for the support and interest they have already shown for our new Commission, and for appointing me the official representative of the IAU to the IAF. I also want to thank my colleagues, F. Drake, G. Gatewood, S. Gulkis, P. Horowitz, and J. Tarter for providing me with useful information which I used in the preparation of this report. In closing I want to thank the nearly 200 fellow IAU members and distinguished colleagues from related disciplines who have enthusiastically joined our new Commission. I hope our new Commission will meet the expectations of all of us.

Prof. MICHAEL D. PAPAGIANNIS *
President, IAU Commission 51

* Dept. of Astronomy, Boston University, Boston, MA 02215, USA.
Tel. (617) 353-2625.

## WORKING GROUP FOR PLANETARY SYSTEM NOMENCLATURE (NOMENCLATURE DU SYSTEME PLANETAIRE)
(Committee of the Executive Committee)

Report of Meetings on 6 June 1980, 16 April 1981, 16-19 August 1982

PRESIDENT: P.M. Millman

A brief summary of the first two meetings listed above has appeared in Trans. IAU vol XVIII A, p. 669, 1982. A detailed listing of action taken at all three meetings, as approved at the General Assembly in Patras, is given herewith.

One new Latin term has been used in addition to the 32 previously listed in Trans. IAU vols. XVI B and XVII B.

COLLIS (Colles) a small hill or knob

The plural form is given in brackets.

The current membership in the WGPSN and in the five nomenclature task groups is listed on page 458, this volume.

### LUNAR NOMENCLATURE

Table 1. 6 names, approved as assigned to craters on the lunar surface. Positions are given in degrees of latitude and longitude.

| | | | |
|---|---|---|---|
| BEALS, Carlyle S. (replaces Riemann A) | 1899-1979 Canadian astronomer | 37.1N | 086.0E |
| BOS, Willem H. van den | 1896-1974 Dutch astronomer | 05.3S | 145.8E |
| JENKINS, Louise (replaces Schubert Z) | 1888-1970 American astronomer | 00.7N | 078.0E |
| KAO, Ping-Tse | 1888-1970 Taiwanese astronomer | 06.7S | 087.8E |
| KELDYSH, Mstislav V. (replaces Hercules A) | 1911-1978 Soviet mathematician | 51.5N | 044.0E |
| RECHT, Albert | 1899-1962 American astronomer | 09.8N | 124.0E |

### MERCURY NOMENCLATURE

Errata: YAKOVLEV, a crater on Mercury, listed at bottom of p. 292, Trans. IAU vol. XVII B, 1980, should be corrected to -

BARMA, Ivan Postnik Yakovlev (16th cen) Russian architect

VENUS NOMENCLATURE

Table 2. 66 names, approved as assigned to features on the surface of Venus. Approximate locations are given in degrees of latitude and longitude.

| | | | |
|---|---|---|---|
| CRATERS | | | |
| COLETTE | (1873-1954) French writer | 65N | 322E |
| EARHART | (1897-1937) American flyer | 72N | 136E |
| NIGHTINGALE | (1820-1910) English nurse | 62N | 132E |
| PAVLOVA | (1881-1931) Russian classical dancer | 14N | 040E |
| SACAJAWEA | (1786-1812) Shoshone Indian guide | 63N | 335E |
| CHASMATA | | | |
| ARTEMIS | Greek goddess of hunt/moon | 30S-42S | 121E-145E |
| DALI | Georgian goddess of hunt | 17S-21S | 155E-180E |
| DEVANA | Czechoslovakian goddess of hunt | 00 | 289E |
| DIANA | Roman goddess of hunt/moon | 15S | 150E |
| GANIS | Western Lapp forest maiden | 19N-10N | 188E-200E |
| HECATE | Greek moon goddess | 15N-18N | 230E-245E |
| HENG-O | Chinese moon goddess | 00 -10N | 350E-000 |
| IX CHEL | Aztec wife of the Sun god | 09S-11S | 071E-078E |
| JUNO | Roman sky goddess, consort of Jupiter | 32S | 102E-120E |
| KUANJA | Mbundu goddess of hunt | 10S-16S | 097E-112E |
| PARGA | Samoyed forest spirit | 08S-41S | 222E-321E |
| QUILLA | Inca moon goddess | 32S-38S | 118E-135E |
| VIRAVA | Mordvin forest mother | 13S-17S | 117E-132E |
| DORSUM | | | |
| JUNO | Roman sky goddess, consort of Jupiter | 32S-35S | 085E-095E |
| LINEAE | | | |
| ANTIOPE | Greek amazon | 40S | 350E |
| GUOR | Northern European valkyrie | 20N | 000 |
| HARIASA | German war goddess | 19N | 015E |
| HIPPOLYTA | Greek amazon | 42S | 345E |
| KARA | Icelandic valkryie, a sweet singer | 44S | 306E |
| LAMPEDO | Scythian amazon queen | 57N | 295E |
| MOLPADIA | Greek amazon | 48S | 359E |
| VIHANSA | German war goddess | 54N | 020E |

Table 2. (Cont'd)

| | | | |
|---|---|---|---|
| MONTES | | | |
| GULA | Babylonian earth mother | 22N | 000 |
| INNINI | Babylonian earth mother, worshipped at Kish | 35S | 329E |
| MAAT | Egyptian goddess of truth and justice | 02S | 194E |
| NOKOMIS | Algonquin earth mother | 20N-18N | 189E-191E |
| OZZA (formerly Gaea) | Persian goddess honored by the Koreishies | 05N | 200E |
| SAPAS (formerly Ceres) | Phoenician goddess | 08N | 188E |
| SIF | Teutonic goddess, Thor's wife | 21N | 352E |
| TEFNUT | Egyptian goddess of dew or rain | 39S | 304E |
| USHAS | Indian goddess of dawn | 25S | 323E |
| PATERAE | | | |
| CLEOPATRA | (69 BC-30 BC) Egyptian queen | 67N | 009E |
| THEODORA | (c.500-548) Byzantine empress, wife of Justinian I | 23N | 280E |
| PLANITIAE | | | |
| AINO | Finnish heroine and water sprite | 45S | 090E |
| ATALANTA | Greek mythology, a swift-footed huntress | 54N | 162E |
| GUINEVERE | British queen, wife of Arthur | 40N | 310E |
| HELEN | Wife of Menalaus in ancient Greece (c.1100 BC) | 55S | 255E |
| LAVINIA | Roman wife of Aeneas | 45S | 350E |
| LEDA | Greek mythology, mother of Helen and Castor | 45N | 065E |
| NAVKA | Small water nymph | 08S | 315E |
| NIOBE | Greek, her 12 children killed by Artemis and Apollo | 38N | 132E |
| RUSALKA | Russian mermaid | 10N-10S | 160E-185E |
| SEDNA | Inuit whose fingers became seals and whales | 40N | 335E |
| REGIONES | | | |
| ASTERIA | Greek titaness | 18N-30N | 228E-270E |
| ATLA | Norse giantess, mother of Heimdall | 20N-05S | 185E-210E |
| BELL | English giantess | 25N-35N | 045E-055E |
| EISILA | Greek giantess | 10N-25N | 350E-050E |
| IMDR | Norse giantess | 42S | 211E |
| METIS | Greek titaness | 72N | 245E-255E |

Table 2. (Cont'd)

| | | | |
|---|---|---|---|
| MNEMOSYNE | Greek titaness | 68N | 275E-285E |
| OVDA | Marijian titaness | 05N-10S | 080E-108E |
| PHOEBE | Greek titaness | 10N-20S | 275E-300E |
| TELLUS | Greek titaness | 35N | 080E |
| TETHUS | Greek titaness | 55N | 100E |
| THEMIS | Greek titaness | 37S-40S | 275E-310E |
| THETIS | Greek titaness | 02S-15S | 118E-140E |
| ULFRUN | Norse giantess | 18N-03S | 220E-230E |
| RUPES | | | |
| HESTIA | Greek hearth goddess | 03N-08N | 059E-075E |
| UT | Siberian, Turco-Tatar goddess of hearth | 48N-53N | 305E-325E |
| VESTA | Roman hearth goddess | 55N-65N | 295E-355E |
| TERRA | | | |
| LADA | Serbian goddess of love | 50S-65S | 285E-110E |

MARS NOMENCLATURE

Table 3. 44 names, approved as assigned to features on the surface of Mars. Positions have been given in degrees of latitude and longitude.

| | | | |
|---|---|---|---|
| CRATERS | | | |
| DOKUCHAEV, Vasily V. | 1846-1903<br>Russian soil scientist | 61S | 127W |
| IBRAGIMOV, Nadir B. | 1932-1977<br>Soviet astronomer | 25.5S | 059W |
| KRISTOFOVICH, Afrikan N. | 1885-1953<br>Soviet paleobotanist | 48S | 262W |
| NORDENSKIOLD, Nils A.E. | 1832-1901<br>Swedish geologist and geographer | 53S | 159W |
| SEMEYKIN, Boris E. | c.1900-1937<br>Soviet astronomer | 42N | 351W |
| SYTINSKAYA, Nadezhda N. | 1906-1974<br>Soviet astronomer | 43N | 052W |
| TIMOSHENKO, Ivan F. | 1918-1941<br>Soviet astronomer | 42N | 064W |
| TOLSTOY, Aleksey | 1882-1945<br>Soviet writer | 48S | 234W |
| COLLES | | | |
| ARIADNES | Classical albedo feature | 35S | 187W |
| DEUTERONILUS | Classical albedo feature | 41N | 338W |

Table 3. (Cont'd)

| | | | |
|---|---|---|---|
| CHAOS | | | |
| ARSINOES | Daughter of Ptolomy Lagus and Bernice, albedo feature | 08S | 028W |
| EOS | Greek name of Aurora, albedo feature | 16S-19S | 043W-050W |
| PYRRHAE | Daughter of Pandora, albedo feature | 11S | 029W |
| DORSUM | | | |
| CERBERUS | Classical albedo feature | 09S-18S | 252W-258W |
| FOSSAE | | | |
| CALYDON | Son of Ares and Astynome | 07.5S | 087W-093W |
| COLOE | Classical albedo feature | 32N-41.5N | 294W-307W |
| ISMENIAE | Classical albedo feature | 37N-42N | 315W-337W |
| NIA | River Gambia, West Africa | 13S-15S | 070W-074W |
| OTI | Classical albedo feature | 07S-12S | 115W-118W |
| TITHONIAE | Classical albedo feature | 01S-03S | 034W-047W |
| LABYRINTHUS | | | |
| ADAMAS | Classical albedo feature | 30N-42N | 247W-257W |
| MENSAE | | | |
| CANDOR | Classical albedo feature | 06S | 028W |
| HEBES | Classical albedo feature | 01S | 032W |
| MONTES | | | |
| EREBUS | Classical albedo feature | 30N-42N | 170W-180W |
| GERYON | Classical albedo feature | 07S-08S | 078W-084W |
| PATERAE | | | |
| MEROE | Classical albedo feature | 07N | 291W |
| NILI | Classical albedo feature | 09N | 293W |
| PLANUM | | | |
| DAEDALIA | Classical albedo feature | 17S-23S | 115W-125W |
| RUPES | | | |
| AMENTHES | Classical albedo feature | 00 -04N | 248W-252W |
| MORPHEOS | Classical albedo feature | 36S-39S | 231W-240W |
| SCOPULUS | | | |
| OENOTRIA | Classical albedo feature | 15S-10S | 270W-292W |
| VALLES | | | |
| ARNUS (replaces Arena Rupes) | River in Tuscany, Italy | 18.5N | 290W |
| BRAZOS | River in Texas, USA | 02.5S-06S | 339W-344W |

Table 3. (Cont'd)

| | | | |
|---|---|---|---|
| DRAVA | River in Yugoslavia | 47S-56S | 192W-197W |
| EVROS | River in Greece | 12S-13S | 343W-349W |
| GRANICUS | Ancient river in Turkey | 27N-30N | 225W-231W |
| HEBRUS | Ancient river in Greece | 17N-22N | 232W-234W |
| INDUS | River in Pakistan | 17N-21N | 317W-322W |
| LOIRE | River in France | 17.5S-22S | 014W-017.5W |
| LICUS | Ancient river in France | 01S-05S | 233W-234W |
| LOUROS | River in Greece | 08S-09.5S | 079W-086W |
| NAKTONG | River in Korea | 00 -09N | 330W-321W |
| REULL | Word for planet in Gaelic | 37S-43S | 248W-265W |
| SCAMANDER | Ancient river near Troy, Turkey | 14N-18N | 330W-331W |

OUTER SOLAR SYSTEM NOMENCLATURE

The following names* were approved for recently discovered (or more completely observed) satellites of Jupiter and Saturn:-

| | Permanent Designation | Temporary Designation |
|---|---|---|
| THEBE | Jupiter XIV | 1979 J2 |
| JANUS | Saturn X | 1980 S1 |
| EPIMETHEUS | Saturn XI | 1980 S3 |
| TELESTO | Saturn XIII | 1980 S13 |
| CALYPSO | Saturn XIV | 1980 S25 |

The following nomenclature was approved for divisions or gaps in the rings of Saturn. The first two names are traditional, as used for divisions seen by earth-based visual observers. The last three are applied to prominent narrow gaps, as revealed by the Voyager 2 photography. (See figure on page 345)

* Following detailed discussions in Patras among members of Commissions 6 and 20 and the WGPSN it was suggested that three names, orginally recommended by the WGPSN (i.e. ADRASTEA for 1979 J1, METIS for 1979 J3, ATLAS for 1980 S28), be considered provisional until further information re the orbital elements of these three objects had been made generally available. These discussions also called attention to the fact that the IAU had not formally assigned the responsibility for the various steps that must be taken following the discovery of new bodies in the solar system. A joint meeting of representatives from the three IAU units noted above was held and a plan for future action was drawn up. This plan was later endorsed by the Executive Committee of the IAU and the General Assembly. It is reproduced in the Astronomer's Handbook, Section 4, p. 383, this volume, and in the IAU Information Bulletin No. 49, January 1983.

| | |
|---|---|
| CASSINI, Giovanni D. (1625-1712) | - the prominent division between the A ring and the B ring, approx. radius 119,000 km. |
| ENCKE, Johann F. (1791-1865) | - the widest division in the A ring, approx. radius 133,500 km. |
| MAXWELL, James C. (1831-1879) | - the widest gap in the C ring, approx. radius 87,500 km. |
| HUYGENS, Christiaan (1629-1695) | - the gap at the inner edge of the Cassini division, approx. radius 117,500 km. |
| KEELER, James E. (1857-1900) | - the gap near the outer edge of the A ring, approx. radius 136,500 km. |

Table 4. 7 names, approved as assigned to features on the surfaces of 2 satellites of Jupiter. Approximate positions are given in degrees of latitude and longitude.

IO NOMENCLATURE

PATERAE

| | | | |
|---|---|---|---|
| HUO SHEN | Chinese god of fire | 15S | 329W |
| KIBERO | Underworld toad who gives mankind fire | 12S | 306W |
| MBALI | Pygmy word for fire | 32S | 007W |
| SHOSHU | Caucasian patron of fire | 19.5S | 324.5W |
| TALOS | A smith, nephew of Daedalus | 26S | 356W |
| TAW | Monguor word for fire or hearth | 33.5S | 000.5W |

GANYMEDE NOMENCLATURE

CRATER

| | | | |
|---|---|---|---|
| ENKIDU | Friend of Gilgamesh | 18S | 322W |

Table 5. 190 names, approved as assigned to features on the surfaces of 9 satellites of Saturn. Approximate positions are given in degrees of latitude and longitude.

JANUS NOMENCLATURE

CRATERS

| | |
|---|---|
| CASTOR | One of the Dioskuroi, a tamer of horses |
| IDAS | Cousin of Gemini |
| LYNCEUS | Cousin of Gemini |
| PHOIBE | Daughter of Leukippos |

EPIMETHEUS NOMENCLATURE

CRATERS

| | |
|---|---|
| HILAIREA | Wife of Pollux |
| POLLUX | Latin name for Polydeukes, Castor's twin |

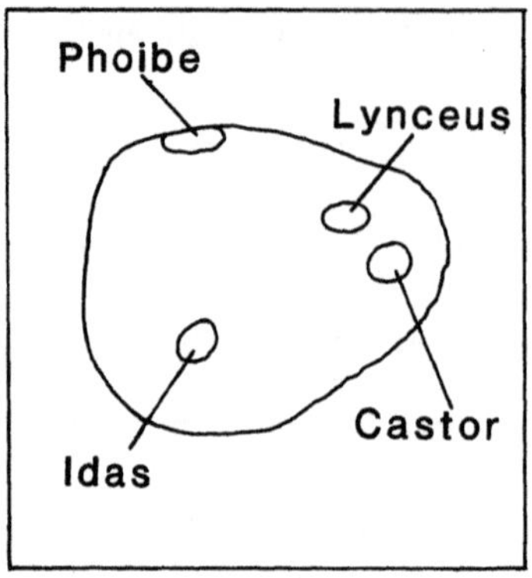


**JANUS**

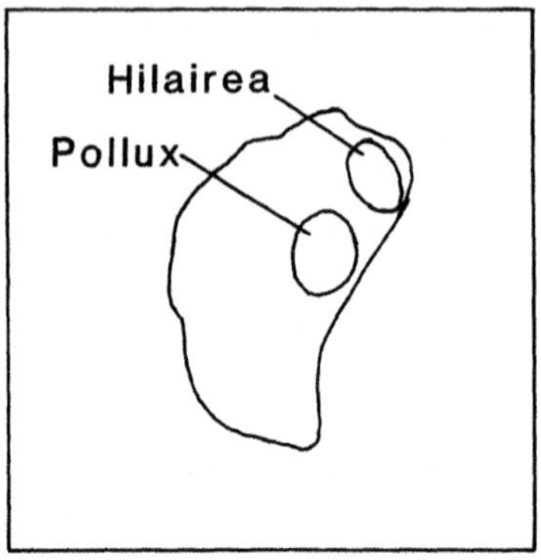


**EPIMETHEUS**

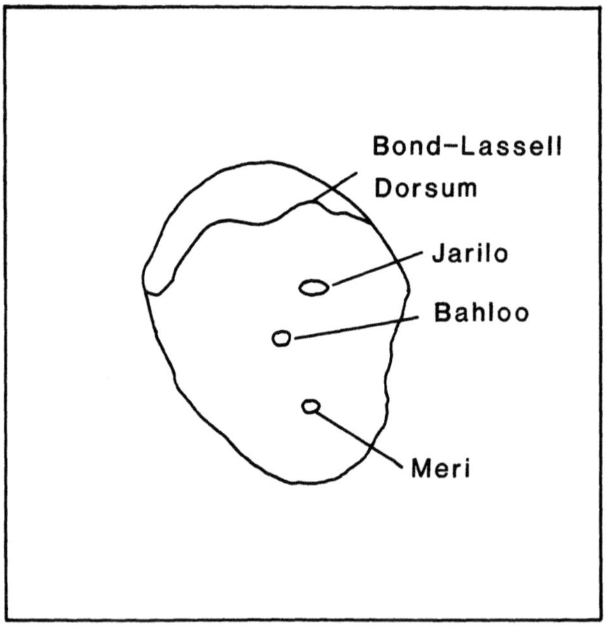


**HYPERION**

Table 5. (Cont'd)

MIMAS NOMENCLATURE

CRATERS

| | | | |
|---|---|---|---|
| ACCOLON | Arthur's companion, jousted with Arthur | 68S | 166W |
| ARTHUR | King of the Round Table assemblage | 35S | 190W |
| BALIN | Knight of "matchless courage and virtue" | 22N | 082W |
| BAN | King of Benwick, father of Sir Launcelot | 47N | 149W |
| BEDIVERE | Knight of the Round Table | 10N | 145W |
| BORS | King of Gaul | 45N | 165W |
| DYNAS | Knight of the Round Table | 08N | 075W |
| ELAINE | Daughter of King Pelles | 44N | 102W |
| GAHERIS | Older son of King Lot | 46S | 287W |
| GALAHAD | Bastard son of Launcelot and Elaine | 47S | 135W |
| GARETH | Youngest son of King Lot | 44S | 280W |
| GAWAIN | Eldest son of King Lot | 60S | 254W |
| GWYNEVERE | Queen, wife of Arthur and lover of Launcelot | 12S | 312W |
| HERSCHEL, William | (1738-1822) Astronomer who discovered Mimas | 00 | 104W |
| IGRAINE | Wife of Uther, mother of Arthur | 40S | 225W |
| ISEULT | Loved by Tristram | 48S | 035W |
| KAY | Royal seneschel at Arthur's court | 46N | 116W |
| LAMEROK | Pellinore's son | 65S | 283W |
| LAUNCELOT | King Arthur's favorite knight | 10S | 317W |
| LOT | Leader of the rebel kings of North and West | 30S | 227W |
| MARK | King of Cornwall | 28S | 297W |
| MERLIN | Magician and prophet, Arthur's mentor | 38S | 215W |
| MODRED | Arthur's bastard son and mortal enemy | 05N | 213W |
| MORGAN | Arthur's half sister, an enchantress | 25N | 240W |
| PALOMIDES | Saracen enemy of Tristram | 04N | 157W |
| PELLINORE | King who pursued the questing beast | 35N | 128W |
| PERCIVALE | Pure Knight, finds the Holy Grail | 01S | 171W |
| TRISTRAM | Saves Iseult, falls in love with her | 58S | 026W |
| UTHER | Ruler of all Britain, Arthur's father | 35S | 244W |

CHASMATA

| | | | |
|---|---|---|---|
| AVALON | Arthurian paradise | 20N-57N | 160W-120W |
| CAMELOT | Home of the Round Table assemblage | 25S-60S | 000 -045W |
| OETA | Shook by Titan in war with Olympians | 10N-35N | 130W-105W |
| OSSA | Mt. Pelion piled on top of it | 10S-30S | 305W-280W |
| PANGEA | Picked up by Titan in war with gods | 25S-55S | 340W-290W |
| PELION | Piled up with Mt. Ossa in war with gods | 20S-25S | 235W-200W |
| TINTAGEL | Home of Igraine, Arthur's mother | 43S-60S | 235W-190W |

Table 5. (Cont'd)

ENCELADUS NOMENCLATURE

CRATERS

| | | | |
|---|---|---|---|
| AHMAD | Youngest son, brings father magic apple | 58N | 304W |
| ALADDIN | Hero of the tale, has the magic lamp | 63N | 017W |
| ALI BABA | Hero of tale, found a great treasure | 55N | 011W |
| DALILAH | Crafty old crone, fools a series of men | 53N | 244W |
| DUBAN | Sage who cured King Yunan of leprosy | 58N | 276W |
| DUNYAZAD | Sister of Shahrazad | 43N | 200W |
| GHARIB | Hero of many tales | 81N | 245W |
| JULNAR | The seaborn heroine | 54N | 340W |
| MUSA | Gets vessels containing Jinni, City of Brass | 73N | 003W |
| PERI-BANU | Genie who marries Ahmad | 63N | 315W |
| SALIH | Brother of Julnar | 06S | 000 |
| SAMAD | Shayk who guides Musa and Talib | 61N | 353W |
| SHAHRAZAD | Daughter of Wazir who tells the tales to king | 49N | 200W |
| SHAHRYAR | King who is beguiled by the tales | 58N | 222W |
| SINDBAD | Voyager who had many marvelous adventures | 66N | 210W |

FOSSAE

| | | | |
|---|---|---|---|
| BASSORAH | Town from which Sindbad embarked | 40N-50N | 023W-345W |
| DARYABAR | Land from which princess Daryabar came | 05N-10N | 020W-335W |
| ISBANIR | Fakir Taj's home | 10S-20N | 000 -350W |

PLANITIAE

| | | | |
|---|---|---|---|
| DIYAR | Country where Khudadad's father ruled | 00 | 250W |
| SARANDIB | Ceylon, island visited by Sindbad | 05N | 300W |

SULCI

| | | | |
|---|---|---|---|
| HARRAN | City where Khudadad's father ruled | 05S-35N | 270W-210W |
| SAMARKAND | Ruled by Zaman, brother of Shahryar | 10S-75N | 340W-300W |

TETHYS NOMENCLATURE

CRATERS

| | | | |
|---|---|---|---|
| AJAX | Greek hero, second only to Achilles | 30S | 285W |
| ANTICLEIA | Mother of Odysseus | 55N | 038W |
| ANTINOUS | Chief of Penelope's wooers | 62S | 275W |
| ARETE | Wife of Alcinous, mother of Nausicaa | 04S | 300W |
| CIRCE | Changed Odysseus' companions into swine | 08S | 049W |
| ELPENOR | Follower of Odysseus | 54N | 268W |
| EUMAEUS | Faithful swineherd who greets Odysseus | 27N | 047W |

Table 5. (Cont'd)

| | | | |
|---|---|---|---|
| EURYCLEIA | Faithful old nurse of Odysseus | 56N | 247W |
| LAERTES | Father of Odysseus | 50S | 060W |
| MELANTHIUS | Disloyal goatherd, insults Odysseus | 62S | 204W |
| MENTOR | Friend of Odysseus | 03N | 039W |
| NAUSICAA | Daughter of Alcinous who advised Odysseus | 80N | 352W |
| NESTOR | Wise old king | 57S | 058W |
| ODYSSEUS | Hero of Odyssey | 30N | 130W |
| PENELOPE | Faithful wife of Odysseus | 10S | 252W |
| PHEMIUS | Minstrel to the wooers | 12N | 290W |
| POLYPHEMUS | Cyclops battled by Odysseus | 05S | 285W |
| TEIRESIAS | Aged prophet, consulted by Odysseus | 62N | 005W |
| TELEMACHUS | Son of Odysseus | 56N | 338W |
| CHASMA | | | |
| ITHACA | Ionian island, home of Odysseus | 80S-50N | 030W-340W |
| DIONE NOMENCLATURE | | | |
| CRATERS | | | |
| ADRASTUS | King of Argos | 64S | 040W |
| AENEAS | Hero of the Aeneid, son of Anchises and Venus | 26N | 047W |
| AMATA | Mother of Lavinia | 07N | 287W |
| ANCHISES | Father of Aeneas | 35S | 063W |
| ANTENOR | Nephew of Priam, escaped fall of Troy | 06S | 008W |
| BUTES | A boxer, defeated by Dares | 68N | 050W |
| CAIETA | Nurse of Aeneas | 25S | 080W |
| CASSANDRA | Daughter of Priam, foretold the future | 42S | 245W |
| CATILLUS | Brother of Tiburtus, twin brother of Coras | 01S | 275W |
| CORAS | Twin brother of Catillus, founder of Tibur | 03N | 268W |
| CREUSA | Daughter of Priam, first wife of Aeneas | 48N | 078W |
| DIDO | Tyrian princess, founded Carthage | 22S | 015W |
| HALYS | A Trojan, defended the camp of Aeneas | 60S | 045W |
| ILIA | Mother by Mars of Romulus and Remus | 03N | 344W |
| ITALUS | Eponymous ancestor of Italians, ancient hero | 20S | 076W |
| LATAGUS | Soldier of Aeneas | 16N | 026W |
| LAUSUS | Son of Mezentius, handsome and brave | 38N | 023W |
| MAGUS | King of the Rutulians, rival of Aeneas | 20N | 024W |
| MASSICUS | Etruscan ally of Aeneas | 36S | 052W |
| PALINURUS | Pilot of the fleet of Aeneas | 05S | 062W |
| REMUS | Brother of Romulus, founder of Rome | 10S | 030W |

Table 5. (Cont'd)

| | | | |
|---|---|---|---|
| RIPHEUS | A Trojan who fought beside Aeneas at Troy | 56S | 029W |
| ROMULUS | Mythical founder of Rome circa 754 BC | 08S | 024W |
| SABINUS | Fabled ancestor of the Sabines | 44S | 190W |
| TURNUS | Rutulian king, rival of Aeneas for Lavinia | 21N | 342W |
| CHASMATA | | | |
| LARISSA | Town in Thessaly, Achilles' native region | 20N-48N | 015W-065W |
| LATIUM | The Trojans' promised land in Italy | 03N-45N | 064W-075W |
| PALATINE | One of the seven hills of Rome | 55S-73S | 075W-320W |
| TIBUR | Ancient town of Italy near Rome | 48N-80N | 060W-080W |
| LINEAE | | | |
| CARTHAGE | An ancient Etruscan city | 20N-10N | 337W-310W |
| PADUA | City in Italy, founded by Antenor | 05N-40S | 245W-190W |
| PALATINE | One of the seven hills of Rome | 10S-55S | 285W-320W |
| RHEA NOMENCLATURE | | | |
| CRATERS | | | |
| AANANIN | Korean god of the heavens | 39N | 330W |
| ADJUA | Mythical heroine, ancestor of Ulci tribe | 46N | 126W |
| AGUNUA | Melanesian god who made sea, land and people | 70N | 065W |
| AMETA | Indonesian ancestor whose blood made Hainuwele | 59N | 014W |
| ARUNAKA | Inca god, creator of all things | 14S | 021W |
| ATUM | Creator god of Heliopolos, son of Ptah | 45S | 000 |
| BULAGAT | Mythological ancestor of Buriat tribe | 35S | 014W |
| BUMBA | Bushongo god, vomited up sun, moon, men, etc. | 70N | 040W |
| BURKHAN | Siberian god who created world | 69N | 288W |
| CON | Coastal creator god | 24S | 010W |
| DJULI | Ukrainian first man, ancestor of the people | 26S | 046W |
| ELLYAY | Yakutian ancestor of the people | 78N | 095W |
| FARO | Mande, sacrificial killing purified Earth | 52N | 121W |
| HAIK | Mythological ancestor of Armenian people | 34S | 027W |
| HAOSO | Manchurian creator of all things | 09N | 008W |
| HELLER | Auracanin creator of men and civilization | 09N | 310W |
| IRACA | Incan creator god who became the moon | 45N | 120W |
| IZANAGI | Japanese mythology, creator god | 49S | 298W |
| IZANAMI | Japanese creator goddess, sister of Izanagi | 46S | 310W |
| JUMO | Marijian sky god | 56N | 065W |
| KARORA | Aranda, Australian ancestor | 07N | 016W |
| KHADO | Nanajan hero who built the world, first shaman | 45N | 349W |

Table 5. (Cont'd)

| | | | |
|---|---|---|---|
| KIHO | Tuamotu, Society Islands, made land and sea | 10S | 354W |
| KUMPARA | Jivaro, Ecuador, creator god | 11N | 321W |
| LEZA | Tonga, originator of conditions of life | 19S | 304W |
| LOWA | Marshall Islands, great creator god | 45N | 009W |
| MALUNGA | Yao, Bantu, creator god and progenitor god | 74N | 049W |
| MANOID | Negrito, Malay Peninsula, progenitor goddess | 33N | 002W |
| MELO | Minyong, India, the original man | 51S | 006W |
| MUBAI | Tibetan heavenly god | 61N | 011W |
| NUM | Samoyed god of heaven | 23N | 093W |
| ORMAZD | Persian progenitor god of light | 62N | 052W |
| PAN KU | Miao creator of all things | 72N | 115W |
| PEDN | Negrito, Malay Peninsula, god, created men | 48N | 340W |
| QAT | New Hebrides, Melanesia, formed men from trees | 23S | 347W |
| SHOLMO | Buriat, Siberia, devil, created evil creatures | 13N | 340W |
| TAAROA | Tahitian creator god | 14N | 099W |
| THUNUPA | Incan god, creator of all things | 51N | 015W |
| TIKA | Supreme Abkhaz being | 25N | 087W |
| TORE | Pygmie lord of world, creator of all things | 01N | 335W |
| TOROM | Ostyah sky god | 68S | 345W |
| UKU | Estonian super-god | 85N | 115W |
| WHANIN | Korean god, creator of all things | 74N | 121W |
| WURAKA | Kakadu, Australia, ancestor of all people | 28N | 357W |
| XAMBA | Supreme Bushman, creator of all things | 04N | 347W |
| XU | Bushman, creator | 61N | 070W |
| YU-TI | Supreme primal Chinese god | 55N | 085W |
| CHASMATA | | | |
| KUN LUN | Chinese mythology, home of the immortals | 37N-50N | 275W-300W |
| PU CHOU | Mountain attacked by Kung Chung, sky fell | 10N-35N | 085W-115W |

HYPERION NOMENCLATURE

| | |
|---|---|
| CRATERS | |
| BAHLOO | The moon, maker of girl babies |
| HELIOS | Son of Hyperion, the Greek sun god |
| JARILO | East-Slavic god of sun, fertility and love |
| MERI | Bororo folk hero, the sun |
| DORSUM | |
| BOND-LASSELL | G.W. Bond and Wm. Lassell, discoverers of Hyperion in 1848 |

Table 5. (Cont'd)

IAPETUS NOMENCLATURE

| | | | |
|---|---|---|---|
| CRATERS | | | |
| ALMERIC | One of 12 peers, killed by Marsilion | 53N | 274W |
| BALIGANT | Emir of Babylon, assisted Marsilion in war | 15N | 225W |
| BASAN | French baron, brother of Basilie | 30N | 197W |
| BERENGER | One of 12 peers, killed by Grandoyne | 59N | 220W |
| BESGUN | Chief cook for Charlemagne's army | 72N | 296W |
| CHARLEMAGNE | King of France who conquered Spanish towns | 54N | 266W |
| GEBOIN | Guarded French dead in Charlemagne's army | 56N | 175W |
| GODEFROY | Charlemagne's defender, bearer of gonfalcon | 78N | 253W |
| GRANDOYNE | Son of Cappadocian king Capuel | 18N | 215W |
| HAMON | Joint commander of Charlemagne's 8th division | 10N | 271W |
| LORANT | French commander against Baligant | 64N | 165W |
| MARSILION | Saracen king of Spain, vassal of Baligant | 41N | 177W |
| MILON | Guarded French dead in Charlemagne's army | 75N | 270W |
| OGIER | Dane in vanguard of Charlemagne's army | 42N | 274W |
| OLIVER | Roland's friend, killed by Marganice | 61N | 203W |
| OTHON | One of 12 peers, guarded French dead | 24N | 344W |
| ROLAND | Charlemagne's nephew, died at Roncevaux | 78N | 030W |
| TURPIN | Archbishop of Rheims | 43N | 000 |
| REGIO | | | |
| CASSINI | G.D. Cassini (1625-1712) discoverer of four satellites of Saturn | 48S-55N | 210W-340W |
| TERRA | | | |
| RONCEVAUX | Where Roland and forces were ambushed by the Spaniards | 30S-90N | 300W-130W |

Ottawa, Ontario

5 November, 1982

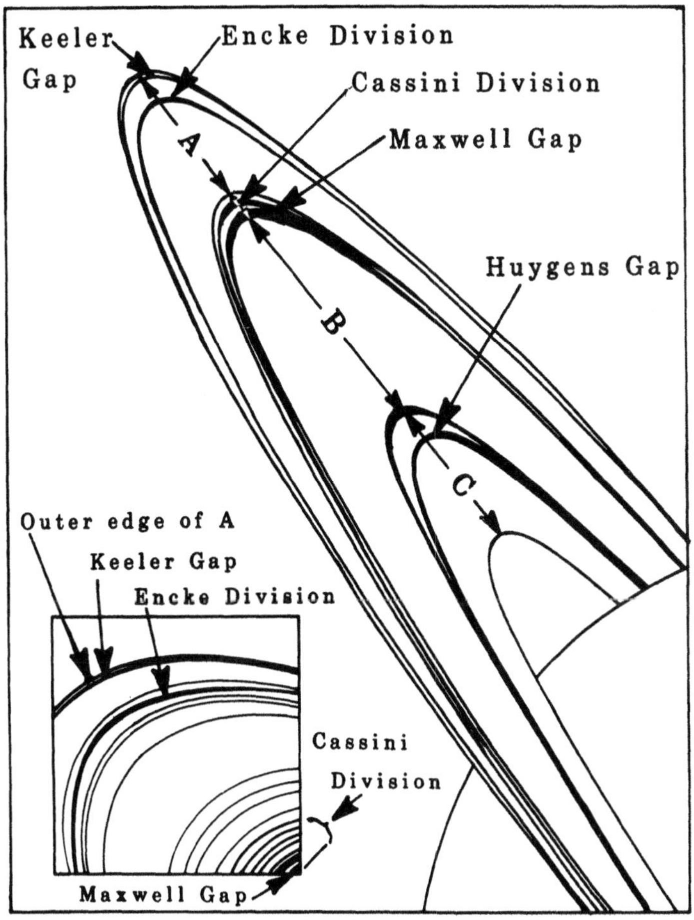


Saturn Ring Nomenclature

# V.
# Report of the Executive Committee (1979-1981)

PRESIDENT: Professor M.K. Vainu Bappu — Indian Institute of Astrophysics, Bangalore 560034, India

VICE-PRESIDENTS:

| | |
|---|---|
| Dr M.W. Feast | South African Astronomical Observatory, PO Box 9, Observatory, Cape 7935, South Africa |
| Dr D.S. Heeschen | National Radio Astronomy Observatory, Edgemont Road, Charlottesville, Virginia 22901, USA |
| Academician E.K. Kharadze | Abastumani Astrophysical Observatory, Mt Canobili, Abastumani 383762, USSR |
| Dr L. Kresak | Astronomical Institute, Slovak Academy of Sciences, 899 30 Bratislava, Czechoslovakia |
| Professor S. van den Bergh | Dominion Astrophysical Observatory, 5071 W. Saanich Road, RR 5, Victoria, British Colombia, V8X 4M6, Canada |
| Professor R. Wilson | Department of Physics & Astronomy, University College London, Gower Street, London WC1E 6BT, UK |

GENERAL SECRETARY: Professor Patrick A. Wayman — Dunsink Observatory, Castleknock, Co. Dublin, Ireland

ASSISTANT GENERAL SECRETARY: Dr Richard M. West — European Southern Observatory, Karl-Schwarzschild Strasse 2, D-8046 Garching bei München, GFR

ADVISERS TO THE EXECUTIVE COMMITTEE:

| | |
|---|---|
| Professor A. Blaauw<br>-former President | Asserstraat 78, 9335 TC Zuidvelde, Netherlands |
| Professor Edith A. Müller<br>-former General Secretary | Observatoire de Genève, CH-1290 Sauverny, Switzerland |

---

INTRODUCTION

The present report covers the period from 1 January 1979 to 31 December 1981. This period includes the first eight months (to 26 August) of the year 1979, which was part of the period in office of the previous Executive Committee; it does not include the first eight months of 1982, still the period of responsibility of the present Executive Committee (to 27 August 1982).

The address of the General Secretary at the XVIIIth General Assembly will include a report for the period 1 January - 16 August 1982.

This report includes a summarized financial report for the calendar years 1979, 1980 and 1981 and a budget proposal for the calendar year 1982, closing the 3-year budgetary period 1979-1982.

## ADMINISTRATION

### Executive Committee

The Executive Committee held the following meetings:

45th meeting: Montreal, Canada, 12, 13 and 21 August 1979

46th meeting: Montreal, Canada, 23 August 1979

47th meeting: Amsterdam, Netherlands, 5-8 August 1980

48th meeting: Abastumani, USSR, 21-25 July 1981

The chairman of the 45th meeting, held with the previous composition of the Executive Committee, was Professor A. Blaauw, President of the IAU. The meeting was attended by all members and advisers, with the exception of Mr J.G. Bolton; Professor M.K.V. Bappu, then President-elect, and Dr R.M. West, Assistant General Secretary elect, participated on 21 August 1979.

The 46th, 47th and 48th meetings were held under the chairmanship of Professor M.K.V. Bappu, President of the IAU. All members and advisers were present, except that Dr L. Kresak was prevented on medical advice from attending the 47th and 48th meetings and Professor S. van den Bergh was unable to attend the 48th meeting.

### Officers' Meetings

The President of the IAU, the General Secretary and the Assistant General Secretary met as follows:

Geneva, Switzerland, 19 May 1979

Montreal, Canada, 10-11 August 1979

Paris, France, 10-11 December 1979

Leiden, Netherlands, 3-4 August 1980

Paris, France, 12-13 January 1981

Tbilisi, Georgia, USSR, 19-20 July 1981

### IAU Secretariat

From 14 September 1979, the secretariat of the IAU occupied a permanent seat in the Pavillon Ouest of l'Observatoire de Paris. The address of the secretariat is:

IAU - UAI
61, avenue de l'Observatoire
F-75014 Paris (France)

Cable address: UNASTRINTER, Paris 014
Telephone: (01) - 325-83-58

It is appropriate to express the thanks of the Executive Committee, on behalf of the Union, to the Conseil de l'Observatoire de Paris and its President, M. Boulon, also to Professor J.-C. Pecker of the Collège de France, former General Secretary of the IAU, and to "IAU France 1976" for the work of preparing accommodation for the IAU Secretariat in 1979, including valuable gifts of furniture and other items, ensuring thereby a smooth transition to the new premises.

IAU Staff

The retirement took place during September 1979 of the IAU Executive Secretary, Dr A. Jappel. Dr Jappel was appointed as Assistant Secretary during 1967 and for twelve years he served the Union devotedly. He used his knowledge of international law and the legal systems of different countries in paying special attention to the financial and legal standing of the Union.

For the major part of his service, Dr Jappel had the support of Mrs Jarka Dankova, Administrative Assistant. In September 1979 Mrs Dankova returned to Czechoslovakia, leaving the service of the IAU. Dr Jappel and Mrs Dankova worked in the four home countries of successive IAU General Secretaries, as follows:

L. Perek, Czechoslovakia, 1967-1970

C. de Jager, Netherlands, 1970-1973

G. Contopoulos, Greece, 1973-1976

E.A. Müller, Switzerland, 1976-1979

Mme Brigitte Manning and Ms Patricia Smiley were engaged respectively as Secretary and Typist at the Paris IAU Secretariat from September 1979.

Adhering Countries

Indonesia was admitted by the XVIIth General Assembly in 1979 as the 49th Adhering Country.

Following discussion at the time of the XVIIth General Assembly between, on one hand, the President and Executive Committee of the IAU, and on the other hand, representatives of the Chinese Astronomical Society, the adherence of China from 1 May 1980 was effected, as a temporary measure, through two adhering organisations. A resolution of the XVIIth General Assembly in 1979 calls for ratification of these arrangements at the XVIIIth General Assembly.

Members of the IAU

The number of IAU members was 3783 on 1 January 1979. 812 new members were admitted at the XVIIth General Assembly and 58 names were deleted by resignation and decease up to December 1979. Since that date there have been recorded 39 further names for deletion. The number of members at 31 December 1981 is therefore estimated to be 4498.

Consultants to IAU Commissions

The 39 IAU Commissions had a total of 227 consultants at 31 December 1981.

## COMMISSIONS OF THE IAU

By resolution of the XVIIth General Assembly, Commission 16 (Physical Study of Planets and Satellites) and Commission 17 (The Moon) were merged to form a new Commission, retaining the title and number of Commission 16.

Reports of each Commission and of the Executive Committee's Working Group for Planetary System Nomenclature are being published during 1982 in IAU Transactions volume XVIIIA, c. 700 pages.

The following remarks concern some of the many Commission activities not otherwise included in subsequent sections of this report:

The Working Group on Project MERIT (Commission 19), in conjunction with the International Association of Geodesy (IUGG), has produced reports on proposed new methods for measuring earth rotation.

The Minor Planet Center at the Smithsonian Institution, Cambridge, Massachusetts, under the direction of Dr B.G. Marsden, has produced Minor Planet Circulars regularly and has rendered annual reports to the Executive Committee.

The IAU Telegram Bureau at the Smithsonian Institution, Cambridge, Massachusetts, under the direction of Dr B.G. Marsden, produced 338 Circulars and despatched telegrams to subscribers over all continents.

The Quarterly Bulletin of Solar Activity continued its work with regular publication of solar activity, as a permanent record, from the Tokyo Astronomical Observatory. The valuable continuation of the Zurich Sunspot numbers has been maintained by the Sunspot Index Data Center under A. Koeckelenbergh at the Observatoire Royal, Brussels, Belgium.

Regular Bulletins on the Motion of the Earth's Pole have been produced by the International Latitude Observatory of Mizusawa, Japan, including publication of the results of the International Latitude Service in a homogeneous system for the years 1900-1979 by S. Yumi and K. Yokoyama.

Operating by agreed rules, Commission 38, under the guidance of Professor J. Delhaye, has maintained financial assistance to individual astronomers to enable them to work in countries other than their own. The demand for these allocations and the rise in air fares have fully stretched the budgetary provisions.

Commission 50, under Professor F. Graham Smith, has co-operated with the Commission Internationale de l'Eclairage to produce a substantial publication giving Guide-Lines for minimizing Urban Sky Glow near Astronomical Observatories. In this way, useful and important precautions worked out in a limited number of countries have been made widely available; financial assistance from UNESCO is acknowledged.

The fruitful discoveries in planetary details by the space vehicles VOYAGER I and II have produced problems of nomenclature, e.g. in Saturn's rings, that are actively under review by the Working Group for Planetary System Nomenclature and Commission 5.

Each important Total Eclipse of the Sun is reviewed by the Organising Committee of Commission 12 and, if appropriate, a co-ordinator has been appointed for the principal country experiencing the eclipse. Examples have been:

| | |
|---|---|
| 31 July 1981 | V.A. Krat for USSR |
| 11 June 1983 | B. Hidayat for Indonesia |

Some of the work of Commission Presidents has been undertaken, due to illness or other circumstances, by Vice-Presidents, as follows

| | |
|---|---|
| Comm. 4 | T. Lederle for A.M. Sinzi |
| Comm. 20 | E. Roemer for G. Sitarski |
| Comm. 33 | R. Wielen for G.G. Kuzmin, with assistance from J. Einasto |

The chief scientific work of Commissions has been in arranging Symposia, Colloquia and Co-sponsored meetings as listed below:

SYMPOSIA

IAU Symposia (held between 1 January 1979 and 31 December 1981)

IAU Symposium No. 85 "Star Clusters", Victoria, British Colombia, Canada 27-31 August 1979

IAU Symposium No. 86 "Radio Physics of the Sun", College Park, Maryland, USA, 7-10 August 1979

IAU Symposium No. 87 "Interstellar Molecules", Mont Tremblant, Québec, Canada, 6-10 August 1979

IAU Symposium No. 88 "Close Binary Stars: Observations and Interpretation", Toronto, Ontario, Canada, 7-10 August 1979

IAU Symposium No. 90 "Solid Particles in the Solar System", Ottawa, Ontario, Canada, 27-30 August 1979

IAU Symposium No. 91 "Solar and Interplanetary Dynamics", Cambridge, Massachusetts, USA, 27-31 August 1979

IAU Symposium No. 92 "Objects of High Redshift", Los Angeles, California, USA, 28-31 August 1979

IAU Symposium No. 93 "Fundamental Problems in the Theory of Stellar Evolution", Kyoto, Japan, 22-25 July 1980

IAU Symposium No. 94 "Origin of Cosmic Rays", Bologna, Italy, 11-14 June 1980

IAU Symposium No. 95 "Pulsars", Bonn, Germany F.R., 26-29 August 1980

IAU Symposium No. 96 "Infrared Astronomy", Kailua-Kona, Hawaii, USA, 23-27 June 1980

IAU Symposium No. 97 "Extragalactic Radio Sources", Albuquerque, New Mexico, USA, 3-7 August 1981

IAU Symposium No. 98 "Be Stars", Munich, Germany F.R., 6-10 April 1981

IAU Symposium No. 99 "Wolf-Rayet Stars: Observations, Physics, and Evolution", Cozumel, Mexico, 16-22 September 1981

SYMPOSIA with IAU Participation

COSPAR/IAU/IUPAP Symposium "Non-Solar Gamma-Rays", Bangalore, India, 30 May - 1 June 1979

SCOSTEP/IAU Symposium "Solar Radio Astronomy, Interplanetary Scintillations and Their Coordination with Spacecraft Observations", Culgoora, N.S.W., Australia, 28-30 November 1979

COSPAR/IAU Symposium "Progress in Planetary Exploration", Budapest, Hungary, 2-4 June 1980

COSPAR/IAU Symposium "Cosmic Rays in the Heliosphere", Budapest, Hungary, 3-4 June 1980

COSPAR/IAU Symposium "Physics of Planetary Magnetospheres", Budapest, Hungary, 4-7 June 1980

COSPAR/IAU Symposium "Theoretical Problems in High-Energy", Budapest, Hungary, 5-7 June 1980

COSPAR/IAMAP/IAU Symposium "Origin and Evolution of Planetary Atmospheres", Hamburg, FRG, August 1981

IUHPS/IAU Symposium "Archaeoastronomy", Oxford, UK, 4-9 September 1981

COLLOQUIA

IAU Colloquia (held between 1 January 1979 and 31 December 1981)

IAU Colloquium No. 51 "Convection and Turbulence in Stellar Atmospheres", London, Ontario, Canada, 27-30 August 1979

IAU Colloquium No. 53 "White Dwarfs and Variable Degenerate Stars", Rochester, New York, USA, 30 July - 2 August 1979

IAU Colloquium No. 54 "Scientific Research with the Space Telescope", Princeton, New Jersey, USA, 8-11 August 1979

IAU Colloquium No. 55 "The Sixth International Colloquium on Ultraviolet and X-Ray Spectroscopy of Astrophysical and Laboratory Plasmas", Toronto, Ontario, Canada, 7-10 July 1980

IAU Colloquium No. 56 "Second International Colloquium on Reference Coordinate Systems for Earth Dynamics", Warsaw, Poland, 8-12 September 1980

IAU Colloquium No. 57 "The Satellites of Jupiter", Kailua-Kona, Hawaii, USA, 13-16 May 1980

IAU Colloquium No. 58 "Stellar Hydrodynamics", Los Alamos, New Mexico, USA, 12-15 August 1980

IAU Colloquium No. 59 "Effects of Mass-Loss on Stellar Evolution", Trieste, Italy, 15-19 September 1980

IAU Colloquium No. 60 "Uranus and the Outer Solar System", Bath, UK, 14-16 April 1981

IAU Colloquium No. 61 "Comets, Gases, Ices, Grains and Plasma", Tucson, Arizona, USA, 11-14 March 1981

IAU Colloquium No. 62 "Current Techniques in Double and Multiple Star Research", Flagstaff, Arizona, USA, 19-21 May 1981

IAU Colloquium No. 63 "High-Precision Earth Rotation and Earth-Moon Dynamics: Lunar Distances and Related Observations", Grasse, France, 22-27 May 1981

IAU Colloquium No. 64 "Automated Data Retrieval in Astronomy", Strasbourg, France, 7-10 July 1981

IAU Colloquium No. 65 "Evolutionary Status of Variable Stars" was cancelled

IAU Colloquium No. 66 "Problems in Solar and Stellar Oscillations", Simferopol, USSR, 1-5 September 1981

IAU Colloquium No. 67 "Instrumentation for Large Telescopes", Zelenchuk, USSR, 8-10 September 1981

IAU Colloquium No. 68 "Astrophysical Parameters for Globular Clusters", Schenectady, New York, USA, 7-10 October 1981

IAU Colloquium No. 69 "Binary and Multiple Stars as Tracers of Stellar Evolution", Bamberg, GFR, 31 August - 3 September 1981

IAU Colloquium No. 70 "The Nature of Symbiotic Stars", Haute-Provence, France, 26-28 August 1981

## REGIONAL ASTRONOMY MEETINGS held under the auspices of the IAU

Fifth European Regional Astronomy Meeting, Liège, Belgium, 28 July - 1 August 1980

Second Latin-American Regional Meeting, Mérida, Venezuela, 19-23 January 1981

Second Asian-Pacific Regional Meeting in Astronomy, Bandung, Indonesia, 24-29 August 1981

Sixth European Regional Astronomy Meeting, Dubrovnik, Yugoslavia, 19-23 October 1981

## YOUNG ASTRONOMERS' SCHOOLS

IAU Commission No. 46 "Teaching of Astronomy" organized three schools (Xth, XIth and XIIth) for young astronomers in the course of 1979 to 1981.

The school at La Laguna, Tenerife, Spain, from 10-23 September 1979 was attended by 63 students from Spain, Portugal, Chile, Argentina, Mexico and Uruguay. The curriculum included Stars and their Evolution, General Astrophysics, Solar Physics, Planetary System, Spectroscopy, Instrumentation and Techniques, Interstellar Matter, and Differential Solar Rotation.

The school at Hvar, Yugoslavia, from 17 September to 8 October 1980 was attended by 35 young astronomers from the Balkan and Mediterranean countries. The curriculum included Astronomical Technics, Astronomical Electrodynamics, Spectroscopy, Stellar Physics, Interstellar Matter, Galactic Astronomy, Solar Physics, Physics of Planets, Meteors and Comets, Structure and Evolution of Stars, Astrophysics from Space, Practical Training.

The school at Cairo, Egypt, from 22 August to 9 September 1981 was attended by 28 students from Egypt, Greece, Netherlands, Hungary, Malta, Nigeria, Portugal, Sudan and China. The curriculum included general astrophysics with emphasis on solar and stellar physics.

## XVIIth GENERAL ASSEMBLY, Montreal, Canada, 14-23 August 1979

Volume XVIIB of IAU Transactions, published in 1980, contains a report of the Proceedings of the XVIIth General Assembly of the Union including the resolutions passed at the Assembly. There follow, in the same volume, the reports of Commission meetings held during the period of the General Assembly and other activities in progress.

Minor revisions of IAU Statutes are included in the publication, in IAU Transactions volume XVIIB, of IAU Statutes, By-laws, Working Rules and IAU Rules for Scientific Meetings (pp. 310-329).

## SALES OF IAU PUBLICATIONS

The Transactions of the IAU sold as follows in the course of the period under report:

| | Number of Copies |
|---|---|
| Vol. XIIIA | 10 |

| | |
|---|---|
| Vol. XIIIB | 6 |
| Vol. XIVA | 8 |
| Vol. XIVB | 12 |
| Vol. XVA | 12 |
| Vol. XVB | 20 |
| Vol. XVIA | 14 |
| Vol. XVIB | 40 |
| Vol. XVIIA part 1 | 682 |
| Vol. XVIIA part 2 | 685 |
| Vol. XVIIA part 3 | 702 |
| Vol. XVIIB | 441 |

The volumes of Highlights of Astronomy were sold as follows over the same period:

| | | Number of Copies |
|---|---|---|
| Highlights I | | 18 |
| Highlights II | | 22 |
| Highlights III | | 26 |
| Highlights IV part 1 | (cloth) | 62 |
| | (paper) | 22 |
| Highlights IV part 2 | (cloth) | 64 |
| | (paper) | 21 |
| Highlights V | (cloth) | 280 |
| | (paper) | 152 |

The sales of Symposium volumes in 1979, 1980, 1981 were:

Symposium volume number, number of copies sold, cloth and paper *

32 - 37
33 - 0
34 - 19
35 - 9
36 - 0
37 - 16
38 - 21
39 - 16
40 - 11
41 - 5
42 - 20
43 - 21
44 - 23
45 - 22
46 - 16
47 - 11
48 - 17
49 - 11 - 11*
50 - 8 - 14*
51 - 11 - 9*
52 - 14 - 7*
53 - 21 - 9*
54 - 13 - 13*
55 - 7 - 16*
56 - 6 - 3*
57 - 11 - 4*
58 - 23 - 19*
59 - 19 - 11*
60 - 11 - 8*
61 - 12 - 10*
62 - 12 - 9*
63 - 29 - 28*
64 - 27 - 20*
65 - 13 - 4*
66 - 18 - 17*
67 - 17 - 18*
68 - 14 - 12*
69 - 14 - 22*
70 - 17 - 26*
71 - 30 - 22*
72 - 10 - 13*
73 - 35 - 32*
74 - 94 - 67*
75 - 91 - 66*
76 - 101 - 69*
77 - 191 - 122*
78 - 461 - 99*
79 - 254 - 142*
80 - 141 - 137*
81 - 551 - 168*
82 - 536 - 171*
83 - 487 - 168*
84 - 489 - 213*
85 - 436 - 266*
86 - 401 - 178*
87 - 397 - 305*
88 - 432 - 217*
89 - 435 - 112*
90 - 381 - 184*
91 - 362 - 167*
92 - 446 - 267*
93 - 399 - 166*
94 - 389 - 184*
95 - 366 - 204*
96 - 434 - 316*
98 - 140 - 18*

## INFORMATION BULLETIN

The Information Bulletin has been prepared, edited and typed for offset reproduction by the General Secretary and the staff of the IAU Secretariat, and printed and distributed by the publisher of the Union. It has appeared twice a year and has been sent, free of charge, to IAU Members, consultants, members of the International Science Writers Association, and to selected international organisations and institutions.

## RELATIONS TO OTHER ORGANISATIONS

The IAU was represented at meetings of the International Council of Scientific Unions (ICSU) as follows:

11 Meeting General Committee, Brussels, July 1979 - E.A. Müller

12/13 Meeting General Committee, Amsterdam, September 1980 - P.A. Wayman

18 General Assembly, Amsterdam, September 1980 - P.A. Wayman

14 Meeting General Committee, Dubrovnik, September 1981 - P.A. Wayman

Professor M.K.V. Bappu, President of the IAU, represents India at ICSU (from 1981) and attended the 14th General Committee Meeting as representative of National delegates to the General Assembly.

The IAU was represented at meetings of the ICSU Committee for Space Research (COSPAR) as follows:

XXIInd Plenary Meeting of COSPAR, Bangalore, May/June 1979 - E.A. Müller

XXIIIrd Plenary Meeting of COSPAR, Budapest, June 1980 - P.A. Wayman

The following representatives to other international (ICSU) bodies have been active since August 1979:

- Federation of Astronomical and Geophysical Services (FAGS): H. Enslin, V. Bumba, under which heading is comprised:

  Quarterly Bulletin of Solar Activity (QBSA): F. Moriyama

  Bureau International de l'Heure (BIH): A. Orte, H.M. Smith

  International Polar Motion Service (IPMS): K. Yokoyama

  International Ursigram and World Day Service (IUWDS): V. Bumba

- ICSU Abstracting Board (ICSU-AB): W.D. Heintz
- Scientific Committee for Solar-Terrestrial Physics (SCOSTEP): M.R. Kundu
- Committee on Science and Technology in Developing Countries (COSTED): L. Houziaux
- Committee for Teaching of Science (CTS): L. Houziaux
- Scientific Committee for Protection of the Environment (SCOPE): F. Graham Smith
- Coordinating Committee for the Moon and Planets (CCMP): P.M. Millman
- Committee on Data in Science and Technology (CODATA): B. Hauck

- Inter-Union Committee on Frequency Allocation for Radio Astronomy and Space Science (IUCAF): R. Schilizzi, G. Westerhout

- Inter-Union Commission on Spectroscopy (IUCS): B. Edlen, J.G. Phillips, M.J. Seaton

The IAU participates in the following international bodies:

-Comité Consultatif pour la Définition de la Seconde (CCDS): W. Markowitz

-Comité Consultatif pour la Définition du Mètre (CCDM): A.H. Cook

-Comité Consultatif International des Radio communications (CCIR): B.J. Robinson, L.H. Doherty, H.M. Smith

-European Physical Society Conference Committee (EPS): R.M. West

An important activity by IAU during 1979-1981 was representation through IUCAF at the World Administrative Radio Conference, Geneva, 1979, where the preservation from artificial interference of bands of frequencies needed in radio astronomy was actively and successfully urged.

Following resolutions passed at the XVIIth General Assembly concerning the establishing of a Giant Equatorial Radio Telescope (GERT), some progress has been made through representation at ICSU and at UNESCO (by national bodies participating in UNESCO) in achieving this ambitious astronomy project for an equatorial country.

The work of BIH continues to be of the greatest importance for definition and maintenance of atomic and astronomical time scales by international agreement.

## FINANCIAL REPORT

In accord with the policy adopted in 1961, the Executive Committee presents to the General Assembly a "Summarized Account of Receipts and Payments" for the years 1979, 1980 and 1981. The account is in Swiss Francs; the figures in the US dollar accounts, the Dutch florin (guilder) accounts and the French franc accounts have been transferred to the summarized (Swiss franc) account at the following ICSU conversion rates:

| | | | |
|---|---|---|---|
| For the year 1979: | US$ = 1.6583 Swfr | FF 1 = 0.3905 Swfr | Dfl 1 = 0.8285 Swfr |
| For the year 1980: | US$ = 1.6683 Swfr | FF 1 = 0.3989 Swfr | Dfl 1 = 0.8433 Swfr |
| For the year 1981: | US$ = 1.9616 Swfr | FF 1 = 0.3666 Swfr | Dfl 1 = 0.7939 Swfr |

The varying conversion rate produces a revaluation (upward) or a devaluation (downward) in the opening total balances at the beginning of a year as compared with the corresponding closing total balances at the end of the previous year. The differences for the years 1979-1981 are as follows (Swiss francs):

Current Accounts plus Savings Accounts, Total Balance

| Closing Balance | | Opening Balance | | Revaluation/ Devaluation |
|---|---|---|---|---|
| 31 December 1978 | 648 627.09 | 1 January 1979 | 629 385.41 | -19 241.68 |
| 31 December 1979 | 551 987.11 | 1 January 1980 | 564 655.98 | +12 668.87 |
| 31 December 1980 | 591 957.90 | 1 January 1981 | 618 373.60 | +26 415.70 |
| | | 1978-81 Net Revaluation | | + 19 842.89 |

Transfers between accounts, due to varying exchange rates not equal to those adopted in producing the summarized account in Swiss Francs, produce gains or losses on exchange that are "fictitious" in the sense that they depend upon the choice of currency used for producing the summarized account. The summarized 1979-81 account has been derived from the detailed accounts prepared for each of the bank accounts and savings accounts held by the Union and certified by the official auditors of the Union. The "Vérifications des Comptes", 1979, 1980 and 1981 produced by R. Bacle, H.E.C., Paris, for the Executive Committee will be available for examination by the Finance Committee of the XVIIIth General Assembly.

It will be noticed in comparison with the Executive Committee Report 1976-78 that the above table of closing and opening balances includes both Savings Accounts and Current Accounts, whereas previous tables have referred to Current Accounts only. There are corresponding changes in the summarized account 1979-81 below:

(i) "Receipts, Item 3" is now "Interest on Accounts", i.e. including Current Accounts and Savings Accounts; this corresponds to the Budget Provision for each triennium, where all interest received is regarded as Income for budgetary purposes.

(ii) Corresponding to a change in the presentation of the Vérification des Comptes, Item 5 - "Other Receipts" and Item 18 - "Other Payments" refer only to miscellaneous items and do not include, as formerly, the "fictitious" sums relating to transfer between accounts; in the Vérification des Comptes these transactions are given separately.

SUMMARIZED ACCOUNT OF RECEIPTS AND PAYMENTS FOR THE 3-YEAR PERIOD 1979-1981 (Swiss Francs)

RECEIPTS

| | | | |
|---|---|---|---|
| 1. | Contributions from Adhering Countries | | |
| | 1979 | 332 860.85 | |
| | 1980 | 316 508.55 | |
| | 1981 | 306 220.69 | 955 590.09 |
| 2. | Revenue from IAU Publications | | |
| | 1979 | 18 489.42 | |
| | 1980 | 23 683.62 | |
| | 1981 | 23 034.34 | 65 207.38 |
| 3. | Interests on Accounts | | |
| | 1979 | 19 948.71 | |
| | 1980 | 24 485.69 | |
| | 1981 | 41 676.08 | 86 110.48 |
| 4. | UNESCO Contribution through ICSU | | |
| | 1979 | 30 237.76 | |
| | 1980 | 27 070.56 | |
| | 1981 | 22 248.00 | 79 556.32 |

| | | | |
|---|---|---|---|
| 5. Other Receipts | | | |
| 1979 | | 13 823.70 | |
| 1980 | | 1 770.35 | |
| 1981 | | - | 15 594.05 |
| | | Total Receipts | 1 202 058.32 |
| | | Excess of Payments over Receipts | 17 254.33 |
| | | | 1 219 312.65 |

PAYMENTS

| | | | |
|---|---|---|---|
| 1. Administrative Office | | | |
| 1979 | | 184 735.19 | |
| 1980 | | 130 190.65 | |
| 1981 | | 134 435.12 | 449 360.96 |
| 2. Contribution to ICSU | | | |
| 1979 | | 8 003.00 | |
| 1980 | | 8 418.00 | |
| 1981 | | 7 977.00 | 24 398.00 |
| 3. Commission Expenses | | | |
| 1979 | | 1 443.10 | |
| 1980 | | 356.60 | |
| 1981 | | 14 023.16 | 15 822.86 |
| 4. Projects of Commissions | | | |
| 4.1 Exchange of Astronomers | | | |
| 1979 | 7 475.48 | | |
| 1980 | 22 268.01 | | |
| 1981 | 27 203.44 | 56 946.93 | |
| 4.2 Teaching of Astronomy | | | |
| 1979 | - | | |
| 1980 | - | | |
| 1981 | 1 176.96 | 1 176.96 | |
| 4.3 Other projects | | | |
| 1979 | 33 047.54 | | |
| 1980 | 6 171.02 | | |
| 1981 | 6 794.22 | 46 012.78 | 104 136.67 |
| 5. General Assembly | | | |
| 1979 | | 159 192.03 | |
| 1980 | | 50 348.35 | |
| 1981 | | - | 209 540.38 |

6. IAU Publications

| | | |
|---|---|---|
| 1979 | 20 032.96 | |
| 1980 | 19 169.08 | |
| 1981 | 24 316.76 | 63 518.80 |

7. Distribution of IAU Publications Free and at Reduced Price

| | | |
|---|---|---|
| 1979 | 7 760.39 | |
| 1980 | 9 339.98 | |
| 1981 | 7 689.32 | 24 789.69 |

8. Executive Committee Meeting

| | | |
|---|---|---|
| 1979 | - | |
| 1980 | 29 081.40 | |
| 1981 | 27 549.12 | 56 630.52 |

9. Officers' Meeting

| | | |
|---|---|---|
| 1979 | 7 677.14 | |
| 1980 | 710.42 | |
| 1981 | 5 399.62 | 13 787.18 |

10. Symposia and Colloquia

| | | |
|---|---|---|
| 1979 | 39 830.53 | |
| 1980 | 37 783.82 | |
| 1981 | 45 777.29 | 123 391.64 |

11. Inter-Union Commissions

| | | |
|---|---|---|
| 1979 | 5 422.64 | |
| 1980 | 14 427.67 | |
| 1981 | 8 120.19 | 27 970.50 |

12. Projects of the Executive Committee

| | | |
|---|---|---|
| 1979 | - | |
| 1980 | - | |
| 1981 | - | - |

13. Representation

| | | |
|---|---|---|
| 1979 | 16 951.90 | |
| 1980 | 5 523.09 | |
| 1981 | 7 466.90 | 29 941.89 |

14. Bank Charges

| | | |
|---|---|---|
| 1979 | 424.99 | |
| 1980 | 315.70 | |
| 1981 | 951.19 | 1 691.88 |

| | | |
|---|---|---|
| 15. Young Astronomers' Schools | | |
| 1979 | 11 248.90 | |
| 1980 | 19 586.85 | |
| 1981 | 21 582.19 | 52 417.94 |
| 16. Regional Meetings | | |
| 1979 | - | |
| 1980 | 12 005.88 | |
| 1981 | 10 057.51 | 22 063.39 |
| 17. Special Nominating Committee | nil | nil |
| 18. Other Payments | | |
| 1979 | 152.55 | |
| 1980 | 100.90 | |
| 1981 | 68.50 | 321.95 |
| | TOTAL | 1 219 784.25 |
| | less REFUNDS (1980) | - 471.60 |
| | NET PAYMENTS | 1 219 312.65 |

BANK ACCOUNTS as on 31 December 1978

| | | Swiss Francs |
|---|---|---|
| Savings Accounts | | 361 740.19 |
| Current Accounts | | 286 886.90 |
| | TOTAL | 648 627.09 |

BANK ACCOUNTS as on 31 December 1981

| | | | |
|---|---|---|---|
| Savings Accounts | | 531 522.58 | |
| Current Accounts | | 132 468.56 | |
| | TOTAL | 663 991.14 | -TOTAL Balance on IAU Accounts as on 31 December 1981 |

Note: The 3-year Summarized Account shows an Excess of Payments over Receipts of SwFr 17 254.33. The relationship between this Excess and the increase in the total balance on IAU Accounts between 31 December 1978 and 31 December 1981 is accounted as follows:

| | | | |
|---|---|---|---|
| 3-year Excess of Payments over Receipts | 17 254.33 | Bank Balances 31.12.78 | 648 627.09 |
| 1980 Loss on Exchange | 891.03 | 1979 Gain on Exchange | 968.30 |
| | | 1981 Gain on Exchange | 3 026.92 |
| Bank Balances 31.12.81 | 663 991.14 | [1]UBS 50 Bank Account | 9 671.30 |
| | | [2]Net Revaluation 1978/81 | 19 842.89 |
| | 682 136.50 | | 682 136.50 |

[1]The UBS 50 Bank Account was entered with Swfr 10 504.27 in 1977. Some payments were made for Commission 50 in 1978; in 1979 the special account was incorporated into the IAU Accounts; since 1979 January 1 all Commission 50 payments have been included as "Commission Expenses".

[2]As given in Table preceding 3-year Accounts.

Auditor's Statement for the Accounts 1979-1981

Messieurs,

Conformément à la mission que vous avez bien voulu nous confier, nous vous présentons le résultat de notre examen des comptes de l'UNION ASTRONOMIQUE INTERNATIONALE pour la période du 1er Janvier 1979 au 31 Décembre 1981.

L'examen a porté notamment sur les documents suivants:

-comptes des recettes et dépenses pour les années 1979 à 1981 vérifiés par nous-mêmes.

Les comptes ont été comparés avec les cahiers de comptes et pièces comptables diverses.

-Le solde des "Refunds" qui n'ont pu être imputés sur des dépenses ou des recettes, correspond à des régularisations sur le compte UBS 50: 471.60 en 1980.

Nous certifions que ces comptes sont en accord complet avec les informations et explications que nous avons demandées pour nous permettre d'effectuer notre examen.

Nous vous donnons ci-après les recettes et dépenses résumant les comptes des années 1979 - 1980 - 1981.

à Paris, le 31 Mars 1982
(signé)
Roger BACLE
HEC
Diplômé Expert-Comptable

Explanatory Notes:

The statement of Receipts and Payments 1979-1981 is arranged under headings that correspond to the Budget approved by the General Assembly, although it is important to note that the triennia of the Budget (e.g. 1980-1982) do not coincide with the triennia of the financial reports. Hence the formation of the Residual Budget for 1982 and the Budget Proposal for 1982 that follows this section of the Report.

The following explanatory notes provide information on each item of the 1979-1981 accounts.

RECEIPTS

1. Contributions from Adhering Organisations

The XVIIth General Assembly, Montreal, Canada, 1979, kept the annual unit of contribution payable by IAU Adhering Organisations unchanged at Swfr 1 465.

The number of adhering countries changed during 1979 from 48 to 49 with the accession of Indonesia to the Union. The total number of units due in each year were:

| | | |
|---|---|---|
| 1979 | 220 | |
| 1980 | 225 | |
| 1981 | 225 | |
| TOTAL | 670 | units, equivalent to Swfr 981 550 |

(See IAU Transactions vol XVIIB, pp. 49 and 526 for the separate financial contribution from two adhering organisations in China).

The following Table shows the number of organisations adhering in the categories specified in the By-laws during 1980 and 1981.

| | | |
|---|---|---|
| 1 unit - | 19 | adhering organisations |
| 2 units - | 14 | " " |
| 4 units - | 7 | " " |
| 6 units - | 2 | " " |
| 10 units - | 2 | " " |
| 14 units - | 2 | " " |
| 20 units - | 3 | " " |
| 30 units - | 1 | adhering organisation |

Total: 50 adhering organisations; 225 units contribution.

At the time of writing this report, 5 adhering organisations were in arrear with contributions up to and including 1981. At the close of the year 1981, however, as covered by the 3-year accounts, there were 36 units outstanding, of which 26 units have been paid since 1 January 1982.

2. Revenue from Publications

The continuing steady level of this item, arising from royalties provided by the Union's publishers, which was maintained through 1979-1981, reflects increased total sales with the introduction of volumes produced at lower cost per page (see earlier section on Sales).

3. Interest from Current and Savings Accounts

During the period 1979-1981 there were, in all, ten accounts holding Savings of the Union; three were closed during the period and three others were opened. At any one time four accounts have been in US$, two in Swiss Francs and one in Guilders. In order to attract interest earnings, the ratio of the sum held in Savings Accounts to that held in Current accounts was increased from 1.26 at the close of 1978 to 4.01 at the end of 1981. Due to fluctuations in exchange rates and interest rates, a balance between interest income and capital value maintenance is not easily procured. Being spread between US$, Swiss Francs, and Dutch florins, reserve funds are not customarily transferred out of their respective categories for speculative purposes. The Executive Committee maintains a conservative policy in respect of its financial reserves, which have been roughly equivalent to 1 1/2 years' expenditure.

Current accounts are held in the United States, Switzerland, the Netherlands and France. Payment of contributions from Adhering Organisations are mainly in Swiss Francs. Transfers of funds are necessary to maintain the various accounts, particularly the account held in Paris, France. The current accounts in Switzerland and the Netherlands bear annual interest.

The change in item 3 from the summarized accounts of the Executive Committee Report for previous triennia, namely the inclusion of interest on Savings Accounts, was explained in the introductory paragraph of this section.

4. Grant from UNESCO subvention to ICSU

The support of scientific Unions by ICSU, using, for the major part of this support, funds provided by UNESCO, has undergone modification during the period of this Report. A lower level of routine allocation of funds is coupled with provision of funds for specific purposes under UNESCO or ICSU contracts. Detailed accounts of expenditure are submitted to ICSU, and thence to UNESCO, each year by the General Secretary.

Sums were allocated during 1979-81 as follows (in US dollars):

| | UNESCO Subvention through ICSU Bank | Contract (Budget items) | ICSU grant | Total (US$) |
|---|---|---|---|---|
| 1979 | 12 000 | 4 000 (15) | 3 000 | 19 000 |
| 1980 | 8 000 | 8 000 (4.1, 7, 15) | - | 16 000 |
| 1981 | 9 600 | 1 200 (4.2) | - | 10 800 |

The amounts received (item 4 of Receipts) are closely similar to the allocations when fluctuating exchange rates are taken into account.

5. Other receipts

As is explained in the introduction to this section, only minor receipts have been included in this item, the former inclusion of transfer sums under this heading being discontinued. Even so, the figure for 1979 includes the donation of UAI France 1976 to the IAU Secretariat.

PAYMENTS

1. Administrative Office

During 1979 the permanent Secretariat of the IAU was established in Paris; two staff were appointed to replace former staff who had retired and premises provided by l'Observatoire de Paris were adapted to the use of the Secretariat. Together with the administrative costs of establishing the secretariat, extra payments were necessary to cover commitments to retiring staff; the cost foreseen in the 1979 Budget Estimate (p.514, EC Report, IAU Transactions XVIIB, 1980) was closely matched by expenditure. During 1980 and 1981 the administrative costs, expressed in Swiss Francs, reverted to the level of 1976-1978.

The expenses of the administrative office include salaries, social charges, health and pension schemes for the staff of the Secretariat (2 persons), audit fees, stationery, electricity (including heating), postal and telephone costs, and attendance costs for the General Secretary. The last four items were previously not a charge on the Union, being covered in other ways. The Union was still indebted in 1979 to the Observatoire de Genève and the Institut d'Astronomie de l'Université de Lausanne for assistance with office needs, removal costs and other resources for the Union. Throughout the period of the Report substantial contributions to the work of the Union have been made by the Dublin Institute for Advanced Studies (Dunsink Observatory) in Ireland and by the European Southern Observatory, Garching, Germany, F.R., (formerly at Geneva, Switzerland) where the General Secretary and Assistant General Secretary respectively have been provided with necessary services. The Executive Committee is grateful to all four institutions for their valuable and vital support, materially reducing the financial burden on the Union.

2. Subscription to ICSU

Adherence to the International Council of Scientific Unions by the IAU is met by a levy of 2.5% of the contributions from Adhering Organisations for the year preceding payment.

3. Commission Expenses

Relatively small sums are allocated from time to time to Commissions to cover non-routine expenditure. Commission 50 expenditures have, however, been included here.

4. Projects of Commissions

Budgetary Estimates were approved by the XVIth and XVIIth General Assemblies for expenditure of Union funds on projects of Commission 38 (Exchange of Astronomers), item 4.1, of Commission 46 (Teaching of Astronomy), item 4.2, and of other, long-established, projects of the IAU, item 4.3. The accompanying Table shows the budgetary allocations and expenditures under these items for the years 1979-1981. Unspent balances of IAU projects have customarily been held over for expenditure in other years. Accounts are not kept of all unspent sums for each project; such sums have simply reverted to the reserves of the Union. An example of release of these reserves during 1979 was expenditure on behalf of the Variable Star Catalogue, as provided for in the Budget Estimate of 1979 in the EC Report 1976-1978 (see p.515, IAU Transactions, vol. XVIIB, 1980). This item contributed to the excess of payments over receipts during the triennium of this report.

5. General Assembly

The expenditure on the XVIIth General Assembly in 1979 was assisted by funds specially provided by the Franklin Mint Corporation donation which was received in 1978 and which therefore does not form part of the Receipts 1979-1981. This is a second cause of the excess of Payments over Receipts in the triennium of this Report. *In toto*, the expenditure on the 1979 General Assembly was below that foreseen in the Budget Estimate for 1979, confirmed by the XVIIth General Assembly. Details of the expenditure will be found in the Accounts of the Union for 1979.

A further cause for the 1979-81 excess of payments over receipts has been the granting by the Executive Committee in 1980 of an exceptional request by the National Organising Committee of the XVIIIth General Assembly for financial support towards the Local Organising Committee's expenses for that Assembly. Funds were required early in 1981 to cover advance hotel deposits in Patras in order to secure the required accommodation. Financial support has also been allowed in respect of student residence grants and of support for IAU Symposia and Colloquia held in Greece at the time of the General Assembly. This financial support was reckoned to be necessary as an exceptional measure in respect of the 1982 General Assembly because the issuing of the invitation was only arranged, with two years' notice, during 1980, following the withdrawal in 1979 of the expected invitation for the 1982 General Assembly to be held in Bulgaria. The Executive Committee has made this exceptional allocation on the basis that no precedent for future General Assemblies is thereby created and that any depletion of IAU reserves caused by this allocation is minimized.

6. IAU Publications

The IAU Information Bulletin was distributed to all adhering bodies, to IAU members and consultants, and to many astronomy libraries, as in previous years. The total number printed was increased from 4 700 in 1979 to 5 400 in 1980 and 1981. The cost of distribution, including postage (in brackets), for the six bulletins was as follows:

| | | pages | | |
|---|---|---|---|---|
| 1979 | No. 41 | 52 | Dfl 11 763.95 | |
| | No. 42 | 64 | Dfl 12 428.35 | |
| 1980 | No. 43 | 56 | Dfl 13 502.16 | (6 828.79) |
| | No. 44 | 36 | Dfl 9 228.88 | (4 976.88) |
| 1981 | No. 45 | 60 | Dfl 19 125.50 | (13 524.50) |
| | No. 46 | 48 | Dfl 11 294.00 | (5 320.59) |
| | Total 1979-1981 | | Dfl 77 342.84 | |

ITEM 4 (Payments): SPECIFIC PROJECTS OF COMMISSIONS (rounded to Swiss Francs)

| Item | Comm. | Project | Allocation* 1979 | Paid 1979 | Paid 1980 | Paid 1981 | Total Paid 1980-1981 | Allocation 1980-1982 | Balance |
|---|---|---|---|---|---|---|---|---|---|
| 4.1 | 38 | Exchange of Astronomers | 25 000 | 7 475.48 | 22 268.01 | 27 203.44 | 49 471.45 | 58 000 | 8 528.55 |
| 4.2 | 46 | Teaching of Astronomy | - | | | 1 176.96 | 1 176.96 | 3 000 | 1 823.04 |
| 4.3 | | Other Projects | | | | | | | |
| | 4 | Ephemerides Bulletin | 830 | 828.34 | - | | - | - | - |
| | 6 | IAU Telegram Bureau | 1 865 | 1 787.70 | 1 944.64 | 2 118.53 | 4 063.17 | 5 700 | 1 636.83 |
| | 16 | Planetary Documentation | 1 765 | 1 763.34 | 1 770.50 | 1 770.00 | 3 540.50 | 5 310 | 1 769.50 |
| | 17 | Lunar Documentation | 1 760 | 1 761.34 | - | - | 1 761.34 | 5 310 | 3 548.66 |
| | 20 | Minor Planet Center | 2 070 | 1 987.40 | 2 456.38 | 2 675.62 | 5 132.00 | 7 200 | 2 068.00 |
| | 27 | Catalogue of Variable Stars | 3 110 | 24 919.42 | | 230.07 | 230.07 | 9 330 | 9 099.93 |
| | | 4.3 Total | 11 400 | 33 047.54 | 6 171.02 | 6 794.22 | 12 965.24 | 32 850 | 19 884.76 |
| | | 4 TOTAL | 36 400 | 40 523.02 | 28 439.03 | 35 174.62 | 63 613.65 | 93 850 | 30 236.35 |

*The allocation to item 4.3 in the Budget for 1979 was inappropriate

The Preliminary Announcement of the XVIIIth General Assembly in 1982 was distributed with IAU Information Bulletin No. 45. Postage charges exceed the cost of printing in some cases.

The publication, in conjunction with D. Reidel Publishing Company, Dordrecht, Netherlands, of the Transactions of the International Astronomical Union (Volume XVIIA in three parts in 1979 and Volume XVIIB in 1980), of Highlights of Astronomy Volume 5 in 1980 and of the Proceedings of IAU Symposia, was carried out under the financial contract with the publisher without expenditure from the IAU accounts.

7. Publications for Developing Countries and for free distribution

Volumes published by D. Reidel on behalf of the IAU are purchased by the Executive Committee for distribution at greatly reduced cost to a small number of libraries in developing countries. Copies are also distributed free of cost to EC members and to Chairmen of Scientific Organising Committees of IAU Symposia. The allocations during 1979-1981 were Swfr 24 789.69 to 9 institutions in developing countries for which payment equivalent to Swfr 4 217.94 was received from these countries.

8. Executive Committee meetings

The costs of the 47th (Amsterdam, 1980) and 48th (Abastumani, 1981) meetings of the Executive Committee comprise travel and subsistence costs under ICSU regulations for the attending members and IAU staff. The amount incurred for 1980 and 1981 was kept within the budget provision of Swfr 60 000 for 1980-82, no expense under this item being allocated during years of a General Assembly.

The Executive Committee greatly appreciates the arrangements made in the Netherlands and in Georgia, USSR, for enabling the meetings of 1980 and 1981 to be held under favorable conditions. The hospitality offered by the Astronomical Council of the Academy of Sciences of the USSR was not limited to the meeting at Abastumani but included many visits to places of scientific and cultural interest. The degree of hospitality offered reduced considerably the total cost to the IAU of the 1981 meeting.

9. Officers' meetings

The expenditure on Officers' meetings other than those held at the time of Executive Committee was kept at an economical level. The total of Swfr 6 110.04 expended in 1980 and 1981 is less than 30% of the 3-year allocation of Swfr 21 000 for 1980-1982.

10. Symposia and Colloquia

This item, which includes expenditure in cosponsored meetings, covered 40 scientific meetings in 1979, 1980 and 1981 as shown below:

| | Symposia | Colloquia | Co-sponsorship |
|---|---|---|---|
| 1979 | 7 | 3 | 2 |
| 1980 | 4 | 5 | 4 |
| 1981 | 3 | 10 | 2 |

Each Chairman of Scientific Organising Committee is empowered to recommend travel grants for participants, the allocation to individuals being scrutinized by the Assistant General Secretary of the Union. With substantial rises in the Swiss-franc cost of air travel, the IAU grants have become increasingly used to cover subsistence costs at the meetings. Participants from some countries can obtain travel tickets but not foreign currency for paying their hotel or hostel charges. Hence the guideline that this money be used principally for travel is being increasingly neglected. Approximately 250 grants averaging Swfr 500 (= $285) were allocated during 1979-1981.

11. Inter-Union Commissions

The annual allocation of the 1980-82 Budget for this item of Swfr 6 700 was exceeded during 1980 due to continually increasing requirements by scientific inter-union commissions (IUCAF, ICSU-AB, CTS, SCOSTEP and the FAGS agencies) for direct financial support from the Unions. The Executive Committee, in 1979, allocated a token amount of Swfr 2 000 p.a. to the FAGS Council out of its 1980-82 Budget, such provision not having been foreseen in the Budget Proposal of 1979.

12. Executive Committee Projects

No Projects were financed within this category during 1979-1981 but the project of Commission 5, involving many other Commissions' activity in an advisory capacity, of C. Jaschek, F. Spite, C. Lortet *et alii* to produce an index of astronomical catalogues and their abbreviated names was adopted in 1981 as an Executive Committee project for completion in 1982.

13. Representation

Part of the cost of IAU representation at bureau and commission meetings of other organisations, including FAGS, CCIR, CCDM, ICSU-AB, EPS, COSPAR, SCOSTEP, SCOPE and COSTED, have been met from these funds.

15. Young Astronomers' Schools

Three Schools were held in the triennium 1979-1981, the amounts allocated, together with the amounts expended and the support allocated through ICSU-UNESCO funds, are shown in note 16 of the Budget proposal 1982 which follows.

The IAU grant usually covers about one third of the costs of a Young Astronomers' School. Funds are used for travel of Lecturers, for transport of elementary teaching equipment, and for travel assistance to some students.

16. Regional Astronomy meetings

Four Regional meetings were supported with an allowance of Swfr 6 000 each to assist attendance of young astronomers. The amounts expended were as follows;

| | SwFr |
|---|---|
| Liège, Belgium, July 1980 | 6 000 |
| Mérida, Venezuela, January 1981 | 4 473.88 |
| Bandung, Indonesia, August 1981 | 5 589.51 |
| Dubrovnik, Yugoslavia, October 1981 | 6 000 |

17. Special Nominating Committee

The work of the Special Nominating Committee (which reports in 1982 to the General Assembly) was transacted without incurring special expenses in 1980 and 1981.

18. Other Payments

As explained in the introductory paragraphs, only minor payments are now included under this heading, substantial financial transfer between accounts being treated separately in the audited annual accounts.

Conclusion

The total balance in the Savings and Current accounts of the Union, valued at Swfr 663 991.14 on 31.12.81 is Swfr 15 364.05 higher than the corresponding figure given for 31.12.78 in the Executive Committee Report for 1976-1978. This is an increase of 2.4% as against a 32% increase for the triennium 1976-1978. The figure of 2.4% is diminished by the causes referred to under the notes on expenditure items 4.3 and 5 above.

RESIDUAL BUDGET FOR THE YEAR 1982 and BUDGET PROPOSAL 1982

1. The figures in the accounts for 1980 and 1981, added together for each Budget item, are subtracted from the figures for the respective Budget items for 1980-1982, as approved by the General Assembly in 1979. In this way a Residual Budget for 1982 is obtained. Negative figures are used to indicate that a particular item of the 1980-1982 Budget, either Receipts or Payments, has already been exceeded.

At Abastumani in July 1981, the Executive Committee approved some changed figures for the triennium, as compared with the adopted Budget 1980-1982. Two sets of figures, rounded in Swiss Francs are therefore given;

(i) Residual Budget for 1982, using 1980-82 Budget figures approved by the XVIIth General Assembly and the income and expenditure figures for 1980 and 1981

(ii) Proposed Budget Estimate for 1982, to be referred to the Finance Committee at the XVIIIth (Patras) General Assembly for presentation to the Assembly, as in IAU Statute 18(b).

2. Residual Budget for 1982

The figures for 1980 and 1981 subtracted from those of the budget approved for 1980-82 give the residual budget for the year 1982 as follows (numbered notes appear in Section 4 below):

Receipts (Swiss francs)

| | | |
|---|---|---|
| 1. | Contributions from Adhering Countries | 344 171[1] |
| 2. | Revenue from publications | 4 282[2] |
| 3. | Interest on current accounts and savings accounts | 15 838[3] |
| 4. | ICSU and ICSU-UNESCO allocations | -2 278[4] |
| 5. | Other receipts | -1 770 |
| | Total receipts | 360 243 |
| | Total payments | 429 665 |
| | Excess of payments over receipts | 69 422[5] |

Payments (Swiss francs)

| | | |
|---|---|---|
| 1. | Administrative Office | |
| | 1.1 Secretariat in Paris, including GS expenses | 118 581[6] |
| | 1.2 Secretarial help to GS and AGS | 30 000[7] |
| | 1.3 Furniture, office equipment, etc. | 16 793[8] |
| 2. | Subscription to ICSU | 7 805 |
| 3. | Commission Expenses | 620[9] |

| | | |
|---|---|---|
| 4. | Projects of Commissions | |
| | 4.1 Exchange of Astronomers | 8 529[10] |
| | 4.2 Teaching of Astronomy | 1 823[11] |
| | 4.3 Other projects | 19 885[12] |
| 5. | General Assembly | 89 652[13] |
| 6. | IAU publications | 11 514 |
| 7. | Publications for EC and to developing countries | 31 971 |
| 8. | Executive Committee meetings 1980, 1981 | 3 369 |
| 9. | Officers' meetings 1980-82 | 14 890 |
| 10. | Symposia and Colloquia (travel grants) | 54 439 |
| 11. | Inter-Union Commissions | -2 548 |
| 12. | Projects of the Executive Committee | 13 000[14] |
| 13. | Representation | 7 010 |
| 14. | Bank Charges | -267[15] |
| 15. | Young Astronomers' Schools | -4 169[16] |
| 16. | Regional Meetings 1980, 1981 | 1 937 |
| 17. | Expenses of Special Nominating Committee | 5 000[17] |
| 18. | Other payments | -169 |
| | | 429 665[18] |

3. Budget Proposal 1982

Based upon careful examination of current trends and figures actually achieved, the Executive Committee proposes the Budget Estimate for 1982 as follows (a note is given wherever an appreciable change occurs, as compared with the Residual Budget, and in many cases the same note serves section 2 and section 3):

| | | |
|---|---|---|
| 1. | Contributions from Adhering Countries | 356 000[1] |
| 2. | Revenue from publications | 18 000[2] |
| 3. | Interest on current accounts and savings accounts | 35 000[3] |
| 4. | ICSU and ICSU-UNESCO allocations | 32 384[4] |
| | Total receipts | 441 384 |
| | Total payments | 527 470 |
| | Excess of payments over receipts | 86 086[5] |

Payments (Swiss Francs)

| | | |
|---|---|---|
| 1. | Administrative Office | |
| | 1.1 Secretariat in Paris, including GS expenses | 140 000[6] |
| | 1.2 Secretarial help to GS and AGS | 5 000[7] |
| | 1.3 Furniture, office equipment, etc. | 20 000[8] |
| 2. | Subscription to ICSU | 7 700 |
| 3. | Commission Expenses | 5 000[9] |

| | | |
|---|---|---|
| 4. | Projects of Commissions | |
| | 4.1 Exchange of Astronomers | 18 500[10] |
| | 4.2 Teaching of Astronomy | 2 800[11] |
| | 4.3 Other projects | 15 370[12] |
| 5. | General Assembly | 169 000[19] |
| 6. | IAU Publications | 24 000[20] |
| 7. | Publications for EC and to developing countries | 10 000[21] |
| 8. | Executive Committee meetings | 0[22] |
| 9. | Officers' Meetings | 8 000 |
| 10. | Symposia and Colloquia (travel grants) | 55 000[23] |
| 11. | Inter-Union Commissions | 7 500 |
| 12. | Projects of the Executive Committee | 10 000[14] |
| 13. | Representation | 7 000[24] |
| 14. | Bank Charges | 600[15] |
| 15. | Young Astronomers' Schools | 22 000[16] |
| 16. | Regional Astronomy Meetings | 0[25] |
| 17. | Expenses of Special Nominating Committee | 0[17] |
| | Total payments | 527 470[18] |

4. Notes of Explanation

Receipts Notes:

1. The residual figure for contributions being greater than one third of the 1980-82 budget figure (Swfr 966 900) indicates that some countries are in arrears for their contributions up to 1981. However, several appreciable contributions were received soon after the 1981 accounts closed on 31 December 1981. The figure estimated for 1982 takes into account a slightly increased total number of units of contribution and makes an allowance for a certain (small) number of late payments.

2. Revenue from publications totalled Swfr 46 718 in 1980-81 making it possible to anticipate additional income from this item even after allowing for the lesser royalties of the new publication contract (see IB 47).

3. Interest rates are always changing, the total figure for 1981 being Swfr 41 676. Thus it is possible to anticipate a substantial increase, in 1982, over the residual budget figure.

4. It was anticipated in the adopted budget that ICSU/UNESCO contributions would substantially exceed the conservative figure provided, which was based on annual payments only without added "contract" payments. The 1982 figure corresponds to the allocation by ICSU for 1982 to the IAU ($18 000).

5. It is to be expected that payments in a closing (General Assembly) year will exceed receipts. The "budgetted" excess for 1982 (Swfr 69 422), incorporating the excesses of receipts over payments actually obtained in 1980 and 1981, produces a net excess of receipts over payments for the triennium 1980-82, of Swfr 890. The Executive Committee has decided that in the present circumstances of steeply rising costs and for other reasons, a net deficit of Swfr 16 000 could be borne for the triennium 1980-82, recognising a substantial surplus against budget for the triennium 1977-79; hence the projected excess of payments over receipts for 1982 has been allowed to rise from 69 422 to 86 086.

Payments Notes:

6. The budgetted payment for the Paris secretariat (item 1.1) maintenance of Swfr 126 667 p.a. was exceeded slightly in 1981 (Swfr 134 435). However, we remark that payments under this item averaged Swfr 134 524 in 1976-78. Therefore, expenditure has been held constant in a period when inflation has affected the Swiss Franc by 4 to 7% p.a. Taking into account certain charges which were not levied in 1980 and 1981 but which are now due, payments under 1.1 in 1982 cannot be less than Swfr 140 000.

7. Item 1.2 has not been drawn on; a figure for 1982 of a reduced amount on a per annum basis is retained in the Budget proposal for 1982.

8. Item 1.3 has scarcely been drawn on, thanks to furniture purchased for the IAU secretariat by "IAU France-1976", which provision is gratefully acknowledged. For 1982 it is proposed to make allowance of Swfr 20 000 towards purchase of word-processing equipment for future modernisation in the Paris secretariat.

9. Items under 3 include undertakings of Commission 25 (Orion atlas) and Commission 50 (IAU/CIE report) in 1981. It is proposed to add a further Swfr 4 380 for this sub-head in 1982 to cover possible proposals of a minor nature.

10. Due to heavy fare increases, demands on funds for Exchange of Astronomers have been considerable. The Executive Committee approved a supplementary allocation of Swfr 10 000 for this item (4.1).

11. Following allocation of ICSU project support, this item (4.2) has been increased by Swfr 1 000 in the proposal for 1982.

12. Under "Other Projects" are included regular specific commitments of IAU Commissions, the allocations in the Residual Budget for 1982 and in the Budget proposal for 1982 being as follows:

| | | | Residual | Proposal |
|---|---|---|---|---|
| (i) | Comm. 6 | Telegram Bureau | 1 637 | 2 000 |
| (ii) | Comm. 16 | Planetary Documentation, Meudon | 1 770 | 1 770 |
| (iii) | Comm. 16 | Lunar Documentation, Meudon | 5 310 | 0 |
| (iv) | Comm. 20 | Minor Planet Center | 2 068 | 2 500 |
| (v) | Comm. 27 | Variable Star Catalogue (USSR) | 9 100 | 9 100 |
| | | Total | 19 885 | 15 370 |

Item (iii) was not called upon for payment in 1980 and 1981 and notice was given by the Executive Committee in 1978 that no monies would be allocated beyond 1982 for items (ii) and (iii). The increases in (i) and (iv) are to accommodate changing dollar-Swfr ratios.

13. As announced in section 1.1(g) on p.14 of Information Bulletin No. 45 (January 1981) financial support of the Greek Local Organising Committee was approved by the Executive Committee in 1980 as an exceptional measure, the sums being utilised for specific purposes within the general provisions of the IAU Budget such as Symposia, Colloquia and travel grants to young astronomers. The imposition of this item renders it virtually impossible to balance the budget for the triennium 1980-82, although, taking into account the surplus of 1977-79, an overall balance in the six-year period 1977-82 is being achieved.

14. The multi-Commission proposal for a comprehensive index of Star Catalogues by Jaschek, Spite, Lortet, et al. is supported to an extent of Swfr 9 500 by the Executive Committee as an EC Project, but no sums have yet been drawn.

15. Due to allocation of some handling charges to this subhead, previously borne under exchange losses, a minor increase is proposed.

16. Young Astronomers' Schools have been held in each year 1979-81 as follows, with the approved allocations shown in US dollars and the payments actually made in Swfr:

| | Allocations($) | | Total Payments |
|---|---|---|---|
| | IAU | ICSU/UNESCO | 1979-81 (Swfr) |
| X-1979<br>La Laguna, Tenerife, Spain | 4 000 | 4 000 | 15 123 |
| XI-1980<br>Hvar, Yugoslavia | 7 000 | 4 000 | 17 904 |
| XII-1981<br>Helwan, Egypt | 11 000 | - | 19 391 |

The total 1980-81 allocation of $22 000 is already in excess of the Swfr 37 000 under item 15 of the 1980-82 Budget expenditure. The XIIIth Young Astronomers School is planned for Nairobi, Kenya, in June 1982. The Executive Committee has agreed to a contribution to this School equivalent to $11 000, of which the ICSU Fund is contributing $4 000. Substantial revision of this budgetary item was anticipated in the notes on the 1980-82 Budget as given in IAU Transactions XVIIB, p.32. The revised amount for 1982 of Swfr 22 000, replacing the "negative" residual figure of Swfr 4 169, represents a remaining small payment to the XIIth (Helwan) School plus the amount required for the XIIIth (Nairobi) Young Astronomers' School. (NB: Since these decisions were made the Nairobi School is cancelled and a modification will be made to this item at the 49th EC meeting in August 1982.)

17. The business of the Special Nominating Committee of 1979-82 was completed during 1981 without drawing on item 17 of the adopted budget and that of the SNC of 1982-85 is unlikely to commence until 1983.

18. An apparent increase in planned payments in 1982 by Swfr 97 805 is matched by an expected increase in receipts of Swfr 81 141 converting the budgetted surplus of Swfr 890 to a deficit of Swfr 15 774; see also Note 5.

19. See Note 13.

20. The IAU Information Bulletin has exceeded budgetary provisions by some 5% and an increased amount is required in 1982. Postage charges are largely responsible for the excess expenditure incurred.

21. Economies in this item have been possible by restricting the distribution slightly in places were the need was less.

22. EC Meeting costs are not incurred in 1982 except under item 5 "General Assembly".

23. Financial support of Symposia and Colloquia is provided to assist travel of participants to scientific meetings on the recommendations of Chairmen of Scientific Organising Committees. A strict limit is applied for each meeting and, in spite of there being an undetermined number of meetings, expenditure was maintained very close to the budget figure.

24. Budgetary support of IAU representatives attendance at meetings of other international bodies is expended on a limited scale, mainly being used to implement IAU interests.

25. No Regional Astronomy Meeting is planned during 1982.

APPENDIX

A. List of Countries Adhering to the IAU

The following is a list of the 49 countries that adhered to the Union in December 1981, giving also the year of admission and the Adhering Organisation.

| Country | Year | Adhering Organisation |
|---|---|---|
| Arab Republic of Egypt | 1935 | Academy of Scientific Research, Dept. Sci. Soc. and International Unions, 101 Kasr El Eini Street, Cairo |
| Argentina | 1927 | Comité Nacional de Astronomia<br>Attn: Ing. J.A. López, Secretary<br>Observatorio Astronomico "Félix Aguilar"<br>Av. Benavidez 8175 Oeste<br>5407 Marquesado, San Juan |
| Australia | 1939 | Australian Academy of Sciences<br>PO Box 783, Canberra City, ACT 2601 |
| Austria | 1955 | Osterreichische Akademie der Wissenschaften<br>Dr Ignaz Seipel-Platz 2, A-1010 Wien |
| Belgium | 1920 | Académie Royale de Belgique, Palais des Académies, 1 rue Ducale, Bruxelles 1 |
| Brazil | 1961 | Dourimar Nunes de Moura, Superintendencia de Cooperacao Internacional - SCI<br>Conselho Nacional de Desenvolvimento Científico e Tecnológico - CNPq<br>Av. W 3 Norte, Q. 507/B,<br>Caixa Postal 11-1142, 70740, Brasilia, DF |
| Bulgaria | 1957 | Bulgarian Academy of Sciences, 7th November Street 1, 1000 Sofia |
| Canada | 1920 | National Research Council of Canada<br>Attn: Dr C. Gauvreau<br>International Relations<br>Ottawa, Ontario K1A OR6 |
| Chile | 1947 | Universidad de Chile, Observatorio Astronomico Nacional, Casilla 36-D, Santiago de Chile |
| China – Nanking | 1935 | Chinese Astronomical Society<br>Attn: Dr Y.C. Chang, Director<br>Purple Mountain Observatory<br>Academia Sinica, Nanking |
| China – Taipei | 1959 | The Astronomy Union located at Taipei, China, Taipei Observatory, Yuan Shan, Taipei 104, Taiwan |
| Colombia | 1967 | Universidad Nacional de Columbia<br>Observatorio Astronomico Nacional<br>Apartado Aereo 2584, Bogota D.E. |
| Cuba | 1970 | Academia de Ciencias de Cuba<br>Attn: Dr B. Sagaro-Delgado<br>Capitolio Nacional, La Habana |

| Country | Year | Adhering Organisation |
|---|---|---|
| Czechoslovakia | 1922 | Czechoslovak Academy of Sciences<br>Narodní 3, Praha 1 |
| Denmark | 1922 | Det Kongelike Danske Videnskabernes Selskab - The Royal Danish Academy of Sciences & Letters, H.C. Andersens Blvd 35<br>DK - 1553 København V |
| Finland | 1948 | Académie des Sciences & Lettres<br>Snellmaninkatu 9-11, Helsinki 17 |
| France | 1920 | Académie des Sciences, Institut de France<br>23, quai de Conti, 75006 Paris |
| Germany, D.R. | 1951 | Akademie der Wissenschaften der DDR<br>Der Generalsekretär, DDR-108 Berlin,<br>Otto-Nuschke-Strasse 22/23 |
| Germany, F.R. | 1951 | Rat Westdeutscher Sternwarte<br>Attn: Prof. Dr R. Kippenhahn<br>MPI für Physik und Astrophysik<br>(Institut für Astrophysik)<br>Karl-Schwarzschild-Str. 1<br>D-8046 Garching bei München |
| Greece | 1920 | Academy of Athens, Panepistimiou Str.,<br>Athens |
| Hungary | 1947 | Hungarian Academy of Sciences<br>Roosevelt Tér 9, Budapest V |
| India | 1964 | Indian National Science Academy<br>Bahadur Shah Zafar Marg, New Delhi 110002 |
| Indonesia | 1979 | National Institute for Sciences<br>Tenku Cik Dithiro 43, Jakarta |
| Iran | 1969 | Institute of Geophysics<br>Attn: Dr M. Sobouti, Tehran University<br>Amirabad Shomali, Tehran |
| Iraq | 1976 | Space & Astronomical Research Centre<br>Council for Scientific Research<br>Attn: Dr N.M. Khalil<br>PO Box 255, Baghdad |
| Ireland | 1947 | The Royal Irish Academy, 19 Dawson Street<br>Dublin 2 |
| Israel | 1954 | The Israel Academy of Sciences & Humanities<br>Albert Einstein Square, Talbieh<br>Jerusalem 91 040, PO Box 4040 |
| Italy | 1921 | Consiglio Nazionale delle Ricerche<br>Piazzale delle Scienze 7, 00100 Rome |
| Japan | 1920 | Science Council of Japan<br>22-34 Roppongi 7 chome, Minato-Ku<br>Tokyo 106 |

| Country | Year | Adhering Organisation |
| --- | --- | --- |
| Korea (DPR) | 1961 | Academy of Sciences, Attn: Dr Sin Mun Gyu<br>General Secretary, Pyongyang |
| Korea (Republic) | 1973 | Korean National Astronomical Observatory<br>Yoksam-Dong, Kangnam-ku, Seoul, 134-03 |
| Mexico | 1921 | Instituto de Astronomia UNAM<br>Apartado Postal 70-264, México D.F. |
| Netherlands | 1922 | Koninklijke Nederlandsche Akademie van<br>Wetenschappen, Kloveniersburgwal 29<br>1011 JV Amsterdam |
| New Zealand | 1964 | The Royal Society of New Zealand<br>Private Bag, Wellington |
| Norway | 1922 | Det Norske Videnskaps-Akademi i Oslo<br>Drammensveinen 78, Oslo 2 |
| Poland | 1924 | Polska Akademia Nauk, Palac Kultury i Nauki<br>Warsaw |
| Portugal | 1924 | Seccáo Portuguesa das Uniões Internacionais<br>Astronómica e Geodésica e Geofísica<br>Praça da Estrela, Lisboa 2 |
| Roumania | 1928 | Roumanian National Committee of Astronomy<br>Astronomical Observatory<br>Str. Cutitul de Argint 5<br>75212 Bucarest, PO 28 |
| South Africa | 1938 | International Relations Division<br>Council for Scientific & Industrial<br>Research Services, PO Box 395, Pretoria 0001 |
| Spain | 1922 | Comision Nacional de Astronomia, Instituto<br>Geografico y Cadastral, General Ibanez 3<br>Madrid (3) |
| Sweden | 1925 | Kungl. Vetenskapsakademien,<br>Attn: Prof. T. Ganelius, Permanent Secretary<br>Box 50005, S-104 05 Stockholm |
| Switzerland | 1923 | Schweizerische Naturforschende Gesellschaft,<br>Zentralsekretariat, Laupenstrasse 10,<br>3000 Bern |
| Turkey | 1961 | Türk Astronomi Dernegi, University<br>Rasathanesi, Beyazit, Istanbul |
| United Kingdom | 1920 | The Royal Society, 6, Carlton House Terrace<br>London SW1Y 5AG |
| Uruguay | 1970 | Ministerio de Educacion y Cultura<br>Comité Nacional de Astronomia<br>8 de Octubre 3255, Montevideo |

| Country | Year | Adhering Organisation |
|---|---|---|
| USA | 1920 | National Academy of Sciences<br>Commission on International Relations<br>2101 Constitution Avenue N.W.<br>Washington DC 20148 |
| USSR | 1935 | Academy of Sciences of the USSR<br>Leninskij Prospekt 14, Moscow 71 |
| Vatican City State | 1932 | Pontificia Academia delle Scienze<br>Citta del Vaticano |
| Venezuela | 1953 | Dr Francisco J. Fuenmayor, Presidente<br>Asociacion Venezolana de Astronomia<br>Apartado 264, Mérida 5101 |
| Yugoslavia | 1935 | Savez društava matematičara, fizičára i astronoma Jugosloslavije, Institut za matematiku i fiziku, Cetinjski put bb, 81000 Titograd |

B. List of Deceased Members

The Executive Committee regrets to record the loss by death of the following 63 Members:

ASHBROOK Dr J. - 4 August 1980
BEALS Dr Carlyle S. - 2 July 1979
BEER Dr Arthur - 20 October 1980
BLOCH Dr M. -
BONEFF Prof. Nicola - May 1979
BONNEAU Prof. M. -
BRKIC Prof. Dr Zaharije - 24 April 1979
BUCHAR Prof. Dr E. - 20 September 1979
BULLARD Sir Edward - 3 April 1980
CARRASCO Dr Garrorena Rafael - 16 December 1981
CECCHINI Prof. G. -
CHIS Prof. Dr Gheorghe - 19 May 1981
COUDER Dr André - 16 January 1979
COUDERC Dr Paul - 5 February 1981
DA SILVA Prof. Dr Inacio Francisco - 9 September 1977
DJURKOVIC Prof. Petar M. - 5 January 1980
DOUGLAS Dr Alex E. - 26 July 1981
FICHERA Prof. E. - February 1978
FINSEN Dr William S. -
FLECKENSTEIN Prof. Dr J.D. - 21 February 1980
GUNTZEL-LINGNER Dr U. - 27 August 1979
GUTH Dr Prof. Vladimir - 25 June 1980
HAGIHARA Prof. Y. - 29 January 1979
HARWOOD M. - August 1979
HERGET Dr P. - 27 August 1981
HIROSE Dr Hideo - 27 October 1981
HUNAERTS Dr Joseph J. - 19 March 1979
IKEDA Dr Tetsuro - 2 October 1981
JOHNSON Dr H.L. - 2 April 1980
KORDYLEWSKI Dr Kazimierz - 11 March 1981

KOZHEVNIKOV Dr N.I. - 16 October 1981
LAUTERBORN Prof. Dr D. - May 1980
LUBECK K. - 11 March 1977
MATSUNAMI Dr N. - 6 September 1979
McCUSKEY Dr S.W. - 22 April 1979
MICHEL H. -
MOHR Prof. Dr J.M. - 16 December 1979
MOREAU Dr F. - 12 November 1979
MYERSCOUGH Dr Valerie P. - 8 November 1980
NØRLUND Prof. Niels Erik - 4 July 1981
PAJDUSAKOVA Dr Ludmila - 6 October 1979
PAYNE-GAPOSCHKIN Dr C.H. - 6 December 1979
PLASKETT Prof. Harry Hemley - 20 January 1980
PLATZECK Dr R. - 4 February 1979
PORTER Dr J.G. - 13 September 1981
RABBEN Dr H.H. - 28 April 1979
REINMUTH Dr K. - 6 May 1979
RICHTER Prof. Dr N.B. - 26 November 1980
ROMANA Dr A. - 13 October 1981
SANDIG Prof. H.-U. - 25 May 1979
SERKOWSKI Dr K. - 7 October 1981
SHAPLEY M.B. - March 1981
STRAUSS Dr Federico M. - 19 March 1981
SMILEY Dr C.H. - 26 July 1977
SWOPE H.H. - 24 November 1980
SYROVATSKIJ S.I. - 26 September 1979
TINSLEY Prof. Beatrice M. - 23 March 1981
UREY, Prof. H.C. -
WEMPE Prof. Dr J. - 29 May 1980
WHELAN Dr John A.J. - 21 December 1981
WYATT Dr S.P., Jr. - 6 December 1980

## C. List of IAU Publications

Transactions of the IAU, Volume XVIIA "Reports on Astronomy, 1979", in set of 3 volumes (Part 1: pp. vii + 223; Part 2: pp. vii + 246; Part 3: pp. vii + 228), price Dfl 168 (Dfl 56 per volume (Members at 33.33% discount). Editor: Edith A. Müller, published by D. Reidel Publishing Company, Dordrecht, Holland, 1979.

Transactions of the IAU, Volume XVIIB: "Proceedings of the Seventeenth General Assembly of the IAU", Montreal, Canada, 1979, pp. x + 527, price Dfl 130 (Members at 33.33% discount). Edited by Patrick A. Wayman, published by D. Reidel Publishing Company, 1980

Highlights of Astronomy, Volume No. 5, as presented at the XVIIth General Assembly of the IAU, 1979, pp. viii + 868, price Dfl 160 (cloth) and Dfl 75 (paper)(Members at 33.33% discount). Edited by Patrick A. Wayman, published by D. Reidel Publishing Company, 1980.

IAU Symposium Volumes

No. 78 "Nutation and the Earth's Rotation", pp. xvi + 266, price Dfl 65, paper Dfl 40 (Members at 33.33% discount)*. Edited by E.P. Fedorov, M.L. Smith and P.L. Bender, published by D. Reidel Publishing Company, 1980.

No. 79 "The Large Scale Structure of the Universe", pp. 520, edited by M.S. Longair and J. Einasto, 1978, Russian version published by "MIIP", Moscow, 1981.

No. 81 "Dynamics of the Solar System", pp. xiv + 347, price Dfl 85, paper Dfl 55 (Members at 33.33% discount)*. Edited by R.L. Duncombe, published by D. Reidel Publishing Company, 1979.

No. 82 "Time and the Earth's Rotation", pp. xix + 332, price Dfl 80, paper Dfl 50 (Members at 33.33% discount)*. Edited by D.D. McCarthy and J.D.H. Pilkington, published by D. Reidel Publishing Company, 1979.

No. 83 "Mass Loss and Evolution of O-Type Stars", pp. xix + 501, price Dfl 120, paper Dfl 60 (Members at 33.33% discount)*. Edited by P.S. Conti and C.W.H. de Loore, published by D. Reidel Publishing Company, 1979.

No. 84 "The Large-Scale Characteristics of the Galaxy", pp. xvi + 606, price Dfl 140, paper Dfl 70 (Members at 33.33% discount)*. Edited by W.B. Burton, published by D. Reidel Publishing Company, 1979.

No. 85 "Star Clusters", pp. xxi + 516, price Dfl 120, paper Dfl 60 (Members at 33.33% discount)*. Edited by J.E. Hesser, published by D. Reidel Publishing Company, 1980.

No. 86 "Radio Physics of the Sun", pp. xx + 474, price Dfl 110, paper Dfl 55 (Members at 33.33% discount)*. Edited by M.R. Kundu and T.E. Gergely, published by D. Reidel Publishing Company, 1980.

No. 87 "Interstellar Molecules", pp. xi + 704, price Dfl 145, paper Dfl 65 (Members at 33.33% discount)*. Edited by B.H. Andrew, published by D. Reidel Publishing Company, 1980.

No. 88 "Close Binary Stars", pp. xviii + 598, price Dfl 130, paper Dfl 60 (Members at 33.33% discount)*. Edited by M. Plavec, D. Popper and R.K. Ulrich, published by D. Reidel Publishing Company, 1980.

No. 89 "Refractional Influences in Astronomy and Geodesy", pp. xxiv + 403, price Dfl 95, paper Dfl 50 (Members at 33.33% discount)*. Edited by G. Teleki and K.-F.E. Tengstrom, published by D. Reidel Publishing Company, 1979.

No. 90 "Solid Particles in the Solar System", pp. xvi + 441, price Dfl 95, paper Dfl 50 (Members at 33.33% discount)*. Edited by I. Halliday and B.A. McIntosh, published by D. Reidel Publishing Company, 1980.

No. 91 "Solar and Interplanetary Dynamics", pp. xx + 558, price Dfl 125, paper Dfl 50 (Members 33.33% discount)*. Edited by M. Dryer and E. Tandberg-Hanssen, published by D. Reidel Publishing Company, 1980.

No. 92 "Objects of High Redshift", pp. xiv + 334, price Dfl 80, paper Dfl 40 (Members at 33.33% discount)*. Edited by G.O. Abell and P.J.E. Peebles, published by D. Reidel Publishing Company, 1980.

No. 93 "Fundamental Problems in the Theory of Stellar Evolution", pp. xv + 347, price Dfl 90, paper Dfl 40 (Members at 33.33% discount)*. Edited by D. Sugimoto, D.Q. Lamb and D.N. Schramm, published by D. Reidel Publishing Company, 1981.

No. 94 "Origin of Cosmic Rays", pp. xv + 409, price Dfl 100, paper Dfl 45 (Members at 33.33% discount)*. Edited by G. Setti, G. Spada and A. Wolfendale, published by D. Reidel Publishing Company, 1981.

No. 95 "Pulsars", pp. xv + 475, price Dfl 115, paper Dfl 55 (Members at 33.33% discount)*. Edited by W. Sieber and R. Wielebinski, published by D. Reidel Publishing Company, 1981.

No. 96 "Infrared Astronomy", pp. xvi + 380, price Dfl 90, paper Dfl 45 (Members at 33.33% discount)*. Edited by C.G. Wynn-Williams and D.P. Cruikshank, published by D. Reidel Publishing Company, 1981.

* Members receive 50% discount on paper edition.

IAU Colloquium Volumes

No. 44 "Physics of Solar Prominences", pp. xiv + 375, edited by E. Jensen, P. Maltby and F.Q. Orrall. Published by Institutt for Teoretisk Astrofysikk, Universitetet I Oslo,

No. 46 "Changing Trends in Variable Star Research", pp. 527, edited by F.M. Bateson, J. Smak and I.H. Urch. Published by the University of Waikato, 1979.

No. 47 "Spectral Classification of the Future, pp. xxii + 575, edited by M.F. McCarthy, S.J., A.G.D. Philip and G.B. Coyne, S.J. Published by the Vatican Observatory, 1979.

No. 48 "Modern Astrometry", pp. xvi + 603, edited by F.V. Prochazka and R.H. Tucker. Published by the Vienna University Observatory, 1979.

No. 49 "Image Formation from Coherence Functions in Astronomy", pp. xxi + 337, edited by C. van Schooneveld. Published by D. Reidel Publishing Company, 1979.

No. 50 "High Angular Resolution Stellar Interferometry", pp. xx + 492, edited by J. Davis and W.J. Tango. Published by Chatterton Astronomy Department, 1979.

No. 51 "Stellar Turbulence", pp. x + 308, edited by D.F. Gray and J.L. Linsky. Published by Springer-Verlag in the series "Lecture Notes in Physics", 1980.

No. 52 "Protostars and Planets, Studies of Star Formation and of the Origin of the Solar System", pp. viii + 756, edited by T. Gehrels with the assistance of M.S. Matthews. Published by the University of Arizona Press, 1978.

No. 53 "White Dwarfs and Variable Degenerate Stars", edited by H.M. van Horn and V. Weidemann. Published by University of Rochester, 1979

No. 54 "Scientific Research with the Space Telescope", pp. 327, edited by M.S. Longair and J.W. Warner. Published by NASA (CP-2111), 1980.

No. 56 "Reference Coordinate Systems for Earth Dynamics", pp. 396, edited by E.M. Gaposchkin and B. Kolaczek. Published by D. Reidel Publishing Company, in the series "Astrophysics and Space Science Library, vol 86, 1981.

No. 57 "The Satellites of Jupiter", in ICARUS, vol. 44, No. 2, November 1980.

No. 58 "Stellar Hydrodynamics", pp. 464, edited by A.N. Cox and D.S. King. Published by D. Reidel Publishing Company, in the series "Space Science Reviews", vol. 27, Nos. 3 and 4, 1980.

No. 59 "Effects of Mass-loss on Stellar Evolution", pp. 566, edited by C. Chiosi and R. Stalio. Published by D. Reidel Publishing Company in the series "Astrophysics and Space Library", vol. 89, 1981.

Other Publications

Proceedings of the Fourth Regional Meeting in Astronomy, held under the auspices of the IAU, abstracts of papers presented at the meeting published by Uppsala Astronomical Observatory, Report No. 12, 1979.

Proceedings of the First Asian-South Pacific Regional Meeting in Astronomy, held under the auspices of the IAU, edited by W. Davidson and B.M. Lewis, published as a special issue of the New Zealand Journal of Science, vol. 22, No. 4, December 1979.

Publications by, or on behalf of, IAU Commissions and by Astronomical Services

Commission No. 6 (Telegram Bureau) circulars Nos. 3318 - 3655, issued by the IAU Central Bureau for Astronomical Telegrams.

Commission No. 10 Cartes Synoptiques de la Chromosphère Solaire, prepared by Observatoire de Meudon, Paris.

Quarterly Bulletin on Solar Activity (QBSA), published by the Tokyo Astronomical Observatory, Mitaka, Tokyo, Japan

Commission No. 10 Sunspot Bulletin No. 1 (January 1980), published by the Sunspot Index Data Center, c/o Dr A. Koeckelenbergh, avenue Circulaire 3, B-1180 Bruxelles, Belgium.

Commission No. 19 Monthly Notes of the IPMS, Nos. 1 (1979) - 12 (1981), published by the International Polar Motion Service, International Latitude Observatory of Mizusawa, Mizusawa-shi, Iwate-ken, Japan.

Commission No. 20 Minor Planet Circulars Nos. 4583-6572 mimeographed circulars issued by the Minor Planet Center, Smithsonian Institution, 60 Garden Street, Cambridge, Mass., 02138, USA.

Commission 25 "A Photometric Atlas of the Orion Nebula Aggregate", published by A.D. Andrews, Armagh Observatory, N. Ireland, UK, 1981.

Commission 26 Circulaires d'Information, Nos. 77-86. President: P. Muller, CERGA, Grasse, France.

Commission 27 Information Bulletin on Variable Stars, prepared by Konkoly Observatory, Budapest, Hungary.

Commission 45 Circular letters, published by the President of the Commission

Commission 46 Newsletter 1 - 10, published by D.G. Wentzel, Astronomy Program, University of Maryland, College Park MD 20742, USA.

Commission 50 "Guidelines for Minimizing Urban Sky Glow near Astronomical Observatories", publication IAU/CIE No. 1, 1980. Can be obtained from the Secretariat.

IAU Information Bulletin Nos. 41 - 46, prepared by the General Secretary.

# VI.
# Astronomer's Handbook

The first "Astronomer's Handbook" was published as IAU Transactions Vol. XII C, 1966. As of November 1, 1982, a few copies are still available at the IAU Secretariat in Paris. Pending publication of a revised, adapted edition of the "Handbook", additions and changes to the sections in the first edition are indicated here.

## 1. SHORT HISTORY OF THE IAU

In continuation of Table 2 in Vol XII C, the information about General Assemblies and Executive Committees, contained in Vols. XIII to XVII B, is summarized and updated in Table 3, shown on the next page.

## 2. ADMINISTRATION AND FINANCES OF THE UNION

### A. Administration

The address of the General Secretary is:

Dr R.M. West
IAU SECRETARIAT
61, avenue de l'Observatoire
F-75014 Paris
France

The Secretariat consists of:

| | |
|---|---|
| Mrs. B. Manning | IAU Secretary |
| Ms. P. Smiley | Typist |

The address of the Assistant General Secretary, to whom all matters related to IAU Symposia and Colloquia should be directed, is:

Dr J.P. Swings
Institut d'Astrophysique
5, avenue de Cointe
B-4200 Cointe-Ougrée
Belgium

TABLE 3

IAU General Assemblies and Executive Committee Members (1967-1985)

| General Assemblies | Union Members (after G.A.) | No. of Adhering Countries | Participants | No. of Commissions | Transactions No. of pages | Highlights No. of pages |
|---|---|---|---|---|---|---|
| 1967 Prague, XIIIth | 2009 | | 1835 | 38 | XIIIA(1156),XIIIB(319) | 1(558) |
| 1970 Brighton, XIVth | 2590 | | 2255 | 39 | XIVA (566), XIVB (378) | 2(805) |
| 1973 Sydney, XVth | 3188 | 47 | 840 | 40 | XVA (768), XVB (343) | 3(584) |
| 1976 Grenoble, XVIth | 3805 | 48 | 2134 | 40 | XVIA (711), XVIB (547) | 4(761) |
| 1979 Montreal, XVIIth | 4513 | 49 | 1965 | 39 | XVIIA(718), XVIIB(538) | 5(876) |
| 1982 Patras, XVIIIth | 5310 | 49 | 1760 | 40 | XVIIIA( ),XVIIIB( ) | 6( ) |

An Extraordinary General Assembly took place in Warsaw, Poland, September 1973, on the occasion of the quincentenary birth of Nicolaus Copernicus.

| Presidents | General Secretaries | Assistant General Secretaries | Vice-Presidents |
|---|---|---|---|
| | | | W.N. Christiansen (1964-1970)<br>M. Schwarzschild (1964-1970)<br>A.B. Severny (1964-1970) |
| O. Heckmann (1967-1970) | L. Perek (1967-1970) | C. de Jager (1967-1970) | M.K.V. Bappu (1967-1973)<br>L. Gratton (1967-1973)<br>J. Sahade (1967-1973) |
| B. Strömgren (1970-1973) | C. de Jager (1970-1973) | G. Contopoulos (1970-1973) | B.J. Bok (1970-1976)<br>B. Lovell (1970 -1976)<br>E.R. Mustel (1970-1976) |
| L. Goldberg (1973-1976) | G. Contopoulos (1973-1976) | E.A. Müller (1973-1976) | J.G. Bolton (1973-1979)<br>Ch. Fehrenbach (1973-1979)<br>W. Iwanowska (1973-1979) |
| A. Blaauw (1976-1979) | E.A. Müller (1976-1979) | P.A. Wayman (1976-1979) | D.S. Heeschen (1976-1982)<br>E.K. Kharadze (1976-1982)<br>S. van den Bergh (1976-1982) |
| M.K.V. Bappu (1979-1982) | P.A. Wayman (1979-1982) | R.M. West (1979-1982) | M.W. Feast (1979-1985)<br>L. Kresak (1979-1985)<br>R. Wilson (1979-1985) |
| R. Hanbury Brown (1982-1985) | R.M. West (1982-1985) | J.P. Swings (1982-1985) | R.P. Kraft (1982-1988)<br>M. Peimbert (1982-1988)<br>Ya.S. Yatskiv (1982-1988) |

### B. Finances

By decision of the XVIIIth General Assembly, the annual unit of contribution, payable by Adhering Bodies to the Union, is fixed at Swiss Francs 1760.00 for the years 1983, 1984 and 1985.

## 3. THE INTERNATIONAL SCIENTIFIC UNIONS

### A. The International Council of Scientific Unions

The ICSU Yearbook contains the most recent information about IAU Participation in ICSU Committees, etc. This Yearbook should also be consulted for all current information on ICSU Unions and Committees (International Council of Scientific Unions, 51, Bd. de Montmorency, F-75016 Paris, France). The IAU is represented in the ICSU General Committee by its General Secretary, Dr Richard M. West, who is also a member of the ICSU Executive Board (1982- ), on behalf of "Earth and Space Science" Unions.

### B. IAU Representation on ICSU Committees

| | |
|---|---|
| 1. COSPAR (Committee on Space Research): | R.M. West |
| 2. ICSU-AB (ICSU Abstracting Board): | W.D. Heintz |
| 3. FAGS (Federation of Astronomical and Geophysical Services): | E. Tandberg Hansen, H. Enslin |
| FAGS Services Representatives: | |
| a) BIH (Bureau International de l'Heure): | S. Iijima, P. Pâquet |
| b) IPMS (International Polar Motion Service): | B. Kolaczek |
| c) IUWDS (International Ursigram and World Day Service): | E. Tandberg Hansen |
| 4. SCOSTEP (Scientific Committee for Solar-Terrestrial Physics): | M.R. Kundu |

### C. IAU Representation in Inter-Union Committees and in ICSU Commissions

| | |
|---|---|
| 1. IUCAF (Inter-Union Committee on Frequency Allocation for Radio Astronomy and Space Sciences): | R. Schilizzi and G. Westerhout |
| 2. CTS (Committee on the Teaching of Science) : | L. Houziaux |
| 3. CODATA (Committee for Data on Science and Technology): | B. Hauck |
| 4. IUCS (Inter-Union Commission on Spectroscopy): | B. Edlén, J.G. Phillips and M.J. Seaton |
| 5. COSTED (Committee on Science and Technology in Developing Countries): | G. Swarup |
| 6. SCOPE (Scientific Committee on Problems of the Environment): | F. Graham Smith |

### D. IAU Representation in Other Organizations

| | |
|---|---|
| 1. IAF (International Astronautical Federation): | M. Papagiannis |
| 2. CCDS (Comité Consultatif pour la Définition de la Seconde): | W. Markowitz and J. Benavente |
| 3. CCDM (Comité Consultatif pour la Définition du Mètre): | A.H. Cook |
| 4. CCIR (International Radio Consultative Committee): | |
| Study Group 7: | H.M. Smith and J. Pilkington |
| Study Group 2: | J. Whiteoak and L.H. Doherty |
| 5. EPS (European Physical Society Conference Committee): | J.P. Swings |

## 4. SERVICES AND FUNCTIONS OF THE IAU

The full text of this section can be found in IAU Transactions Vol. XIIC, pp. 17-38. Changes are indicated as follows: XIIIB (pp. 235-236); XIVB (p. 253); XVB (pp. 187-191); XVIB (p. 375) and XVIIB (p. 308).

The following changes are now made:

§B.4 The Minor Planet Center (project supported by Commission 20)

Address: IAU Minor Planet Center
Smithsonian Astrophysical Observatory
60, Garden Street
Cambridge, MA 02138
USA

TWX: 710-320-6842 ASTROGRAM CAM

Telephone: (617) 864-5758

Director : Dr B.G. Marsden

Assistant Directors: Dr C. Bardwell
Mr. D. Green

The file of observations now contains more than 250 000 positions of minor planets and comets, and is available on computer-readable tape.

§D Central Bureau for Astronomical Telegrams

The address, telex and telephone numbers are as for the Minor Planet Center (cf. above). A 24-hour service is maintained.

A revised Astronomical Telegram Code was issued in 1980, cf. IAU Information Bulletin No. 46, pp. 12-16.

## IAU Telegramme Code (1980)

### First groups for all telegrams

DISCOVERER OBJECT OBSERVER AAAAB

(1) The name of the discoverer(s) and/or other designation, such as year and letter for a comet, constellation for a nova, galaxy for a supernova (N = NGC, I = IC, M = Messier), etc.

(2) The type of object (COMET, OBJECT, NOVA, SUPERNOVA, VSTAR, etc.).

(3) The name of the observer(s) and/or computer(s).

(4) AAAA = equinox (assumed to be the mean equinox of the beginning of the year specified)

B = 1 for an approximate position;
= 2 for an accurate position;
= 3 for orbital elements of an object moving around the sun;
= 4 for an ephemeris.

Note: when an ephemeris follows the elements, group 4 is replaced by the word EPHEMERIS, and elements and ephemeris are to be referred to the same equinox.

### Middle groups for a position

Approximate: CDDEE FFFGH IIJJJ LMMNN PQRRS TUUUU VWWXX
Accurate: CDDEE FFFGH IIJJK KKKLM NNNPP PQRRS TUUUU VWWXX

(5) Date of observation:

C = final digit of year;
DD = month (01 = January, ...., 12 = December);
EE = day (UT).

(6) FFFGH = time of observation in decimals of a day (UT). This group may be omitted entirely in the case of observations of stationary objects.

(7) For an approximate position:
Right ascension $II^hJJ^m.J$
Declination L MM°NN'
P = 0

For an accurate position:
Right ascension $II^hJJ^mKK^s.KK$
Declination L MM°NN'PP".P

L = sign: 2 = positive, 1 = negative.

for comets:

Q = 1 for "total" magnitude ($m_1$);
= 2 for "nuclear" magnitude ($m_2$).

RR = magnitude (nearest integer; if negative, add 100).

S = appearance, according to the following:

| | Nothing reported about tail | Tail < 1° | Tail > 1° |
|---|---|---|---|
| Stellar appearance | 0 | | |
| Nothing reported about appearance of object | 1 | 2 | 3 |
| Object diffuse, without central condensation | 4 | 5 | 6 |
| Object diffuse, with central condensation | 7 | 8 | 9 |

For minor planets, novae, supernovae and variable stars:

Q = 3 for visual magnitude ($m_v$);
= 4 for range 0.38-0.50 μm (includes $m_{pg}$ and standard B);
= 5 for range 0.48-0.65 μm (includes $m_{pv}$ and standard V);
= 6 for range 0.3-0.4 μm (includes standard U);
= 7 for range 0.6-1.0 μm (includes standard R and I);
= 8 for range 1.0-5.0 μm (includes standard J,H, K and L);
= 9 for range 5.0- 20 μm (includes standard N and Q).

RR.S = magnitude (nearest 0.1; if negative, add 100). The group PQRRS may be repeated if magnitudes for different colors are being communicated.

(8) For comets and minor planets, the daily motion:

In right ascension T $UU^m.UU$
In declination V WW°XX'

For supernovae, the offset from the nucleus of the parent galaxy:

In right ascension T UUUU"
In declination V WWXX"

T, V = sign: 2 = positive (east, north), 1 = negative (west, south).

Group 8 may be omitted if unknown or irrelevant.

Middle groups for orbital elements

CDDEE FFFGH IIIII JJJJJ KKKKK TTTTT UUUUU

(5) Date of perihelion passage:

C = final digit of year;
DD = month (01 = January, ...., 12 = December);
EE = day (ET).

(6) FFF = time of perihelion passage in decimals of a day (ET).

G = time interval in days (nearest integer) between the first and last positions used for the computation; 0 = 10 days or more.

H = the number and quality of the observations on which the computation is based, or by which the elements have been checked, according to the table:

| | Maximum residuals > 5" | 1"-5" | < 1" |
|---|---|---|---|
| Fewer than three accurate positions* | 1 | 2 | 3 |
| Three accurate positions | 4 | 5 | 6 |
| More than three accurate positions | 7 | 8 | 9 |

*or other factors decreasing the reliability of the orbit, such as approximate, doubtful or unsatisfactorily distributed positions.

(7) III°II = $\omega$;
JJJ°JJ = $\Omega$;
KKK°KK = i.

(8) T.TTTT = q (in AU)
U.UUUU = e (omitted if the orbit is parabolic; e = 1).

Middle groups for ephemerides

CDDEE (IIJJJ LMMNN 9TTTT 8UUUU) cddee

(5) Date of the first position in the ephemeris:

C = final digit of year;
DD = month (01 = January, ...., 12 = December);
EE = day (ET)

(6) It is understood that ephemerides are given for $0^h$ ET.

(7) Right ascension $II^hJJ^m.J$
Declination L MM°NN'

L = sign: 2 = positive, 1 = negative

(8) T.TTT = geocentric distance in AU, preceded by a 9;
U.UUU = heliocentric distance in AU, preceded by an 8.

Groups 7 and 8 are repeated as often as is necessary to complete the ephemeris. Positions should be given at uniform intervals (10, 5, 2 or 1 days). Group 8 nee not be given for every line in the ephemeris.

cddee is the date of the last position in the ephemeris, in the same form as before.

Final groups for all telegrams

YYYYY ZZZZZ REMARKS COMMUNICATOR

(9) YYYYY = the last five figures of the sum of all the groups of digits including and following group 4 (containing the equinox).

ZZZZZ = the last five figures of the sum of the groups giving the right ascension, declination and magnitude only (or $\omega$, $\Omega$ and i in the case of orbital elements). In practice, this "second checksum" includes only the items under group 7 (or involving the letters I through S).

Note: any digit that is unknown, not significant, or otherwise to be withhel should be replaced by a slash (/); this is to count as zero in the summations. A grou of five slashes (/////) should be avoided. If two or more observations are being telegraphed in the same message, checksums should follow each one. A checksum of all numerical information is useful even if the telegram is not sent in code.

(10) Any additional remarks in qualification of the observation or the computation. In the case of an accurate position the location of the observing station could be specified.

(11) The name of the communicator(s). In the case of messages sent out by th Central Bureau, this is followed by a code giving the telegram category and the UT date and time of transmission. For example, 3DEC01/2315Z indicates a category 3 messa transmitted on Dec. $1^d23^h15^m$ UT.

Examples

(1) MEIER COMET MEIER 19501 80427 08/// 07187 25347 01108 20119 10017 71706 33642 MEIER

Meier reports that he has discovered and observed a comet as follows:

| 1978 UT | $\alpha_{1950}$ | $\delta_{1950}$ | $m_1$ |
|---|---|---|---|
| Apr. 27.08 | $7^h18^m.7$ | +53°47' | 10 |

Object diffuse with condensation, tail < 1°. Daily motion: $\Delta\alpha = +1^m.19$, $\Delta\delta = -17'$.

(2) 79VA OBJECT GICLAS 19502 91116 22014 01381 82522 50402 34120 01057 68425 19502 91117 21181 01453 10222 45530 5//// 39005 07205 LOWELLOBS

The Lowell Observatory communicates the following precise positions, by Giclas, of the object 1979 VA:

| 1979 UT | $\alpha_{1950}$ | $\delta_{1950}$ | $m_{pg}$ |
|---|---|---|---|
| Nov. 16.22014 | $1^h38^m18^s.25$ | +25°04'02".3 | 12.0 |
| 17.21181 | 1 45 31.02 | +24 55 30.5 | |

(3) 1979C COMET CANDY 19503 90723 40547 05191 16475 13489 03925 89853 35155 EPHEMERIS 90626 08380 20033 91180 80794 08393 20613 08393 21135 91293 80602 08372 21647 90711 42172 16966 PERTHOBS

The Perth Observatory communicates the following parabolic orbital elements and ephemeris, by Candy, for comet 1979c. The elements are based on more than three accurate observations, covering an arc of 4 days, and leave residuals in excess of 5".

T = 1979 July 23.405 ET    $\omega$ = 51°.91)
                                              $\Omega$ = 164.75) 1950.0
q = 0.3925 AU                         i = 134.89)

| 1979 ET | $\alpha_{1950}$ | $\delta_{1950}$ | $\Delta$ | r |
|---|---|---|---|---|
| June 26.0 | $8^h38^m.0$ | + 0°33' | 1.180 | 0.794 |
| July 1.0 | 8 39.3 | + 6 13 | | |
| 6.0 | 8 39.3 | +11 35 | 1.293 | 0.602 |
| 11.0 | 8 37.2 | +16 47 | | |

(4) CHERNYKH OBJECT MARSDEN 19504 81004 00119 11514 23570 11336 90495 81464 23470 11214 23403 11103 23360 11001 23336 10904 90856 81711 81029 10393 84330 APOLLO TYPE ASTEROID MAGNITUDE FOURTEEN MARSDEN

Marsden communicates the following ephemeris, calculated by him for the object discovered by Chernykh. The object is an Apollo-type asteroid.

| 1978 ET | $\alpha_{1950}$ | $\delta_{1950}$ | $\Delta$ | r | Mag. |
|---|---|---|---|---|---|
| Oct. 4.0 | $0^h11^m.9$ | -15°14' | | | |
| 9.0 | 23 57.0 | -13 36 | 0.495 | 1.464 | 14 |
| 14.0 | 23 47.0 | -12 14 | | | |
| 19.0 | 23 40.3 | -11 03 | | | |
| 24.0 | 23 36.0 | -10 01 | | | |
| 29.0 | 23 33.6 | - 9 04 | 0.856 | 1.711 | |

(5) SERPENS NOVA HONDA 19001 80301 817// 17461 11444 03087 12994 31992 KOZAI

Kozai reports that Honda has observed a nova in Serpens, as follows:

| 1978 UT | $\alpha_{1900}$ | $\delta_{1900}$ | $m_V$ |
|---|---|---|---|
| Mar. 1.817 | $17^h46^m.1$ | -14°44' | 8.7 |

(6) JOHNSON N4321 SUPERNOVA RAFANELLI 19501 90421 12204 21606 0512/ 20056 10087 78995 38930 ROSINO

Rosino reports an observation by Rafanelli of the supernova discovered by Johnson in NGC 4321 ($\alpha = 12^h20^m.4$, $\delta = +16°06'$, equinox 1950.0). On 1979 Apr. 21 UT the photovisual magnitude was 12, and the measured offset from the nucleus of the galaxy 56" east and 87" south.

(7) V1668 CYGNUS VSTAR BALDINELLI JONES 19751 80915 21416 24355 04072 05067 06068 07062 08047 09992 86745 86079 FARNSBARN

Farnsbarn reports the following magnitudes and colors of V1668 Cyg ($\alpha = 21^h41^m.6$, $\delta = +43°55'$, equinox 1975.0), observed on 1978 Sept. 15 UT by Baldinelli and Jones: V = 6.7, B-V = +0.5, U-B = -0.4; average infrared magnitudes were: 0.6-1.0 μm 6.2; 1-5 μm, 4.7; 5-20 μm, -0.8.

## §E Assignment of Designations and Names to Newly-Discovered Objects in the Solar System

During the XVIIIth General Assembly, representatives of IAU Commissions 6 and 20, and the EC Working Group on Planetary System Nomenclature decided to establish the scheme shown below for the assignation of designation and names. The scheme was endorsed by the IAU Executive Committee in its 50th meeting on 26 August 1982

Notes to Table:

C6 = Commission 6 (Astronomical Telegrams)
President: M. Candy (1982-85)

C20 = Commission 20 (Positions and Motions of Minor Planets, Comets and Satellites)
President: E. Roemer (1982-85)

WGPSN = Working Group on Planetary System Nomenclature
Chairman: H. Masursky (1982-85)

CBAT = Central Bureau for Astronomical Telegrams
Director: B.G. Marsden

MPC = Minor Planet Center
Director B.G. Marsden

P = President

VP = Vice-President

| Newly-discovered | A<br>Principal Planet or satellite | B<br>Newly discovered minor planet | C<br>Newly discovered comet | D<br>Rings |
|---|---|---|---|---|
| 1. Who assigns temporary designations to | Planet: C6 - CBAT<br>Satellite: C6 - CBAT | C20 - MPC | C6 - CBAT (or C20 for discoveries on old plates) | C6 - CBAT |
| 2. Who decides when a permanent designation can be given to | C20 committee (P,VP, Chm. + 2 from WG on Satellites - meet annually or as required) | C20 - MPC | C20 (published in Minor Planet/Comet Circular) | WGPSN |
| 3. Who assigns a permanent designation to | C20 | C20 (published in Minor Planet/Comet Circular) | C20 (published in Minor Planet/Comet Circular) | WGPSN |
| 4. Who decides when a name can be given to | C20 (when 3A is implemented, 4A becomes operational) | C20 - MPC | C6 - CBAT (or C20 in case of discoveries on old plates) | WGPSN |
| 5. Who assigns a name to | WGPSN (consult discoverers; advice on conflicting discovery claims from C20) | C20 - MPC | C6 - CBAT (or C20 in case of discoveries on old plates) | WGPSN |
| 6. Who names topographic/ albedo features on | WGPSN | WGPSN | WGPSN | WGPSN |

## 5. PUBLICATIONS OF THE IAU

Lists of IAU Publications were published in the IAU Information Bulletin as follows: No. 43 (January 1980, p. 49); No. 44 (June 1980, pp. 26-27); No. 45 (January 1981, pp. 49-50); No. 46 (June 1981, pp. 37-38); No. 47 (January 1982, p. 26) and No. 48 (June 1982, p. 61). A full list is included in the 1979-81 Report of the Executive Committee, Chapter 5 of this volume, on page 377.

## 6. MEETINGS OF THE INTERNATIONAL ASTRONOMICAL UNION

Details of IAU Symposia, Colloquia and other scientific meetings are reported in successive issues of the IAU Information Bulletin and a full list is included in the 1979-81 Report of the Executive Committee, Chapter 5 of this volume, on page 351.

## 7. EXECUTIVE COMMITTEE 1982 - 1985

President

Professor R. Hanbury Brown
School of Physics
University of Sydney
Sydney, NSW 2006
Australia

Vice-Presidents

Dr M.W. Feast
South African Astronomical Observatory
P.O. Box 9, Observatory, Cape 7935
South Africa

Professor R.P. Kraft
Lick Observatory
University of California
Santa Cruz, CA 95064
USA

Dr L. Kresak
Astronomical Institute
Slovak Academy of Sciences
842 28 Bratislava
Czechoslovakia

Professor M. Peimbert
Inst. de Astronomia
Apdo. Postal 70264
04510 Mexico, D.F.
Mexico

Professor R. Wilson
Department of Physics and Astronomy
University College London
Gower Street
London WC1E 6BT
UK

Professor Ya. S. Yatskiv
Main Astronomical Observatory
Ukrainian Academy of Sciences
252127 Kiev
USSR

General Secretary

Dr R.M. West
European Southern Observatory
Karl-Schwarzschild-Str. 2
D-8046 Garching bei München
Germany, F.R.

Assistant General Secretary

Dr J.P. Swings
Institut d'Astrophysique
5, avenue de Cointe
B-4200 Cointe-Ougrée
Belgium

Adviser to the Executive Committee

Professor P.A. Wayman
Dunsink Observatory
Castlenock, Co. Dublin
Ireland

# 8. STATUTES AND BY-LAWS

# UNION ASTRONOMIQUE INTERNATIONALE

## STATUTS

### I. DENOMINATION, BUTS ET DOMICILE

1. L'Union Astronomique Internationale (ci-après dénommée l'Union) est une organisation non-gouvernementale, qui a pour buts de:

   (a) développer l'astronomie par la coopération internationale,
   (b) encourager l'étude et le développement de l'astronomie sous tous ses aspects,
   (c) servir et sauvegarder les intérêts de l'astronomie.

2. L'Union a son siège légal à Bruxelles.

### II. AFFILIATION DE L'UNION

3. L'Union adhère au Conseil International des Unions Scientifiques.

### III. MEMBRES DE L'UNION

4. L'Union a pour membres:

   (a) des personnes morales (Pays adhérents)
   (b) des membres individuels (Membres).

### IV. ORGANISATIONS AFFILIEES

5. L'Union peut accepter l'affiliation d'organisations internationales non-gouvernementales qui contribuent au développement de l'astronomie.

### V. PAYS ADHERENTS

6. Les pays adhèrent à l'Union

   soit:
   (a) par l'intermédiaire de l'organisation par laquelle ils adhèrent au Conseil International des Unions Scientifiques, ou par l'intermédiaire d'un Comité National d'Astronomie approuvé par cette organisation,

   soit:
   (b) s'ils n'adhèrent pas au Conseil International des Unions Scientifiques, par l'intermédiaire d'un Comité National d'Astronomie reconnu par le Comité Exécutif de l'Union.

   (c) Les Organisations ou Comités mentionnés à l'article 6(a) et les Comités Nationaux d'Astronomie mentionnés à l'article 6(b) sont dénommés ci-après organismes adhérents.

7. L'adhésion d'un pays à l'Union est proposée par le Comité Exécutif et approuvée par l'Assemblée Générale; elle prend fin si le pays se retire de l'Union.

8. Les Pays Adhérents sont répartis en catégories. Le nombre des catégories est fixé par le Règlement. Un pays qui sollicite son adhésion indique la catégorie dans laquelle il désire être classé. La proposition peut être refusée par le Comité Exécutif si la catégorie est manifestement inadéquate.

# INTERNATIONAL ASTRONOMICAL UNION

## STATUTES

### I. DENOMINATION, OBJECTS AND DOMICILE

1. The International Astronomical Union (referred to as the Union) is a non-governmental organization, whose objects are:

   (a) to develop astronomy through international co-operation,
   (b) to promote the study and development of astronomy in all aspects,
   (c) to further and safeguard the interests of astronomy.

2. The legal domicile of the Union is Brussels.

### II. ADHERENCE TO THE UNION

3. The Union adheres to the International Council of Scientific Unions.

### III. COMPOSITION OF THE UNION

4. The Union is composed of:

   (a) corporate members (Adhering Countries)
   (b) individual members (Members).

### IV. AFFILIATED ORGANIZATIONS

5. The Union may admit the affiliation of international non-governmental organizations which contribute to the development of astronomy.

### V. ADHERING COUNTRIES

6. Countries adhere to the Union

   either:
   (a) through the organization by which they adhere to the International Council of Scientific Unions, or through a National Committee of Astronomy approved by that organization,

   or
   (b) if they do not adhere to the International Council of Scientific Unions, through a National Committee of Astronomy recognized by the Executive Committee of the Union.

   (c) The Adhering Organizations and National Committees of Astronomy are referred to as adhering bodies.

7. Adherence of a country to the Union is approved, on the proposal of the Executive Committee, by the General Assembly; it terminates if the country withdraws from the Union.

8. Adhering Countries are classified in categories. The number of categories shall be specified in the By-laws.
   A country requesting adherence shall specify the category in which it desires to be classed. The specification may be declined by the Executive Committee if the category proposed is manifestly inadequate.

## VI. MEMBRES

9. Les Membres sont admis dans l'Union par le Comité Exécutif, sur proposition de l'un des organismes adhérents mentionnés à l'article 6, en considération de leur activité dans une branche de l'astronomie.

## VII. ASSEMBLEE GENERALE

10. (a) L'activité de l'Union est dirigée par l'Assemblée Générale des représentants des Pays Adhérents et des Membres. Chaque Pays Adhérent nomme un représentant autorisé à voter en son nom.
    (b) L'Assemblée Générale rédige un Règlement qui précise les modalités d'application des Statuts.
    (c) Elle nomme un Comité Exécutif chargé d'exécuter les décisions de l'Assemblée Générale, et d'administrer l'Union pendant la période séparant les réunions de deux Assemblées Générales ordinaires successives. Le Comité Exécutif rend compte de sa gestion à l'Assemblée Générale. L'Assemblée Générale, en acceptant le rapport du Comité Exécutif, le décharge de sa responsabilité.

11. (a) Sur les questions concernant l'administration de l'Union, sans implication budgétaire, le vote à l'Assemblée Générale a lieu par Pays Adhérent, chaque pays disposant d'une voix. Les Pays Adhérents qui ne sont pas à jour de leurs cotisations annuelles au 31 décembre de l'année précédant l'Assemblée Générale ne peuvent pas participer aux votes.
    (b) Sur les questions engageant le budget de l'Union, le vote a lieu de même par Pays Adhérent, dans les conditions et avec les réserves prévues à l'article 11(a), le nombre de voix de chaque Pays Adhérent étant égal à l'indice de sa catégorie, définie conformément à l'article 8, augmenté d'une unité.
    (c) Les Pays Adhérents peuvent voter par correspondance sur les questions figurant à l'ordre du jour de l'Assemblée Générale.
    (d) Un scrutin n'est valable que si au moins deux tiers des Pays Adhérents disposant du droit de vote en vertu de l'article 11(a) y prennent part.

12. Sur les questions scientifiques n'engageant pas le budget de l'Union, les Membres de l'Union disposent chacun d'une voix.

13. Sur toutes les questions prévues aux articles 11 et 12, les décisions sont prises à la majorité absolue des suffrages. Cependant, une décision de modification des Statuts n'est valable que si elle a été prise à la majorité des deux tiers des voix des Pays Adhérents qui disposent du droit de vote en vertu de l'article 11(a).

14. Une proposition de modification des Statuts ne peut être discutée que si elle figure, en tant que telle, à l'ordre du jour de l'Assemblée Générale.

## VII. COMITE EXECUTIF

15. Le Comité Exécutif se compose du Président de l'Union, de six Vice-Présidents, du Secrétaire Général et du Secrétaire Général Adjoint, élus par l'Assemblée Générale sur la proposition du Comité Spécial des Nominations.

## IX. COMMISSIONS DE L'UNION

16. L'Assemblée Générale crée des Commissions en vue d'assurer la réalisation des buts qu'elle se propose.

## VI. MEMBERS

9. Members are admitted to the Union by the Executive Committee, on the proposal of an adhering body referred to in article 6, with regard to their achievements in some branch of astronomy.

## VII. GENERAL ASSEMBLY

10. (a) The work of the Union is directed by the General Assembly of representatives of Adhering Countries and of Members. Each Adhering Country appoints a representative authorized to vote in its name.
(b) The General Assembly draws up By-laws governing the application of the Statutes.
(c) It appoints an Executive Committee to implement the decisions of the General Assembly, and to direct the affairs of the Union in the interval between meetings of two successive ordinary General Assemblies. The Executive Committee reports to the General Assembly. The General Assembly, in accepting the report of the Executive Committee, discharges it of liability.

11. (a) On questions concerning the administration of the Union, not involving its budget, voting at the General Assembly is by Adhering Country, each country having one vote. Adhering Countries which have not paid their annual contributions up to 31 December of the year preceding the General Assembly may not participate in the voting.

(b) On questions involving the budget of the Union, voting is similarly by Adhering Country, under the same conditions and with the same reservations as in article 11(a), the number of votes for each Adhering Country being one greater than the number of its category, as defined in article 8.
(c) Adhering Countries may vote by correspondence on questions on the agenda for the General Assembly.
(d) A vote is valid only if at least two thirds of the Adhering Countries having the right to vote by virtue of article 11(a) participate in it.

12. On scientific questions not involving the budget of the Union the Members of the Union each have one vote.

13. On all questions in articles 11 and 12, decisions are taken by an absolute majority of the votes cast. However, a decision to change the Statutes is only valid if taken with the approval of at least two thirds of the votes of the Adhering Countries having the right to vote by virtue of article 11(a).

14. A motion to change the Statutes can only be discussed if it appears, in specific terms, on the agenda for the General Assembly.

## VIII. EXECUTIVE COMMITTEE

15. The Executive Committee consists of the President of the Union, six Vice-Presidents, the General Secretary and the Assistant General Secretary elected by the General Assembly on the proposal of a Special Nominating Committee.

## IX. COMMISSIONS OF THE UNION

16. The General Assembly forms Commissions for such purposes as it may decide.

## X. REPRESENTATION LEGALE DE L'UNION

17. Le Secrétaire Général est le représentant légal de l'Union.

## XI. BUDGET ET COTISATIONS

18. (a) Pour chaque Assemblée Générale ordinaire, le Comité Exécutif prépare un projet de budget pour la période à courir jusqu'à l'Assemblée Générale ordinaire suivante, ainsi que les comptes de l'Union pour la période précédente. Il les soumet au Comité des Finances pour examen; ce Comité des Finances est composé de membres nommés par les organismes adhérents, à raison d'un membre par organisme, et il est approuvé par l'Assemblée Générale. Lors de sa première séance pendant l'Assemblée Générale, le Comité des Finances élit un Président parmi ses membres.

(b) Le Comité des Finances examine les comptes de l'Union pour voir si les dépenses engagées ont été conformes aux voeux émis lors de la précédente réunion de l'Assemblée Générale et il s'assure que le budget proposé vise à la poursuite de la politique de l'Assemblée Générale, telle qu'elle est interprétée par le Comité Exécutif. Il présente des rapports sur ces questions qu'il soumet à l'Assemblée Générale pour approbation des comptes, et pour décision sur le budget.

(c) Chaque Pays Adhérent verse annuellement à l'Union un nombre d'unités de cotisation qui est fonction de sa catégorie. Le nombre d'unités de cotisation pour chaque catégorie est fixé par le Règlement.

(d) Le montant de l'unité de cotisation est fixé par l'Assemblée Générale, sur la proposition du Comité Exécutif et avec l'avis du Comité des Finances.

(e) Le paiement des cotisations est à la charge des organismes adhérents. La responsabilité de chaque Pays Adhérent envers l'Union est limitée au montant des cotisations dues par ce pays à l'Union.

(f) Un Pays Adhérent qui cesse d'adhérer à l'Union renonce de ce fait à ses droits sur l'actif de l'Union.

## XII. DISSOLUTION DE L'UNION

19. La décision de dissoudre l'Union n'est valable que si elle est prise à la majorité des trois quarts des voix des Pays Adhérents qui disposent du droit de vote en vertu de l'article 11(a).

## XIII. DEVOLUTION DE L'AUTORITE EN CAS DE FORCE MAJEURE

20. Si, par suite d'événements indépendants de la volonté de l'Union, des circonstances apparaissent qui rendent impossible le respect des clauses de ces Statuts et du Règlement établi par l'Assemblée Générale, les organes et membres du Comité Exécutif de l'Union, dans l'ordre fixé ci-dessous, prendront toutes dispositions qu'ils jugeront nécessaires pour la continuation du fonctionnement de l'Union. Ces dispositions devront être soumises à une autorité supérieure dès que cela deviendra possible, jusqu'à ce qu'une Assemblée Générale extraordinaire puisse être réunie. L'autorité est dévolue dans l'ordre ci-dessous: l'Assemblée Générale; une Assemblée Générale extraordinaire; le Comité Exécutif, réuni ou par correspondance; le Président de l'Union; le Secrétaire Général; ou, à défaut de la possibilité de recourir à l'une de ces autorités ou de leur disponibilité, un des Vice-Présidents.

## X. LEGAL REPRESENTATION OF THE UNION

17. The General Secretary is the legal representative of the Union.

## XI. BUDGET AND DUES

18. (a) For each ordinary General Assembly the Executive Committee prepares a budget proposal covering the period to the next ordinary General Assembly, together with the accounts of the Union for the preceding period. It submits these to the Finance Committee for consideration; this Finance Committee consists of one member nominated by each adhering body and approved by the General Assembly. At its first meeting during the General Assembly, the Finance Committee elects a Chairman from among its members.

(b) The Finance Committee examines the accounts of the Union from the point of view of responsible expenditure within the intent of the previous General Assembly, and it considers whether the proposed budget is adequate to implement the policy of the General Assembly, as interpreted by the Executive Committee. It submits reports on these matters to the General Assembly for approval of the account and decision on the budget.

(c) Each Adhering Country pays annually to the Union a number of units of contribution according to its category. The number of units of contribution for each category shall be specified in the By-laws.

(d) The amount of the unit of contribution is determined by the General Assembly, on the proposal of the Executive Committee and with the advice of the Finance Committee.

(e) The payment of contributions is the responsibility of the adhering bodies. The liability of each Adhering Country in respect of the Union is limited to the amount of that country's dues to the Union.

(f) An Adhering Country that ceases to adhere to the Union resigns at the same time its rights to a share in the assets of the Union.

## XII. DISSOLUTION OF THE UNION

19. The decision to dissolve the Union is only valid if taken with the approval of three quarters of the votes of the Adhering Countries having the right to vote by virtue of article 11(a).

## XIII. EMERGENCY POWERS

20. If, through events outside the control of the Union, circumstances arise in which it is inpracticable to comply with the provisions of these Statutes and of the By-laws drawn up by the General Assembly, the organs and officers of the Union, in the order specified below, shall take such actions as they deem necessary for the continued operation of the Union. Such action shall be reported to a higher authority immediately this becomes practicable until such time as an extraordinary General Assembly can be convened. The following is the order of authority:
The General Assembly; an extraordinary General Assembly; the Executive Committee in meeting or by correspondence; the President of the Union; the General Secretary; or failing the practicability or availability of any of the above, one of the Vice-Presidents.

XIV. CLAUSES FINALES

21. Ces Statuts entrent en vigueur le 1er Septembre 1970.

22. Les présents Statuts sont publiés en versions française et anglaise. En cas d'incertitude, la version française fait seule autorité.

## REGLEMENT

### I. LES MEMBRES DE L'UNION

1. Les demandes d'adhésion des pays à l'Union Astronomique Internationale (ci-après dénommée l'Union) sont examinées par le Comité Exécutif et soumises à l'approbation de l'Assemblée Générale.

2. Les propositions de modifications de la liste des Membres sont, après examen attentif des suggestions des Présidents de Commissions, soumises pour avis au Comité des Nominations, composé d'un représentant de chaque Pays Adhérent désigné par l'organisme adhérent habilité, avant la décision du Comité Exécutif.

3. Les Commissions peuvent, avec l'approbation du Comité Exécutif, coopter des consultants qu'elles jugent en mesure d'apporter une contribution utile à leur travail. L'adhésion des consultants a pour terme le dernier jour de la première Assemblée Générale ordinaire qui suit leur admission, à moins qu'elle ne soit renouvelée.

4. Une organisation affiliée peut participer au travail de l'Union dans les conditions fixées par accord entre l'organisation et le Comité Exécutif.

### II. L'ASSEMBLEE GENERALE

5. L'Union se réunit en Assemblée Générale ordinaire régulièrement une fois tous les trois ans. Si le lieu et la date de l'Assemblée Générale ordinaire n'ont pas été décidés lors de la précédente Assemblée Générale, ils sont fixés par le Comité Exécutif et communiqués aux organismes adhérents au moins six mois à l'avance.

6. Le Président peut convoquer, avec l'accord du Comité Exécutif, une Assemblée Générale extraordinaire. Il est tenu de le faire à la demande du tiers des Pays Adhérents.

7. L'Ordre du Jour de chaque Assemblée Générale ordinaire est arrêté par le Comité Exécutif et communiqué aux Organismes Adhérents au moins quatre mois avant le premier jour de la réunion. Il devra inclure la proposition du Comité Exécutif concernant le montant de l'unité de cotisation qui permet l'application de l'article 24.

8. (a) L'Ordre du Jour doit inclure toute motion ou proposition reçue par le Secrétaire Général au moins cinq mois avant le premier jour d'une Assemblée Générale ordinaire, qu'elle émane d'un organisme adhérent, d'une Commission de l'Union, ou d'une Commission mixte dans laquelle l'Union est représentée.

   (b) Une motion ou proposition concernant l'administration ou le budget de l'Union qui ne figure pas à l'Ordre du Jour, préparé par le Comité

## XIV. FINAL CLAUSES

21. These Statutes enter into force on 1 September 1970.

22. The present Statutes are being published in French and English version. In case of doubt, the French version is the only authority.

# BY-LAWS

## I. MEMBERSHIP

1. Applications of countries for adherence to the International Astronomical Union (referred to as the Union) are examined by the Executive Committee and submitted to the General Assembly for approval.

2. Proposed changes in the list of Members are, with due regard to the suggestions of the Presidents of Commissions, submitted for advice to the Nominating Committee, consisting of one representative of each Adhering Country designated by the appropriate adhering body, before decision by the Executive Commitee.

3. Commissions may, with the approval of the Executive Committee, co-opt consultants whom they consider may contribute to their work. The adherence of consultants expires on the last day of the ordinary General Assembly next following their admission, unless renewed.

4. An affiliated organization may participate in the work of the Union as mutually agreed between the organization and the Executive Committee.

## II. GENERAL ASSEMBLY

5. The Union meets in ordinary General Assembly, as a rule, once every three years. The place and date of the ordinary General Assembly, unless determined by the General Assembly at its previous meeting, shall be fixed by the Executive Committee and communicated to the adhering bodies at least six months beforehand.

6. The President, with the consent of the Executive Committee, may summon an extraordinary General Assembly. He must do so at the request of one third of the Adhering Countries.

7. The agenda of business for each ordinary General Assembly is determined by the Executive Committee and is communicated to the adhering bodies at least four months before the first day of the meeting. It shall include the proposal of the Executive Committee in regard to the unit of contribution as called for in article 24.

8. (a) Any motion or proposal received by the General Secretary at least five months before the first day of an ordinary General Assembly, whether from an adhering body, from a Commission of the Union, or from an Inter-Union Commission on which the Union is represented, must be placed on the agenda.

   (b) A motion or proposal concerning the administration or budget of the Union which does not appear on the agenda prepared by the Executive

Exécutif, ou tout amendement à une motion qui figure à l'Ordre du Jour, ne peut être discuté qu'avec l'accord prélable des deux tiers au moins des voix des Pays Adhérents représentés à l'Assemblée Générale et disposant du droit de vote en vertu de l'article 11(a) des Statuts.

9. S'il y a doute sur le caractère administratif ou scientifique d'une question donnant lieu à un vote, l'avis du Président est prépondérant.

10. En cas de partage égal des voix, le Président a voix prépondérante.

11. Le Président peut inviter des représentants d'autres organisations, des scientifiques et de jeunes astronomes à participer à l'Assemblée Générale. Avec l'accord du Comité Exécutif, il peut déléguer ce privilège au Secrétaire Général en ce qui concerne les représentants d'autres organisations, aux organismes adhérents en ce qui concerne les scientifiques et les jeunes astronomes.

III. LE COMITE SPECIAL DES NOMINATIONS

12. (a) Les propositions pour les élections du Président de l'Union, des six Vice-Présidents, du Secrétaire Général et du Secrétaire Général Adjoint sont soumises à l'Assemblée Générale par le Comité Spécial des Nominations. Ce Comité se compose du Président en fonction et du Président sortant, d'un membre proposé par le Comité Exécutif sortant et n'appartenant ni au Comité Exécutif actuel ni au Comité Exécutif précédent, et de quatre membres élus par le Comité des Nominations parmi 12 membres proposés par les Présidents de Commissions.
A l'exception du Président en fonction et du Président sortant, les membres actuels et les anciens membres du Comité Exécutif ne doivent pas faire partie du Comité Spécial des Nominations. Les membres du Comité Spécial des Nominations doivent tous appartenir à des pays différents.

(b) Le Secrétaire Général et le Secrétaire Général Adjoint participent au travail du Comité Spécial des Nominations à titre consultatif.

(c) Le Comité Spécial des Nominations est nommé par l'Assemblée Générale et est responsable directement devant elle. Il reste en fonction jusqu'à la fin de l'Assemblée Générale ordinaire qui suit immédiatement sa nomination, et il peut combler toute vacance survenant parmi ses membres.

IV. LE COMITE EXECUTIF ET SES MEMBRES

13. (a) Le Président de l'Union reste en fonction jusqu'à la fin de l'Assemblée Générale ordinaire qui suit immédiatement celle de son élection; les Vice-Présidents restent en fonction jusqu'à la fin de la deuxième Assemblée Générale ordinaire qui suit celle de leur élection. Ils ne sont pas rééligibles immédiatement pour les mêmes fonctions.

(b) Le Secrétaire Général et le Secrétaire Général Adjoint restent en fonction jusqu'à la fin de l'Assemblée Générale ordinaire qui suit immédiatement celle de leur élection. Normalement, le Secrétaire Général Adjoint succède au Secrétaire Général, mais l'un et l'autre peuvent être réélus aux mêmes fonctions pour une seconde période consécutive.

(c) Les élections ont lieu au cours de la dernière réunion de l'Assemblée Générale, les noms des candidats proposés ayant été annoncés au cours d'une réunion antérieure.

Committee, or any amendment to a motion that appears on the agenda, shall only be discussed with the prior approval of at least two thirds of the votes of Adhering Countries represented at the General Assembly and having the right to vote by virtue of Statute 11(a).

9. If there is doubt as to the administrative or scientific character of a question giving rise to a vote, the President determines the issue.

10. Where there is an equal division of votes, the President determines the issue.

11. The President may invite representatives of other organizations, scientists and young astronomers to participate in the General Assembly. Subject to the agreement of the Executive Committee he may delegate this privilege concerning representatives of other organizations to the General Secretary, and concerning scientists and young astronomers to the adhering bodies.

## III. SPECIAL NOMINATING COMMITTEE

12. (a) Proposals for elections to the President of the Union, six Vice-Presidents, the General Secretary and the Assistant General Secretary are submitted to the General Assembly by the Special Nominating Committee. This consists of the President and past President of the Union, a member proposed by the retiring Executive Committee, and four members elected by the Nominating Committee from among twelve Members proposed by Presidents of Commissions. Other than the President and immediate past President, present and former members of the Executive Committee shall not serve on the Special Nominating Committee. No two members of the Special Nominating Committee shall belong to the same country.

(b) The General Secretary and the Assistant General Secretary participate in the work of the Special Nominating Committee in an advisory capacity.

(c) The Special Nominating Committee is appointed by the General Assembly to which it reports direct. It remains in office until the end of the ordinary General Assembly next following that of its appointment, and it may fill any vacancy occurring among its members.

## IV. OFFICERS AND EXECUTIVE COMMITTEE

13. (a) The President of the Union remains in office until the end of the ordinary General Assembly next following that of his election; the Vice-Presidents remain in office until the end of the second ordinary General Assembly following that of their election. They may not be re-elected immediately to the same offices.

(b) The General Secretary and the Assistant General Secretary remain in office until the end of the ordinary General Assembly next following that of their election. Normally the Assistant General Secretary succeeds the General Secretary though both officers may be re-elected for another term.

(c) The election takes place at the last session of the General Assembly, the names of the candidates proposed having been announced at a previous session.

14. Le Président sortant et le Secrétaire Général sortant deviennent conseillers du Comité Exécutif jusqu'à la fin de l'Assemblée Générale ordinaire qui suit immédiatement celle de la fin de leur mandat. Ils participent au travail du Comité Exécutif et assistent à ses réunions sans droit de vote.

15. Le Comité Exécutif peut combler toute vacance survenant en son sein. Toute personne ainsi nommée reste en fonction jusqu'à l'Assemblée Générale ordinaire suivante.

16. Le Comité Exécutif peut rédiger et publier des Directives pour expliciter les Statuts et le Règlement.

17. Le Comité Exécutif nomme le représentant de l'Union qui doit siéger au sein du Comité Exécutif du Conseil International des Unions Scientifiques; si ce répresentant n'est pas déjà un membre élu du Comité Exécutif, il devient conseiller.

18. (a) Le Secrétaire Général est responsable auprès du Comité Exécutif des dépenses qu'il engage, qui ne doivent pas dépasser le montant des fonds mis à sa disposition.

(b) Un bureau administratif, sous la direction du Secrétaire Général, est chargé de la correspondance, de la gestion des fonds de l'Union, et de la conservation des archives.

## V. COMMISSIONS

19. (a) Les Commissions de l'Union poursuivent les buts scientifiques de l'Union par des moyens tels que l'étude de domaines particuliers de l'Astronomie, l'encouragement de recherches collectives et la discussion de questions relatives aux accords internationaux et à la standardisation.

(b) Les Commissions de l'Union établissent des rapports sur les sujets qui leur ont été confiés.

20. Chaque Commission se compose de:

(a) un Président et au moins un Vice-Président élus par l'Assemblée Générale sur la proposition du Comité Exécutif. Ils demeurent en fonction jusqu'à la fin de l'Assemblée Générale ordinaire qui suit immédiatement celle de leur élection. Ils ne sont pas normalement rééligibles.

(b) un Comité d'Organisation, dont les membres sont désignés par la Commission sous réserve de l'approbation du Comité Exécutif. Le Comité d'Organisation assiste le Président et le(s) Vice-Président(s) dans leur tâche. Une Commission peut décider qu'elle n'a pas besoin de Comité d'Organisation.

(c) des membres de l'Union, nommés par les Présidents, Vice-Président(s) et Comité d'Organisation, en considération de leurs spécialités; leur désignation est soumise à confirmation par le Comité Exécutif.

21. Entre deux Assemblées ordinaires, les Présidents de Commissions peuvent coopter, parmi les Membres de l'Union, de nouveaux membres des Comités d'Organisation et des Commissions elles-mêmes.

22. Les Commissions rédigent leur propre règlement. Les décisions sont prises, à l'intérieur des Commissions, par un vote de leurs membres.

14. The retiring President and the retiring General Secretary become advisers to the Executive Committee until the end of the ordinary General Assembly next following that of their retirement. They participate in the work of the Executive Committee and attend its meetings without voting right.

15. The Executive Committee may fill any vacancy occurring among its members. Any person so appointed remains in office until the next ordinary General Assembly.

16. The Executive Committe may draw up and publish Working Rules to implement the Statutes and By-laws.

17. The Executive Committee appoints the Union's representative on the Executive Committee of the International Council of Scientific Unions; if not already an elected member of the Executive Committee, this representative will become its adviser.

18. (a) The General Secretary is responsible to the Executive Committee for not incurring expenditure in excess of the funds at his disposal.

(b) An administrative office, under the direction of the General Secretary, conducts the correspondence, administers the funds, and preserves the archives of the Union.

V. COMMISSIONS

19. (a) The Commissions of the Union shall pursue the scientific objects of the Union by activities such as the study of special branches of astronomy, the encouragement of collective investigations, and the discussion of questions relating to international agreements or to standardization.

(b) The Commissions of the Union shall prepare reports on the work with which they are concerned.

20. Each Commission consists of:

(a) A President and at least one Vice-President elected by the General Assembly on the proposal of the Executive Committee. They remain in office until the end of the ordinary General Assembly next following that of their election. They are not normally re-eligible.

(b) An Organizing Committee, whose members are appointed by the Commission subject to the approval by the Executive Committee. The Organizing Committee assists the President and Vice-President(s) in their duties. A Commission may decide that it needs no Organizing Committee.

(c) Members of the Union, appointed by the President, Vice-President(s) and the Organizing Committee, in consideration of their special interests; their appointment is subject to the confirmation by the Executive Committee.

21. Between two ordinary General Assemblies, Presidents of Commissions may co-opt, from among Members of the Union, new members to the Organizing Committees and to the Commissions themselves.

22. Commissions draw up their own rules. Decisions within Commissions are taken according to the vote of their members.

## VI. ORGANISMES ADHERENTS

23. Le rôle des organismes adhérents est d'encourager et de coordonner, sur leurs territoires respectifs, l'étude des diverses branches de l'astronomie, particulièrement en ce qui concerne leurs besoins sur le plan international. Ils ont le droit de soumettre au Comité Exécutif des propositions pour discussion par l'Assemblée Générale.

## VII. FINANCES

24. Chaque Pays Adhérent verse à l'Union une cotisation annuelle, qui est un multiple de l'unité de cotisation en fonction de sa catégorie, comme suit:

| | | | | | | | | |
|---|---|---|---|---|---|---|---|---|
| Catégories définies conformément à l'article 8 des Statuts: | 1 | 2 | 3 | 4 | 5 | 6 | 7 | 8 |
| Nombre respectif d'unités de cotisations: | 1 | 2 | 4 | 6 | 10 | 14 | 20 | 30 |

25. Les ressources de l'Union sont consacrées à la poursuite de ses buts, y compris:

(a) les frais de publication et les dépenses administratives;

(b) l'encouragement des activités astronomiques qui nécessitent la coopération internationale.

(c) la cotisation due par l'Union au Conseil International des Unions Scientifiques.

26. Les ressources provenant de dons sont utilisées par l'Union en tenant compte des voeux exprimés par les donateurs.

## VIII. PUBLICATIONS

27. L'Union a la propriété littéraire de tous les textes imprimés dans ses publications, sauf accord différent.

28. Les Membres de l'Union ont le droit de recevoir les publications de l'Union gratuitement ou à prix réduit , à la discrétion du Comité Exécutif qui décide en fonction de la situation financière de l'Union.

## IX. CLAUSES FINALES

29. Ce règlement entre en vigueur le 1er septembre 1970.
Il peut être modifié avec l'approbation de la majorité absolue des voix des Pays Adhérents qui disposent du droit de vote en vertu de l'article 11(a) des Statuts.

30. Le présent règlement est publié en versions française et anglaise. En cas d'incertitude, la version française fait seule autorité.

## VI. ADHERING BODIES

23. The functions of the adhering bodies are to promote and co-ordinate, in their respective territories, the study of the various branches of astronomy, more especially in relation to their international requirements. They are entitled to submit to the Executive Committee motions for discussion by the General Assembly.

## VII. FINANCE

24. Each Adhering Country pays annually to the Union a number of units of contribution according to its category as follows:

| | | | | | | | | |
|---|---|---|---|---|---|---|---|---|
| Category as defined in Statute 8: | 1 | 2 | 3 | 4 | 5 | 6 | 7 | 8 |
| Number of units of contribution: | 1 | 2 | 4 | 6 | 10 | 14 | 20 | 30 |

25. The income of the Union is to be devoted to its objects, included

    (a) costs of publication and expenses of administration

    (b) the promotion of astronomical enterprises requiring international co-operation

    (c) the contribution due from the Union to the International Council of Scientific Unions.

26. Funds derived from donations are used by the Union in accordance with the wishes expressed by the donors.

## VIII. PUBLICATIONS

27. The Union has the copyright to all materials printed in its publications, unless otherwise arranged.

28. Members of the Union are entitled to receive the publications of the Union free of charge or at reduced prices at the discretion of the Executive Committee taking due regard of the financial situation of the Union.

## IX. FINAL CLAUSES

29. These By-laws enter into force on 1 September 1970.
They can be changed with the approval of an absolute majority of the votes of the Adhering Countries having the right to vote by virtue of Statute 11(a).

30. The present By-laws are being published in French and English version. In case of doubt, the French version is the only authority.

# 9. WORKING RULES

# UNION ASTRONOMIQUE INTERNATIONALE

## DIRECTIVES

### I. PUBLICATIONS

1. Les publications de l'Union Astronomique Internationale approuvées, dans le budget, par l'Assemblée Générale, sont préparées par le Bureau Administratif de l'Union.

2. Les Commissions de l'Union peuvent, avec l'approbation du Comité Exécutif, avoir leurs propres publications.

3. Le Comité Exécutif décide sur la proposition du Secrétaire Général, des modalités de distribution des publications de l'Union.

4. Les Membres de l'Union peuvent acquérir les publications de l'Union à un prix réduit.

### II. APPARTENANCE A L'UNION

#### A. *Pays Adhérents*

6. Les demandes d'adhésion à l'Union formulées par les pays sont examinées par le Comité Exécutif compte tenu des points suivants:

(a) justesse du choix de la catégorie dans laquelle le pays souhaite être classé
(b) situation actuelle de l'astronomie dans le pays formulant la demande, et ses possibilités de développement
(c) mesure dans laquelle le futur organisme adhérent est représentatif de l'activité astronomique de son pays.

7. Les demandes proposant une contribution annuelle appropriée seront soumises pour décision à l'Assemblée Générale, avec la recommandation du Comité Exécutif.

#### B. *Membres*

8. Les personnes proposées pour devenir Membres de l'Union doivent en principe être choisies parmi des astronomes et des chercheurs dont les activités sont liées à l'astronomie, compte tenu de
(a) la qualité de leur œuvre scientifique
(b) la mesure dans laquelle leur activité scientifique implique des recherches astronomiques
(c) leur désir de contribuer à la poursuite des buts de l'Union.

9. Les jeunes astronomes doivent être considérés comme pouvant devenir Membres de l'Union dès qu'ils ont fait la preuve de leur capacité (en principe par une thèse de doctorat ou son équivalent) et de leur aptitude (quelques années d'activité fructueuse) à mener une recherche personnelle.

10. Pour les astronomes professionnels, leur contribution à l'astronomie peut consister soit en des recherches personnelles, soit en une collaboration assidue à des programmes importants d'observations.

# INTERNATIONAL ASTRONOMICAL UNION

## WORKING RULES

### I. PUBLICATIONS

1. The publications of the International Astronomical Union approved, in the budget, by the General Assembly are prepared by the Administrative Office of the Union.

2. Commissions of the Union may, with the approval of the Executive Committee, issue their publications independently.

3. The distribution of publications of the Union is decided, on the proposal of the General Secretary, by the Executive Committee.

4. Members may purchase the publications of the Union at reduced prices.

### II. MEMBERSHIP

#### A. *Adhering Countries*

6. Applications of countries for adherence to the Union are examined by the Executive Committee for

(a) the adequacy of the category in which the country wishes to be classified
(b) the present state and expected development of astronomy in the applying country
(c) the degree to which the prospective adhering body is representative of its country's astronomical activity.

7. Applications proposing an adequate annual contribution to the Union shall, with the recommendation of the Executive Committee, be submitted to the General Assembly for decision.

#### B. *Members*

8. Individuals proposed for Union membership should, as a rule, be chosen from among astronomers and scientists, whose activity is closely linked with astronomy taking into account

(a) the standard of their scientific achievement
(b) the extent to which their scientific activity involves research in astronomy
(c) their desire to assist in the fulfilment of the aims of the Union.

9. Young astronomers should be considered eligible for membership after they have shown their capability (as a rule Ph.D. or equivalent) of and experience (some years of successful activity) in conducting original research.

10. For full-time professional astronomers the achievement in astronomy may consist either of original research or of substantial contributions to major observational programmes.

11. Les autres personnes ne peuvent devenir Membres de l'Union que si certains de leurs travaux originaux concernent étroitement la recherche astronomique.

12. Huit mois avant une Assemblée Générale ordinaire, il sera demandé aux organismes adhérents de proposer de nouveaux Membres. Les propositions devront parvenir au Secrétaire Général au moins cinq mois avant la première session de l'Assemblée Générale. Les propositions reçues après cette date limite ne seront prises en considération que si des circonstances exceptionnelles justifient le retard.

13. Chaque proposition du nouveau Membre doit être présentée séparément et indiquer le nom, les prénoms et l'adresse postale du candidat (de préférence celle de son Institut ou Observatoire), ses date et lieu de naissance, l'Université devant laquelle il a soutenu sa thèse ou le diplôme équivalent, la date de soutenance, la situation actuelle du candidat, les titres et renseignements bibliographiques de deux ou trois de ses articles ou publications les plus significatifs et, s'il y a lieu, tous les renseignements susceptibles d'être pris en considération par le Comité des Nominations.

14. (a) Les Présidents de Commissions qui désirent suggérer de nouveaux Membres doivent adresser leurs suggestions au Secrétaire Général au moins cinq mois avant la première session d'une Assemblée Générale ordinaire. Les propositions devront fournir les mêmes renseignements que ceux mentionnés à l'article 13.
(b) Le Secrétaire Général fait part de ces suggestions aux organismes adhérents intéressés.

15. Le Secrétaire Général préparera deux listes pour le Comité des Nominations
(a) l'une contenant les noms des candidats proposés par les organismes adhérents
(b) l'autre contenant les noms des candidats proposés par les Présidents de Commissions, mais qui ne sont pas déjà inclus dans les propositions des organismes adhérents.

16. A partir des deux listes mentionnées à l'article 15, le Comité des Nominations prépare les propositions définitives de nouveaux Membres de l'Union.

17. Les organismes adhérents peuvent proposer la radiation de Membres ayant abandonné le domaine de l'astronomie pour d'autres activités, à moins qu'ils ne continuent à apporter une contribution à l'astronomie. Ces propositions doivent être portées à la connaissance du Secrétaire Général et du Membre concerné.

18. Le Secrétaire Général publiera la liste alphabétique des Membres de l'Union dans les Transactions de chaque Assemblée Générale ordinaire.

## III. MEMBRES DES COMMISSIONS

19. Les membres des Commissions de l'Union sont cooptés par les Commissions. Cette procédure est régie par des règles établies par les Commissions elles-mêmes.

20. Les Commissions devraient choisir, ou approuver, la liste des membres de leurs commissions compte tenu de la spécialité de ces personnes, en particulier de leur activité scientifique dans le domaine de recherches de la Commission, et de leur contribution au travail de la Commission. Elles peuvent

(a) inviter des Membres de l'Union à devenir membres de la Commission
(b) radier les membres de la Commission qui n'ont pas contribué à son activité
(c) accepter ou refuser les demandes présentées par des Membres de l'Union, ou par des personnes proposées comme tels, en vue d'appartenir à la Commission
(d) suggérer l'élection comme Membres de l'Union de personnes n'y appartenant pas, ce qui leur permettrait alors de devenir membres de la Commission.

11. Others are eligible for membership only if they are making original contributions closely linked with astronomical research.

12. Eight months before an ordinary General Assembly, adhering bodies will be asked to propose new Members. The proposals should reach the General Secretary not later than five months before the first session of the General Assembly. Proposals received after the closing date will only be taken into consideration if the delay is justified by exceptional circumstances.

13. Each proposal shall be written separately. It should include the name, first names and postal address of the candidate, preferably that of his Institute or Observatory, his place and date of birth, the University and the year of his Ph.D. or equivalent title, his present occupation, titles and bibliographic data of two or three of his more important papers or publications, and details, if any, worthy to be considered by the Nominating Committee.

14. (a) Presidents of Union Commissions wishing to suggest new Members for admittance should address their suggestions to the General Secretary five months before the first session of an ordinary General Assembly. The proposals should contain particulars as in article 13.

(b) The General Secretary notifies the adhering bodies in questions of such suggestions.

15. The General Secretary shall prepare two lists for the Nominating Committee

(a) one containing the candidates proposed by the adhering bodies,
(b) the other containing those suggested by Presidents of Commissions, but not included among the proposals of adhering bodies.

16. The Nominating Committee prepares the final proposals for Union membership from the two lists as mentioned in article 15.

17. Adhering bodies should propose cancellation of Members who have left the field of astronomy for other interests, unless they continue to contribute to astronomy. Such proposals should be announced to the Member concerned and to the General Secretary.

18. The alphabetical list of Union Members will be published by the General Secretary in the Transactions of each ordinary General Assembly.

## III. COMMISSION MEMBERSHIP

19. Members of Union Commissions are co-opted by Commissions. The rules governing the procedure of such co-option are drawn up by the Commissions themselves.

20. Commissions should choose, or approve of, Commission members taking into account their special interests, in particular their scientific activity in the appropriate fields of research and their contribution to the work of the Commission. They may

(a) invite Members to become members of their Commission
(b) remove members who have not contributed to the work of the Commission
(c) accept or reject applications for membership from existing or proposed Members
(d) suggest non-Members for election as Members, thus enabling them to become members of the Commission.

21. Les Membres de l'Union ne peuvent pas, en règle générale, appartenir à plus de trois Commissions.

22. Les Membres de l'Union peuvent demander à être admis dans une Commission en écrivant au Président de cette Commission. Ils ne devraient faire cette demande que si leur propre activité rentre dans le cadre des recherches de la Commission et s'ils sont décidés à contribuer au travail de la Commission.

23. Les membres des Commissions peuvent se retirer d'une Commission en écrivant à son Président.

24. En envoyant leurs propositions de nouveaux Membres, les organismes adhérents peuvent également suggérer le choix d'une Commission pour chaque candidat.

25. Le Secrétaire Général enregistera et analysera la liste des membres des Commissions; si cela est nécessaire, il tentera de trouver une solution aux anomalies évidentes.

26. Le Secrétaire Général publiera la liste des membres des Commissions dans les Transactions de chaque Assemblée Générale ordinaire.

## IV. CONSULTANTS

27. Peuvent être élus Consultants des personnes qui ne sont pas astronomes, mais qui sont susceptibles de servir les intérêts de l'astronomie.

28. Les Commissions doivent en principe envoyer, pour approbation, leurs propositions de consultants au Secrétaire Général au moins cinq mois avant la première session d'une Assemblée Générale ordinaire.

29. Le Secrétaire Général préparera une liste des personnes proposées comme consultants et la soumettra pour approbation au Comité Exécutif.

30. Le Bureau Administratif établira une liste alphabétique des consultants.

31. Les consultants peuvent participer aux réunions de l'Union. Ils peuvent avoir droit de vote dans leurs Commissions respectives. Ils reçoivent gratuitement le Bulletin d'Information de l'Union.

## V. REUNIONS SCIENTIFIQUES

32. Le Secrétaire Général publiera un règlement pour les réunions scientifiques organisées ou parrainées par l'Union.

## VI. CONTACTS EXTERIEURS

33. Aucune relation avec des tiers, imputable à l'Union, ne sera entreprise par quiconque membre de l'Union, si ce n'est sous l'autorité du Secrétaire Général.

34. Les représentants de l'Union dans d'autres organisations, en particulier les Comités de l'ICSU et les Commissions Inter-Unions, seront désignés par le Comité Exécutif. Les noms sont proposés par les Présidents des Commissions concernées.

35. Les dépenses encourues par les représentants de l'Union dans d'autres organisations seront remboursées à la discrétion du Secrétaire Général, dans les limites du Budget adopté par l' Assemblée Générale. Les représentants sont priés d'obtenir l'accord préalable du Secrétaire Général avant d'engager ces dépenses.

21. Members may not, as a rule, be members of more than three Commissions.

22. Members may apply for Commission membership by writing to the President of the Commission concerned. Such applications should only be made if the Member is actively engaged in the appropriate field of research and is prepared to contribute to the work of the Commission.

23. Members of Commissions may resign from a Commission by writing to its President.

24. Adhering bodies, in sending in their proposals for new Members, may also suggest one Commission for each candidate.

25. The General Secretary will record and analyse the list of members of Commissions; if necessary he will try to resolve any outstanding anomalies.

26. The list of Commission members will be published by the General Secretary in the Transactions of each ordinary General Assembly.

## IV. CONSULTANTS

27. Eligible as Consultants are non-astronomers in a position to further the interest in astronomy.

28. Proposals of Commissions for the approval of consultants should, as a rule, reach the General Secretary not later than five months before the first session of an ordinary General Assembly.

29. The General Secretary shall prepare a list of those proposed for admission as consultants and submit it to the Executive Committee for approval.

30. The Administrative Office will maintain an alphabetical list of consultants.

31. Consultants may participate in the meetings of the Union. They may have voting right in the respective Commission. They receive, free of charge, the Information Bulletin of the Union.

## V. SCIENTIFIC MEETINGS

32. The General Secretary shall publish rules for scientific meetings organized or sponsored by the Union.

## VI. EXTERNAL CONTACTS

33. No dealings with third parties, attributable to the Union, shall be undertaken by any Member of the Union except on the authority of the General Secretary.

34. Representatives of the Union in other bodies, especially ICSU Committees and ICSU Inter-Union Committees, shall be appointed by the Executive Committee. Nominations are sought from Presidents of appropriate Commissions.

35. Expenses incurred by Representatives of the Union in other bodies will be reimbursed at the discretion of the General Secretary, within the provisions of the Budget Estimate adopted by the General Assembly. Representatives are required to obtain prior approval of the General Secretary before incurring such expenses.

## VII. ASSEMBLEES GENERALES

36. Huit mois avant l'Assemblée Générale, le Secrétaire Général envoie aux Comités Nationaux d'Astronomie et aux organisations adhérentes le budget préparé par le Comité Exécutif, pour commentaires.

## VII. GENERAL ASSEMBLIES

36. The General Secretary distributes the budget prepared by the Executive Committee to National Committees of Astronomy and/or Adhering Organizations for comments eight months before the General Assembly.

## 10. RULES FOR SCIENTIFIC MEETINGS

These rules are published in Transactions Vol. XVIIB, pp. 330-337.

## 11. LIST OF COUNTRIES ADHERING TO THE UNION

The following is the list of 49 countries that adhere to the Union, giving the Adhering Organization, the year of admission, and the approximate number of individual members.

| Country | Adhering Organization | Year | Members |
|---|---|---|---|
| ARAB REPUBLIC OF EGYPT | Academy of Scientific Research and Technology<br>Dept. of Scientific Societies and International Unions<br>101, Kast El-Einy Street<br>Cairo | 1925 | 33 |
| ARGENTINA | Comité Nacional de Astronomia<br>Observatorio Astronomico<br>Laprida 854<br>5.000 Cordoba | 1927 | 41 |
| AUSTRALIA | Australian Academy of Sciences<br>P.O. Box 783<br>Canberra City, ACT 2601 | 1939 | 107 |
| AUSTRIA | Bundesministerium für Unterricht und Kunst in Wien<br>Minoritenplatz 5<br>A-1014 Wien | 1955 | 26 |
| BELGIUM | Académie royale de Belgique<br>Palais des Académies<br>Rue Ducale 1<br>B-1000 Bruxelles | 1920 | 82 |
| BRAZIL | Conselho Nacional de Desenvolvimento Científico e Tecnológico - CNPq<br>Av. W 3 Norte, Q. 507/B<br>Caixa Postal 11-1142<br>70740 Brasilia, DF | 1961 | 29 |
| BULGARIA | Bulgarian Academy of Sciences<br>7th November Street 1<br>1000 Sofia | 1957 | 27 |
| CANADA | National Research Council of Canada<br>International Relations<br>Ottawa, Ontario, K1A OR6 | 1920 | 162 |
| CHILE | Universidad de Chile<br>Observatorio Astronomico Nacional<br>Casilla 36-D<br>Santiago de Chile | 1947 | 16 |

| Country | Adhering Organization | Year | Members |
|---|---|---|---|
| CHINA | Chinese Astronomical Society<br>Purple Mountain Observatory<br>Academia Sinica<br>Nanking | 1935 | 93 |
| | Astronomy Union located at Taipei, China<br>Taipei Observatory, Taipei, Taiwan | 1959 | 16 |
| COLOMBIA | Universidad Nacional de Columbia<br>Observatorio Astronomico Nacional<br>Apartado Aereo 2584<br>Bogota D.E. | 1967 | 3 |
| CUBA | Academia de Ciencias de Cuba<br>Capitolio Nacional<br>La Habana | 1970 | 0 |
| CZECHOSLOVAKIA | Czechoslovak Academy of Sciences<br>Narodnî 3<br>Praha 1 | 1922 | 62 |
| DENMARK | Det Kongelige Danske Videnskabernes Selskab<br>The Royal Danish Academy of Sciences and Letters<br>H.C. Andersens Boulevard 35<br>DK-1553 København V | 1922 | 40 |
| FINLAND | Académie des Sciences et Lettres<br>Snellmaninkatu 9-11<br>Helsinki 17 | 1948 | 22 |
| FRANCE | Académie des Sciences<br>Institut de France<br>23, quai de Conti<br>F-75006 Paris | 1920 | 389 |
| GERMANY, D.R. | Akademie der Wissenschaften der DDR<br>Otto-Nuschke-Straße 22/23<br>DDR-108 Berlin | 1951 | 43 |
| GERMANY, F.R. | Rat Westdeutscher Sternwarten<br>c/o Prof. Dr. R. Kippenhahn<br>MPI für Physik und Astrophysik<br>Institut für Astrophysik<br>Karl-Schwarzschild-Straße 1<br>D-8046 Garching bei München | 1951 | 260 |
| GREECE | Academy of Athens<br>Panepistimiou Str.<br>Athens | 1920 | 61 |
| HUNGARY | Hungarian Academy of Sciences<br>Roosevelt Tér 9<br>Budapest V | 1947 | 29 |

| Country | Adhering Organization | Year | Members |
|---|---|---|---|
| INDIA | Indian National Science Academy<br>Bahadur Shah Zafar Marg<br>New Delhi 110002 | 1964 | 100 |
| INDONESIA | National Institute for Sciences<br>Tenku Cik Dithiro 43<br>Jakarta | 1979 | 2 |
| IRAN | Institute of Geophysics<br>Tehran University<br>Amirabad Shomali<br>Tehran | 1969 | 3 |
| IRAQ | Space & Astronomical Research Centre<br>Council for Scientific Research<br>P.O. Box 255<br>Baghdad | 1976 | 3 |
| IRELAND | National Committee of Astronomy<br>The Royal Irish Academy<br>19 Dawson Street<br>Dublin 2 | 1947 | 23 |
| ISRAEL | The Israel Academy of Sciences<br>and Humanities<br>Albert Einstein Square, Talbieh<br>P.O. Box 4040<br>Jerusalem 91 040 | 1954 | 36 |
| ITALY | Consiglio Nazionale delle Ricerche<br>Piazale delle Scienze 7<br>I-00100 Rome | 1921 | 233 |
| JAPAN | Science Council of Japan<br>22-34 Roppongi 7 chome<br>Minato-Ku<br>Tokyo 106 | 1920 | 245 |
| KOREA (DPR) | Academy of Sciences<br>Pyongyang | 1961 | 3 |
| KOREA (REPUBLIC) | Korean National Astronomical<br>Observatory<br>Yoksam-Dong<br>Kangnam-ku<br>Seoul, 134-03 | 1973 | 8 |
| MEXICO | Instituto de Astronomía UNAM<br>Apartado Postal 70-264<br>Mexico D.F. | 1921 | 32 |
| NETHERLANDS | Koninklijke Nederlandse Akademie<br>van Wetenschappen<br>Kloveniersburgwal 29<br>1011 JV Amsterdam | 1922 | 134 |

| Country | Adhering Organization | Year | Members |
|---|---|---|---|
| NEW ZEALAND | The Royal Society of New Zealand<br>Private Bag<br>Wellington | 1964 | 18 |
| NORWAY | Det Norske Videnskaps-Akademi i Oslo<br>Drammensveien 78<br>Oslo 2 | 1922 | 18 |
| POLAND | Polska Akademia Nauk<br>Palac Kultury i Nauki<br>Warsaw | 1922 | 70 |
| PORTUGAL | Secção Portuguesa das Uniões Inter-<br>nacionais Astronómica e Geodésica<br>e Geofîsica<br>Praça da Estrela<br>Lisboa 2 | 1924 | 15 |
| ROUMANIA | Roumanian National Committee of Astronomy<br>Astronomical Observatory<br>Str. Cutitul de Argint 5<br>75212 Bucarest, P.O. 28 | 1928 | 17 |
| SOUTH AFRICA | International Relations Division<br>Council for Scientific and<br>Industrial Research Services<br>P.O. Box 395<br>Pretoria 0001 | 1938 | 19 |
| SPAIN | Comision Nacional de Astronomía<br>Instituto Geografico y Cadastral<br>General Ibanez 3<br>Madrid (3) | 1922 | 43 |
| SWEDEN | Kungl. Vetenskapsakademien<br>Box 50005<br>S-104 05 Stockholm | 1925 | 66 |
| SWITZERLAND | Schweizerische Naturforschende Gesellschaft<br>Zentralsekretariat<br>Laupenstraße 10<br>CH-3000 Bern | 1923 | 46 |
| TURKEY | Türk Astronomi Dernegi<br>University Rasathanesi<br>Beyazit, Istanbul | 1961 | 43 |
| UNITED KINGDOM | The Royal Society<br>6, Carlton House Terrace<br>London SW1Y 5AG | 1920 | 461 |
| URUGUAY | Ministerio de Educacion y Cultura<br>8 de Octubre 3255<br>Montevideo | 1970 | 0 |

| Country | Adhering Organization | Year | Members |
|---|---|---|---|
| U.S.A. | National Academy of Sciences<br>Commission on International Relations<br>2101 Constitution Avenue N.W.<br>Washington DC 20148 | 1920 | 1527 |
| U.S.S.R. | Academy of Sciences of the USSR<br>Leninskij Prospekt 14<br>Moscow 71 | 1935 | 418 |
| VATICAN CITY STATE | Pontificia Academia delle Scienze<br>Citta del Vaticano | 1932 | 6 |
| VENEZUELA | Asociacion Venezolana de Astronomia<br>Apartado 264<br>Mérida 5101 | 1953 | 10 |
| YUGOSLAVIA | Savez društva matematičara,<br>fizičara i astronoma Jugoslavije<br>Institut za matematiku i fiziku<br>Cetinjski put bb<br>81000 Titograd | 1935 | 30 |

# VII.
# Membership

## 1. MEMBERSHIP OF COMMISSIONS

COMMISSION NO 4.

EPHEMERIDES (EPHEMERIDES)

PRESIDENT: T. Lederle

VICE-PRESIDENT: B.Morando

COMITE D'ORGANISATION: V.K.Abalakin, S.Aoki, R.L.Duncombe, J.H.Lieske, A.Orte, P.K.Seidelmann, A.M.Sinzi, G.A.Wilkins, B.D.Yallop.

MEMBRES: J.Arias-de-Greiff, A.Bandyopadhyay, A.Bec-Borsenberger, P.Bretagnon, J.Chapront, F.Chollet, A.Deprit, D.W.Dunham, B.Emerson, A.D.Fiala, W.Fricke, M.A.Fursenko, N.I.Glebova, F.Gondolatsch, R.F.Haupt, J.Henrard, P.M.Janiczek, K.J.Johnston, R.W.King Jr., H.Kinoshita, W.J.Klepczynski, B.Kolaczek, Y.Kubo, N.C.Lahiri, L.V.Morrison, I.I.Mueller, C.Oesterwinter, D.A.O'Handley, R.D.Reasenberg, H.Schwan, I.I.Shapiro, A.S.Sochilina, E.M.Standish Jr., G.E.Taylor, Y.T.Ting, T.C.Van Flandern, H.B.Wackernagel, J.G.Williams, G.M.R.Winkler, A.Yamazaki, X.-Z.Zhao.

COMMISSION NO 5.

DOCUMENTATION ET DONNEES ASTRONOMIQUES (DOCUMENTATION AND ASTRONOMICAL DATA)

PRESIDENT: W.D.Heintz

VICE-PRESIDENT: G.A.Wilkins

COMITE D'ORGANISATION: O.B.Dluzhnevskaya, B.Hauck, C.Jaschek, P.Lantos, J.-C.Pecker, L.Schmadel, C.E.Worley.

MEMBRES: H.A.Abt, N.Baker, W.P. Bidelman, J.Bouska, R.J.Davis, M.S.Davis, D.W.Dewhirst, R.S.Dixon, R.L.Duncombe, W.Fricke, R.H.Garstang, R.F.Griffin, A.Heck, I.Heinrich, V.V.Ivanov, Z.Kadla, J.Kleczek, T.Lederle, J.Lequeux, M.C.Lortet, G.Lynga, D.Ya.Martynov, A.Maxwell, D.N.McNamara, J.M.Mead, A.J.Meadows, P.Mein, A.F.J.Moffat, S.A.Mitton, S.Nishimura, F.Ochsenbein, K.F.Ogorodnikov, L.E.Pasinetti, A.G.D.Philip, L.N.Radlova, L.Remy-Battiau, P.F.Renson, J.Sahade, K.H.Schmidt, J.R.Shakeshaft, I.S.Shcherbina-Samojlova, F.Spite, Y.Terashita, S.B.Tritton, A.G.Velghe, H.H.Voigt, W.H.Warren Jr., P.T.Wallace, P.A.Wayman, V.Weidemann, G.Westerhout.

COMMISSION NO 6.

TELEGRAMMES ASTRONOMIQUES (ASTRONOMICAL TELEGRAMS)

PRESIDENT: M.P.Candy

VICE-PRESIDENT: A.Mrkos

Sans Comite d'Organisation

MEMBRES: F.Biraud, C.U.Cesco, L.E.Cunningham, E.Everhart, J.E.Grindlay, J.Hers, Y.Kozai, B.G.Marsden, D.YA.Martynov, K.A.Pounds, E.Roemer, L.Rosino, P.Simon.

COMMISSION NO 7.

MECANIQUE CELESTE (CELESTIAL MECHANICS)

PRESIDENT: J.Kovalevsky

VICE-PRESIDENT: V.A.Brumberg

COMITE D'ORGANISATION: E.P.Aksenov, J.Chapront, S.Ferraz-Mello, J.Hadjidemetriou, J.Henrard, W.H.Jeffreys, Y.Kozai, H.Scholl, P.K.Seidelmann, A.Sinclair.

MEMBRES: V.K.Abalakin, K.Aksnes, C.A.Altavista, G.Antonacopoulos, S.Aoki, B>B>Baghos, G.Balmino, J.M.Batrakov, A.Bec-Borsenberger, D.Benest, D.G.Bettis, K.B.Bhatnagar, G.Bozis, E.Brieva, C.J.Brookes, R.Broucke, O.Calame, M.P.Candy, P.J.Cefola, R.P.Cid, G.Contopoulos, A.H.Cook, C.C.Counselman, L.E.Cunningham, J.M.A.Danby, M.S.Davis, S.Debarbat, V.G.Demin, J.De Orus, A.Deprit, J.R.Dormand, G.N.Duboshin, R.L.Duncombe, R.Dvorak, H.K.Eichhorn, P.E.Elyasberg, N.V.Emelianov, B.Erdi, E.Everhart, H.Fabre, P.Farinella, A.Fiala, C.Froeschle, I.V.Galibna, E.M.Gaposchkin, B.Garfinkel, S.Gaska, G.E.O.Giacaglia, P.Goldreich, C.L.Goudas, E.A.Grebenikov, R.J.Greenberg, N.P.Grushinskij, S.E.Hamid, D.C.Heggie, Y.E.Helali, M.Henon, G.Hori, V.A.Izvekov, P.M.Janiczek, A.H.Jupp, D.M.Katsis, W.H.Kaula, K.V.Kholshevnikov, D.G.King-Hele, H.Kinoshitta, P.Kustaanheimo, P.Lala, J.P.Lazovic, J.H.Lieske, C.A.Lundquist, N.C.Magnaradze, C.Marchal, V.Markellos, B.G.Marsden, V.Matas, A.G.Mavraganis, J.Meffroy, R.Meire, W.G.Melbourne, G.A.Merman, J.Message, F.Mignard, A.Milani, O.C.Mohler, M.B.Moons, B.Morando, J.D.Mulholland, P.Musen, V.F.Myachin, P.E.Nacozy, F.Nahon, B.N.Noskov, V.S.Novoselov, D.A.O'Handley, C.Oesterwinter, T.B.Omarov, A.A.Orlov, S.J.Peale, M.S.Petrovskaya, D.A.Pierce, B.Popovic, W.J.Robinson, A.E.Roy, J.A.Ryabov, J.L.Sagnier, P.Sconzo, J.Schubart, L.Sehnal, W.Sessin, I.I.Shapiro, S.G.Sharaf, K.A.Shteins, M.Sidlichovsky, C.L.Siegel, Z.Sima, A.T.Sinclair, J.W.Siry, E.M.Standish, I.Stellmacher, G.F.Sultanov, V.Szebehely, S.K.Tatevian, M.Y.Tawadrous, D.B.Taylor, Y.Thiry, F.Tong, CH.Varvoglis, S.N.Vashkov'yak, R.Vilhena de Moraes, J.P.Vinti, C.A.Williams, M.S.Yarov-yarovoj, Z.-H.Yi, J.Yoshida, K.Zare, X.-Z.Zhao.

COMMISSION NO 8.

ASTRONOMIE DE POSITION (POSITIONAL ASTRONOMY)

PRESIDENT: J.A.Hughes

VICE-PRESIDENT: Y.Requieme

COMITE D'ORGANISATION: G.Billaud, G.Carrasco, W.Fricke, V.S.Gubanov, E.Hoeg, B.L.Klock, J.A.Lopez, D.-J.Luo, M.Miyamoto, L.V.Morrison, I.Nikoloff, G.Teleki, Y.S.Yatskiv.

MEMBRES: C.A.Anguita, J.G.Argyracos, P.Bacchus, D.C.Backer, B.K.Bagildinskij, J.Bem, P.Benevides-Soares, R.Bien, R.L.Branham, W.N.Brouw, M.F.Bykov, R.A.Carestia, J.M.Chamberlain, N.A.Chernega, F.Chlietovsky, F.Chollet, T.E.Corbin, C.C.Counselman(III), T.Dambara, S.Debarbat, C.De Vegt, R.J.Dejaiffe, A.F.Dravskikh, D.P.Duma, R.L.Duncombe, H.K.Eichhorn, R.T.Fedorova, M.Feissel, H.-J.Fogh Olsen, V.A.Fomin, F.S.Gauss, W.Gliese, P.Grudler, A.P.Gulyaev, W.D.Heintz, L.Helmer, P.D.Hemenway, N.-S.Hu, K.Hurukawa, P.Jackson, K.Johnston, A.S.Kharin, Y.L.Kokurin, V.V.Konin, G.S.Kosin, J.Kovalevsky, F.Laclare, P.Lacroute, S.Laustsen, T.Lederle, D.-M.Li, N.-Y.Li, L.Lindegren, P.Loyola, W.T.Manrique, E.Marcus, L.N.Mavridis, P.Melchior, L.A.Mitic, C.A.Murray, A.I.Nefed'eva, A.A.Nemiro, F.Noel, J.J.S.P.Osorio, G.M.Petrov, J.Pham-Van, K.Pilowski, G.I.Pinigin, V.V.Podobed, G.Polnitzky, D.D.Polozhentsev, A.Poma, E.Proverbio, L.Quijano, E.Raimond, A.Reiz, W.H.Robertson,

S.Roeser, J.-M.Rousseau, J.L.Russell, I.Rusu, S.Sadzakov, M.Sanchez, K.Sato, F.Schmeidler, H.Schwan, B.M.Sevarlic, K.P.Sims, C.A.Smith, L.Stange, R.Stone, D.V.Thomas, R.H.Tucker, P.T.Wallace, H.G.Walter, H.W.Wood, T.-Q.Xu, A.Yamazaki, H.Yasuda, S.-H.Ye, M.Yoshizawa, L.Yu, M.S.Zverev.

COMMISSION NO 9.

INSTRUMENTS ASTRONOMIQUES (ASTRONOMICAL INSTRUMENTS)

PRESIDENT: W.C.Livingston

VICE-PRESIDENT: C.M.Humphries

COMITE D'ORGANISATION: J.C.Bhattacharyya, J.Davis, J.L.Heudier, I.M.Kopylov, G.LeLievre, J.P.Maillard, A.B.Meinel, N.N.Mikhel'son, E.H.Richardson, J.Ring, J.Rosch, N.Steshenko, R.G.Tull, M.F.Walker, W.L.Wilcock, G.Wlerick.

MEMBRES: H.D.Ables, C.Aime, T.R.Albrecht, P.D.Atherton, H.W.Babcock, A.Baranne, W.A.Baum, R.Beer, R.G.Bingham, J.Borgnino, J.Brault, J.B.Breckinridge, I.I.Breido, W.B.Burton, B.Campbell, P.Charvin, D.Clarke, D.L.Crawford, D.G.Currie, J.N.Desai, O.D.Dokuchaeva, D.Dravins, M.Duchesne, T.Dunham Jr., L.Dunkelman, D.Enard, O.Engvold, C.Fehrenbach, J.M.Fletcher, A.F.Fomenko, W.K.Ford Jr., F.S.Gauss, J.Gay, P.R.Gillingham, I.S.Glass, V.M.Grigorjev, W.Grundmann, D.Gutke, C.F.W.Harmer, D.L.Harmer, H.M.Heckathorn, A.Hewitt, N.-S.Hu, E.J.Hysom, B.K.Ioannisiani, A.P.Jayarajan, S.Jeffers, J.V.Jelley, H.Jenkner, B.Jones, R.D.Joseph, I.D.Karachentsev, T.A.Kipper, K.E.Kissell, B.L.Klock, P.Koehler, Y.P.Korovyakovskij, B.J.Kovachev, T.J.N.Kreidel, C.Kuehne, A.Labeyrie, B.E.Lasker, M.Lelyer, G.R.Lemaitre, D.H.Martins, J.D.Mcgee, D.Mcmullan, L.Mertz, A.G.Millikan, B.L.Morgan, D.C.Morton, S.S.Murray, Y.Nakai, J.Nelson, V.B.Nikonov, S.Nishimura, R.S.Nunes, C.R.O'Dell, G.Odgers, A.J.Penny, A.D.Petford, J.D.Pope, C.J.Pritchet, V.V.Prokof'eva, A.Th.Purgathofer, R.Racine, K.D.Rakos, N.K.Reay, L.B.Robinson, C.Roddier, F.Roddier, L.Rusconi, V.S.Rylov, C.Sanchez Magro, G.V.Schultz, J.D.Schumann, B.Servan, P.V.Shcheglov, K.Shivanandan, H.J.Smith, M.J.Smyth, J.W.Storey, D.-G.Su, J.-P.Swings, H.Tueg, B.L.Ulich, B.Valnicek, G.W.Van Citters, F.J.Vrba, A.R.Walker, G.A.H.Walker, P.T.Wallace, D.C.Wells, R.M.West, J.A.Westphal, H.Woehl, S.P.Worden, X.-L.Wu, A.Wyller, Ch.G.Wynne, S.-J.Yang, B.Zealey.

COMMISSION NO 10.

ACTIVITE SOLAIRE (SOLAR ACTIVITY)

PRESIDENT: E.Tandberg Hanssen

VICE-PRESIDENT: M.Pick

COMITE D'ORGANISATION: A.Bhatnagar, V.Gaizauskas, E.A.Gurtovenko, E.Machado, D.B.Melrose, E.Priest, K.Tanaka, H.Zirin.

MEMBRES: A.R.Abbasov, A.Abrami, H.S.Ahluwalia, C.Alissandrakis, R.C.Altrock, M.D.Altschuler, M.K.M.Aly, P.Ambroz, K.A.Anderson, A.Antalova, S.K.Antiochos, R.G.Athay, Y.Avignon, E.Balli, V.Banin, C.H.Barrow, J.R.Bates, J.M.Beckers, B.Bell, G.Belvedere, A.O.Benz, D.E.Billings, A.D.Bonov, J.-L.Bougeret, R.Bray, J.C.Brown, G.Brueckner, E.Bruener, A.Bruzek, V.Bumba, C.J.Cannon, P.Carlqvist, G.A.Chapman, B.Chen, Chungchien Cheng, V.E.Chertoprud, E.L.Chupp, M.Cimino, H.E.Coffey, R.Coutrez,

A.E.Covington, C.J.Crannell, J.L.Culhane, I.K.Czada, K.P.Dare, A.Datlowe, T.de Groot, C.de Jager, B.R.Dennis, K.P.Dere, F.L.Deubner, L.Dezso, Y.-J.Ding, M.Dizer, H.Dodson-Prince, A.Dollfus, M.Dryer, E.E.Dubov, G.A.Dulk, R.A.Duncan, R.B.Dunn, J.Eddy, G.H.Elste, O.Engvold, W.Erickson, R.Falciani, F.Farnik, W.A.Feibelman, K.-J.Feng, A.D.Fokker, T.Fortini, E.Fossat, H.Friedman, L.Fritzova-Svestkova, A.H.Gabriel, D.Galloway, T.E.Gergely, G.B.Gel'frejkh, R.L.Gilliland, P.A.Gilman, R.G.Giovanelli, W.Gleissberg, R.S.Gnevysheva, G.Godoli, M.H.Gokhale, S.T.Gopasjuk, R.E.Guseinov, J.P.Hagen, J.Hanasz, R.T.Hansen, J.Harvey, E.Haug, J.Hayward, E.R.Hedeman, E.Hiei, E.Hildner, T.Hirayama, T.E.Holzer, R.Howard, G.J.Hurford, C.Hyder, B.Ioshpa, V.I.Ivanchuk, F.W.Jaeger, J.Jakimiec, E.Jensen, K.Jockers, H.P.Jones, O.Kaburaki, K.Kai, B.Kalman, M.Kanno, M.O.Kjeldseth, J.Kleczek, M.Klvana, S.Knoska, A.Koeckelenberg, M.Kopecky, S.Koutchmy, A.Kovacs, N.I.Kozhevnikov, V.A.Krat, F.Krause, L.Krivsky, A.Krueger, J.Kubota, G.V.Kuklin, M.R.Kundu, H.Kunzel, M.Kuperus, H.Kurokawa, L.N.Kurochka, P.B.Landecker, D.A.Landman, J.Leibacher, J.-L.Leroy, X.-Q.Li, J.V.Lincoln, M.A.Livshits, R.E.Loughhead, B.-C.Low, R.MacQueen, C.J.Macris, M.Makita, H.H.Malitson, P.Maltby, J.M.Malville, S.L.Mandelshtam, M.J.Martres, O.T.Matsuura, W.Mattig, A.Maxwell, M.McCabe, S.M.P.McKenna Lawlor, D.McLean, P.Mein, J.Mergentaler, A.G.Michalitsianos, R.Michard, E.I.Mogilevskij, O.C.Mohler, I.G.Moiseev, G.E.Moreton, F.Moriyama, N.N.Morozhenko, S.Motta, R.Muller, E.R.Mustel, S.Nagasawa, Y.Nakagawa, O.Namba, G.J.Nelson, W.M.Neupert, G.Newkirk Jr., K.Nishi, R.W.Noyes, H.Nussbaumer, V.A..Obridko, K.Ohki, F.Q.Orrall, J.H.Parkinson, W.H.Parkinson, L.Paterno, B.M.Pedersen, K.Pflug, K.J.H.Phillips, J.-P.Picat, J.H.Piddington, G.W.Pneuman, P.N.Polupan, M.A.Raadu, A.P.Rao, J.Rayole, D.E.Rees, E.M.Reeves, E.J.Rhodes Jr., M.Rigutti, B.Roberts, R.D.Robinson, M.Roemer, P.R.Romanchuk, B.Rompolt, J.Rosch, I.W.Roxburgh, J.P.Rozelot, B.M.Rubashev, V.Rusin, D.Rust, V.Ruzdjak, B.Ruzickova-Topolova, M.Rybanski, K.Saito, K.Sakurai, C.Sawyer, K.H.Schatten, A.Schluter, E.J.Schmahl, B.Schmieder, H.U.Schmidt, S.A.Schoolman, E.H.Schroeter, M.Semel, R.Servajean, A.B.Severny, A.H.Shapley, M.A.Shea, N.R.Sheeley Jr., R.A.Shine, R.Silberberg, G.F.Sitnik, E.M.Slonim, D.F.Smith, E.V.P.Smith, Henry.J.Smith, G.Smolkov, D.S.Spicer, H.C.Spruit, G.Stellmacher, J.O.Stenflo, N.N.Stepanian, V.E.Stepanov, N.V.Steshenko, M.Stix, P.A.Sturrock, J.Suda, Z.Suemoto, Z.Svestka, J.Sykora, T.Takakura, R.Talon, T.Tamenaga, H.Tanaka, R.G.Teske, E.Tifrea, A.Tlamicha, M.Trellis, B.Tritakis, R.Treumann, J.Tuominen, Y.Uchida, J.H.Underwood, B.Valnicek, J.A.Van Allen, G.Van Hoven, F.Van't Veer, A.A.Vaughan, D.Venkatesan, S.K.Vinod, J.Vorpahl, J.C.Vial, Ju.I.Vitinskij, G.F.Vyalshin, M.Waldmeier, D.G.Wentzel, E.Wiehr, J.Wilcox, J.P.Wild, P.Wilson, A.Wittman, H.Wohl, L.Woltjer, Shi-Tsan Wu, J.Xanthakis, N.A.Yakovkin, S.-H.Ye, H.Yoshimura, R.A.Zappala, A.Zelenka, B.-R.Zhang, H.-Q.Zhang, Z.-D.Zhang, Y.D.Zhugzhda, P.Zlobec, Y.-X.Zou, C.Zwaan.

## COMMISSION NO 12.

## RADIATION ET STRUCTURE DE L'ATMOSPHERE SOLAIRE
## (RADIATION AND STRUCTURE OF THE SOLAR ATMOSPHERE)

PRESIDENT: R.W.Noyes

VICE-PRESIDENT: M.Kuperus

COMITE D'ORGANISATION: B.Chen, F.L.Deubner, E.Fossat, J.W.Harvey, V.A.Kotov, J.Leibacher, K.R.Sivaraman, J.O.Stenflo, Y.Uchida, P.R.Wilson.

MEMBRES: L.Acton, M.G.Adam, C.Aime, C.E.Alissandrakis, C.W.Allen, R.C.Altrock, M.Altschuler, H.Ando, G.M.R.Ansari, R.G.Athay, D.B.Beard, J.M.Beckers, J.E.Beckman, N.Bel, G.Benford, A.Bhatnagar, J.C.Bhattacharyya, D.E.Billings, D.E.Blackwell, J.E.Blamont, K.M.Bohm, E.Bohm-Vitense, R.Bonnet, D.L.Book, J.W.Brault, R.J.Bray,

J.B.Breckenridge, G.E.Brueckner, E.C.Bruener Jr., A.Bruzek, V.Bumba, C.Cannon, G.Ceppatelli, G.A.Chapman, C.C.Cheng, V.F.Chistyakov, E.Chvojkova, T.A.Clark, J.W.Cook, A.N.Cox, I.J.D.Craig, L.E.Cram, P.Delache, L.Delbouille, C.de Jager, L.Dezso, D.Dravins, E.E.Dubov, S.Dumont, L.Dunkelman, R.B.Dunn, F.N.Edmonds, I.Elliott, G.Elste, G.L.Epstein, J.V.Evans, R.Falciani, U.Feldman, A.Fiala, P.Foukal, E.M.Frazier, H.Friedman, A.Gabriel, V.Gaizauskas, V.P.Gaur, O.J.Gingerich, R.G.Giovanelli, M.N.Gnevyshev, G.Godoli, N.Gokdogan, L.Goldberg, M.V.Goldman, M.T.Gomez, S.I.Gopaschuk, C.Gordon, N.Grevesse, E.A.Gurtovenko, S.S.Hasan, R.F.Haupt, E.Hiei, T.Hirayama, H.Holweger, B.H.Horton, M.Hotinli, L.House, R.Howard, R.M.E.Illing, J.T.Jefferies, H.P.Jones, C.Jordan, S.Jordan, B.Kalman, M.Kanno, S.Kato, P.Kaufmann, I.Kawaguchi, S.L.Keil, T.S.Khetsuriani, F.Kneer, E.V.Kononovich, M.Kopecky, S.Koutchmy, G.G.Kraemer, V.A.Krat, A.Kubicela, G.V.Kuklin, M.R.Kundu, G.Laborde, D.Labs, B.J.La Bonte, D.A.Landman, R.B.Leighton, J.-L.Leroy, J.L.Linsky, W.C.Livingston, J.L.Locke, M.Lopez-Arroyo, R.E.Loughhead, R.Luest, E.A.Makarova, M.Makita, M.Marik, E.Marilli, J.T.Mariska, S.Matsushima, W.Mattig, S.M.P.McKenna Lawlor, P.Mein, J.Mergentaler, R.Mewe, R.Michard, M.V.Migeotte, D.Mihalas, R.Milkey, CH.Moore-Sitterly, F.Moriyama, Z.Mouradian, E.A.Mueller, R.H.Munro, O.Namba, L.Neven, G.A.Newkirk, K.R.Nicolas, M.Nicolet, G.M.Nikolsky, K.Nishi, A.Nordlund, F.Q.Orral, Y.Osaki, L.Oster, M.C.Pande, D.Papathanasoglou, W.H.Parkinson, J.M.Pasachoff, J.-C.Pecker, R.Peyturaux, K.Pflug, A.K.Pierce, R.E.Rees, E.M.Reeves, G.Righini, M.Rigutti, G.Roberti, F.Roddier, G.Roland, V.Rusin, R.J.Rutten, M.Rybansky, J.I.Sakai, D.Samain, T.Sakurai, E.J.Schmahl, S.J.Schwartz, M.J.Seaton, M.Semel, A.B.Severny, M.J.Shallis, R.A.Shine, G.F.Sitnik, A.Skumanich, V.M.Sobolev, P.Souffrin, D.S.Spicer, J.Staude, R.Stebbins, G.Stellmacher, V.E.Stepanov, M.Stix, Z.Suemoto, Z.Svestka, J.W.Swensson, K.Tanaka, E.Tandberg-Hanssen, R.B.Teplitskaya, J.H.Thomas, R.N.Thomas, R.Tousey, T.T.Tsap, W.Unno, U.K.Uus, A.Unsold, G.Van Hoven, G.Va.Vasil'eva, A.H.Vaughan, J.C.Vial, S.Volonte, M.Vukicevic-Karabin, M.Waldmeier, J.W.Warwick, D.G.Wentzel, A.D.Wittman, G.Wlerick, H.Wohl, S.P.Worden, L.-X.Wu, A.Wyller, H.Yoshimura, N.H.Youssef, A.Zelenka, Y.N.Zhugzhda, H.Zirin, J.B.Zirker, C.Zwaan.

## COMMISSION NO 14.

DONNEES ATOMIQUES ET MOLECULAIRES (ATOMIC AND MOLECULAR DATA)

PRESIDENT: A.H.Gabriel

VICE-PRESIDENT: R.W.Nicholls

COMITE D'ORGANISATION: D.R.Johnson, S.L.Mandel'shtam, H.Nussbaumer, J.G.Phillips, S.Sahal, K.Takayanagi, W.Wiese.

MEMBRES: S.J.Adelman, C.W.Allen, K.L.Andrew, M.-Chr.Artru, K.M.Baird, H.A.G.Barnard, R.Barrow, D.R.Bates, O.Bely, F.Bely-Dubau, K.A.Berrington, E.Biemont, J.H.Black, M.Blaha, L.M.Branscomb, J.W.Brault, G.E.Bromage, A.Burgess, D.F.Carbon, P.K.Carroll, J.H.Carver, A.H.Cook, C.H.Corliss, S.J.Czyzak, A.Dalgarno, A.H.Delsemme, J.Desesquelles, G.H.F.Diercksen, P.P.Dobronravin, K.Dressler, J.Dubau, M.Dufay, B.Edlen, E.Engelhard, G.L.Epstein, P.Faucher, N.Feautrier, P.Felenbok, U.Fink, D.Flower, R.H.Garstang, W.R.S.Garton, Yu.V.Glagolevskij, C.Goldbach, I.P.Grant, L.C.Green, D.W.O.Heddle, R.A.Hefferlin, R.Herman, G.Herzberg, J.E.Hesser, L.L.House, M.C.E.Huber, W.F.Huebner, C.J.Humphreys, A.W.Irwin, P.Jacquinot, F.M.Johnson, F.Joly, C.Jordan, H.L.Jordan, J.Junkes, T.Kato, K.G.Kessler, R.B.King, A.E.Kingston, T.A.Kipper, J.L.Kohl, A.Lagerqvist, D.A.Landman, J.Lang, J.-M.Launay, G.M.Lawrence, D.Layzer, W.Lochte-Holtgreven, M.Loulergue, F.J.Lovas, B.L.Lutz, J.P.Maillard, M.Malinovsky, W.C.Martin, H.E.Mason, R.W.P.Mcwhirter, O.A.Mel'nikov, R.Mewe, M.Migeotte, O.C.Mohler, A.Monfils, CH.Moore-Sitterly, T.E.Nevin, G.H.Newsom,

G.Nollez, S.Obi, L.Oetken, T.Oka, A.Omont, G.S.Orton, W.H.Parkinson, G.Peach, D.Petrini, B.Petropoulos, H.Pfennig, V.K.Prokof'ev, F.Querci, J.Richter, J.E.Ross, E.Roueff, A.Schadee, J.Schrijver, M.J.Seaton, B.W.Shore, M.Simoda, J.A.Smit, G.Smith, P.L.Smith, W.H.Smith, W.B.Somerville, N.G.Sorensen, L.Steenman-Clark, V.S.Strelnitsky, H.P.Summers, J.P.Swings, P.Swings, J.B.Tatum, J.Tech, R.Tousey, E.Trefftz, H.Van Regemorter, W.Van Rensbergen, C.M.Varsavsky, D.A.Varshalovich, H.Voelk, S.Volonte, V.Vujnovic, G.W.Wares, J.C.Weisheit, S.Weniger, R.Wilson, G.Winnewisser, L.G.Young, C.J.Zeippen, H.Zirin.

## COMMISSION NO 15.

## ETUDE PHYSIQUE DES COMETES, DES PETITES PLANETES ET DES METEORITES

## (PHYSICAL STUDY OF COMETS, MINOR PLANETS AND METEORITES)

PRESIDENT: C.R.Chapman

VICE-PRESIDENT: L.Kresak

COMITE D'ORGANISATION: B.D.Donn, F.Dossin, H.Fechtig, M.Festou, D.L.Matson, H.U.Schmidt, Z.Sekanina, L.M.Shulman, J.Wasson, G.W.Wetherill, L.Wilkening.

MEMBRES: M.F.A'hearn, C.Allegre, E.Anders, D.A.Andrienko, C.Arpigny, E.Barker, D.B.Beard, Ch.Bertaud, M.Beyer, L.Biermann, J.E.Blamont, N.T.Bobrovnikoff, J.Bouska, E.L.G.Bowell, J.C.Brandt, A.Brecher, W.Brunk, L.F.Burlaga, J.A.Burns, M.P.Candy, Z.Ceplecha, V.I.Cherednichenko, W.Cochran, C.B.Cosmovici, C.Cristescu, J.Crovisier, D.P.Cruikshank, A.Danks, C.Degewij, A.H.Delsemme, A.A.Demenko, I.C.Deqewij, W.A.Deutschman, O.V.Dobrovsky, A.E.Douglas, M.Dryer, A.T.Ershkovich, E.Everhart, A.Eviatar, P.Farinella, P.Feldman, J.A.Fernandez, M.Gaffey, T.Gehrels, E.Gerard, D.M.Gibson, R.H.Giese, J.Gradie, J.M.Greenberg, L.Grossman, S.Grudzinska, E.Grun, I.Halliday, M.S.Hanner, B.W.Hapke, A.W.Harris, W.K.Hartmann, M.O.Harwit, L.Haser, H.Haupt, G.Herzberg, W.F.Huebner, D.W.Hughes, R.Hutchinson, K.Ibadinov, W.Irvine, S.Isobe, W.M.Jackson, K.Jockers, T.V.Johnson, H.U.Keller, R.F.Knack, L.Kohoutek, V.P.Konopleva, C.T.Kowal, E.L.Krinov, P.Lamy, P.Lancaster Brown, H.P.Larson, L.A.Lebovsky, A.C.Levasseur-Regourd, B.J.Levin, W.Liller, C.F.Lillie, C.A.Lindsey, R.Luest, D.F.Lupishko, R.A.Lyttleton, D.J.Malaise, S.P.Maran, B.G.Marsden, M.T.Martel-Chossat, O.T.Matsuura, T.B.McCord, R.E.McCroskey, S.M.P.McKenna-Lawlor, D.Meisel, D.A.Mendis, B.Milet, F.D.Miller, P.Millman, E.Moore, D.D.Morrison, A.Mrkos, T.Nakamura, J.S.Neff, N.F.Ness, R.L.Newburn, C.R.O'Dell, J.A.O'Keefe, E.J.Opik, P.Pellas, H.A.Perez-de-Tejada, C.B.Pilcher, E.M.Pittich, P.Proisy, J.Rahe, L.Remy-Battiau, D.Revelle, E.Roemer, F.Scaltriti, M.V.Schmidt, M.Shimizu, A.M.Simonenko, K.R.Sivaraman, L.E.Snyder, P.Swings, H.Takeda, H.Tanabe, E.F.Tedesco, K.Tomita, V.Vanysek, G.J.Veeder, J.F.Veverka, S.K.Vsekhsvjatskij, M.Wallis, J.Wasson, R.L.Waterfield, P.Wehinger, P.R.Weissman, F.L.Whipple, L.L.Wilkening, W.Wisniewski, J.A.Wood, A.Woszczyk, S.Wyckoff, A.A.Yavnel', D.Yeomans, V.Zappala, B.H.Zellner.

## COMMISSION NO 16.

## ETUDE PHYSIQUE DES PLANETES ET DES SATELLITES

## (PHYSICAL STUDY OF PLANETS AND SATELLITES)

PRESIDENT: V.G.Teijfel'

VICE-PRESIDENT: G.E.Hunt

COMITE D'ORGANISATION: R.F.Beebe, J.L.Bertaux, A.Brahic, J.L.Elliot, D.Gautier, L.V.Ksanfomaliti, M.Y.Marov, H.Masursky, G.Neukum, T.C.Owen, V.V.Shevchenko, B.A.Smith.

MEMBRES: T.Akabane, E.Anders, D.W.G.Arthur, C.H.Barrow, W.A.Baum, R.Beer, M.S.Belton, P.Bender, G.L.Berge, J.E.Blamont, M.S.Bobrov, L.N.Bondarenko, P.B.Boyce, A.Brecher, A.L.Broadfoot, W.E.Brunk, K.E.Bullen, J.A.Burns, N.F.Bystrov, O.Calame, W.S.Cameron, H.Camichel, D.B.Campbell, S.Catalano, J.W.Chamberlain, C.R.Chapman, E.H.Collinson, G.Colombo, A.F.Cook, J.Connes, C.C.Counselman, D.P.Cruikshank, M.E.Davies, J.Degewij, W.C.De Marcus, G.De Mottoni y Palacios, J.R.Dickel, A.Dollfus, F.D.Drake, V.P.Dzhapiashvili, F.El Baz, W.E.Elston, Von R. Eshleman, L.W.Esposito, D.S.Evans, P.Farinella, G.Fielder, U.Fink, K.P.Florenskij, K.Fox, L.S.Galkin, I.V.Gavrilov, J.E.Geake, T.Gehrels, J.Geiss, H.L.Giclas, P.Gierasch, T.Gold, P.Goldreich, R.M.Goldstein, R.M.Goody, P.Gorenstein, C.L.Goudas, J.Green, L.Grossman, P.Guerin, J.E.Guest, S.Gulkis, S.T.Habibullin, T.Hagfors, J.S.Hall, I.Halliday, G.Herzberg, G.P.Horedt, J.W.Hovenier, W.B.Hubbard, D.M.Hunten, H.C.Ingrao, W.M.Irvine, K.Iwasaki, H.Jeffreys, T.V.Johnson, R.F.Jurgens, R.V.Karandikar, W.M.Kaula, R.I.Kiladze, V.S.Kislyuk, Y.Kondo, Z.Kopal, N.A.Kozyrev, S.S.Kumar, V.G.Kurt, A.D.Kuzmin, A.L.Lane, H.P.Larson, G.A.Lejkin, B.J.Levin, J.S.Lewis, F.Link, G.W.Lockwood, B.L.Lutz, D.Ya.Martynov, D.L.Matson, C.H.Mayer, T.B.McCord, M.B.McElroy, A.J.Meadows, B.M.Middlehurst, J.S.Mikhail, R.L.Millis, P.M.Millman, S.Miyamoto, P.Moore, V.I.Morov, A.V.Morozhenko, D.Morrison, L.V.Morrison, M.Moutsoulas, J.D.Mulholland, R.E.Murphy, Y.Nakagawa, N.F.Ness, J.A.O'Keefe, E.J.Opik, I.Ottelet, K.Pang, B.Petropoulous, G.H.Pettengill, J.B.Pollack, J.Rosch, S.K.Runcorn, E.L.Ruskol, V.Safronov, C.Sagan, M.U.Sagitov, J.Saissac, I.I.Shapiro, T.Shimizu, E.Shoemaker, W.Sinton, H.J.Smith, R.Smoluchowski, L.A.Soderblom, C.P.Sonett, E.C.Stone, D.F.Strobel, R.G.Strom, J.D.Strong, R.J.Terrile, C.Tombaugh, T.W.Thompson, L.Trafton, V.S.Troitsky, G.L.Tyler Jr., J.A.Van Allen, T.C.Van Flandern, J.Veverka, R.M.Walker, L.V.Wallace, L.H.Wasserman, J.T.Wasson, T.Weimer, G.W.Wetherill, E.A.Whitaker, R.L.Wildey, I.P.Williams, J.G.Williams, J.A.Wood, A.Woszczyk, C.F.Yoder, A.T.Young, L.G.Young.

## COMMISSION NO 19.

## ROTATION DE LA TERRE (ROTATION OF THE EARTH)

PRESIDENT: Ya.S.Yatskiv

VICE-PRESIDENT: W.Klepczynski

COMITE D'ORGANISATION: F.Barlier, H.Enslin, C.Kakuta, B.Kolaczek, I.I.Mueller, V.Naumov, P.Paquet, E.Proverbio, M.G.Rochester, S.-Y.Ye, K.Yokoyama.

MEMBRES: H.J.M.Abraham, D.Arabelos, P.L.Bender, N.S.Blinov, G.Billaud, J.Bonanomi, P.Brosche, O.Calame, W.Cannon, M.Catalan, D.Currie, W.E.Carter, J.G.Davies, S.V.Debarbat, R.Dejaiffe, D.Djurovic, C.Dramba, B.Elsmore, J.L.Fanselow, E.P.Fedorov, M.Feissel, H.F.Fliegel, J.Gaignebet, E.M.Gaposhkin, E.Groten, B.Guinot, R.G.Hall, H.Hellwig, G.Hemmleb, K.Hosoyama, K.Hurukawa, S.Iijima, W.Jaks, H.Jeffreys, A.M.Kalmykov, R.King, S.H.Knowles, Y.Kokurin, L.D.Kostina, Y.Kozai, K.Lambeck, T.Lederle, M.Lefebvre, J.J.Levallois, J.H.Lieske, D.-J.Luo, S.Manabe, Wm.Markowitz, O.N.Matsakis, D.D.McCarthy, M.Meinig, W.Melbourne, P.J.Melchior, J.Mietelski, A.A.Mikhailov, V.Milovanovic, M.Miyadi, J.Moczko, P.Morgan, L.V.Morrison, T.Nicolini, A.M.Nobili, N.P.J.O'Hora, I.Okamoto, S.Okazaki, T.Okuda, M.Ooe, W.Opalski, A.Orte, L.Oterma, N.N.Parijskij, N.N.Pavlov, J.D.H.Pilkington, A.Poma, J.Popelar, N.A.Popov, Y.I.Prodan, L.Randic, J.-P.Ren, D.S.Robertson, S.K.Runcorn, L.V.Rykhlova, V.I.Sakharov, M.Sanchez, T.Sasao, K.Sato, B.E.Schutz, N.Sekiguchi, B.M.Sevarlic,

M.Sevilla, I.I.Shapiro, E.Silverberg, M.Slade, F.G.Smith, H.M.Smith, D.E.Smylie, G.Stanila, F.R.Stephenson, A.Stoyko, C.Sugawa, S.Takagi, B.D.Tapley, V.K.Tarady, G.Teleki, D.V.Thomas, M.Torao, M.Tsao, I.Tsubokawa, G.Veis, R.Vicente, J.Vondrak, T.-S.Wan, K.Wako, W.Ward, G.Wilkins, J.G.Williams, P.Wilson, G.Winkler, S.-X.Wu, T.-Q.Xu, S.Yumi.

COMMISSION NO 20.

POSITIONS ET MOUVEMENTS DES PETITES PLANETES, DES COMETES ET DES SATELLITES

(POSITIONS AND MOTIONS OF MINOR PLANETS, COMETS AND SATELLITES)

PRESIDENMT: E.Roemer

VICE-PRESIDENT: C.J.Van Houten

COMITE D'ORGANISATION: V.Yu.Batrakov, M.P.Candy, Y.Kozai, L.Kresak, L.K.Kristensen, B.G.Marsden, H.Scholl, G.E.Taylor, D.K.Yeomans, Y.Z.Zhang,

MEMBRES: V.K.Abalakin, M.F.A'hearn, K.Aksnes, S.J.V.Arend, P.B.Babadzhanov, A.Bec-Borsenberger, N.A.Belyaev, D.Benest, J.C.Bennett, M.Bielicki, R.Bien, F.Boerngen, E.L.G.Bowell, R.L.Branham, J.A.Bruwer, J.A.Burns, O.Calame, A.Carusi, N.S.Chernykh, C.Cristescu, L.E.Cunningham, M.De Pascual Martinez, H.Debehogne, A.H.Delsemme, M.A.Dirikis, D.W.Dunham, R.Dvorak, F.K.Edmondson, J.L.Elliot, Yu.V.Evdokimov, E.Everhart, S.Ferraz-Mello, F.A.Franklin, R.R.Freitas Mourao, C.Froeschle, I.V.Galibina, B.Garfinkel, T.Gehrels, H.L.Giclas, A.C.Gilmore, R.J.Greenberg, R.S.Harrington, A.W.Harris, I.Hasegawa, E.F.Helin, P.D.Hemenway, J.Hers, H.Hurnik, K.Hurukawa, P.A.Ianna, V.A.Izvekov, E.I.Kazimirchak-Polonskaya, A.S.Khatsiashvili, T.Kiang, A.R.Klemola, L.Kohoutek, J.Kovalevsky, C.T.Kowal, C.I.Lagerkvist, J.H.Lieske, B.A.Lindblad, N.R.Lomb, M.Lovas, R.E.McCrosky, B.Milet, R.Millis, B.Mintz-Blanco, B.Morando, J.D.Mulholland, P.E.Nacozy, T.Nakamura, A.M.Nobili, L.Oterma, D.Pascu, D.A.Pierce, E.M.Pittich, B.Popovic, M.B.Protitch, L.Quijano, M.Rapaport, H.Rickman, W.H.Robertson, J.L.Sagnier, L.D.Schmadel, G.Schrutka, J.Schubart, H.-E.Schuster, P.K.Seidelmann, Z.Sekanina, P.J.Shelus, V.A.Shor, K.A.Shtejns, A.T.Sinclair, G.Sitarski, W.Strobel, G.F.Sultanov, Ko.Tomita, C.Torres, S.Vaghi, T.C.Van Flandern, I.Van Houten-Groeneveld, S.K.Vsekhsvyatskij, D.T.Vu, L.H.Wasserman, P.R.Weissman, R.M.West, F.L.Whipple, P.-Wild, I.P.Williams, J.G.Williams, N.Yakhontova-samojlova, P.E.Zadunaisky, V.Zappala, J.-X.Zhang, K.Ziolkowski.

COMMISSION NO 21.

LA LUMIERE DU CIEL NOCTURNE (THE LIGHT OF THE NIGHT SKY)

PRESIDENT: R.H.Giese

VICE-PRESIDENT: K.Mattila

COMITE D'ORGANISATION: R.Dumont, Y.I.Galperin, M.S.Hanner, A.C.Levasseur-Regourd, F.Sanchez-Mmartinez, H.Tanabe, J.L.Weinberg.

MEMBRES: P.Alvarez, K.Anderson, R.Angione, A.S.Asaad, W.J.Baggaley, C.J.Banos, D.R.Bates, D.E.Blackwell, J.E.Blamont, A.L.Broadfoot, J.W.Chamberlain, A.F.Cook, J.Dachs, N.B.Divari, M.Dufay, L.Dunkelman, H.Elsasser, H.Fechtig, P.D.Feldman,

L.M.Fishkova, M.Fracassini, J.M.Greenberg, I.Halliday, M.Harwit, U.Haug, R.C.Henry, W.Hofmann, M.Huruhata, G.S.Ivanov-Kholodny, J.F.James, A.H.Jarrett, J.Kaplan, R.V.Karandikar, Z.V.Karygina, A.Kastler, H.B.Knaflich, S.Koutchmy, P.V.Kulkarni, P.Lamy, C.Leinert, C.F.Lillie, F.Link, T.Matsumoto, M.Maucherat-Joubert, T.G.Megrelishvili, N.Y.Misconi, D.H.Morgan, S.Mukai, T.Mukai, S.Nawar, L.Neuzil, E.P.Ney, M.Nicolet, V.B.Nikonov, T.Nishimura, L.E.Pasinetti, R.W.B.Pearce, J.Pfleiderer, E.Pitz, M.Rapaport, H.W.Ripken, F.E.Roach, R.Robley, R.G.Roosen, D.A.Rozhkovskij, C.Sanchez-Magro, P.P.Saxena, M.Schuessler, A.S.Sharov, R.K.Soberman, J.G.Sparrow, H.J.Staude, T.I.Toroshlidze, Y.L.Trutse, J.A.Tyson, J.A.Van Allen, H.C.Van de Hulst, L.V.Wallace, G.Weill, S.Weniger, A.N.Witt, R.D.Wolstencroft, R.H.Zerull.

COMMISSION NO 22.

METEORES ET LA POUSSIERE INTERPLANETAIRE (METEORS AND INTERPLANETARY DUST)

PRESIDENT: O.I.Belkovich

VICE-PRESIDENT: D.E.Brownlee

COMITE D'ORGANISATION: P.B.Babadzahnov, W.J.Baggaley, W.G.Elford, H.Fechtig, M.S.Hanner, J.A.M.McDonnell, J.Stohl, K.Tomita.

MEMBRES: W.N.Abbott, E.Anders, D.B.Beard, R.S.Bibarsov, A.T.Blackwell, Z.Ceplecha, K.S.Clifton, A.F.Cook, J.G.Davies, J.Davis, E.L.Fireman, R.H.Giese, B.P.Glass, E.Grun, M.Hajdukova, A.Hajduk, M.S.Halliday, G.A.Harvey, I.Hasegawa, G.S.Hawkins, C.L.Hemenway, J.S.Hey, P.W.Hodge, S.S.Hong, J.A.Hoppe, D.W.Hughes, L.G.Jacchia, R.C.Jennison, J.Jones, T.R.Kaiser, B.L.Kashscheev, L.A.Katasev, C.S.L.Keay, A.Kizilirmak, K.V.Kostylev, Kh.N.Kramer, L.Kresak, M.Kresakova, E.L.Krinov, Z.Kviz, P.Lamy, V.N.Lebedinets, B.J.Levin, B.A.Lindblad, A.C.B.Lovell, R.E.McCrosky, B.A.McIntosh, H.Miles, P.M.Millman, N.Y.Misconi, R.L.Newburn, J.A.O'Keefe, E.J.Opik, V.Padevet, Z.Plavcova, G.Polnitzky, V.Porubcan, J.Rajchl, D.O.Revelle, H.W.Ripken, R.G.Roosen, J.A.Russell, Z.Sekanina, C.Y.Shao, M.Simek, A.N.Simonenko, R.K.Soberman, R.B.Southworth, J.Stohl, E.P.Tedesco, A.K.Terenteva, F.F.Verniani, J.L.Weinberg, G.W.Wetherill, I.P.Williams, F.L.Whipple, J.A.Wood, D.K.Yeomans, A.A.Yavnel'.

COMMISSION NO 24.

ASTROMETRIE PHOTOGRAPHIQUE (PHOTOGRAPHIC ASTROMETRY)

PRESIDENT: W.Gliese

VICE-PRESIDENT: A.Upgren

COMITE D'ORGANISATION: T.E.Corbin, H.K.Eichhorn, G.Gatewood, I.I.Kanaev, T.E.Lutz, J.D.Stock, W.F.Van Altena,

MEMBRES: K.D.Abhyankar, N.Argue, G.F.Benedict, A.Blaauw, R.Bouigue, R.Branham, P.Brosche, Liang-Tia G.Chiu, J.W.Christy, J.Churms, S.V.M.Clube, P.Connes, M.Creze, K.Cudworth, C.C.Dahn, J.Delhaye, A.N.Deutsch, CH.De Vegt, W.Dieckvoss, J.Dommanget, B.Elsmore, F.W.Fallon, J.Fanselow, N.V.Fatchikhin, F.Firneis, M.G.Firneis, M.G.Fracastoro, O.G.Franz, L.W.Fredrick, A.Fresneau, L.Gallouet, I.V.Gavrilov, H.L.Giclas, A.N.Goyal, R.B.Hanson, R.S.Harrington, W.D.Heintz, P.D.Hemenway,

J.L.Hershey, J.-L.Heudier, G.Hill, E.D.Hoffleit, J.Hughes, P.A.Ianna, H.Jahreiss, W.H.Jefferys, K.Johnston, B.F.Jones, D.H.P.Jones, A.R.Klemola, B.L.Klock, I.G.Kolchinskij, P.Lacroute, K.K.Lapushka, A.A.Latypov, V.V.Lavdovskij, R.S.Le poole, S.L.Lippincott, A.M.Lozinskij, P.K.Lu, W.J.Luyten, V.S.Matiagin, H.A.Mcalister, M.O.Menessier, J.Meurers, C.A.Murray, W.Nicholson, J.K.O'Connell, T.Oja, A.B.Onegina, V.V.Podobed, D.D.Polozhentsev, H.I.Potter, F.V.Prochazka, L.Quijano, N.G.Rizvanov, W.H.Robertson, E.Roemer, J.L.Russell, W.L.Sanders, K.Sims, C.A.Smith, L.Stange, J.W.Stein, R.C.Stone, K.Aa.Strand, D.V.Thomas, C.Turon, A.Valbousquet, P.Van de Kamp, S.Vasilevskis, N.E.Wagman, H.G.Walter, L.Wan, L.H.Wasserman, A.J.Wesselink, G.Westerhout.

COMMISSION NO 25.

PHOTOMETRIE ET POLARIMETRIE STELLAIRE (STELLAR PHOTOMETRY AND POLARIMETRY)

PRESIDENT: J.Tindbergen

VICE-PRESIDENT: F.G.Rufener

COMITE D'ORGANISATION: M.S.Bessell, P.F.Chugajnov, J.Dachs, P.Edwards, I.S.Glass, J.A.Graham, B.Hauck, D.Kilkenny, A.U.Landolt, I.S.McLean, V.Piirola, N.M.Shakhovskoj.

MEMBRES: H.D.Ables, S.Adelman, R.Albrecht, A.N.Argue, J.D.R.Bahng, L.Baldinelli, W.Becker, A.Behr, J.H.Bigay, B.M.Blanco, B.J.Bok, B.Bookmyer, J.Borgman, M.Breger, H.A.Bruck, B.W.Carney, G.Chis, A.W.J.Cousins, G.V.Coyne, D.L.Crawford, C.C.Dahn, M.K.Das, J.K.Denoyelle, H.T.Eelsalu, A.Feinstein, J.D.Fernie, W.S.Fitch, J.C.Forte, L.Gallouet, T.Gehrels, R.D.Gehrz, M.Golay, G.Goy, A.D.Grauer, M.Grenon, A.Gutierrez-Moreno, D.S.Hall, R.H.Hardie, D.Hayes, A.Heck, H.Hensberge, P.W.Hill, R.Hilditch, W.A.Hiltner, E.B.Holmberg, A.R.Hyland, J.B.Irwin, M.Jerzykiewicz, S.C.Joshi, G.E.Kron, W.Kunkel, Z.Kviz, P.G.Laskarides, B.M.Lasker, R.Lenzen, G.W.Lockwood, H.G.Marraco, T.Markkanen, A.Masani, P.Mayer, M.F.McCarthy, E.E.Mendoza, J.Menzies, P.Mianes, E.Milone, R.I.Mitchell, T.Moffett, H.Moreno, S.C.Morris, A.B.Muller, G.S.Mumford, V.B.Nikonov, P.Notni, K.Osawa, V.S.Oskanyan, J.W.Pel, A.Penny, C.L.Perry, R.J.Pfeiffer, A.G.D.Philip, E.L.Robinson, C.Roslund, E.Rybka, A.E.Rydgren, M.B.K.Sarma, E.G.Schmidt, H.Schmidt, W.Schoneich, S.J.Shawl, M.J.Smyth, U.W.Steinlin, J.Stock, H.S.Stockman, R.P.S.Stone, V.L.Straizys, W.Strohmeier, D.J.Sullivan, P.Szkody, J.A.Thomas, I.Todoran, C.R.Tolbert, H.Trodahl, B.Ulrich, V.Ureche, R.A.Vardanian, A.G.Velghe, F.J.Vrba, A.A.E.Wallenquist, Th.Walraven, W.H.Warren Jr., D.Weistrop, A.Wesselink, N.M.White, R.V.Willstrop, W.Wisniewski, S.S.O.Wramdemark, Y.Yamashita, A.T.Young.

COMMISSION NO 26.

ETOILES DOUBLES ET MULTIPLES (DOUBLE AND MULTIPLE STARS)

PRESIDENT: M.Fracastoro

VICE-PRESIDENT: K.D.Rakos

COMITE D'ORGANISATION: O.G.Franz, J.Dommanget, W.D.Heintz, H.A.McAlister, A.Poveda, C.D.Scarfe.

MEMBRES: H.A.Abt, C.Allen, S.Arend, P.Baize, A.H.Batten, E.Cabrita, B.Cester, P.Couteau, R.B.Culver, A.N.Dadaev, A.N.Deutsch, D.W.Dunham, F.C.Fekel, O.E.Ferrer,

R.R.Freitas Mourao, G.Gatewood, E.H.Geyer, R.S.Harrington, J.L.Hershey, C.A.Hernandez, B.Hidajat, F.Holden, G.Ishida, A.A.Kiselev, P.G.Kulikovskij, M.I.Kumsiashvili, S.L.Lippincott, N.L.Magalashvili, C.Meyer, C.L.Morbey, P.Muller, R.Pannunzio, G.M.Popovic, J.L.Russell, G.N.Salukvadze, A.M.Shulberg, K.Aa.Strand, M.Tapia, A.Valbousquet, P.Van de Kamp, K.A.Van der Hucht, E.Van Dessel, R.L.Walker, E.W.Weiss, N.P.Wieth-Knudsen, R.H.Wilson, C.E.Worley.

COMMISSION NO 27.

ETOILES VARIABLES (VARIABLE STARS)

PRESIDENT: N.H.Baker

VICE-PRESIDENT: B.Szeidl

COMITE D'ORGANISATION: M.Breger, A.N.Cox, J.D.Fernie, R.E.Gershberg, M.Jerzykiewicz, L.N.Mavridis, L.N.Mirzoyan, J.R.Percy, A.M.Van Genderen, B.Warner.

MEMBRES: M.L.Aizenman, I.F.Alanija, H.Ando, A.D.Andrews, E.Antonello, V.P.Arkhipova, G.Asteriadis, A.Baglin, G.A.Bakos, T.G.Barnes, P.Bastien, F.M.Bateson, G.T.Bath, E.P.Belserene, G.Belvedere, Ch.Bertaud, G.Berthomieu, M.S.Bessell, D.R.Bochonko, C.T.Bolton, H.E.Bond, B.V.Bopp, J.Boulon, A.A.Boyarchuk, M.E.Boyarchuk, C.J.Butler, D.Butler, P.B.Byrne, R.M.Catchpole, E.Chavira, A.M.Cherepaschuk, G.Chis, R.F.Christy, P.F.Chugajnov, C.M.Coutts-Clement, B.C.Cogan, M.Cohen, M.E.Contadakis, M.De Groot, S.Demers, R.G.Deupree, R.J.Dickens, D.Dupuy, W.Dziembowski, P.J.Edwards, Yu.N.Efremov, A.N.Eskioglu, A.Evans, D.S.Evans, N.R.Evans, M.W.Feast, W.A.Feibelman, G.J.Ferland, W.S.Fitch, M.Friedjung, M.S.Frolov, G.Gahm, J.S.Gallagher, S.Gaposchkin, S.C.B.Gascoigne, E.H.Geyer, D.M.Gibson, W.Gieren, Yu.V.Glagolevskij, G.Godoli, V.G.Gorbatsky, J.Graham, G.L.Grasdalen, J.Grygar, G.Guerrero, E.F.Guinan, H.Gursky, J.A.Hackwell, D.S.Hall, C.J.Hansen, G.Haro, A.Heiser, G.H.Herbig, R.B.Herr, J.Hers, J.E.Hesser, F.J.Heyden, H.A.Hill, P.W.Hill, E.D.Hoffleit, N.Houk, M.Huruhata, J.B.Hutchings, I.Iben, T.Jarzebowski, A.F.Jones, S.Kanyo, A.H.Karp, P.N.Kholopov, R.Kippenhahn, I.M.Kopylov, K.Kordylewski, R.P.Kraft, J.Krautter, J.M.Kreiner, W.Krzeminski, M.Kubiak, L.V.Kuhi, J.I.Kumsiashvili, W.E.Kunkel, D.W.Kurtz, K.K.Kwee, M.T.V.T.Lago, A.U.Landolt, P.G.Laskarides, P.Ledoux, P.Leite Scheid, K.C.Leung, G.W.Lockwood, M.C.Lortet-Zuckermann, L.S.Luud, B.F.Madore, P.Maffei, F.M.Mahmoud, E.N.Makarenko, G.Mannino, T.E.Margrave, W.L.Martin, A.Masani, J.A.Mattei, M.W.Mayall, D.H.McNamara, M.-O.Menessier, W.J.Miller, E.F.Milone, L.A.Milone, T.J.Moffett, N.Morguleff, N.D.Morrison, G.S.Mumford, P.G.Murdin, R.E.Nather, J.S.Neff, P.G.Niarchos, N.S.Nicolov, G.J.Odgers, A.Opolski, Y.Osaki, K.Osawa, V.Oskanyan, J.Papaloizou, M.Parthasarathy, L.Paterno, L.C.Peltier, J.O.Petersen, V.Piirola, S.L.Piotrowski, L.Plaut, M.Popova, J.E.Pringle, J.Provost, Yu.P.Pskovskij, K.D.Rakos, P.Renson, G.Richter, E.L.Robinson, A.W.Rodgers, M.Rodomo, G.Romano, Yu.S.Romanov, L.Rosino, J.Rountree, A.R.Sadik, N.N.Samus, W.H.Sandmann, A.Sanyal, J.P.Sareyan, M.B.K.Sarma, H.B.Sawyer Hogg, E.G.Schmidt, R.Schoembs, P.R.Schwartz, R.Scuflaire, M.Shara, W.A.Sherwood, R.R.Shobbrook, S.D.Sinvhal, J.Smak, P.Smeyers, Harlan J. Smith M.A.Smith, S.Starrfield, R.F.Stellingwerf, K.Stepien, C.L.Sterken, M.J.Stift, R.S.Stobie, W.Strohmeier, L.Szabados, P.Szkody, M.Takeuti, G.A.Tammann, P.Tempesti, A.Terzan, H.R.E.Tjin a Djie, C.A.Torres, J.Tremko, V.P.Tsesevich, A.Tsioumis, A.V.Tutukov, R.Tylenda, P.D.Usher, J.C.Valtier, S.Van Agt, A.Van Hoof, R.Viotti, N.Vogt, A.A.Wachmann, E.N.Walker, M.F.Walker, G.Wallerstein, T.Walraven, R.F.Webbink, A.Wehlau, W.Wenzel, A.J.Wesselink, R.M.Williamon, L.A.Willson, L.A.Wilson, R.F.Wing, W.Wisniewski, P.R.Wood, F.W.Wright, D.-R.Xiong, B.Zuckerman.

COMMISSION NO 28.

GALAXIES (GALAXIES)

PRESIDENT: V.C.Rubin

VICE-PRESIDENT: P.C.Van der Kruit

COMITE D'ORGANISATION: S.D'Odorico, J.E.Einasto, J.A.Graham, P.W.Hodge, I.D.Karachentsev, D.Lynden-Bell, A.Toomre, M.H.Ulrich, K.Wakamatsu, B.E.Westerlund.

MEMBRES: M.Aaronson, G.O.Abell, H.D.Ables, E.L.Agvero, H.Alfven, R.J.Allen, V.A.Ambartsumian, Y.Andrillat, M.A.Arakelian, A.L.Ardeberg, H.C.Arp, E.O.Athanassoula, M.Azzopardi, M.K.Babadzhanianc, J.N.Bahcall, M.E.Bailey, C.Barbieri, R.Barbon, W.A.Baum, G.F.Benedict, N.Bergvall, R.H.Berman, F.Bertola, J.H.Bigay, J.J.Binney, C.P.Blackman, G.O.Boeshaar, B.J.Bok, A.Boksenberg, H.Bondi, F.Boerngen, T.M.Borchkhadze, A.Bosma, L.Bottinelli, A.Braccesi, K.Brecher, P.Brosche, M.T.Bruck, P.Bruston, E.M.Burbidge, G.R.Burbidge, J.O'N.Burns, D.Burnstein, G.G.Byrd, R.D.Cannon, M.Capaccioli, L.Carrasco, G.J.Carranza, R.R.Carswell, C.Casini, P.Chamaraux, G.L.Chincarini, K.K.Chuvaev, G.Contopoulos, H.G.Corwin, G.Courtes, A.C.Danks, A.F.Davidsen, R.D.Davies, J.De la Moe, G.De Vaucouleurs, E.K.Denisyuk, E.A.Dibay, A.S.Di Serego, G.Di Tullio, K.J.Donner, L.L.Dressel, P.Dubois, M.C.Edmunds, G.Efstathiou, R.D.Ekers, D.M.Elmegreen, A.M.Elvius, V.F.Esipov, D.S.Evans, A.P.Fairall, M.W.Feast, C.Fehrenbach, J.Feitzinger, G.B.Field, A.Florsch, C.Foltz, A.F.Fomenko, H.Ford, W.K.Ford, C.W.Fraser, K.C.Freeman, J.W.Fried, J.S.Gallagher, G.Galletta, S.C.B.Gascoigne, F.D.Ghigo, R.Giovanelli, K.Gopal, W.M.Goss, S.T.Gottesman, L.Gouguenheim, S.A.Grandi, G.L.Grasdalen, R.E.Griffiths, T.Hara, E.Hardy, R.J.Harms, G.Haro, O.Heckmann, D.S.Heeschen, J.Heidmann, P.Hickson, P.M.N.Hintzen, E.B.Holmberg, F.Hoyle, J.P.Huchra, R.W.Hunstead, J.H.Hunter Jr., G.D.Illingworth, I.A.Issa, H.M.Johnson, Th.W.Jones, J.Jugaku, M.Kafatos, M.Kalinkov, A.T.Kalloglian, M.Kaufman, K.I.Kellerman, R.C.Kennicutt, E.Y.Khachikian, T.Kiang, I.R.King, T.D.Kinman, R.P.Kirshner, K.Kodaira, G.Kojoian, I.M.Kopylov, J.Kormendy, Yu.P.Korovyakovskij, R.G.Kron, N.A.Krumm, C.K.Kumar, M.R.Kundu, P.Kustaanheimo, J.P.J.Lafon, R.B.Larson, B.M.Lasker, A.Lauberts, A.Lausberg, D.Layzer, R.J.Leacock, G.Lelievre, J.Lequewx, J.Li, Q.-B.Li, C.C.Lin, P.O.Lindblad, K.Y.Lo, R.V.E.Lovelace, F.J.Low, B.T.Lynds, R.C.Lynds, V.M.Lyuty, H.T.Macgillivray, C.D.Mackay, B.F.Madore, P.Magnenat, M.L.Malagnini, J.W.-K.Mark, B.E.Markarian, L.Martinet, J.Materne, E.Maurice, N.U.Mayall, B.P.McBreen, R.D.McClure, W.H.McCrea, B.Q.McGimsey Jr., G.C.McVittie, T.K.Menon, H.R.Miller, J.S.Miller, R.H.Miller, B.Y.Mills, I.F.Mirabel, C.L.Morbey, W.W.Morgan, C.Moss, J.R.Mould, H.S.Murdoch, S.S.Murray, J.V.Narlikar, G.Neugebauer, T.W.Noonan, A.Oemler jr., S.Okamura, J.B.Oke, H.Oleak, G.C.Omer, J.H.Oort, A.M.Osman, D.E.Osterbrock, F.N.Owen, J.Pachner, A.G.Pacholczyk, T.L.Page, M.Peimbert, W.D.Pence, W.L.Peters, B.M.Peterson, Ch.J.Peterson, M.Petrou, P.Pismis de Recillas, V.S.Popov, A.Poveda, T.P.Prabhu, K.H.Prendergast, W.H.Press, M.L.Prevot-Burnichon, CH.J.Pritchet, I.I.Pronik, V.I.Pronik, H.Quintana, G.Reaves, V.C.Reddish, M.Reinhardt, H.B.Richer, L.J.Rickard, W.Rindler, M.S.Roberts, W.W.Roberts, J.G.Robertson, I.Robinson, H.J.Rood, J.A.Rose, H.J.Roeser, L.Rosino, A.Rots, K.Rudnicki, R.Sancisi, R.H.Sanders, C.L.Sarazin, W.L.W.Sargent, W.C.Saslaw, S.K.Sastry, M.Schmidt, E.L.Schucking, G.V.Schultz, H.Schulz, H.E.Schuster, U.J.Schwarz, F.Schweizer, D.W.Sciama, E.L.Scott, L.Searle, I.E.Segal, A.Serrano, J.L.Sersic, G.Setti, R.K.Shahbazian, J.R.Shakeshaft, C.D.Shane, P.A.Shaver, B.S.P.Shen, W.A.Sherwood, G.A.Shields, G.S.Shostak, B.F.Smith, H.C.Smith, H.E.Smith Jr., M.G.Smith, Y.Sobouti, A.M.Soltan, H.Spinrad, W.A.Stein, D.W.N.Stibbs, R.P.S.Stone, R.G.Strom, J.W.Sulenti, K.Takarada, B.Takase, R.J.Talbot, G.A.Tammann, V.YU.Terebizh, R.J.Terlevich, Y.Terzian, L.A.Thompson, N.Thonnard, T.X.Thuan, W.G.Tifft, H.M.Tovmassian, H.J.Treder, R.B.Tully, E.L.Turner, J.A.Tyson, M.Urbanik, T.S.Van Albada, S.Van der Bergh, J.M.Van der Hulst, H.Van der Laan, H.Van Woerden, R.K.Varma, M.P.Veron, N.Visvanathan, B.A.Vorontsov-Velyaminov, M.J.Ward, J.W.Warner, D.W.Weedman, P.A.Wehinger, G.A.Welch, A.E.Whitford, R.Wielen, J.P.Wiita, P.Wild, A.Wilkinson, R.F.Williams, TH.B.Williams, B.J.Wills, A.G.Wilson, K.H.Winkler, G.Wlerick, R.Wood, J.D.Wray, C.G.Wynn-Williams, J.S.Young, A.V.Zasov, Ya.B.Zel'dovich, R.J.Zinn.

COMMISSION NO 29.

SPECTRES STELLAIRES (STELLAR SPECTRA)

PRESIDENT: J.Jugaku

VICE-PRESIDENT: G.Cayrel-de-Strobel

COMITE D'ORGANISATION: Y.Andrillat, W.K.Bonsack, P.S.Conti, C.R.Cowley, C.O.R.Jaschek, O.H.Levato, M.A.Smith, R.Viotti, S.C.Wolff.

MEMBRES: K.D.Abhyankar, H.A.Abt, S.J.Adelman, G.C.L.Aikman, L.H.Aller, H.Andrillat, A.S.Asaad, I.A.Aslanov, G.B.Baratta, D.C.Barry, J.Berger, CH.Bertaud, M.S.Bessel, W.P.Bidelman, A.M.Boesgaard, A.Boggess, R.Bouigue, J.Boulon, A.A.Boyarchuk, E.S.Brandi, F.C.Bruhweiler, W.Buscombe, S.Catalano, A.Cassatella, F.Castelli, R.Cayrel, J.L.Climenhaga, A.D.Code, A.P.Cowley, M.De Groot, L.Divan, V.Doazan, P.P.Dobronravin, M.V.Dolidze, D.Dravins, T.Dunham Jr., M.M.Dworetsky, F.N.Edmonds Jr., D.S.Evans, R.Faraggiana, M.W.Feast, M.J.Fernandez-Figueroa, R.Freire, M.Friedjung, A.M.Fringant, Y.Fujita, R.F.Garrison, T.Gehren, R.E.Gershberg, D.P.Gilra, Ju.V.Glagolevskij, I.N.Glushneva, J.Goebel, L.Goldberg, V.G.Gorbatskij, L.Gratton, D.F.Gray, J.L.Greenstein, R.E.M.Griffin, R.F.Griffin, H.G.Groth, B.N.G.Guthrie, M.Hack, W.A.Hagen, P.Harmanec, D.L.Harmer, G.Haro, J.R.W.Heintze, K.G.Henlze, G.H.Herbig, R.Herman, L.Houziaux, A.R.Hyland, M.Jaschek, H.R.Johnson, P.C.Keenan, A.V.Kharitonov, V.Khokhlova, R.B.King, T.A.Kipper, K.Kodaira, T.Kogure, N.S.Komarov, I.M.Kopylov, P.Koubsky, R.J.Kovachev, R.P.Kraft, J.Krempec-Krygier, R.N.Kumajgorodskaya, D.Labs, D.L.Lambert, G.Larsson-Leander, D.S.Leckrone, J.B.Lester, D.D.Locanthi, L.S.Luud, H.M.Maitzen, S.Malaroda, D.H.Mcnamara, C.Megessier, O.A.Mel'nikov, J.E.Milligan, A.F.J.Moffat, H.W.Moos, N.Morguleff, Morrison, D.C.Morton, E.R.Mustel, R.W.Nicholls, A.A.Nikitin, S.Nishimura, L.Oetken, J.B.Oke, M.Y.Orlov, K.Osawa, B.E.J.Pagel, S.B.Parsons, L.E.Pasinetti, F.Paterson-Beeckmans, D.W.Peat, A.Pedoussaut, B.F.Peery, G.J.Peters, M.N.Perrin, R.C.Peterson, C.A.Pilachowski, M.Plavec, R.H.Poeckert, G.W.Preston, A.Przybylski, F.Querci, M.Rego Fernandez, D.Reimers, A.E.Ringuelet, A.W.Rodgers, K.Sadakane, J.Sahade, R.E.Schild, G.Scholz, W.Seggewiss, D.E.Shchegolev, S.N.Shore, U.Sinnerstad, A.Slettebak, J.Smak, J.Smolinski, Th.Snow, F.Spite, M.Spite, R.Stalio, A.Stawikowski, T.P.Stecher, R.E.Stencel, S.N.Svolopoulos, J.W.Swensson, J.-P.Swings, P.Swings, S.Taffara, M.Takada, J.Tech, G.I.Thompson, T.Tsuji, I.Tuominen, A.B.Underhill, K.Utsumi, K.A.Van der Hucht, C.Van't Veer, O.Vilhu, H.H.Voigt, S.S.Vogt, J.M.Vreux, G.N.Wallerstein, M.D.Waterworth, G.A.Wegner, P.Wehinger, W.H.Wehlau, W.W.Weiss, P.Wellmann, S.Weniger, P.M.Williams, O.C.Wilson, R.Wilson, R.F.Wing, B.Wolf, H.J.Wood, K.O.Wright, S.Wyckoff, A.A.Wyller, Y.Yamashita.

COMMISSION NO 30.

VITESSES RADIALES (RADIAL VELOCITIES)

PRESIDENT: A.G.Davis Philip

VICE-PRESIDENT: J.Andersen

COMITE D'ORGANISATION: M.Duflot, D.P.Hube, E.Maurice, M.Mayor, J.Sahade.

MEMBRES: H.A.Abt, M.Azzopardi, L.A.Balona, M.Barbier-Brossat, A.H.Batten, W.R.Beardsley, W.I.Beavers, F.C.Bertiau, C.T.Bolton, R.Bouigue, J.Boulon, D.Crampton, J.K.de Jonge, G.De Vaucouleurs, F.K.Edmondson, H.T.Eelsalu, C.Fehrenbach, J.M.Fletcher, A.Florsch, Y.Georgelin, F.Gieseking, R.F.Griffin, J.R.W.Heintze,

G.Hill, C.C.Huang, M.Imbert, I.D.Karachentsev, R.P.Kraft, D.W.Latham, N.Martin, C.L.Morbey, B.Nordstrom, L.Oetken, A.Pedoussaut, C.L.Perry, G.W.Preston, L.Prevot, E.Rebeirot, V.C.Rubin, N.B.Sanwal, C.D.Scarfe, K.Serkowski, J.Stock, E.Van Dessel, R.V.Willstrop.

COMMISSION NO 31.

L'HEURE (TIME)

PRESIDENT: G.Hemmleb

VICE-PRESIDENT: D.D.McCarthy

COMITE D'ORGANISATION: N.S.Blinov, C.C.Costain, H.F.Fliegel, B.Guinot, S.Iijima, Y.-R.Miao, A.Orte, P.Paquet, J.Pilkington, E.Proverbio.

MEMBRES: M.K.Abele, P.M.Afanasjeva, D.W.Allan, S.Aoki, D.J.Belotserkovski, J.Benavente, P.L.Bender, G.Billaud, J.Bonanomi, G.Caprioli, J.M.Chamberlain, I.Dominski, K.Dorenwendt, C.Dramba, H.Enslin, M.-K.Fujimoto, J.Gaignebet, T.Gokmen, H.-R.Gong, R.G.Hall, H.Hellwig, J.Hers, C.Kakuta, W.J.Klepczynski, T.Lederle, J.H.Lieske, A.M.Lozinskij, W.Markowitz, D.N.Matsakis, W.G.Melbourne, P.Melchior, A.A.Mikhailov, P.Morgan, I.I.Mueller, V.A.Naumov, F.Noel, W.Opalski, P.Parcelier, N.N.Pavlov, S.B.Pushkin, L.Randic, D.S.Robertson, W.Schuler, V.P.Shcheglov, A.M.Sinzi, H.M.Smith, D.E.Smylie, G.Stanila, G.V.Starytsin, J.M.Steele, A.Stoyko, S.Takagi, M.Tsao, A.Tsuchiya, R.O.Vicente, H.B.Wackernagel, L.Webrova, N.P.Wieth-knudsen, G.A.Wilkins, G.M.R.Winkler, S.-X.Wu, S.-H.Ye, S.Yumi.

COMMISSION NO 33.

STRUCTURE ET DYNAMIQUE DU SYSTEME GALACTIQUE

(STRUCTURE AND DYNAMICS OF THE GALACTIC SYSTEM)

PRESIDENT: R.Wielen

VICE-PRESIDENT: W.B.Burton

COMITE D'ORGANISATION: L.Blitz, W.Iwanowska, E.K.Kharadze, G.G.Kuzmin, D.Lynden-Bell, G.Lynga, M.Mayor, M.Miyamoto.

MEMBRES: S.Aarseth, T.A.Agekjan, W.J.Altenhoff, V.A.Ambartsumian, P.Andrle, V.A.Antonov, S.Aoki, A.L.Ardeberg, G.Asteriadis, E.Athanassoula, L.Balazs, J.E.Baldwin, B.Barbanis, N.F.Bash, B.Baud, W.Becker, E.M.Berkhuijsen, J.H.Bigay, J.J.Binney, A.Blaauw, V.M.Blanco, B.J.Bok, J.Boulon, B.F.Burke, H.V.Cane, L.Carrasco, J.Y.Caswell, E.B.Churchwell, S.V.M.Clube, N.F.Comins, G.Contopoulos, G.Courtes, D.Crampton, D.L.Crawford, M.Creze, K.M.Cudworth, S.Cuperman, R.D.Davies, T.De Jong, J.Delhaye, J.K.Denoyelle, H.R.Dickel, J.R.Dickel, N.H.Dieter, D.Downes, J.S.Drilling, R.M.Dzigvashvili, F.K.Edmondson, Y.N.Efremov, J.Einasto, B.G.Elmegreen, H.Elsasser, T.Elvius, E.Evangelidis, S.M.Faber, M.W.Feast, Ch.Fehrenbach, J.V.Feitzinger, R.P.Fenkart, M.P.Fitzgerald, K.C.Freeman, W.Fricke, M.Fujimoto, I.L.Genkin, Y.M.Georgelin, Y.P.Georgelin, G.F.Gilmore, W.Gliese, P.Goldreich, M.A.Gordon, E.J.Grayzeck, K.Gyldenkerne, H.J.Habing, K.Hamajima, U.Haug, A.Hayli, O.Heckmann, C.E.Heiles, M.Henon, W.Herbst, R.W.Hobbs, F.Hohl, G.Hori, V.A.Hughes, A.N.M.Hulsbosch, R.M.Humphreys, C.Hunter, S.Ikeuchi, S.Inagaki, K.A.Innanen,

J.B.Irwin, S.Isobe, C.Iwaniszewska-Lubienska, M.Iye, P.D.Jackson, H.Jahreiss, C.O.R.Jaschek, H.M.Johnson, D.H.P.Jones, M.Kaburaki, N.B.Kalandadze, A.J.Kalnajs, S.Kato, F.J.Kerr, P.N.Kholopov, I.R.King, T.D.Kinman, G.Klare, G.R.Knapp, L.N.Kolesnik, J.Kormendy, R.Kulsrud, R.Kurth, S.A.Kutuzov, J.P.Lafon, R.B.Larson, M.Lecar, J.Li, C.C.Lin, P.O.Lindblad, K.Loden, L.O.Loden, M.Lunel, W.J.Luyten, D.J.MacConnell, D.A.MacRae, K.G.Malmquist, R.N.Manchester, J.W.K.Mark, L.Marochnik, L.Martinet, D.S.Mathewson, L.N.Mavridis, M.F.McCarthy, M.O.Menessier, P.G.Mezger, R.H.Miller, I.F.Mirabel, L.N.Mirzoyan, A.F.J.Moffat, G.Monnet, W.W.Morgan, M.R.Morris, G.Munch, C.A.Murray, J.C.Muzzio, F.Nahon, Th.Neckel, A.H.Nelson, H.Niimi, K.F.Ogorodnikov, T.Oja, H.Okuda, A.Ollongren, J.H.Oort, J.P.Ostriker, M.W.Ovenden, P.E.Palmer, Th.Papayannopoulos, T.A.Pauls, E.D.Pavlovskaya, M.Peimbert, L.Perek, C.L.Perry, P.Pesch, A.G.Davis Philip, K.Pilowski, P.Pismis, L.Plaut, R.M.Price, W.Priester, J.M.Ramberg, K.W.Riegel, M.S.Roberts, W.W.Roberts, B.J.Robinson, K.Rohlfs, N.G.Roman, V.C.Rubin, M.T.Ruiz, G.B.Rybicki, E.Saar, Aa.Sandqvist, N.Sanduleak, H.Schmidt, K.H.Schmidt, M.Schmidt, Th.Schmidt-Kaler, H.M.Schwerdtfeger, W.Seggewiss, J.A.Sellwood, W.W.Shane, A.S.Sharov, D.Sher, T.Shimizu, F.Shu, W.L.H.Shuter, S.C.Simonson, A.Slettebak, P.M.Solomon, E.A.Spiegel, F.W.Stecker, A.E.Stefanovitch-Gomez, U.W.Steinlin, C.B.Stephenson, D.W.N.Stibbs, A.Stroebel, C.Sturch, S.N.Svolopoulos, V.Szebehely, G.A.Tammann, CH.Terzides, Pik Sin The, K.O.Thielheim, W.Tobin, A.Toomre, M.Tosa, C.F.Trefzger, A.Tsioumis, P.Turon-Lacarrieu, A.R.Upgren, P.C.Van der Kruit, A.Van Hoff, H.Van Woerden, H.L.Vanderlinden, P.O.Vandervoort, C.M.Varsavksy, A.G.Velghe, V.R.Venugopal, G.L.Verschuur, M.Vetesnik, V.I.Voroshilov, P.A.Wayman, H.F.Weaver, D.Weistrop, G.Westerhout, B.E.Westerlund, R.E.White, J.B.Whiteoak, D.C.B.Whittet, R.Wielebinski, Th.L.Wilson, L.Woltjer, P.R.Woodward, Sir Richard Woolley, S.S.O.Wramdemark, Y.Yoshii, C.Yuan.

## COMMISSION NO 34.

## MATIERE INTERSTELLAIRE (INTERSTELLAR MATTER)

PRESIDENT: M.Peimbert

VICE-PRESIDENT: J.Lequeux

COMITE D'ORGANISATION: S.D'Odorico, B.G.Elmegreen, E.B.Kostyakova, U.Mebold, J.S.Mathis, V.Radhakrishnan, P.A.Shaver, Y.Terzian, P.Wannier.

MEMBRES: P.Aanestad, A.Acker, D.K.Aitken, K.Akabane, L.H.Aller, W.Altenhoff, B.H.Andrew, C.D.Andriesse, H.Andrillat, Y.Andrillat, V.P.Arkhipova, T.T.Arny, L.Avery, W.I.Axford, J.W.M.Baars, J.E.Baldwin, T.M.Bania, M.Barlow, A.H.Barrett, A.Barnes, F.N.Bash, A.Baudry, E.Becklin, J.Beckman, S.V.W.Beckwith, N.Bel, J.Bergeron, E.M.Berkhuijsen, A.Bernat, J.Bieging, L.Biermann, K.Birkle, J.Black, J.Blades, G.Blair, R.C.Bless, N.G.Bochkarev, G.Boeshaar, A.Boggess, R.C.Bohlin, B.J.Bok, J.Borgman, P.W.J.L.Brand, E.Braunsfurth, G.Bromage, R.Brown, M.T.Bruck, F.C.Bruhweiler, A.Bunner, A.Burgess, B.F.Burke, W.B.Burton, N.V.Bystrova, E.R.Capriotti, R.G.Carruthers, J.L.Caswell, C.Cesarsky, D.Cesarsky, R.Chevalier, M.Chopinet, E.B.Churchwell, F.Clark, A.D.Code, M.Cohen, S.Collin-Souffrin, R.Costero, G.Courtes, L.Cowie, D.P.Cox, G.V.Coyne, J.Crovisier, P.Cruvellier, D.D.Cudaback, P.Cugnon, S.J.Czyzak, C.Dahn, A.Dalgarno, A.C.Danks, R.D.Davies, K.S.De Boer, P.Deen Singh, S.Deguchi, T.de Jong, J.De la Noe, E.A.Dibaj, H.R.Dickel, M.J.Disney, O.D.Dokutchaeva, M.A.Dopita, J.Dorschner, D.Downes, S.Drapatz, T.Dunham, A.Dupree, R.J.Dufour, E.Dwek, J.E.Dyson, K.Elliott, M.Elitzur, A.M.Elvius, J.Emerson, P.Encrenaz, V.F.Esipov, N.J.Evans II, S.Falk, S.A.E.G.Falle, D.J.Faulkner, W.Feibelman, S.R.Federman, M.Felli, J.E.Felten, D.Field, G.B.Field, B.Flannery, D.Flower, H.Ford, J.R.Forster, C.Friedemann, I.Furniss, P.Frisch, F.F.Gardner,

J.E.Gaustad, J.Gay, T.Gehrels, Y.Georgelin, E.Gerard, H.Gerola, D.Gezari, D.P.Gilra, R.Giovannelli, P.D.Godfrey, J.Goebel, P.Goldreich, D.W.Goldsmith, P.Goldsmith, S.J.Goldstein, F.A.Goldsworthy, C.P.Gordon, M.A.Gordon, I.V.Gosachinsky, W.M.Goss, G.Grasdalen, J.M.Greenberg, M.Grewing, M.Guelin, T.R.Gull, G.A.Gurzadjan, H.J.Habing, J.Hackwell, J.S.Hall, S.Harris, E.G.Hardebeck, J.Harrington, R.H.Harten, T.W.Hartquist, P.Harvey, R.Haynes, H.Hefele, C.E.Heiles, H.L.Helfer, K.G.Henize, G.H.Herbig, G.Herzberg, B.Hidajat, L.A.Higgs, W.A.Hiltner, R.H.Hildebrand, H.H.Hippelein, A.Hjalmarson, R.M.Hjellming, L.M.Hobbs, B.Hoglund, D.Hollenbach, J.Hollis, S.S.Hong, L.Houziaux, C.T.Hua, V.A.Hughes, A.N.M.Hulsbosch, D.Hummer, J.B.Hutchings, W.M.Irvine, S.Isobe, I.A.Issa, H.Itoh, P.Jackson, E.B.Jenkins, R.E.Jennings, F.M.Johnson, H.M.Johnson, K.J.Johnston, F.Jones, M.Jura, M.Kafatos, M.A.Kaftan-Kassin, F.D.Kahn, N.Kaifu, J.B.Kaler, F.Kamijo, I.Kazes, W.H.Kegel, F.J.Kerr, E.K.Kharadze, G.S.Khromov, H.Kimura, R.C.Kirkpatrick, R.Kirshner, R.Knack, G.R.Knapp, J.Knude, L.Kohoutek, I.G.Kolesnik, M.M.Komesaroff, J.Koornneef, Y.Kondon, J.Koppen, K.S.Krishna Swamy, C.Kumar, M.R.Kundu, M.Kutner, N.D.Kylafis, K.B.Kwitter, S.Kwok, C.Lada, H.Lambrecht, W.Langer, B.M.Lasker, C.Laurent, A.M.Le Squeren, T.J.Lee, C.M.Leung, W.Liller, C.C.Lin, R.Linke, H.S.Liszt, K.Y.Lo, L.O.Loden, R.Loren, M.C.Lortet-Zuckerman, F.Lovas, F.J.Low, T.A.Lozinskaya, R.Lucas, B.L.Lutz, B.T.Lynds, T.Maihara, D.C.V.Mallik, R.N.Manchester, J.Manfroid, R.Martin, W.G.Mathews, D.S.Mathewson, J.S.Mathis, K.Mattila, R.A.McCray, W.H.McCrea, R.X.McGee, C.McKee, C.McKeith, J.McLeod, D.McNally, J.Meaburn, R.R.Meier, R.Mendez, T.K.Menon, P.Meszaros, P.G.Mezger, A.Michalitsianos, K.W.Michel, T.Millar, J.S.Miller, D.K.Milne, Y.Minn, D.H.Morgan, M.Morimoto, M.Morris, D.C.Morton, T.P.Mouschovias, S.Mufson, G.Munch, P.Myers, Y.Nakada, K.Nandy, G.Neugebauer, H.Nussbaumer, J.Nittman, C.R.O'Dell, H.Ohtani, H.Okuda, J.H.Oort, T.Osaki, J.Osborne, D.E.Osterbrock, L.M.Ozernoy, P.E.Palmer, N.Panagia, V.Pankonin, E.N.Parker, T.Pauls, J.-C.Pecker, M.V.Penston, A.Penzias, M.Perinotto, W.Peters, V.Petrosian, A.P.Phillips, T.Phillips, P.Pismis, W.Poeppel, S.R.Pottasch, R.M.Price, I.I.Pronik, YU.P.Pskovskij, E.Raimond, J.C.Raymond, R.J.Reynolds, L.J.Rickard, G.Righini-Cohen, R.R.Robbins, W.W.Roberts, B.J.Robinson, A.E.E.Rodgers, L.F.Rodriguez, R.Roger, K.Rohlfs, W.K.Rose, L.Rosino, I.Roxburgh, D.A.Rozhkovskij, R.H.Rubin, V.C.Rubin, Y.Sabano, F.Sabbadin, E.E.Salpeter, R.Sancisi, A.Sandqvist, C.Sarazin, F.Sato, S.Sato, B.Savage, M.P.Savedoff, J.Scalo, C.Schalen, E.Schatzman, F.Scherb, P.A.G.Scheuer, J.Schmid-Burgk, Th.Schmidt, Th.Schmidt-Kaler, G.V.Schultz, P.Schwartz, U.J.Schwarz, R.D.Schwartz, E.H.Scott, M.J.Seaton, M.Seki, G.A.Shah, W.W.Shane, C.Y.Shao, M.M.Shapiro, S.L.Sharpless, S.J.Shawl, P.V.Shcheglov, W.Sherwood, G.Shields, I.S.Shklovskij, F.Shu, J.M.Shull, B.M.Shustov, W.L.Shuter, R.Silberberg, J.Silk, S.Simons, J.P.Sivan, J.Skilling, B.W.Smith, P.Smith, R.L.Snell, T.P.Snow, V.V.Sobolev, S.Sofia, P.M.Solomon, S.Tarafdar, K.N.Taylor, G.Tenorio-Tagle, P.Thaddeus, P.S.The, A.R.Thompson, N.Thonnard, H.A.Thronson, S.Torres-Peimbert, C.H.Townes, R.R.Treffers, B.E.Turner, K.C.Turner, M.H.Ulrich, W.Unno, H.C.Van de Hulst, P.Van den Bout, H.Van der Laan, H.Van Woerden, V.Vanysek, D.A.Varshalovich, G.L.Verschuur, J.L.Vidal, A.Vidal-Madjar, M.Viner, N.Visvanathan, B.A.Vorontsov-Vel'yaminov, F.Vrba, G.A.H.Walker, G.D.Watt, H.F.Weaver, B.L.Webster, K.W.Weiler, J.C.Weisheit, H.J.Wendker, P.R.Wesselius, G.Westerhout, R.J.Weymann, G.J.White, P.A.Whitelock, J.B.Whiteoak, D.Whittet, A.P.Whitworth, N.C.Wickramasinghe, D.Williams, D.A.Williams, I.P.Williams, R.E.Williams, A.Willis, S.Willner, R.W.Wilson, A.Winnberg, A.N.Witt, R.D.Wolstencroft, L.Woltjer, P.Woodward, N.J.Woolf, H.A.Wootten, E.Wright, C.C.Wu, G.Wynn-Williams, J.Xing, S.Yabushita, D.G.York, H.Yorke, W.Zealey, M.Zeilik, H.Zimmerman, B.Zuckerman.

COMMISSION NO 35.

CONSTITUTION DES ETOILES (STELLAR CONSTITUTION)

PRESIDENT: A.N.Cox

VICE-PRESIDENT: D.Sugimoto

COMITE D'ORGANISATION: P.Bodenheimer, C.Chiosi, D.J.Faulkner, A.Maeder, J.-L.Tassoul, R.J.Tayler, A.Tutukov, V.Weidemann, J.P.Zahn.

MEMBRES: M.Aizenman, S.P.S.Anand, I.Appenzeller, K.Arai, N.Arimoto, W.D.Arnett, J.Audouze, A.Baglin, G.Baym, G.Beaudet, G.Berthomieu, L.Biermann, G.S.Bisnovatyi-Kogan, S.A.Bludman, K.H.Boehm, H.Bondi, A.Boury, R.R.Brownlee, G.R.Burbidge, A.G.W.Cameron, T.R.Carson, V.Castellani, J.I.Castor, G.R.Caughlan, S.Chandrasekhar, C.Chevalier, H.Y.Chiu, J.Christensen-Dalsgaard, R.F.Christy, J.M.Cohen, J.J.Cowan, T.G.Cowling, J.P.Cox, F.D'Antona, C.G.Davis, D.S.P.Dearborn, J.-P.De Greve, W.Deinzer, P.R.Demarque, K.H.Despain, R.G.Deupree, P.S.A.Dingens, O.B.Dluzhnevskaya, R.H.Durisen, W.A.Dziembowski, A.C.Edwards, T.W.Edwards, P.P.Eggleton, T.A.Eminzade, A.S.Endal, I.Epstein, E.V.Ergma, Y.Eriguchi, D.Ezer, J.Faulkner, B.P.Flannery, G.Fontaine, J.E.Forbes, E.Fossat, W.A.Fowler, Yu.L.Frantsman, K.J.Fricke, M.Fujimoto, M.Gabriel, B.S.Geroyannis, P.Giannone, R.A.Gingold, S.-M.Gong, M.Goossens, D.O.Gough, E.Graham, T.Hamada, C.Hayashi, P.M.Heilesen, E.R.H.Hilf, J.Hitotsuyangai, R.Hoshi, F.Hoyle, R.-G.Huang, I.Iben, V.S.Imshennik, G.R.Isaak, T.Ishizuka, N.Itoh, R.A.James, H.Kaehler, M.Kiguchi, D.S.King, R.Kippenhahn, R.K.Kochar, L.N.Kolesnik, D.Koester, D.S.Kothari, A.Kovetz, M.Kozlowski, M.Krook, S.S.Kumar, R.Kushwaha, D.Q.Lamb Jr., S.A.Lamb, R.B.Larson, P.G.Laskarides, J.Latour, N.R.Lebovitz, P.Ledoux, J.R.D.Lepine, H.Li, J.E.Littleton, W.J.Maciel, W.H.Mccrea, G.Marx, A.Masani, A.G.Massevich, T.J.Mazurek, I.Mazzitelli, Yu.-K.Melik-Alaverdian, L.Mestel, E.Meyer-Hofmeister, G.Michaud, R.Mitalas, S.Miyaji, J.J.Monaghan, D.R.Moore, S.C.Morris, J.A.Morgan, D.L.Moss, C.Moellenhoff, D.K.Nadyozhin, T.Nakano, T.Nakamura, K.Nakazawa, K.Nariai, S.Narita, M.J.Newman, M.Nishida, K.Nomoto, A.P.Odell, N.Ohyama, I.Okamoto, Y.Osaki, J.P.Ostriker, B.Paczynski, A.A.Pamjatn'ikh, G.C.Pande, J.C.B.Papaloizou, D.Pines, A.D.Pinotsis, M.Plavec, M.D.Popova, V.V.Porfir'ev, A.Poveda, A.J.R.Prentice, J.Provost, Q.-Y.Qu, K.-H.Raedler, S.Ramadurai, H.Reeves, A.Reiz, A.Renzini, R.Rood, S.Rosseland, C.A.Rouse, I.Roxburgh, G.Ruben, H.Saio, S.Sakashita, T.Sakurai, E.E.Salpeter, K.Sato, M.P.Savedoff, J.M.Scalo, K.H.Schatten, E.Schatzman, D.N.Schramm, B.F.Schutz, M.Schwarzschild, R.Scuflaire, R.L.Sears, Z.F.Seidov, K.V.Sengbusch, G.Shaviv, H.Shibahashi., Y.Shibata, R.Sienkiewicz, E.M.Sion, P.Smeyers, R.C.Smith, Y.Sobouti, P.Souffrin, W.M.Sparks, E.A.Spiegel, S.S.Starrfield, R.F.Stellingwerf, D.W.N.Stibbs, P.A.Strittmatter, B.Stromgren, K.Suda, P.A.Sweet, A.V.Sweigart, M.Tassoul, S.Temesvary, H.C.Thomas, H.R.E.Tjin-a-Djie, J.E.Tohline, J.Toomre, V.L.Trimble-Weber, J.W.Truran, I.V.Tuominen, J.V.Tuominen, J.Uchida, R.Ulrich, W.Unno, U.Uus, D.Vandenberg, R.Van der Borght, H.B.Van der Raay, H.M.Van Horn, K.A.Van Riper, M.S.Vardya, G.Vauclair, S.C.Vila, O.Vilhu, R.A.Ward, T.A.Weaver, R.F.Webbink, V.Weidemann, A.Weigert, N.O.Weiss, J.C.Wheeler, R.E.Wilson, K.H.Winkler, P.R.Wood, G.A.E.Wright, D.-R.Xiong, H.W.Yorke, S.A.Zhevakin, J.Ziolkowski.

COMMISSION NO 36.

LA THEORIE DES ATMOSPHERES STELLAIRES (THE THEORY OF STELLAR ATMOSPHERES)

PRESIDENT: B.Gustafsson

VICE-PRESIDENT: K.Kodaira

COMITE D'ORGANISATION: D.F.Gray, A.G.Hearn, W.Kalkofen, R.P.Kudritski, D.Mihalas, A.A.Sapar, M.J.Seaton, G.Traving, A.B.Underhill.

MEMBRES: K.D.Abhyankar, L.H.Aller, R.C.Altrock, C.Arpigny, R.G.Athay, L.H.Auer, J.R.Auman, E.H.Avrett, B.Baschek, R.A.Bell, A.P.Bernat, C.Bertoud, L.Biermann, C.Blanco, R.C.Bless, A.M.Boesgaard, K.M.Boehm, E.Boehm-Vitense, I.D.Bues, C.J.Cannon,

D.Carbon, T.R.Carson, J.Cassinelli, J.I.Castor, R.Cayrel, G.Cayrel de Strobel, R.E.S.Clegg, P.S.Conti, C.R.Cowley, L.E.Cram, L.Crivellari, Y.Cuny, C.G.Davis, C.de Jager, Ph.Delache, H.Domke, P.L.Dufton, S.Dumont, A.K.Dupree, F.N.Edmonds, G.H.Elste, K.Eriksson, E.Evangelidis, R.Faraggiana, G.D.Finn, R.Foy, H.Frisch, U.Frisch, C.Froeschle, H.Gail, K.B.Gebbie, O.Gingerich, N.Gokdogan, Ch.Gordon, D.Gough, I.P.Grant, J.L.Greenstein, N.Grevesse, V.P.Grinin, H.G.Groth, E.A.Gussman, M.H.Hack, B.M.Haisch, J.Hardorp, J.N.Heasley, L.W.Hartmann, J.Hekela, J.Hitotuyangai, H.Holweger, T.E.Holzer, M.Hotinli, L.L.House, J.Houtgast, I.Hubeny, D.G.Hummer, K.Hunger, J.B.Hutchings, V.V.Ivanov, J.T.Jefferies, H.R.Johnson, L.W.Kamp, R.S.Kandel, A.H.Karp, V.L.Khokhlova, R.I.Klein, D.Koester, A.K.Kolesov, E.Kontizas, K.S.Krishna Swamy, R.P.Kudritzki, L.V.Kuhi, S.S.Kumar, R.L.Kurucz, R.S.Kushwaha, D.L.Lambert, J.Leibacher, J.L.Linsky, J.M.Marlborough, M.Matsumoto, S.Matsushima, G.Michaud, S.Miyamoto, M.Mnatsakanian, S.Mukai, E.A.Muller, G.Munch, E.R.Mustel, J.P.Mutschlecner, D.I.Nagirner, K.Nariai, J.S.Neff, L.Neven, B.J.O'Mara, F.Q.Orrall, J.Oxenius, B.E.J.Pagel, R.J.Panek, L.Pasinetti, J.-C.Pecker, A.Peraiah, G.J.Peters, J.G.Phillips, S.R.Pottasch, F.Praderie, D.N.Rachkovsky, L.W.Ramsey, D.Reimers, J.E.R.Ross, R.J.Rutten, G.B.Rybicki, S.Saito, N.A.Sakhibullin, D.C.Schmalberger, J.Schmid-Burgk, M.Scholz, D.Schoenberner, E.Sedlmayer, R.A.Shine, H.L.Shipman, E.Simmonneau, T.Simon, G.Sitnik, A.Skumanich, V.V.Sobolev, Y.Sobouti, P.Souffrin, M.A.J.Snijders, E.A.Spiegel, F.Spite, M.Spite, R.Stalio, R.F.Stein, R.E.Stencel, K.Stepien, D.W.N.Stibbs, S.E.Strom, T.L.Swihart, S.P.Tarafdar, R.N.Thomas, J.Toomre, T.Tsuji, S.Ueno, A.Uesugi, W.Unno, H.Van Regemorter, C.Van't Veer, F.Van't Veer, I.Vardavas, M.Vardya, T.Viik, S.V.Weber, R.Wehrse, V.Weidemann, P.Wellmann, D.T.Wickramsinghe, P.R.Wilson, S.J.Wilson, K.O.Wright, A.A.Wyller, E.G.Yanovitsky, H.W.Yorke, J.P.Zahn, J.B.Zirker, C.Zwaan.

COMMISSION NO 37.

AMAS STELLAIRES ET ASSOCIATIONS (STAR CLUSTERS AND ASSOCIATIONS)

PRESIDENT: K.C.Freeman

VICE-PRESIDENT: D.C.Heggie

COMITE D'ORGANISATION: G.L.H.Harris, G.Lynga, P.E.Nissen, C.A.Pilachovski, G.M.Salukvadze.

MEMBRES: S.Aarseth, T.A.Agekjan, A.K.Alksnis, C.P.Allen, B.Balazs, K.A.Barkhatova, W.Becker, A.Bijaoui, A.Blaauw, P.B.Bouvier, M.Breger, J.Brodie, M.S.Burkhead, D.Butler, G.G.Byrd, R.D.Cannon, B.Carney, C.A.Christian, M.S.Chun, J.J.Claria, K.M.Cudworth, J.Cuffey, G.S.Da Costa, I.A.Daube, P.Demarque, S.Demers, R.J.Dickens, Q.B.Dluzhnevskaya, Yu.N.Efremov, J.Einasto, M.S.A.Ella, B.G.Elmegreen, M.W.Feast, A.Feinstein, M.P.Fitzgerald, J.C.Forte, S.C.B.Gascoigne, M.Golay, J.E.Grindlay, C.Grubissich, D.A.Hanes, W.E.Harris, C.A.Harvel, S.M.Hassan, T.G.Hawarden, H.Henon, W.Herbst, J.E.Hesser, J.G.Hills, I.Iben, G.D.Illingworth, K.Ischida, K.A.Janes, D.H.P.Jones, Z.Kadla, L.W.Kamp, P.N.Kholopov, I.R.King, M.Kontizas, G.E.Kron, C.J.Lada, G.Larsson-Leander, A.Laval, V.V.Lavdovskij, M.H.Liller, T.Lloyd Evans, L.O.Loden, W.Lohmann, P.K.Lu, D.Lynden-Bell, A.Maeder, T.Markkanen, H.G.Marraco, D.H.Martins, T.K.Menon, J.W.Menzies, J.C.Mermilliod, J.Meurers, A.F.J.Moffat, W.W.Morgan, J.Mould, C.A.Murray, E.B.Newell, K.Ogura, W.Osborn, A.M.I.Oosman, E.S.Parsamian, A.J.Penny, C.J.Peterson, M.S.Petrovskaya, A.G.D.Philip, M.D.Popova, A.Poveda, R.Rajamohan, V.C.Reddish, L.Rosino, M.L.Roth, J.Rountree, J.Ruprecht, N.N.Samus, W.L.Sanders, H.B.Sawyer Hogg, H.E.Schuster, A.Sharov, S.J.Shawl, D.Sher, M.Shimoda, A.Terzan, P.S.The, D.G.Turner, A.Upgren, W.F.Van Altena, S.Van den Bergh, D.Vandenbergh, Yu.K.Veltmann, N.Vogt, S.Von Hoerner, G.A.H.Walker, M.F.Walker, A.A.E.Wallenquist, L.Wan, W.H.Warren, H.F.Weaver, A.Wehlau, R.E.White, R.Wielen, R.Woolley, S.S.O.Wramdemark, H.H.Wu, R.J.Zinn.

COMMISSION NO 38.

ECHANGE DES ASTRONOMES (EXCHANGE OF ASTRONOMERS)

PRESIDENT: F.B.Wood

VICE-PRESIDENT: E.A.Mueller

COMITE D'ORGANISATION: A.A.Boyarchuk, J.Delhaye, D.A.MacRae, P.M.Routly, J.Sahade, F.G.Smith, C.R.Tolbert, H.H.Voigt, S.-H.Ye.

MEMBRES: A.W.Al-Sabti, B.J.Bok, A.Florsch, G.Godoli, H.Haupt, G.Keller, K.-C.Leung, H.Li, M.Marik, S.Miyamoto, I.S.Nha, S.E.Okoye, A.Opolski, T.L.Page, A.Reiz, G.Ruben, E.V.P.Smith, G.Teleki, J.P.Wild.

COMMISSION NO 40.

RADIOASTRONOMIE (RADIO ASTRONOMY)

PRESIDENT: K.I.Kellerman

VICE-PRESIDENT: J.E.Baldwin

COMITE D'ORGANISATION: J.G.Ables, N.W.Broten, G.A.Dulk, B.Hoeglund, N.S.Kardashev, M.Morimoto, S.Okoye, G.Setti, R.G.Strom, G.Swarup, S.-G.Wang, L.Weliachew, T.L.Wilson.

MEMBRES: A.Abrami, P.A.R.Ade, K.Akabane, J.K.Alexander, R.J.Allen, H.Aller, M.F.Aller, W.J.Altenhoff, B.H.Andrew, P.E.Argyle, M.E.Arnal, G.E.Assousa, L.W.Avery, Y.Avignon, J.W.M.Baars, E.E.Baart, L.B.Baath, D.C.Backer, D.S.Bagri, A.Balklavs, A.H.Barret, C.Barrow, N.H.Bartel, D.Basu, R.H.T.Bates, A.O.Benz, G.L.Berge, E.Berkhuijsen, J.H.Bieging, F.Biraud, R.D.Blandfort, E.J.Blum, A.Boehme, A.Boischot, J.G.Bolton, R.S.Booth, V.Boriakoff, P.F.Bowers, R.N.Bracewell, S.Ya.Braude, J.D.Bregman, A.H.Bridle, W.N.Brouw, I.W.A.Browne, G.R.Burbidge, B.F.Burke, C.Caroubalos, T.D.Carr, J.L.Caswell, M.Ceccarelli, Y.Chikada, W.A.Christiansen, W.N.Christiansen, B.G.Clark, D.H.Clark, F.O.Clark, M.H.Cohen, R.J.Cohen, T.Cole, J.J.Condon, E.K.Conklin, R.G.Conway, B.F.C.Cooper, J.M.Cordes, C.H.Costain, R.A.J.Coutrez, A.E.Covington, P.C.Crane, D.L.Croom, J.Crovisier, D.D.Cudaback, E.J.Daintree, T.Daishido, J.G.Davies, R.D.Davies, M.M.Davis, R.J.Davis, T.De groot, C.de Jager, J.Delannoy, J.De la Noe, J.F.Denisse, B.K.Dennison, W.A.Dent, P.E.F.Dewdney, H.R.Dickel, J.R.Dickel, N.H.Dieter, R.Dixon, J.N.Douglas, D.Downes, G.S.Downs, F.D.Drake, A.F.Dravskikh, B.A.Dubinskij, P.J.Duffett-Smith, F.J.Dyson, R.D.Ekers, O.Elgaroy, G.R.A.Ellis, B.Elsmore, G.Elwert, E.E.Epstein, W.C.Erickson, G.Eriksen, Von R.Eshleman, K.D.Evans, M.Ewing, R.Fanti, G.Feix, P.A.Feldman, M.Felli, J.E.Felten, G.B.Field, J.W.Findlay, R.Fleischer, A.D.Fokker, E.B.Fomalont, D.N.Fort, R.H.Frater, H.Friedman, Y.Fukui, E.Fuerst, J.Galt, F.F.Gardner, K.H.Gebler, G.B.Gel'frejkh, H.Gent, R.Genzel, T.E.Gergely, F.D.Ghigo, D.M.Gibson, V.L.Ginzburg, I.M.Gioia, T.Gold, P.F.Goldsmith, S.J.Goldstein, H.C.Goldwire, R.Gonze, U.V.Gopala Rao, M.A.Gordon, S.Gorgolewski, I.V.Gosachinskij, W.M.Goss, J.F.R.Gower, P.C.Gregory, M.Grewing, M.Guelin, S.Gulkis, S.F.Gull, O.Hachenberg, F.T.Haddock, J.P.Hagen, P.A.Hamilton, J.Hanasz, R.Hanbury Brown, T.H.Hankins, P.Hardee, D.E.Harris, R.H.Harten, T.R.Hartz, A.Haschick, C.G.T.Haslam, M.P.Haynes, R.F.Haynes, C.Hazard, D.S.Heeschen, J.Heidmann, C.Heiles, A.Hewish, J.S.Hey, L.A.Higgs, E.R.Hilf, R.E.Hills, H.Hirabayashi, R.M.Hjellming, P.T.P.Ho, D.Hoang Binh, R.W.Hobbs, J.A.Hoegbom, D.E.Hogg, J.M.Hollis, B.G.Hooghoudt, W.E.Howard, W.K.-H.Huchtmeier, V.A.Hughes, A.N.M.Hulsbosch, R.W.Hunstead, R.N.Ikhsavov, V.N.Ikhsanova, M.Inoue, W.J.Jaffe, M.A.Janssen, D.L.Jauncey, R.C.Jennison, L.E.B.Johansson, D.R.Johnson,

K.Johnston, F.Joly, M.N.Joshi, M.Kaftan-Kassim, F.D.Kahn, K.Kai, N.Kaifu, T.Kakinuma, V.K.Kapahi, P.Kaufmann, Kin-Aki Kawabata, I.Kazes, S.Kenderdine, F.J.Kerr, M.L.Kesteven, A.G.Kislyakov, Hsien-Ching Ko, M.M.Kobrin, G.Kojoian, M.Komesaroff, D.V.Korolkov, V.A.Kotelnikov, J.D.Kraus, T.Krishnan, P.P.Kronberg, E.Kruegel, J.M.E.Kuijpers, W.Kundt, M.R.Kundu, V.N.Kuril'chik, A.J.Kus, M.L.Kutner, A.D.Kuz'min, C.J.Lada, M.Laffineur, K.R.Lang, W.D.Langer, P.Lantos, M.I.Large, Y.Leblanc, J.R.D.Lepine, J.Lequeux, A.M.Le Squeren, C.M.Leung, A.E.Lilley, R.A.Linke, A.G.Little, L.Th.Little, K.-Y.Lo, J.L.Locke, M.S.Longair, R.B.Loren, T.Lozinskaya, Sir Bernard Lovell, A.G.Lyne, J.M.MacLeod, D.A.MacRae, G.H.MacDonald, J.Machalski, R.N.Manchester, S.P.Maran, A.P.Marscher, R.N.Martin, D.N.Matheson, K.Mattila, D.N.Matsakis, L.I.Matveyenko, A.Maxwell, C.H.Mayer, W.B.McAdam, P.M.McCulloch, R.X.McGee, S.McKenna-Lawlor, D.J.McLean, U.Mebold, M.L.Meeks, T.K.Menon, P.G.Mezger, G.K.Miley, B.Y.Mills, D.K.Milne, I.F.Mirabel, A.T.Moffet, I.G.Moiseev, A.P.Molchanov, J.Moran, I.Morison, F.Moriyama, R.Morras, D.Morris, M.R.Morris, C.A.Muller, H.S.Murdoch, R.L.Mutel, P.Myers, G.D.Nicolson, J.D.O'Sullivan, J.H.Oort, D.E.Osterbrock, F.N.Owen, A.G.Pacholczyk, P.Palmer, V.L.Pankonin, M.Papagiannis, Yu.N.Parijskij, E.A.Parker, A.Parrish, J.M.Pasachoff, I.K.K.Pauliny-Toth, T.A.Pauls, A.Pedlar, A.A.Penzias, R.A.Perley, W.L.Peters, G.H.Pettengill, T.G.Phillips, M.Pick, J.E.B.Ponsonby, G.G.Pooley, R.Porcas, R.A.Preston, E.Preuss, R.M.Price, W.Priester, J.J.Puschell, V.Radhakrishnan, E.Raimond, R.Ramaty, A.P.Rao, V.A.Razin, A.C.S.Readhead, G.Reber, W.Reich, M.J.Reid, J.-C.Ribes, L.J.Rickard, B.J.Rickett, J.M.Riley, J.A.Roberts, M.S.Roberts, D.S.Robertson, J.G.Robertson, B.J.Robinson, R.D.Robinson, R.C.Roeder, R.S.Roger, A.E.E.Rogers, D.H.Rogstad, K.Rohlfs, J.D.Romney, B.Rowson, L.Rudnick, G.H.B.Rydbeck, O.E.H.Rydbeck, Sir Martin Ryle, N.F.Ryzhkov, A.E.Salomonovich, E.E.Salpeter, V.A.Sanamjan, Ch.V.Sastry, F.Sato, A.Savage, E.Scalise Jr., P.A.G.Scheuer, R.T.Schilizzi, R.Schlickeiser, M.Schmidt, G.V.Schultz, P.R.Schwartz, U.J.Schwarz, J.S.Scott, P.F.Scott, E.R.Seaquist, G.A.Seielstad, J.H.Seiradakis, D.Shaffer, J.R.Shakeshaft, P.Shaver, K.V.Sheridan, A.J.Shimmins, I.S.Shklovsky, G.B.Sholomitsky, W.L.H.Shuter, W.Sieber, P.A.Simon, R.P.Sinha, M.A.Slade, O.B.Slee, V.I.Slysh, A.G.Smith, D.F.Smith, F.G.Smith, N.S.Soboleva, Y.Sofue, R.L.Sorochenko, J.H.Spencer, R.E.Spencer, R.Sramek, G.J.Stanley, D.Stannard, J.L.Steinberg, P.Stewart, R.G.Stone, W.T.Sullivan, G.W.Swenson Jr., H.Tabara, K.Takagi, T.Takakura, H.Tanaka, R.Tanaka, J.C.Tarter, Y.Terzian, P.Thomasson, A.R.Thompson, C.Thum, A.Tlamicha, C.R.Tolbert, P.Tomasi, H.M.Tovmassian, C.H.Townes, K.P.Tritton, V.S.Troitsky, T.H.Troland, S.Tsuruta, Z.Turlo, B.E.Turner, K.C.Turner, A.J.Turtle, B.T.Ulrich, M.H.Ulrich, J.P.Vallee, H.C.Van de Hulst, T.Van der Hulst, P.C.Van der Kruit, H.Van der Laan, J.Van Nieuwkoop, H.Van Woerden, P.Vanden Bout, C.M.Varsavsky, T.Velusamy, V.R.Venugopal, P.Veron, G.L.Verschuur, M.R.Viner, C.M.Wade, J.V.Wall, M.C.Walmsley, D.Walsh, T.-S.Wan, S.-G.Wang, P.G.Wannier, J.F.C.Wardle, P.J.Warner, J.W.Warwick, H.F.Weaver, H.J.Wendker, G.Westerhout, K.C.Westfold, J.B.Whiteoak, N.C.Wickramasinghe, R.Wielebinski, J.P.Wild, P.N.Wilkinson, D.Williams, A.G.Willis, B.Wills, D.Wills, R.F.Willson, A.S.Wilson, R.W.Wilson, W.J.Wilson, J.E.Wink, A.Winnberg, G.Winnewisser, A.Witzel, L.Woltjer, A.W.Woodsworth, A.E.Wright, P.-Y.Xu, V.V.Zaitsev, V.V.Zhelezniakov, S.Zieba, P.Zlobec, B.M.Zuckerman.

COMMISSION NO 41.

HISTOIRE D'ASTRONOMIE (HISTORY OF ASTRONOMY)

PRESIDENT: O.Pedersen

VICE-PRESIDENT: J.A.Eddy

COMITE D'ORGANISATION: H.Eelsalu, E.Forbes, M.A.Hoskin, Verdet, K.Yabuuti.

MEMBRES: S.M.R.Ansari, J.G.Argyracos, R.Berendzen, R.L.Bishop, J.P.Brunet,

J.B.Carlson, Z.G.Chen, M.Cimino, Z.Dadic, J.Darius, T.L.Deeming, E.Dekker, D.W.Dewhirst, J.Dobrzycki, A.V.Douglas, N.P.Erpylev, D.S.Evans, D.Fernie, D.K.Ferrari-d'Occhieppo, J.Filliozat, M.G.Firneis, P.S.Florides, H.C.Freiesleben, R.R.Freitas-Mourao, O.Gingerich, G.S.Hawkins, A.Hayli, D.Heggie, D.Herrmann, H.Hirose, E.J.Hysom, Z.Horsky, H.D.Howse, G.M.Idlis, G.Jackisch, J.E.Kennedy, G.S.Khromov, T.Kiang, H.C.King, D.Kotsakis, P.G.Kulikovsky, K.R.Lang, J.R.Levy, F.Link, S.M.P.McKenna-Lawlor, A.J.Meadows, J.Merleau-Ponty, K.P.Moesgaard, S.Nakayama, J.North, G.C.Omer, D.Osterbrock, J.Pelseneer, C.J.Peterson, W.Petri, D.Pingree, A.Pogo, N.A.Porter, E.Poulle, C.A.Ronan, E.Rosen, E.Rybka, P.Rybka, V.P.Shcheglov, K.S.Shukla, F.R.Stephenson, W.Sullivan, A.Sundman, S.N.Svolopoulos, N.Swerdlov, R.Taton, V.Thoren, D.Wattenberg, E.A.Whitaker, G.J.Whitrow, C.A.Wilson, H.Wright, Z.Z.Xi, D.K.Yeomans, Q.M.Zhao.

## COMMISSION NO 42.

## ETOILES DOUBLES SERREES (CLOSE BINARY STARS)

PRESIDENT: A.H.Batten

VICE-PRESIDENT: J.Smak

COMITE D'ORGANISATION: K.D.Abhyankar, J.Andersen, A.M.Cherepashchuk, M.Kitamura, R.H.Koch, Y.Kondo, K.-Ch.Leung, M.Rodono, G.Shaviv, E.P.J.Van den Heuvel, B.Warner.

MEMBRES: E.Antonopoulou, B.M.Barber, G.T.Bath, L.Binnendijk, W.Blitzstein, C.T.Bolton, B.B.Bookmyer, S.Bonnazzola, B.W.Bopp, R.Breinhorst, P.Broglia, R.R.Brownlee, F.Bruhweiler, E.Budding, A.N.Bunner, S.Catalano, B.Cester, R.C.Chambliss, K.-Y.Chen, G.Chanmugam, G.G.Cillie, J.V.Clausen, D.Chochol, G.W.Collins, A.P.Cowley, S.Cristaldi, A.N.Dadaev, J.-P.De Greve, M.J.H.De Groot, J.De Kort, O.Demircan, N.A.Doughty, H.Drechsel, H.W.Duerbeck, J.A.Eaton, E.G.Ebbighausen, P.P.Eggleton, J.Faulkner, F.Fekel, K.Ferrari d'Occhieppo, C.Firmani, B.P.Flannery, M.Fracassini, M.G.Fracastoro, Yu.L.Frantsman, L.W.Fredrick, A.Fresa, M.Friedjung, S.Gaposhkin, C.D.Garmany, E.H.Geyer, P.Giannone, D.M.Gibson, G.Giuricin, J.Grygar, E.F.Guinan, A.F.Gulliver, H.Gursky, K.Gyldenkerne, D.S.Hall, G.Hammerschlag-Hensberge, P.Harmanec, J.Hazlehurst, T.J.Herczeg, R.W.Hilditch, G.Hill, R.M.Hjellming, M.Hoffmann, T.Horak, D.P.Hube, J.Hutchings, M.Imbert, J.B.Irwin, C.Jaschek, P.C.Joss, I.Jurkevich, C.D.Kandpal, V.G.Karetnikov, S.Kawabata, A.Kizilirmak, Z.Kopal, P.Koubsky, R.P.Kraft, V.A.Krat, Z.Kraicheva, J.Krautter, J.M.Kreiner, S.Kriz, K.G.Kron, A.Kruszewski, W.Krzeminski, M.I.Kumsiashvily, M.Kurpinska-Vinierska, Z.Kviz, K.K.Kwee, C.H.Lacy, D.Q.Lamb, A.U.Landolt, G.Larsson-Leander, M.I.Lavrov, A.P.Linndnell, L.B.Lucy, V.M.Lyuty, N.L.Magalashvily, A.Mammano, F.Mardirossian, E.Marilli, B.F.Marino, D.Ya.Martynov, J.A.Mattei, H.Mauder, P.Mayer, G.E.Mc Cluskey, J.McDonald, J.E.Merrill, E.Meyer-Hofmeister, M.Mezzetti, E.F.Milone, S.W.Mochnacki, T.H.Morgan, G.S.Mumford, H.M.K.Naimy, K.Nariai, R.E.Nather, B.Nelson, IL-S.Nha, B.Nordstrom, A.Okazaki, J.P.Oliver, M.V.Ovenden, B.Paczynski, G.J.Peters, V.Piirola, S.Piotrowski, M.Plavec, R.S.Polidan, D.M.Popper, J.Pringle, A.Purgathofer, J.B.Rafert, J.Rahe, T.Rahunen, K.D.Rakos, S.Refsdal, E.G.Reuning, H.Ritter, J.A.Robertson, E.L.Robinson, P.Rovithis, I.W.Roxburgh, S.Rucinski, A.R.Sadik, J.Sahade, N.B.Sanwal, A.Sanyal, F.Scaltriti, C.D.Scarfe, H.Schmidt, H.J.Schober, E.F.Schoeffel, W.Seggewiss, I.Semeniuk, N.I.Shakura, L.Z.Shen, F.H.Shu, A.R.Shulberg, Z.Sima, J.F.L.Simmons, S.D.Sinvhal, R.C.Smith, S.Sobieski, S.Soderhjelm, J.-E.Solheim, W.M.Sparks, S.Starrfield, R.E.Stencel, W.Strohmeier, D.Sugimoto, A.Sundman, M.A.Svechnikov, R.Szafraniec, P.Szkody, I.Todoran, J.Tremko, V.Trimble, V.P.Tsesevich, A.V.Tutukov, V.Ureche, W.Van Hamme, A.Van Hoof, J.Van Paradijs, F.Van't Veer, M.Vetesnik, O.Vilhu, W.S.G.Walker, K.Walter, M.J.Ward, R.F.Webbink, W.H.Wehlau, A.Weigert, E.J.Weiler, P.Wellmann, A.J.Wesselink, C.Whyte, T.Widorn, R.M.Williamon, R.E.Williams, R.E.Wilson, D.B.Wood, F.B.Wood, K.O.Wright, A.Yamasaki, D.S.Zhai, J.Ziolkowski, E.J.Zuiderwijk.

COMMISSION NO 44.

L'ASTRONOMIE A PARTIR DE L'ESPACE (ASTRONOMY FROM SPACE)

PRESIDENT: M.Oda

VICE-PRESIDENT: Y.Kondo

COMITE D'ORGANISATION: A.A.Boyarchuk, K.Fredga, M.Grewing, D.C.Morton, L.Peterson, E.Peytremann, A.B.Underhill, B.Valnicek, K.A.Van der Hucht, R.J.Van Duinen, A.J.Willis.

MEMBRES: L.W.Acton, J.K.Alexander, B.Aschenbach, G.F.Benedict, J.E.Blamont, R.C.Bless, A.Boggess, R.C.Bohlin, A.Boksenberg, R.M.Bonnet, R.L.F.Boyd, J.C.Brandt, B.C.Brinkman, G.E.Brueckner, F.C.Bruhweiler, E.C.Bruner, V.Bumba, M.Burger, W.M.Burton, M.F.Campbell, P.K.Carroll, J.H.Carver, R.C.Catura, G.Clark, P.A.Charles, D.Chochol, T.A.Chubb, T.A.Clark, A.D.Code, G.Courtes, C.J.Crannell, L.Culhane, A.F.Davidsen, R.J.Davis, C.de Jager, B.R.Dennis, J.F.Dolan, L.Dunkelman, A.Dupree, R.Faraggiana, G.G.Fazio, C.E.Fichtel, D.C.Fisher, G.J.Fishman, B.Fitton, I.Furniss, H.Friedman, A.H.Gabriel, D.Y.Gezari, R.Giacconi, D.P.Gilra, H.Glaser, T.Gold, L.Goldberg, P.Gondhalekar, H.D.Greyber, R.E.Griffiths, T.R.Gull, H.Gursky, M.Hack, F.Haddock, K.L.Hallam, R.J.Harms, T.R.Hartz, C.C.Harvey, P.M.Harvey, S.Hayakawa, A.G.Hearn, H.M.Heckathorn, J.Heise, H.Helmken, K.G.Henize, J.C.Henoux, H.Hensberge, H.E.Hinteregger, J.A.Hoffman, L.Houziaux, M.C.E.Hueber, C.Jamar, E.B.Jenkins, C.Jordan, M.Kafatos, K.Kasturirangan, G.G.Kramer, W.L.Kraushaar, V.G.Kurt, H.J.Lamers, D.S.Leckrone, P.Lemaire, W.Lewin, B.A.Lindblad, J.Linsky, Sir Bernard Lovell, P.Luest, F.Machetto, D.J.Malaise, H.M.Malitson, A.Manara, S.L.Mandelstam, S.P.Maran, T.M.K.Marar, M.Y.Marov, M.Matsuoka, G.E.McCluskey, K.G.McCracken, R.W.P.McWhirter, J.P.Mehltretter, R.Mewe, A.A.Mikhailov, A.Michalitsianos, J.L.Modisette, A.Monfils, H.W.Moos, T.H.Morgan, D.C.Morton, E.A.Muller, T.L.Murdock, N.F.Ness, W.H.Neupert, G.A.Newkirk, R.Novick, V.Novotny, R.Noyes, H.Olthof, E.O'Mongain, T.Owen, F.Pacini, M.Papagiannis, J.H.Parkinson, W.H.Parkinson, G.Peters, K.Pinkau, R.Poliden, K.Pounds, S.D.Price, V.D.Prokof'ev, J.Rahe, R.V.Rao, M.Rees, E.M.Reeves, W.A.Rense, G.Righini-Cohen, N.G.Roman, J.D.Rosendhal, G.Ruben, A.E.Salomonovich, K.Sato, B.Savage, W.Schoneich, G.V.Schultz, S.J.Schwartz, P.Selvelli, A.B.Severny, K.Shivanandan, G.B.Sholomitsky, P.Simon, P.C.Simon, B.A.Smith, H.A.Smith, H.J.Smith, T.P.Snow, S.Sofia, R.Speer, T.P.Stecher, J.L.Steinberg, R.E.Stencel, P.A.Stern, H.S.Stockman, R.G.Stone, T.Takakura, Y.Tanaka, W.A.Traub, J.Truemper, W.L.Upson, H.F.Van Beek, H.C.Van de Hulst, L.P.Van Speybroeck, J.C.Vial, A.Vidal-Madjar, R.Viotti, D.Walsh, J.W.Warner, E.J.Weiler, J.L.Weinberg, P.Wesselius, J.A.Westphal, S.P.Willner, R.Wilson, J.D.Wray, C.C.Wu, K.Yamashita, J.C.Zarneki, M.V.Zombeck, H.C.Zou.

COMMISSION NO 45.

CLASSIFICATION STELLAIRE (STELLAR CLASSIFICATION)

PRESIDENT: V.L.Straizys

VICE-PRESIDENT: R.F.Garrison

COMITE D'ORGANISATION: R.A.Bell, M.Golay, A.Heck, M.Jaschek, D.J.MacConnell, A.G.D.Philip, A.Slettebak.

MEMBRES: H.Albers, A.L.Ardeberg, J.D.R.Bahng, M.Barbier, D.C.Barry, R.A.Bartaya, W.P.Bidelman, V.M.Blanco, P.B.Boyce, W.Buscombe, R.Buser, B.Cester, A.M.Cherepashchuk, J.Christy, J.J.Claria, A.P.Cowley, D.Crampton, D.L.Crawford,

L.Divan, M.Duflot, T.Elvius, D.S.Evans, M.W.Feast, Ch.Fehrenbach, M.Fracassini, M.Gerbaldi, E.H.Geyer, Ju.V.Glagolevskij, H.H.Guetter, G.Gurzadyan, M.H.Hack, K.L.Hallam, B.Hauck, D.S.Hayes, A.Heck, K.G.Henize, R.Herman, P.W.Hill, W.A.Hiltner, A.A.Hoag, N.Houk, L.Huang, R.M.Humphreys, C.Jaschek, P.C.Keenan, E.K.Kharadze, G.E.Kron, J.Landi-Dessy, O.H.Levato, T.Lloyd-Evans, K.Loden, F.J.Low, G.Lynga, J.Lutz, A.Maeder, H.Maehara, S.Malaroda, M.F.McCarthy, R.D.McClure, D.H.McNamara, J.Mead, E.E.Mendoza, W.W.Morgan, N.Morguleff, K.Nandy, V.B.Nikonov, P.Notni, T.Oja, E.H.Olsen, K.Osawa, W.Osborn, S.Parsons, L.E.Pasinetti, C.L.Perry, G.W.Preston, V.C.Reddish, H.B.Richer, J.Rountree, M.Rudkjobing, N.Sanduleak, N.B.Sanwal, R.E.Schild, TH.Schmidt-Kaler, W.C.Seitter, S.L.Sharpless, U.Sinnerstad, S.D.Sinvhal, U.W.Steinlin, C.B.Stephenson, J.Stock, A.Strobel, B.Stromgren, A.R.Upgren, N.R.Walborn, G.A.H.Walker, B.Warner, W.Warren, R.M.West, B.Westerlund, P.R.Wesselius, J.A.Williams, O.C.Wilson, R.Wing, H.H.Wu, S.Wyckoff, K.M.Yoss, K.Zdanavicius.

## COMMISSION NO 46.

## L'ENSEIGNEMENT DE L'ASTRONOMIE (THE TEACHING OF ASTRONOMY)

PRESIDENT: L.N.Houziaux

VICE-PRESIDENTS: M.Rigutti, C.Ivaniszewska

COMITE D'ORGANISATION: S.Ferraz-Mello, M.Gerbalbi, J.Kleczek, E.V.Kononovich, D.McNally, S.E.Okoye, R.R.Robbins, Aa.Sandqvist, D.G.Wentzel.

MEMBRES: A.Acker, A.Z.Aiad, S.M.Alladin, H.L.Andrillat, S.H.R.Ansari, A.S.Asaad, I.Atanasijevic, V.Barocas, L.Bottinelli, W.Buscombe, E.Brieva, P.B.Byrne, J.Casanovas, A.Catala-Poch, J.M.Chamberlain, D.Clarke, S.Codina, E.A.Dibay, N.Dinulescu, F.Dominko, N.A.Doughty, D.L.Dupuy, O.Elgaroy, K.J.Feng, I.Ferrin, F.Fuenmayor, L.Gouguenheim, H.S.Gurm, B.Hauck, H.F.Haupt, S.Hazer, B.Hidayat, D.B.Hoff, S.Isobe, V.V.Ivanov, A.H.Jarrett, H.E.Jorgensen, M.Karabin, J.E.Kennedy, V.Kourganoff, Loh Yu Kyung, K.Lumme, R.C.Maddison, M.Marik, J.C.D.Marsh, L.Mavridis, H.G.Miles, H.Moreno, E.A.Muller, N.S.Nicolov, B.Onderlicka, W.Osborn, J.Osorio, N.Owaki, J.M.Pasachoff, B.F.Peery, J.Percy, E.Proverbio, Q.-Y.Qu, M.Rego-Fernandez, A.Ringuelet, L.J.Robinson, A.W.Rodgers, L.F.Rodriguez, C.Roslund, A.E.Roy, A.R.Sadik, J.L.Safko, H.Scheffler, T.Schmidt, P.A.H.Seymour, Chung Shan Shen, H.L.Shipman, R.Steinitz, K.N.Taylor, K.P.Tritton, J.A.Van Paradijs, V.Vujvovik, W.Zealey, M.Zeilik, H.Zimmermann.

## COMMISSION NO 47.

## COSMOLOGIE (COSMOLOGY)

PRESIDENT: J.Audouze

VICE-PRESIDENT: G.C.Setti

COMITE D'ORGANISATION: G.O.Abell, L.P.Grischuk, J.Gunn, S.Hayakawa, B.J.T.Jones, M.J.Rees, R.A.Sunyaev, H.Van der Laan.

MEMBRES: K.Aizu, H.Alfven, H.Andrillat, J.E.Baldwin, J.M.Bardeen, J.M.Barnothy, J.D.Barrow, D.Basu, J.E.Beckman, N.Bel, F.Bertola, S.P.Bhavsar, G.V.Bicknell, S.A.Bludman, A.F.Bogorodsky, J.R.Bond, H.Bondi, W.B.Bonnor, K.Brecher, G.R.Burbidge, M.Calvani, A.Cavaliere, G.Chinearini, D.M.Chitre, J.M.Cohen, J.J.Condon, W.Davidson,

P.C.W.Davies, M.M.Davis, A.Dekel, G.De Zotti, J.Demaret, G.De Vaucouleurs, R.H.Dicke, D.Dionysiou, A.G.Doroshkevich, C.C.Dyer, J.Ehlers, J.Einasto, G.F.R.Ellis, S.M.Faber, S.W.Falk, S.M.Fall, L.-Z.Fang, J.E.Felten, G.B.Field, P.S.Florides, R.Fong, H.C.Ford, W.R.Forman, M.Fujimoto, T.Fukui, M.G.Geller, O.Godart, T.Gold, D.W.Goldsmith, P.Goret, L.Gratton, S.A.Gregory, H.D.Greyber, S.Hacyan, K.Hara, E.Hardy, R.J.Harms, E.R.Harrison, S.W.Hawking, C.Hayashi, X.-T.He, O.Heckmann, J.Heidmann, M.Heller, F.Hoyle, J.P.Huchra, V.Icke, D.L.Jauncey, R.C.Kapoor, I.D.Karachentsev, U.Kasper, S.Kato, K.Kawabata, K.I.Kellerman, A.K.Kembhavi, J.Kormendy, A.Kovetz, B.Z.Kozlovsky, P.E.Kustaanheimo, K.W.Lake, J.P.Lasota, A.Lausberg, D.Layzer, J.Lequeux, D.E.Liebscher, M.S.Longair, T.Lu, M.A.H.MacCallum, F.Mardirossian, J.Marek, J.Materne, J.C.Mather, J.D.McCrea, W.H.McCrea, G.C.McVittie, F.Melchiorri, P.Meszaros, M.Mezzetti, C.W.Misner, R.A.Muller, H.Nariai, J.V.Narlikar, Y.Ne'eman, W.T.Ni, M.Nishida, P.D.Noerdlinger, T.W.Noonan, I.Novikov, R.F.O'Connell, A.Oemler, S.E.Okoye, G.C.Omer, R.Omnes, J.H.Oort, L.M.Ozernoi, G.Paal, J.Pachner, D.N.Page, J.C.Pecker, P.J.E.Peebles, A.Penzias, S.Persides, B.A.Peterson, V.Petrosian, W.H.Press, Q.-Y.Qu, H.Reeves, M.Reinhardt, W.Rindler, D.H.Roberts, I.Robinson, R.C.Roeder, I.W.Roxburgh, V.C.Rubin, V.Ruddy, L.Rudnick, M.Ryle, H.Saar, A.A.Sapar, W.L.W.Sargent, K.Sato, A.Savage, E.Schatzman, P.A.G.Scheuer, M.Schmidt, D.N.Schramm, E.Schucking, V.F.Schwartzman, D.W.Sciama, E.L.Scott, I.E.Segal, G.A.Seielstad, J.L.Sersic, P.A.Shaver, S.F.Shandarin, G.Shaviv, K.Shivanandan, R.L.E.Simon, H.E.Smith, G.F.Smoot, N.Spyrou, F.W.Stecker, G.Steigman, J.M.Stewart, W.R.Stoeger, M.F.Struble, C.R.Subrahmanya, J.M.G.Surdej, S.Szalay, G.A.Tammann, J.C.Tarter, G.E.Tauber, R.J.Tayler, L.A.Thompson, X.Thuan, W.G.Tifft, H.J.Treder, S.D.Tremaine, V.L.Trimble, R.B.Tully, E.L.Turner, M.S.Turner, P.C.Vaidya, V.Vanysek, C.V.Vishveswara, H.H.Von Borzeszkowski, A.S.Webster, S.Weinberg, P.S.Wesson, J.A.Wheeler, S.D.M.White, G.J.Whitrow, D.Wilkinson, C.M.Will, A.G.Wilson, E.L.Wright, B.C.Xanthopoulos, YA.B.Zel'dovich, A.L.Zel'manov, J.L.Zhang, A.Zieba, S.Zieba, E.J.Zuiderwijk,

## COMMISSION NO 48.

## ASTROPHYSIQUE DE GRANDE ENERGIE (HIGH-ENERGY ASTROPHYSICS)

PRESIDENT: R.Giacconi

VICE-PRESIDENT: C.Cesarsky

COMITE D'ORGANISATION: G.Clark, F.Pacini, P.Scheuer, E.Salpeter, R.A.Sunyaev, J.Truemper, L.Woltjer, Q.-Y.Qu.

MEMBRES: M.Abramowicz, D.Adams, H.S.Ahluwalia, H.Alfven, L.Alvarez, K.M.V.Apparao, J.Arons, B.Aschenbach, E.Asseo, J.Audouze, Y.Avni, W.I.Axford, W.A.Baan, J.Barnothy, D.Basu, G.A.Baym, R.H.Becker, M.C.Begelman, G.Benford, J.Bergeron, G.V.Bicknell, S.Biswas, S.Bladman, R.D.Blandford, S.A.Bludman, S.Bonazzola, S.Bonometto, D.J.Bond, R.L.F.Boyd, K.Brecher, A.N.Bunner, G.R.Burbidge, A.S.Burrows, A.G.W.Cameron, W.C.Cash, M.Casse, R.C.Catura, G.K.Caughlan, A.Cavaliere, S.Chandrasekhar, S.M.Chitre, T.A.Chubb, E.L.Chupp, J.M.Cohen, S.Collin-Souffrin, J.J.Condon, L.L.Cowie, R.Cowsik, M.Cruise, J.L.Culhane, J.M.Da Costa, G.Dautcourt, A.F.Davidsen, W.Davidson, L.Davis Jr., M.M.Davis, F.De Felice, T.De Graaff, D.S.De Young, H.Debrunner, B.R.Dennis, B.S.Dewitt, R.H.Dicke, M.Disney, J.F.Dolan, F.D.Drake, J.G.Duthie, P.J.Edwards, D.Eichler, J.Eilek, W.D.Evans, A.C.Fabian, L.-Z.Fang, G.Fazio, J.E.Felten, K.B.Fenton, A.Ferrari, C.Fichtel, G.B.Field, P.C.Fisher, W.R.Forman, W.A.Fowler, H.Friedman, G.P.Garmire, V.L.Ginzburg, T.Gold, D.W.Goldsmith, W.D.Gonzalez-a, K.I.Greisen, H.Greyber, M.Grewing, R.E.Griffiths, J.Grindlay, J.E.Gunn, H.Gursky, O.Guseinov, A.N.Hall, M.O.Harwit, S.W.Hawking, S.Hayakawa, J.Heise, D.J.Helfand, R.N.Henriksen, R.C.Henry, D.Hoang Binh, J.A.Hoffman, N.J.Holloway, F.Hoyle, J.Ipser, W.Israel, K.Ito, J.C.Jackson, W.J.Jaffe, J.V.Jelley,

J.R.Jokipii, F.C.Jones, T.W.Jones, P.C.Joss, P.Kafka, F.D.Kahn, J.I.Katz, K.Kellerman, E.Kellogg, A.K.Kembhavi, J.Kirk, G.Kocharov, M.Kondo, B.Z.Kozlovsky, E.Kreisel, K.G.B.Kristiansson, R.M.Kulsrud, W.Kundt, V.G.Kurt, D.Q.Lamb Jr., F.Lamb, S.A.Lamb, M.Lampton, G.J.Lasher, J.M.Lattimer, S.M.Lea, Q.-B.Li, E.Liang, J.Linsley, M.S.Longair, R.V.E.Lovelace, T.Lu, R.Luest, D.Lynden-Bell, T.Maccacaro, I.M.Martin, E.P.Masetz, K.O.Mason, M.Matsuoka, T.J.Mazurek, B.P.McBreen, R.A.McCray, D.B.Melrose, L.Mestel, P.Meszaros, F.Meyer, J.-P.Meyer, F.C.Michel, J.C.Miller, S.Miyamoto, P.Morrison, Y.Ne'eman, R.Novick, R.F.O'Connell, M.Oda, S.E.Okoye, J.P.Ostriker, D.F.O'Sullivan, L.N.Ozernoy, A.G.Pacholczyk, C.G.Page, G.G.C.Palumbo, E.Parker, J.H.Parkinson, I.Pauliny-Toth, C.Perola, B.A.Peterson, L.E.Peterson, V.Petrosian, J.H.Piddington, K.Pinkau, N.A.Porter, K.A.Pounds, E.Preuss, H.Quintana, V.Radhakrishnan, S.Ramadurai, M.J.Rees, H.Reeves, M.Reinhardt, T.N.Rengarajan, B.Rossi, M.Salvati, W.T.Sanders III, L.Sartori, W.C.Saslaw, M.P.Savedoff, J.D.Scargle, E.T.Scharlemann, K.H.Schatten, E.Schatzman, R.Schilizzi, D.N.Schramm, V.F.Schwartzman, D.W.Sciama, J.S.Scott, G.Seielstad, G.Setti, J.Shaham, N.Shakura, M.M.Shapiro, P.Shaver, G.Shaviv, G.A.Shields, I.Shklovsky, C.S.Shukre, R.Silberberg, J.Skilling, B.W.Smith, F.G.Smith, S.Sofia, L.Spitzer, G.Steigman, A.A.Stepanian, H.S.Stockman Jr., P.A.Sturrock, J.H.Swank, F.Takahara, R.J.Tayler, N.J.Terrell, K.S.Thorne, V.L.Trimble, J.W.Truran, S.Tsuruta, K.A.Van Riper, N.Vidal, S.-G.Wang, T.A.Weaver, A.S.Webster, M.C.Weisskopf, D.G.Wentzel, K.C.Westfold, J.A.Wheeler, C.M.Will, J.R.Wilson, R.Wolstencroft, D.M.Worrall, YA.B.Zeldovich, H.-Q.Zhang, J.-L.Zhang, M.V.Zombeck.

## COMMISSION NO 49.

### LA PLASMA INTERPLANETAIRE ET L'HELIOSPHERE (THE INTERPLANETARY PLASMA AND THE HELIOSPHERE)

PRESIDENT: I.W.Roxburgh

VICE-PRESIDENT: S.Grzedzielski

COMITE D'ORGANISATION: S.Cuperman, A.Z.Dolginov, H.J.Fahr, H.U.Keller, F.Paresce, G.Setti, S.T.Suess, C.S.Weller.

MEMBRES: H.S.Ahluwalia, S.Ananthakrishnan, K.A.Anderson, S.M.R.Ansari, E.Antonucci, A.Barnes, C.H.Barrow, C.A.Barth, J.L.Bertaux, D.E.Blackwell, P.W.Blum, R.M.Bonnet, C.S.Bowyer, J.C.Brandt, L.F.Burlaga, J.W.Chamberlain, B.Chen, F.Crifo, C.de Jager, P.Delache, D.S.De young, M.Dryer, B.Durney, J.E.Dyson, E.V.Ergma, Von R. Eshleman, A.Eviatar, J.Feynman, G.B.Field, J.T.Gosling, S.R.Habbal, C.C.Harvey, J.Heyvaerts, J.V.Hollweg, T.E.Holzer, J.A.Ionson, J.R.Jokipii, T.Kakinuma, S.A.Kaplan, J.P.Lafon, E.Levy, N.A.Lotova, R.Luest, R.MacQueen, A.Mangeney, O.T.Matsuura, H.Mavromichaliki, D.A.Mendis, L.Mestel, F.C.Michel, X.Moussas, Y.Nakagawa, G.A.Newkirk, E.N.Parker, F.Perkins Jr., K.Pflug, G.Pneuman, M.A.Raadu, A.C.S.Readhead, N.K.Reay, B.J.Rickett, A.C.Riddle, H.W.Ripken, F.E.Roach, R.Rosner, C.T.Russel, E.P.Sarris, C.Sawyer, E.Schatzman, F.Scherb, K.Schindler, H.U.Schmidt, A.B.Severny, S.D.Shawhan, M.A.Shea, D.F.Smith, C.P.Sonett, R.G.Stone, P.A.Sturrock, H.Tanaka, L.A.Vainshtein, J.A.Van Allen, M.K.Wallis, T.Watanabe, J.M.Wilcox, J.P.Wild, S.T.Wu.

## COMMISSION NO 50.

### PROTECTION DES SITES D'OBSERVATOIRES EXISTANTS ET POTENTIELS (PROTECTION OF EXISTING AND POTENTIAL OBSERVATORY SITES)

PRESIDENT: A.A.Hoag

VICE-PRESIDENT: S.Van den Bergh

COMITE D'ORGANISATION: C.Anguita, A.Ardeberg, F.Bertola, V.Blanco, R.Cayrel, P.J.Edwards, J.T.Jefferies, W.Mattig, B.E.J.Oke, F.G.Smith, M.F.Walker.

MEMBRES: J.Arias de Grieff, A.S.Asaad, J.C.Bhattacharyya, C.Blanco, A.A.Boyarchuk, J.Casanovas, D.L.Crawford, J.Davis, J.Dommanget, L.Dunkelman, H.Elsaesser, E.Goebel, S.-M.Gong, K.Gyldenkerne, L.Huang, A.Kubicela, E.M.Leibowitz, T.K.J.Markkanen, S.Marx, L.N.Mavridis, E.Mendoza, S.Newar, S.B.Novikov, K.Osawa, J.Osorio, F.Sanchez, R.T.Schilizzi, P.V.Shcheglov, L.Z.Shen, C.Torres, J.Tremko, P.A.Wayman, W.Woolf, A.Woszczyk, B.-R.Zhang.

COMMISSION NO 51.

RECHERCHE DE LA VIE DANS L'UNIVERS (SEARCH FOR EXTRATERRESTRIAL LIFE)

PRESIDENT: M.D.Papagiannis

VICE-PRESIDENTS: N.S.Kardashev, F.D.Drake

COMITE D'ORGANISATION: R.D.Brown, P.Connes, G.D.Gatewood, L.Goldberg, J.Jugaku, G.Marx, F.Pacini, M.J.Rees, V.S.Troitsky.

MEMBRES: H.M.K.Al-Naimy, A.W.Al-Sabti, I.F.Almar, V.A.Ambartsumian, H.Ando, G.A.Bakos, B.A.Balazs, J.A.Ball, D.Basu, W.A.Baum, J.E.Beckman, R.F.Beebe, D.Benest, R.Berendzen, P.L.Bernacca, F.Biraud, R.C.Bless, S.C.Bowyer, P.B.Boyce, R.N.Bracewell, B.Campbell, J.B.Carlson, T.D.Carr, E.J.Chaisson, T.A.Clark, D.G.Currie, A.H.Delsemme, R.S.Dixon, G.S.Downs, F.J.Dyson, E.E.Epstein, N.J.Evans II, P.A.Feldman, G.B.Field, F.J.Firneis, M.G.Firneis, P.C.Fisher, M.Fracassini, L.W.Fredrick, M.Fujimoto, T.Gehrels, V.L.Ginzburg, F.Giovanelli, G.Godoli, D.W.Goldsmith, R.J.Gott III, C.Goudis, J.L.Greenstein, P.C.Gregory, S.Gulkis, J.E.Gunn, F.T.Haddock, B.M.Haisch, A.Hajduk, R.S.Harrington, M.H.Hart, D.S.Heeschen, J.Heidmann, G.H.Herbig, T.j.Herczeg, J.L.Hershey, J.-L.Heudier, H.Hirabayashi, J.M.Hollis, D.M.Hunten, J.H.Hunter, E.J.Hyson, G.M.Idlis, W.M.Irvine, R.Jastrow, S.Jeffers, R.C.Jennison, E.M.Jones, P.Kafka, P.Kaufmann, H.-U.Keller, K.I.Kellerman, M.J.Klein, S.H.Knowles, D.Kocer, D.Kotsakis, J.D.Krauss, L.V.Ksanfomaliti, T.B.H.Kuiper, A.D.Kuzmin, E.A.Lilley, S.L.Lippincott, L.O.Loden, Sir Bernard Lovell, P.Maffei, B.E.Markarian, M.Ya.Marov, T.Matsuda, L.N.Mavridis, H.A.McAlister, E.E.Mendoza, B.L.Milet, Y.K.Minn, I.F.Mirabel, M.Morimoto, V.I.Moroz, M.R.Morris, P.Morrison, M.Moutsoulas, N.Nakagawa, P.G.Niarchos, M.Oda, J.P.Ostriker, T.C.Owen, Yu.N.Parijskij, L.E.Pasinetti, L.Perek, J.B.Pollack, J.E.B.Ponsonby, F.V.Prochazka, E.Proverbio, H.Quintana, J.M.Quintana, R.T.Rood, M.Rowan-Robinson, R.H.Rubin, J.L.Russell, C.Sagan, K.Sakurai, R.Sancisi, J.D.Scargle, H.j.Schober, C.L.Seeger III, Chun-Shan Shen, M.Shimizu, I.S.Shklovsky G.S.Shostak, W.L.H.Shuter, V.I.Slysh, Harlan J. Smith, H.A.Smith, L.E.Snyder, R.Stalio, V.Straizys, P.A.Sturrock, W.T.Sullivan III, J.C.Tarter, R.K.Tavakol, P.Thaddeus, C.R.Tolbert, H.M.Tovmassian, C.H.Townes, V.L.Trimble, E.L.Turner, A.Valbousquet, P. Van den Kamp, T.C. Van Flandern, D.A.Varshalovich, M.Vazquez, V.R.Venugopal, S. Von Hoerner, M.K.Wallis, F.G.Watson, W.J.Welch, G.Westerhout, G.W.Wetherill, I.P.Williams, R.F.Willson, B.M.Zuckerman.

GROUPE DE TRAVAIL POUR LA NOMENCLATURE DU SYSTEME PLANETAIRE

(WORKING GROUP FOR PLANETARY SYSTEM NOMENCLATURE)

PRESIDENT: H.Masursky

MEMBRES: K.Aksnes, G.E.Hunt, M.Ya.Marov, P.M.Millman, D.Morrison, T.C.Owen, V.V.Shevchenko, B.A.Smith, V.G.Tejfel'.

TASK GROUPS:

1) Lunar Nomenclature:

V.V.Shevchenko (Chairman), A.Dollfus, F.El-Baz, H.Masursky, P.M.Millman, S.Runcorn, E.A.Whitaker.

2) Mercury Nomenclature:

D.Morrison (Chairman), D.B.Campbell, M.E.Davies, A.Dollfus, N.P.Erpylev, J.E.Guest.

3) Venus Nomenclature:

M.Ya.Marov (Chairman), A.T.Basilevsky, D.B.Campbell, R.M.Goldstein, R.F.Jurgens, H.Masursky, G.H.Pettengill.

4) Mars Nomenclature:

B.A.Smith (Chairman), A.Dollfus, M.Ya.Marov, D.Ya.Martynov, H.Masursky, S.Miyamoto, C.Sagan.

5) Outer Solar System Nomenclature:

T.C.Owen (Chairman), K.Aksnes, M.S.Bobrov, A.Brahic, M.E.Davies, N.P.Erpylev, H.Masursky, B.A.Smith, V.G. Tejfel'.

## 2. ALPHABETICAL LIST OF MEMBERS

On 26 August 1982, the total number of individual members was about 5200.

Reprints of this list are available from the IAU Publisher, D. Reidel Publishing Company:

P.O. Box 17
3300 AA Dordrecht
Holland

A'HEARN MICHAEL F DR
ASTRONOMY PROGRAM
UNIV OF MARYLAND
COLLEGE PK MD 20742
U.S.A.

AANESTAD PER ARNE DR
PHYSICS DEPT
ARIZONA STATE UNIV
TEMPE AZ 85281
U.S.A.

AARONSON MARC
STEWARD OBSERVATORY
UNIV OF ARIZONA
TUCSON AZ 85721
U.S.A.

AARSETH SVERRE J DR
INST OF ASTRONOMY
MADINGLEY RD
CAMBRIDGE CB3 OHA
U.K.

ABALAKIN VICTOR K DR
INSTITUTE FOR THEORE
TICAL ASTRONOMY
10 KUTUZOV QUAY
192187 LENINGRAD
U.S.S.R.

ABBASOV ALIK R DR
SHEMAKHA ASTRO-
PHYSICAL OBSERVATORY
373243 SHEMAKHA AZER
U.S.S.R.

ABBOTT WILLIAM N DR
MICHALACOPOULOU 42
ATHENS 612
GREECE

ABDALA J DR
AV SOLANO
EDIFICIO ARAGUANEY
PLANTA BAJA NO 1
CHACAITO CARACAS 106
VENEZUELA

ABELE M K DR
LATVIAN STATE UNIVER
SITY ASTRONOMICAL
OBSERVATORY
226098 RIGA
U.S.S.R.

ABELL GEORGE O PROF
DEPT OF ASTRONOMY
UNIVERSITY OF
CALIFORNIA
LOS ANGELES CA 90024
U.S.A.

ABHYANKAR KRISHNA D PROF
DEPT OF ASTRONOMY
OSMANIA UNIVERSITY
HYDERABAD 500 007
INDIA

ABLES HAROLD D DR
FLAGSTAFF STATION
US NAVAL OBSERVATORY
P O BOX 1149
FLAGSTAFF AZ 86001
U.S.A.

ABLES J G DR
CSIRO
DIV OF RADIOPHYSICS
P O BOX 76
EPPING NSW 2121
AUSTRALIA

ABOU-EL-ELLA MOHAMED S DR
HELWAN OBSERVATORY
HELWAN CAIRO
ARAB REP. OF EGYPT

ABRAHAM HENRY J M
7 CASTLEREAGH
CRESCENT
MACQUARIE ACT 2614
AUSTRALIA

ABRAHAM ZULEMA DR
INST PESQ ESPACIAIS
CRAAM RADIO OBS
C P 515
12200 S JOSE CAMPOS
BRAZIL

ABRAMI ALBERTO PROF
OSSERVATORIO ASTRONO
MICO DI TRIESTE
I-34131 TRIESTE
ITALY

ABRAMOWICZ MAREK DR
N COPERNICUS ASTRON
CENTER
BARTYCKA 18
00-716 WARSAW
POLAND

ABREU DE BRITO E J C DR
OBSERVATORIO ASTRO-
NOMICO DE LISBOA
TAPADA
LISBON
PORTUGAL

ABT HELMUT A DR
KITT PEAK NATL OBS
BOX 26732
TUCSON AZ 85726
U.S.A.

ABU EL ATA NABIL DR
BUREAU D LONGITUDES
77 AVENUE DENFERT-
ROCHEREAU
F-75014 PARIS
FRANCE

ACKER AGNES DR
OBSERVATOIRE
11 RUE DE
L UNIVERSITE
F-67000 STRASBOURG
FRANCE

ACTON LOREN W DR
LOCKHEED PALO ALTO
RESEARCH LAB
3251 HANOVER ST
PALO ALTO CA 94304
U.S.A.

ADAM MADGE G DR
DEPT OF ASTROPHYSICS
SOUTH PARKS RD
OXFORD OX1 3RQ
U.K.

ADAMS A N MR
6549 N 35TH ROAD
ARLINGTON VA 22213
U.S.A.

ADAMS DAVID J DR
ASTRONOMY DEPT
THE UNIVERSITY
LEICESTER LE1 7RH
U.K.

ADAMS THOMAS F DR
G-7 MS 329 LOS ALA-
MOS SCIENTIFIC LAB
P O BOX 1663
LOS ALAMOS NM 87545
U.S.A.

ADE PETER A R DR
PHYSICS DEPT
QUEEN MARY COLLEGE
MILE END ROAD
LONDON E1 4NS
U.K.

ADEL ARTHUR F PROF EMER
NORTHERN ARIZONA
UNIVERSITY
BOX 5679
FLAGSTAFF AZ 86001
U.S.A.

ADEL SHARAF MOHAMED DR
DEPT OF ASTRONOMY
CAIRO UNIVERSITY
CAIRO
ARAB REP. OF EGYPT

ADELMAN SAUL J DR
DEPT OF PHYSICS
THE CITADEL
CHARLESTON SC 29409
U.S.A.

ADOLFSSON TORD DR
KRAGEHOLMSGATAN 12
S-216 19 MALMO
SWEDEN

AFANASJEVA PRASKOVYA M DR
PULKOVO OBSERVATORY
196140 LENINGRAD
U.S.S.R.

AGEKJAN TATEOS A PROF
UNIVERSITY OF
LENINGRAD
OBSERVATORY
199178 LENINGRAD
U.S.S.R.

AGRINIER BERNARD L MR
C E N SACLAY
B P NO 2
F-91190 GIF-YVETTE
FRANCE

AGUERO ESTELA L DR
OBS ASTRONOMICO
LAPRIDA NO 854
5000 CORDOBA
ARGENTINA

AHLUWALIA HARJIT SINGH DR
DEPT PHYSICS ASTRON
UNIV OF NEW MEXICO
800 YALE BLVD N E
ALBUQUERQUE NM 87131
U.S.A.

AHMED IMAM IBRAHIM
HELWAN OBSERVATORY
HELWAN NEAR CAIRO
ARAB REP. OF EGYPT

AHNERT P DR
ZENTRAL INST FUER
ASTROPHYSIK
STERNWARTE SONNEBERG
DDR-64 SONNEBERG
G.D.R.

AIAD A Z DR
CAIRO UNIVERSITY
FACULTY OF SCIENCE
ASTRONOMY DEPT
GEZA ORMAN
ARAB REP. OF EGYPT

AIKAWA TOSHIKI DR
ASTRONOMICAL INST
TOHOKU UNIVERSITY
AOBAYAMA SENDAI
JAPAN

AIKMAN G CHRIS L DR
DOMINION ASTROPH OBS
5071 W SAANICH ROAD
RR 5
VICTORIA BC V8X 4M6
CANADA

AIME C DR
DEPT D'ASTROPHYSIQUE
UNIVERSITE DE NICE
PARC VALROSE
F-06034 NICE CEDEX
FRANCE

AITKEN DAVID K DR
DEPT PHYS AND ASTRO
UNIV COLLEGE LONDON
GOWER STREET
LONDON WC1E 6BT
U.K.

AIZENMAN MORRIS L DR
DIV ASTRONOMICAL SC
NAT SC FOUNDATION
RM 615 1800 G ST NW
WASHINGTON DC 20550
U.S.A.

AIZU KO PROF
PHYSICS DEPT
RIKKYO UNIVERSITY
NISHI-IKEBUKURO 3
TOSHIMAKU TOKYO 171
JAPAN

AKABANE KENJI A PROF
TOKYO ASTRONOMICAL
OBSERVATORY
OSAWA MITAKA
TOKYO 181
JAPAN

AKABANE TOKUHIDE DR
HIDA OBSERVATORY
506-13 KAMITAKARA
GIFU
JAPAN

AKCAYLI MELEK M A DR
EGE UNIVERSITESI
FEN FAKUTESI
GOK BILIMLERI ENST
BORNOVA IZMIR
TURKEY

AKSENOV E P DR
STERNBERG STATE AS-
TRONOMICAL INSTITUTE
117234 MOSCOW
U.S.S.R.

AKSNES KAARE DR.
NORWEGIAN DEFENCE
RESEARCH ESTABLISHM
P O BOX 25
2007 KJELLER
NORWAY

AKYOL MUSTAFA U PROF
EGE UNIVERSITY
SCIENCE FACULTY
DEPT OF ASTRONOMY
BORNOVA IZMIR
TURKEY

AL-NAIMY HAMID M K DR
ASTRONOMY & SPACE
RES CTR JADIRIYA
P O BOX 255
BAGHDAD
REP. OF IRAQ

AL-SABTI ABDUL ADIM DR
PHYSICS DEPT
SCIENCE COLLEGE
BAGHDAD UNIVERSITY
BAGHDAD
REP. OF IRAQ

ALANIJA I F DR
ABASTUMANI ASTROPHY-
SICAL OBSERVATORY
383762 ABASTUMANI
GEORGIA
U.S.S.R.

ALBANO J DR
OBS ASTRONOMICO
LA PLATA
ARGENTINA

ALBERS HENRY PROF
VASSAR COLLEGE
OBSERVATORY
POUGHKEEPSIE NY12601
U.S.A.

ALBRECHT RUDOLF DR
INS F ASTRONOMIE
UNIVERSITAET WIEN
TUERKENSCHANZSTR 17
A-1180 WIEN
AUSTRIA

ALDROVANDI RUBEN DR
INSTITUTO DI FISICA
TEORICA
RUA PAMPLONA 145
01405 SAO PAULO SP
BRAZIL

ALDROVANDI SUELI M V DR
INSTITUTO ASTRONOMI
CO E GEOFISICO
CP 30627
01000 SAO PAULO SP
BRAZIL

ALECIAN GEORGES
OBSERVATOIRE PARIS
DEPT D'ASTROPHYSIQUE
FONDAMENTALE
F-92190 MEUDON
FRANCE

ALEXANDER JOHN B
ROYAL GREENWICH OBS
HAILSHAM
EAST SUSSEX BN27 1RP
U.K.

ALEXANDER JOSEPH K
CODE 693
GODDARD SPACE FLIGHT
CENTER
GREENBELT MD 20771
U.S.A.

ALFVEN HANNES PROF
DEPT OF PLASMA
PHYSICS
ROYAL INST OF TECHN
S-100 44 STOCKHOLM
SWEDEN

ALISSANDRAKIS C PH D
LAB OF ASTROPHYSICS
UNIV OF ATHENS
PANEPISTIMIOPOLIS
ATHENS 621
GREECE

ALKSNE ZENTA DR
RADIOASTROPHYSICAL
OBSERVATORY
TURGENEVA 19
226524 RIGA LATVIA
U.S.S.R.

ALKSNIS ANDREJS DR
RADIOASTROPHYSICAL
OBSERVATORY
226524 RIGA LATVIA
U.S.S.R.

ALLADIN SALEH MOHAMED DR
DEPT OF ASTRONOMY
OSMANIA UNIVERSITY
HYDERABAD 500 007
INDIA

ALLAMANDOLA LOUIS JOHN DR
STERREWACHT
HUYGENS LAB
POSTBUS 9513
2300 RA LEIDEN
THE NETHERLANDS

ALLAN DAVID W DR
NATIONAL BUREAU OF
STANDARDS
BOULDER CO 80302
U.S.A.

ALLEGRE CLAUDE PROF
INST DE PHYSIQUE DU
GLOBE
4 PLACE JUSSIEU
F-75005 PARIS
FRANCE

ALLEN C W PROF
MT STROMLO
OBSERVATORY
WODEN PO ACT 2606
AUSTRALIA

ALLEN CHRISTINE
INSTITUTO DE ASTRONO
MIA UNAM
APDO POSTAL 70-264
04510 MEXICO DF
MEXICO

ALLEN DAVID A DR
ANGLO AUSTRALIAN
OBSERVATORY
PO BOX 296
EPPING NSW 2121
AUSTRALIA

ALLEN RONALD J DR
KAPTEYN LABORATORIUM
POSTBUS 800
9700 AV GRONINGEN
THE NETHERLANDS

ALLER HUGH D DR
DEPT OF ASTRONOMY
UNIVERSITY MICHIGAN
953 PHYS-ASTRO BLDG
ANN ARBOR MI 48109
U.S.A.

ALLER LAWRENCE HUGH
ASTRONOMY DEPT
UNIVERSITY OF CALIF
MATH-SCIENCES BLDG
LOS ANGELES CA 90024
U.S.A.

ALLER MARGO F DR
ASTRONOMY DEPT
UNIVERSITY MICHIGAN
ANN ARBOR MI 48109
U.S.A.

ALLOIN DANIELLE DR
DAPHE
OBS DE MEUDON
F-92190 MEUDON
FRANCE

ALMAR IVAN DR
SATELLITE GEODETIC
OBSERVATORY
POB 546
1373 BUDAPEST
HUNGARY

ALTAMORE ALDO
OSS ASTRONOMICO
VIALE DEL PARCO
MELLINI 84
I-00136 ROMA
ITALY

ALTAVISTA CARLOS A DR
OBS ASTRONOMICO
PASEO DEL BOSQUE
1900 LA PLATA
ARGENTINA

ALTENHOFF WILHELM J DR
MPI FUER
RADIOASTRONOMIE
AUF DEM HUEGEL 69
D-5300 BONN
GERMANY, F.R.

ALTROCK RICHARD C DR
AIR FORCE GEOPHYSICS
LABORATORY
SACRAMENTO PEAK.OBS
SUNSPOT NM 88349
U.S.A.

ALTSCHULER MARTIN D PROF
DEPT RAD THER/BX 522
HOSP UNIV OF PENNA
3400 SPRUCE ST
PHILADELPHIA PA19104
U.S.A.

ALURKAR S K DR
PHYSICAL RES LAB
NAVRANGPURA
AHMEDABAD 380 009
INDIA

ALVAREZ HECTOR DR.
DEPTO DE ASTRONOMIA
UNIVERSIDAD DE CHILE
CASILLA 36-D
SANTIAGO
CHILE

ALVAREZ LUIS DR
LAWRENCE RADIATION
LABORATORY
UNIVERSITY OF CALIF
BERKELEY CA 94720
U.S.A.

ALVAREZ P
INSTITUTO ASTROFISI-
CA CANARIAS
UNIVERSIDAD LAGUNA
TENERIFE CANARIAS
SPAIN

ALY M K PROF DIRECTOR
RABITATUL AALAM AL-
ISLAMI
ISLAMIC OBSERVATORY
MECCA ALMUKARRAMA
SAUDI ARABIA

AMBARTSUMIAN V A PROF DR
BYURAKAN ASTROPHYSI
CAL OBSERVATORY
378433 ARMENIA
U.S.S.R.

AMBROZ PAVEL DR
ASTRONOMICAL INST OF
THE CZECHOSLOVAK
ACADEMY OF SCIENCES
25165 ONDREJOV
CZECHOSLOVAKIA

ANAND S P S DR
APPLIED RESEARCH AND
SYSTEMS
1998 GENERALS HWY
ANNAPOLIS MD 21401
U.S.A.

ANANTHAKRISHNAN S
RADIOASTRONOMY CTR
TATA INSTITUTE OF
FUNDAMENTAL RESEARCH
OOTACAMUND 643 001
INDIA

ANDERS EDWARD PROF
UNIVERSITY OF
CHICAGO
5640 S ELLIS AVE
CHICAGO IL 60637
U.S.A.

ANDERSEN JOHANNES LEKTOR
BRORFELDE OBS
COPENHAGEN UNIV
DK-4340 TOLLOSE
DENMARK

ANDERSEN TORBEN BRENDER
COPENHAGEN UNIV
OBSERVATORY
OESTER VOLDGADE 3
DK-1350 COPENHAGEN K
DENMARK

ANDERSON BRYAN DR
UNIV OF MANCHESTER
JODRELL BANK
MACCLESFIELD
CHESHIRE SK11 9DL
U.K.

ANDERSON CHRISTOPHER M DR
WASHBURN OBSERVATORY
UNIVERSITY OF
WISCONSIN
MADISON WI 53706
U.S.A.

ANDERSON KINSEY A PROF
SPACE SCIENCES LAB
UNIVERSITY OF CALIF
BERKELEY CA 94720
U.S.A.

ANDO HIROYASU DR
TOKYO ASTRONOMICAL
OBSERVATORY
OSAWA 2-21-1 MITAKA
TOKYO 181
JAPAN

ANDREW BRYAN H DR
HERZBERG INST OF
ASTROPHYSICS
NAT RES COUNCIL
OTTAWA ONT K1A 0R6
CANADA

ANDREW KENNETH L PROF
DEPT OF PHYSICS
PURDUE UNIVERSITY
W LAFAYETTE IN 47907
U.S.A.

ANDREWS DAVID A DR
ARMAGH OBSERVATORY
ARMAGH BT61 9DG
U.K.

ANDREWS DONALD H
DOMINION ASTROPH OBS
5071 W SAANICH ROAD
RR 5
VICTORIA BC V8X 4M6
CANADA

ANDREWS PETER J DR
ROYAL GREENWICH
OBSERVATORY
HAILSHAM
EAST SUSSEX BN27 1RP
U.K.

ANDRIENKO DMITRY A DR
VLADIMIRSKAYA 51 53
FLAT 75
252003 KIEV
U.S.S.R.

ANDRIESSE CORNELIS D DR
VALERIUSLAAN 15
6865 JA DOORWERTH
THE NETHERLANDS

ANDRILLAT HENRI L PROF
LAB D'ASTRONOMIE
UNIV DES SCIENCES ET
TECHN DU LANGUEDOC
F-34060 MONTPELLIER
FRANCE

ANDRILLAT YVETTE DR
OBSERVATOIRE
DE HAUTE-PROVENCE
ST-MICHEL L OBS
F-04300 FORCALQUIER
FRANCE

ANDRLE P DR
ASTRONOMICAL INST
CZECH ACAD OF SC
BUDECSKA 6
12023 PRAHA 2
CZECHOSLOVAKIA

ANGEL J ROGER P PROF
STEWARD OBSERVATORY
UNIVERSITY OF
ARIZONA
TUCSON AZ 85721
U.S.A.

ANGELETTI LUCIO DR
OSSERVATORIO ASTRO-
NOMICO DI ROMA
I-00100 ROMA
ITALY

ANGELOV TRAJKO
INST OF ASTRONOMY
UNIV OF BELGRADE
STUDENTSKI TRG 16
YU-11000 BELGRADE
YUGOSLAVIA

ANGIONE RONALD J DR
ASTRONOMY DEPT
SAN DIEGO STATE
UNIVERSITY
SAN DIEGO CA 92182
U.S.A.

ANGUITA C A DR
UNIVERSIDAD DE CHILE
OBS ASTRON NACIONAL
CASILLA 36-D
SANTIAGO
CHILE

ANILE ANGELO M
OSSERVATORIO
ASTROFISICO
CITTA UNIVERSITARIA
I-95123 CATANIA
ITALY

ANSARI S M RAZAULLAH DR
PHYSICS DEPT
ALIGARH MUSLIM
UNIVERSITY
ALIGARH UP 202001
INDIA

ANTALOVA ANNA
ASTRONOMICAL INST
SLOVAK ACADEMY OF SC
05960 TATRANSKA
LOMNICA
CZECHOSLOVAKIA

ANTIOCHOS SPIRO KOSTA
INST FOR PLASMA RES
ERL 262
STANFORD UNIVERSITY
STANFORD CA 94305
U.S.A.

ANTONACOPOULOS GREG PROF
DEPT OF ASTRONOMY
UNIVERSITY OF PATRAS
PATRAS
GREECE

ANTONELLO ELIO
OSS ASTRONOMICO
DI MERATE
VIA BIANCHI 46
I-22055 MERATE
ITALY

ANTONOPOULOU E DR
DEPT OF ASTRONOMY
UNIVERSITY OF ATHENS
PANEPISTIMIOPOLIS
ATHENS 621
GREECE

ANTONOV V A DR
LENINGRAD STATE UNI-
VERSITY ASTRONOMICAL
OBSERVATORY
199178 LENINGRAD
U.S.S.R.

ANTONUCCI ESTER DR
ISTITUTO DI FISICA
UNIVERSITA DI TORINO
C D AZEGLIO 46
I-10125 TORINO
ITALY

ANZER ULRICH DR
MPI FUER PHYSIK UND
ASTROPHYSIK
FOEHRINGER RING 6
D-8000 MUENCHEN 6
GERMANY, F.R.

AOKI SHINKO PROF
TOKYO ASTRONOMICAL
OBSERVATORY
MITAKA
TOKYO 181
JAPAN

APPARAO K M V DR
TATA INST FUND RES
HOMI BHABH ROAD
BOMBAY 400 005
INDIA

APPENZELLER IMMO PROF
LANDESSTERNWARTE
KOENIGSTUHL
D-6900 HEIDELBERG 1
GERMANY, F.R.

ARABELOS D DR
DEPT OF GEODETIC
ASTRONOMY
UNIV OF THESSALONIKI
THESSALONIKI
GREECE

ARAI KENZO DR
DEPT OF PHYSICS
KUMAMOTO UNIVERSITY
KUMAMOTO 860
JAPAN

ARAKELIAN MARAT A DR
BYURAKAN ASTROPHYSI-
CAL OBSERVATORY
378433 BYURAKAN
ARMENIA
U.S.S.R.

ARDAVAN H DR
INST OF ASTRONOMY
MADINGLEY ROAD
CAMBRIDGE
U.K.

ARDEBERG ARNE L DR
ESO
K-SCHWARZSCHILD-ST 2
D-8046 GARCHING B M
GERMANY, F.R.

AREND S DR
AVENUE DE SATURNE 11
B-1180 BRUXELLES
BELGIUM

ARGUE A NOEL MR
INST OF ASTRONOMY
THE OBSERVATORIES
MADINGLEY RD
CAMBRIDGE CB3 OHA
U.K.

ARGYLE P E DR
DOMINION RADIO
ASTROPHYSICAL OBS
BOX 248
PENTICTON BC V2A 6K3
CANADA

ARGYRAKOS JEAN PROF DR
UNIV TECHNIQUE
NATIONALE D'ATHENES
193 PATISSION ST
ATHENS 816
GREECE

ARIAS DE GREIF JORGE PROF
OBSERVATORIO
NACIONAL
APARTADO 2584
BOGOTA
COLOMBIA

ARIMOTO NOBUO DR
ASTRONOMICAL INST
TOHOKU UNIVERSITY
SENDAI 980
JAPAN

ARKHIPOVA V P DR
STERNBERG STATE AS-
TRONOMICAL INSTITUTE
117234 MOSCOW
U.S.S.R.

ARLOT JEAN-EUDES
BUREAU LONGITUDES
77 AVENUE DENFERT-
ROCHEREAU
F-75014 PARIS
FRANCE

ARNAL MARCELO EDMUNDO DR
INSTITUTO ARGENTINO
DE RADIOASTRONOMIA
CASILLA DE CORREO 5
1894-VILLA ELISA
ARGENTINA

ARNETT W DAVID PROF
ENRICO FERMI INST
933 E 56TH STREET
CHICAGO IL 60637
U.S.A.

ARNOULD MARCEL L DR
INST D'ASTRONOMIE
UNIV LIB BRUXELLES
AVE F D ROOSEVELT 50
B-1050 BRUXELLES
BELGIUM

ARNQUIST WARREN N DR
8127 DELGANY AVE
PLAYA DEL REY
CA 90291
U.S.A.

ARNY THOMAS T DR
DEPT PHYSICS ASTR
UNIVERSITY OF MA
GRC TOWER B
AMHERST MA 01002
U.S.A.

ARONS JONATHAN
DEPT OF ASTRONOMY
UNIVERSITY OF CALIF
BERKELEY CA 94720
U.S.A.

ARP HALTON DR
HALE OBSERVATORIES
813 SANTA BARBARA ST
PASADENA CA 91101
U.S.A.

ARPIGNY CLAUDE PROF
INST D ASTROPHYSIQUE
UNIVERSITE DE LIEGE
B-4200 COINTE OUGREE
BELGIUM

ARSENIJEVIC JELISAVETA
ASTRONOMICAL OBS
VOLGINA 7
YU-11050 BEOGRAD
YUGOSLAVIA

ARTHUR DAVID W G
US GEOLOGICAL SURVEY
FLAGSTAFF ARIZONA
FLAGSTAFF AZ 86001
U.S.A.

ARTRU M C DR
OBS DE MEUDON
F-92190 MEUDON
FRANCE

ARTZNER GUY
LPSP
B P 10
F-91370 VERRIERES-
LE-BUISSON
FRANCE

ASAAD ADLY S PROF
OBSERVATORY OF
HELWAN
HELWAN
ARAB REP. OF EGYPT

ASCHENBACH BERND PH D
HEITERWANGERSTR 54
D-8000 MUENCHEN 70
GERMANY, F.R.

ASLAN ZEKI DR
ASTRONOMY DEPT
UNIV OF ANKARA
FEN FAKULTESI
ANKARA
TURKEY

ASLANOV I A DR
SHEMAKHA ASTROPHYSI-
CAL OBSERVATORY
373243 AZERBAIDZAN
U.S.S.R.

ASSEO ESTELLE DR
OBSERVATOIRE DE
MEUDON DAF
F-92190 MEUDON
FRANCE

ASSOUSA GEORGE ELIAS DR
DEPT TERR MAGNETISM
CARNEGIE INST WASH
5241 BROAD BR RD N W
WASHINGTON DC 20015
U.S.A.

ASTERIADIS GEORGIOS DR
DEPT OF GEODETIC
ASTRONOMY
UNIV THESSALONIKI
THESSALONIKI
GREECE

ATANASIJEVIC IVAN DR
FACULTY OF SCIENCES
6500 GL NIJMEGEN
THE NETHERLANDS

ATHANASSOULA E O DR
OBSERVATOIRE
DE BESANCON
41B AVE OBSERVATOIRE
F-25000 BESANCON
FRANCE

ATHAY R GRANT DR
HIGH ALTITUDE
OBSERVATORY
P O BOX 1470
BOULDER CO 80302
U.S.A.

ATHERTON PAUL DAVID
ASTRONOMY GROUP
BLACKETT LABORATORY
IMPERIAL COLLEGE
LONDON SW7
U.K.

ATKINSON ROBERT D E DR
SWAIN HALL WEST 319
ASTRONOMY DEPT
INDIANA UNIVERSITY
BLOOMINGTON IN 47401
U.S.A.

AUBIER MONIQUE G DR
OBSERVATOIRE DE
MEUDON
F-92190 MEUDON
FRANCE

AUDOUZE JEAN DR
INST D'ASTROPHYSIQUE
98 BIS BLVD ARAGO
F-75014 PARIS
FRANCE

AUER LAWRENCE H DR
HIGH ALTITUDE
OBSERVATORY
P O BOX 1470
BOULDER CO 80302
U.S.A.

AULUCK F C PROF
DEPT OF PHYSICS AND
ASTROPHYSICS
UNIVERSITY OF DELHI
DELHI 110007
INDIA

AUMAN JASON R PROF
DEPT OF GEOPHYSICS
AND ASTRONOMY
UNIVERSITY OF BC
VANCOUVER BC V6T 1W5
CANADA

AUNER GERHARD DR
INST FUER ASTRONOMIE
UNIVERSITAETSSTR 4
A-6020 INNSBRUCK
AUSTRIA

AURIEMMA GIULIO DR
CNR ASTROFISICO
SPAZIALE C P 67
I-00044 FRASCATI
ROMA
ITALY

AUVERGNE MICHEL
OBS DE NICE
B.P 252
F-06007 NICE CEDEX
FRANCE

AVERY LORNE W DR
HERZBERG INST OF
ASTROPHYSICS
NAT RES COUNC CANADA
OTTAWA ONT K1A 0R6
CANADA

AVIGNON YVETTE DR
OBSERVATOIRE
DE MEUDON
F-92190 MEUDON
FRANCE

AVNI YORAM DR
WEIZMANN INSTITUTE
REHOVOT
ISRAEL

AVRETT EUGENE H DR
CENTER FOR
ASTROPHYSICS
60 GARDEN ST
CAMBRIDGE MA 02138
U.S.A.

AWAD MERVAT EL-SAID DR
ASTRONOMY DEPT
FACULTY OF SCIENCE
CAIRO UNIVERSITY
CAIRO
ARAB REP. OF EGYPT

AWADALLA NABIL SHOUKRY
HELWAN OBSERVATORY
HELWAN NEAR CAIRO
ARAB REP. OF EGYPT

AXFORD W IAN PROF
MPI FUER AERONOMIE
POSTFACH 20
D-3411 KATLENBURG-
LINDAU 3
GERMANY, F.R.

AYDIN CEMAL ASSOC PROF
AU FEN FAKULTESI
ASTRONOMI BOLUMU
BESEVLER ANKARA
TURKEY

AZZOPARDI M MR
OBS DE TOULOUSE
1 AV C FLAMMARION
F-31510 TOULOUSE
FRANCE

BAAN WILLEM A
DEPT OF ASTRONOMY
525 DAVEY LABORATORY
PENNSYLVANIA S U
UNIV PARK PA 16802
U.S.A.

BAARS JAAP W M DR
MPI FUER
RADIOASTRONOMIE
AUF DEM HUEGEL 69
D-5300 BONN
GERMANY, F.R.

BAART EDWARD E PROF
DEPT OF PHYSICS
RHODES UNIVERSITY
P O BOX 94
GRAHAMSTOWN 6140
SOUTH AFRICA

BAATH LARS B DR
PL 8481
S-439 00 ONSALA
SWEDEN

BABADZHANIANC MICHAIL DR
ASTRONOMICAL OBS
LENINGRAD UNIVERSITY
LENINGRAD
U.S.S.R.

BABADZHANOV P B DR
ASTROPHYSICAL INSTI-
TUTE OF TADJIKIAN
ACADEMY OF SCIENCES
734670 DUSHANBE
U.S.S.R.

BABCOCK HORRACE W DR
HALE OBSERVATORIES
813 SANTA BARBARA ST
PASADENA CA 91101
U.S.A.

BACCHUS PIERRE PROF
1 IMPASSE DE
L-OBSERVATOIRE
F-59000 LILLE
FRANCE

BACKER DONALD CH DR
601 CAMPBELL HALL
UNIV OF CALIFORNIA
BERKELEY CA 94720
U.S.A.

BAECK NICOLE A L DR
195 E-3 PLEIN
B-9218 LEDEBERG
BELGIUM

BAERENTZEN JORN
ELMEHOJVEJ 66
DK-8270 HOJBJERG
DENMARK

BAGGALEY WILLIAM JACK DR
PHYSICS DEPT
UNIV OF CANTERBURY
CHRISTCHURCH
NEW ZEALAND

BAGHOS BALEGH B DR
HELWAN OBSERVATORY
HELWAN CAIRO
ARAB REP. OF EGYPT

BAGILDINSKIJ BRONISLAV
K DR
PULKOVO OBSERVATORY
196140 LENINGRAD
U.S.S.R.

BAGLIN ANNIE DR
OBSERVATOIRE DE NICE
BP 252
F-06007 NICE CEDEX
FRANCE

BAGRI DURGADAS S
RADIOASTRONOMY CTR
TATA INSTITUTE OF
FUNDAMENTAL RESEARCH
OOTACAMUND 643 001
INDIA

BAHCALL JOHN N PROF
INSTITUTE FOR
ADVANCED STUDY
PRINCETON NJ 08540
U.S.A.

BAHNER KLAUS DR
LANDESSTERNWARTE
KOENIGSTUHL
D-6900 HEIDELBERG 1
GERMANY, F.R.

BAHNG JOHN D R PROF
DEARBORN OBSERVATORY
NORTHWESTERN
UNIVERSITY
EVANSTON IL 60201
U.S.A.

BAILEY MARK EDWARD
ASTRONOMY CTR
UNIV OF SUSSEX
FALMER
BRIGHTON BN1 6JQ
U.K.

BAIRD GEORGE A DR
PHYSICS DEPT
UNIVERSITY COLLEGE
BELFIELD
DUBLIN 4
IRELAND

BAIRD KENNETH M DR
DIVISION OF PHYSICS
NAT RES COUNC CANADA
MONTREAL ROAD
OTTAWA ONT K1A OR6
CANADA

BAIZE PAUL DR
6 RUE DAUBIGNY
F-75017 PARIS
FRANCE

BAJAJA E DR
INSTITUTO ARGENTINO
DE RADIOASTRONOMIA
CASILLA NO 5
VILLA ELISA PR BS AS
ARGENTINA

BAKER JAMES GILBERT DR
14 FRENCH DRIVE
BEDFORD NH 03102
U.S.A.

BAKER NORMAN H PROF
ASTRONOMY DEPT
COLUMBIA UNIVERSITY
PUPIN HALL 538 W 120
NEW YORK N Y 10027
U.S.A.

BAKOS GUSTAV A PROF
DEPT OF PHYSICS
UNIVERSITY WATERLOO
WATERLOO ONT N2L 3G1
CANADA

BALAZS BELA A DR
DEPT OF ASTRONOMY
EOTVOS UNIVERSITY
KUN BELA TER 2
H-1083 BUDAPEST
HUNGARY

BALAZS LAJOS G DR
KONKOLY OBSERVATORY
BOX 67
H-1525 BUDAPEST
HUNGARY

BALDINELLI LUIGI DR
VIA MONTEBELLO 4
I-40141 BOLOGNA
ITALY

BALDWIN JACK A DR
CERRO TOLOLO
INTERAMERICAN OBS
CASILLA 603
LA SERENA
CHILE

BALDWIN JOHN E DR
CAVENDISH LABORATORY
MADINGLEY RD
CAMBRIDGE CB3 OHE
U.K.

BALDWIN RALPH B
6190 GATEHOUSE DR SE
GRAND RAPIDS MI49506
U.S.A.

BALICK BRUCE PROF
ASTRONOMY DEPT FM-20
UNIV OF WASHINGTON
SEATTLE WA 98195
U.S.A.

BALKLAVS A E DR
RADIOASTROPHYSICAL
OBSERVATORY LATVIAN
ACADEMY OF SCIENCES
226026 RIGA
U.S.S.R.

BALL JOHN A DR
HARVARD COLLEGE
OBSERVATORY
HARVARD MA 01451
U.S.A.

BALLARIO M C PROF
OSSERVATORIO ASTRONO
MICO DI ARCETRI
VIA S LEONARDO
I-50100 FIRENZE
ITALY

BALLI EDIBE PROF
UNIVERSITE
RASATHANESI
ISTANBUL
TURKEY

BALMINO GEORGES G DR
CTR NAT D'ETUDES SPA
GROUP RECH GEOD SPAT
18 AVE EDOUARD BELIN
F-31400 TOULOUSE
FRANCE

BALONA LUIS ANTERO DR
S A ASTRON OBS
P O BOX 9
OBSERVATORY 7935
CAPE
SOUTH AFRICA

BANDERMANN L W DR
21131 GRENOLA DRIVE
CUPERTINO CA 95014
U.S.A.

BANDYOPADHYAY A DR
INDIA METEO DEPT
INDIAN ASTRON EPH UN
P546 BK N NW ALIPORE
CALCUTTA 700 053
INDIA

BANIA THOMAS MICHAEL
DEPT OF ASTRONOMY
BOSTON UNIVERSITY
725 COMMONWEALTH AVE
BOSTON MA 02215
U.S.A.

BANIN V G DR
SIBIZMIR
P B 4
664697 IRKUTSK 33
U.S.S.R.

BANOS COSMAS J DR
OBSERVATOIRE NATIO-
NAL D ATHENES
ATHENS 306
GREECE

BANOS GEORGE J PROF
UNIVERSITY OF
IOANNINA
ASTRONOMY DEPT
IOANNINA
GREECE

BARANNE A DR
OBSERVATOIRE
DE MARSEILLE
2 PLACE LE VERRIER
F-13004 MARSEILLE
FRANCE

BARATTA G B DR
OSSERVATORIO ASTRO-
NOMICO DI ROMA
I-00100 ROMA
ITALY

BARBANIS BASIL PROF
DEPT OF ASTRONOMY
UNIV OF THESSALONIKI
THESSALONIKI
GREECE

BARBARO G DR
OSSERVATORIO ASTRON
VICOLO DELL
OSSERVATORIO
PADOVA
ITALY

BARBIER MADELEINE AST ADJ
OBSERVATOIRE
DE MARSEILLE
2 PLACE LE VERRIER
F-13004 MARSEILLE
FRANCE

BARBIERI CESARE PROF
INST DI ASTRONOMIA
UNIVERSITA DI PADOVA
I-35100 PADOVA
ITALY

BARBON ROBERTO PROF
OSSERVATORIO
ASTROFISICO
I-36012 ASIAGO VI
ITALY

BARCZA SZABOLCS DR
KONKOLY OBSERVATORY
P O BOX 67
H-1525 BUDAPEST
HUNGARY

BARDEEN JAMES M PROF
PHYSICS DEPT FM-15
UNIVERSITY OF
WASHINGTON
SEATTLE WA 98195
U.S.A.

BARKAT ZALMAN PROF
THEORETICAL PHYSICS
HEBREW UNIVERSITY
JERUSALEM
ISRAEL

BARKER BRUCE MICHAEL
DEPT OF PHYSICS AND
ASTRONOMY
UNIV OF ALABAMA
UNIVERSITY AL 35486
U.S.A.

BARKER EDWIN S DR
ASTRONOMY DEPT
RLM 15 220
UNIV OF TEXAS
AUSTIN TX 78712
U.S.A.

BARKHATOVA KLAUDIA PROF
STATE UNIVERSITY
LENIN ST
620083 SVERDLOVSK
U.S.S.R.

BARLAI KATALIN DR
KONKOLY OBSERVATORY
BOX 67
H-1525 BUDAPEST
HUNGARY

BARLETTI RAFFAELE ENG
OSSERVATORIO ARCETRI
LARGO FERMI 5
I-50125 FIRENZE
ITALY

BARLIER FRANCOIS E DR
CERGA
AV COPERNIC
F-06130 GRASSE
FRANCE

BARLOW MICHAEL J DR
DEPT PHYSICS/ASTRON
UNIV COLLEGE LONDON
GOWER STREET
LONDON WC1E 6BT
U.K.

BARNARD HANNES A J DR
PHYSICS DEPT
UNIVERSITY OF B C
2075 WESBROOK MALL
VANCOUVER V6T 1W6
CANADA

BARNES AARON DR
NASA AMES RES CTR
CODE 245-3
MOFFETT FLD CA 94035
U.S.A.

BARNES III THOMAS G DR
DEPT OF ASTRONOMY
UNIVERSITY OF TEXAS
AUSTIN TX 78712
U.S.A.

BARNOTHY JENO DR PROF
833 LINCOLN ST
EVANSTON IL 60201
U.S.A.

BAROCAS VINICIO DR
JEREMIAH HORROCKS
AND WILFRED HALL OBS
MOOR PARK
PRESTON PR1 6AD
U.K.

BARRETO LUIZ MUNIZ PROF
OBSERVATORIO
NACIONAL
RUA GEN BRUCE 586
20000 RIO DE JANEIRO
BRAZIL

BARRETT ALAN H PROF
ROOM 26-331
MASSACHUSETT INST
OF TECHNOLOGY
CAMBRIDGE MA 02139
U.S.A.

BARROW RICHARD F DR
PHYSICAL CHEMISTRY
LABORATORY
SOUTH PARKS RD
OXFORD OX1 3QZ
U.K.

BARTEL NORBERT HARALD DR
MPI FUER
RADIOASTRONOMIE
AUF DEM HUEGEL 69
D-5300 BONN
GERMANY, F.R.

BARTOLINI CORRADO
OSSERVATORIO
ASTRONOMICO
VIA ZAMBONI 33
I-40126 BOLOGNA
ITALY

BASH FRANK N PROF
ASTRONOMY DEPT
UNIVERSITY OF TEXAS
AUSTIN TX 78712
U.S.A.

BASU BAIDYANATH READER
APPLIED MATHEMATICS
DEPARTMENT C U
92 A P C ROAD
CALCUTTA-700009
INDIA

BATES DAVID R PROF
DEPT OF APPLIED MATH
AND THEORETICAL PHYS
QUEENS UNIVERSITY
BELFAST BT7 1NN
U.K.

BATESON FRANK M OBE
P O BOX 3093
GREERTON TAURANGA
NEW ZEALAND

BATTANER EDUARDO DR
INSTITUTO DE ASTROFI
SICA DE ANDALUCIA
APDO 2144
GRANADA
SPAIN

BAUD BOUDEWIJN DR
KAPTEYN LABORATORIUM
HOOGBOUW WSN
POSTBUS 800
9700 AV GRONINGEN
THE NETHERLANDS

BAUM WILLIAM A DR
LOWELL OBSERVATORY
P O BOX 1269
FLAGSTAFF AZ 86002
U.S.A.

BAYM GORDON ALAN DR
DEPT OF PHYSICS
UNIV OF ILLINOIS
URBANA IL 61801
U.S.A.

BARROW COLIN H DR
SWISS FED INST TECHN
RADIO ASTRONOMY GR
HOCHSTRASSE 58
CH-8044 ZUERICH
SWITZERLAND

BARRY DON C DR
DEPT OF ASTRONOMY
UNIVERSITY OF
SOUTHERN CALIFORNIA
LOS ANGELES CA 90007
U.S.A.

BARTH CHARLES A PROF
LASP UNIVERSITY
OF COLORADO
BOULDER CO 80309
U.S.A.

BASART JOHN P
COOVER HALL
IOWA STATE UNIV
AMES IA 50011
U.S.A.

BASTIEN PIERRE DR
ASTRON INSTITUT
STERNWARTE
AUF DEM HUEGEL 71
D-5300 BONN 1
GERMANY, F.R.

BASU DIPAK DR
DEPT OF PHYSICS
UNIVERSITY OF THE
WEST INDIES
ST AUGUSTINE
TRINIDAD AND TOBAGO

BATES J RAPHAEL
METEOROLOGICAL
SERVICE
GLASNEVIN HILL
DUBLIN 9
IRELAND

BATH GEOFFREY T DR
DEPT OF ASTROPHYSICS
UNIVERSITY OF OXFORD
OXFORD OX1 3RQ
U.K.

BATTEN ALAN H DR
HERZBERG INST ASTRO
DOMINION ASTRO OBS
5071 W SAANICH RD
VICTORIA BC V8X 4M6
CANADA

BAUDRY ALAIN DR
OBSERVATOIRE
DE BORDEAUX
F-33270 FLOIRAC
FRANCE

BAUSTIAN W W MR
KITT PEAK NAT OBS
950 N CHERRY AVE
P O BOX 26732
TUCSON AZ 85726
U.S.A.

BEALE JOHN S DR
ROYAL GREENWICH
OBSERVATORY
HAILSHAM
EAST SUSSEX BN27 1RP
U.K.

BARROW JOHN DAVID
ASTRONOMY CENTRE
UNIVERSITY OF SUSSEX
FALMER
BRIGHTON BN1 9QH
U.K.

BARTAYA R A DR
ABASTUMANI ASTROPHY-
SICAL OBSERVATORY
383762 ABASTUMANI
GEORGIA
U.S.S.R.

BARTHOLDI PAUL DR
OBSERVATOIRE
DE GENEVE
CH-1290 SAUVERNY GE
SWITZERLAND

BASCHEK BODO PROF
INST F THEORETISCHE
ASTROPHYSIK
IM NEUENHEIMER F 294
D-6900 HEIDELBERG
GERMANY, F.R.

BASTIN J A PROF
PHYSICS DEPT
QUEEN MARY COLLEGE
MILE END ROAD
LONDON E1 4NS
U.K.

BATES BRIAN DR
DEPT OF PURE AND
APPLIED PHYSICS
QUEENS UNIVERSITY
BELFAST BT7 1NN
U.K.

BATES RICHARD HEATON T DR
ELECTRICAL ENG DEPT
UNIV OF CANTERBURY
CHRISTCHURCH
NEW ZEALAND

BATRAKOV J M DR
INSTITUTE OF THEORE-
TICAL ASTRONOMY
192187 LENINGRAD
U.S.S.R.

BATTISTINI PIERLUIGI DR
OSSERVATORIO
ASTRONOMICO
VIA ZAMBONI 33
I-40126 BOLOGNA
ITALY

BAUER CARL A DR
506 DAVEY
PENNA STATE UNIV
UNIV PK PA 16802
U.S.A.

BAUTZ LAURA P DR
1325 18TH ST NW
APT 506
WASHINGTON DC 20036
U.S.A.

BEARD DAVID B DR
DEPT PHYSICS ASTRON
UNIV OF KANSAS
LAWRENCE KS 66045
U.S.A.

BEARDSLEY WALLACE R DR
DEPT PHYS AND ASTRON
UNIV OF PITTSBURGH
ALLEGHENY OBS B 7658
PITTSBURGH PA 15214
U.S.A.

BEAUDET GILLES DR
DEPT DE PHYSIQUE
UNIV DE MONTREAL
CP 6128
MONTREAL PQ H3C 3J7
CANADA

BEAVERS WILLET I DR
ERWIN W FICK OBS
IOWA STATE
UNIVERSITY
AMES IA 50010
U.S.A.

BEC-BORSENBERGER ANNICKDR
BUREAU LONGITUDES
77 AVE DENFERT-
ROCHEREAU
F-75014 PARIS
FRANCE

BECK H G
VEB CARL ZEISS
FORSCHUNGSZENTRUM
CARL ZEISS STR 1
DDR 69 JENA
G.D.R.

BECKER FRIEDRICH DR PROF
WEITLSTR 66-2109
D-8000 MUENCHEN 45
GERMANY, F.R.

BECKER ROBERT A DR
P O BOX 4609
CARMEL CA 93921
U.S.A.

BECKER ROBERT HOWARD
PHYSICS DEPT
VIRGINIA POLYTECHNIC
INST & STATE UNIV
BLACKSBURG VA 24061
U.S.A.

BECKER WILHELM PROF
ASTRON INST
UNIVERSITAET BASEL
VENUSSTR 7
CH- 4102 BINNINGEN
SWITZERLAND

BECKERS JACQUES M DR
MULTIPLE MIRROR
TELESCOPE OBS
UNIV OF ARIZONA
TUCSON AZ 85718
U.S.A.

BECKLIN ERIC E DR
INST FOR ASTRONOMY
2680 WOODLAWN DR
HONOLULU HI 96822
U.S.A.

BECKMAN JOHN E DR
DEPT OF PHYSICS
QUEEN MARY COLLEGE
MILE END ROAD
LONDON E1 4NS
U.K.

BECKWITH S V W
DEPT OF ASTRONOMY
SPACE SCIENCES BLDG
CORNELL UNIVERSITY
ITHACA NY 14853
U.S.A.

BEEBE RETA FAYE DR
NEW MEXICO ST UNIV
DEPT OF ASTRONOMY
BOX 4500
LAS CRUCES NM 88003
U.S.A.

BEER REINHARD DR
JET PROPULSION LAB
CALIF INST OF TECH
4800 OAK GROVE DR
PASADENA CA 91103
U.S.A.

BEGELMAN MITCHELL CRAIG
INST OF ASTRONOMY
MADINGLEY ROAD
CAMBRIDGE CB3 0HA
U.K.

BEGGS DENIS W MR
INSTITUTE OF
ASTRONOMY
MADINGLEY ROAD
CAMBRIDGE CB3 0HA
U.K.

BEHR ALFRED PROF
HAMBURGER STERNWARTE
GOJENBERGSWEG 112
D-2050 HAMBURG 80
GERMANY, F.R.

BEINTEMA DOUWE A DR
SPACE RESEARCH DEPT
UNIV OF GRONINGEN
P O BOX 800
9700 AV GRONINGEN
THE NETHERLANDS

BEKENSTEIN JACOB D DR
PHYSICS DEPT
BEN GURION UNIV
P O B 653
BEERSHEVA
ISRAEL

BEL NICOLE J DR
OBSERVATOIRE
DE MEUDON
F-92190 MEUDON
FRANCE

BELIAYEV NIKOLAJ A DR
INST FOR THEORETICAL
ASTRONOMY
10 KUTUZOV QUAY
192187 LENINGRAD
U.S.S.R.

BELKOVICH O I DR
ENGELHARDT ASTRONOMI
CAL OBSERVATORY
422526 KAZAN
U.S.S.R.

BELL BARBARA DR
CENTER FOR
ASTROPHYSICS
60 GARDEN STREET
CAMBRIDGE MA 02138
U.S.A.

BELL KENNETH LLOYD DR
DEPT OF APPLIED MATH
AND THEORET PHYSICS
QUEEN'S UNIVERSITY
BELFAST BT7 1NN
U.K.

BELL MORLEY B
HERZBERG INST OF
ASTROPHYSICS
NAT RES COUNCIL
OTTAWA ONT K1A 0R6
CANADA

BELL ROGER A DR
ASTRONOMY PROGRAM
UNIVERSITY OF
MARYLAND
COLLEGE PARK MD20742
U.S.A.

BELOTSERKOVSKIJ
DAVID J DR
GOSSTANDART USSR
LENINSKY PROSPECT 9
117049 MOSCOW
U.S.S.R.

BELSERENE EMILIA P
HARVARD CLG OBS
C/O PLATE STACKS
60 GARDEN ST
CAMBRIDGE MA 02138
U.S.A.

BELTON MICHAEL J S DR
SOLAR SYSTEM PROG
KITT PEAK NAT OBS
950 N CHERRY AVE
TUCSON AZ 85726
U.S.A.

BELVEDERE GAETANO DR
IST DI ASTRONOMIA
CITTA UNIVERSITARIA
I-95125 CATANIA
ITALY

BELY OLEG DR
OBSERVATOIRE DE NICE
B P 252
F-06007 NICE CEDEX
FRANCE

BELY-DUBAU FRANCOISE
OBSERVATOIRE DE NICE
B P 252
F-06007 NICE CEDEX
FRANCE

BEM J DR
OBSERVATORIUM ASTR
UL KOPERNIKA 11
51-622 WROCLAW
POLAND

BENACCHIO LEOPOLDO
OSS ASTRONOMICO
VICOLO DELL
OSSERVATARIO 5
PADOVA
ITALY

BENAVENTE JOSE
INST Y OBSERVATORIO
DE MARINA
SAN FERNANDO CADIZ
SPAIN

BENDER PETER L DR
JILA
UNIVERSITY OF
COLORADO
BOULDER CO 80309
U.S.A.

BENEDICT G F DR
DEPT OF ASTRONOMY
UNIV OF TEXAS
AUSTIN TX 78712
U.S.A.

BENEST D DR
OBS DE NICE
B P 252
F-06007 NICE CEDEX
FRANCE

BENEVIDES SOARES P DR
INSTITUTO ASTRONOMI-
CO E GEOFISICO
CAIXA POSTAL 30 627
SAO PAULO 00 SP
BRAZIL

BENFORD GREGORY DR
PHYSICS DEPT
UNIV OF CALIFORNIA
IRVINE CA 92717
U.S.A.

BENNETT JOHN CAISTER DR
90 MALAN STREET
RIVIERA
PRETORIA 0084
SOUTH AFRICA

BENSAMMAR S DR
OBS DE MEUDON
F-92190 MEUDON
FRANCE

BENVENUTI P DR
OSSERVATORIO ASTRO-
FISICO
I-36012 ASIAGO
ITALY

BENZ ARNOLD DR
RADIO ASTRONOMY GR
INST FUER ASTRONOMIE
ETH ZENTRUM
CH-8092 ZUERICH
SWITZERLAND

BERENDZEN RICHARD DR
PROVOST
THE AMERICAN
UNIVERSITY
WASHINGTON DC 20016
U.S.A.

BERG RICHARD A DR
DEPT PHYS AND ASTRON
UNIV OF ROCHESTER
419 SPACE SCI CTR
ROCHESTER NY 14627
U.S.A.

BERGE GLENN L DR
OWENS VALLEY RADIO
OBSERVATORY
CALTECH 102 24
PASADENA CA 91125
U.S.A.

BERGEAT JACQUES G DR
OBSERVATOIRE DE LYON
F-69230 SAINT-GENIS-
LAVAL
FRANCE

BERGER CHRISTIANE DR
CERGA
AVENUE COPERNIC
F-06130 GRASSE
FRANCE

BERGER JACQUES G DR
OBSERVATOIRE
DE PARIS
61 AV OBSERVATOIRE
F-75014 PARIS
FRANCE

BERGER XAVIER DR
CERGA
AVENUE COPERNIC
F-6130 GRASSE
FRANCE

BERGERON JACQUELINE A DR
INST D'ASTROPHYSIQUE
98 BIS BD ARAGO
F-75014 PARIS
FRANCE

BERGSTRALH JAY T DR
JPL
M/X 1833 - 365
4800 OAK GROVE DRIVE
PASADENA CA 91103
U.S.A.

BERGVALL NILS AKE SIGVARD
ASTRONOMISKA
OBSERVATORIET
BOX 515
S-751 20 UPPSALA
SWEDEN

BERKHUIJSEN ELLY M DR
MPI FUER
RADIOASTRONOMIE
AUF DEM HUEGEL 69
D-5300 BONN
GERMANY, F.R.

BERMAN ROBERT HIRAM DR
MIT
RM 36-227
77 MASSACHUSETTS AVE
CAMBRIDGE MA 02139
U.S.A.

BERNACCA P DR
OSSERVATORIO ASTRO-
FISICO
ASIAGO VICENZA
ITALY

BERNAT ANDREW PLOUS DR
INST FOR ASTRONOMY
UNIV OF HAWAII
P O BOX 209
KULA MAUI HI 96790
U.S.A.

BERRINGTON KEITH ADRIAN
DEPT OF APPLIED MATH
& THEORETICAL PHYS
QUEEN'S UNIVERSITY
BELFAST BT7 1NN
U.K.

BERTAUX J L DR
SERVICE D AERONOMIE
B P NO 3
F-91370 VERRIERES-
LE-BUISSON
FRANCE

BERTHOMIEU GABRIELLE DR
OBSERVATOIRE DE NICE
BP 252
F-06007 NICE CEDEX
FRANCE

BERTIAU FLOR C PROF
WAVERSEBAAN 220
B-3030 HEVERLEE
BELGIUM

BERTIN GIUSEPPE DR
SCUOLA NORMALE SUP
PIAZZA DEI CAVALIERI
I-56100 PISA
ITALY

BERTOLA FRANCESCO PROF
OSSERVATORIO
ASTRONOMICO
I-35100 PADOVA
ITALY

BERTOUT CLAUDE PH D
LANDESSTERNWARTE
KOENIGSTUHL
D-6900 HEIDELBERG 1
GERMANY, F.R.

BESSELL MICHAEL S DR
MT STROMLO
OBSERVATORY
WODEN PO ACT 2606
AUSTRALIA

BETTIS DALE G PROF
TICOM
UNIVERSITY OF TEXAS
AT AUSTIN
AUSTIN TX 78712
U.S.A.

BEUERMANN KLAUS P PROF
INST ASTRON ASTROPH
TU BERLIN
ERNST-REUTER-PL 7
D-1000 BERLIN
GERMANY, F.R.

BHATIA PREM K DR
DEPT OF MATHEMATICS
UNIVERSITY OF
JODHPUR
JODHPUR-342001
INDIA

BHATIA V B DR
DEPT OF PHYSICS AND
ASTROPHYSICS
DELHI UNIVERSITY
DELHI-110007
INDIA

BHATNAGAR ARVIND DR
VEDHSHALA UDAIPUR
SOLAR OBS
11 VIDYA MARG
UDAIPUR RAJAS 313001
INDIA

BHATNAGAR K B DR
ZAKIR HUSSAIN COLL
UNIV OF DELHI
AJMERI GATE
NEW DELHI 110 006
INDIA

BHATTACHARYYA J C PROF
INDIAN INSTITUTE OF
ASTROPHYSICS
BANGALORE 560034
INDIA

BHAVSAR SUKETU P
ASTRONOMY CENTRE
UNIVERSITY OF SUSSEX
FALMER
BRIGHTON BN1 9QH
U.K.

BHONSLE RAJARAM V PROF
PHYSICAL RESEARCH
LABORATORY
NAVRANGPURA
AHMEDABAD 380009
INDIA

BIANCHINI ANTONIO DR
OSSERVATORIO ASTRO-
FISICO
I-36012 ASIAGO
ITALY

BIBARSOV RAVIL'SH DR
ASTROPHYSICAL INST
ACAD OF SCS OF THE
TADJIK SSR
734670 DUSHANBE
U.S.S.R.

BICAK JIRI DR
DEPT OF MATH PHYSICS
CHARLES UNIVERSITY
POVLTAVSKA 8
180 00 PRAHA 8
CZECHOSLOVAKIA

BICKNELL GEOFFREY V DR
MOUNT STROMLO &
SIDING SPRING OBS
PRIVATE BAG
WODEN P.O. ACT 2606
AUSTRALIA

BIDELMAN WILLIAM P PROF
WARNER AND SWASEY
OBSERVATORY
CASE WESTERN RES UN
CLEVELAND OH 44106
U.S.A.

BIEGING JOHN HAROLD DR
RADIO ASTRONOMY LAB
CAMPBELL HALL
UNIV OF CALIFORNIA
BERKELEY CA 94720
U.S.A.

BIELICKI MACIEJ DR
ASTRONOMICAL
OBSERVATORY
AL UJAZDOWSKIE 4
00-478 WARSZAWA
POLAND

BIEMONT EMILE DR
RUE A LECLEROS 39
B-4652 XHENDELESSE
HERVE
BELGIUM

BIEN REINHOLD DR
ASTRONOMISCHES
RECHEN-INSTITUT
MOENCHHOFSTR 12-14
D-6900 HEIDELBERG
GERMANY, F.R.

BIERMANN LUDWIG PROF DR
MPI F ASTROPHYSIK
FOEHRINGER RING 6
D-8000 MUENCHEN 40
GERMANY, F.R.

BIERMANN PETER L DR
MPI FUR RADIOASTRON
AUF DEM HUEGEL 69
D-5300 BONN
GERMANY, F.R.

BIGNELL R C DR
NRAO-VLA
SOCORRO NM 87801
U.S.A.

BIJAOUI ALBERT DR
OBSERVATOIRE DE NICE
B P 252
F-06007 NICE CEDEX
FRANCE

BILLAUD G J
CERGA
AVE COPERNIC
F-6130 GRASSE
FRANCE

BILLINGS DONALD E PROF
UNIVERSITY COLORADO
DEPT OF ASTRO-
GEOPHYSICS
BOULDER CO 80309
U.S.A.

BINGHAM RICHARD G DR
ROYAL GREENWICH
OBSERVATORY
HAILSHAM
EAST SUSSEX BN27 1RP
U.K.

BINNENDIJK L PROF DR
FLOWER AND COOK
OBSERVATORY
UNIV OF PENNSYLVANIA
PHILADELPHIA PA19104
U.S.A.

BINNEY JAMES J DR
DEPT THEOR PHYSICS
1 KEBLE ROAD
OXFORD OX1 3NP
U.K.

BIRAUD FRANCOIS DR
OBSERVATOIRE
DE MEUDON
F-92190 MEUDON
FRANCE

BIRKLE KURT PH D
MPI FUER ASTRONOMIE
KOENIGSTUHL
D-6900 HEIDELBERG 1
GERMANY, F.R.

BISHOP ROY L DR
DEPT OF PHYSICS
ACADIA UNIVERSITY
WOLFVILLE NS B0P 1X0
CANADA

BISIACCHI GIANFRANCO DR
INST DE ASTRONOMIA
UNAM
APDO POSTAL 70-264
04510 MEXICO DF
MEXICO

BISNOVATYI-KOGAN G S DR
SPACE RESEARCH INSTI
TUTE USSR ACADEMY OF
SCIENCES
117810 MOSCOW
U.S.S.R.

BISWAS SUKUMAR DR
SOLAR PHYS GROUP
TATA INST FUND RES
HOMI BHABHA ROAD
BOMBAY 400 005
INDIA

BLAAUW A PROF DR
ASSERSTRAAT 78
9335 TC ZUIDVELDE
THE NETHERLANDS

BLACK JOHN HARRY DR
HARVARD COLLEGE OBS
CTR FOR ASTROPHYSICS
60 GARDEN ST
CAMBRIDGE MA 02138
U.S.A.

BLACKMAN CLINTON PAUL DR
UNIV OF EDINBURGH
ROYAL OBSERVATORY
BLACKFORD HILL
EDINBURGH EH9 3HJ
U.K.

BLACKWELL ALAN TREVOR
METEORITE PROJECT
NAT RES COUNCIL
UNIV SASKATCHEWAN
SASKATOON S7N 0W0
CANADA

BLACKWELL DONALD E PROF
DEPT OF ASTROPHYSICS
SOUTH PARKS ROAD
OXFORD OX1 3RQ
U.K.

BLADES J CHRIS DR
ESA SATEL TR STATION
V FRANCA DEL CASTILL
P O BOX 54065
MADRID
SPAIN

BLAHA MILAN DR
DEPT OF PHYSICS AND
ASTRONOMY
UNIVERSITY MARYLAND
COLLEGE PARK MD20742
U.S.A.

BLAIR GUY NORMAN DR
ASTRONOMY PROGRAM
DEPT EARTH SPACE SCS
SUNY
STONY BROOK NY 11794
U.S.A.

BLAMONT JACQUES E PROF
SERVICE D AERONOMIE
DU CNRS B P 3
F-91370 VERRIERES
LE BUISSON
FRANCE

BLANDFORD R D DR
THEORET ASTROPHYSICS
130-33
CALTECH
PASADENA CA 91125
U.S.A.

BLEEKER JOHAN A M DR IR
HUYGENS LABORATORY
WASSENAARSEWEG 78
2300 RA LEIDEN
THE NETHERLANDS

BLITZ LEO
UNIV OF MARYLAND
ASTRONOMY PROGRAM
COLLEGE PARK MD20742
U.S.A.

BLUM EMILE-JACQUES DR
IRAM - VOIE 10
DOMAINE UNIVERSITAIR
F-38406 ST MARTIN
D'HERES
FRANCE

BOBROV M S DR
ASTRONOMICAL COUNCIL
USSR ACADEMY SCI
PYATNITSKAYA UL 48
109017 MOSCOW
U.S.S.R.

BOCHAREV NIKOLAJ G DR
STERNBERG STATE
ASTRONOMICAL INST
117234 MOSCOW
U.S.S.R.

BOEHM KARL-HEINZ PROF
ASTRONOMY DEPT
UNIVERSITY OF
WASHINGTON
SEATTLE WA 98195
U.S.A.

BOEHME SIEGFRIED DR
ASTRONOMISCHES
RECHEN INSTITUT
MOENCHHOFSTR 12-14
D-6900 HEIDELBERG
GERMANY, F.R.

BOESGAARD ANN M PROF
INST FOR ASTRONOMY
2680 WOODLAWN DR
HONOLULU HI 96822
U.S.A.

BOGGESS NANCY W DR
NASA HEADQUARTERS
WASHINGTON DC 20546
U.S.A.

BOHLIN J DAVID DR
CODE ST
NASA HEADQUARTERS
WASHINGTON DC 20546
U.S.A.

BLANCO CARLO DR
OSSERVATORIO
ASTROFISICO
CITTA UNIVERSITARIA
I-95125 CATANIA
ITALY

BLASIUS KARL RICHARD DR
PLANETARY SC INST
283 S LAKE
SUITE 218
PASADENA CA 91101
U.S.A.

BLESS ROBERT C PROF
ASTRONOMY DEPT
UNIVERSITY WISCONSIN
475 N CHARTER ST
MADISON WI 53706
U.S.A.

BLITZSTEIN WILLIAM DR
DEPT OF ASTRONOMY
UNIV OF PENNSYLVANIA
D RITTENHOUSE LAB E1
PHILADELPHIA PA19174
U.S.A.

BLUM PETER PROF
INST F ASTROPHYSIK
UNIVERSITAET BONN
AUF DEM HUEGEL 71
D-5300 BONN 1
GERMANY, F.R.

BOBROVNIKOFF NICHOLAS T
DR
1623 VISALIA AVE
BERKELEY CA 94707
U.S.A.

BOCHONKO DONALD R DR
DEPT MATH ASTRONOMY
UNIV OF MANITOBA
WINNIPEG MB R3T 2N2
CANADA

BOEHM-VITENSE ERIKA PROF
ASTRONOMY DEPT
UNIVERSITY OF
WASHINGTON
SEATTLE WA 98195
U.S.A.

BOERNER GERHARD DR
MPI FUER PHYSIK UND
ASTROPHYSIK
FOEHRINGER RING 6
D-8000 MUENCHEN
GERMANY, F.R.

BOESHAAR GREGORY ORTH DR
DEPT PHYSICS/ASTRON
FREE UNIVERSITY
DE BOELELAAN 1081
1081 HV AMSTERDAM
THE NETHERLANDS

BOGORODSKIJ A F DR
KIEV STATE UNIVERSI-
TY ASTRONOMICAL
OBSERVATORY
252053 KIEV
U.S.S.R.

BOHLIN RALPH C DR
CODE 672
GODDARD SPACE
FLIGHT CENTER
GREENBELT MD 20771
U.S.A.

BLANCO VICTOR M DR
CERRO TOLOLO
INTERAMERICAN OBS
CASILLA 63 D
LA SERENA
CHILE

BLECHA ANDRE BORIS G DR
16 RUE ET-DUMONT
CH-1204 GENEVA
SWITZERLAND

BLINOV N S DR
STERNBERG STATE AS-
TRONOMICAL INSTITUTE
117234 MOSCOW
U.S.S.R.

BLUDMAN SIDNEY A PROF
DEPT OF PHYSICS
UNIVERSITY OF
PENNSYLVANIA
PHILADELPHIA PA19104
U.S.A.

BO SHU-REN
NO 1 GONG YUAN WEST
ROAD
BEIJING
PEOPLE'S REPUBLIC CHINA

BOCCHIA ROMEO DR
OBS DE BORDEAUX
AVENUE P SEMIROT
33270 FLOIRAC
FRANCE

BODENHEIMER PETER PROF
LICK OBSERVATORY
UNIVERSITY OF CALIF
SANTA CRUZ CA 95064
U.S.A.

BOEHME A DR
HEINRICH HERTZ INST
SOLAR TERRESTR PHYS
TELEGRAFENBERG
DDR 15 POTSDAM
G.D.R.

BOERNGEN FREIMUT DR PH
ZENTR INST ASTROPHYS
K SCHWARZSCHILD OBS
DDR-6901 TAUTENBURG
G.D.R.

BOGGESS ALBERT DR
CODE 673
GODDARD SPACE
FLIGHT CENTER
GREENBELT MD 20771
U.S.A.

BOHANNAN BRUCE EDWARD
SOMMERS-BAUSCH OBS
UNIV OF COLORADO
BOX 391
BOULDER CO 80309
U.S.A.

BOHRMANN ALFRED PROF
SCHAERSTRASSE 23
D-2050 HAMBURG 80
GERMANY, F.R.

BOISCHOT ANDRE DR
OBSERVATOIRE DE
MEUDON
F-92190 MEUDON
FRANCE

BOK BART J PROF
STEWARD OBSERVATORY
UNIVERSITY ARIZONA
TUCSON AZ 85721
U.S.A.

BOKSENBERG ALEC PROF
ROYAL GREENWICH OBS
HERSTMONCEUX CASTLE
HAILSHAM
E SUSSEX BN27 1RP
U.K.

BOLCAL C DR
UNIV OBSERVATORY
UNIVERSITE
ISTANBUL
TURKEY

BOLDT ELIHU DR
CODE 661
GODDARD SPACE FLIGHT
CENTER
GREENBELT MD 20771
U.S.A.

BOLEY FORREST
WILDER LABORATORY
DARTMOUTH COLLEGE
HANOVER NH 03755
U.S.A.

BOLTON C THOMAS PROF
DAVID DUNLAP OBS
P O BOX 360
RICHMOND HILL
ONT L4C 4Y6
CANADA

BOLTON JOHN G
39 PANORAMA CRESCENT
BODERIM QLD 4556
AUSTRALIA

BONANOMI J DR
OBSERVATOIRE
DE NEUCHATEL
CH-2000 NEUCHATEL
SWITZERLAND

BONAZZOLA S DR
OBS DE MEUDON
F-92190 MEUDON
FRANCE

BOND HOWARD E DR
DEPT PHYSICS AND AST
LOUISIANA STATE
UNIVERSITY
BATON ROUGE LA 70803
U.S.A.

BOND JOHN RICHARD
DEPT OF PHYSICS
STANFORD UNIVERSITY
STANFORD CA 94305
U.S.A.

BONDARENKO L N DR
STERNBERG STATE AS-
TRONOMICAL INSTITUTE
117234 MOSCOW
U.S.S.R.

BONDI HERMANN PROF SIR
DEPT OF ENERGY
THAMES HOUSE SOUTH
MILLBANK
LONDON SW1P 4QJ
U.K.

BONET JOSE A
URBANIZACION CERCADO
MESA
VILLAS DEL PRADO 7
LA LAGUNA TENERIFE
SPAIN

BONEV T MR
OBS POPULAIRE
STARA-ZAGORA
BULGARIA

BONIFAZI ANGELO DR
OSSERVATORIO ASTRO-
NOMICO
I-40100 BOLOGNA
ITALY

BONNET ROGER M DR
LPSP CNRS
BP 10
F-91370 VERRIERES
LE BUISSON
FRANCE

BONNOR W B PROF
QUEEN ELIZABETH
COLLEGE
CAMPDEN HILL ROAD
LONDON W8 7AH
U.K.

BONOLI FABRIZIO
OSS ASTRONOMICO
UNIVERSITARIO
CASELLA POSTALE 596
I-40100 BOLOGNA
ITALY

BONOMETTO SILVIO A DR
ISTITUTO DI FISICA
G GALILEI 8
VIA MARZOLO
I-35100 PADOVA
ITALY

BONOV ANGEL DR
ASTRONOMICAL
OBSERVATORY
P O BOX 36
1504 SOFIA 4
BULGARIA

BONSACK WALTER K PROF
INST FOR ASTRONOMY
2680 WOODLAWN DR
HONOLULU HI 96822
U.S.A.

BOOK DAVID L
CODE 4040
US NAVAL RES LAB
WASHINGTON DC 20375
U.S.A.

BOOKMYER BEVERLY B DR
DEPT OF PHYSICS AND
ASTRONOMY
CLEMSON UNIVERSITY
CLEMSON SC 29631
U.S.A.

BOOTH ROY S DR
NUFFIELD RADIO ASTRO
NOMY LABORATORIES
JODRELL BANK
MACCLESFIELD SK119DL
U.K.

BOPP BERNARD W DR
DEPT PHYSICS ASTRON
UNIV OF TOLEDO
TOLEDO OH 43606
U.S.A.

BORCHKHADZE TANGIZ M DR
ABASTUMANI ASTROPHY-
SICAL OBSERVATORY
383762 ABASTUMANI
GEORGIA
U.S.S.R.

BORD DONALD JOHN
DEPT OF PHYSICS
BENEDICTINE COLLEGE
ATCHISON KS 66002
U.S.A.

BORGMAN JAN DR PROF
KAPTEYN OBSERVATORY
MENSINGHEWEG 20
9301 KA RODEN DR
THE NETHERLANDS

BORGNINO J DR
DEPT D'ASTROPHYSIQUE
UNIVERSITE DE NICE
PARC VALROSE
F-06034 NICE CEDEX
FRANCE

BORIAKOFF VALENTIN
NAIC
420 SPACE SC BLDG
CORNELL UNIVERSITY
ITHACA NY 14853
U.S.A.

BORRA ERMANNO F DR
DEPT DE PHYSIQUE
UNIVERSITE LAVAL
STE FOY
QUEBEC PQ G1K 7P4
CANADA

BOSMA ALBERT DR
STERREWACHT
POSTBUS 9513
2300 RA LEIDEN
THE NETHERLANDS

BOSMA PIETER B DR
DEPT PHYSICS/ASTRON
FREE UNIVERSITY
DE BOELELAAN 1081
1081 HV AMSTERDAM
THE NETHERLANDS

BOSMAN-CRESPIN DENISE
BLVD D AVROY 68
BTE 093
B-4000 LIEGE
BELGIUM

BOTELHEIRO A DR
OBS ASTRONOMICO
LISBOA
PORTUGAL

BOTEZ ELVIRA DR
INSTITUT D ENSEIGNE-
MENT SUPERIEUR
5800 SUCEAVA
ROUMANIA

BOTTINELLI LUCETTE DR
OBSERVATOIRE
DE MEUDON
RADIOASTRONOMIE
F-92190 MEUDON
FRANCE

BOUGERET J L DR
OBS DE MEUDON
F-92190 MEUDON
FRANCE

BOUIGUE ROGER PROF
LAB D'ASTRONOMIE
UNIVERSITE
PAUL SABATIER
F-31077 TOULOUSE CDX
FRANCE

BOULESTEIX JAQUES
OBS DE MARSEILLE
2 PLACE LE VERRIER
F-13248 MARSEILLE
FRANCE

BOULON JACQUES J DR
OBSERVATOIRE PARIS
61 AVENUE DE
L OBSERVATOIRE
F-75014 PARIS
FRANCE

BOUSKA JIRI DR
DEPT OF ASTRONOMY
CHARLES UNIVERSITY
SVEDSKA 8
150 00 PRAHA
CZECHOSLOVAKIA

BOUVIER PIERRE PROF
OBSERVATOIRE
DE GENEVE
CH-1290 SAUVERNY GE
SWITZERLAND

BOWELL E L G DR
LOWELL OBS
P O BOX 1269
FLAGSTAFF AZ 86002
U.S.A.

BOWEN EDWARD G DR
5010 MAXWELL AVENUE
WEST RIVER MD 20881
U.S.A.

BOWERS PHILLIP F
CODE 4134
NAVAL RESEARCH LAB
WASHINGTON DC 20375
U.S.A.

BOWYER C STUART PROF
UNIVERSITY
OF CALIFORNIA
ASTRONOMY DEPT
BERKELEY CA 94720
U.S.A.

BOYARCHUK A A DR
CRIMEAN ASTROPHYSI-
CAL OBSERVATORY USSR
ACAD OF SCI NAUCHNY
334413 CRIMEA
U.S.S.R.

BOYARCHUK MARGARITA E DR
CRIMEAN ASTROPHYS
OBSERVATORY
P/O NAUCHNYJ
334413 CRIMEA
U.S.S.R.

BOYCE PETER B DR
AMERICAN ASTRON SOC
1816 JEFFERSON PL NW
WASHINGTON DC 20036
U.S.A.

BOYD ROBERT L F PROF
MULLARD SPACE
SCIENCE LAB UCL
HOLMBURY ST MARY
DORKING RH5 6NS
U.K.

BOYDAG-YILDIZDOGDU F S
ACADEMY OF ISTANBUL
ENG & ARCHITECTURE
DEPT OF PHYSICS
ISTANBUL
TURKEY

BOYTON P E DR
ASTRONOMY DEPT
UNIV OF WASHINGTON
SEATTLE WA 98195
U.S.A.

BOZIS GEORGE PROF
DEPT THEOR MECHANICS
UNIVERSITY
OF THESSALONIKI
THESSALONIKI
GREECE

BOZKURT SUKRU DR
EGE UNIVERSITY
OBSERVATORY
PK 21
BORNOVA IZMIR
TURKEY

BRACCESI ALESSANDRO PROF
OSSERVATORIO
ASTRONOMICO
VIA ZAMBONI 33
I-40126 BOLOGNA
ITALY

BRACEWELL RONALD N PROF
DURAND 329
STANFORD UNIVERSITY
STANFORD CA 94305
U.S.A.

BRAES L L E DR
STERREWACHT
POSTBUS 9513
2300 RA LEIDEN
THE NETHERLANDS

BRAHDE ROLF DR
INSTITUTE OF THEORE-
TICAL ASTROPHYSICS
UNIVERSITY OF OSLO
BLINDERN OSLO 3
NORWAY

BRAHIC ANDRE DR
OBS DE MEUDON
F-92190 MEUDON
FRANCE

BRANCH DAVID DR
DEPT OF PHYSICS AND
ASTRONOMY
UNIVERSITY OKLAHOMA
NORMAN OK 73019
U.S.A.

BRAND PETER W J L DR
DEPT OF ASTRONOMY
UNIVERSITY
OF EDINBURGH
EDINBURGH EH8 9YL
U.K.

BRANDI ELISANDE STELLA DR
OBS ASTRONOMICO
PASEO DE BOSQUE
1900-LA PLATA
ARGENTINA

BRANDIE GEORGE W DR
ENVIRONMENTAL ENG
DUPUIS HALL
QUEENS UNIVERSITY
KINGSTON ONT K7L 3N6
CANADA

BRANDT JOHN C DR
CODE 680
GODDARD SPACE
FLIGHT CENTER
GREENBELT MD 20771
U.S.A.

BRANDUARDI-RAYMONT G
MULLARD SPACE
SCIENCE LAB
HOLMBURY ST MARY
DORKING SURREY
U.K.

BRANHAM RICHARD L JR
US NAVAL OBSERVATORY
TRANSIT CIRCLE DIV
WASHINGTON DC 20390
U.S.A.

BRANSCOMB L M DR
NATIONAL BUREAU OF
STANDARDS
WASHINGTON DC 20025
U.S.A.

BRANSON NICHOLAS J B A DR
GENERAL BOARD OF THE
FACULTIES
THE OLD SCHOOLS
CAMBRIDGE CB2 1TT
U.K.

BRAUDE SEMION YA PROF AG
INST RADIOPHY ELECTR
ACADEMY OF SCIENCES
THE UKRAINIAN SSR
310085 KHARKOV
U.S.S.R.

BRAULT JAMES W DR
KITT PEAK NATIONAL
OBSERVATORY
PO BOX 26732
TUCSON AZ 85726
U.S.A.

BRAUNINGER H DR
MPI FUER PHYSIK UND
ASTROPHYSIK
INST EXTRATERR PHYS
D-8046 GARCHING
GERMANY, F.R.

BRAUNSFURTH EDWARD PH D
ASTRONOMISCHES INST
RUHR UNIVERSITAET
POSTFACH 102148
D-4630 BOCHUM
GERMANY, F.R.

BRAY ROBERT J DR
CSIRO NATIONAL MEASU
REMENT LABORATORY
PO BOX 218
LINDFIELD NSW 2070
AUSTRALIA

BREALEY G A P ENG
DOMINION ASTROPH OBS
5071 W SAANICH ROAD
RR 5
VICTORIA BC V8X 4M6
CANADA

BRECHER AVIVA DR PROF
ARTHUR D LITTLE INC
ACORN PARK
CAMBRIDGE MA 02140
U.S.A.

BRECHER KENNETH PROF
DEPT OF ASTRONOMY
BOSTON UNIVERSITY
725 COMMONWEALTH AVE
BOSTON MA 02215
U.S.A.

BRECKINRIDGE JAMES B DR
J P L CALTECH
MAIL STOP 183-301
4800 OAK GROVE DR
PASADENA CA 91103
U.S.A.

BREGER MICHEL DR
DEPT OF ASTRONOMY
UNIVERSITY OF TEXAS
AUSTIN TX 78712
U.S.A.

BREGMAN JACOB D IR
NETHERLANDS FOUNDA-
TION RADIOASTRONOMY
POSTBUS 2
7990 AA DWINGELOO
THE NETHERLANDS

BREINHORST ROBERT A DR
ASTRON INSTITUT
STERNWARTE
AUF DEM HUEGEL 71
D-5300 BONN 1
GERMANY, F.R.

BREJDO IZABELLA I DR
PULKOVO OBSERVATORY
196140 LENINGRAD
U.S.S.R.

BRETAGNON P DR
BUREAU D LONGITUDES
77 AVENUE DENFERT-
ROCHEREAU
F-75014 PARIS
FRANCE

BRIDLE ALAN H PROF
DEPT OF PHYSICS
QUEENS UNIVERSITY
KINGSTON ONT K7L 3N6
CANADA

BRIEVA EDUARDO PROF
OBSERVATORIO
NACIONAL
APARTADO 2584
BOGOTA
COLOMBIA

BRIHAYE CHARLES C A DR
SCHEMERINGLAAN 14A
1900 OVERIJSE
BELGIUM

BRINI D PROF
LABORATORIO TESRE
VIA CASTAGNOLI 1
I-40100 BOLOGNA
ITALY

BRINKMAN BERT C DR
SPACE RESEARCH LAB
BENELUXLAAN 21
3527 HS UTRECHT
THE NETHERLANDS

BRIOT DANIELLE
OBS DE PARIS
61 AVE DE L'OBS
F-75014 PARIS
FRANCE

BROADFOOT A LYLE DR
SPACE SCS INST USC
TUCSON LAB
3625 EAST AJO WAY
TUCSON AZ 85173
U.S.A.

BRODERICK JOHN DR
PHYSICS DEPT
VPI AND SU
BLACKSBURG VA 24061
U.S.A.

BROGLIA PIETRO DR
OSSERVATORIO
ASTRONOMICO
I-22055 MERATE COMO
ITALY

BROMAGE GORDON E DR
ASTROPHYSICS RES DIV
S R C APPLETON LAB
CULHAM LAB
ABINGDON OXON
U.K.

BROOKES CLIVE J DR
1 RYEFIELD CLOSE
WEST HAGLEY
STOURBRIDGE
W MIDLANDS DY9 0JS
U.K.

BROSCHE PETER PROF
OBSERVATORIUM
HOHER LIST
D-5568 DAUN
GERMANY, F.R.

BROSTERHUS E B F DR
DOMINION ASTROPH OBS
5071 W SAANICH ROAD
RR 5
VICTORIA BC V8X 4M6
CANADA

BROTEN NORMAN W
HERZBERG INST OF
ASTROPHYSICS
NAT RES COUNC CANADA
OTTAWA ONT K1A 0R6
CANADA

BROUCKE ROGER DR
7203 RUNNING ROPE
CIRCLE
AUSTIN TX 78731
U.S.A.

BROUW W N DR
RADIOSTERRENWACHT
POSTBUS 2
7990 AA DWINGELOO
THE NETHERLANDS

BROWN ALEXANDER
DEPT OF PHYSICS
QUEEN MARY COLLEGE
MILE END ROAD
LONDON E1 4NS
U.K.

BROWN HARRISON DR
DIV OF GEOLOGICAL
SCIENCES
CALTECH
PASADENA CA 91103
U.S.A.

BROWN JOHN C DR
DEPT OF ASTRONOMY
GLASGOW UNIVERSITY
GLASGOW G12 8QQ
U.K.

BROWN ROBERT L DR
NATIONAL RADIO
ASTRONOMY'OBS
BOX 2
GREEN BANK WV 24944
U.S.A.

BROWN RONALD D PROF
CHEMISTRY DEPT
MONASH UNIVERSITY
WELLINGTON ROAD
CLAYTON VICT 3168
AUSTRALIA

BROWNE IAN W A DR
NUFFIELD RADIO ASTRO
NOMY LABORATORIES
JODRELL BANK
MACCLESFIELD SK119DL
U.K.

BROWNLEE DONALD E PROF
DEPT OF ASTRONOMY
UNIV OF WASHINGTON
SEATTLE WA 98195
U.S.A.

BROWNLEE ROBERT R DR
G-DO MS 570
LOS ALAMOS SCI
LABORATORY
LOS ALAMOS NM 87544
U.S.A.

BRUCK HERMANN A PROF
CRAIGOWER
PENICUIK EH26 9LA
U.K.

BRUCK MARY T DR
ROYAL OBSERVATORY
EDINBURGH EH9 3HJ
U.K.

BRUECKNER GUENTER E DR
CODE 7142
NAVAL RESEARCH
LABORATORY
WASHINGTON DC 20375
U.S.A.

BRUHWEILER F C JR
IVE OBSERVATORY
CODE 685/CSC
NASA GSFC
GREENBELT MD 20771
U.S.A.

BRUMBERG VICTOR A DR
INSTITUTE FOR THEORE
TICAL ASTRONOMY
192187 LENINGRAD
U.S.S.R.

BRUNER EL C JR DR
LOCKHEED P A RES LAB
DEPT 52-12 BLDG 202
3251 HANOVER ST
PALO ALTO CA 94304
U.S.A.

BRUNET JEAN-PIERRE DR
OBSERVATOIRES
1 AVE C FLAMMARION
F-31500 TOULOUSE
FRANCE

BRUNK WILLIAM E DR
OFF SPACE SC CODE SL
NASA HEADQUARTERS
400 MARYLAND AVE S W
WASHINGTON DC 20546
U.S.A.

BRUSTON P DR
L P S P
B P NO 10
F-91370 VERRIERES-
LE-BUISSON
FRANCE

BRUWER J A
SOUTH AFRICAN ASTRO-
NOMICAL OBSERVATORY
P O BOX 3718
JOHANNESBURG 2000
SOUTH AFRICA

BRUZEK ANTON DR
KIEPENHEUER INSTITUT
FUER SONNENPHYSIK
SCHOENECKSTRASSE 6
D-7800 FREIBURG
GERMANY, F.R.

BRUZUAL GUSTAVO
CENTRO DE INVEST DE
ASTRONOMIA CIDA
AP P 264
MERIDA 5101-A
VENEZUELA

BUCHLER J ROBERT PROF
DEPT OF PHYSICS AND
ASTRONOMY
UNIVERSITY FLORIDA
GAINESVILLE FL 32611
U.S.A.

BUDDING EDWIN DR
CARTER OBSERVATORY
P O BOX 2909
WELLINGTON 1
NEW ZEALAND

BUES IRMELA D DR
REMEIS STERNWARTE
STERNWARTSTRASSE 7
D-8600 BAMBERG
GERMANY, F.R.

BUFF JAMES S DR
DEPT PHYSICS ASTRON
DARTMOUTH COLLEGE
HANOVER NH 03755
U.S.A.

BUHL DAVID DR
INFR RADIO ASTRO BR
NASA GODDARD SPACE
FLIGHT CTR CODE 693
GREENBELT MD 20771
U.S.A.

BULLEN K E PROF
DEPT OF APPLIED MATH
UNIVERSITY OF SYDNEY
SYDNEY NSW 2006
AUSTRALIA

BUMBA VACLAV DR
ASTRONOMICAL INST OF
THE CZECHOSLOVAK
ACADEMY OF SCIENCES
251 65 ONDREJOV
CZECHOSLOVAKIA

BUNNER A N DR
PERKIN-ELMER CORP
M S 846
100 WOOSTER HTS RD
DANBURY CT 06810
U.S.A.

BURBIDGE E MARGARET PROF
DEPT OF PHYSICS
UNIV OF CALIFORNIA
SAN DIEGO MCCO-11
LA JOLLA CA 92093
U.S.A.

BURBIDGE GEOFFREY R PROF
OFF OF THE DIRECTOR
KITT PEAK NAT OBS
P O BOX 26732
TUCSON AZ 85726
U.S.A.

BURCH CECIL R DR
PHYSICS DEPT
THE UNIVERSITY
BRISTOL 8
U.K.

BURGER J J DR IR
WERKGROEP KOSMISCHE
STRALING K ONNES LAB
NIEUWSTEEG 18
2311 SB LEIDEN
THE NETHERLANDS

BURGER MARIJKE DR
SPARRENSTRAAT 58
1170 BRUSSEL
BELGIUM

BURGESS ALAN DR
DEPT OF APPLIED MATH
AND THEORETICAL PHYS
SILVER STREET
CAMBRIDGE CB3 9EW
U.K.

BURGESS DAVID D DR
BLACKETT LABORATORY
IMPERIAL COLLEGE OF
SCI AND TECHNOLOGY
LONDON SW7 2BZ
U.K.

BURKE BERNARD F DR
DEPT OF PHYSICS
M I T RM 26-335
CAMBRIDGE MA 02139
U.S.A.

BURKE J ANTHONY DR
DEPT OF PHYSICS
UNIVERSITY OF
VICTORIA
VICTORIA BC V8W 2Y2
CANADA

BURKHART C DR
OBS DE LYON
69230 ST-GENIS-LAVAL
FRANCE

BURKHEAD MARTIN S
ASTRONOMY DEPT
INDIANA UNIVERSITY
SWAIN HALL WEST
BLOOMINGTON IN 47405
U.S.A.

BURKI GILBERT DR
OBS DE GENEVE
CH-1290 SAUVERNY
SWITZERLAND

BURLAGA L F DR
NASA GSFC
CODE 692
GREENBELT MD 20771
U.S.A.

BURNELL SUSAN J B DR
MULLARD SPACE SC LAB
UNIV COLLEGE LONDON
LONDON
U.K.

BURNS JACK O'NEIL JR
DEPT OF PHYSICS &
ASTRONOMY
UNIV OF NEW MEXICO
ALBUQUERQUE NM 97131
U.S.A.

BURNS JOSEPH A PROF
CENTER RADIOPHYSICS
SPACE RESEARCH
CORNELL UNIVERSITY
ITHACA NY 14850
U.S.A.

BURROWS ADAM SETH
DEPT OF PHYSICS
SUNY
STONY BROOK NY 11794
U.S.A.

BURSA MILAN DR
ASTRONOMICAL INST
CSAV
BUDECSKA 6
120 23 PRAHA 2
CZECHOSLOVAKIA

BURSTEIN DAVID
NRAO
EDGEMONT ROAD
CHARLOTTESVILLE
VA 22903
U.S.A.

BURTON W BUTLER DR
STERREWACHT
POSTBUS 9513
2300 RA LEIDEN
THE NETHERLANDS

BURTON WILLIAM M
SPACE ASTROPH R25
RUTHERFORD APPL LAB
CHILTON DIDCOT
OXON OX11 0QX
U.K.

BUSCOMBE WILLIAM PROF
ASTRONOMY DEPT
NORTHWESTERN
UNIVERSITY
EVANSTON IL 60201
U.S.A.

BUSER ROLAND DR
ASTRONOMICAL INST
UNIV OF BASEL
VENUSSTRASSE 7
CH-4102 BINNINGEN
SWITZERLAND

BUTI B PROF
PHYSICAL RESEARCH
NAVRANGPURA
AHMEDABAD 380009
INDIA

BUTLER C JOHN DR
ARMAGH OBSERVATORY
ARMAGH BT61 9DG
U.K.

BUTLER DENNIS DR
ASTRONOMY DEPT
YALE UNIV
BOX 2023 YALE STAT
NEW HAVEN CT 06520
U.S.A.

BYARD P L DR
DEPT OF ASTRONOMY
OHIO STATE UNIV
174 W 18TH AVE
COLUMBUS OH 43210
U.S.A.

BYKOV M F DR
ASTRONOMICAL INST
UZBEKIAN ACADEMY OF
SCIENCES
700000 TASHKENT
U.S.S.R.

BYRD GENE G DR
DEPT PHYSICS ASTRON
UNIV OF ALABAMA
BOX 1921
UNIVERSITY AL 35486
U.S.A.

BYRNE PATRICK B DR
ARMAGH OBSERVATORY
ARMAGH BT61 9DG
U.K.

BYSTROV NIKOLAI F DR
PULKOVO OBSERVATORY
196140 LENINGRAD
U.S.S.R.

BYSTROVA NATALIJA V DR
SPECIAL ASTROPHYSI-
CAL OBSERVATORY
LENINGRAD BRANCH
196140 LENINGRAD
U.S.S.R.

CABRITA EZEQUIEL DR
OBSERVATORIO
DE LISBOA
TAPADA LISBOA
PORTUGAL

CACCIANI ALESSANDRO DR
OSSERVATORIO
ASTRONOMICO DI ROMA
I-00100 ROMA
ITALY

CACCIANI CARLA DR
IST DI ASTRONOMIA
I-40100 BOLOGNA
ITALY

CACCIN BRUNO
OSSERVATORIO ASTRONO
MICO DI CAPODIMONTE
VIA MOIARIELLO 16
I-80131 NAPOLI
ITALY

CADEZ ANDREJ DR
DEPT OF PHYSICS
UNIV OF LJUBLJANA
JADRANSKA 19
61000 LJUBLJANA
YUGOSLAVIA

CAHN JULIUS H ASSOC PROF
UNIVERSITY OF
ILLINOIS OBSERVATORY
URBANA IL 61801
U.S.A.

CALAMAI G PROF
OSSERVATORIO ASTRO
FISICO DI ARCETRI
FIRENZE
ITALY

CALAME ODILE DR
CERGA
AVENUE COPERNIC
F-06130 GRASSE
FRANCE

CALDWELL JOHN JAMES
ESS DEPT SUNY
STONY BROOK NY 11794
U.S.A.

CALLEBAUT DIRK K DR
DEPT OF PHYSICS UIA
UNIVERSITEITSPLEIN 1
B 2610 WILRIJK-ANTW
BELGIUM

CALOI VITTORIA DR
CNR
ASTROFISICA SPAZIALE
C P 67 ROMA
I-00044 FRASCATI
ITALY

CALVANI MASSIMO DR
OSSERVATORIO ASTRO-
NOMICO
I-35100 PADOVA
ITALY

CALVET NURIA DR
CENTRO DE INVEST DE
ASTRONOMIA CIDA
AP P 264
MERIDA 5101-A
VENEZUELA

CALVO MANUEL
DEPTO DE ASTRONOMIA
UNIVERSIDAD ZARAGOZA
ZARAGOZA 9
SPAIN

CAMARENA BADIA V ADJ PROF
DPTO FIS TIERRA COS
FACULTAD DE CIENCIAS
UNIV DE ZARAGOZA
ZARAGOZA
SPAIN

CAMERON ALASTAIR G W PROF
HARVARD COLLEGE
OBSERVATORY
60 GARDEN STREET
CAMBRIDGE MA 02138
U.S.A.

CAMERON WINIFRED S MRS
CODE 601
GODDARD SPACE FLIGHT
CENTER
GREENBELT MD 20771
U.S.A.

CAMICHEL HENRI DR
OBSERVATOIRE
DE TOULOUSE
1 CAMILLE FLAMMARION
F-31500 TOULOUSE
FRANCE

CAMPBELL BRUCE DR
CFHT CORPORATION
BOX 1597
KAMUELA HI 96743
U.S.A.

CAMPBELL DONALD B
ARECIBO OBSERVATORY
BOX 995
ARECIBO PR 00612
U.S.A.

CAMPBELL JAMES W
ROYAL OBSERVATORY
BLACKFORD HILL
EDINBURGH EH9 3HJ
U.K.

CAMPBELL MURRAY F
DEPT OF PHYSICS &
ASTRONOMY
COLBY COLLEGE
WATERVILLE ME 04901
U.S.A.

CAMPUSANO L E
DEPTO DE ASTRONOMIA
UNIVERSIDAD DE CHILE
CASILLA 36-D
SANTIAGO
CHILE

CANAL RAMON M DR
DPTO FIS TIERRA COS
UNIVERSIDAD
DE BARCELONA
BARCELONA 14
SPAIN

CANAVAGGIA RENEE DR
OBSERVATOIRE
DE PARIS
61 AV L OBSERVATOIRE
F-75014 PARIS
FRANCE

CANDY MICHAEL P MR
PERTH OBSERVATORY
WALNUT ROAD
BICKLEY 6076
AUSTRALIA

CANE HILARY VIVIEN
CODE 690
NASA GSFC
GREENBELT MD 20771
U.S.A.

CANFIELD RICHARD C DR
PHYSICS C-011 UCSD
LA JOLLA CA 92093
U.S.A.

CANIZARES CLAUDE R PROF
MIT
RM 37-501
CAMBRIDGE MA 02139
U.S.A.

CANNON C J DR
DEPT OF APPLIED
MATHEMATICS
UNIVERSITY OF SYDNEY
SYDNEY NSW 2006
AUSTRALIA

CANNON RUSSELL D DR
ROYAL OBSERVATORY
BLACKFORD HILL
EDINBURGH EH9 3HJ
U.K.

CANNON WAYNE H DR
DEPT OF PHYSICS
YORK UNIVERSITY
4700 KEELE STREET
DOWNSVIEW ONT M3J1P3
CANADA

CANTO JORGE DR
INSTITUTO DE
ASTRONOMIA UNAM
APDO POSTAL 70264
MEXICO 04510 DF
MEXICO

CANTU ALBERTO M DR
OSSERVATORIO ARCETRI
LARGO FERMI 5
I-50125 FIRENZE
ITALY

CAPACCIOLI MASSIMO DR
ASTRONOMICAL OBS
UNIVERSITY OF PADOVA
I-35100 PADOVA
ITALY

CAPEN CHARLES F
LOWELL OBSERVATORY
BOX 1269
FLAGSTAFF AZ 86001
U.S.A.

CAPITAINE NICOLE
OBS DE PARIS
61 AV DE L'OBS
F-75014 PARIS
FRANCE

CAPLAN JAMES
OBSERVATOIRE DE
MARSEILLE
PLACE LE VERRIER
F-13004 MARSEILLE
FRANCE

CAPRIOLI GIUSEPPE PROF
OSSERVATORIO ASTRONO
MICO DI ROMA
VIA TRIONFALE 204
ROMA
ITALY

CAPRIOTTI EUGENE R DR
DEPT OF ASTRONOMY
OHIO STATE UNIV
5058 A SMITH LAB
COLUMBUS OH 43210
U.S.A.

CAPUTO FILIPPINA DR
CNR ASTROFISICA
SPAZIALE C P 67
ROMA
I-00044 FRASCATI
ITALY

CAPUZZO DOLCETTA ROBERTO
OSS ASTRONOMICO
VIALE DEL PARCO
MELLINI 84
I-00136 ROMA
ITALY

CARBON DUANE F DR
KITT PEAK NAT OBS
P O BOX 26732
TUCSON AZ 85726
U.S.A.

CARDONA OCTAVIO DR
INST NAC DE ASTROFIS
OPTICA Y ELECTRONICA
APDO POSTAL 216
PUEBLA PUE
MEXICO

CARDUS ALMEDA J DR
OBSERVATORIO DEL
EBRO
ROQUETAS
TORTOSA BARCELONA
SPAIN

CARESTIA REINALDO A DR
OBS ASTRON F AGUILAR
AV BENAVIDEZ 8175
OESTE RIVADAVIA
5400 SAN JUAN
ARGENTINA

CARLBERG RAYMOND GARY DR
INST OF ASTRONOMY
MADINGLEY ROAD
CAMBRIDGE CB3 0HA
U.K.

CARLETON NATHANIEL P DR
SMITHSONIAN ASTROPHY
SICAL OBSERVATORY
60 GARDEN ST
CAMBRIDGE MA 02138
U.S.A.

CARLQVIST PER A DR
DEPT PLASMA PHYSICS
ROYAL INST OF
TECHNOLOGY
S-10044 STOCKHOLM 70
SWEDEN

CARLSON JOHN B
ASTRONOMY PROGRAM
UNIV OF MARYLAND
COLLEGE PARK MD20742
U.S.A.

CARM LAWRENCE E DR
SACRAMENTO PEAK OBS
SUNSPOT NM 88349
U.S.A.

CARNEY BRUCE WILLIAM
DEPT OF PHYS & ASTRO
PHILLIPS HALL 039A
UNIV NORTH CAROLINA
CHAPEL HILL NC 27514
U.S.A.

CAROFF LAWRENCE J
SPACE SC DIV NASA
AMES RES CTR MS245-3
MOFFETT FIELD
CA 94035
U.S.A.

CAROUBALOS C A PROF
LAB ELECTRONIC PHYS
UNIVERSITY OF ATHENS
104 SOLONOS STREET
ATHENS 144
GREECE

CARPENTER LLOYD DR
GODDARD SPACE FLIGHT
CENTER
CODE 921
GREENBELT MD 20771
U.S.A.

CARR BERNARD JOHN
INST OF ASTRONOMY
MADINGLEY ROAD
CAMBRIDGE CB3 OHA
U.K.

CARR THOMAS D PROF
DEPT OF PHYSICS AND
ASTRONOMY
UNIVERSITY FLORIDA
GAINESVILLE FL 32611
U.S.A.

CARRANZA G J DR
LAPRIDA 880
CORDOBA
ARGENTINA

CARRASCO GUILLERMO DR
OBS ASTRON NATIONAL
UNIVERSIDAD DE CHILE
CASILLA 36-D
SANTIAGO
CHILE

CARRASCO LUIS DR
INSTITUTO DE
ASTRONOMIA
APDO POSTAL 70264
04510 MEXICO DF
MEXICO

CARROLL P KEVIN PROF
PHYSICS DEPT
UNIVERSITY COLLEGE
BELFIELD
DUBLIN 4
IRELAND

CARRUTHERS GEORGE R DR
SPACE SCIENCE DIV
US NAVAL RES LAB
CODE 7123
WASHINGTON DC 20375
U.S.A.

CARSON T R DR
UNIVERSITY
OBSERVATORY
BUCHANAN GARDENS
ST ANDREWS FIFE
U.K.

CARSWELL ROBERT F DR
INST OF ASTRONOMY
MADINGLEY ROAD
CAMBRIDGE CB3 OHA
U.K.

CARTER WILLIAM EUGENE
19004 OXCART PLACE
GAITHERSBURG MD20879
U.S.A.

CARUSI ANDREA
IST ASTROF SPAZIALE
REP PLANETOLOGIA
VIALE DELL'UNIV 11
I-00185 ROMA
ITALY

CARVER JOHN H PROF
RESEARCH SCHOOL OF
PHYSICAL SCIENCES
AUSTRALIAN NAT UNIV
CANBERRA A C T 2601
AUSTRALIA

CASANOVAS JUAN DR
VATICAN OBSERVATORY
I-00120 VATICAN CITY
VATICAN CITY STATE

CASH WEBSTER C JR
LASP BOX 392
UNIV OF COLORADO
BOULDER CO 80309
U.S.A.

CASINI CATERINA DR
ISTITUTO DI
ASTRONOMIA
VIA BRERA 28
I-20121 MILANO
ITALY

CASSATELLA ANGELO DR
LAB ASTROFISICA SPAZ
P O B 67
I-00044 FRASCATI
ITALY

CASSE M DR
SECT D'ASTROPHYSIQUE
CEN SACLAY
B P NO 2
F-91190 GIF-S-YVETTE
FRANCE

CASSINELLI JOSEPH P DR
ASTRONOMY DEPT
UNIVERSITY WISCONSIN
MADISON WI 53706
U.S.A.

CASTELLANI VITTORIO PROF
LAB ASTROFISICA
C P 67
I-00044 FRASCATI
ITALY

CASTELLI FIORELLA DR
OSSERVATORIO ASTRO-
NOMICO
VIA TIEPOLO 11
I-34131 TRIESTE
ITALY

CASTELLI JOHN P
AFGL-PHP
HANSCOM AFB MA 01731
U.S.A.

CASTOR JOHN I DR
L-18 LAWRENCE LIVER
MORE NATIONAL LAB
P O BOX 808
LIVERMORE CA 94550
U.S.A.

CASWELL J L DR
CSIRO RADIOPHYSICS
P O BOX 76
EPPING NSW 2121
AUSTRALIA

CATALA ASUNCION P DR
DPTO FIS TIERRA COS
UNIVERSIDAD
DE BARCELONA
BARCELONA 28
SPAIN

CATALAN MANUEL DR
INSTITUTO Y
OBSERVATORIO MARINA
SAN FERNANDO CADIZ
SPAIN

CATALANO FRANCESCO A DR
OSSERVATORIO
ASTROFISICO
CITTA UNIVERSITARIA
I-95125 CATANIA
ITALY

CATALANO SANTO DR
IST DI ASTRONOMIA
CITTA UNIVERSITARIA
I-95125 CATANIA
ITALY

CATCHPOLE ROBIN MICHAEL
SAAO
P O BOX 9
OBSERVATORY 7935
CAPE
SOUTH AFRICA

CATO TORGNY DR
RODDBAATSGATAN 3
S-421 71 VAESTRA
FROELUNDA
SWEDEN

CATURA RICHARD C DR
LOCKHEED RES LAB
DEPT 52 12 BLDG 202
3251 HANOVER ST
PALO ALTO CA 94304
U.S.A.

CAUGHLAN GEORGEANNE R
PROF
DEPT OF PHYSICS
MONTANA STATE UNIV
BOZEMAN MT 59715
U.S.A.

CAVALIERE ALFONSO G PROF
ISTITUTO ASTRONOMICO
UNIVERSITA ROMA
P A MORO 2
I-00185 ROMA
ITALY

CAYREL DE STROBEL GIUSADR
OBS DE MEUDON
F-92190 MEUDON
FRANCE

CAYREL ROGER DR
OBS DE PARIS
61 AV L'OBSERVATOIRE
F-75014 PARIS
FRANCE

CAZENAVE ANNY DR
CNES-GRGS
18 AVE EDOUARD BELIN
F-31400 TOULOUSE
FRANCE

CAZZOLA P DR
OSSERVATORIO ASTRON
VICOLA DELL
OSSERVATORIO
PADOVA
ITALY

CECCARELLI M PROF
ISTITUTO DI FISICA
GENERALE
UNIV DI BOLOGNA
I-40100 BOLOGNA
ITALY

CEFOLA PAUL J DR
MAIL STATION 64
C S DRAPER LAB
555 TECHNOLOGY SQ
CAMBRIDGE MA 02139
U.S.A.

CELNIKIER LUDWIK DR
OBSERVATOIRE
DE MEUDON
F-92190 MEUDON
FRANCE

CEPLECHA ZDENEK DR
ASTRONOMICAL INST
CZECHO ACAD SCIENCES
251 65 ONDREJOV
OBSERVATORY
CZECHOSLOVAKIA

CEPPATELLI GUIDO DR
OSSERVATORIO ASTRO-
FISICO DI ARCETRI
I-50125 FIRENZE
ITALY

CESARSKY CATHERINE J DR
INST RECHERCHE FOND
CEN SACLAY
DPHG/SAP-BAT 28
F-91191 GIF-YVETTE
FRANCE

CESARSKY DIEGO A DR
INST D'ASTROPHYSIQUE
98 BIS BD ARAGO
F-75014 PARIS
FRANCE

CESCO CARLOS DR
OBS ASTR F AGUILAR
BENAVIDEZ 8175 OESTE
5407 MARQUESADO
SAN JUAN
ARGENTINA

CESTER BRUNO PROF
OSSERVATORIO
ASTRONOMICO
VIA TIEPOLO 11
I-34131 TRIESTE
ITALY

CHAFFEE FREDERIC H DR
MT HOPKINS
OBSERVATORY
P O BOX 97
AMADO AZ 85640
U.S.A.

CHAISSON ERIC J PROF
HAVERFORD COLLEGE
HAVERFORD PA 19041
U.S.A.

CHAMARAJX P DR
OBS DE MEUDON
F-92190 MEUDON
FRANCE

CHAMBERLAIN JOSEPH M DR
ADLER PLANETARIUM
1300 S LAKE SHORE DR
CHICAGO IL 60505
U.S.A.

CHAMBERLAIN JOSEPH W PROF
DEPT SPACE PHYSICS
AND ASTRONOMY
RICE UNIVERSITY
HOUSTON TX 77001
U.S.A.

CHAMBLISS CARLSON R DR
KUTZTOWN STATE
COLLEGE
KUTZTOWN PA 19530
U.S.A.

CHANDRA SUBHASH
MIS PHILIPS LABS
345 SCARBOROUGH RD
BRIAR CLIFF NY 10510
U.S.A.

CHANDRASEKHAR S PROF
933 E 56 STREET
CHICAGO IL 60637
U.S.A.

CHANMUGAN GABESAR PROF
DEPT OF PHYSICS
AND ASTRONOMY
LA STATE UNIVERSITY
BATON ROUGE LA 70803
U.S.A.

CHAPMAN CLARK R DR
PLANETARY SC INST
2030 E SPEEDWAY
SUITE 201
TUCSON AZ 85719
U.S.A.

CHAPMAN GARY A DR
DEPT PHYSICS ASTRON
CALIF STATE UNIV
18111 NORDHOFF ST
NORTHRIDGE CA 91324
U.S.A.

CHAPMAN ROBERT D DR
CODE 680
GODDARD SPACE FLIGHT
CENTER
GREENBELT MD 20771
U.S.A.

CHAPRONT J DR
BUREAU D LONGITUDES
77 AVE DENFERT-
ROCHEREAU
F-75014 PARIS
FRANCE

CHAPRONT-TOUZE M DR
BUREAU D LONGITUDES
77 AVENUE DENFERT-
ROCHEREAU
F-75014 PARIS
FRANCE

CHARLES PHILIP ALLAN
DEPT OF ASTROPHYSICS
SOUTH PARKS ROAD
OXFORD OX1 3RQ
U.K.

CHARVIN PIERRE DR
INAG
77 AVENUE DENFERT
ROCHEREAU
F-75014 PARIS
FRANCE

CHAU WAI Y PROF
PHYSICS DEPT
QUEEN S UNIVERSITY
KINGSTON ONT K7L 3N6
CANADA

CHAVIRA ENRIQUE SR
INST NAC ASTROPHYS
OPTICA Y ELECTRONICA
AP POSTAL 216 Y 51
PUEBLA PUE
MEXICO

CHEN BIAO
PURPLE MOUNTAIN
OBSERVATORY
NANJING
PEOPLE'S REPUBLIC CHINA

CHEN DAOHAN
PURPLE MOUNTAIN OBS
NANJING
PEOPLE'S REPUBLIC CHINA

CHEN JIAN-SHENG
PEKING ASTRONOMICAL
OBSERVATORY
ACADEMIA SINICA
PEKING
PEOPLE'S REPUBLIC CHINA

CHEN KWAN-YU PROF
DEPT OF PHYSICS AND
ASTRONOMY
UNIVERSITY FLORIDA
GAINESVILLE FL 32611
U.S.A.

CHEN ZUN-GUI
BEIJING PLANETARIUM
BEIJING
PEOPLE'S REPUBLIC CHINA

CHENG CHUNG-CHIEH DR
NAVAL RESEARCH LAB
CODE 7174C
WASHINGTON DC 20375
U.S.A.

CHEREDNICHENKO V I DR
KIEV POLYTECHNICAL
INSTITUTE
252056 KIEV
U.S.S.R.

CHEREPASHCHUK A M DR
STERNBERG STATE
ASTRONOMICAL INST
NOMICAL INSTITUTE
117234 MOSCOW
U.S.S.R.

CHERNEGA N A A DR
OBSERVATORY OF THE
KIEV UNIVERSITY
OBSERVATORNAYA 3
252053 KIEV
U.S.S.R.

CHERNYKH N S DR
CRIMEAN ASTRO-
PHYSICAL OBSERVATORY
NAUCHNY
334413 CRIMEA
U.S.S.R.

CHERTOPRUD V E DR
HYDROMETEOROLOGICAL
CENTER OF THE USSR
123376 MOSCOW
U.S.S.R.

CHEVALIER CLAUDE DR
OBSERVATOIRE
DE MEUDON
F-92190 MEUDON
FRANCE

CHEVALIER ROGER A DR
DEPT OF ASTRONOMY
UNIV OF VIRGINIA
P O BOX 3818
CHARLOTTESVL VA22903
U.S.A.

CHIKADA YOSHIHIRO DR
NOBEYAMA RADIO OBS
TOKYO ASTRON OBS
NOBEYAMA
NAGANO PREF 384-13
JAPAN

CHINCARINI GUIDO L DR
DEPT PHYS AND ASTRON
UNIV OF OKLAHOMA NEL
SON BD 440 W BROOKS
NORMAN OK 73069
U.S.A.

CHIOSI CESARE S DR
ISTITUTO DI
ASTRONOMIA
UNIVERSITA DI PADOVA
I-35100 PADOVA
ITALY

CHISTYAKOV V E DR
SOLAR STATION
PRIMORSKY KRAY
692533 GORNOTAEZHNOE
U.S.S.R.

CHITRE D M DR
COMPUTER SCS CORP.
8728 COLESVILLE RD
SILVER SPG MD 20910
U.S.A.

CHITRE SHASHIKUMAR M DR
TATA INSTITUTE
HOMI BHABHA ROAD
COLABA
BOMBAY 400005
INDIA

CHIU HONG-YEE DR
MUDD BLDG RM 828
COLUMBIA UNIV
NEW YORK NY 10027
U.S.A.

CHIU LIANG-TIA GEORGE
T J WATSON RES CTR
BOX 218
YORKTOWN HEIGHTS
NY 10598
U.S.A.

CHIUDERI CLAUDIO PROF
OSSERVATORIO ARCETRI
LARGO FERMI 5
I-50125 FIRENZE
ITALY

CHIUDERI-DRAGO F PROF
OSSERVATORIO ASTROFI
SICO DI ARCETRI
LARGO E FERMI 5
I-50125 FIRENZE
ITALY

CHLIETOVSKY FRANCA DR
OSSERVATORIO ASTRO-
NOMICO
BRERA
MILANO
ITALY

CHMIELEWSKI YVES DR
OBS DE GENEVE
51 CHEMIN MAILLETTES
CH-1290 SAUVERNY
SWITZERLAND

CHOCHOL DRAHOMIR
ASTRONOMICAL INST
SLOVAK ACADEMY OF SC
05960 TATRANSKA
LOMNICA
CZECHOSLOVAKIA

CHOI BON CHOL
ACADEMY OF SCIENCES
PYONGYANG ASTRONOMI-
CAL OBSERVATORY
PYONGYANG
DEM PEOPLE'S REP OF KOREA

CHOLLET FERNAND
OBSERVATOIRE PARIS
61 AVENUE DE
L OBSERVATOIRE
F-75014 PARIS
FRANCE

CHOPINET MARGUERITE DR
OBSERVATOIRE
DE BORDEAUX
F-33270 FLOIRAC
FRANCE

CHOU KYONG CHOL PROF
KYUNG HEE UNIVERSITY
SEOUL
REPUBLIC OF KOREA

CHRISTENSEN-DALSGAARD J
ADVANCED STUDY PROGR
NCAR
P O BOX 1470
BOULDER CO 80302
U.S.A.

CHRISTIAN CAROL ANN
CANADA-FRANCE-HAWAII
TELESCOPE CORP
BOX 1597
KAMUELA HI 96743
U.S.A.

CHRISTIANSEN WAYNE A
DEPT OF PHYSICS AND
ASTRONOMY
UNIV NORTH CAROLINA
CHAPEL HILL NC 27514
U.S.A.

CHRISTIANSEN WILBUR PROF
MOUNT STROMLO OBS
PRIVATE BAG
WODEN P O ACT 2606
AUSTRALIA

CHRISTOPHE-GLAUME J DR
SERVICE D'AERONOMIE
B P NO 3
F-91370 VERRIERES-LE
BUISSON
FRANCE

CHRISTY JAMES WALTER DR
EXPLORATORY DEV ST
U S NAVAL OBS
WASHINGTON DC 20390
U.S.A.

CHRISTY ROBERT F DR
PROVOST OFFICE
CALIFORNIA INSTITUTE
OF TECHNOLOGY
PASADENA CA 91125
U.S.A.

CHUBB TALBOT A DR
5023 N 38TH ST
ARLINGTON VA 22207
U.S.A.

CHUGAJNOV P F DR
CRIMEAN ASTROPHYSI-
CAL OBSERVATORY USSR
ACAD OF SCI NAUCHNIY
334413 CRIMEA
U.S.S.R.

CHUN MUN-SUK DR
DEPT ASTRO/METEOROL
COLLEGE OF SCIENCE
YONSEI UNIVERSITY
SEOUL
REPUBLIC OF KOREA

CHUPP EDWARD L DR
PHYSICS DEPT
DEMERITT HALL
UNIV NEW HAMPSHIRE
DURHAM NH 03824
U.S.A.

CHURCHWELL EDWARD B DR
WASHBURN OBSERVATORY
UNIVERSITY WISCONSIN
475 N CHARTER ST
MADISON WI 53706
U.S.A.

CHURMS JOSEPH
S A ASTRONOMICAL OBS
P O BOX 9
OBSERVATORY
CAPE 7935
SOUTH AFRICA

CHUVAEV K K DR
CRIMEAN ASTROPHYSI-
CAL OBSERVATORY USSR
ACAD OF SCI NAUCHNIY
334413 CRIMEA
U.S.S.R.

CHVOJKOVA WOYK E DR
ASTRONOMICAL INST
CZECHO ACAD OF SC
BUDECSKA 6
12023 PRAHA 2
CZECHOSLOVAKIA

CIATTI FRANCO DR
OSSERVATORIO
ASTROFISICO
I-36012 ASIAGO VI
ITALY

CID PALACIOS RAFAEL PROF
FACULTAD DE CIENCIAS
CIUDAD UNIVERSITARIA
ZARAGOZA
SPAIN

CILLIE G G PROF
4 MINSERIE STREET
STELLENBOSCH 7600
SOUTH AFRICA

CIMINO MASSIMO A PROF
VIA A CADLOLO 19
I-00136 ROMA
ITALY

CLARIA J DR
OBSERVATORIO
ASTRONOMICO
LAPRIDA 854
5000 CORDOBA
ARGENTINA

CLARK ALFRED JR PROF
DEPT MECH AND AERO
SCIENCES
UNIVERSITY ROCHESTER
ROCHESTER NY 14627
U.S.A.

CLARK BARRY G DR
NATIONAL RADIO
ASTRONOMY OBS
P O BOX 0
SOCORRO NM 87801
U.S.A.

CLARK DAVID H DR
SPACE & ASTROPH DIV
RUTHERFORD APPL LAB
CHILTON DIDCOT
OXON OX11 0QX
U.K.

CLARK FRANK OLIVER DR
DEPT PHYSICS ASTRON
UNIV OF KENTUCKY
LEXINGTON KY 40506
U.S.A.

CLARK GEORGE W PROF
MIT ROOM 37-611
CAMBRIDGE MA 02139
U.S.A.

CLARK THOMAS A DR
RADIO ASTRO BRANCH
NASA GODDARD SPACE
FLIGHT CTR CODE 693
GREENBELT MD 20771
U.S.A.

CLARK THOMAS ALAN DR
PHYSICS DEPT
UNIV OF CALGARY
2920-24 AVENUE N W
CALGARY ALB T2N 1N4
CANADA

CLARKE DAVID DR
DEPT OF ASTRONOMY
THE UNIVERSITY
GLASGOW G12 8QQ
U.K.

CLARKE THOMAS R DR
MCLAUGHLIN PLANETAR
ROYAL ONTARIO MUSEUM
100 QUEENS PK CRESC
TORONTO ONT M5S 2C6
CANADA

CLAUSEN JENS VIGGO LEKTOR
COPENHAGEN UNIVER-
SITY OBSERVATORY
BRORFELDE
DK-4340 TOLLOSE
DENMARK

CLAYTON DONALD D PROF
DEPT SPACE PHYSICS
AND ASTRONOMY
RICE UNIVERSITY
HOUSTON TX 77001
U.S.A.

CLEGG PETER E DR
QUEEN MARY COLLEGE
MILE END RD
LONDON E1 4NS
U.K.

CLEGG ROBIN E S DR
DEPT PHYSICS/ASTRON
UNIV COLLEGE LONDON
GOWER STREET
LONDON WC1E 6BT
U.K.

CLEMENT MAURICE J PROF
DEPT OF ASTRONOMY
UNIVERSITY OF
TORONTO
TORONTO ONT M5S 1A7
CANADA

CLIFTON KENNETH ST DR
NASA E S 64
MARSHALL SFC AL35812
U.S.A.

CLIMENHAGA JOHN L PROF
DEPT OF PHYSICS
UNIVERSITY
OF VICTORIA
VICTORIA BC V8W 2Y2
CANADA

CLUBE S V M DR
ROYAL OBSERVATORY
EDINBURGH EH9 3HJ
U.K.

CLUTTON-BROCK MARTIN DR
DEPT MATH ASTRONOMY
UNIV OF MANITOBA
WINNIPEG MB R3T 2N2
CANADA

COCHRAN WILLIAM DAVID DR
DEPT OF ASTRONOMY
UNIV OF TEXAS
AUSTIN TX 78712
U.S.A.

COCKE W JOHN PROF
STEWARD OBSERVATORY
UNIVERSITY ARIZONA
TUCSON AZ 85721
U.S.A.

CODE ARTHUR D
WASHBURN OBSERVATORY
UNIV OF WISCONSIN
MADISON WI 53706
U.S.A.

CODINA LANDABERRY S J DR
INSTITUTO ASTRONOMI-
CO I GEOFISICO USP
CAIXA POSTAL 30627
01000 SAO PAULO SP
BRAZIL

CODINA VIDAL J M DR
AV JOSE ANTONIO 679
BARCELONA 13
SPAIN

COELHO BALSA M C DR
OBS ASTRONOMICO
UNIV DE COIMBRA
COIMBRA
PORTUGAL

COFFEEN DAVID L DR
INST FOR SPACE STUDY
2880 BROADWAY
NEW YORK NY 10025
U.S.A.

COFFEY HELEN E DR
NOAA
NGSDC D63
BOULDER CO 80303
U.S.A.

COGAN BRUCE C DR
MT STROMLO OBS
PRIVATE BAG
WODEN P O ACT 2606
AUSTRALIA

COHEN JEFFREY M DR
PHYSICS DEPT
UNIV OF PENNSYLVANIA
PHILADELPHIA PA19174
U.S.A.

COHEN JUDITH DR
KITT PEAK NATIONAL
OBSERVATORY
P O BOX 26732
TUCSON AZ 85726
U.S.A.

COHEN LEON PROF
DEPT OF PHYSICS
HUNTER COLLEGE
695 PARK AVE
NEW YORK NY 10021
U.S.A.

COHEN MARSHALL H PROF
CALTECH 102-24
PASADENA CA 91125
U.S.A.

COHEN MARTIN DR
ASTRONOMY DEPT
UNIV OF CALIFORNIA
BERKELEY CA 94720
U.S.A.

COHEN RAYMOND J DR
NRAL
JODRELL BANK
MACCLESFIELD
CHESHIRE SK11 9DL
U.K.

COLBURN DAVID S DR
245-3 NASA
AMES RESEARCH CENTER
MOFFETT FIELD
CA 94035
U.S.A.

COLE TREVOR W DR
ELECTRICAL ENGIN
UNIV OF SYDNEY
P O BOX 76
SYDNEY NSW 2006
AUSTRALIA

COLGATE STIRLING A DR
THEORETICAL DIVISION
LOS ALAMOS SCIENTI-
FIC LAB MS 210
LOS ALAMOS NM 87845
U.S.A.

COLIN JAQUES DR
OBS DE BESANCON
41 BIS AVE DE L'OBS
F-25000 BESANCON
FRANCE

COLLIN-SOUFFRIN SUZY DR
OBSERVATOIRE
DE MEUDON
F-92190 MEUDON
FRANCE

COLLINS GEORGE W II PROF
174 WEST 18TH AVE
COLUMBUS OH 43210
U.S.A.

COLLINSON EDWARD H
THE COPSE
CHURCH LANE PLAYFORD
IPSWICH
SUFFOLK IP6 9DR
U.K.

COLOMB FERNANDO R DR
INST ARGENTINO DE
RADIOASTRONOMIA
CAS DE CORREO NO 5
1894 VILLA ELISA BA
ARGENTINA

COLOMBO G PROF DR
IST MECCANICA APPL
UNIV DI PADOVA
VIA F MARZOLO 9
I-35100 PADOVA
ITALY

COMBES FRANCOISE DR
OBS DE PARIS
DEPT RADIOASTRONOMIE
F-92190 MEUDON
FRANCE

COMBES M MR
OBS DE MEUDON
F-92190 MEUDON
FRANCE

COMINS NEIL FRANCIS
BENNETT HALL
UNIVERSITY OF MAINE
ORONO ME 04401
U.S.A.

CONCONI PAOLO DR
OSSERVATORIO ASTRO-
NOMICO
BRERA
MILANO
ITALY

CONDON JAMES J DR
PHYSICS DEPT VA POLY
TECHN INST STATE UNI
325 ROBESON HALL
BLACKSBURG VA 24061
U.S.A.

CONKLIN EDWARD K
C O FORTH INC
2309 PACIFIC CST HWY
HERMOSA BCH CA 90254
U.S.A.

CONNES JANINE DR
CIRCE
BAT 506
F-91440 ORSAY CEDEX
FRANCE

CONNES PIERRE DR
SERVICE D'ASTRONOMIE
B P 3
F-91370 VERRIERES LE
BUISSON
FRANCE

CONTADAKIS M E DR
DEPT OF GEODETIC
ASTRONOMY
UNIV OF THESSALONIKI
THESSALONIKI
GREECE

CONTI PETER S DR
JILA
UNIVERSITY COLORADO
BOULDER CO 80309
U.S.A.

CONTOPOULOS GEORGE PROF
ASTRONOMY DEPT
PANEPISTIMIOPOLIS
ATHENS 621
GREECE

CONWAY ROBIN G DR
NRAL
JODRELL BANK
CHESHIRE SK11 9DL
U.K.

COOK ALAN H PROF
CAVENDISH LABORATORY
MADINGLEY RD
CAMBRIDGE CB3 OHE
U.K.

COOK ALLAN F DR
CENTER FOR ASTROPHY-
SICS HCO-SAO
60 GARDEN ST
CAMBRIDGE MA 02138
U.S.A.

COOK JOHN W
8032 SLEEPY VIEW LN
SPRINGFIELD VA 22153
U.S.A.

COOKE B A DR
X-RAY ASTRONOMY
GROUP PHYSICS DEPT
THE UNIVERSITY
LEICESTER LE1 7RH
U.K.

CORADINI ANGIOLETTA
IST ASTROF SPAZIALE
REP PLANETOLOGIA
VIALE UNIVERSITA 11
I-00185 ROMA
ITALY

CORBIN THOMAS ELBERT DR
U S NAVAL OBS
TRANSIT CIRCLE DIV
WASHINGTON DC 20390
U.S.A.

CORDES JAMES M
SPACE SCIENCES BLDG
CORNELL UNIVERSITY
ITHACA NY 14853
U.S.A.

CORLISS C H DR
NATIONAL BUREAU OF
STANDARDS
WASHINGTON DC 20234
U.S.A.

CORNEJO ALEJANDRO A MSS
INAOE
APDO POSTAL 216 Y 51
PUEBLA PUE
MEXICO

CORWIN HAROLD G JR DR
DEPT OF ASTRONOMY
RLM 15-212
UNIV OF TEXAS
AUSTIN TX 78712
U.S.A.

COSMOVICI CRISTIANO B DR
INST F OPTOELEKTRON
DFVLR
D-8031 WESSLING
GERMANY, F.R.

COSTA ENRICO
IST ASTROF SPAZIALE
CASELLA POSTALE 67
I-00044 FRASCATI
ITALY

COSTA VICTOR DR
CORTIJO CARDENETE
BELICENA GRANADA
SPAIN

COSTAIN C C DR
DIV OF PHYSICS
NAT RES COUNCIL
OTTAWA K1A 0R6
CANADA

COSTAIN CARMAN H DR
DOMINION RADIO
ASTROPHYSICAL OBS
BOX 248
PENTICTON BC V2A 6K3
CANADA

COSTERO RAFAEL MC
INST DE ASTRONOMIA
APDO POSTAL 70-264
04510 MEXICO DF
MEXICO

COUNSELMAN CHARLES C PROF
MIT DEPT EARTH AND
PLANETARY SCIENCES
CAMBRIDGE MA 02139
U.S.A.

COUPINOT GERARD DR
OBSERVATOIRE
DE TOULOUSE
1 CAMILLE FLAMMARION
F-31500 TOULOUSE
FRANCE

COURTES GEORGES PROF
OBSERVATOIRE DE
MARSEILLE
2 PLACE LE VERRIER
F-13004 MARSEILLE
FRANCE

COUSINS A W J DR
S A ASTRONOMICAL OBS
P O BOX 9
OBSERVATORY
CAPE 7935
SOUTH AFRICA

COUTEAU PAUL PROF
OBSERVATOIRE DE NICE
B P 252
F-06007 NICE
FRANCE

COUTREZ RAYMOND A J PROF
UNIVERSITE LIBRE DE
BRUXELLES
50 AV F D ROOSEVELT
B-1050 BRUXELLES
BELGIUM

COUTTS-CLEMENT CHRISTINE
M PROF
DEPT OF ASTRONOMY
UNIVERSITY TORONTO
TORONTO ONT M5S 1A7
CANADA

COVINGTON ARTHUR E
269 PLEASANT PARK RD
OTTAWA ONT K1H 5M7
CANADA

COWAN JOHN J DR
DEPT PHYSICS/ASTRON
UNIV OF OKLAHOMA
NORMAN OK 73019
U.S.A.

COWIE LENNOX LAUCHLAN DR
PRINCETON UNIV OBS
PEYTON HALL
PRINCETON NJ 08540
U.S.A.

COWLEY ANNE P DR
ASTRONOMY DEPT
UNIVERSITY
OF MICHIGAN
ANN ARBOR MI 48109
U.S.A.

COWLEY CHARLES R PROF
ASTRONOMY DEPT
UNIVERSITY
OF MICHIGAN
ANN ARBOR MI 48109
U.S.A.

COWLING THOMAS G PROF
19 HOLLIN GARDENS
HEADINGLY
LEEDS LS16 5NL
U.K.

COWSIK R DR
TATA INST FUND RES
HOMI BHABHA ROAD
BOMBAY 400 005
INDIA

COX ARTHUR N DR
LOS ALAMOS
SCIENTIFIC LAB
P O BOX 1663
LOS ALAMOS NM 87545
U.S.A.

COX DONALD P PROF
DEPT OF PHYSICS
UW MADISON
1150 UNIVERSITY AVE
MADISON WI 53706
U.S.A.

COX JOHN P PROF
JILA
UNIVERSITY OF
COLORADO
BOULDER CO 80309
U.S.A.

COYNE GEORGE V DR
SPECOLA VATICANA
I-00120 CITTA DEL
VATICANO
ITALY

CRAIG IAN JONATHAN D DR
DEPT OF APPLIED MATH
UNIV OF WAIKATO
HAMILTON
NEW ZEALAND

CRAINE ERIC RICHARD DR
STEWARD OBS
UNIV OF ARIZONA
TUCSON AZ 85721
U.S.A.

CRAM LAWRENCE EDWARD DR
CSIRO
DIV OF APPLIED PHYS
P O BOX 218
LINDFIELD NSW 2070
AUSTRALIA

CRAMPTON DAVID DR
DOMINION ASTROPH OBS
5071 W SAANICH ROAD
RR 5
VICTORIA BC V8X 4M6
CANADA

CRANE PATRICK C
NRAO
P O BOX O
SOCORRO NM 87801
U.S.A.

CRANNELL CAROL JO DR
NASA-GSFC
CODE 684
GREENBELT MD 20771
U.S.A.

CRAWFORD DAVID L
KITT PEAK NATIONAL
OBSERVATORY
BOX 26732
TUCSON AZ 85726
U.S.A.

CREZE MICHEL DR
OBSERVATOIRE DE
BESANCON
41 AV L OBSERVATOIRE
F-25000 BESANCON
FRANCE

CRIFO FRANCOISE DR
DAPHE
OBSERVATOIRE DE
MEUDON
F-92190 MEUDON
FRANCE

CRISTALDI SALVATORE DR
OSSERVATORIO
ASTROFISICO
CITTA UNIVERSITARIA
I-95125 CATANIA
ITALY

CRISTESCU CORNELIA G DR
OBSERVATOIRE
DE BUCAREST
75212 BUCAREST
ROUMANIA

CRIVELLARI LUCIO
OSS ASTRONOMICO
DI TRIESTE
VIA G B TIEPOLO 11
I-34131 TRIESTE
ITALY

CROOM DAVID L DR
RUTHERFORD & APPLE-
TON LABS
CHILTON DIDCOT
OXFORDSHIRE OX11 0QX
U.K.

CROVISIER J DR
OBS DE MEUDON
F-92190 MEUDON
FRANCE

CRUIKSHANK DALE P DR
INST FOR ASTRONOMY
2680 WOODLAWN DR
HONOLULU HI 96822
U.S.A.

CRUISE ADRIAN MICHAEL DR
MULLARD SPACE SC LAB
HOLMBURY ST MARY
DORKING
SURREY RH5 6NS
U.K.

CRUTCHER RICHARD M DR
UNIVERSITY OF
ILLINOIS OBSERVATORY
URBANA IL 61801
U.S.A.

CRUVELLIER PAUL E DR
LAB ASTRO SPATIALE
TRAVERSE DU SIPHON
LES TROIS LUCS
F-13012 MARSEILLE
FRANCE

CSADA IMRE K DR
KONKOLY OBSERVATORY
BOX 67
H-1525 BUDAPEST
HUNGARY

CUDABACK DAVID D DR
RADIO ASTRONOMY LAB
UNIVERSITY
OF CALIFORNIA
BERKELEY CA 94720
U.S.A.

CUDWORTH KYLE MCCABE DR
YERKES OBS
WILLIAMS BAY WI53191
U.S.A.

CUFFEY J MR
N M STATE UNIV
DEPT OF EARTH SC AND
ASTRONOMY
UNIV PK NM 88001
U.S.A.

CUGNON PIERRE DR
OBSERVATOIRE ROYAL
DE BELGIQUE
AVE CIRCULAIRE 3
B-1180 BRUXELLES
BELGIUM

CUGUSI LEONINO DR
IST DI ASTRONOMIA
UNIVERSITA
CAGLIARI
ITALY

CULHANE LEONARD DR
MULLARD SPACE SC LAB
HOLMBURY ST MARY
DORKING
SURREY RH5 6NS
U.K.

CULVER ROGER BRUCE DR
DEPT OF PHYSICS
COLORADO STATE UNIV
FT COLLINS CO 80523
U.S.A.

CUNNINGHAM LELAND E PROF
DEPT OF ASTRONOMY
UNIVERSITY OF
CALIFORNIA
BERKELEY CA 94720
U.S.A.

CUNY YVETTE J DR
OBSERVATOIRE DE
MEUDON
PLACE JANSSEN
F-92190 MEUDON
FRANCE

CUPERMAN SAMI PROF
DEPT OF PHYSICS AND
ASTRONOMY
TEL-AVIV UNIVERSITY
RAMAT-AVIV
ISRAEL

CURIR ANNA
OSS ASTRONOMICO DI
TORINO
STRADA OSS 20
10025 PINO TORINESE
ITALY

CURRIE DOUGLAS G DR
DEPT PHYSICS ASTRON
UNIV OF MARYLAND
COLLEGE PK MD 20742
U.S.A.

CZYZAK STANLEY J DR
ASTRONOMY DEPT
OHIO STATE UNIV
174 W 18TH ST
COLUMBUS OH 43210
U.S.A.

D'ANTONA FRANCESCA DR
OSSERVATORIO ASTRO-
NOMICO
I-00100 ROMA
ITALY

D'ODORICO SANDRO DR
ESO
K SCHWARZSCHILD ST 2
D-8046 GARCHING B M
GERMANY, F.R.

DA COSTA GARY STEWART
DEPT OF ASTRONOMY
YALE UNIVERSITY
BOX 6666
NEW HAVEN CT 06511
U.S.A.

DA COSTA JOSE MARQUES DR
INST PESQUISAS
ESPACIAIS
INPE C P 515
12200 S JOSE CAMPOS
BRAZIL

DA ROCHA VIEIRA E DR
INT DE FISICA
UNIV FEDERAL DA
RIO GRANDE DO SUL
90000 PORTO ALLEGRE
BRAZIL

DA SILVA A V C S
OBS ASTRONOMICO
UNIV SANTA CLARA
COIMBRA
PORTUGAL

DA SILVA LICIO DR
OBSERVATORIO NACION
R GEN BRUCE 586
S CRISTOVAO
20000 RIO DE JANEIRO
BRAZIL

DACHS JOACHIM PROF DR
ASTRONOMISCHES INST
RUHR-UNIVERSITAET
POSTFACH 102148
D-4630 BOCHUM 1
GERMANY, F.R.

DADAEV ALEKSANDR N DR
PULKOVO OBSERVATORY
196140 LENINGRAD
U.S.S.R.

DADIC ZARKO DR
INST SCIENCE HISTORY
YUGOS ACAD SC ART
DEMETROVA 18
41000 ZAGREB
YUGOSLAVIA

DAEPPEN WERNER
INST OF ASTRONOMY
CAMBRIDGE CB3 OHA
U.K.

DAHN CONARD CURTIS DR
U S NAVAL OBS
P O BOX 1149
FLAGSTAFF AZ 86002
U.S.A.

DAIGNE GERARD G
OBS DE MEUDON
F-92190 MEUDON
FRANCE

DAINTREE EDWARD J DR
NUFFIELD RADIO ASTRO
NOMY LABS
JODRELL BANK
MACCLESFIELD SK119DL
U.K.

DAISHIDO TSUNEAKI DR
DEPT OF SCIENCE
SCHOOL OF EDUCATION
WASEDA UNIVERSITY
SHINJUKU-KU TOKYO
JAPAN

DALGARNO ALEXANDER PROF
HARVARD COLLEGE OBS
60 GARDEN STREET
CAMBRIDGE MA 02138
U.S.A.

DALLAPORTA N PROF
ISTITUTO ASTRONOMIA
UNIVERSITA
VICOLO OSSERVATORIO
I-35100 PADOVA
ITALY

DALTABUIT ENRIQUE DR
INSTITUTO DE
ASTRONOMIA UNAM
APDO POSTAL 70264
MEXICO DF
MEXICO

DAMBARA TAKESHI PROF
SHIZUOKA UNIVERSITY
OTANI SHIZUOKA 422
JAPAN

DANBY J M ANTHONY DR
DEPT OF MATHEMATICS
NORTH CAROLINA STATE
UNIVERSITY
RALEIGH NC 27607
U.S.A.

DANESE LUIGI DR
OSSERVATORIO ASTRO-
NOMICO
I-35100 PADOVA
ITALY

DANKS ANTHONY C R
ESO
SCHLEISSHEIMER ST 17
D-8046 GARCHING B M
GERMANY, F.R.

DANZIGER I JOHN DR
ESO
K SCHWARZSCHILD ST 2
D-8046 GARCHING B M
GERMANY, F.R.

DARIUS JON
UNIVERSITY COLLEGE
DEPT OF PHYSICS &
ASTRONOMY
LONDON WC1E 6BT
U.K.

DAS MRINAL KANTI
DEPT OF PHYSICS
UNIVERSITY OF DELHI
DHAULA KUAN
NEW DELHI 110 021
INDIA

DATLOWE AYTON DR
LOCKHEED P A RES LAB
DEPT 52-12 BLDG 202
3251 HANOVER STREET
PALO ALTO CA 94304
U.S.A.

DAUBE-KURZEMNIECE I A DR
RADIOASTROPHYSICAL
OBSERVATORY LATVIAN
ACADEMY OF SCIENCES
226026 RIGA
U.S.S.R.

DAUTCOURT G DR
ZENTRINST ASTROPHYS
STERNWARTE BABELSB
R LUXEMBURG STR 17A
DDR-1502 POTSDAM
G.D.R.

DAVIDSEN ARTHUR FALNES DR
DEPT OF PHYSICS
JOHNS HOPKINS UNIV
CHARLES AND 34TH STS
BALTIMORE MD 21218
U.S.A.

DAVIDSON KRIS DR
SCHOOL PHYS AND ASTR
UNIV OF MINNESOTA
116 CHURCH ST S E
MINNEAPOLIS MN 55455
U.S.A.

DAVIDSON WILLIAM PROF
MATHEMATICS DEPT
UNIVERSITY OF OTAGO
P O BOX 56
DUNEDIN
NEW ZEALAND

DAVIES JOHN G DR
NRAL
JODRELL BANK
MACCLESFIELD
CHESHIRE SK11 9DC
U.K.

DAVIES MERTON E MR
1414 SAN REMO DRIVE
PACIFIC PALISADES
CA 90272
U.S.A.

DAVIES PAUL C W
SCHOOL OF PHYSICS
THE UNIVERSITY
NEWCASTLE
-UPON-TYNE NE1 7RU
U.K.

DAVIES RODNEY D PROF
NUFFIELD RADIO
ASTRONOMY LABS
JODRELL BANK
MACCLESFIELD SK119DL
U.K.

DAVIS CECIL G JR
UNIVERSITY OF
CALIFORNIA LASL
BOX 1663 MS 420
LOS ALAMOS NM 87545
U.S.A.

DAVIS JOHN DR
SCHOOL OF PHYSICS
UNIVERSITY OF SYDNEY
SYDNEY NSW 2006
AUSTRALIA

DAVIS LEVERETT JR PROF
405-47 CALTECH
PASADENA CA 91125
U.S.A.

DAVIS MARC DR
DEPT OF ASTRONOMY
HARVARD UNIVERSITY
60 GARDEN ST
CAMBRIDGE MA 02138
U.S.A.

DAVIS MICHAEL M DR
ARECIBO OBSERVATORY
P O BOX 995
ARECIBO PR 00612
U.S.A.

DAVIS MORRIS S PROF
DEPT PHYS ASTRONOMY
UNIV NORTH CAROLINA
284 PHILLIPS H 039A
CHAPEL HILL NC 27514
U.S.A.

DAVIS RICHARD J DR
NUFFIELD RADIO
ASTRONOMY LABS
JODRELL BANK
MACCLESFIELD SK119DL
U.K.

DAVIS ROBERT J DR
CENTER FOR ASTRO
PHYSICS
60 GARDEN ST
CAMBRIDGE MA 02138
U.S.A.

DAWE JOHN ALAN DR
U K S T U
PRIVATE BAG
SIDING SPRING OBS
COONABARABRAN 2857
AUSTRALIA

DE BERGH C DR
GROUPE PLANETES
SECT D'ASTROPHYSIQUE
OBS DE MEUDON
F-92190 MEUDON
FRANCE

DE BIASE G A DR
OSS ASTRONOMICO
VIALE DEL PARCO
MELLINI 84
ROMA
ITALY

DE BOER KLAAS S DR
ASTRONOMISCHES INST
UNIV TUEBINGEN
WALDHAUSER STR 64
D-7400 TUEBINGEN
GERMANY, F.R.

DE BRUYN ANTONIUS G DR
RADIOSTERREWACHT
POSTBUS 2
7990 AA DWINGELOO
THE NETHERLANDS

DE CASTRO A DR
OBSERVATORIO
ASTRONOMICO
ALFONSO X11 5
MADRID 7
SPAIN

DE FELICE FERNANDO DR
IST FISICA GALILEO
GALILEI
UNIVERSITA DI PADOVA
I-35100 PADOVA
ITALY

DE FREITAS PACHECO J A DR
OBSERVATORIO DE SAO
PAULO USP
CAIXA POSTAL 30627
SAO PAULO
BRAZIL

DE GRAAF T DR
INST VOOR FONETISCHE
WETENSCHAPPEN
GROTE ROZENSTRAAT 31
9712 TG GRONINGEN
THE NETHERLANDS

DE GRAAFF W DR
STERREKUNDIG INST
ZONNENBURG 2
3512 NL UTRECHT
THE NETHERLANDS

DE GRAAUW TH DR
ESA
SPACE SCIENCE DEPT
POSTBUS 299
2200 AG NOORDWIJK
THE NETHERLANDS

DE GREVE JEAN PIERRE DR
ASTROPHYSICAL INST
V U B PLEINLAAN 2
B-1050 BRUSSELS
BELGIUM

DE GROOT MART DR
ARMAGH OBSERVATORY
ARMAGH BT61 9DG
U.K.

DE GROOT T DR
STERREKUNDIG
INSTITUUT
ZONNENBURG 2
3512 NL UTRECHT
THE NETHERLANDS

DE JAGER CORNELIS PROF
THE ASTRONOMICAL
INSTITUTE
BENELUXLAAN 21
3527 HS UTRECHT
THE NETHERLANDS

DE JAGER G DR
DEPT OF PHYSICS
RHODES UNIVERSITY
P O BOX 94
GRAHAMSTOWN 6140
SOUTH AFRICA

DE JONG T DR
ASTRONOMICAL INST
ROETERSTRAAT 15
1018 WB AMSTERDAM
THE NETHERLANDS

DE JONGE J K DR
DEPT OF ASTRONOMY
UNIV OF PITTSBURGH
RIVERVIEW PARK
PITTSBURGH PA 15214
U.S.A.

DE KORT JULES J DR
HOUTLAAN 4
6500 GV NIJMEGEN
THE NETHERLANDS

DE KORTE P A J DR
COSMIC RAY WORKING
GROUP HUYGENS LAB
WASSENAARSEWEG 78
2300 RA LEIDEN
THE NETHERLANDS

DE LA HERRAN V JOSE ENG
INSTITUTO DE
ASTRONOMIA MEXICO
AP POSTAL 971
MEXICO 1 DF
MEXICO

DE LA NOE JEROME DR
OBS DE BORDEAUX
AV PIERRE SEMIROT
F-33270 FLOIRAC
FRANCE

DE LA REZA RAMIRO DR
OBSERVATORIO
NACIONAL
R GEN BRUCE 586 ZC08
20000 RIO DE JANEIRO
BRAZIL

DE LOORE CAMIEL PROF
ASTROPHYSICAL INST
VRIJE UNIVERSITEIT
VAN BRUSSEL
B-1050 BRUSSELS
BELGIUM

DE MARCUS WENDELL C PROF
UNIVERSITY OF
KENTUCKY-LEXINGTON
LEXINGTON KY 40506
U.S.A.

DE MOTTONI Y PALACIOS DR
PRIVAT OBSERVATORY
VIA ROSSELLI 15-23
I-16145 GENOVA
ITALY

DE PASCUAL MARTINEZ M DR
OBSERVATORIO
ASTRONOMICO
ALFONSO XII 5
MADRID 7
SPAIN

DE SABBATA V PROF DR
IST DI FISICA
UNIV DI BOLOGNA
VIA IRNERIO 46
I-40100 BOLOGNA
ITALY

DE VAUCOULEURS G PROF
DEPT OF ASTRONOMY
UNIVERSITY OF TEXAS
AUSTIN TX 78712
U.S.A.

DE VEGT CHR PROF DR
HAMBURGER STERNWARTE
GOJENBERGSWEG 112
D-2050 HAMBURG 80
GERMANY, F.R.

DE YOUNG DAVID S DR
KITT PEAK NAT OBS
P O BOX 26732
TUCSON AZ 85726
U.S.A.

DE ZOTTI GIANFRANCO DR
OSSERVATORIO ASTRO-
NOMICO
I-35100 PADOVA
ITALY

DEARBORN DAVID S P DR
STEWARD OBS
UNIV OF ARIZONA
TUCSON AZ 85721
U.S.A.

DEBARBAT SUZANNE V DR
OBS DE PARIS
61 AV L'OBSERVATOIRE
F- 75014 PARIS
FRANCE

DEBEHOGNE HENRI DR SC
OBSERVATOIRE ROYAL
DE BELGIQUE
3 AVE CIRCULAIRE
B-1180 BRUXELLES
BELGIUM

DEBRUNNER HERMANN DR
PHYSIKALISCHES INST
UNIVERSITAET BERN
SIDLERSTRASSE 5
CH-3000 BERN
SWITZERLAND

DECAUX BERNARD
20 RUE BOIS LE VENT
F-75016 PARIS
FRANCE

DEEMING T J DR
DIGITAL GEOPH CORP
3701 KIRBY DR
SUITE 1144
HOUSTON TX 77098
U.S.A.

DEERENBERG AD J DR
COSMIC RAY WORKING
GROUP HUYGENS LAB
WASSENAARSEWEG 78
2300 RA LEIDEN
THE NETHERLANDS

DEGEWIJ JOHAN DR
MPI FUER AERONOMIE
POSTFACH 20
D-3411 KATLENBURG-
LINDAU 3
GERMANY, F.R.

DEGUCHI SHUJI DR
C/O DEPT ASTRONOMY
FACULTY OF SCIENCE
TOKYO UNIVERSITY
BUNKYO-KU TOKYO 113
JAPAN

DEHARVENG JEAN MICHEL DR
LAS TRAVERSE DU SIP-
HON LES TROIS LUCS
F-13012 MARSEILLE
FRANCE

DEHARVENG LISE DR
OBS DE MARSEILLE
2 PLACE LE VERRIER
F-13248 MARSEILLE C4
FRANCE

DEINZER W PROF DR
UNIVERSITAETS-
STERNWARTE
GEISMARLANDSTR 11
D-3400 GOETTINGEN
GERMANY, F.R.

DEJAIFFE RENE J DR
OBSERVATOIRE ROYAL
DE BELGIQUE
AVE CIRCULAIRE 3
B-1180 BRUXELLES
BELGIUM

DEKKER E DR
MUSEUM BOERHAVE
STEENSTRAAT 1A
2312 BS LEIDEN
THE NETHERLANDS

DEL RIO GERARDO DR
NATIONAL ASTRON OBS
ALFONSO XII 3
MADRID 7
SPAIN

DELABOUDINIERE J P
LPSP
B P 10
F-91370 VERRIERES-LE
-BUISSON
FRANCE

DELACHE PHILIPPE J DR
OBSERVATOIRE DE NICE
BP 252
F-06007 NICE CEDEX
FRANCE

DELANNOY JEAN DR
IRAM - VOIE 10
DOMAINE UNIVERSITAIR
F-38406 ST MARTIN
D'HERES
FRANCE

DELBOUILLE LUC PROF
INST D ASTROPHYSIQUE
UNIVERSITE DE LIEGE
B-4200 COINTE OUGREE
BELGIUM

DELCROIX ANDRE J S DR
19A R E VANDERVELDE
B-7230 FRAMERIES
BELGIUM

DELHAYE JEAN PROF
OBSERVATOIRE PARIS
61 AVE DE
L OBSERVATOIRE
F-75014 PARIS
FRANCE

DELLI SANTI SAVERIO
OSS ASTRONOMICO
UNIVERSITARIO
CASELLA POSTALE 596
I-40100 BOLOGNA
ITALY

DELSEMME ARMAND H PROF DR
DEPT OF PHYSICS AND
ASTRONOMY
UNIVERSITY OF TOLEDO
TOLEDO OH 43606
U.S.A.

DEMARCQ JEAN L ING
OBSERVATOIRE DE NICE
B P 252
F-06007 NICE CEDEX
FRANCE

DEMARET JACQUES DR
158/033 AVENUE DE
L'OBSERVATOIRE
B-4000 LIEGE
BELGIUM

DEMARQUE P PROF
YALE UNIV OBS
260 WHITNEY AVE
BOX 6666
NEW HAVEN CT 06511
U.S.A.

DEMENKO A A DR
KIEV STATE UNIVERSI-
TY ASTRONOMICAL OB-
SERVATORY
252053 KIEV
U.S.S.R.

DEMERS SERGE DR
DEPT DE PHYSIQUE
UNIV DE MONTREAL
CP 6128 SUCC A
MONTREAL PQ H3C 3J7
CANADA

DEMIN V G PROF DR
STERNBERG STATE AS-
TRONOMICAL INSTITUTE
117234 MOSCOW
U.S.S.R.

DEMIRCAN OSMAN DR
EGE UNIVERSITY OBS
BORNOVA-IZMIR
TURKEY

DENIS CARLO DR
INST D'ASTROPHYSIQUE
UNIVERSITE DE LIEGE
AVENUE DE COINTE 5
B-4200 COINTE-OUGREE
BELGIUM

DENISSE JEAN-FRANCOIS DR
48 RUE M.LE PRINCE
F-75006 PARIS
FRANCE

DENISYUK EDVARD K DR
ASTROPHYSICAL INST
480068 ALMA ATA
U.S.S.R.

DENNIS BRIAN ROY DR
NASA-GSFC
CODE 684
GREENBELT MD 20771
U.S.A.

DENNISON BRIAN KENNETH
PHYSICS DEPT
VIRGINIA POLYTECHNIC
INSTITUTE
BLACKSBURG VA 24061
U.S.A.

DENNISON E W DR
ELECTRONIC VISION CO
11526 SORRENTO
VALLEY ROAD
SAN DIEGO CA 92121
U.S.A.

DENNISON P A DR
DEPT OF PHYSICS
UNIV OF ADELAIDE
ADELAIDE SA 5001
AUSTRALIA

DENOYELLE JOZEF KIC
VICTOR ALLARDSTR 126
B-1180 BRUXELLES
BELGIUM

DENT WILLIAM A PROF
PHYSICS ASTRONOMY
DEPT GRC TOWER B
UNIVERSITY OF MASS
AMHERST MA 01002
U.S.A.

DEPRIT ANDRE PROF
CTR FOR APPLIED MATH
NATIONAL BUREAU OF
STANDARDS
WASHINGTON DC 20234
U.S.A.

DERE KENNETH PAUL
CODE 4163
NAVAL RESEARCH LAB
WASHINGTON DC 20375
U.S.A.

DERMAN I ETHEM DR
ANKARA UNIVERSITY
FACULTY OF SCIENCE
ANKARA
TURKEY

DERMENDJIEV VLADIMIR DR
DEPT OF ASTRONOMY
BULGARIAN ACAD OF SC
7TH NOVEMBER STR 1
SOFIA
BULGARIA

DESAI JYOTINDRA N
PHYS RESEARCH LAB
AHMEDABAD 380 009
INDIA

DESESQUELLES JEAN DR
UNIVERSITE LYON I
CAMPUS LA DOUA
F-69621 VILLEURBANNE
FRANCE

DESPAIN KEITH HOWARD DR
DEPT OF ASTRONOMY
HAVERFORD COLLEGE
HAVERFORD PA 19041
U.S.A.

DEUBNER FRANZ-LUDWIG DR
INST F ASTRONOMIE U
ASTROPHYSIK DER UNIV
AM HUBLAND
D-8700 WUERZBURG
GERMANY, F.R.

DEUPREE ROBERT G DR
DEPT OF ASTRONOMY
BOSTON UNIV
725 COMMONWEALTH AVE
BOSTON MA 02215
U.S.A.

DEUTSCH ALEKSANDR N
PROF DR
PULKOVO OBSERVATORY
196140 LENINGRAD
U.S.S.R.

DEUTSCHMAN W.A. DR.
DEPT.PHYSICS ASTRON.
ROSE-HULMAN INST.
5500 WABASH AVE.
TERRE HAUTE IN 47803
U.S.A.

DEVINNEY E J DR
100 UNION AVE
DELANCO NJ 08075
U.S.A.

DEWDNEY P E F DR
186 MATSON PL
PENTICTON BC
CANADA

DEWHIRST DAVID W DR
INST OF ASTRONOMY
THE OBSERVATORIES
MADINGLEY RD
CAMBRIDGE CB3 OHA
U.K.

DEWITT B S DR
DEPT OF PHYSICS
UNIV OF TEXAS
AUSTIN TX 78712
U.S.A.

DEWITT JOHN H JR
3602 HOODS HILLS RD
NASHVILLE TN 37215
U.S.A.

DEWITT-MORETTE C PROF
DEPT OF ASTRONOMY
UNIVERSITY OF TEXAS
AUSTIN TX 78712
U.S.A.

DEZSO LORANT PROF
HELIOPHYSICAL
OBSERVATORY
H-4010 DEBRECEN
HUNGARY

DI SEREGO ALIGHIERI S DR
ASTRONOMY DIV OF ESA
ESTEC
P O BOX 299
2200 AG NOORDWIJK
THE NETHERLANDS

DI TULLIO GRAZIELLA DR
OSSERVATORIO ASTRO-
NOMICO
I-35100 PADOVA
ITALY

DIBAY E A DR
STERNBERG STATE AS-
TRONOMICAL INSTITUTE
117234 MOSCOW
U.S.S.R.

DICKE ROBERT H PROF
JOSEPH HENRY LABS
PHYSICS DEPT
PRINCETON UNIVERSITY
PRINCETON NJ 08540
U.S.A.

DICKEL HELENE R DR
UNIVERSITY OF
ILLINOIS OBSERVATORY
URBANA IL 61801
U.S.A.

DICKEL JOHN R ASSOC PROF
UNIVERSITY OF
ILLINOIS OBSERVATORY
URBANA IL 61801
U.S.A.

DICKENS ROBERT J DR
ROYAL GREENWICH
OBSERVATORY
HAILSHAM
EAST SUSSEX BN27 1RP
U.K.

DICKINSON DALE F DR
JPL
183-301
4800 OAK GROVE DRIVE
PASADENA CA 91103
U.S.A.

DIECKVOSS WILHELM KE PROF
HAMBURGER STERNWARTE
D-2050 HAMBURG 80
GERMANY, F.R.

DIERCKSEN GEERD H F PH D
MPI FUER PHYSIK UND
ASTROPHYSIK
FOEHRINGER RING 6
D-8000 MUENCHEN 40
GERMANY, F.R.

DIETER N H DR
RADIO ASTRONOMY LAB
UNIV OF CALIFORNIA
BERKELEY CA 94720
U.S.A.

DIMITRIJEVIC MILAN
INST ZA PRIMENJENU
FIZIKU MARSALA-BIR
JUZOVA 14 A POSTF 24
YU-11001 BEOGRAD
YUGOSLAVIA

DINESCU A DR
INST GEODESIE PHOTO-
GRAMMETRIE CARTOGRAP
1A BD DE LEXPOSITION
78334 BUCAREST
ROUMANIA

DING YOU-JI
YUNNAN OBSERVATORY
KUNMING
YUNNAN PROVINCE
PEOPLE'S REPUBLIC CHINA

DINGENS P PROF DR
STERRENKUNDIG INST
RIJKSUNIVERSITEIT
KORTRIJKSE STWEG 523
B-9000 GENT
BELGIUM

DINULESCU NICOLAE I PROF
SOSEANA KISELEFF 13
72168 BUCURESTI
SECTOR 1
ROUMANIA

DIONYSIOU DEMETRIOS D DR
DEPT OF ASTRONOMY
UNIVERSITY OF ATHENS
ATHENS 621
GREECE

DIRIKIS M A DR
LATVIAN STATE UNIVER
SITY ASTRONOMICAL
OBSERVATORY
226098 RIGA
U.S.S.R.

DISNEY MICHAEL J DR
UNIVERSITY COLLEGE
DEPT OF APPLIED MATH
& THEOR PHYS
CARDIFF CF1 1XL
U.K.

DIVAN LUCIENNE DR
INSTITUT
ASTROPHYSIQUE
98 BIS BD ARAGO
F-75014 PARIS
FRANCE

DIVARI N B DR
ODESSA POLYTECHNICAL
INSTITUTE
270044 ODESSA
U.S.S.R.

DIXON ROBERTS S DR
OHIO STATE UNIVER-
SITY RADIO OBS
2015 NEIL AVE
COLUMBUS OH 43210
U.S.A.

DIZER MUAMMER PROF
KANDILLI OBSERVATORY
CENGELKOY
ISTANBUL
TURKEY

DJUROVIC DRAGUTIN M DR
DEPT OF ASTRONOMY
FAC OF SCIENCES
STUDENTSKI TRG 16
11000 BEOGRAD
YUGOSLAVIA

DLUZHNEVSKAYA O B DR
ASTRONOMICAL COUNCIL
USSR ACAD OF SCI
PYATNITSKAYA UL 48
109017 MOSCOW
U.S.S.R.

DOAN H H DR
OBSERVATOIRE DE LYON
F-69230 SAINT-GENIS
LAVAL
FRANCE

DOAZAN VERA DR
OBSERVATOIRE
DE PARIS
61 AVE L OBSERVATOIR
F-75014 PARIS
FRANCE

DOBRITSCHEV V M MR
DEPT OF ASTRONOMY
7TH NOVEMBER ST 1
SOFIA
BULGARIA

DOBRONRAVIN PETER P DR
CRIMEAN ASTROPHYSI-
CAL OBSERVATORY
334413 NAUCHNYJ
CRIMEA
U.S.S.R.

DOBROVOLSKY OLEG V PROF
INST OF ASTROPHYSICS
734670 DUSHANBE
U.S.S.R.

DOBRZYCKI JERZY PROF
GWIAZDZISTA 27/169
01-814 WARSZAWA
POLAND

DODD RICHARD J DR
CARTER OBSERVATORY
P O BOX 2909
WELLINGTON 1
NEW ZEALAND

DOGAN NADIR PROF
A U FEN FAKULTESI
ANKARA
TURKEY

DOHERTY LORNE H DR
HERZBERG INST OF
ASTROPHYSICS
NAT RES COUNC CANADA
OTTAWA ONT K1A OR6
CANADA

DOHERTY LOWELL R PROF
ASTRONOMY DEPT
UNIVERSITY WISCONSIN
475 N CHARTER ST
MADISON WI 53706
U.S.A.

DOKUCHAEVA O D DR
STERNBERG STATE AS-
TRONOMICAL INSTITUTE
117234 MOSCOW
U.S.S.R.

DOLAN JOSEPH F DR
CODE 681
NASA GODDARD SPACE
FLIGHT CENTER
GREENBELT MD 20771
U.S.A.

DOLGINOV ARKADY Z PROF DR
IOFFE PHYSICAL
TECHNICAL INSTITUTE
194021 LENINGRAD
U.S.S.R.

DOLIDZE M V DR
ABASTUMANI ASTROPHY-
SICAL OBSERVATORY
383762 ABASTUMANI
GEORGIA
U.S.S.R.

DOLLFUS AUDOUIN PROF
OBSERVATOIRE DE
MEUDON
F-92190 MEUDON
FRANCE

DOMINKO FRAN PROF DR
SARANOVICEVA 11
61000 LJUBLJANA
YUGOSLAVIA

DOMINSKI IRENEUSZ DR
BOROWIEC
63-120 KORNIK
POLAND

DOMKE HELMUT PH D
ZENTR INST ASTROPHYS
DER ADW DER DDR
ROSA-LUXEMBURGST 17A
DDR-1502 POTSDAM
G.D.R.

DOMMANGET J DR
OBSERVATOIRE ROYAL
DE BELGIQUE
AVE CIRCULAIRE 3
B-1180 BRUXELLES
BELGIUM

DONN BERTRAM D
GODDARD SPACE FLIGHT
CENTER
CODE 691
GREENBELT MD 20771
U.S.A.

DONNER KARL JOHN DR
OBS & ASTROPHYS LAB
UNIV OF HELSINKI
TAEHTITORNINMAEKI
SF-00130 HELSINKI 13
FINLAND

DOPITA MICHAEL A DR
MT STROMLO
OBSERVATORY
PRIVATE BAG
WODEN PO ACT 2606
AUSTRALIA

DORENWENDT KLAUS DR
PHYSIKALISCH-TECHN.
BUNDESANSTALT
BUNDESALLEE 100
D-3300 BRAUNSCHWEIG
GERMANY, F.R.

DORMAND JOHN R DR
MATHEMATICS DEPT
TEESSIDE POLYTECHNIC
MIDDLESBROUGH
CLEVELAND TS1 3BA
U.K.

DOROSHKEVICH A G DR
INSTITUTE F APPLIED
MATHEMATICS
USSR ACADEMY OF SCI
125047 MOSCOW
U.S.S.R.

DORSCHNER JOHANN DR
UNIVERSITAETS
STERNWARTE JENA
SCHILLERGAESSCHEN 2
DDR-69 JENA
G.D.R.

DOS REIS M PROF
DIRECTOR OF THE
OBSERVATORY
COIMBRA
PORTUGAL

DOSCHEK GEORGE A DR
NAVAL RESEARCH
LABORATORY
CODE 7125.8
WASHINGTON DC 20375
U.S.A.

DOSSIN F DR
UNIVERSITE DE LIEGE
INST D ASTROPHYSIQUE
AVE DE COINTE 5
B-4200 COINTE OUGREE
BELGIUM

DOUGHTY NOEL A DR
PHYSICS DEPT
UNIVERSITY OF
CANTERBURY
CHRISTCHURCH 1
NEW ZEALAND

DOUGLAS A VIBERT DR
402 67 SYDENHAM ST
KINGSTON ONT K7L 3H2
CANADA

DOUGLAS JAMES N PROF
DEPT OF ASTRONOMY
UNIV OF TEXAS
AUSTIN TX 78712
U.S.A.

DOWNES ANN JULIET B
MULLARD RADIO
ASTRONOMY OBS
MADINGLEY ROAD
CAMBRIDGE CB3 OHE
U.K.

DOWNES DENNIS D DR
IRAM - VOIE 10
DOMAINE UNIVERSITAIR
F-38406 ST MARTIN
D'HERES
FRANCE

DOWNS GEORGE S DR
JPL
MAIL STOP 238/420
4800 OAK GROVE DRIVE
PASADENA CA 91103
U.S.A.

DRAKE FRANK D PROF
N A I C
402 SPACE SCI BLDG
CORNELL UNIVERSITY
ITHACA NY 14853
U.S.A.

DRAMBA C PROF
OBS ASTRONOMIQUE
5 CUTITUL DE ARGINT
P.O.BOX 28
75212 BUCAREST
ROUMANIA

DRAPATZ SIEGFRIED W DR
MPI F EXTRATERRESTRI
SCHE PHYSIK
D-8046 GARCHING
GERMANY, F.R.

DRAVINS DAINIS DR
LUND OBSERVATORY
SVANEGATAN 9
S-222 24 LUND
SWEDEN

DRAVSKIKH A F DR
SPECIAL ASTROPHYSI-
CAL OBS USSR ACAD SC
LENINGRAD BRANCH
196140 LENINGRAD
U.S.S.R.

DRECHSEL HORST DR
REMEIS-STERNWARTE
ASTR INST UNIV ERL-N
STERNWARTSTR 7
D-8600 BAMBERG
GERMANY, F.R.

DRESSEL LINDA L
NASA GSFC
CODE 661
GREENBELT MD 20771
U.S.A.

DRESSLER KURT PROF
LAB PHYSIK CHEMIE
ETH ZENTRUM
CH-8092 ZUERICH
SWITZERLAND

DREVER RONALD W P DR
DEPT OF NATURAL PHIL
GLASGOW UNIVERSITY
GLASGOW G12 8QQ
U.K.

DRILLING JOHN S
DEPT OF PHYSICS AND
ASTRONOMY
LOUISIANA STATE UNIV
BATON ROUGE LA 70803
U.S.A.

DROFA VASILIY K DR
ASTRONOMICAL OBS
KIEV UNIVERSITY
252053 KIEV
U.S.S.R.

DRYER MURRAY DR
SPACE ENVIRONMENT
LAB
NOAA ERL
BOULDER CO 80302
U.S.A.

DUBAU J DR
OBS DE MEUDON
F-92190 MEUDON
FRANCE

DUBINSKIJ B A DR
INSTITUTE OF RADIO
AND ELECTRONICS USSR
ACADEMY OF SCIENCES
103907 MOSCOW
U.S.S.R.

DUBOIS PASCAL DR
OBS DE STRASBOURG
11 RUE DE L'UNIV
F-67000 STRASBOURG
FRANCE

DUBOSHIN G N PROF DR
STERNBERG STATE AS-
TRONOMICAL INSTITUTE
117234 MOSCOW
U.S.S.R.

DUBOV EMIL E PROF
WDC-B2
MOLODEZHNAYA 3
117296 MOSCOW
U.S.S.R.

DUCHESNE MAURICE DR
OBSERVATOIRE
DE PARIS
61 AVE OBSERVATOIRE
F-75014 PARIS
FRANCE

DUERBECK HILMAR W DR
OBS HOHER LIST
UNIV STERNWARTE BONN
D-5568 DAUN
GERMANY, F.R.

DUERST JOHANNES DR
SWISS FEDERAL OBS
SCHMELZBERGSTR 25
CH 8006 ZUERICH
SWITZERLAND

DUFAY MAURICE PROF
UNIVERSITE CLAUDE-
BERNARD-LYON I
43 BD DU 11 NOVEMBRE
F-69621 VILLEURBANNE
FRANCE

DUFFETT-SMITH P J
MULLARD RADIO
ASTRONOMY OBS
CAVENDISH ROAD
CAMBRIDGE CB3 OHE
U.K.

DUFLOT MARCELLE DR
OBSERVATOIRE DE
MARSEILLE
2 PLACE LE VERRIER
F-13248 MARSEILLE C4
FRANCE

DUFLOT RENEE DR
OBSERVATOIRE
DE MARSEILLE
F-13004 MARSEILLE
FRANCE

DUFOUR REGINALD JAMES
SPACE PHYSICS DEPT
RICE UNIVERSITY
HOUSTON TX 77001
U.S.A.

DUFTON PHILIP L DR
DEPT OF APPLIED MATH
QUEEN'S UNIVERSITY
BELFAST BT7 1NN
U.K.

DULEY WALTER W PROF
PHYSICS DEPT
YORK UNIVERSITY
DOWNSVIEW ONT M3J1P3
CANADA

DULK GEORGE A PROF
DEPT ASTRO-GEOPHYS
UNIV OF COLORADO
BOULDER CO 80302
U.S.A.

DULTZIN DEBORAH DR
INSTITUTO DE
ASTRONOMIA UNAM
APDO POSTAL 70264
MEXICO 04510 DF
MEXICO

DUMA DMITRIJ P DR
MAIN ASTRONOMICAL OB
UKRAINIAN ACAD OF SC
252127 KIEV
U.S.S.R.

DUMONT RENE DR
OBSERVATOIRE
DE BORDEAUX
F-33270 FLOIRAC
FRANCE

DUMONT SIMONE DR
INST D'ASTROPHYSIQUE
98 BIS BLVD ARAGO
F-75014 PARIS
FRANCE

DUNCAN ROBERT A PROF
CSIRO
DIV OF RADIOPHYSICS
P O BOX 76
EPPING NSW 2121
AUSTRALIA

DUNCOMBE RAYNOR L DR
DEPT OF AEROSPACE
ENGINEERING
UNIVERSITY OF TEXAS
AUSTIN TX 78712
U.S.A.

DUNHAM DAVID W
COMPUTER SCS CORP
8728 COLESVILLE RD
SILVER SPG MD 20910
U.S.A.

DUNHAM THEODORE JR DR
P O BOX D
CHOCORUA NH 03817
U.S.A.

DUNKELMAN LAWRENCE
CODE 685 LAB F ASTRO
AND SOLAR PHYSICS
GODDARD SPACE FL CTR
GREENBELT MD 20771
U.S.A.

DUNN RICHARD B DR
SACRAMENTO PEAK
OBSERVATORY
SUNSPOT NM 88349
U.S.A.

DUPREE ANDREA K DR
CTR F ASTROPHYSICS
60 GARDEN ST
CAMBRIDGE MA 02138
U.S.A.

DUPUY DAVID L DR
DEPT OF ASTRONOMY
ST MARY S UNIVERSITY
HALIFAX NS B3H 3C3
CANADA

DURISEN RICHARD H DR
DEPT OF ASTRONOMY
SWAIN WEST 319
INDIANA UNIV
BLOOMINGTON IN 47401
U.S.A.

DURNEY BERNARD R DR
SACRAMENTO PEAK OBS
SUNSPOT NM 88349
U.S.A.

DURRANT CHRISTOPHER J DR
KIEPENHEUER INST
FUER SONNENPHYSIK
SCHOENECKSTR 6
D-7800 FREIBURG
GERMANY, F.R.

DUTHIE JOSEPH G PROF
UNIVERSITY ROCHESTER
DEPT OF PHYSICS AND
ASTRONOMY
ROCHESTER NY 14627
U.S.A.

DVORAK RUDOLF DR
ASTRONOMISCHES INST
DER UNIVERSITAET
A-8010 GRAZ
AUSTRIA

DWEK ELIAHU
CODE 693.2
NASA GSFC
GREENBELT MD 20771
U.S.A.

DWORETSKY MICHAEL M DR
DEPT OF PHYSICS AND
ASTRONOMY
UNIV COLLEGE LONDON
LONDON WC1E 6BT
U.K.

DYCK M DR
KITT PEAK NATIONAL
OBSERVATORY
P O BOX 26732
TUCSON AZ 85726
U.S.A.

DYER CHARLES CHESTER DR
PHYS SCS GR RM R4227
SCARBOROUGH COLLEGE
UNIV OF TORONTO
TORONTO ONT M1G 3S7
CANADA

DYER EDWARD R DR
NATIONAL ACADEMY OF
SCIENCES
2101 CONSTITUTION AV
WASHINGTON DC 20418
U.S.A.

DYSON F J DR
INSTITUTE FOR
ADVANCED STUDIES
PRINCETON NJ 08540
U.S.A.

DYSON JOHN E DR
ASTRONOMY DEPT
UNIVERSITY OF
MANCHESTER
MANCHESTER M13 9PL
U.K.

DZHAPIASHVILI V P DR
ABASTUMANI ASTROPHY-
SICAL OBSERVATORY
383762 ABASTUMANI
GEORGIA
U.S.S.R.

DZIEMBOWSKI W DR
ASTRONOMICAL CENTER
UL BARTYCKA 18
00-716 WARSAW
POLAND

DZIGVASHVILI R M DR
ABASTUMANI ASTROPHY-
SICAL OBSERVATORY
383762 ABASTUMANI
U.S.S.R.

EATON JOEL A DR
DYER OBS
BOX 1803 STA B
NASHVILLE TN 37235
U.S.A.

EBBIGHAUSEN E G
PHYSICS DEPT
UNIVERSITY OF OREGON
EUGENE OR 97403
U.S.A.

ECCLES MICHAEL J DR
29 BUCKSTONE LOAN E
EDINBURGH 10
U.K.

EDDY JOHN A DR
HIGH ALTITUDE
OBSERVATORY
BOX 3000
BOULDER CO 80307
U.S.A.

EDLEN BENGT PROF
DEPT OF PHYSICS
SOELVEGATAN 14
S-223 62 LUND
SWEDEN

EDMONDS FRANK N JR DR
RLM 15.212
UNIVERSITY OF TEXAS
AUSTIN TX 78712
U.S.A.

EDMONDSON FRANK K PROF
GOETHE LINK OBS
319A SWAIN HALL WEST
INDIANA UNIVERSITY
BLOOMINGTON IN 47405
U.S.A.

EDMUNDS MICHAEL GEOFFREY
DEPT OF APPLIED MATH
& ASTRONOMY
UNIV CLG P O BOX 78
CARDIFF CF1 1XL
U.K.

EDWARDS ALAN CH DR
DEPT OF ASTROPHYSICS
SOUTH PARKS ROAD
OXFORD OX1 3RQ
U.K.

EDWARDS P J PROF
DEPT OF PHYSICS
UNIV OF OTAGO
P O BOX 56
DUNEDIN
NEW ZEALAND

EDWARDS TERRY W
DEPT OF PHYSICS
AND ASTRONOMY
UNIV OF ROCHESTER
ROCHESTER NY 14627
U.S.A.

EELSALU HEINO DR
TARTU OBSERVATORY
202-444 TORAVERE
ESTONIA
U.S.S.R.

EFREMOV YU I DR
INSTITUTE F APPLIED
MATHEMATICS
USSR ACADEMY OF SCI
125047 MOSCOW
U.S.S.R.

EFREMOV YURY N DR
STERNBERG
ASTRONOMICAL INST
117234 MOSCOW
U.S.S.R.

EFSTATHIOU GEORGE
INST OF ASTRONOMY
MADINGLEY ROAD
CAMBRIDGE CB3 OHA
U.K.

EGGLETON PETER P DR
INSTITUTE OF
ASTRONOMY
MADINGLEY RD
CAMBRIDGE CB3 OHE
U.K.

EGRET DANIEL DR
OBS DE STRASBOURG
11 RUE DE L'UNIV
F-67000 STRASBOURG
FRANCE

EHLERS JURGEN PROF
MPI FUER PHYSIK UND
ASTROPHYSIK MUENCHEN
D-8000 MUENCHEN 40
GERMANY, F.R.

EHRGMA E V DR
ASTRONOMICAL COUNCIL
USSR ACADEMY OF SCI
PYATNITSKAYA UL 48
109017 MOSCOW
U.S.S.R.

EICHHORN HEINRICH K DR
DEPT OF ASTRONOMY
UNIV OF FLORIDA
231 SPACE SC RES BLD
GAINESVILLE FL 32611
U.S.A.

EICHLER DAVID DR
ASTRONOMY PROGRAM
UNIV OF MARYLAND
COLLEGE PK MD 20742
U.S.A.

EILEK JEAN
PHYSICS DEPT
NEW MEXICO TECH
SOCORRO NM 87801
U.S.A.

EINICKE O H LEKTOR
UNIVERSITY
OBSERVATORY
OSTERVOLDGADE 3
DK-1350 COPENHAGEN K
DENMARK

EINSATO JAAN DR
TARTU OBSERVATORY
202444 TORAVERE
ESTONIA
U.S.S.R.

EKERS RONALD J DR
KAPTEYN LABORATORIUM
POSTBUS 800
9700 AV GRONINGEN
THE NETHERLANDS

EL EID MOUNIB DR
UNIVERSITAETSSTERN-
WARTE
GEISMARLANDSTR 11
D-3400 GOETTINGEN
GERMANY, F.R.

EL-BAZ F DR
CTR EARTH AND PLANE-
TARY STUDIES
SMITHSONIAN INST.
WASHINGTON DC 20560
U.S.A.

EL-RAEY MOHAMED E DR
DEPT OF PHYSICS
FACULTY OF SCIENCE
UNIV OF ALEXANDRIA
ALEXANDRIA
ARAB REP. OF EGYPT

EL-SHAARAWY M B DR
HELWAN OBSERVATORY
HELWAN NEAR CAIRO
ARAB REP. OF EGYPT

ELFORD W G DR
DEPT OF PHYSICS
UNIV OF ADELAIDE
ADELAIDE
AUSTRALIA

ELGAROY OYSTEIN PROF
INST THEOR ASTROPHYS
UNIVERSITY OF OSLO
P O BOX 1029
BLINDERN OSLO 3
NORWAY

ELITZUR MOSHE
DEPT OF PHYS & ASTRO
UNIV OF KENTUCKY
LEXINGTON
KY 40506-0055
U.S.A.

ELLDER JOEL DR
ONSALA SPACE
OBSERVATORY
S-430 34 ONSALA
SWEDEN

ELLIOT HARRY PROF
BLACKETT LABORATORY
IMPERIAL COLLEGE
LONDON SW7 2BZ
U.K.

ELLIOT JAMES L DR
DEPT EARTH/PLANET SC
MIT BLDG 54-422A
CAMBRIDGE MA 02139
U.S.A.

ELLIOTT IAN DR
DUNSINK OBSERVATORY
CASTLEKNOCK
CO DUBLIN
IRELAND

ELLIOTT KENNETH H DR
DEPT OF SPACE RES
UNIV OF BIRMINGHAM
P O BOX 363
BIRMINGHAM B15 2TT
U.K.

ELLIS G R A PROF
UNIV OF TASMANIA
P O BOX 252D
HOBART TASMANIA
AUSTRALIA

ELLIS GEORGE F R PROF
DEPT OF APPLIED MATH
UNIV OF CAPE TOWN
RONDEBOSCH CAPE 7700
SOUTH AFRICA

ELLIS RICHARD S
PHYSICS DEPT
DURHAM UNIVERSITY
SOUTH ROAD
DURHAM DH1 3LE
U.K.

ELMEGREEN BRUCE GORDON DR
DEPT OF ASTRONOMY
PUPIN LABS
COLUMBIA UNIV
NEW YORK NY 10027
U.S.A.

ELMEGREEN DEBRA MELOY
DEPT OF ASTRONOMY
COLUMBIA UNIVERSITY
NEW YORK NY 10027
U.S.A.

ELSAESSER HANS PROF
MPI FUER ASTRONOMIE
D-6900 HEIDELBERG-
KOENIGSTUHL
GERMANY, F.R.

ELSMORE BRUCE DR
CAVENDISH LABORATORY
MADINGLEY ROAD
CAMBRIDGE CB3 OHE
U.K.

ELST E DR
KONINGLIJKE STERREN-
WACHT VAN BELGIE
RINGLAAN 3
B-1180 BRUSSEL
BELGIUM

ELSTE GUNTHER H DR
DEPT OF ASTRONOMY
UNIVERSITY OF
MICHIGAN
ANN ARBOR MI 48109
U.S.A.

ELSTON WOLFGANG E PROF
DEPT OF GEOLOGY
UNIVERSITY OF
NEW MEXICO
ALBUQUERQUE NM 87131
U.S.A.

ELVIS MARTIN S DR
CTR FOR ASTROPHYSICS
60 GARDEN ST
CAMBRIDGE MA 02138
U.S.A.

ELVIUS AINA M PROF
STOCKHOLM OBS
S-133 00
SALTSJOEBADEN
SWEDEN

ELVIUS TORD PROF
NORRLANDSG 34 F
S-752 29 UPPSALA
SWEDEN

ELWERT GERHARD PROF
LEHRSTUHL F THEORE-
TISCHE ASTROPHYSIK
UNIV TUEBINGEN
D-7400 TUEBINGEN
GERMANY, F.R.

ELYASBERG P E PROF DR
SPACE RESEARCH
INSTITUTE
USSR ACADEMY OF SCI
117810 MOSCOW
U.S.S.R.

EMELIANOV NIKOLAJ V DR
STERNBERG STATE
ASTRONOMICAL INST
117234 MOSCOW
U.S.S.R.

EMERSON BRIAN MR
ROYAL GREENWICH OBS
HERSTMONCEUX CASTLE
HAILSHAM
E SUSSEX BN27 1RP
U.K.

EMERSON JAMES P R
DEPT OF PHYSICS
QUEEN MARY COLLEGE
MILE END ROAD
LONDON E1 4NS
U.K.

EMINZADE T A DR
SHEMAKHA ASTROPHYSI-
CAL OBSERVATORY
373243 AZERBAIDZAN
U.S.S.R.

ENCRENAZ PIERRE J DR
ECOLE NORMALE
SUPERIEURE
24 RUE LHOMOND
F-75005 PARIS
FRANCE

ENCRENAZ T DR
GROUPE PLANETES
SECT D'ASTROPHYSIQUE
OBS DE MEUDON
F-92190 MEUDON
FRANCE

ENDAL ANDREW S DR
DEPT PHYSICS ASTRON
LOUISIANA STATE UNIV
BATON ROUGE LA 70803
U.S.A.

ENGELHARD E J G PROF DR
SACKRING 34
D-3300 BRAUNSCHWEIG
GERMANY, F.R.

ENGIN SEMANUR ASSOC PROF
DEPT OF ASTRONOMY
UNIVERSITY OF ANKARA
FEN FAKULTESI
ANKARA
TURKEY

ENSLIN HEINZ DR
DEUTSCHES HYDROGRA-
PHISCHES INSTITUT
POSTFACH 220
D-2000 HAMBURG 4
GERMANY, F.R.

EPSTEIN GABRIEL LEO DR
NASA-GSFC
CODE 682
GREENBELT MD 20771
U.S.A.

ERDI B DR
ASTRONOMICAL DEPT
LORAND EOTVOS UNIV
KUN BELA TER 2
H-1083 BUDAPEST
HUNGARY

ERIKSEN GUNNAR PROF
INST THEOR ASTROPHYS
UNIVERSITY OF OSLO
P O BOX 1029
BLINDERN OSLO 3
NORWAY

ERSHKOVICH ALEXANDER DR
58/1 HANKIN ST
RAANANA
ISRAEL

ESHLEMAN VON R PROF
DURAND 221
STANFORD UNIVERSITY
STANFORD CA 94305
U.S.A.

ESPOSITO F PAUL PROF
DEPT OF PHYSICS 11
UNIVERSITY OF
CINCINNATI
CINCINNATI OH 45221
U.S.A.

EVANGELIDIS E DR
UNIVERSITY OF PATRAS
DEPT OF ASTRONOMY
PATRAS
GREECE

EVANS J V DR
MIT LINCOLN LAB
MILLSTONE/HAYSTACK
OBSERVATORY
LEXINGTON MA 02173
U.S.A.

EVANS NANCY R DR
183 BRIAR HILL AVE
TORONTO ONT M4R 1H8
CANADA

EVANS W DOYLE
390 EL CONEJO
LOS ALAMOS NM 87544
U.S.A.

ENGVOLD ODDBJOERN DR
INST THEOR ASTROPHYS
UNIVERSITY OF OSLO
P O BOX 1029
BLINDERN OSLO 3
NORWAY

EPPS HARLAND WARREN PROF
DEPT OF ASTRONOMY
UNIVERSITY OF CALIF
MATH SCI RM 8983
LOS ANGELES CA 90024
U.S.A.

EPSTEIN ISADORE PROF
ASTRONOMY DEPT
COLUMBIA UNIVERSITY
NEW YORK NY 10027
U.S.A.

ERICKSON WILLIAM C DR
ASTRONOMY PROGRAM
UNIVERSITY MARYLAND
COLLEGE PARK
MD 20742
U.S.A.

ERIKSSON KJELL DR
ASTRONOMISKA
OBSERVATORIET
BOX 515
S-751 20 UPPSALA
SWEDEN

ERTAN A YENER DR
EGE UNIVERSITY
FEN FAKUELTESI
ASTRONOMI BOELUEMUE
BORNOVA-IZMIR
TURKEY

ESIPOV V F DR
STERNBERG STATE AS-
TRONOMICAL INSTITUTE
117234 MOSCOW
U.S.S.R.

ESPOSITO LARRY W
LASP
CAMPUS BOX 392
UNIV OF COLORADO
BOULDER CO 80309
U.S.A.

EVANS ANEURIN
DEPT OF PHYSICS
UNIV OF KEELE
KEELE ST5 5BG
U.K.

EVANS JOHN W DR
SACRAMENTO PEAK
OBSERVATORY
SUNSPOT NM 88349
U.S.A.

EVANS NEAL J II ASST PROF
DEPT OF ASTRONOMY
UNIVERSITY OF TEXAS
AUSTIN TX 78712
U.S.A.

EVDOKIMOV YU V DR
ENGELHARDT ASTRONOMI
CAL OBSERVATORY
422526 KAZAN
U.S.S.R.

ENOME SHINZO ASSOC PROF
RESEARCH INSTITUTE
OF ATMOSPHERICS
NAGOYA UNIVERSITY
TOYOKAWA 442
JAPAN

EPSTEIN EUGENE E DR
I A GETTING LABS
130-422 AERO CORP
BOX 92957
LOS ANGELES CA 90009
U.S.A.

EPSTEIN RICHARD I DR
NORDITA
BLEGDAMSVEJ 17
DK-2100 COPENHAGEN O
DENMARK

ERIGUCHI YOSHIHARU DR
DEPT EART SC & ASTRO
COLL GEN EDUCATION
UNIV TOKYO KOMABA
MEGURO TOKYO 153
JAPAN

ERPYLEV N P DR
ASTRONOMICAL COUNCIL
USSR ACADEMY OF SCI
PYATNITSKAYA UL 48
109017 MOSCOW
U.S.S.R.

ERUSHEV N N DR
CRIMEAN ASTROPHYSI-
CAL OBSERVATORY USSR
ACAD OF SCI NAUCHNIY
334413 CRIMEA
U.S.S.R.

ESKIOGLU A NIHAT
DEVLET MUHENDISLIK
MIMARLIK AKADEMISI
ADAPAZARI - SAKARYA
TURKEY

ESTALELLA ROBERTO DR
JOAN GUELL 139-145
7-4A
BARCELONA-28
SPAIN

EVANS DAVID S PROF
DEPT OF ASTRONOMY
UNIVERSITY OF TEXAS
AUSTIN TX 78712
U.S.A.

EVANS KENTON D DR
PHYSICS DEPT
THE UNIVERSITY
LEICESTER LE1 7RN
U.K.

EVANS ROGER G DR
RUTHERFORD
LABORATORY
DIDCOT OX11 0QX
U.K.

EVERHART EDGAR DR
PHYSICS DEPT AND
CHAMBERLIN OB
UNIVERSITY OF DENVER
DENVER CO 80208
U.S.A.

EVIATAR AHARON PROF
DEPT OF GEOPHYSICS
AND PLANETARY SC
TEL-AVIV UNIVERSITY
TEL-AVIV
ISRAEL

EWEN HAROLD I DR
60 BEAVER ROAD
WESTON MA 02193
U.S.A.

EWING MARTIN S
CALTECH 102-24
PASADENA CA 91125
U.S.A.

EZER DILHAN PROF
MIDDLE EAST TECH-
NICAL UNIVERSITY
ANKARA
TURKEY

FABER SANDRA M PROF
LICK OBSERVATORY
UNIVERSITY OF
CALIFORNIA
SANTA CRUZ CA 95064
U.S.A.

FABIAN ANDREW C DR
INSTITUTE OF
ASTRONOMY
MADINGLEY ROAD
CAMBRIDGE CB3 OHE
U.K.

FABRE HERVE DR
2 AVENUE MARECHAL
FOCH
F-06310 BEAULIEU MER
FRANCE

FAELTHAMMAR CARL GUNNE PR
DEPT PLASMA PHYSICS
ROYAL INST OF
TECHNOLOGY
S-100 44 STOCKHOLM70
SWEDEN

FAHLMAN GREGORY G DR
DEPT GEOPHYS AND AST
UNIVERSITY OF B C
2075 WESBROOK PLACE
VANCOUVER BC V6T 1W5
CANADA

FAHR HANS JOERG PROF DR
INST ASTROPHY EXTRA-
TERRESTR F UNIV BONN
AUF DEM HUEGEL 71
D-5300 BONN
GERMANY, F.R.

FAHY EDWARD F PROF
PHYSICS DEPT
UNIVERSITY COLLEGE
CORK
IRELAND

FAIRALL ANTHONY P DR
DEPT OF ASTRONOMY
UNIV OF CAPE TOWN
RONDEBOSCH 7700
CAPE
SOUTH AFRICA

FALCIANI ROBERTO DR
ARCETRI ASTROPHYSI-
CAL OBS
LARGO E FERMI 5
I-50125 FIRENZE
ITALY

FALK SYDNEY W JR DR
DEPT OF ASTRONOMY
UNIV OF TEXAS
AUSTIN TX 78712
U.S.A.

FALL S M DR
INST OF ASTRONOMY
MADINGLEY ROAD
CAMBRIDGE CB3 OHA
U.K.

FALLE SAMUEL A DR
UNIVERSITY OF LEEDS
LEEDS LS2 9JT
U.K.

FALLER JAMES E PROF
JILA NBS
UNIVERSITY COLORADO
BOULDER CO 80309
U.S.A.

FALLON FREDERICK W DR
DEPT OF ASTRONOMY
UNIV OF S FLORIDA
TAMPA FL 33620
U.S.A.

FANG CHENG
DEPT OF ASTRONOMY
NANJING UNIVERSITY
NANJING
PEOPLE'S REPUBLIC CHINA

FANG LI-ZHI
ASTROPHYSICS RES.DIV
UNIV OF SCIENCE&TECH
HEFEI
ANHUI PROVINCE
PEOPLE'S REPUBLIC CHINA

FANSELOW JOHN LYMAN
JET PROPULSION LAB
M/S 264-748
4800 OAK GROVE DR
PASADENA CA 91109
U.S.A.

FANTI CARLA GIOVANNINI
LAB DI RADIOASTRONO-
MIA
VIA IRNERIO 46
I-40126 BOLOGNA
ITALY

FANTI R DR
PHYSICAL INST
UNIV OF BOLOGNA
VIA IRNERIO 46
I-40126 BOLOGNA
ITALY

FARAGGIANA ROSANNA PROF
OSSERVATORIO
ASTRONOMICO
VIA TIEPOLO 11
I-34131 TRIESTE
ITALY

FARINELLA PAOLO DR
IST DI MATEMATICA
L TONELLI UN DI PISA
PIAZZA D CAVALIERI 2
I-56100 PISA
ITALY

FARNIK FRANTISEK
ASTRONOMICAL INST
CZECH ACADEMY OF SC
ONDREJOV OBSERVATORY
25165 ONDREJOV
CZECHOSLOVAKIA

FATCHIKHIN NIKOLAJ V DR
PULKOVO OBSERVATORY
196140 LENINGRAD
U.S.S.R.

FAUCHER P DR
OBS DE NICE
B P 252
F-06007 NICE CEDEX
FRANCE

FAULKNER DONALD J DR
MT STROMLO
OBSERVATORY
WODEN PO ACT 2606
AUSTRALIA

FAULKNER JOHN PROF
UNIVERSITY OF
CALIFORNIA
SANTA CRUZ CA 95064
U.S.A.

FAWELL DEREK R DR
UNIVERSITY OF LONDON
OBSERVATORY
MILL HILL PARK
LONDON NW7 2QS
U.K.

FAY THEODORE D DR
MAIL STOP 19
TELEDYNE BROWN ENG
CUMMINGS RES PARK
HUNTSVILLE AL 35807
U.S.A.

FAZIO GIOVANNI G DR
CENTER FOR
ASTROPHYSICS HCO SAO
60 GARDEN ST
CAMBRIDGE MA 02138
U.S.A.

FEAST MICHAEL W DR
S A A O
P O BOX 9
OBSERVATORY
CAPE 7935
SOUTH AFRICA

FEAUTRIER NICOLE DR
OBSERVATOIRE
DE MEUDON
F-92190 MEUDON
FRANCE

FECHTIG HUGO DR
SANSERWEG 3
D-6906 LEIMEN
GERMANY, F.R.

FEDERMAN STEVEN ROBERT
DEPT OF ASTRONOMY
UNIV OF TEXAS
AUSTIN TX 78712
U.S.A.

FEDOROV E P DR
MAIN ASTRON OBS
UKRAINIAN ACAD OF SC
GOLOSEEVO
252127 KIEV
U.S.S.R.

FEDOROVA RIMMA T DR
ASTRONOMICAL OBS
327000 NIKOLAEV
U.S.S.R.

FEGAN DAVID J DR
PHYSICS DEPT
UNIV COLLEGE
BELFIELD
DUBLIN 4
IRELAND

FEHRENBACH CHARLES PROF
OBSERVATOIRE DE
HAUTE PROVENCE
ST MICHEL L OBS
F-04300 FORCALQUIER
FRANCE

FEIBELMAN WALTER A DR
NASA-GSFC
CODE 685
GREENBELT MD 20771
U.S.A.

FEINSTEIN A DR
OBS ASTRONOMICO
1900 LA PLATA
ARGENTINA

FEISSEL MARTINE
OBSERVATOIRE PARIS
61 AVE DE
L-OBSERVATOIRE
F-75014 PARIS
FRANCE

FEITZINGER JOHANNES V PHD
ASTRONOMISCHES INST
RUHR UNIVERSITAET
POSTFACH 102148
D-4630 BOCHUM
GERMANY, F.R.

FEIX GERHARD DR
RUHR UNIVERSITAET
DEPT XII
POSTFACH 102 148
D-4630 BOCHUM
GERMANY, F.R.

FEJES I TVAN DR
FOMI SATELLITE
GEODETIC OBSERVATORY
PO BOX 546
1373 BUDAPEST
HUNGARY

FEKEL FRANK JR
6949 KNIGHTHOOD LANE
COLUMBIA MD 21045
U.S.A.

FELDMAN PAUL A DR
HERZBERG INST ASTROD
NAT RES COUNC CANADA
100 SUSSEX DR
OTTAWA ONT K1A 0R6
CANADA

FELDMAN PAUL DONALD DR
PHYSICS DEPT
JOHNS HOPKINS UNIV
BALTIMORE MD 21218
U.S.A.

FELDMAN U DR
DEPT OF PHYSICS AND
ASTRONOMY
TEL-AVIV UNIV
RAMAT-AVIV
ISRAEL

FELDMAN URI
CODE 4174
NAVAL RESEARCH LAB
WASHINGTON DC 20375
U.S.A.

FELENBOK PAUL DR
OBS DE MEUDON
F-92190 MEUDON
FRANCE

FELLGETT PETER B PROF
DEPT OF CYBERNETICS
3 EARLY GATE
WHITEKNIGHTS
READING RG6 2AL
U.K.

FELLI MARCELLO DR
OSSERVATORIO ARCETRI
LARGO FERMI 5
I-50125 FIRENZE
ITALY

FELTEN JAMES E DR
CODE 660
GODDARD SPACE FL CTR
GREENBELT MD 20771
U.S.A.

FENG KE-JIA
DEPT OF ASTRONOMY
BEIJING NORMAL
UNIVERSITY
BEIJING
PEOPLE'S REPUBLIC CHINA

FENKART ROLF P DR
ASTRONOMISCHES INST
UNIVERSITAET BASEL
VENUSSTR 7
CH-4102 BINNINGEN
SWITZERLAND

FENTON K B DR
DEPT OF PHYSICS
UNIV OF TASMANIA
HOBART TASM 7001
AUSTRALIA

FERETTI LUIGINA
ISTITUTO DI
RADIOASTRONOMIA
VIA IRNERIO 46
I-40126 BOLOGNA
ITALY

FERLAND GARY JOSEPH
DEPT OF PHYS & ASTRO
UNIV OF KENTUCKY
LEXINGTON KY 40506
U.S.A.

FERNANDEZ JEAN-CLAUDE DR
OBSERVATOIRE DE NICE
B P 252
F-06007 NICE CEDEX
FRANCE

FERNANDEZ-FIGUEROA M J DR
ASTROFIS FAC DE FIS
UNIV COMPLUTENSE
CIUDAD UNIVERSITARIA
MADRID 3
SPAIN

FERNIE J DONALD PROF
DAVID DUNLAP OBS
P O BOX 360
RICHMOND HILL
ONT L4C 4Y6
CANADA

FERRARI ATTILIO DR
ISTITUTO DI FISICA
GENERALE DELL.UNIV.
CR M D AZEGLIO 46
I-10125 TORINO
ITALY

FERRARI D OCCHIEPPO K DR
OESTERREICHISCHE AK
DER WISSENSCHAFTEN
DR-I-SEIPEL-PLATZ2
A-1110 WIEN
AUSTRIA

FERRARI TONIOLO MARCO
IST ASTROFISICA
SPAZIALE
CASELLA POSTALE 67
I-00044 FRASCATI
ITALY

FERRAZ-MELLO S PROF DR
UNIV DE SAO PAULO
DEPT ASTRONOMIA
CAIXA POSTAL 30627
01000-SAO PAULO SP
BRAZIL

FERRER OSVALDO EDUARDO DR
OBS ASTRONOMICO
PASEO DEL BOSQUE
1900-LA PLATA
ARGENTINA

FERRIN IGNACIO
UNIV DE LOS ANDES
FACULDAD DE CIENCIAS
DEPTO DE FISICA
MERIDA
VENEZUELA

FERRINI FEDERICO
IST DI ASTRONOMIA
UNIVERSITA DI PISA
VIA BUONARROTI 2
I-56100 PISA
ITALY

FESTOU M DR
SERVICE D'AERONOMIE
CNRS B P NO 3
F-91370 VERRIERES
LE BUISSON
FRANCE

FEYNMAN JOAN DR
DEPT OF PHYSICS
BOSTON COLLEGE
CHESTNUT HL MA 02167
U.S.A.

FIALA ALAN D DR
U S NAVAL OBS
WASHINGTON DC 20390
U.S.A.

FICARRA A MR
LAB NAZIONALE DI
RADIOASTRONOMIA
VIA IRNERIO 46
I-40126 BOLOGNA
ITALY

FICHTEL CARL E DR
GODDARD SPACE FLIGHT
CENTER
CODE 662
GREENBELT MD 20771
U.S.A.

FIELD DAVID
SCHOOL OF CHEMISTRY
CANTOCKS CLOSE
BRISTOL BS8 1TS
U.K.

FIELD GEORGE B PROF
CENTER FOR
ASTROPHYSICS
60 GARDEN ST
CAMBRIDGE MA 02138
U.S.A.

FIELDER GILBERT DR
LUNAR AND PLANETARY
UNIT E S DEPT
LANCASTER UNIVERSITY
LANCASTER LA1 4YR
U.K.

FINDLAY JOHN W DR
NAT RADIO ASTR OBS
EDGEMONT RD
CHARLOTTESVILLE
VA 22901
U.S.A.

FINK U DR
LUNAR AND PLANETARY
LABORATORY
UNIVERSITY ARIZONA
TUCSON AZ 85721
U.S.A.

FINN G D DR
INST FOR ASTRONOMY
2525 CORREA ROAD
HONOLULU HI 96822
U.S.A.

FINZI ARRIGO DR
DEPT OF MATHEMATICS
TECHNION I I T
HAIFA
ISRAEL

FIREMAN EDWARD L
ASTROPHYSICAL OBS
SMITHSONIAN INST
60 GARDEN ST
CAMBRIDGE MA 02138
U.S.A.

FIRMANI CLAUDIO A PROF
INSTITUTO DE
ASTRONOMIA UNAM
A P 70-264
04510 MEXICO DF
MEXICO

FIRNEIS FRIEDRICH J DR
OESTERR AKADEMIE DER
WISSENSCHAFTEN
FLEISCHMARKT 20/1/1
A-1010 WIEN
AUSTRIA

FIRNEIS MARIA G DR
INST FUR ASTRONOMIE
TUERKENSCHANZSTR 17
A-1180 WIEN
AUSTRIA

FIROR JOHN W DR
NATIONAL CTR FOR AT-
MOSPHERIC RESEARCH
P O BOX 3000
BOULDER CO 80703
U.S.A.

FISCHEL DAVID DR
CODE 671
GODDARD SPACE FLIGHT
CENTER
GREENBELT MD 20771
U.S.A.

FISCHER P LEANDER PROF
FREYUNG 6
A-1010 WIEN
AUSTRIA

FISCHER STANISLAV DR
ASTRONOMICAL INST
CSAV
BUDECSKA 6
120 23 PRAHA 2
CZECHOSLOVAKIA

FISHER J RICHARD
NRAO
P O BOX 2
GREEN BANK WV 24944
U.S.A.

FISHER PHILIP C
RUFFNER ASSOCIATES
P O BOX 7070
MENLO PARK CA 94025
U.S.A.

FISHER RICHARD R DR
HIGH ALTITUDE OBS
P O BOX 3000
BOULDER CO 80307
U.S.A.

FISHKOVA L M PROF
ABASTUMANI ASTROPHY-
SICAL OBSERVATORY
383762 ABASTUMANI
GEORGIA
U.S.S.R.

FISHMAN GERALD J
MSFC ASTROPHYSICS BR
SPACE SCIENCES LAB
MARSHALL SFC AL35812
U.S.A.

FITCH WALTER S DR
STEWARD OBSERVATORY
UNIV OF ARIZONA
TUCSON AZ 85721
U.S.A.

FITTON BRIAN DR
ASTRONOMY DIV
ESTEC
POSTBUS 299
2200 AG NOORDWYK
THE NETHERLANDS

FITZGERALD M PIM PROF
DEPT OF PHYSICS
UNIVERSITY WATERLOO
WATERLOO ONT N2L 3G1
CANADA

FIX JOHN D DR
DEPT OF PHYSICS AND
ASTRONOMY
UNIVERSITY OF IOWA
IOWA CITY IA 52240
U.S.A.

FLANNERY BRIAN PAUL DR
CTR FOR ASTROPHYSICS
60 GARDEN STREET
CAMBRIDGE MA 02138
U.S.A.

FLEISCHER ROBERT DR
1733 CHURCH ST N W
WASHINGTON DC 20036
U.S.A.

FLETCHER J MURRAY
DOMINION ASTROPH OBS
5071 W SAANICH ROAD
RR 5
VICTORIA BC V8X 4M6
CANADA

FLIEGEL HENRY F
JPL
4800 OAK GROVE DRIVE
PASADENA CA 91103
U.S.A.

FLOQUET MICHELE DR
OBS DE PARIS
DEPEG
F-92190 MEUDON
FRANCE

FLORENTIN-NIELSEN R M SC
COPENHAGEN UNIVERSI-
TY OBSERVATORY
BRORFELDE
DK-4340 TOLLOSE
DENMARK

FLORIDES PETROS S PROF
SCHOOL OF MATHEMA-
TICS TRINITY COLLEGE
DUBLIN 2
IRELAND

FLORSCH ALPHONSE DR
OBSERVATOIRE
DE STRASBOURG
11 RUE L UNIVERSITE
F-67000 STRASBOURG
FRANCE

FLOWER DAVID R DR
DEPT OF PHYSICS
UNIVERSITY OF DURHAM
DURHAM DH1 3LE
U.K.

FOGARTY WILLIAM G DR
2511 PLOVER RD
WISCONSIN RAPIDS
WI 54494
U.S.A.

FOGH OLSEN H J
COPENHAGEN UNIV
OBSERVATORY
BRORFELDE
DK-4340 TOLLOSE
DENMARK

FOKKER AAD D DR
STERREWACHT
SONNENBORGH
ZONNENBURG 2
3512 NL UTRECHT
THE NETHERLANDS

FOLTZ CRAIG
DEPT OF ASTRONOMY
174 WEST 18TH AVE
COLUMBUS OH 43210
U.S.A.

FOMALONT EDWARD B DR
NATIONAL RADIO
ASTRONOMY OBS
P O BOX O
SOCORRO NM 87801
U.S.A.

FOMENKO ALEXANDER F DR
SPECIAL ASTROPHYS OB
NIZHNIJ ARKHYZ
357140 STAVROPOLSKIJ
KRAJ
U.S.S.R.

FOMIN VALERY A DR
PULKOVO OBSERVATORY
196140 LENINGRAD
U.S.S.R.

FOMINOV ALEXANDR M DR
INST OF THEORETICAL
ASTRONOMY
10 KUTUZOV QUAY
192187 LENINGRAD
U.S.S.R.

FONG RICHARD
DEPT OF PHYSICS
UNIV OF DURHAM
SOUTH ROAD
DURHAM DH1 3LE
U.K.

FONTAINE GILLES DR
DEPT OF PHYSICS
UNIV OF MONTREAL
P O BOX 6128
MONTREAL PQ H3C 3J7
CANADA

FORBES ERIC G DR
HISTORY DEPT
EDINBURGH UNIVERSITY
50 GEORGE SQUARE
EDINBURGH EH8 9JY
U.K.

FORBES J E DR
P O BOX 88120
INDIANAPOLIS IN46208
U.S.A.

FORD HOLLAND C ASSOC PROF
SPACE TELESCOPE SC
HOMEWOOD CAMPUS
THE JOHNS HOPKINS UN
BALTIMORE MD 21218
U.S.A.

FORD W K JR DR
TERRESTRIAL MAGNET
CARNEGIE INSTITUTION
5241 BROAD BR RD N W
WASHINGTON DC 20015
U.S.A.

FORMAN WILLIAM RICHARD DR
SMITHSONIAN OBS
60 GARDEN STREET
CAMBRIDGE MA 02138
U.S.A.

FORREST WILLIAM JOHN
DEPT OF PHYSICS
AND ASTRONOMY
UNIV OF ROCHESTER
ROCHESTER NY 14627
U.S.A.

FORSTER JAMES RICHARD DR
CSIRO DIVISION OF
RADIOPHYSICS
P O BOX 76
EPPING NSW 2121
AUSTRALIA

FORT BERNARD P DR
OBSERVATOIRE
DE MEUDON
F-92190 MEUDON
FRANCE

FORT DAVID NORMAN DR
733 LONSDALE ROAD
OTTAWA K1K 0J9
CANADA

FORTE JUAN CARLOS DR
OBS ASTRONOMICO
PASEO DEL BOSQUE
1900-LA PLATA
ARGENTINA

FORTI GIUSEPPE DR
OSSERVATORIO ARCETRI
LARGO FERMI 5
I-50125 FIRENZE
ITALY

FORTINI TERESA DR
OSSERVATORIO
ASTRONOMICO
I-00136 ROMA
ITALY

FOSBURY ROBERT A E DR
ROYAL GREENWICH
OBSERVATORY
HAILSHAM
E SUSSEX BN27 1RP
U.K.

FOSSAT ERIC DR
OBSERVATOIRE DE NICE
BP 252
F-06007 NICE CEDEX
FRANCE

FOUKAL PETER V DR
CENTER FOR
ASTROPHYSICS
60 GARDEN ST
CAMBRIDGE MA 02138
U.S.A.

FOWLER WILLIAM A PROF
CALTECH 106 38
PASADENA CA 91125
U.S.A.

FOX KENNETH DR
DEPT PHYSICS ASTRON
UNIV OF TENNESSEE
503 PHYSICS
KNOXVILLE TN 37916
U.S.A.

FOX W E MR
40 WINDSOR ROAD
NEWARK NOTTINGHAMS
U.K.

FOY RENAUD DR
OBSERVATOIRE
DE MEUDON
F-92190 MEUDON
FRANCE

FRACASSINI MASSIMO PROF
OSSERVATORIO
ASTRONOMICO DI BRERA
VIA BRERA 28
I-20121 MILANO
ITALY

FRACASTORO MARIO G PROF
OSSERVATORIO ASTRONO
MICO DI TORINO
PINO TORINESE
I-10025 TORINO
ITALY

FRANDSEN SOEREN PROF
INST OF ASTRONOMY
UNIVERSITY OF AARHUS
DK-8000 AARHUS C
DENMARK

FRANKLIN FRED A DR
PLANETARY SC DIV
CTR F ASTROPHYSICS
60 GARDEN ST
CAMBRIDGE MA 02138
U.S.A.

FRANTSMAN YU L DR
RADIOASTROPHYSICAL
OBSERVATORY LATVIAN
ACADEMY OF SCIENCES
226026 RIGA
U.S.S.R.

FRANZ OTTO G DR
LOWELL OBSERVATORY
P O BOX 1269
FLAGSTAFF AZ 86002
U.S.A.

FRASER C W DR
UNIV OBSERVATORY
BUCHANAN GARDENS
ST ANDREWS
FIFE KY16 9LZ
U.K.

FRATER ROBERT H DR
UNIVERSITY OF SYDNEY
SYDNEY N S W 2006
AUSTRALIA

FRAZIER EDWARD N DR
AEROSPACE
CORPORATION
P O BOX 92957
LOS ANGELES CA 90009
U.S.A.

FREDGA KERSTIN PROF
SWEDISH BOARD FOR
SPACE ACTIVITIES
TRITONVAEGEN 27
S-171 54 SOLNA
SWEDEN

FREDRICK LAURENCE W PROF
LEANDER MCCORMICK
OBSERVATORY BOX 3818
CHARLOTTESVILLE
VA 22903
U.S.A.

FREEMAN KENNETH C DR
MT STROMLO
OBSERVATORY
WODEN PO ACT 2606
AUSTRALIA

FREIESLEBEN H C DR
FLORENTINER STR 20
APP 6120
D-7000 STUTTGART 75
RIEDENBERG
GERMANY, F.R.

FREIRE FERRERO R G DR
OBS ASTRONOMIQUE
11 R DE L'UNIVERSITE
F-67000 STRASBOURG
FRANCE

FREITAS MOURAO R R DR
OBS NACIONAL
R GENERAL BRUCE 586
SAO CRISTOVAO
RIO DE JANEIRO
BRAZIL

FRESA A PROF
VIA PORTO 23
I-84100 SALERNO
ITALY

FRESNEAU A DR
OBS DE STRASBOURG
11 R DE L'UNIVERSITE
F-67000 STRASBOURG
FRANCE

FRICKE KLAUS DR
UNIVERSITAETSSTERNW
UNIV GOETTINGEN
GEISMARLANDSTR 11
D-3400 GOETTINGEN
GERMANY, F.R.

FRICKE WALTER PROF DR
DIRECTOR ASTRONOMI-
SCHES RECHENINSTITUT
12-14 MOENCHHOFSTR
D-6900 HEIDELBERG
GERMANY, F.R.

FRIED JOSEF WILHELM DR
MAX-PLANCK-INSTITUT
FUER ASTRONOMIE
KOENIGSTUHL
D-6900 HEIDELBERG 1
GERMANY, F.R.

FRIEDEMANN CHRISTIAN DR
UNIVERSITAETS
STERNWARTE JENA
SCHILLERGAESSCHEN 2
DDR-69 JENA
G.D.R.

FRIEDJUNG MICHAEL DR
INST D ASTROPHYSIQUE
98 BIS BLVD ARAGO
F-75014 PARIS
FRANCE

FRIEDLANDER MICHAEL PROF
DEPT OF PHYSICS
WASHINGTON
UNIVERSITY
ST LOUIS MO 63130
U.S.A.

FRIEDMAN HERBERT DR
US NAVAL RESEARCH
LAB CODE 7100
WASHINGTON DC 20375
U.S.A.

FRINGANT ANNE-MARIE DR
OBS DE PARIS
BAT 8
61 AV L'OBSERVATOIRE
F-75014 PARIS
FRANCE

FRISCH HELENE DR
OBSERVATOIRE DE NICE
BP 252
F-06007 NICE-CEDEX
FRANCE

FRISCH PRISCILLA
LAB FOR ASTROPHYSICS
& SPACE RESEARCH
933 E 56TH STREET
CHICAGO IL 60637
U.S.A.

FRISCH URIEL DR
OBSERVATOIRE DE NICE
B P 252
F-06007 NICE-CEDEX
FRANCE

FRITZOVA-SVESTKA L DR
147 DOPPERSTRAAT
3752 JC BUNSCHOTEN
THE NETHERLANDS

FROESCHLE CHRISTIANE D DR
OBSERVATOIRE DE NICE
BP 252
F-06007 NICE CEDEX
FRANCE

FROESCHLE CLAUDE DR
OBSERVATOIRE DE NICE
B P 252
F-06007 NICE-CEDEX
FRANCE

FROESCHLE MICHEL DR
CERGA
AVENUE COPERNIC
F-06130 GRASSE
FRANCE

FROGEL JAY A DR
CERRO TOLOLO INTER-
AMERIC OBSERVATORY
CASILLA 63 D
LA SERENA
CHILE

FROLOV M S DR
ASTRONOMICAL COUNCIL
USSR ACADEMY OF SCI
PYATNITSKAYA UL 48
109017 MOSCOW
U.S.S.R.

FROST KENNETH J DR
NASA GSFC
CODE 684
GREENBELT MD 20771
U.S.A.

FRYE GLENN M PROF
PHYS DEPT ROCK BLDG
CASE WESTERN RESERVE
UNIVERSITY
CLEVELAND OH 44106
U.S.A.

FU DELIAN
BEIJING ASTRONOMICAL
OBSERVATORY
ACADEMIA SINICA
BEIJING
PEOPLE'S REPUBLIC CHINA

FU-SHONG KUO
DEPT OF PHYSICS
INST PHYS ASTRONOMY
NAT CENTRAL UNIV
CHUNG LI
TAIWAN

FUCHS JOSEPH PROF DR
GOLDSCHLAGSTR 185-9
A-1140 WIEN
AUSTRIA

FUENMAYOR FRANCISCO J DR
UNIV DE LOS ANDES
FACULDAD DE CIENCIAS
DEPTO DE FISICA
MERIDA
VENEZUELA

FUERST ERNST DR
MPI FUER
RADIOASTRONOMIE
AUF DEM HUEGEL 69
D-5300 BONN
GERMANY, F.R.

FUJIMOTO MASA-KATSU DR
TOKYO ASTRONOMICAL
OBSERVATORY
TOKYO 181
JAPAN

FUJIMOTO MASAYUKI DR
NIIGATA UNIVERSITY
1-7-2 SAIJYO-MACHI
JYOETSU
NIIGATA 943
JAPAN

FUJIMOTO MITSUAKI DR
DEPT OF PHYSICS
NAGOYA UNIVERSITY
NAGOYA 464
JAPAN

FUJITA YOSHIO PROF
DEPT OF ASTRONOMY
UNIVERSITY OF TOKYO
BUNKYO KU
TOKYO 113
JAPAN

FUJIWARA AKIRA DR
DEPT OF PHYSICS
KYOTO UNIVERSITY
KYOTO 606
JAPAN

FUKUI TAKAO DR
DEPT OF LIBERAL ARTS
DOKKYO UNIVERSITY
SAKAE-MACHI 600
SOKA SAITAMA
JAPAN

FUKUI YASUO DR
DEPT OF PHYSICS
NAGOYA UNIVERSITY
FUROCHO CHIKUSAKU
NAGOYA 464
JAPAN

FULCHIGNONI MARCELLO PROF
LABORATORIO DI
ASTROFISICA SPAZIALE
E.N.R.REP PLANETOL
I-00044 FRASCATI
ITALY

FURENLID K INGEMAR DR
DEPT PHYS AND ASTRON
GEORGIA STATE UNIV
ATLANTA GA 30303
U.S.A.

FURNISS IAN
DEPT OF PHYS & ASTRO
UNIV COLLEGE LONDON
GOWER STREET
LONDON WC1E 6BT
U.K.

FURSENKO M A DR
INSTITUTE OF THEORE-
TICAL ASTRONOMY
192187 LENINGRAD
U.S.S.R.

FUSI PECCI FLAVIO
OSS ASTRONOMICO
UNIVERSATORIO
CASELLA POSTALE 596
I-40100 BOLOGNA
ITALY

GABRIEL ALAN H
RUTHERFORD APPLETON
LAB SPACE/ASTROPHYS
CHILTON DIDCOT
OXFORDSHIRE OX11 0QX
U.K.

GABRIEL MAURICE R DR
INST D ASTROPHYSIQUE
UNIVERSITE DE LIEGE
B-4200 COINTE OUGREE
BELGIUM

GAHM GOESTA F DR
STOCKHOLM
OBSERVATORY
S-133 00
SALTSJOEBADEN
SWEDEN

GAIGNEBET J DR
CERGA
AVE COPERNIC
F-06130 GRASSE
FRANCE

GAIL H P DR
INST THEOR ASTROPHYS
DER UNIVERSITAET
IM NEUENHEIMER F 294
D-6900 HEIDELBERG 1
GERMANY, F.R.

GAIZAUSKAS VICTOR DR
HERZBERG INST OF
ASTROPHYSICS
NAT RES COUNC CANADA
OTTAWA ONT K1A 0R6
CANADA

GALAL A A DR
HELWAN OBSERVATORY
HELWAN NEAR CAIRO
ARAB REP. OF EGYPT

GALEOTTI PIERO PROF
LABORATORIO DI
COSMO-GEOFISICA
CORSO FIUME 4
I-10133 TORINO
ITALY

GALIBINA I V DR
INSTITUTE OF THEORE-
TICAL ASTRONOMY
192187 LENINGRAD
U.S.S.R.

GALKIN LEONID S A DR
CRIMEAN ASTROPHYSI-
CAL OBSERVATORY
334413 PO NAUCHNY
CRIMEA
U.S.S.R.

GALLAGHER III JOHN S DR
OBSERVATORY
UNIVERSITY OF
ILLINOIS
URBANA IL 61801
U.S.A.

GALLET R M
964 7TH STREET
BOULDER CO 80302
U.S.A.

GALLETTA GIUSEPPE DR
OSSERVATORIO ASTRO-
NOMICO
I-35100 PADOVA
ITALY

GALLOUET LOUIS DR
OBSERVATOIRE
DE PARIS
61 AVE OBSERVATOIRE
F-75014 PARIS
FRANCE

GALLOWAY DAVID DR
MPI FUER ASTROPHYSIK
KARL-SCHWARZSCHILD-
STR 1
D-8046 GARCHING B M
GERMANY, F.R.

GALPERIN YU I DR
SPACE RESEARCH
INSTITUTE
USSR ACADEMY OF SCI
117810 MOSCOW
U.S.S.R.

GALT JOHN A DR
DOMINION RADIO
ASTROPHYSICAL OBS
BOX 248
PENTICTON BC V2A 6K3
CANADA

GAMALELDIN ABDULLA I DR
HELWAN OBSERVATORY
HELWAN-CAIRO
ARAB REP. OF EGYPT

GAMMELGAARD PETER PROF
ASTRONOMISK INSTITUT
AARHUS UNIVERSITET
DK-8000 AARHUS
DENMARK

GAPOSCHKIN E M DR
55 FARMCREST AVE
LEXINGTON MA 02173
U.S.A.

GAPOSHKIN SERGEI ILLAR DR
HARVARD COLLEGE
OBSERVATORY
60 GARDEN ST
CAMBRIDGE MA 02138
U.S.A.

GARCIA-PELAYO JOSE DR
PASEO DE LA RONDA
92 6C
GRANADA
SPAIN

GARDNER FRANCIS F DR
CSIRO DIVISION OF
RADIOPHYSICS
P O 76
EPPING NSW 2121
AUSTRALIA

GARFINKEL BORIS DR
YALE UNIVERSITY
OBSERVATORY
NEW HAVEN CT 06520
U.S.A.

GARLICK GEORGE F DR
267 SOUTH BELOIT AVE
LOS ANGELES CA 90049
U.S.A.

GARMANY CATHERINE D DR
JILA
UNIV OF COLORADO
BOULDER CO 80309
U.S.A.

GARMIRE GORDON P PROF
320-47 CALTECH
PASADENA CA 91125
U.S.A.

GARNIER ROBERT ING
OBSERVATOIRE DE LYON
F-69230 ST-GENIS-LAVAL
FRANCE

GARRISON ROBERT F PROF
DAVID DUNLAP OBS
P O BOX 360
RICHMOND HILL
ONT L4C 4Y6
CANADA

GARSTANG ROY H PROF
JILA UNIVERSITY
COLORADO
BOULDER CO 80309
U.S.A.

GARTON W R S PROF
BLACKETT LABORATORY
IMPERIAL COLLEGE
LONDON SW7 2BZ
U.K.

GARZOLI S L DR
M T DE ALVEAR 2412
4 H
BUENOS AIRES
ARGENTINA

GASCOIGNE S C B DR
MT STROMLO
OBSERVATORY
WODEN PO ACT 2606
AUSTRALIA

GASKA STANISLAW DR
INSTITUTE ASTRONOMY
UL CHOPINA 12-18
87-100 TORUN
POLAND

GATEWOOD GEORGE DIRECTOR
ALLEGHENY OBS
OBSERVATORY STATION
PITTSBURG PA 15214
U.S.A.

GAUR V P
UTTAR PRADESH
STATE OBSERVATORY
NAINITAL UP
INDIA

GAUSS F STEPHEN
US NAVAL OBSERVATORY
WASHINGTON DC 20390
U.S.A.

GAUSTAD JOHN E PROF
DEPT OF ASTRONOMY
UNIVERSITY OF
CALIFORNIA
BERKELEY CA 94720
U.S.A.

GAUTIER DANIEL
OBSERVATOIRE
DE MEUDON
F-92190 MEUDON
FRANCE

GAVIOLA ENRIQUE PROF DR
INSTITUTO BALSEIRO
8400 BARILOCHE
ARGENTINA

GAVRILOV I V DR
MAIN ASTRONOMICAL
OBS OF THE UKRAINIAN
ACAD OF SCIENCES
252127 KIEV
U.S.S.R.

GAY JEAN DR
CERGA
AVE COPERNIC
F-06130 GRASSE
FRANCE

GEAKE JOHN E DR
PHYSICS DEPT
U M I S T
MANCHESTER M60 1QD
U.K.

GEBBIE KATHARINE B DR
JILA
UNIVERSITY OF
COLORADO
BOULDER CO 80302
U.S.A.

GEBLER KARL-HEINZ DR
RADIOASTRONOMISCHES
INST DER UNIV BONN
AUF DEM HUEGEL 71
D-5300 BONN 1
GERMANY, F.R.

GEHRELS TOM PROF
LUNAR LABORATORY
UNIVERSITY OF
ARIZONA
TUCSON AZ 85721
U.S.A.

GEHREN THOMAS PH D
INST ASTRON/ASTROPH
UNIV MUENCHEN
SCHEINERSTRASSE 1
D-8000 MUENCHEN 80
GERMANY, F.R.

GEHRZ ROBERT DOUGLAS DR
DEPT PHYSICS ASTRON
UNIV OF WYOMING
UNIV STA BOX 3905
LARAMIE WY 82071
U.S.A.

GEISS JOHANNES PROF
PHYS INST
UNIVERSITAET BERN
SIDLERSTR 5
CH-3012 BERN
SWITZERLAND

GELFREJKH GEORGIJ B DR
PULKOVO OBSERVATORY
196140 LENINGRAD
U.S.S.R.

GELLER MARGARET JOAN
CENTER FOR ASTROPHYS
60,GARDEN STREET
CAMBRIDGE MA 02138
U.S.A.

GENKIN IGOR L PROF DR
PHYSICS FACULTY
KAZAKH STATE UNIV
KOMSOMOLSKAYA 96
480012 ALMA ATA
U.S.S.R.

GENT HUBERT MR
PROSPECT HOUSE
SHERWOOD LANE
LOWERWICK
WORCS WR2 4NX
U.K.

GENZEL REINHARD DR
DEPT OF PHYSICS
UNIV OF CALIFORNIA
563 BIRGE HALL
BERKELEY CA 94720
U.S.A.

GEORGELIN YVON P DR
OBSERVATOIRE
DE MARSEILLE
2 PLACE LE VERRIER
F-13004 MARSEILLE
FRANCE

GEORGELIN YVONNE M DR
OBS DE MARSEILLE
2 PLACE LE VERRIER
F-13004 MARSEILLE
FRANCE

GERARD ERIC DR
OBSERVATOIRE DE
MEUDON DEPT DE RADIO
ASTRONOMIE
F-92190 MEUDON
FRANCE

GERBAL D DR
OBS DE MEUDON
F-92190 MEUDON
FRANCE

GERBALDI MICHELE DR
INST D ASTROPHYSIQUE
98 BIS BD ARAGO
F-75014 PARIS
FRANCE

GERGELY TOMAS ESTEBAN DR
ASTRONOMY PROGRAM
UNIV OF MARYLAND
COLLEGE PK MD 20742
U.S.A.

GERLEI OTTO
HELIOPHYSICAL
OBSERVATORY
4010 DEBRECEN
HUNGARY

GEROLA HUMBERTO DR
IBM THOMAS J.WATSON
RESEARCH CENTER
P O BOX 218
YORKTWN HTS NY 10598
U.S.A.

GEROYANNIS B S DR
DEPT OF ASTRONOMY
UNIV OF PATRAS
PATRAS
GREECE

GERSHBERG R E DR
CRIMEAN ASTROPHYSI-
CAL OBSERVATORY
NAUCHNY
334413 CRIMEA
U.S.S.R.

GETMANTSEV G G DR
RADIOPHYSICAL
RESEARCH INSTITUTE
603600 GORKIJ
U.S.S.R.

GEYER EDWARD H PROF DR
OBSERVATORIUM
HOHER LIST
D-5568 DAUN EIFEL
GERMANY, F.R.

GEZARI DANIEL YSA DR
NASA-GSFC
CODE 693 2
GREENBELT MD 20771
U.S.A.

GHIGO FRANCIS D DR
DEPT OF ASTRONOMY
UNIV OF MINNESOTA
116 CHURCH ST SE
MINNEAPOLIS MN 55455
U.S.A.

GHOBROS RUSHDY AZER DR
HELWAN OBSERVATORY
HELWAN NR CAIRO
ARAB REP. OF EGYPT

GIACAGLIA G E O PROF
ESCOLA POLITECHNICA
U S P
CAIXA POSTAL 8174
SAO PAULO
BRAZIL

GIACCONI RICCARDO PROF
DIRECTOR SPACE TELE
SCOP SCIENCE INST
HOMEWOOD CAMPUS
BALTIMORE MD 21218
U.S.A.

GIACHETTI RICCARDO DR
VIA BOLOGNESE 98
I-50100 FIRENZE
ITALY

GIANNONE PIETRO PROF
OSSERVATORIO
ASTRONOMICO
V D PARCO MELLINI 84
I-00136 ROMA
ITALY

GIANNUZZI MARIA A DR
PIAZZA GRAMSCI 4
I-00041 ALBANO
LAZIALE
ITALY

GIBSON DAVID MICHAEL DR
PHYSICS DEPT
NM INST MINING TECHN
CAMPUS STATION
SOCORRO NM 87801
U.S.A.

GICLAS H L MR
LOWELL OBSERVATORY
BOX 1269
FLAGSTAFF AZ 86002
U.S.A.

GIERASCH PETER J DR
318 SPACE SC BLDG
CORNELL UNIVERSITY
ITHACA NY 14853
U.S.A.

GIEREN WOLFGANG P DR
DEPTO DE FISICA
UNIV DE LOS ANDES
APARTADO AEREO 4976
BOGOTA
COLOMBIA

GIESE RICHARD H PROF
RUHR UNIVERSITAET
BEREICH EXTRATERRE-
STRISCHE PHYSIK
D-4630 BOCHUM
GERMANY, F.R.

GIESEKING FRANK DR
OBS HOHER LIST
UNIV STERNWARTE BONN
D-5568 DAUN
GERMANY, F.R.

GIETZEN JOSEPH W
ROYAL GREENWICH
OBSERVATORY
HAILSHAM
EAST SUSSEX BN27 1RP
U.K.

GILLILAND RONALD LYNN
HIGH ALTITUDE OBS
NCAR P O BOX 3000
BOULDER CO 80307
U.S.A.

GILLINGHAM PETER DR
ANGLO AUSTRALIAN OBS
PRIVATE BAG
COONABARABRAN
NSW 2857
AUSTRALIA

GILMAN PETER A DR
ADVANCED STUDY PROGR
NAT CTR FOR ATM RES
BOX 3000
BOULDER CO 80307
U.S.A.

GILMORE ALAN C MR
P O BOX 20
LAKE TEKAPO
NEW ZEALAND

GILMORE GERARD FRANCIS
ROYAL OBSERVATORY
BLACKFORD HILL
EDINBURGH EH9 3HJ
U.K.

GILRA DAYA P DR
KAPTEYN ASTRON INST
UNIVERSITY GRONINGEN
P O BOX 800
9700 AV GRONINGEN
THE NETHERLANDS

GINGERICH OWEN PROF
CTR FOR ASTROPHYSICS
CAMBRIDGE MA 02138
U.S.A.

GINGOLD ROBERT ARTHUR DR
MOUNT STROMLO &
SIDING SPRING OBS
PRIVATE BAG
WODEN P O ACT 2606
AUSTRALIA

GINZBURG VITALY L PROF
P N LEBEDEV PHY INST
LENINSKY PROSPECT 53
117924 GSP
MOSCOW B 333
U.S.S.R.

GIOIA ISABELLA M DR
LAB RADIOASTRONOMIA
VIA IRNERIO 46
I-40126 BOLOGNA
ITALY

GIOVANELLI RONALD G DR
NATIONAL MEASUREMENT
LABORATORY
PO BOX 218
LINDFIELD NSW 2070
AUSTRALIA

GIOVANNELLI FRANCO DR
LAB ASTROFISICO SPAZ
P O B 67
I-00044 FRASCATI
ITALY

GIOVANNELLI RICCARDO DR
ARECIBO OBS
P O BOX 995
ARECIBO PR 00612
U.S.A.

GIOVANNINI FONTI C
OSS ASTRONOMICO
VIA ZAMBONI 33
BOLOGNA
ITALY

GIOVANNINI GABRIELE
ISTITUTO DI
RADIOASTRONOMIA
VIA IRNERIO 46
I-40126 BOLOGNA
ITALY

GIURICIN GIULIANO
VIA COLOGNA 45
I-34127 TRIESTE
ITALY

GLAGOLEVSKIJ JU V DR
SPECIAL ASTROPHYSI-
CAL OBSERVATORY
USSR ACADEMY OF SCI
357140 N ARKHYZ
U.S.S.R.

GLASER HAROLD DR
775 UNIVERSITY HALL
UNIV OF CALIFORNIA
SYSTEMWIDE ADMINISTR
BERKELEY CA 94720
U.S.A.

GLASPEY JOHN W DR
DEPT DE PHYSIQUE
UNIV DE MONTREAL
CP 6128 SUCC A
MONTREAL QUE H3C 3J7
CANADA

GLASS BILLY PRICE DR
DEPT OF GEOLOGY
UNIV OF DELAWARE
NEWARK DE 19711
U.S.A.

GLASS IAN S DR
S A ASTRONOMICAL OBS
P O BOX 9
OBSERVATORY CP 7935
SOUTH AFRICA

GLASSGOLD ALFRED E PROF
NEW YORK UNIVERSITY
PHYSICS DEPT
NEW YORK NY 10003
U.S.A.

GLEBOCKI ROBERT DR
INSTITUTE OF PHYSICS
UL WITA STWOSZA 57
80-308 GDANSK
POLAND

GLEBOVA NINA I DR
INST THEORET ASTRON
USSR ACAD OF SCS
10 KUTUZOV QUAY
192187 LENINGRAD
U.S.S.R.

GLEDHILL JOHN A PROF
DEPT OF PHYSICS
RHODES UNIVERSITY
P O BOX 94
GRAHAMSTOWN 6140
SOUTH AFRICA

GLEISSBERG WOLFGANG
PROF DR
BUCHENWEG 12
D-6374 OBERURSEL
GERMANY, F.R.

GLENCROSS WILLIAM M DR
DEPT OF PHYSICS-ASTR
UNIVERSITY COLLEGE
GOWER STREET
LONDON WC1E 6BT
U.K.

GLIESE WILHELM DR
ASTRONOMISCHES
RECHEN INSTITUT
MOENCHHOFSTR 12 14
D-6901 HEIDELBERG
GERMANY, F.R.

GLUSHNEVA I N DR
STERNBERG STATE AS-
TRONOMICAL INSTITUTE
117234 MOSCOW
U.S.S.R.

GNEDIN YURIJ N DR
LENINGRAD PHYSICAL
TECHNICAL INST
194021 LENINGRAD
U.S.S.R.

GNEVYSHEV MSTISLAV N DR
PULKOVO OBSERVATORY
196140 LENINGRAD
U.S.S.R.

GNEVYSHEVA RAISA S DR
PULKOVO OBSERVATORY
196140 LENINGRAD
U.S.S.R.

GODART ODON PROF
RUE DE CHATEAU 96
B-1288 ROUSVAL
BELGIUM

GODFREY PETER DOUGLAS DR
CHEMISTRY DEPT
MONASH UNIVERSITY
WELLINGTON ROAD
CLAYTON VIC 3168
AUSTRALIA

GODOLI GIOVANNI PROF
OSSERVATORIO ASTRO-
FISICO DI ARCETRI
LARGO ENRICO FERMI 5
I-50125 FIRENZE
ITALY

GODWIN J G DR
UNIV OBSERVATORY
SOUTH PARKS ROAD
OXFORD OX1 3RQ
U.K.

GOEBEL ERNST DR
INST F ASTRONOMIE
UNIVERSITAET WIEN
TUERKENSCHANZSTR 17
A-1180 WIEN
AUSTRIA

GOEBEL JOHN H DR
SPACE SCIENCE DIV
ASTROPHYS EXP BRANCH
NASA-AMES RES CTR
MOFFETT FLD CA 94035
U.S.A.

GOELBASI ORHAN DR
ISTANBUL UNIVERSITY
FACULTY OF SCIENCE
ISTANBUL
TURKEY

GOKDOGAN NUZHET PROF
UNIVERSITY
OBSERVATORY
UNIV ISTANBUL
ISTANBUL
TURKEY

GOKHALE MORESHWAR HARI DR
INDIAN INSTITUTE OF
ASTROPHYSICS
BANGALORE 560034
INDIA

GOKMEN TARIK ASSOC PROF
UNIVERSITY
OBSERVATORY
34 ISTANBUL
TURKEY

GOLAY MARCEL PROF
OBSERVATOIRE
DE GENEVE
CH-1290 SAUVERNY GE
SWITZERLAND

GOLD THOMAS PROF
SPACE SCIENCE BLDG
CORNELL UNIVERSITY
ITHACA NY 14853
U.S.A.

GOLDBACH C MME
INST D'ASTROPHYSIQUE
98 BIS BD ARAGO
F-75014 PARIS
FRANCE

GOLDBERG LEO PROF
KITT PEAK NATIONAL
OBSERVATORY
P O BOX 26732
TUCSON AZ 85726
U.S.A.

GOLDMAN MARTIN V
DEPT ASTRO-GEOPHYS
CAMPUS BOX 391
UNIV OF COLORADO
BOULDER CO 80309
U.S.A.

GOLDREICH P DR
CALTECH
PASADENA CA 91109
U.S.A.

GOLDSMITH DONALD W. DR.
2153 RUSSELL STREET
BERKELEY CA 94705
U.S.A.

GOLDSMITH PAUL F DR
DEPT PHYSICS ASTRON
GRC TOWER B RM 626
UNIV OF MASS
AMHERST MA 01003
U.S.A.

GOLDSMITH S. DR.
DEPT.OF PHYSICS AND
ASTRONOMY
TEL AVIV UNIVERSITY
TEL AVIV
ISRAEL

GOLDSTEIN RICHARD M DR
JET PROPULSION
LABORATORY
4800 OAK GROVE DRIVE
PASADENA CA 91011
U.S.A.

GOLDSTEIN SAMUEL J PROF
UNIVERSITY VIRGINIA
PO BOX 3818
CHARLOTTESVILLE
VA 22903
U.S.A.

GOLDSWORTHY FREDERICK A
PROF
SCHOOL MATHEMATICS
UNIVERSITY OF LEEDS
LEEDS LS2 9JT
U.K.

GOLDWIRE HENRY C JR
UNIVERSITY OF
CALIFORNIA LASL
BOX 1663 MS 420
LOS ALAMOS NM 87545
U.S.A.

GOLLNOW H DR
MT STROMLO OBS
CANBERRA ACT
AUSTRALIA

GOLUB LEON DR
HARVARD COLLEGE
OBSERVATORY
60 GARDEN ST
CAMBRIDGE MA 02138
U.S.A.

GOMES ALERCIO M DR
R GAVIAO PEIXOTO 13
AP 1401
ICARAI 24000
NITEROJ ERJ
BRAZIL

GOMEZ GONZALEZ J DR
PASEO IMPERIAL 29 6H
MADRID 5
SPAIN

GOMEZ MARIA THERESA DR
OSSERVATORIO ASTRO-
NOMICO
NAPOLI
ITALY

GONDHALEKAR PRABHAKAR DR
RUTHERFORD & APPLE-
TON LABS
CHILTON DIDCOT
OXFORDSHIRE OX11 0QX
U.K.

GONDOLATSCH FRIEDRICH PRF
ASTRONOMISCHES
RECHENINSTITUUT
MOENCHHOFSTR 12-14
D-6900 HEIDELBERG
GERMANY, F.R.

GONG HUI-REN
SHANGHAI OBSERVATORY
SHANGHAI
PEOPLE'S REPUBLIC CHINA

GONG SHU-MO
PURPLE MOUNTAIN
OBSERVATORY
NANJING
PEOPLE'S REPUBLIC CHINA

GONZALES-A WALTER D DR
INST PESQUISAS
ESPACIAIS
INPE C P 515
12200 S JOSE CAMPOS
BRAZIL

GONZALEZ G
INST NAC DE ASTROFIS
OPT Y ELECTRONICA
APT POSTALES 216 51
PUEBLA PUE
MEXICO

GONZE ROGER F J IR
OBSERVATOIRE ROYAL
DE BELGIQUE
AVE CIRCULAIRE 3
B-1180 BRUXELLES
BELGIUM

GOODY R M
CEPP
PIERCE HALL
29 OXFORD STREET
CAMBRIDGE MA 02138
U.S.A.

GOOSSENS MARCEL DR
LOZENHOEKSTRAAT 127
B-2850 KEERBERGEN
BELGIUM

GOPAL KRISHNA
RADIO ASTRONOMY GR
TIFR CENTRE
INDIAN INST OF SC
BANGALORE 560 012
INDIA

GOPALA RAO U V MR
ASTROPHYSICAL OBS
KODAIKANAL 3
INDIA

GOPASYUK S I DR
CRIMEAN ASTROPHYSI-
CAL OBSERVATORY USSR
ACAD OF SCI NAUCHNIY
334413 CRIMEA
U.S.S.R.

GORBATSKY VITALIJ G PROF
LENINGRAD UNIVERSITY
OBSERVATORY
199178 LENINGRAD
U.S.S.R.

GORDON CHARLOTTE PROF
11 RUE TOURNEFORT
359 FO BATIMENT C
F-75005 PARIS
FRANCE

GORDON COURTNEY P PROF
HAMPSHIRE COLLEGE
AMHERST MA 01002
U.S.A.

GORDON I M DR
INSTITUTE OF RADIO
PHYSICS AND
ELECTRONICS
310085 KHARKOV
U.S.S.R.

GORDON KURTISS J PROF
HAMPSHIRE COLLEGE
AMHERST MA 01002
U.S.A.

GORDON MARK A DR
NAT RADIO ASTRONOMY
OBS SUITE 100
2010 N FORBES BLVD
TUCSON AZ 85705
U.S.A.

GORENSTEIN PAUL DR
CENTER FOR ASTRO-
PHYSICS
60 GARDEN STREET
CAMBRIDGE MA 02138
U.S.A.

GORET P DR
SECT D'ASTROPHYSIQUE
CEN SACLAY
B P NO 2
F-91190 GIF-S-YVETTE
FRANCE

GORGOLEWSKI STANISLAW DR
INST OF ASTRONOMY
NIC COPERNICUS UNIV
CHOPINA 12-18
87100 TORUN
POLAND

GOSACHINSKIJ I V DR
LENINGRAD BRANCH OF
SAO
196140 LENINGRAD
U.S.S.R.

GOSLING JOHN T DR
P4- MS436
LOS ALAMOS SCIENTI-
FIC LABORATORY
LOS ALAMOS NM 87545
U.S.A.

GOSS W MILLER DR
KAPTEYN ASTRONOMICAL
INSTITUTE
POSTBUS 800
9700 AV GRONINGEN
THE NETHERLANDS

GOSSNER SIMONE D MRS
415 E 50TH STREET
NEW YORK NY 10022
U.S.A.

GOTT III J RICHARD
ASST PROF
DEPT ASTROPHYS SC
PRINCETON UNIVERSITY
PRINCETON NJ 08540
U.S.A.

GOTTESMAN STEPHEN T DR
PHYSICS AND
ASTRONOMY
UNIVERSITY FLORIDA
GAINESVILLE FL 32611
U.S.A.

GOTTLIEB CARL A DR
GODDARD INST FOR
SPACE STUDIES
2880 BROAD WAY
NEW YORK NY 10025
U.S.A.

GOTTLIEB KURT
46 JENNINGS STREET
CURTIN ACT 2605
AUSTRALIA

GOUDAS CONSTANTINE L PROF
DEPT OF MECHANICS
UNIVERSITY OF PATRAS
PATRAS
GREECE

GOUDIS CHRISTOS D PROF
UNIVERSITY OF PATRAS
DEPT.OF ASTRONOMY
PATRAS
GREECE

GOUGH DOUGLAS O DR
INST OF ASTRONOMY
MADINGLEY RD
CAMBRIDGE CB3 OHA
U.K.

GOUGUENHEIM LUCIENNE
OBSERVATOIRE
DE MEUDON
RADIOASTRONOMIE
F-92190 MEUDON
FRANCE

GOULD ROBERT J PROF
PHYSICS DEPT
UCSD
LA JOLLA CA 92093
U.S.A.

GOUTTEBROZE PIERRE DR
LPSP
B.P.10
F-91370 VERRIERES-LE
-BUISSON
FRANCE

GOWER ANN C DR
UNIVERSITY OF VICTO-
RIA OBSERVATORY
VICTORIA BC V8W 2Y2
CANADA

GOWER J F R DR
1615 MCTAVISH ROAD
R R 2
SIDNEY B C V8L 3S1
CANADA

GOY GERALD DR
OBSERVATOIRE
DE GENEVE
CH-1290 SAUVERNY
SWITZERLAND

GOYAL A N DR
DEPT OF MATHEMATICS
UNIV OF RAJASTHAN
JAIPUR
INDIA

GRABOSKE HAROLD C JR
LAWRENCE LIVERMORE
LABORATORY
P O BOX 808
LIVERMORE CA 94550
U.S.A.

GRABOWSKI BOLESLAW DR
INSTITUTE OF PHYSICS
UL OLESKA 48
45951 OPOLE
POLAND

GRADSZTAJN E DR
DEPT PHYSICS/ASTRON
TEL-AVIV UNIVERSITY
RAMAT-AVIV 69 978
ISRAEL

GRAHAM ERIC DR
HIGH ALTITUDE OBS
NATL CENTER ATMO RES
PO BOX 3000
BOULDER CO 80307
U.S.A.

GRAHAM JOHN A DR
CERRO TOLOLO
INTERAMERICAN OBS
CASILLA 63 D
LA SERENA
CHILE

GRAHL BERND H DR
MPI FUER RADIOASTRO-
NOMIE
AUF DEM HUEGEL 69
D-5300 BONN
GERMANY, F.R.

GRAINGER J F DR
PHYSICS DEPT
UMIST
MANCHESTER M60 1QD
U.K.

GRANDI STEVEN ALDRIDGE DR
DEPT OF ASTRONOMY
MATH SCS BLDG
UNIV OF CALIFORNIA
LOS ANGELES CA 90024
U.S.A.

GRANT IAN P DR
PEMBROKE COLLEGE
OXFORD OX1 1DW
U.K.

GRASDALEN GARY L DR
DEPT PHYSICS ASTRON
UNIV OF WYOMING
P O BOX 3905 UN STA
LARAMIE WY 82071
U.S.A.

GRATTON LIVIO PROF
LABORATORIO DI
ASTROFISICA SPAZIALE
CAS POST 67
I-00044 FRASCATI
ITALY

GRAUER ALBERT D
DEPT OF PHYS & ASTRO
UALR
33RD & UNIVERSITY
LITTLE ROCK AR 72204
U.S.A.

GRAY DAVID F PROF
DEPT OF ASTRONOMY
UNIVERSITY OF
WESTERN ONTARIO
LONDON ONT N6A 5B9
CANADA

GRAYZECK EDWIN J DR
PHYSICS DEPT
UNIV OF NEVADA
4505 S MARYLAND PKY
LAS VEGAS NV 89154
U.S.A.

GREBENIKOV E A PROF DR
INSTITUTE OF THEORE-
TICAL AND EXPERIMEN-
TAL PHYSICS
117259 MOSCOW
U.S.S.R.

GREC GERARD
DEPT D'ASTROPHYSIQUE
UNIVERSITE DE NICE
PARC VALROSE
F-06034 NICE
FRANCE

GREEN JACK PROF
DEPT OF GEOLOGY
CALIF STATE UNIVER-
SITY-LONG BEACH
LONG BEACH CA 90840
U.S.A.

GREEN LOUIS C PROF
HAVERFORD COLLEGE
HAVERFORD PA 19041
U.S.A.

GREEN ROBIN M DR
DEPT OF ASTRONOMY
GLASGOW UNIVERSITY
GLASGOW G12 8QQ
U.K.

GREENBERG J MAYO DR
HUYGENS LABORATORY
UNIVERSITY OF LEIDEN
WASSENAARSEWEG 78
2300 RA LEIDEN
THE NETHERLANDS

GREENBERG RICHARD DR
PLANETARY SCIENCE
INSTITUTE
2030 E SPEEDWAY
TUCSON AZ 85719
U.S.A.

GREENSTEIN GEORGE PROF
AMHERST COLLEGE
ASTRONOMY DEPT
AMHERST MA 01002
U.S.A.

GREENSTEIN J L PROF
HALE OBSERVATORIES
CALIF INST OF TECHN
1201 E CALIFORNIA ST
PASADENA CA 91125
U.S.A.

GREGORINI LORETTA
ISTITUTO DI
RADIOASTRONOMIA
VIA IRNERIO 46
I-40126 BOLOGNA
ITALY

GREGORY PHILIP C DR
PHYSICS DEPT
UNIVERSITY OF B C
2075 WESBROOK MALL
VANCOUVER BC V6T 1W5
CANADA

GREGORY STEPHEN ALBERT DR
PHYSICS DEPT
BOWLING GREEN STATE
UNIVERSITY
BOWLING GRN OH 43403
U.S.A.

GREGUL A YA DR
OBSERVATORY OF THE
KIEV UNIVERSITY
OBSERVATORNAYA 3
252053 KIEV
U.S.S.R.

GREISEN KENNETH I PROF
DEPT OF ASTRONOMY
228 SPACE SC BLDG
CORNELL UNIVERSITY
ITHACA NY 14853
U.S.A.

GRENON MICHEL DR
OBS DE GENEVE
CH-1290 SAUVERNY
SWITZERLAND

GREVESSE N DR
INSTITUT
D ASTROPHYSIQUE
B-4200 COINTE-OUGREE
BELGIUM

GREWING MICHAEL PROF
ASTRONOMISCHES INST
UNIVERSITAET
WALDHAUSERSTR 64
D-7400 TUEBINGEN
GERMANY, F.R.

GREYBER HOWARD D DR
10123 FALLS ROAD
POTOMAC MD 20854
U.S.A.

GRIFFIN RITA E M DR
THE OBSERVATORIES
MADINGLEY ROAD
CAMBRIDGE CB3 0HA
U.K.

GRIFFIN ROGER F DR
THE OBSERVATORIES
MADINGLEY ROAD
CAMBRIDGE CB3 0HA
U.K.

GRIFFITH JOHN S PROF
DEPT OF MATH SC
LAKEHEAD UNIVERSITY
THUNDER BAY
ONT P7B 5E1
CANADA

GRIFFITHS RICHARD E DR
HARVARD SMITHSONIAN
CTR FOR ASTROPHYSICS
60 GARDEN STREET
CAMBRIDGE MA 02138
U.S.A.

GRIGORJEV VICTOR M DR
SIBERIAN INST OF TER
MAGN,IONOSPH,RAD W P
P BOX 4
664697 IRKUTSK
U.S.S.R.

GRINDLAY JONATHAN E DR
HARVARD SMITHSONIAN
CENTER FOR ASTROPHYS
60 GARDEN ST
CAMBRIDGE MA 02138
U.S.A.

GRININ V P DR
CRIMEAN ASTROPHYSI-
CAL OBSERVATORY USSR
ACAD OF SCI NAUCHNIY
334413 CRIMEA
U.S.S.R.

GRISHCHJK L P DR
STERNBERG STATE AS-
TRONOMICAL INSTITUTE
117234 MOSCOW
U.S.S.R.

GROSBOL PREBEN JOHNSON DR
ESO
K-SCHWARZSCHILD-ST 2
D-8046 GARCHING B M
GERMANY, F.R.

GROSS PETER G PROF
WARNER SWASEY
OBSERVATORY
1975 TAYLOR RD
CLEVELAND OH 44112
U.S.A.

GROSSMAN ALLEN S PROF
ERWIN FICK
OBSERVATORY
IOWA ST UNIVERSITY
AMES IA 50011
U.S.A.

GROSSMAN LAWRENCE PROF
DEPT GEOPHYSICAL SC
UNIVERSITY CHICAGO
5734 SOUTH ELLIS AVE
CHICAGO IL 60637
U.S.A.

GROSSMANN-DOERTH U DR
KIEPENHEUER INST
FUER SONNENPHYSIK
SCHOENECKSTR 6
D-7800 FREIBURG
GERMANY, F.R.

GROTEN ERWIN PROF
INST F PHYSIKALISCHE
GEODAESIE
PETERSENSTR 13
D-6100 DARMSTADT
GERMANY, F.R.

GROTH EDWARD J III
PHYSICS DEPT
JADWIN HALL
PRINCETON UNIVERSITY
PRINCETON NJ 08540
U.S.A.

GROTH HANS G PROF DR
INST ASTRON/ASTROPH
UNIV MUENCHEN
SCHEINERSTRASSE 1
D-8000 MUENCHEN 80
GERMANY, F.R.

GRUBISSICH C PROF DR
VIA AOSTA 34/5
I-35100 PADOVA
ITALY

GRUDLER PIERRE
C E R G A
AVE COPERNIC
F-06130 GRASSE
FRANCE

GRUDZINSKA STEFANIA DR
INST OF ASTRONOMY
UL CHOPINA 12-18
MIKOLAJA KOPERNIKA
87100 TORUN
POLAND

GRUEFF GAVRIL DR
LABORATORIO DI RADIO
ASTRONOMIA C N R
VIA IRNERIO 46
I-40126 BOLOGNA
ITALY

GRUEN EBERHARD DR
MPI FUER KERNPHYSIK
POSTFACH 102 980
D-6900 HEIDELBERG
GERMANY, F.R.

GRUNDMANN WALTER
DOMINION ASTROPHYSI-
CAL OBSERVATORY
5071 W SAANICH ROAD
VICTORIA BC V8X 4M6
CANADA

GRUSHINSKIJ N P PROF DR
STERNBERG STATE AS-
TRONOMICAL INSTITUTE
117234 MOSCOW
U.S.S.R.

GRYGAR JIRI DR
INST OF PHYSICS
CZECHOSLOVAK ACAD SC
CS-250 68 REZ
CZECHOSLOVAKIA

GRZEDZIELSKI STANISLAW DR
POLISH ACAD OF SC
SPACE RESEARCH CTR
ORDONA 21
01-237 WARSZAWA
POLAND

GUARNIERI ADRIANO DR
OSSERVATORIO
ASTRONOMICO
VIA ZAMBONI 33
I-40126 BOLOGNA
ITALY

GUBANOV VADIM S DR
PULKOVO OBSERVATORY
196140 LENINGRAD
U.S.S.R.

GUDUR N DR
EGE UNIVERSITY
OBSERVATORY
P.K.21 BORNOVA
IZMIR
TURKEY

GUELIN MICHEL DR
OBSERVATOIRE DE
MEUDON
F-92190 MEUDON
FRANCE

GUERIN PIERRE DR
INST D ASTROPHYSIQUE
98 BIS BLVD ARAGO
F-75014 PARIS
FRANCE

GUETTER HARRY HENDRICK
US NAVAL OBSERVATORY
P O BOX 1149
FLAGSTAFF AZ 86002
U.S.A.

GUINAN EDWARD FRANCIS DR
DEPT OF ASTRONOMY
VILLANOVA UNIV
VILLANOVA PA 19085
U.S.A.

GULL STEPHEN F DR
ST JOHN'S COLLEGE
CAMBRIDGE
U.K.

GULMEN OMUR DR
EGE UNIVERSITY
OBSERVATORY
PK 21
BORNOVA IZMIR
TURKEY

GUNN JAMES E PROF
DEPT ASTROPHYS SCS
PRINCETON UNIVERSITY
PRINCETON NJ 08544
U.S.A.

GURSHTEIN A A DR
INST HIST SC & TECHN
USSR ACAD OF SCIENCE
STAROPANSKY 1/5
103012 MOSCOW
U.S.S.R.

GURZADIAN G A PROF DR
BYURAKAN ASTROPHYSI-
CAL OBSERVATORY
378433 ARMENIA
U.S.S.R.

GUSSMANN E A DR
ZENTRALINST ASTROPHY
ASTROPHYSIKALIS OBS
TELEGRAFENBERG
DDR-15 POTSDAM
G.D.R.

GUTHRIE BRUCE N G DR
ROYAL OBSERVATORY
BLACKFORD HILL
EDINBURGH
EDINBURGH EH9 3HJ
U.K.

HABBAL SHADIA RIFAI
HARVARD COLL OBS
60 GARDEN STREET
CAMBRIDGE MA 02138
U.S.A.

HACHENBERG OTTO PROF DR
RADIOASTRON INST DER
UNIV BONN
AUF DEM HUEGEL 71
D-5300 BONN
GERMANY, F.R.

GUERRERO GIANANTONIO DR
OSSERVATORIO ASTRO-
NOMICO
MERATE
MILANO
ITALY

GUIBERT JEAN DR
OBSERVATOIRE
DE MEUDON
F-92190 MEUDON
FRANCE

GUINOT BERNARD R PROF
BUREAU INTERNATIONAL
DE L HEURE
61 AV L OBSERVATOIRE
F-75014 PARIS
FRANCE

GULL THEODRE R DR
CODE 683
GODDARD
SPACE FLIGHT CENTER
GREENBELT MD 20771
U.S.A.

GULYAEV A P DR
STERNBERG ASTRO INST
UNIVERSITETSKIJ
PROSPEKT 13
117234 MOSCOW
U.S.S.R.

GUO QUAN SHI
PURPLE MOUNTAIN OBS
ACADEMIA SINICA
NANJING
PEOPLE'S REPUBLIC CHINA

GURSKY HERBERT DR
CENTER FOR
ASTROPHYSICS
60 GARDEN STREET
CAMBRIDGE MA 02138
U.S.A.

GUSEINOV O KH DR
SHEMAKHA ASTROPHYSI-
CAL OBSERVATORY
373243 AZERBAIDZAN
U.S.S.R.

GUSTAFSSON BENGT DR
ASTRONOMICAL
OBSERVATORY
BOX 515
S-751 20 UPPSALA
SWEDEN

GUTIERREZ-MORENO A DR MRS
DEP DE ASTRONOMIA
UNIVERSIDAD DE CHILE
CASILLA 36-D
SANTIAGO
CHILE

HABIBULLIN SH T PROF DR
ENGELGARDT ASTRONO-
MICAL OBSERVATORY
422526 KAZAN
U.S.S.R.

HACHISU IZUMI DR
DEPT OF AERONAUTICAL
ENGINEERING
KYOTO UNIVERSITY
KYOTO 606
JAPAN

GUEST JOHN E DR
UNIVERSITY OF LONDON
OBSERVATORY
MILL HILL PARK
LONDON NW7 2QS
U.K.

GUIDICE DONALD A DR
A F GEOPHYSICS
LABORATORY
HANSCOM AFB MA 01731
U.S.A.

GULKIS SAMUEL DR
JET PROPULSION
LABORATORY
4800 OAK GROVE DRIVE
PASADENA CA 91103
U.S.A.

GULLIVER AUSTIN FRASER DR
DEPT OF PHYSICS
UNIV OF ALBERTA
EDMONTON ALB T6G 2J1
CANADA

GULYAEV RUDOLF A DR
IZMIRAN
AKADEMGORODOK
142092 MOSCOW REGION
U.S.S.R.

GURM HARDEN S DR
DEPT OF ASTRONOMY
AND SPACE SCIENCES
PANJABI UNIVERSITY
PATIALA 147002
INDIA

GURTOVENKO E A DR
MAIN ASTRONOMICAL
OBS UKRAINIAN ACAD
OF SCIENCES
252127 KIEV
U.S.S.R.

GUSEJNOV R EH DR
SHEMAKHA ASTROPHYSI-
CAL OBSERVATORY
373243 AZERBAIDZAN
U.S.S.R.

GUTCKE DIETRICH
CARL-ZEISS STR 1
DDR-6900 JENA
G.D.R.

GYLDENKERNE KJELD DR
COPENHAGEN UNIVER-
SITY OBSERVATORY
BRORFELDE
DK-4340 TOLLOSE
DENMARK

HABING H J DR
STERREWACHT
POSTBUS 9513
2300 RA LEIDEN
THE NETHERLANDS

HACK MARGHERITA PROF
OSSERVATORIO
ASTRONOMICO
VIA TIEPOLO 11
I-34131 TRIESTE
ITALY

HACKWELL JOHN A DR
DEPT PHYSICS ASTRON
UNIV OF WYOMING
LARAMIE WY 82070
U.S.A.

HACYAN SHAHEN DR.
INST DE ASTRONOMIA
UNAM
APDO POSTAL 70-264
04510 MEXICO DF
MEXICO

HADDOCK FRED T DR
UNIV OF MICHIGAN
937 PHYSICS-ASTRONO-
MY BLDG
ANN ARBOR MI 48104
U.S.A.

HADJIDEMETRIOU JOHN DPROF
DEPT OF THEORETICAL
MECHANICS
UNIV OF THESSALONIKI
THESSALONIKI
GREECE

HAEFNER REINHOLD DR
UNIV STERNWARTE
SCHEINERSTRASSE 1
D-8000 MUENCHEN 80
GERMANY, F.R.

HAENSEL PAWEL DR
N COPERNICUS ASTRON
CENTER
BARTYCKA 18
00-716 WARSAW
POLAND

HAERENDEL G DR
MPI FUER PHYSIK UND
ASTROPHYSIK
INST EXTRATERR PHYS
D-8046 GARCHING
GERMANY, F.R.

HAGEN JOHN P
613 W PARK AVE
STATE COLLEGE
PA 16801
U.S.A.

HAGEN WENDY ANNE
WHITIN OBSERVATORY
WELLESLEY COLLEGE
WELLESLEY MA 02181
U.S.A.

HAGEN-THORN V A DR
LENINGRAD UNIVERSITY
10 LINIJA 33
199178 LENINGRAD
U.S.S.R.

HAGFORS T DR
MIT LINCOLN LAB
MILLSTON/HAYSTACK
LEXINGTON MA 02173
U.S.A.

HAISCH BERNARD MICHAEL
LOCKHEED PALO ALTO
RESEARCH LABORATORY
3251 HANOVER STREET
PALO ALTO CA 94304
U.S.A.

HAJDUK ANTON DR
ASTRONOMICAL INST
SAV
DUBRAVSKA CESTA A/11
89930 BRATISLAVA
CZECHOSLOVAKIA

HALL ANDREW NORMAN
DEPT OF ASTROPHYSICS
SOUTH PARKS ROAD
OXFORD OX1 3RQ
U.K.

HALL DONALD N DR
KITT PEAK NATIONAL
OBSERVATORY
P O 26732
TUCSON AZ 85726
U.S.A.

HALL DOUGLAS S DR
DYER OBSERVATORY
VANDERBILT
UNIVERSITY
NASHVILLE TN 37235
U.S.A.

HALL JOHN S DR
STAR ROUTE 1
BOX 97
SEDONA AZ 86336
U.S.A.

HALL R GLENN DR
3612 SPRING STREET
CHEVY CHASE MD 20015
U.S.A.

HALLAM KENNETH L DR
CODE 685 LAB FOR
ASTRON SOLAR PHYSICS
GSFC
GREENBELT MD 20771
U.S.A.

HALLIDAY I DR
HERZBERG INSTITUTE
OF ASTROPHYSICS
NATIONAL RES COUNCIL
OTTAWA ONT K1A OR6
CANADA

HAMADA TETSUO PROF
DEPT OF PHYSICS
IBARAKI UNIVERSITY
310 MITO
JAPAN

HAMAJIMA KIYOTOSHI DR
KISO BRANCH OF THE
TOKYO ASTRON OBS
MITAKEMURA KISOGUN
NAGANOKEN 397-01
JAPAN

HAMDY M A M DR
HELWAN OBSERVATORY
HELWAN
ARAB REP. OF EGYPT

HAMEEN ANTILLA KAARLE A
PROF DR
ASTRONOMY DEPT
UNIVERSITY OF OULU
SF-90100 OULU 10
FINLAND

HAMID S EL DIN DR
DEPT OF ASTRONOMY
FACULTY OF SCIENCE
UNIVERSITY FOJAD
GIZA CAIRO
ARAB REP. OF EGYPT

HAMILTON P A DR
UNIV OF TASMANIA
HOBART TASMANIA
AUSTRALIA

HAMMERSCHLAG ROBERT H
DR IR
ASTRONOMICAL INST
SERVAAS BOLWERK 13
3512 NK UTRECHT
THE NETHERLANDS

HAMMERSCHLAG-HENSBERGE
G DR
ASTRONOMICAL INST
ROETERSTRAAT 15
1018 WB AMSTERDAM
THE NETHERLANDS

HAMZAOGLU ESAT E H DR
EGE UNIVERSITESI
FEN FAKULTESI
GOK BILIMLERI ENST
BORNOVA IZMIR
TURKEY

HAN TIANQI
INST OF GEODESY &
GEOPHYSICS
CHIN ACADEMY OF SCS
WUCHANG HUBEI
PEOPLE'S REPUBLIC CHINA

HANASZ JAN DR
ASTRONOMICAL CENTER
UL CHOPINA 12/13
MICAL CENTRE P A N
87100-TORUN
POLAND

HANBURY BROWN R PROF
SCHOOL OF PHYSICS
UNIVERSITY OF SYDNEY
SYDNEY NSW 2006
AUSTRALIA

HANES DAVID A DR
INST OF ASTRONOMY
UNIV OF CAMBRIDGE
CAMBRIDGE
U.K.

HANG HENG-RONG
PURPLE MOUNTAIN OBS
NANJING
PEOPLE'S REPUBLIC CHINA

HANKINS TIMOTHY HAMILTON
DARTMOUTH COLLEGE
HANOVER NH 03755
U.S.A.

HANNER MARTHA S DR
JPL
MS 183-301
PASADENA CA 91103
U.S.A.

HANSEN CARL J PROF
JILA
UNIVERSITY OF
COLORADO
BOULDER CO 80309
U.S.A.

HANSON ROBERT B DR
LICK OBSERVATORY
UNIV OF CALIFORNIA
SANTA CRUZ CA 95064
U.S.A.

HAPKE BRUCE W DR
DEPT EARTH PLANET SC
321 OLD ENGIN HALL
UNIV OF PITTSBURGH
PITTSBURGH PA 15235
U.S.A.

HARDEBECK ELLEN G DR
3106 TUMBLEWEED RD
BISHOP CA 93514
U.S.A.

HARDORP JOHANNES PROF
DEPT EARTH SPACE SC
STATE UNIVERSITY
STONY BROOK NY 11794
U.S.A.

HARMER CHARLES F W MR
ROYAL GREENWICH OBS
HERSTMONCEUX CASTLE
HAILSHAM
E SUSSEX BN27 1RP
U.K.

HARO GUILLERMO DR
INST NAC ASTROFISICA
OPTICA Y ELECTRONICA
A P 216 Y 51
PUEBLA PUE
MEXICO

HARRIS ALAN WILLIAM DR
JPL
MS 264-720
4800 OAK GROVE DRIVE
PASADENA CA 91103
U.S.A.

HARRIS STELLA
DEPT OF PHYSICS
QUEEN MARY COLLEGE
MILE END ROAD
LONDON E1 4NS
U.K.

HARROWER GEORGE A DR
LAKEHEAD UNIVERSITY
THUNDER BAY
ONT P7B 5E1
CANADA

HARTL HERBERT DR
INST F ASTRONOMIE
UNIVERSITAETSSTR 4
A-6020 INNSBRUCK
AUSTRIA

HARTMANN WILLIAM K
PLANETARY SC INST
2030 E SPEEDWAY
SUITE 201
TUCSON AZ 85719
U.S.A.

HANSEN LEIF LECTURER
COPENHAGEN UNIVERSI-
TY OBSERVATORY
OSTERVOLDGADE 3
DK-1350 COPENHAGEN K
DENMARK

HANSSON NILS DR
LUND OBSERVATORY
SVANEGATAN 9
S-222 24 LUND
SWEDEN

HARA KEN NOSUKE DR
SENDAI-SHIRITSU-JOSH
SENIOR HIGH SCHOOL
KASHIWAGI 3-3-1
SENDAI 980
JAPAN

HARDEE PHILIP
DEPT OF PHYS & ASTRO
UNIV OF ALABAMA
BOX 1921
UNIVERSITY AL 35486
U.S.A.

HARDY EDUARDO DR PH
DEPT DE PHYSIQUE
UNIVERSITE LAVAL
FAC DES SCS GENIE
QUEBEC G1K 7P4
CANADA

HARMER DIANNE L MRS
ROYAL GREENWICH OBS
HERSTMONCEUX CASTLE
HAILSHAM
E SUSSEX BN27 1RP
U.K.

HARRINGTON J PATRICK DR
ASTRONOMY PROGRAM
UNIVERSITY MARYLAND
COLLEGE PK MD 20742
U.S.A.

HARRIS DANIEL E DR
CTR FOR ASTROPHYSICS
60 GARDEN STREET
CAMBRIDGE MA 02138
U.S.A.

HARRIS WILLIAM E DR
DEPT OF PHYSICS
MCMASTER UNIV
HAMILTON ONT L8S 4M1
CANADA

HART MICHAEL H DR
PHYSICS DEPT
TRINITY UNIV
SAN ANTONIO TX 78284
U.S.A.

HARTLEY KENNETH F DR
ROYAL GREENWICH
OBSERVATORY
HAILSHAM
EAST SUSSEX BN27 1RP
U.K.

HARTOOG MARK RICHARD DR
LICK OBS
UNIV OF CALIFORNIA
SANTA CRUZ CA 95064
U.S.A.

HANSEN RICHARD T MR
1693 OAK CIRCLE
LAKE CITY FL 32055
U.S.A.

HAO YUN-XIANG
DEPT OF ASTRONOMY
BEIJING NORMAL UNIV
BEIJING
PEOPLE'S REPUBLIC CHINA

HARA TETSUYA DR
DEPT OF PHYSICS
KYOTO SANGYO UNIV
KYOTO 603
JAPAN

HARDIE R PROF
DYER OBSERVATORY
VANDERBILT
UNIVERSITY
NASHVILLE TN 37235
U.S.A.

HARMANEC PETR DR
ASTRONOMICAL INST
CZECHOSLOVAK ACAD
OF SCIENCES
251 65 ONDREJOV
CZECHOSLOVAKIA

HARMS RICHARD JAMES DR
PHYSICS DEPT B-019
UNIV OF CALIFORNIA
SAN DIEGO
LA JOLLA CA 92093
U.S.A.

HARRINGTON ROBERT S DR
U S NAVAL
OBSERVATORY
WASHINGTON DC 20390
U.S.A.

HARRIS GRETCHEN L H DR PH
DEPT OF PHYSICS
UNIV OF WATERLOO
WATERLOO ONT
CANADA

HARRISON EDWARD R PROF
DEPT OF PHYSICS AND
ASTRONOMY
UNIVERSITY OF MASS
AMHERST MA 01003
U.S.A.

HARTEN RONALD H DR
RADIOSTERRENWACHT
POSTBUS 2
7990 AA DWINGELOO
THE NETHERLANDS

HARTMANN LEE WILLIAM
CTR FOR ASTROPHYSICS
60,GARDEN STREET
CAMBRIDGE MA 02138
U.S.A.

HARTQUIST THOMAS WILBUR
DEPT OF PHYSICS &
ASTRONOMY
UNIVERSITY COLLEGE
LONDON WC1E 6BT
U.K.

HARTWICK F DAVID A DR
UNIVERSITY VICTORIA
DEPT OF PHYSICS
VICTORIA BC V8W 2Y2
CANADA

HARTZ THEODORE R DR
COMMUNICATIONS
RESEARCH CENTER
POB 11490 STATION H
OTTAWA ONT K2H 8S2
CANADA

HARVEL CHRISTOPHER ALVIN
6161 STEPHEN'S
FOREST ROAD
COLUMBIA MD 21045
U.S.A.

HARVEY CHRISTOPHER C DR
OBSERVATOIRE
DE MEUDON
F-92190 MEUDON
FRANCE

HARVEY GALE A DR
INST F GESCHICHTE D
NATURWISSENSCHAFTEN
GOETHE UNIVERSITAET
D-6000 FRANKFURT
GERMANY, F.R.

HARVEY J W DR
KITT PEAK NATIONAL
OBSERVATORY
950 N CHERRY AVE
TUCSON AZ 85726
U.S.A.

HARVEY PAUL MICHAEL DR
STEWARD OBS
UNIV OF ARIZONA
TUCSON AZ 85721
U.S.A.

HARWIT MARTIN PROF
SPACE SCIENCE BLDG
CORNELL UNIVERSITY
ITHACA NY 14853
U.S.A.

HASAN SAIYID STRAJUL
INDIAN INSTITITE OF
ASTROPHYSICS
BANGALORE 560 034
INDIA

HASCHICK AUBREY
HAYSTACK OBSERVATORY
WESTFORD MA 01886
U.S.A.

HASEGAWA HIROICHI DR
DEPT OF PHYSICS
KYOTO UNIVERSITY
SAKYO-KU
KYOTO 606
JAPAN

HASEGAWA ICHIRO DR
2-3-11 SAIDAIJINO-
GAMI
NARA 631
JAPAN

HASER LEO N K DR
MPI FUER EXTRATERRE-
STRISCHE PHYSIK
D-8046 GARCHING
GERMANY, F.R.

HASLAM C GLYN T DR
MPI F RADIO
ASTRONOMIE
AUF DEM HUEGEL 69
D-5300 BONN
GERMANY, F.R.

HASSAN S M DR
HELWAN INST OF ASTRO
NOMY AND GEOPHYSICS
HELWAN NEAR CAIRO
ARAB REP. OF EGYPT

HATTORI AKIRA A PROF
HIDA OBSERVATORY
KAMITAKARA
GIFU KEN 506 13
JAPAN

HAUCK BERNARD PROF
INST D ASTRONOMIE
UNIVERSITE LAUSANNE
CH-1290 CHAVANNES
DES BOIS
SWITZERLAND

HAUG EBERHARD DOZ DR
MOZARTSTRASSE 20
D-7418 METZINGEN
GERMANY, F.R.

HAUG ULRICH PROF
HAMBURGER STERNWARTE
GOJENBERGSWEG 112
D-2050 HAMBURG 80
GERMANY, F.R.

HAUGE OIVIND DR
INSTITUTE OF THEORE-
TICAL ASTROPHYSICS
UNIVERSITY OF OSLO
BLINDERN OSLO 3
NORWAY

HAUPT HERMANN F PROF
ASTRONOMISCHES INST
DER UNIVERSITAET
UNIVERSTAETSPLATZ 5
A-8010 GRAZ
AUSTRIA

HAUPT RALPH F
3701 DULWICK DRIVE
SILVER SPG MD 20906
U.S.A.

HAUPT WOLFGANG DR
GIRONDELLE 105
D-4630 BOCHUM
GERMANY, F.R.

HAVLEN ROBERT J DR
NAT RADIO ASTRON OBS
EDGEMONT ROAD
CHARLOTTESVL VA22901
U.S.A.

HAVNES OVE DR
AURORAL OBSERVATORY
UNIVERSITY OF TROMSO
N-9000 TROMSO
NORWAY

HAWARDEN T G DR
UK 48 INCH SCHMIDT
TELESCOPE UNIT
PRIVATE BAG
COONABARAN NSW 2857
AUSTRALIA

HAWKING STEPHEN W PROF
DEPT APPLIED MATHS
AND THEOR PHYSICS
SILVER STREET
CAMBRIDGE CB3 9EW
U.K.

HAWKINS G S DR
16 GARDEN ROAD
WELLESLEY MA 02181
U.S.A.

HAYAKAWA SACHIO PROF
DEPT OF PHYSICS
NAGOYA UNIVERSITY
FUROKO CHIKUSAKU
NAGOYA 464
JAPAN

HAYASHI CHUSHIRO PROF
PHYSICS DEPT
FACULTY OF SCIENCE
KYOTO UNIVERSITY
KYOTO 606
JAPAN

HAYES DONALD S DR
KITT PEAK NAT OBS
P O BOX 26732
TUCSON AZ 85726
U.S.A.

HAYLI AVRAM PROF
OBSERVATOIRE DE LYON
F-69230 ST GENIS
LAVAL
FRANCE

HAYMES ROBERT C PROF
DEPT OF SPACE PHY-
SICS AND ASTRONOMY
RICE UNIVERSITY
HOUSTON TX 77001
U.S.A.

HAYNES MARTHA P
NRAO
P O BOX 2
GREEN BANK WV 24944
U.S.A.

HAYNES RAYMOND F PROF
CSIRO
DIV OF RADIOPHYSICS
P O BOX 76
EPPING N S W 2121
AUSTRALIA

HAYWARD JOHN
DEPT OF ASTRONOMY
THE UNIVERSITY
GLASGOW G12 8QQ
U.K.

HAZARD CYRIL DR
INST OF ASTRONOMY
MADINGLEY ROAD
CAMBRIDGE CB3 OHE
U.K.

HAZER S DR
FACULTY OF SCIENCE
DEPT OF ASTRONOMY
P K 21
BORNOVA IZMIR
TURKEY

HAZLEHURST JOHN DR
HAMBURGER STERNWARTE
GOJENBERGSWEG 112
D-2050 HAMBURG 80
GERMANY, F.R.

HE MIAO-FU
SHANGHAI OBSERVATORY
ACADEMIA SINICA
SHANGHAI
PEOPLE'S REPUBLIC CHINA

HE XIANG-TAO
ROYAL OBSERVATORY
BLACKFORD HILL
EDINBURGH EH9 3HJ
U.K.

HEAP SARA R DR
CODE 672 GODDARD
SPACE FLIGHT CENTER
GREENBELT MD 20771
U.S.A.

HEARN ANTHONY G DR
STERREKUNDIG INST
SERVAAS BOLWERK 13
3512 NK UTRECHT
THE NETHERLANDS

HEARNSHAW JOHN B DR
DEPT OF PHYSICS
UNIVERSITY OF
CANTERBURY
CHRISTCHURCH
NEW ZEALAND

HEASLEY JAMES NORTON
INST FOR ASTRONOMY
2680 WOODLAWN DRIVE
HONOLULU HI 96822
U.S.A.

HECK ANDRE DR
VILLAFRANCA SATEL-
LITE TRACKING STAT
APARTADO 54065
MADRID
SPAIN

HECKATHORN HARRY M
US NAVAL RESEARCH
LABORATORY
CODE 4143 2
WASHINGTON DC 20375
U.S.A.

HECKMANN OTTO PROF DR
WOHNSTIFT GOETTINGEN
APP B 1403
CHARLOTTENB.STR 29
D-3400 GOETTINGEN
GERMANY, F.R.

HEDDLE DOUGLAS W O PROF
DEPT OF PHYSICS
ROYAL HOLLOWAY
COLLEGE
EGHAM SURREY TW200EX
U.K.

HEDEMAN E RUTH
14 N MAIN ST
NO 201
CLARKSTON MI 48016
U.S.A.

HEESCHEN DAVID S DR DIR
NAT RADIO ASTR OBS
EDGEMONT RD
CHARLOTTESVILLE
VA 22901
U.S.A.

HEFELE HERBERT PH D
MPI FUER ASTRONOMIE
KONIGSTUHL
D-6900 HEIDELBERG 1
GERMANY, F.R.

HEFFERLIN RAY A PROF
PHYSICS DEPT
S MISSISSIPPI COLL
DRAWER H
COLLEGEDALE TN 37315
U.S.A.

HEGGIE DOUGLAS C DR
UNIV OF EDINBURGH
EDINBURGH EH9 3JZ
U.K.

HEGYI DENNIS J ASS PROF
RANDALL LABORATORY
UNIVERSITY OF
MICHIGAN
ANN ARBOR MI 48109
U.S.A.

HEIDMANN JEAN DR
OBS DE MEUDON
F-92190 MEUDON
FRANCE

HEILES CARL PROF
ASTRONOMY DEPT
UNIVERSITY OF
CALIFORNIA
BERKELEY CA 94720
U.S.A.

HEILESEN POUL MARTIN DR
BOULEVARDEN 14 2 16
DK-2800 LYNGBY
DENMARK

HEINRICH INGE DR
ASTRONOMISCHES
RECHEN-INSTITUT
MOENCHHOFSTR 12-14
D-6900 HEIDELBERG 1
GERMANY, F.R.

HEINTZ WULFF D DR
DEPT OF ASTRONOMY
SWARTHMORE COLLEGE
SWARTHMORE PA 19081
U.S.A.

HEINTZE J R W DR
STERREWACHT
SONNENBORGH
SERVAAS BOLWERK 13
3512 NK UTRECHT
THE NETHERLANDS

HEISE JOHN DR
SPACE RESEARCH LAB
BENELUXLAAN 21
3527 HS UTRECHT
THE NETHERLANDS

HEISER ARNOLD M DR
DYER OBSERVATORY
VANDERBILT UNIV
BOX 1803-STA B
NASHVILLE TN 37235
U.S.A.

HEKELA JAN DR
ASTRONOMICAL INST
CZECHOSLOVAK ACAD
OF SCIENCES
251 65 ONDREJOV
CZECHOSLOVAKIA

HELALI YHYA E DR
HELWAN OBSERVATORY
HELWAN-CAIRO
ARAB REP. OF EGYPT

HELFAND DAVID JOHN
COLUMBIA ASTROPHYS
LABORATORY
538 WEST 120TH ST
NEW YORK NY 10027
U.S.A.

HELFER H LAWRENCE PROF
DEPT OF PHYSICS AND
ASTRONOMY
UNIVERSITY ROCHESTER
ROCHESTER NY 14627
U.S.A.

HELIN ELEANOR FRANCIS
JET PROPULSION LAB
4800 OAK GROVE DRIVE
PASADENA CA 91109
U.S.A.

HELLER MICHAEL DR
ASTRONOMICAL OBS
UL ORLA 171
30-244 KRAKOW
POLAND

HELLWIG HELMUT WILHELM DR
NAT BUREAU STANDARDS
TIME & FREQ DIV
325 BROADWAY RM 2031
BOULDER CO 80303
U.S.A.

HELMER LEIF
COPENHAGEN UNIV
OBSERVATORY
BRORFELDE
DK-4340 TOLLOSE
DENMARK

HELMKEN HENRY F DR
SMITHSONIAN
ASTROPHYSICAL OBS
60 GARDEN ST
CAMBRIDGE MA 02138
U.S.A.

HELT BODIL E
UNIVERSITY
OBSERVATORY
OSTER VOLDGADE 3
DK-1350 COPENHAGEN
DENMARK

HEMENWAY CURTIS L PROF
DUDLEY OBS PLAZA 7
1202 TROY
SCHENECTADY RD
LATHAM NY 12110
U.S.A.

HEMENWAY P D
ASTRONOMY DEPT
UNIVERSITY OF TEXAS
AUSTIN TX 78712
U.S.A.

HEMMLEB GERHARD DR
ZENTRALINSTITUT
PHYSIK DER ERDE
TELEGRAFENBERG A 17
DDR-15 POTSDAM
G.D.R.

HENIZE KARL G ASTRONAUT
CODE CB
JOHNSON SPACE CENTER
HOUSTON TX 77058
U.S.A.

HENON MICHEL C DR
OBSERVATOIRE DE NICE
B P 252
F-06007 NICE CEDEX
FRANCE

HENOUX JEAN-CLAUDE DR
OBSERVATOIRE
DE MEUDON
F-92190 MEUDON
FRANCE

HENRARD JACQUES PROF
FAC UNIV DE NAMUR
61 RUE DE BRUXELLES
B-5000 NAMUR
BELGIUM

HENRIKSEN RICHARD N DR
ASTRONOMY GROUP
DEPT OF PHYSICS
QUEENS UNIVERSITY
KINGSTON ONT K7L 3N6
CANADA

HENRY RICHARD C. PROF.
PHYSICS DEPARTMENT
JOHNS HOPKINS UNIV.
BALTIMORE MD 21218
U.S.A.

HENSBERGE HERMAN
ADELAARSTRAAT 23
B-2060 MERKSEM
BELGIUM

HERBIG GEORGE H DR
LICK OBSERVATORY
UNIVERSITY OF
CALIFORNIA
SANTA CRUZ CA 95064
U.S.A.

HERBST ERIC DR
DEPT OF PHYSICS
DUKE UNIVERSITY
DURHAM NC 27706
U.S.A.

HERBST WILLIAM DR
ASTRONOMY DEPT
WESLEYAN UNIV
MIDDLETOWN CT 06457
U.S.A.

HERCZEG TIBOR J PROF DR
DEPT OF PHYSICS AND
ASTRONOMY
UNIVERSITY OKLAHOMA
NORMAN OK 73069
U.S.A.

HERMAN RENEE DR
7 BIS RUE TRUDON
F-92160 ANTONY
FRANCE

HERNANDEZ CARLOS ALBERTO
OBS ASTRONOMICO
PASEO DEL BOSQUE
1900-LA PLATA
ARGENTINA

HERR RICHARD B DR
PHYSICS DEPT
UNIVERSITY DELAWARE
NEWARK DE 19711
U.S.A.

HERRMANN DIETER DR
ARCHENHOLD OBS
AND PUBLIC OBS
1193 BERLIN-TREPTOW
G.D.R.

HERS JAN MR
P.O.BOX 48
SEDGEFIELD 6573
SOUTH AFRICA

HERSHEY JOHN L DR
SPROUL OBSERVATORY
SWARTHMORE COLLEGE
SWARTHMORE PA 19081
U.S.A.

HERZBERG GERHARD DR
HERZBERG INSTITUTE
OF ASTROPHYSICS
NAT RES COUNCIL
OTTAWA ONT K1A 0R6
CANADA

HESSER JAMES E DR
DOMINION ASTROPH OBS
5071 W SAANICH ROAD
RR 5
VICTORIA BC V8X 4M6
CANADA

HEUDIER J L MR
CERGA
AVENUE COPERNIC
F-06130 GRASSE
FRANCE

HEWISH ANTONY PROF
CAVENDISH LABORATORY
MADINGLEY RD
CAMBRIDGE CB3 0HE
U.K.

HEWITT A V DR
US NAVAL OBSERVATORY
P O BOX 1149
FLAGSTAFF AZ 86001
U.S.A.

HEY JAMES STANLEY DR
4 SHORTLANDS CLOSE
WILLINGDON
EASTBOURNE
SUSSEX BN22 0JE
U.K.

HEYDEN FRANCIS J SJ DR
MANILA OBSERVATORY
P O BOX 1231
MANILA
REPUBLIC OF PHILIPPINES

HEYVAERTS JEAN DR
OBSERVATOIRE
DE MEUDON
F-92190 MEUDON
FRANCE

HIBBS ALBERT R MGR PLANS
JET PROPULSION
LABORATORY
4800 OAK GROVE DRIVE
PASADENA CA 91103
U.S.A.

HICKSON PAUL DR
DEPT GEOPHYS ASTRON
UNIVERSITY OF B C
VANCOUVER BC V6T 1W5
CANADA

HIDAJAT BAMBANG PROF DR
BOSSCHA OBSERVATORY
LEMBANG JAVA
INDONESIA

HIDALGO MIGUEL A DR
FACULTAD DE CIENCIAS
CIUDAD UNIVERSITARIA
ZARAGOZA
SPAIN

HIDE RAYMOND PROF
GEOPHYSICAL FLUID
DYNAMICS LABORATORY
METEOROLOGICAL OFF
BRACKNELL BERKSHIRE
U.K.

HIEI EIJIRO DR
TOKYO ASTRONOMICAL
OBSERVATORY
OSAWA MITAKA
TOKYO 181
JAPAN

HIGGS LLOYD A DR
HERZBERG INSTITUTE
OF ASTROPHYSICS
NAT RES COUNC CANADA
OTTAWA ONT K1A 0R6
CANADA

HILDEBRAND ROGER H
UNIV OF CHICAGO
ENRICO FERMI INST
5630 S ELLIS AVE
CHICAGO IL 60637
U.S.A.

HILDITCH RONALD W DR
UNIVERSITY
OBSERVATORY
BUCHANAN GARDENS
ST ANDREWS KY16 9LZ
U.K.

HILDNER ERNEST DR
CODE ES 52
MARSHALL SFC
HUNTSVILLE AL 35812
U.S.A.

HILF EBERHARD R H PH D
PAUL-WAGNERSTR 56
D-6100 DARMSTADT
GERMANY, F.R.

HILL GRAHAM DR
DOMINION ASTROPH OBS
5071 W SAANICH ROAD
RR 5
VICTORIA BC V8X 4M6
CANADA

HILL HENRY ALLEN DR
DEPT OF PHYSICS
BLDG 81
UNIV OF ARIZONA
TUCSON AZ 85721
U.S.A.

HILL PHILIP W DR
UNIVERSITY
OBSERVATORY
ST ANDREWS
FIFE KY16 9LZ
U.K.

HILLEBRANDT WOLFGANG PH D
MPI FUER PHYSIK UND
ASTROPHYSIK
FOEHRINGER RING 6
D-8000 MUENCHEN 40
GERMANY, F.R.

HILLIARD R DR
STEWARD OBSERVATORY
UNIV OF ARIZONA
TUCSON AZ 85721
U.S.A.

HILLS JACK G DR
DEPT ASTRON ASTROPHY
MICHIGAN STATE UNIV
PHYS-ASTROPHYS BLDG
E LANSING MI 48824
U.S.A.

HILLS RICHARD E DR
CAVENDISH LABORATORY
MADINGLEY RD
CAMBRIDGE CB3 OHE
U.K.

HILTNER W ALBERT PROF
DEPT OF ASTRONOMY
UNIVERSITY MICHIGAN
ANN ARBOR MI 48109
U.S.A.

HINKLE KENNETH H
950 N CHERRY AVE
P O BOX 26732
TUCSON AZ 85726
U.S.A.

HINTEREGGER HANS E DR
AIR FORCE GEOPHYSICS
LABORATORY
HANSCOM AFB MA 01731
U.S.A.

HINTZEN PAUL MICHAEL N DR
NASA-GSFC
CODE 681
GREENBELT MD 20771
U.S.A.

HIPPELEIN HANS H DR
MPI FUER ASTRONOMIE
D-6900 HEIDELBERG
GERMANY, F.R.

HIRABAYASHI HISASHI DR
TOKYO ASTRONOMICAL
OBSERVATORY
OSAWA MITAKA
TOKYO 181
JAPAN

HIRAI MASANORI DR
FUKUOKA UNIVERSITY
OF EDUCATION
729 MUNAKATA
FUKUOKA 811-41
JAPAN

HIRAYAMA TADASHI DR
TOKYO ASTRONOMICAL
OBSERVATORY
OSAWA MITAKA
TOKYO 181
JAPAN

HIRST WILLIAM P
1 CLIFFORD CRESCENT
BERGVLIET CP 7800
SOUTH AFRICA

HIRTH WOLFGANG ERNST PH D
THEODOR-HEUSSTR 18
D-5354 WEILERSWIST
GERMANY, F.R.

HITOTSUYANAGI JUICHI PROF
DEPT OF ASTRONOMY
TOHOKU UNIVERSITY
ARAMAKI SENDAI 980
JAPAN

HJALMARSON AKE G DR
ONSALA SPACE
OBSERVATORY
S-430 34 ONSALA
SWEDEN

HJELLMING ROBERT M DR
NATIONAL RADIO ASTRO
NOMY OBSERVATORY
P O BOX 0
SOCORRO NM 87801
U.S.A.

HO PAUL T P
601 CAMPBELL HALL
UNIV OF CALIFORNIA
BERKELEY CA 94720
U.S.A.

HOAG ARTHUR A DR
LOWELL OBSERVATORY
BOX 1269
FLAGSTAFF AZ 86002
U.S.A.

HOANG BINH DY DR
OBSERVATOIRE
DE MEUDON
F-92190 MEUDON
FRANCE

HOBBS LEWIS M DR
YERKES OBSERVATORY
UNIVERSITY OF
CHICAGO
WILLIAMS BAY WI53191
U.S.A.

HOBBS R W DR
NASA-GODDARD SPACE
FLIGHT CENTER
CODE 685
GREENBELT MD 20771
U.S.A.

HODGE PAUL W PROF
ASTRONOMY FM20
UNIVERSITY OF
WASHINGTON
SEATTLE WA 98195
U.S.A.

HOEG ERIK DR
COPENHAGEN OBS
OSTERVOLDGADE 3
DK-1350 COPENHAGEN
DENMARK

HOEGBOM JAN A DR
STOCKHOLM
OBSERVATORY
S-133 00
SALTSJOEBADEN
SWEDEN

HOEGLUND BERTIL PROF
ONSALA SPACE
OBSERVATORY
S-430 34 ONSALA
SWEDEN

HOEKSTRA ROEL DR
SPACE RESEARCH
LABORATORY
3527 HS UTRECHT
THE NETHERLANDS

HOEY MICHAEL J DR
PHYSICS DEPT
UNIVERSITY COLLEGE
BELFIELD
DUBLIN 4
IRELAND

HOFF DARREL BARTON
ASTRONOMY COORDINA-
TOR
UNIV NORTHERN IOWA
CEDAR FALLS IA 50614
U.S.A.

HOFFLEIT E DORRIT DR
DEPT OF ASTRONOMY
BOX 2023
YALE STATION
NEW HAVEN CT 06520
U.S.A.

HOFFMAN JEFFREY ALAN DR
NASA-JSC
CODE CB-4
HOUSTON TX 77058
U.S.A.

HOFFMANN MARTIN DR
OBS HOHER LIST
STERNWARTE DER
UNIVERSITAET BONN
D-5568 DAUN
GERMANY, F.R.

HOFMANN WILFRED DR
ASTRONOMISCHES
RECHEN-INSTITUT
MOENCHHOFSTR 12-14
D-6900 HEIDELBERG 1
GERMANY, F.R.

HOGG DAVID E DR
NAT RADIO ASTR OBS
EDGEMONT RD
CHARLOTTESVILLE
VA 22901
U.S.A.

HOHL FRANK DR
NASA MS-160
LANGLEY RESEARCH
CENTER
HAMPTON VA 23665
U.S.A.

HOLDEN FRANK
2 COLWICH CRESCENT
STAFFORD ST16 3XP
U.K.

HOLLENBACH DAVID JOHN DR
MAIL STOP 245-6
NASA/AMES RES CTR
MOFFETT FLD CA 94035
U.S.A.

HOLLIS JAN MICHAEL DR
NASA GSFC
CODE 685
GREENBELT MD 20771
U.S.A.

HOLLOWAY NIGEL J DR
ASTRONOMY CENTRE
UNIV OF SUSSEX
FALMER
BRIGHTON BN1 9QH
U.K.

HOLLWEG JOSEPH V
DEPT OF PHYSICS
UNIV OF
NEW HAMPSHIRE
DURHAM NH 03824
U.S.A.

HOLMBERG ERIK B PROF
ENELIDEN 2
S-433 00 PARTILLE
SWEDEN

HOLWEGER HARTMUT PROF DR
INST THEOR PHY STERN
NEUE UNIV PHYS ZENTR
OLSHAUSENST GEBN 61C
D-2300 KIEL 1
GERMANY, F.R.

HOLZER THOMAS EDWARD DR
HIGH ALTITUDE OBS
NCAR
P O BOX 3000
BOULDER CO 80307
U.S.A.

HONEYCUTT R KENT PROF
ASTRONOMY DEPT
SWAIN HALL WEST
INDIANA UNIVERSITY
BLOOMINGTON IN 47405
U.S.A.

HONG SEUNG SOO DR
DEPT OF ASTRONOMY
COLLEGE NATURAL SCS
SEOUL NATIONAL UNIV
SEOUL 150
REPUBLIC OF KOREA

HOOGHOUDT B G IR
PRINSENLAAN 10
2341 KT OEGSTGEEST
THE NETHERLANDS

HOPPE J A PROF DR
SONNENBERGSTR 12
DDR 69 JENA
G.D.R.

HORAK TOMAS B DR
GEOFYZIKA NP
JECNA 29A
61200 BRNO
CZECHOSLOVAKIA

HORAK ZDENEK PROF DR
VIETNAMSKA 2
16000 PRAHA 6
CZECHOSLOVAKIA

HOREDT GEORG PAUL DR
LUNAR & PLANETARY
LABORATORY
UNIVERSITY ARIZONA
TUCSON AZ 85721
U.S.A.

HORI GENICHIRO PROF
DEPT OF ASTRONOMY
UNIVERSITY OF TOKYO
BUNKYO TOKYO 113
JAPAN

HOROWITZ GERALD PROF
RACAH INST OF PHYS
HEBREW UNIVERSITY
JERUSALEM
ISRAEL

HORSKY JAN DR
DEPT OF THEOR PHYS
PURKYNE UNIVERSITY
KOTLARSKA 2
611 37 BRNO
CZECHOSLOVAKIA

HORSKY Z DR
ASTRONOMICAL INST
CSAV
BUDECSKA 6
120 23 PRAHA
CZECHOSLOVAKIA

HORTON BRIAN H DR
DEPT OF PHYSICS
UNIV OF ADELAIDE
GPO BOX 498
ADELAIDE S A 5001
AUSTRALIA

HORVATH ANDRAS DR
TIT PLANETARIUM &
URANIA OBSERVATORY
BOX 46
H-1476 BUDAPEST
HUNGARY

HOSHI REIUN DR
DEPT OF PHYSICS
RIKKYO UNIVERSITY
NISHIIKEBUKURO 3-CH
TOSHIMA TOKYO 171
JAPAN

HOSKIN MICHAEL A DR
CHURCHILL COLLEGE
CAMBRIDGE CB3 ODS
U.K.

HOSKING ROGER J PROF
MATHEMATICS DEPT
UNIVERSITY WAIKATO
HAMILTON
NEW ZEALAND

HOSOKAWA YOSHIMASA H PROF
SAKIGAOKA 3-4-9
FUNABASHI-CITY
CHIBA PREFECTURE 274
JAPAN

HOSOYAMA KENNOSHUKE DR
INTERNATIONAL LATITU
DE OBSERVATORY
HOSHIGAOKA 2-CHOME
MIZUSAWA IWATE 023
JAPAN

HOTINLI METIN DR
UNIVERSITE
RASATHANESI
UNIVERSITE ISTANBUL
TURKEY

HOUCK JAMES R
ASTRONOMY DEPT
CORNELL UNIVERSITY
220 SPACE SCIENCE BD
ITHACA NY 14853
U.S.A.

HOUK NANCY DR
DEPT OF ASTRONOMY
UNIV OF MICHIGAN
1045 PHYS-ASTRO BLDG
ANN ARBOR MI 48109
U.S.A.

HOUSE FRANKLIN C DR
ESA-ESOC
ROBERT-BOSCH-STR 5
D-6100 DARMSTADT
GERMANY, F.R.

HOUSE LEWIS L DR
HIGH ALTITUDE OBS
NAT CTR ATMOSPH RES
P O BOX 3000
BOULDER CO 80303
U.S.A.

HOUTGAST JAKOB DR
STERREWACHT
ZONNENBURG 2
3512 NK UTRECHT
THE NETHERLANDS

HOUZIAUX L PROF
INST D'ASTROPHYSIQUE
5 AV DE COINTE
B-4200 COINTE-OUGREE
BELGIUM

HOVENIER J W DR
FREE UNIVERSITY
DEPT PHYSICS/ASTRON
DE BOELELAAN 1081
1081 HV AMSTERDAM
THE NETHERLANDS

HOWARD ROBERT F DR
HALE OBSERVATORIES
813 SANTA BARBARA ST
PASADENA CA 91101
U.S.A.

HOWARD W MICHAEL DR
LAWRENCE LIVERMORE
LABORATORY L 71
LIVERMORE CA 94550
U.S.A.

HOWARD WILLIAM E III DR
N STUART ST
ARLINGTON VA 22207
U.S.A.

HOWSE H DEREK
12 BARNFIELD ROAD
RIVERHEAD
SEVENOAKS
KENT TN13 2AY
U.K.

HOYLE FRED SIR
COCKLEY MOOR
DOCKRAY
PENRITH
CUMBRIA CA11 CLG
U.K.

HOYNG PETER DR
SPACE RESEARCH LAB
BENELUXLAAN 21
3527 HS UTRECHT
THE NETHERLANDS

HSIANG-KUANG TSENG
INST PHYS ASTRONOMY
NAT CENTRAL UNIV
CHUNG LI
TAIWAN

HU JING-YAO
BEIJING OBSERVATORY
BEIJING
PEOPLE'S REPUBLIC CHINA

HU NING-SHENG
NANJING ASTRONOMICAL
INSTRUMENTS FACTORY
NANJING
JIANGSU PROVINCE
PEOPLE'S REPUBLIC CHINA

HUA CHON TRUNG DR
VILLA LA MAJORICA
LES EMBUS
F-13190 ALLAUCH
FRANCE

HUANG CHANG-CHUN
PURPLE MOUNTAIN
OBSERVATORY
NANJING
PEOPLE'S REPUBLIC CHINA

HUANG KE-LIANG
ASTRONOMY DEPARTMENT
NANJING UNIVERSITY
NANJING
PEOPLE'S REPUBLIC CHINA

HUANG KUN-YI
PURPLE MOUNTAIN
OBSERVATORY
NANJING
PEOPLE'S REPUBLIC CHINA

HUANG LIN
BEIJING OBSERVATORY
BEIJING
PEOPLE'S REPUBLIC CHINA

HUANG RUN-QIAN
YUNNAN OBSERVATORY
KUNMING
YUNNAN PROVINCE
PEOPLE'S REPUBLIC CHINA

HUANG YINLIANG
YUNNAN OBSERVATORY
KUNMING
YUNNAN PROVINCE
PEOPLE'S REPUBLIC CHINA

HUBBARD WILLIAM B PROF
DEPT OF PLANETARY
SCIENCES
UNIVERSITY ARIZONA
TUCSON AZ 85721
U.S.A.

HUBE DOUGLAS P DR
DEPT OF PHYSICS
UNIVERSITY ALBERTA
EDMONTON ALTA T6G2J1
CANADA

HUBENET HENRI DR
STERRENWACHT
SONNENBORGH
ZONNENBURG 2
UTRECHT 2500
THE NETHERLANDS

HUBENY IVAN
ASTRONOMICAL INST
CZECH ACADEMY OF SC.
ONDREJOV OBSERVATORY
25165 ONDREJOV
CZECHOSLOVAKIA

HUBER MARTIN C DR
ATOM & ASTROPHYSICS
GROUP
ETH ZENTRUM
CH 8092 ZUERICH
SWITZERLAND

HUBERT-DELPLACE
ANNE-MARIE DR
OBSERVATOIRE
DE MEUDON
F-92190 MEUDON
FRANCE

HUCHRA JOHN PETER DR
CTR FOR ASTROPHYSICS
60 GARDEN STREET
CAMBRIDGE MA 02138
U.S.A.

HUCHTMEIER WALTER K DR
MPI FUER
RADIOASTRONOMIE
AUF DEM HUEGEL 69
D-5300 BONN 1
GERMANY, F.R.

HUDSON HUGH S DR
PHYSICS C-011
UCSD
LA JOLLA CA 92093
U.S.A.

HUEBNER WALTER F DR
T-4
LOS ALAMOS SCIENTI-
FIC LABORATORY
LOS ALAMOS NM 87545
U.S.A.

HUGHES DAVID W DR
DEPT OF PHYSICS
THE UNIVERSITY
SHEFFIELD S3 7RH
U.K.

HUGHES JAMES A DR
US NAVAL OBSERVATORY
WASHINGTON DC 20390
U.S.A.

HUGHES VICTOR A PROF
DEPT OF PHYSICS
QUEENS UNIVERSITY
KINGSTON ONT K7L 3N6
CANADA

HUGUENIN G RICHARD
DEPT PHYS AND ASTRON
UNIV OF MASS
GRADUATE RES CTR
AMHERST MA 01002
U.S.A.

HULSBOSCH A N M DR
STERRENKUNDIG INST
KATH UNIV
TOERNOOIVELD
6525 ED NIJMEGEN
THE NETHERLANDS

HUMMER DAVID G DR
JOINT INST F LABORA-
TORY ASTROPHYSICS
BOULDER CO 80309
U.S.A.

HUMPHREYS CURTIS JUDSON DR
6065 RIVERSIDE AVE
RIVERSIDE CA 92506
U.S.A.

HUNDHAUSEN A DR
HIGH ALTITUDE OBS
BOULDER CO 80302
U.S.A.

HUNT GARRY E DR
ATMOSPH PHYS GROUP
BLACKETT LAB
IMPERIAL COLLEGE
LONDON SW7 2BZ
U.K.

HUNTER CHRISTOPHER PROF
MATHEMATICS DEPT
FLORIDA STATE
UNIVERSITY
TALLAHASSEE FL 32306
U.S.A.

HURNIK HIERONIM PROF
OBSERWATORIUM
ASTRONOMICZNE UAM
SLONECZNA 36
60-286 POZNAN
POLAND

HUTCHEON RICHARD J DR
X-RAY ASTRONOMY GR
PHYSICS DEPT
UNIV OF LEICESTER
LEICESTER LE1 7RH
U.K.

HYLAND A R HARRY DR
MT STROMLO
OBSERVATORY
WODEN PO ACT 2606
AUSTRALIA

HYUN JONG-JUNE PROF
SEOUL NAT UNIVERSITY
SINLIM-DONG
KWANAK-KU
SEOUL 151
REPUBLIC OF KOREA

IBADINOV KHURSANDKUL DR
INST OF ASTROPHYSICS
ACAD OF SCS OF THE
TADJIK SSR
734670 DUSHANBE
U.S.S.R.

IBBETSON PETER AARON DR
WISE OBSERVATORY
TEL AVIV UNIVERSITY
TEL AVIV
ISRAEL

ICKE V DR
DEPT OF ASTRONOMY
UNIV OF MINNESOTA
116 CHURCH ST S E
MINNEAPOLIS MN 55455
U.S.A.

IKEUCHI SATORU DR
DEPT OF PHYSICS
HOKKAIDO UNIVERSITY
SAPPORO 060
JAPAN

HUMPHREYS ROBERTA M PROF
ASTRONOMY DEPT
UNIVERSITY OF
MINNESOTA
MINNEAPOLIS MN 55455
U.S.A.

HUNGER KURT PROF
INST THEOR PHY STERN
NEUE UNIV PHYS ZENTR
OLSHAUSENST CB N 61C
D-2300 KIEL 1
GERMANY, F.R.

HUNT ROLAND DR
DEPT OF MATHEMATICS
UNIV OF STRATHCLYDE
26 RICHMOND STREET
GLASGOW G1 1XH
U.K.

HUNTER JAMES H PROF
DEPT OF ASTRONOMY
T W BRYANT SP SC BDG
UNIV OF FLORIDA
GAINESVILLE FL 32611
U.S.A.

HURUHATA MASAAKI PROF
TOKYO ASTRONOMICAL
OBSERVATORY
OSAWA 2-CHOME
MITAKA TOKYO 181
JAPAN

HUTCHINGS JOHN B DR
DOMINION ASTROPH OBS
5071 W SAANICH ROAD
RR 5
VICTORIA BC V8X 4M6
CANADA

HYNEK J ALLEN PROF
DEARBORN OBSERVATORY
NORTHWESTERN
UNIVERSITY
EVANSTON IL 60201
U.S.A.

IANNA PHILIP A
PO BOX 3818
CHARLOTTESVILLE
VA 22903
U.S.A.

IBANEZ MIGUEL DR
UNIV DE LOS ANDES
FACULDAD DE CIENCIAS
DEPTO DE FISICA
MERIDA
VENEZUELA

IBEN ICKO JR PROF
ASTRONOMY DEPT
UNIVERSITY OF
ILLINOIS
URBANA IL 61801
U.S.A.

IDLIS G M DR
INSTITUTE F HISTORY
OF SCI AND TECHNOLO-
GY USSR ACAD OF SCI
103012 MOSCOW
U.S.S.R.

IKHSANOV ROBERT N DR
PULKOVO OBSERVATORY
196140 LENINGRAD
U.S.S.R.

HUMPHRIES C M DR
ROYAL OBSERVATORY
BLACKFORD HILL
EDINBURGH EH9 3HJ
U.K.

HUNSTEAD RICHARD W DR
SCHOOL OF PHYSICS
UNIVERSITY OF SYDNEY
SYDNEY N S W 2006
AUSTRALIA

HUNTEN DONALD M PROF
LUNAR AND PLANETARY
LABORATORY
UNIVERSITY ARIZONA
TUCSON AZ 85721
U.S.A.

HURFORD GORDON JAMES
264-33 CALTECH
PASADENA CA 91125
U.S.A.

HURUKAWA KIICHIRO DR
TOKYO ASTRONOMICAL
OBSERVATORY
OSAWA MITAKA
181 TOKYO
JAPAN

HYDER C L DR
HIGH ALTITUDE OBS
P O BOX 3000
BOULDER CO 80307
U.S.A.

HYSOM E J
65 DALKEITH RD
HARPENDEN
HERTS AL5 5PP
U.K.

IANNINI GUALBERTO DR
OBSERVATORIO ASTRONO
MICO DE CORDOBA
LAPRIDA 854
5000 CORDOBA
ARGENTINA

IBANOGLU C DR
EGE UNIVERSITY
OBSERVATORY
BORNOVA IZMIR
TURKEY

ICHIMARU SETSUO DR
DEPT OF PHYSICS
UNIVERSITY OF TOKYO
BUNKYO-KU
TOKYO 113
JAPAN

IIJIMA SHIGETAKA PROF
TOKYO ASTRONOMICAL
OBSERVATORY
OSAWA MITAKA TOK 181
JAPAN

IKHSANOVA VERA N DR
PULKOVO OBSERVATORY
196140 LENINGRAD
U.S.S.R.

ILL MARTON J DR
TOTH KALMAN U 19
OBSERVATORY
6501 BAJA
HUNGARY

ILLES ALMAR ERZSEBET DR
KONKOLY OBSERVATORY
BOX 67
1525 BUDAPEST
HUNGARY

ILLING R M E
HIGH ALTITUDE OBS
NCAR
P O BOX 3000
BOULDER CO 80307
U.S.A.

ILLINGWORTH GARTH D DR
KITT PEAK NAT OBS
P O BOX 26732
TUCSON AZ 85726
U.S.A.

IMBERT MAURICE DR
OBSERVATOIRE
DE MARSEILLE
2 PLACE LE VERRIER
F-13004 MARSEILLE
FRANCE

IMSHENNIK V S DR
INST THEOR EXP PHYS
CHEREMUSHKINSKAJA 25
117259 MOSCOW
U.S.S.R.

INAGAKI SYOGO DR
INST OF ASTRONOMY
MADINGLEY ROAD
CAMBRIDGE CB3 OHA
U.K.

INGRAO HECTOR C
58 HUNDREDS RD
WELLESLEY HILLS
MA 02181
U.S.A.

INNANEN KIMMO A PROF
CRESS PHYSICS DEPT
YORK UNIVERSITY
4700 KEELE STREET
DOWNSVIEW ONT M3J1P3
CANADA

INOUE HAJIME DR
INST OF SPACE &
ASTRONAUTICAL SC
KOMABA
TOKYO 153
JAPAN

INOUE MAKOTO DR
NOBEYAMA RADIO OBS
NOBEYAMA MINAMI-MAKI
MURA MINAMI-SAKU
NAGANO 384-13
JAPAN

INOUE TAKESHI PROF
KYOTO SANGYO
UNIVERSITY
KYOTO 603
JAPAN

IOANNISIANI B K DR
MAIN ASTRONOMICAL
OBSERVATORY PULKOVO
196140 LENINGRAD
U.S.S.R.

IONSON JAMES ALBERT
NASA GSFC
CODE 682
GREENBELT MD 20771
U.S.A.

IOSHPA B A DR
INSTITUTE OF TERRE-
STRIAL MAGNETISM AND
IONOSPHERE
142092 AKADEMGORODOK
U.S.S.R.

IPSER JAMES R PROF
DEPT OF PHYSICS
WILLIAMSON HALL
UNIV OF FLORIDA
GAINESVILLE FL 32611
U.S.A.

IRELAND JOHN G DR
C-O 13 GORDON RD
BELVEDERE
KENT DA17 6EA
U.K.

IRIARTE B MR
INST NAC DE ASTROFIS
OPT Y ELECTRONICA
APT POSTALES 216 51
PUEBLA PUE
MEXICO

IRVINE WILLIAM M PROF
ASTRONOMY GRC
TOWER B
UNIVERSITY MASS
AMHERST MA 01003
U.S.A.

IRWIN ALAN W DR
DEPT OF PHYSICS
UNIV OF VICTORIA
P O BOX 1700
VICTORIA V8W 2Y2
CANADA

IRWIN JOHN B PROF
2744 TYNDALL AVE
TUCSON AZ 85719
U.S.A.

ISAAK GEORGE R DR
DEPT OF PHYSICS
UNIV OF BIRMINGHAM
P O BOX 363
BIRMINGHAM B15 2TT
U.K.

ISERN JORGE DR
C/SEPULVEDA 83-6-3A
BARCELONA-15
SPAIN

ISHIDA GORO DR
KAMOGATA-2039
ASAGUCHI
OKAYAMA
JAPAN

ISHIDA KEIICHI DR
TOKYO ASTRONOMICAL
OBSERVATORY
2-21-1 OSAWA MITAKA
JAPAN

ISHIZAWA TOSHISUKE A PROF
DEPT OF ASTRONOMY
UNIVERSITY OF KYOTO
KYOTO 606
JAPAN

ISHIZUKA TOSHIHISA DR
DEPT OF PHYSICS
IBARAKI UNIVERSITY
MITO 310
JAPAN

ISOBE SHUZO DR
TOKYO ASTRONOMICAL
OBSERVATORY
181 MITAKA TOKYO
JAPAN

ISRAEL FRANK P DR
SP/ESTEC
POSTBUS 289
2200 AG NOORDWIJK
THE NETHERLANDS

ISRAEL GUY MARCEL DR
LAB D'AERONOMIE
B.P.3
F-91370 VERRIERES-LE
-BUISSON
FRANCE

ISRAEL WERNER PROF
PHYSICS DEPT
UNIVERSITY OF
ALBERTA
EDMONTON ALTA T6G2E1
CANADA

ISSA ALI DR
HELWAN OBSERVATORY
HELWAN-CAIRO
ARAB REP. OF EGYPT

ISSERSTEDT JOERG DR
INSTITUT ASTRONOMIE
UND ASTROPHYSIK
AM HUBLAND
D-7800 WUERZBURG
GERMANY, F.R.

ITO KENSUKE A PROF
RIKKYO UNIVERSITY
NISHI-IKEBUKURO
TOKYO 171
JAPAN

ITOH HIROSHI DR
DEPT OF ASTRONOMY
UNIVERSITY OF KYOTO
KYOTO 606
JAPAN

ITOH NAOKI DR
DEPT OF PHYSICS
SOPHIA UNIVERSITY
7 KIOICHO CHIYODA-KU
TOKYO 102
JAPAN

IVANCHUK VICTOR I DR
KIEV UNIVERSITY
OBSERVATORNAYA 3
252053 KIEV
U.S.S.R.

IVANOV G R DR
DEPT OF ASTRONOMY
ANTON IVANOV ST 5
P O 120
1126 SOFIA
BULGARIA

IVANOV VSEVOLOD V DR
ASTRONOMICAL OBS
LENINGRAD UNIVERSITY
10TH LINIJA 33
199178 LENINGRAD
U.S.S.R.

IVANOV-KHOLODNY G S DR
INSTITUTE OF APPLIED
GEOPHYSICS USSR ACA-
DEMY OF SCIENCES
107150 MOSCOW
U.S.S.R.

IVES JOHN CH DR
MULLARD SPACE SC LAB
UNIV COLLEGE LONDON
LONDON
U.K.

IWANISZEWSKA CECYLIA DR
INST OF ASTRONOMY
N COPERNICUS UNIV
CHOPINA 12 18
87-100 TORUN
POLAND

IWANOWSKA WILHELMINA DR
INST OF ASTRONOMY
CHOPINA 12 18
87-100 TORUN
POLAND

IWASAKI KYOSUKE DR
KWASAN OBSERVATORY
YAMASHINA KYOTO 607
JAPAN

IYE MASANORI DR
TOKYO ASTRONOMICAL
OBSERVATORY
UNIV OF TOKYO
MITAKA 181
JAPAN

IZVEKOV V A DR
INST OF THEORETICAL
ASTRONOMY
10 KUTUZOV QUAY
192187 LENINGRAD
U.S.S.R.

JAAKKOLA TOIVO S
OBSERVATORY
TAHTITORNINMAKI
SF-00130 HELSINKI
FINLAND

JACCHIA LUIGI G DR
CENTER FOR
ASTROPHYSICS
60 GARDEN ST
CAMBRIDGE MA 02138
U.S.A.

JACKISCH G DR
ZENTRALINSTITUT FUER
ASTROPHYSIK
STERNWARTE SONNEBERG
DDR-64 SONNEBERG
G.D.R.

JACKSON J C DR
16 THE PARK
NEWARK
NOTTINHAMS NG24 1SO
U.K.

JACKSON PAUL DR
INST FUER ASTRONOMY
DER UNIV WIEN
TUERKENSCHANZSTR 17
A-1180 WIEN
AUSTRIA

JACKSON PETER DOUGLAS DR
ASTRONOMY PROGRAM
UNIV OF MARYLAND
COLLEGE PK MD 20742
U.S.A.

JACKSON WILLIAM M DR
DEPT OF CHEMISTRY
HOWARD UNIVERSITY
WASHINGTON DC 20059
U.S.A.

JACOBS KENNETH C DR
CENTER 344
BELL LABORATORIES
600 MOUNTAIN AVENUE
MURRAY HILL NJ 07974
U.S.A.

JACOBSEN THEODOR S PROF
6205 17TH AVE NE
SEATTLE WA 98115
U.S.A.

JACQUINOT P DR
LAB AIME COTTON
BAT 505
UNIVERSITE PARIS SUD
F-91405 ORSAY CEDEX
FRANCE

JAEGER FRIEDRICH W PROF
ZENTRALINSTITUT FUER
SOLARTERR PHYSIK
DDR-15 POTSDAM
G.D.R.

JAFFE WALTER JOSEPH DR
NAT RADIO ASTRONOMY
OBSERVATORY
EDGEMONT ROAD
CHARLOTTESVL VA22901
U.S.A.

JAHREISS HARTMUT DR
ASTRONOMISCHES
RECHEN INSTITUT
MOENCHHOFSTR 12-14
D-6900 HEIDELBERG 1
GERMANY, F.R.

JAKIMIEC JERZY DR
ASTRONOMICAL INST
KOPERNIKA 11
51 622 WROCLAW
POLAND

JAKS WALDEMAR DR
ASTRON LATITUDE OBS
SPACE RES CTR PAS
BOROWIEC
63-120 KORNIK
POLAND

JAMAR CLAUDE A J DR
8 RUE DE BATTY
B-4200 COINTE-OUGREE
BELGIUM

JAMES JOHN F MR
SCHUSTER LABORATORY
THE UNIVERSITY
MANCHESTER M13 9PL
U.K.

JAMES RICHARD A DR
DEPT OF ASTRONOMY
THE UNIVERSITY
MANCHESTER M13 9PL
U.K.

JAMESON RICHARD F DR
ASTRONOMY DEPT
LEICESTER UNIVERSITY
LEICESTER LE1 7RH
U.K.

JANES KENNETH A DR
ASTRONOMY DEPT
BOSTON UNIVERSITY
725 COMMONWEALTH AVE
BOSTON MA 02215
U.S.A.

JANICZEK PAUL M DR
US NAVAL OBSERVATORY
WASHINGTON DC 20390
U.S.A.

JANKOVICS ISTVAN DR
KONKOLY OBSERVATORY
BOX 67
1525 BUDAPEST
HUNGARY

JANSSEN MICHAEL ALLEN
JET PROPULSION LAB
MAIL STOP 183-301
4800 OAK GROVE DRIVE
PASADENA CA 91109
U.S.A.

JARNEFELT GUSTAF J PROF
OHJAAJANTIE 3 B 18
SF-00400 HELSINKI 40
FINLAND

JARRETT ALAN H PROF
BOYDEN OBSERVATORY
P O BOX 334
BLOEMFONTEIN 9300
SOUTH AFRICA

JARZEBOWSKI TADEUSZ DR
ASTRONOMICAL
INSTITUTE
KOPERNIKA 11
51-622 WROCLAW
POLAND

JASCHEK CARLOS O R PROF
OBSERVATOIRE
11 R DE L UNIVERSITE
F-67000 STRASBOURG
FRANCE

JASCHEK MERCEDES DR
OBSERVATOIRE
11 R DE L UNIVERSITE
F-67000 STRASBOURG
FRANCE

JASTROW ROBERT
INST F SPACE STUDIES
2880 BROADWAY
NEW YORK NY 10025
U.S.A.

JAUNCEY DAVID L DR
CSIRO DIVISION OF
RADIOPHYSICS
PO BOX 76
EPPING NSW 2121
AUSTRALIA

JAVET PIERRE PROF
AV DE BEAUMONT 36
CH-1012 LAUSANNE
SWITZERLAND

JAYARAJAN A P MR
ASTROPHYSICAL OBS
KODAIKANAL 3
INDIA

JEFFERIES JOHN T DR
INST FOR ASTRONOMY
2680 WOODLAWN DR
HONOLULU HI 96822
U.S.A.

JEFFERS STANLEY DR
CRESS PHYSICS DEPT
YORK UNIVERSITY
4700 KEELE ST
DOWNSVIEW M3J 1P3
CANADA

JEFFERYS WILLIAM H DR
ASTRONOMY DEPT
UNIVERSITY OF TEXAS
AUSTIN TX 78712
U.S.A.

JEFFREYS HAROLD PROF SIR
160 HUNTINGDON ROAD
CAMBRIDGE CB3 OLB
U.K.

JELLEY JOHN V PHD FRAS
NP-DIV BLDG 8 AERE
HARWELL OXON OX110RA
U.K.

JENKINS EDWARD B DR
PRINCETON UNIVERSITY
OBSERVATORY
PRINCETON NJ 08540
U.S.A.

JENKINS L F MS
YALE UNIV OBS
BOX 2023
YALE STATION
NEW HAVEN CT 06520
U.S.A.

JENKNER HELMUT DR
SPACE TELESC SC INST
JOHNS HOPKINS UNIV
HOMEWOOD CAMPUS
BALTIMORE MD 21218
U.S.A.

JENNER DAVID C PROF
DEPT OF ASTRONOMY
UNIV.OF WASHINGTON
SEATTLE, WA 98195
U.S.A.

JENNINGS R E DR
DEPT PHYSICS AND AST
UNIVERSITY COLLEGE
GOWER STREET
LONDON WC1E 6BT
U.K.

JENNISON ROGER C PROF
ELECTRONICS LAB
UNIVERSITY OF KENT
CANTERBURY CT2 7NT
U.K.

JENSCH A
PESTALOZZISTR 9
DDR 69 JENA
G.D.R.

JENSEN EBERHART PROF
INSTITUTE OF THEORE-
TICAL ASTROPHYSICS
UNIVERSITY OF OSLO
BLINDERN OSLO 3
NORWAY

JERZYKIEWICZ MIKOLAJ DR
ASTRONOMICAL INST
KOPERNIKA 11
51-622 WROCLAW
POLAND

JIANG SHI-YANG
BEIJING ASTRONOMICAL
OBSERVATORY
BEIJING
PEOPLE'S REPUBLIC CHINA

JIN WEN-JING
SHANGHAI OBSERVATORY
SHANGHAI
PEOPLE'S REPUBLIC CHINA

JOCKERS KLAUS DR
MPI FUER AERONOMIE
POSTFACH 20
D-3411 KATLENBURG-L
GERMANY, F.R.

JOHANSEN KAREN T LEKTOR
IGELSO
DK-4300 HOLBAEK
DENMARK

JOHANSSON L E B DR
RAGAKERN 18
S-417 26 GOETEBORG
SWEDEN

JOHNSON DONALD R DR
716 TIFFANY COURT
GAITHERSBURG MD20760
U.S.A.

JOHNSON FRED M PROF DR
DEPT PHYS AND ASTRON
CALIF STATE UNIV
2308 E AMERIGE AVE
FULLERTON CA 92634
U.S.A.

JOHNSON HOLLIS R PROF
ASTRONOMY DEPT
INDIANA UNIVERSITY
BLOOMINGTON IN 47405
U.S.A.

JOHNSON HUGH M DR
DEPT 52-12 BLDG 255
LOCKHEED MISSILES
3170 PORTER DRIVE
PALO ALTO CA 94304
U.S.A.

JOHNSON MARTIN DR
PHYSICS DEPT
UNIVERSITY
BIRMINGHAM B15 2TT
U.K.

JOHNSON TORRENCE V DR
JET PROPULSION
LABORATORY
4800 OAK GROVE DRIVE
PASADENA CA 91103
U.S.A.

JOHNSTON KENNETH
CODE 7134
NAVAL RESEARCH
LABORATORY
WASHINGTON DC 20375
U.S.A.

JOKIPII J R PROF
DEPT OF PLANETARY SC
UNIVERSITY OF
ARIZONA
TUCSON AZ 85721
U.S.A.

JOLY FRANCOIS DR
UNIVERSITE DE
BORDEAUX I
40 RUE LAMARTINE
F-33400 TALENCE
FRANCE

JONES ALBERT F MR
31 RANUI ROAD
STOKE
NELSON
NEW ZEALAND

JONES BARBARA
CASS C-011
UNIV OF CALIFORNIA
AT SAN DIEGO
LA JOLLA CA 92093
U.S.A.

JONES BERNARD J T DR
INST OF ASTRONOMY
MADINGLEY ROAD
CAMBRIDGE
U.K.

JONES BURTON DR
LICK OBSERVATORY
UNIVERSITY
OF CALIFORNIA
SANTA CRUZ CA 95064
U.S.A.

JONES DEREK H P DR
ROYAL GREENWICH
OBSERVATORY
HAILSHAM
EAST SUSSEX BN27 1RP
U.K.

JONES ERIC M
LOS ALAMOS SCIENTI-
FIC LABORATORY
LOS ALAMOS NM 87545
U.S.A.

JONES FRANK CULVER DR
NASA-GSFC
CODE 660
GREENBELT MD 20771
U.S.A.

JONES HARRISON PRICE DR
KITT PEAK NAT OBS
SOLAR STATION
900 N CHERRY AVE
TUCSON AZ 85726
U.S.A.

JONES JAMES DR
DEPT OF PHYSICS
UNIV WESTERN ONTARIO
LONDON ONT N6A 5B9
CANADA

JONES JANET E DR
INST OF ASTRONOMY
MADINGLEY ROAD
CAMBRIDGE
U.K.

JONES THOMAS WALTER DR
DEPT OF ASTRONOMY
UNIV OF MINNESOTA
116 CHURCH ST SE
MINNEAPOLIS MN 55455
U.S.A.

JORDAN CAROLE DR
DEPT THEORETICAL PHY
1 KEBLE ROAD
OXFORD UNIVERSITY
OXFORD OX1 3NP
U.K.

JORDAN H L DR DIREKTOR
INST PLASMAPHYSIK
KERNFORSCHUNGSANLAGE
JUELICH GMBH PF 365
D-5170 JUELICH 1
GERMANY, F.R.

JORDAN STUART D DR
LAB ASTRO SOLAR PHYS
NASA-GODDARD SPACE
FLIGHT CTR CODE 682
GREENBELT MD 20771
U.S.A.

JORDEN PAUL RICHARD
ROYAL GREENWICH OBS
HERSTMONCEUX CASTLE
HAILSHAM
E SUSSEX BN27 1RP
U.K.

JORGENSEN HENNING E LEKT
UNIVERSITY
OBSERVATORY
OSTERVOLDGADE 3
DK-1350 COPENHAGEN K
DENMARK

JOSEPH J H DR
DEPT OF GEOPHYSICS
AND PLANETARY SC
TEL-AVIV UNIVERSITY
RAMAT AVIV
ISRAEL

JOSEPH ROBERT D DR
BLACKETT LAB
ASTRONOMY GROUP
IMPERIAL COLLEGE
LONDON SW7 2BZ
U.K.

JOSHI MOHAN N DR
RADIO ASTRONOMY
CENTRE OF TIFR
POST BOX 8
OOTACAMUND 643001
INDIA

JOSHI SURESH CHANDRA DR
UP STATE OBSERVATORY
MANORA PEAK
NAINI TAL 263129
INDIA

JOSS PAUL CHRISTOPHER DR
MIT
ROOM 6-203
CAMBRIDGE MA 02139
U.S.A.

JOVANOVIC BOZIDAR
FAC OF AGRICULTURE
INST WATERRANGING
VELJKA VLAHOVICA
YU-21000 NOVI SAD
YUGOSLAVIA

JUGAKU JUN DR
TOKYO ASTRONOMICAL
OBSERVATORY
OSAWA MITAKA
TOKYO 181
JAPAN

JUNG JEAN DR
AMBASS FRANCE WASH
FRENCH SC MISSION
2011 EYE STREET NW
WASHINGTON DC 20006
U.S.A.

JUNKES JOSEF DR
12 HOHENBRUNNERSTR
D-8014 NEUBIBERG
GERMANY, F.R.

JUPP ALAN A DR
APPL MATH THEOR PHYS
UNIVERSITY LIVERPOOL
P O BOX 147
LIVERPOOL L69 3BX
U.K.

JURA MICHAEL DR
DEPT OF ASTRONOMY
UCLA
LOS ANGELES CA 90024
U.S.A.

JURGENS RAYMOND F
M S 238/420
JET PROPULSION LAB
4800 OAK GROVE DRIVE
PASADENA CA 91109
U.S.A.

JURKEVICH IGOR DR
3130 PORT WAY
ANNAPOLIS MD 21403
U.S.A.

KABURAKI MASAKI PROF
4-24-9 KICHIJYOJI
MINAMI
MUSASHINO TOKYO 180
JAPAN

KABURAKI OSAMU DR
ASTRONOMICAL INST
TOHOKU UNIVERSITY
SENDAI 980
JAPAN

KADLA ZDENKA I DR
PULKOVO OBSERVATORY
196140 LENINGRAD
U.S.S.R.

KAEHLER HELMUTH DR
HAMBURGER STERNWARTE
GOJENBERGSWEG 112
D-2050 HAMBURG 80
GERMANY, F.R.

KAFATOS MINEAS DR
PHYSICS DEPT
GEORGE MASON UNIV
FAIRFAX VA 22030
U.S.A.

KAFKA PETER
MPI FUER PHYSIK UND
ASTROPHYSIK
8000 MUENCHEN 40
GERMANY, F.R.

KAFTAN MAY A DR
ASTRONOMICAL OBS
FOUND OF SCIENTIFIC
RESEARCH P O BOX 255
JADIRIYAH BAGHDAD
REP. OF IRAQ

KAHLER STEPHEN W DR
AMERICAN SCIENCE AND
ENGINEERING INC
955 MASSACHUSETTS AV
CAMBRIDGE MA 02139
U.S.A.

KAHN FRANZ D PROF
DEPT OF ASTRONOMY
THE UNIVERSITY
MANCHESTER M13 9PL
U.K.

KAI KEIZO DR
NOBEYAMA SOLAR RADIO
OBSERVATORY
NOBEYAMA MINAMIMAKI
NAGANO 384-13
JAPAN

KAKINUMA TAKAKIYO T PROF
RESEARCH INSTITUTE
OF ATMOSPHERICS
NAGOYA UNIVERSITY
TOYOKAWA AICHI 442
JAPAN

KALER JAMES B PROF
UNIVERSITY OF
ILLINOIS OBSERVATORY
URBANA IL 61801
U.S.A.

KALLOGLIAN ARSEN T DR
BYURAKAN ASTROPHYSI-
CAL OBSERVATORY
378433 BYURAKAN
ARMENIA
U.S.S.R.

KALNAJS AGRIS J DR
MT STROMLO
OBSERVATORY
WODEN PO ACT 2606
AUSTRALIA

KAMINISHI KEISUKE PROF
FACULTY OF SCIENCE
KUMAMOTO UNIVERSITY
KUROKAMI 5 CHOME
860 KUMAMOTO
JAPAN

KANAEV IVAN I DR
PULKOVO OBSERVATORY
196140 LENINGRAD
U.S.S.R.

KANDPAL C D
UTTER PRADESH
STATE OBSERVATORY
NAINITAL UP
INDIA

KANNO MITSUO DR
HIDA OBSERVATORY
KAMITAKARA
GIFU-KEN 506 13
JAPAN

KAPLAN J DR
DEPT OF PHYSICS
UCLA
LOS ANGELES CA 90024
U.S.A.

KARAALI SALIH DR
ISTANBUL UNIVERSITY
FACULTY OF SCIENCE
ISTANBUL
TURKEY

KARDASHEV N S DR
SPACE RESEARCH
INSTITUTE
USSR ACADEMY OF SCI
117810 MOSCOW
U.S.S.R.

KAIFU NORIO DR
TOKYO ASTRONOMICAL
OBSERVATORY
OSAWA MITAKA
TOKYO 181
JAPAN

KAKUTA CHUICHI DR
INTERNATIONAL LATITU
DE OBSERVATORY
HOSHIGAOKA 2-CHOME
MIZUSAWA IWATE 023
JAPAN

KALINKOV MARIN P DR
DEPT OF ASTRONOMY
BULGARIAN AC OF SC
7 NOVEMBER ST 1
1000 SOFIA
BULGARIA

KALMAN BELA DR
HELIOPHYSICAL OBS
P O BOX 30
4010 DEBRECEN
HUNGARY

KAMEL OSMAN M DR
FACULTY OF SCIENCES
ASTRONOMY DEPT
CAIRO UNIVERSITY
GIZA
ARAB REP. OF EGYPT

KAMP LUCAS WILLEM DR
DEPT OF ASTRONOMY
BOSTON UNIV
725 COMMONWEALTH AVE
BOSTON MA 02215
U.S.A.

KANDEL ROBERT S DR
SERV AERONOMIE CNRS
B P NO 3
F-91370 VERRIERES-LE
BUISSON
FRANCE

KANE SHARAD R DR
SPACE SCIENCES LAB
UNIVERSITY OF CALI-
FORNIA-BERKELEY
BERKELEY CA 94720
U.S.A.

KANYO SANDOR DR
KONKOLY OBSERVATORY
BOX 67
1525 BUDAPEST
HUNGARY

KAPLAN LEWIS D DR
ATMOSPHERIC E ENVI-
RONMENTAL RES INC
840 MEMORIAL DRIVE
CAMBRIDGE MA 02139
U.S.A.

KARACHENTSEV I D DR
SPECIAL ASTROPHYSI-
CAL OBSERVATORY
USSR ACADEMY OF SCI
357140 N ARKHYZ
U.S.S.R.

KARETNIKOV VALENTIN G R
ODESSA STATE UNIV
270014 ODESSA
U.S.S.R.

KAISER T R PROF
DEPT OF PHYSICS
THE UNIVERSITY
SHEFFIELD S3 7RH
U.K.

KALANDADZE N B DR
ABASTUMANY ASTROPHY-
SICAL OBSERVATORY
383762 ABASTUMANY
U.S.S.R.

KALKOFEN W DR
SMITHSONIAN ASTROPHY
SICAL OBSERVATORY
60 GARDEN ST
CAMBRIDGE MA 02138
U.S.A.

KALMYKOV A M DR
ASTRONOMICAL INSTITU
TE UZBEKIAN ACADEMY
OF SCIENCES
700000 TASHKENT
U.S.S.R.

KAMIJO FUMIO PROF DR
DEPT OF ASTRONOMY
UNIVERSITY OF TOKYO
YAYOI BUNKYO KU
TOKYO 113
JAPAN

KAMPER KARL W DR
DAVID DUNLAP
OBSERVATORY
RICHMOND HILL
ONT L4C 4Y6
CANADA

KANDEMIR GUELCIN
ITUMMF
MAT-FIZIK
MACKA
ISTANBUL
TURKEY

KANEKO NOBORU DR
DEPT OF PHYSICS
FACULTY OF SCIENCE
HOKKAIDO UNIVERSITY
060 SAPPORO
JAPAN

KAPAHI V K DR
TIFR CENTRE
INDIAN INSTITUTE OF
SCIENCE
BANGALORE 560 012
INDIA

KAPOOR RAMESH CHANDER
INDIAN INSTITUTE OF
ASTROPHYSICS
BANGALORE 560 034
INDIA

KARANDIKAR R V PROF
DEPT OF ASTRONOMY
OSMANIA UNIVERSITY
HYDERABAD 500 007
INDIA

KARP ALAN HERSH DR
IBM PALO ALTO SC CTR
1530 PAGE MILL RD
PALO ALTO CA 94304
U.S.A.

KARPINSKIJ VADIM N DR
PULKOVO OBSERVATORY
196140 LENINGRAD
U.S.S.R.

KARYGINA ZOYA V DR
ASTROPHYSICAL INST
480068 ALMA-ATA
U.S.S.R.

KASHSCHEEV B L PROF DR
KHARKOV INSTITUTE
FOR RADIOELECTRONICS
310059 KHARKOV
U.S.S.R.

KASPER U DR
ZNTRINST ASTROPHYSIK
STERNW BABELSBERG
R LUXEMBURG STR 17A
DDR-1502 POTSDAM
G.D.R.

KASTLER ALFRED PROF
ECOLE NORMALE
SUPERIEURE
45 RUE D ULM
F-75005 PARIS
FRANCE

KASTURIRANGAN K DR
ISRO
STAELLITE CENTRE
BANGALORE 560 058
INDIA

KATASEV L A DR
ASTRONOMICAL COUNCIL
USSR ACADEMY OF SCI
PYATNITSKAYA UL 48
109017 MOSCOW
U.S.S.R.

KATGERT PETER DR
STERREWACHT
POSTBUS 9513
2300 RA LEIDEN
THE NETHERLANDS

KATGERT-MERKELIJN J K DR
STERREWACHT
HUYGENS LAB
WASSENAARSEWEG 78
2300 RA LEIDEN
THE NETHERLANDS

KATO SHOJI PROF
DEPT OF ASTRONOMY
UNIVERSITY OF KYOTO
KITASHIRAKAWA OIWAKE
SAKYOKU KYOTO 606
JAPAN

KATO TAKAKO DR
INST OF PLASMA PHYS
NAGOYA UNIVERSITY
RURO-CHO CHIKUSA-KU
NAGOYA 464
JAPAN

KATSIS DEM DR
12 RUE VARNIS
NEA SMYRNE
ATHENS
GREECE

KATZ JONATHAN I
DEPT OF PHYSICS
WASHINGTON UNIV
ST LOUIS MI 63105
U.S.A.

KATZ JOSEPH DR
RACAH INST OF PHYS
HEBREW UNIV
JERUSALEM
ISRAEL

KAUFMAN MICHELE DR
PHYSICS DEPT
OHIO ST UNIVERSITY
174 W 18TH AVE
COLUMBUS OH 43210
U.S.A.

KAUFMANN JENS PETER DR
INST F ASTROPHYSIK
T U
ERNST REUTER PLATZ 7
D-1000 BERLIN 10
GERMANY, F.R.

KAUFMANN PIERRE PROF
INST PESQ ESPACIAIS
CRAAM RADIO OBS
C P 515
12200 S JOSE CAMPOS
BRAZIL

KAULA WILLIAM M PROF
INST OF GEOPHYSICS
UNIVERSITY
OF CALIFORNIA
LOS ANGELES CA 90024
U.S.A.

KAWABATA KINAKI PROF
DEPT OF PHYSICS
NAGOYA UNIVERSITY
FUROCHO CHIKUSAKU
NAGOYA 464
JAPAN

KAWABATA SHUSAKU PROF
KYOTO GAKUEN
UNIVERSITY
SOGABE MACHI KAMEOKA
KYOTO 621
JAPAN

KAWAGUCHI ICHIRO PROF
DEPT OF ASTRONOMY
FACULTY OF SCIENCE
UNIVERSITY OF KYOTO
606 KYOTO
JAPAN

KAWATA YOSHIYUKI DR
KANAZAWA INSTITUTE
OF TECHNOLOGY
NONOICHO
KANAZAWA MINAMI 921
JAPAN

KAZANTIS PANAYOTIS DR
DEPT OF ASTRONOMY
UNIV OF GLASGOW
GLASGOW G12 8QQ
U.K.

KAZES ILYA DR
OBSERVATOIRE
DE MEUDON
F-92190 MEUDON
FRANCE

KEAY COLIN S L PROF
PHYSICS DEPT
NEWCASTLE UNIVERSITY
NEWCASTLE NSW 2308
AUSTRALIA

KEENAN PHILIP C PROF EMER
PERKINS OBSERVATORY
BOX 449
DELAWARE OH 43015
U.S.A.

KEGEL WILHELM H PROF
INST THEOR PHYSIK
UNIV FRANKFURT
R-MAYER-STR 8-10
D-6000 FRANKFURT/M 1
GERMANY, F.R.

KEIL STEPHEN L
SACRAMENTO PEAK OBS
SUNSPOT NM 88349
U.S.A.

KELLER CHARLES F
UNIVERSITY OF
CALIFORNIA LASL
BOX 1663 MS 420
LOS ALAMOS NM 87545
U.S.A.

KELLER GEOFFREY
410 HIGHGATE AVE
WORTHINGTON OH 43085
U.S.A.

KELLER H UWE DR
MPI FUER AERONOMIE
D-3411 KATLENBURG
LINDAU 3
GERMANY, F.R.

KELLER HANS U DR
OBSERVATORY AND
PLANETARIUM
POB 161 NECKARST 47
D-7000 STUTTGART
GERMANY, F.R.

KELLERMANN KENNETH I DR
NATIONAL RADIO ASTRO
NOMY OBSERVATORY
P O BOX 2
GREEN BANK WV 24944
U.S.A.

KELLOGG EDWIN M DR
17 GRAY ST
CAMBRIDGE MA 02138
U.S.A.

KELLOGG WILLIAM W DR
NATIONAL CTR FOR
ATMOSPHERIC RESEARCH
BOULDER CO 80302
U.S.A.

KEMBHAVI AJIT K
INST OF ASTRONOMY
MADINGLEY ROAD
CAMBRIDGE CB3 OHA
U.K.

KENDERDINE SIDNEY DR
MULLARD RADIO ASTRO-
NOMY LABORATORY
MADINGLEY ROAD
CAMBRIDGE CB3 OHE
U.K.

KENNEDY JOHN E PROF
UNIV OF SASKATCHEWAN
SASKATOON
SASKATCHEWAN S7N OWO
CANADA

KENNICUTT ROBERT C JR
DEPT OF ASTRONOMY
UNIV OF MINNESOTA
116 CHURCH ST S E
MINNEAPOLIS MN 55455
U.S.A.

KERES KH P PROF DR
IAPHA ESTONIAN ACADE
MY OF SCIENCES
TORAVERE
202444 TARTJ
U.S.S.R.

KERR FRANK J DR
PROVOST
MATH PHYS SCS EN DIV
UN OF MARYLAND
COLLEGE PK MD 20742
U.S.A.

KERR ROY P PROF
UNIV OF CANTERBURY
PRIVATE BAG
CHRISTCHURCH
NEW ZEALAND

KESSLER KARL G DR
CTR ABS PHYS QUANTIT
NAT BUREAU STANDARDS
B-164 PHYSICS BLDG
WASHINGTON DC 20234
U.S.A.

KESTEVEN MICHAEL J L DR
QUEEN S UNIVERSITY
KINGSTON ONT K7L 3N6
CANADA

KHACHIKIAN E Y DR
BYURAKAN ASTROPHYSI-
CAL OBSERVATORY
378433 ARMENIA
U.S.S.R.

KHARADZE E K PROF
ABASTUMANY ASTROPHY-
SICAL OBSERVATORY
383762 ABASTUMANY
GEORGIA
U.S.S.R.

KHARE BISHUN N DR
306 SPACE SC BLDG
CORNELL UNIVERSITY
ITHACA NY 14853
U.S.A.

KHARIN A S DR
MAIN ASTRON OBS
UKRAINIAN ACAD OF SC
GOLOSEEVO
252127 KIEV
U.S.S.R.

KHARITONOV ANDREJ V DR
ASTROPHYSICAL INST
480068 ALMA-ATA
U.S.S.R.

KHATSIASHVILI ALFEZ SH DR
ABASTUMANY ASTRO-
PHYSICAL OBSERVATORY
383762 ABASTUMANY
U.S.S.R.

KHETSURIANI T S DR
ABASTUMANY ASTROPHY-
SICAL OBSERVATORY
383762 ABASTUMANY
U.S.S.R.

KHOKHLOVA V L DR
ASTRONOMICAL COUNCIL
USSR ACADEMY OF SCI
PYATNITSKAYA UL 48
109017 MOSCOW
U.S.S.R.

KHOLOPOV P N DR
STERNBERG STATE AS-
TRONOMICAL INSTITUTE
117234 MOSCOW
U.S.S.R.

KHOLSHEVNIKOV K V DR
LENINGRAD STATE UNI-
VERSITY ASTRONOMICAL
OBSERVATORY
199178 LENINGRAD
U.S.S.R.

KHROMOV G S DR
ASTRONOMICAL COUNCIL
USSR ACADEMY OF SCI
PYATNITSKAYA UL 48
109017 MOSCOW
U.S.S.R.

KIANG TAO PROF
DUNSINK OBSERVATORY
CASTLEKNOCK
CO DUBLIN
IRELAND

KIASATPOOR AHMAD PROF
PHYSICS DEPT
UNIVERSITY OF
ISFAHAN
DANESHGAH E ISFAHAN
IRAN

KIBBLEWHITE EDWARD J DR
INSTITUTE OF
ASTRONOMY
MADINGLEY RD
CAMBRIDGE CB3 OHE
U.K.

KIGUCHI MASAYOSHI DR
DEPT OF PHYSICS
KYOTO UNIVERSITY
KYOTO 606
JAPAN

KIKUCHI SADAEMON PROF
ASTRONOMICAL OBS
TOHOKU UNIVERSITY
ARAMAKI
SENDAI 980
JAPAN

KILADZE R I DR
ABASTUMANI ASTROPHY-
CAL OBSERVATORY
383762 ABASTUMANY
U.S.S.R.

KILAR BOGDAN DR
FACULTY OF GEODESY
LJUBLJANA UNIVERSITY
JAMOVA 2
61000 LJUBLJANA
YUGOSLAVIA

KILKENNY DAVID DR
SOUTH AFRICAN ASTRO-
NOMICAL OBSERVATORY
P O BOX 9
OBSERVATORY 7935
SOUTH AFRICA

KIM YONG HEK DR
ACADEMY OF SCIENCES
PYONGYANG ASTRONOMI-
CAL OBSERVATORY
PYONGYANG
DEM PEOPLE'S REP OF KOREA

KIMURA HIROSHI DR
PURPLE MOUNTAIN OBS
ACADEMIA SINICA
NANKING
PEOPLE'S REPUBLIC CHINA

KING ANDREW R DR
ASTRONOMY DEPT
UNIV OF LEICESTER
LEICESTER
U.K.

KING DAVID S PROF
DEPT OF PHYSICS AND
ASTRONOMY
UNIVERSITY N MEXICO
ALBUQUERQUE NM 87131
U.S.A.

KING HENRY C DR
TRILLIUM 5 HILLBURY
CLOSE CHESHAM BOIS
AMERSHAM
BUCKS HP6 5LB
U.K.

KING IVAN R PROF
ASTRONOMY DEPT
UNIVERSITY OF
CALIFORNIA
BERKELEY CA 94720
U.S.A.

KING R B DR
P O BOX 725
MEDOCINO CA 95460
U.S.A.

KING ROBERT WILSON JR DR
DEPT EARTH PLANET SC
MIT 54-627
CAMBRIDGE MA 02139
U.S.A.

KING-HELE DESMOND G MR
ROYAL AIRCRAFT EST
FARNBOROUGH HANTS
U.K.

KINGSTON ARTHUR E DR
DEPT OF APPLIED MATH
AND THEORET PHYSICS
QUEEN'S UNIVERSITY
BELFAST BT7 1NN
U.K.

KINMAN THOMAS D DR
KITT PEAK NATIONAL
OBSERVATORY
P O BOX 26732
TUCSON AZ 85726
U.S.A.

KINOSHITA HIROSHI DR
TOKYO ASTRONOMICAL
OBSERVATORY
OSAWA MITAKA
181 TOKYO
JAPAN

KIPPENHAHN RUDOLF PROF
MPI FUER PHYSIK UND
ASTROPHYSIK
K SCHWARZSCHILD ST 1
D-8046 GARCHING B M
GERMANY, F.R.

KIPPER AKSEL PROF
IAPHA
ESTONIAN ACAD OF SC
TORAVERE
20244 TARTU
U.S.S.R.

KIPPER TONUA DR
IAPHA
ESTONIAN ACAD OF SC
TORAVERE
20244 TARTU
U.S.S.R.

KIRAL ADNAN PROF
UNIV OBSERVATORY
BEYAZIT
ISTANBUL
TURKEY

KIRBIYIK HALIL DR
MIDDLE EAST TECHNI-
CAL UNIVERSITY
PHYSICS DEPT
ANKARA
TURKEY

KIRK JOHN DR
MPI FUER ASTROPHYSIK
KARL-SCHWARZSCHILD-
STR 1
D-8046 GARCHING B M
GERMANY, F.R.

KIRKPATRICK RONALD C DR
MS 220
LOS ALAMOS SCIENTI-
FIC LABORATORY
LOS ALAMOS NM 87545
U.S.A.

KIRSHNER ROBERT PAUL DR
DEPT OF ASTRONOMY
UNIV OF MICHIGAN
ANN ARBOR MI 48103
U.S.A.

KISELEV ALEXEJ A DR
PULKOVO OBSERVATORY
196140 LENINGRAD
U.S.S.R.

KISLYAKOV ALBERT G DR
APPLIED PHYSICS INST
ACADEMY OF SCIENCE
ULYANOV STREET 46
603600 GORKY
U.S.S.R.

KISLYUK V S DR
MAIN ASTRON OBS
UKRAINIAN ACAD OF SC
GOLOSEEVO
252127 KIEV
U.S.S.R.

KISSELL KENNETH E DR
2202 MALBON AVE
SIMI VALLEY CA 93063
U.S.A.

KITAMURA M PROF
TOKYO ASTRONOMICAL
OBSERVATORY
MITAKA TOKYO 181
JAPAN

KITAMURA SEIICHI DR
SHIGA UNIVERSITY
OTSU SHIGA 520
JAPAN

KITCHIN CHRISTOPHER R DR
HATFIELD POLYTECHNIC
OBSERVATORY
BAYFORDBURY
HERTFORD HERTS
U.K.

KIZILIRMAK ABDULLAH PROF
EGE UNIVERSITY
SCIENCE FACULTY
ASTRONOMY DEPT.
BORNOVA IZMIR
TURKEY

KJAERGAARD PER DR
COPENHAGEN UNIVERSI-
TY OBSERVATORY
OSTER VOLDGADE 3
DK-1350 COPENHAGEN
DENMARK

KJELDSETH MOE OLAF DR
INSTITUTE OF THEORE-
TICAL ASTROPHYSICS
UNIVERSITY OF OSLO
BLINDERN OSLO 3
NORWAY

KLARE GERHARD DR
LANDESSTERNWARTE
KOENIGSTUHL
D-6900 HEIDELBERG 1
GERMANY, F.R.

KLARMANN JOSEPH PROF
DEPT OF PHYSICS
WASHINGTON
UNIVERSITY
ST LOUIS MO 63130
U.S.A.

KLECZEK JOSIP DR
ASTRONOMICAL
INSTITUTE
25165 ONDREJOV
CZECHOSLOVAKIA

KLEIN MICHAEL J DR
SPACE SCIENCES DIV
JET PROPULSION LAB
BLDG 183B-365
PASADENA CA 91103
U.S.A.

KLEIN RICHARD I DR
KITT PEAK NAT OBS
P O BOX 26732
TUCSON AZ 85726
U.S.A.

KLEINMANN DOUGLAS E DR
CENTER FOR ASTROPHYS
60 GARDEN ST
CAMBRIDGE MA 02138
U.S.A.

KLEMOLA ARNOLD R DR
LICK OBSERVATORY
UCSC
SANTA CRUZ CA 95064
U.S.A.

KLEMPERER W K DR
DIV ASTRONOMICAL SC
NAT SCIENCE FOUND
1800 G STREET N W
WASHINGTON DC 20550
U.S.A.

KLEPCZYNSKI WILLIAM J DR
US NAVAL OBSERVATORY
WASHINGTON DC 20390
U.S.A.

KLIMISHIN I A DR
PEDAGOGIC INSTITUTE
284000 IVANO-
FRANKOV SK
U.S.S.R.

KLINGLESMITH DANIEL A DR
GSFC
CODE 671
GREENBELT MD 20771
U.S.A.

KLIORE ARVYDAS JOSEPH DR
JPL
4800 OAK GROVE DRIVE
PASADENA CA 91103
U.S.A.

KLOCK B L DR
US NAVAL OBSERVATORY
WASHINGTON DC 20390
U.S.A.

KLVANA MIROSLAV
ASTRONOMICAL INST
CZECH ACADEMY OF SC
ONDREJOV OBSERVATORY
25165 ONDREJOV
CZECHOSLOVAKIA

KNACK ROGER F DR
DEPT EARTH SPACE SCS
SUNY
STONY BROOK NY 11794
U.S.A.

KNAFLICH H B MS
BOEING COMPANY
AEROSPACE GROUP
M S 88-21 POB 3999
SEATTLE WA 98124
U.S.A.

KNAPP GILLIAN R DR
DEPT ASTROPHYS SCS
PRINCETON UNIVERSITY
PRINCETON NJ 08544
U.S.A.

KNEER F DR
KIEPENHEUER INST
FUER SONNENPHYSIK
SCHOENECKSTR 6
D-7800 FREIBURG-BR
GERMANY, F.R.

KNIFFEN DONALD A DR
NASA-GSFC
CODE 662
GREENBELT MD 20771
U.S.A.

KNOSKA STEFAN
ASTRONOMICAL INST
SLOVAK ACADEMY OF SC
05960 TATRANSKA
LOMNICA
CZECHOSLOVAKIA

KNOWLES STEPHEN H DR
CODE 7134
NAVAL RESEARCH
LABORATORY
WASHINGTON DC 20375
U.S.A.

KNUDE JENS KIRKESKOV DR
COPENHAGEN UNIV
OBSERVATORY
OSTER VOLDGADE 3
DK-1350 COPENHAGEN-K
DENMARK

KO HSIEN C PROF
DEPT OF ELECTRICAL
ENGINEERING
OHIO ST UNIVERSITY
COLUMBUS OH 43210
U.S.A.

KOBAYASHI EISUKE DR
DEPT OF EARTH SCS
SC ED INST OSAKA PR
4-15 KARITA-CHO SUMI
YOSHI-KU OSAKA 558
JAPAN

KOBRIN M M PROF DR
INSTITUTE OF APPLIED
PHYSICS
603600 GORKIJ
U.S.S.R.

KOCER D DR
UNIV OBSERVATORY
UNIVERSITY
ISTANBUL
TURKEY

KOCH LYDIE DR
DPH EP ES BAT 28
CEN SACLAY
B P 2
F-91190 GIF S YVETTE
FRANCE

KOCH ROBERT H DR
DEPT OF ASTRONOMY
UNIV OF PENNA DRL
33RD AND WALNUT STS
PHILADELPHIA PA19104
U.S.A.

KOCHAR R K DR
INDIA INSTITUTE OF
ASTROPHYSICS
BANGALORE 560 034
INDIA

KOCHAROV GRANT E PROF
PHYSICAL
TECHN INSTITUTE
ACAD SCIENCES USSR
194021 LENINGRAD
U.S.S.R.

KODAIRA KEIICHI A PROF
DEPT OF ASTRONOMY
FACULTY OF SCIENCE
UNIVERSITY OF TOKYO
113 TOKYO BUNKYO-KU
JAPAN

KOECKELENBERGH ANDRE DR
OBSERVATOIRE ROYAL
DE BELGIQUE
AVE CIRCULAIRE 3
B-1180 BRUXELLES
BELGIUM

KOEHLER H PROF DR
SAUERBRUCHSTRASSE 6
D-7920 HEIDENHEIM
A-D BRENZ
GERMANY, F.R.

KOEHLER JAMES A PROF
PHYSICS DEPT
UNIV OF SASKATOON
SASKATOON S7N OWO
CANADA

KOEHLER PETER
CARL-ZEISS STR 1
DDR-6900 JENA
G.D.R.

KOEPPEN JOACHIM DR
INST THEOR ASTROPH
UNIV HEIDELBERG
IM NEUENHEIMER F 294
D-6900 HEIDELBERG
GERMANY, F.R.

KOESTER DETLEV DR
INST THEOR PHY STERN
NEUE UNIV PHYS ZENTR
OLSHAUSENST GEBN 61C
D-2300 KIEL 1
GERMANY, F.R.

KOGURE TOMOKAZU DR
DEPT OF ASTRONOMY
FACULTY OF SCIENCE
UNIVERSITY OF KYOTO
KYOTO 606
JAPAN

KOHL JOHN L DR
CENTER FOR
ASTROPHYSICS
60 GARDEN ST
CAMBRIDGE MA 02138
U.S.A.

KOHOUTEK LUBOS DR
HAMBURGER STERNWARTE
GOJENSBERGSWEG
D-2050 HAMBURG 80
GERMANY, F.R.

KOJOIAN GABRIEL DR
DEPT OF PHYSICS
UNIV OF WISCONSIN
EAU CLAIRE WI 54701
U.S.A.

KOKURIN YURIJ L DR
LEBEDEV PHYS INST
USSR ACAD OF SCS
LENINSKY PROSPECT 53
117924 MOSCOW
U.S.S.R.

KOLACZEK BARBARA DR
PLAN. GEODESY DEPT
UL BARTYCKA 18
00-716 WARSAW
POLAND

KOLCHINSKIJ I G DR
MAIN ASTRONOMICAL
OBS UKRAINIAN ACAD
OF SCIENCES
252127 KIEV
U.S.S.R.

KOLESNIK IGOR G DR
MAIN ASTRON OBS
UKRAINIAN AC OF SC
GOLOSEEVO
252127 KIEV
U.S.S.R.

KOLESNIK L N DR
MAIN ASTRONOMICAL
OBS UKRAINIAN ACAD
OF SCIENCES
252127 KIEV
U.S.S.R.

KOLESOV A K DR
LENINGRAD STATE UNI-
VERSITY ASTRONOMICAL
OBSERVATORY
199178 LENINGRAD
U.S.S.R.

KOLEV DIMITAR ZDRAVKOV
BLOCK D Z-20
SMOLJAN
BULGARIA

KOLLBERG ERIK L DR
RES LAB ELECTRONICS
CHALMERS UNIVERSITY
OF TECHNOLOGY FACK
S-402 20 GOETEBORG
SWEDEN

KOMAROV N S DR
ODESSA STATE UNIVER-
SITY ASTRONOMICAL
OBSERVATORY
270014 ODESSA
U.S.S.R.

KOMESAROFF MAX M D SC
CSIRO DIVISION OF
RADIOPHYSICS
BOX 76
EPPING NSW 2121
AUSTRALIA

KONDO MASAAKI DR
FAC GENERAL EDUC
UNIVERSITY OF TOKYO
KOMABA MEGURO
TOKYO 153
JAPAN

KONDO MASAYUKI DR
TOKYO ASTRONOMICAL
OBSERVATORY
UNIVERSITY OF TOKYO
MITAKA TOKYO 181
JAPAN

KONDO YOJI DR
CODE 683
GODDARD SPACE FL CTR
GREENBELT MD 20771
U.S.A.

KONIN V V DR
NIKOLAEV BRANCH OF
THE MAIN ASTRONOMI-
CAL OBSERVATORY
327000 NIKOLAEV
U.S.S.R.

KONONOVICH EDWARD V DR
STERNBERG ASTRONOMI-
CAL INSTITUTE
117234 MOSCOW
U.S.S.R.

KONOPLEVA V P DR
MAIN ASTRONOMICAL
OBS UKRAINIAN ACAD
OF SCIENCES
252127 KIEV
U.S.S.R.

KONTIZAS EVANGELOS PH D
NAT OBS OF ATHENS
ASTRONOMICAL INST
THISSION
ATHENS 305
GREECE

KONTIZAS MARY PH D
DEPT OF ASTRONOMY
UNIV OF ATHENS
PANEPISTIMIOPOLIS
ATHENS 621
GREECE

KOORNNEEF JAN DR
EUROPEAN SOUTHERN
OBSERVATORY
CASILLA 16317
SANTIAGO 9
CHILE

KOPAL ZDENEK PROF
DEPT OF ASTRONOMY
UNIVERSITY OF
MANCHESTER
MANCHESTER M13 9PL
U.K.

KOPECKY MILOSLAV DR
ASTRONOMICAL INST
CZECHOSLOVAK ACADEMY
OF SCIENCES
251 65 ONDREJOV OBS
CZECHOSLOVAKIA

KOPP ROGER A DR
LOS ALAMOS SCIENTI-
FIC LABORATORY
M S 531
LOS ALAMOS NM 87545
U.S.A.

KOPYLOV I M DR
SPECIAL ASTROPHYSI-
CAL OBSERVATORY
USSR ACADEMY OF SCI
357140 N ARKHYZ
U.S.S.R.

KORCHAK A A DR
INSTITUTE OF TERRE-
STRIAL MAGNETISM AND
IONOSPHERE
142092 ACADEMGORODOK
U.S.S.R.

KORMENDY JOHN DR
DOMINION ASTROPH OBS
5071 W SAANICH ROAD
RR 5
VICTORIA BC V8X 4M6
CANADA

KOROLKOV DMITRY V PROF DR
PULKOVO OBSERVATORY
196140 LENINGRAD
U.S.S.R.

KOROVYAKOVSKIJ YURIJ P DR
SPECIAL ASTROPHYS OB
NIZHNEIJ ARKHYZ
357140 STAVROPOLSKIJ
KRAJ
U.S.S.R.

KOSIN GENNADIJ S DR
PULKOVO OBSERVATORY
196140 LENINGRAD
U.S.S.R.

KOSTINA LIDIJA D DR
PULKOVO OBSERVATORY
196140 LENINGRAD
U.S.S.R.

KOSTYAKOVA ELENA B DR
STERNBERG STATE
ASTRONOMICAL INST
117234 MOSCOW
U.S.S.R.

KOSTYLEV K V DR
ENGELHARDT ASTRONO-
MICAL OBSERVATORY
422526 KAZAN
U.S.S.R.

KOTELNIKOV V A ACAD
INSTITUTE OF RADIO
AND ELECTRONICS
USSR ACADEMY OF SCI
103907 MOSCOW
U.S.S.R.

KOTHARI D.S. DR.
DEPT.OF PHYSICS
UNIVERSITY OF DELHI
NEW DELHI
INDIA

KOTOV V A DR
CRIMEAN ASTROPHYSI-
CAL OBSERVATORY USSR
ACAD OF SCI NAUCHNIY
334413 CRIMEA
U.S.S.R.

KOTSAKIS DEMETRIUS PROF
HIPPOCRATES STR 189
ATHENS-708
GREECE

KOUBSKY PAVEL
ASTRONOMICAL INST
CZECH ACADEMY OF SC
ONDREJOV OBSERVATORY
25165 ONDREJOV
CZECHOSLOVAKIA

KOURGANOFF VLADIMIR PROF
20 AVE PAUL APPELL
F-75014 PARIS
FRANCE

KOUTCHMY SERGE DR
INSTITUT D ASTROPHY-
SIQUE CNRS
98 BIS BD ARAGO
F-75014 PARIS
FRANCE

KOVACHEV B J DR
ACADEMIE BULGARE
SECTION ASTRONOMIE
RUE 7 NOVEMBER NO 1
SOFIA
BULGARIA

KOVACS AGNES DR
HELIOPHYSICAL OBS
P O BOX 30
H-4010 DEBRECEN
HUNGARY

KOVACS GEZA DR
KONKOLY OBSERVATORY
HUNG ACADEMY OF SC
BOX 67
H-1525 BUDAPEST
HUNGARY

KOVAL I K DR
MAIN ASTRONOMICAL
OBSERVATORY
GOLOSEEVO
252127 KIEV
U.S.S.R.

KOVALEVSKY JEAN DR
CERGA
AV COPERNIC
F-06130 GRASSE
FRANCE

KOVAR N S DR
PHYSICS DEPT
UNIV OF HOUSTON
HOUSTON TX 77004
U.S.A.

KOVAR ROBERT P DR
9666 E ORCHARD DR
ENGLEWOOD CO 80111
U.S.A.

KOVETZ ATTAY PROF
DEPT OF PHYSICS AND
ASTRONOMY
TEL AVIV UNIVERSITY
TEL-AVIV
ISRAEL

KOWAL CHARLES THOMAS DR
DEPT OF ASTROPHYSICS
CALTECH
PASADENA CA 91125
U.S.A.

KOYAMA SHIN PROF DR
KAGAWA UNIVERSITY
SAIWAI CHO
TAKAMATSU 760
JAPAN

KOZAI YOSHIHIDE PROF
TOKYO ASTRONOMICAL
OBSERVATORY
OSAWA MITAKA
TOKYO 181
JAPAN

KOZIEL KAROL PROF DR
UL 22 LIPCA 16
43-460 WISLA
POLAND

KOZLOVSKY B Z DR
DEPT OF PHYSICS AND
ASTRONOMY
TEL-AVIV UNIVERSITY
RAMAT AVIV
ISRAEL

KOZLOWSKI MACIEJ DR
ASTRONOMICAL
OBSERVATORY
AL UJAZDOWSKIE 4
00-478 WARSZAWA
POLAND

KOZYREV NIKOLAJ A DR
PULKOVO OBSERVATORY
196140 LENINGRAD
U.S.S.R.

KRAEMER GERHARD DR
ERLENWEG 24
D-7400 TUEBINGEN
GERMANY, F.R.

KRAFT ROBERT P PROF
LICK OBSERVATORY
UNIVERSITY
OF CALIFORNIA
SANTA CRUZ CA 95064
U.S.A.

KRAICHEVA Z DR
DEPT OF ASTRONOMY
BULGARIAN AC OF SCS
7TH NOVEMBER ST 1
1000 SOFIA
BULGARIA

KRAMER KH N DR
ODESSA STATE UNIVER-
SITY ASTRONOMICAL
OBSERVATORY
270014 ODESSA
U.S.S.R.

KRANJC A DR
OSS ASTRONOMICO DI
BRERA
VIA BRERA 28
I-20121 MILANO
ITALY

KRASINSKY GEORGE A DR
INST THEORET ASTRON
USSR ACAD OF SCS
10 KUTUZOV QUAY
192187 LENINGRAD
U.S.S.R.

KRAT VLADIMIR A PROF DR
PULKOVO OBSERVATORY
196140 LENINGRAD
U.S.S.R.

KRAUS JOHN D PROF
RADIO OBSERVATORY
OHIO ST UNIVERSITY
2015 NEIL AVE
COLUMBUS OH 43210
U.S.A.

KRAUSE F DR
ZNTRINST ASTROPHYSIK
ASTROPHYS OBS
TELEGRAFENBERG
DDR-15 POTSDAM
G.D.R.

KRAUSHAAR WILLIAM L PROF
DEPT OF PHYSICS
UNIV OF WISCONSIN
1150 UNIVERSITY AVE
MADISON WI 53706
U.S.A.

KRAUTTER JOACHIM DR
EUROP SOUTHERN OBS
KARL-SCHWARZSCHILD-
STR 2
D-8046 GARCHING
GERMANY, F.R.

KREIDL TOBIAS J N
LOWELL OBSERVATORY
P O BOX 1269
FLAGSTAFF AZ 86002
U.S.A.

KREINER JERZY MAREK DR
UL SPOLDZIELCOW 8/52
30-682 KRAKOW 47
POLAND

KREISEL E DR
ZNTRINST ASTROPHYSIK
STERNW BABELSBERG
ROSA LUXEMBURGST 17A
DDR-1502 POTSDAM
G.D.R.

KRELOWSKI JACEK DR
INST OF ASTRONOMY
N COPERNICUS UNIV
CHOPINA 12/18
PL-87-100 TORUN
POLAND

KREMPEC JANINA DR
N.COPERNICUS AST CTR
ASTROPHYSICAL LAB
CHOPINA 12/18
87-100 TORUN
POLAND

KRESAK LUBOR DR
ASTRONOMICAL INST
SLOVAK ACADEMY OF SC
84228 BRATISLAVA
CZECHOSLOVAKIA

KRESAKOVA MARGITA DR
ASTRONOMICAL INST
SAV
DUBRAVSKA CESTA 11/A
89930 BRATISLAVA
CZECHOSLOVAKIA

KRIEGER ALLEN S DR
AMERICAN SCIENCE AND
ENGINEERING
955 MASSACHUSETTS AV
CAMBRIDGE MA 02139
U.S.A.

KRINOV EUGEN L PROF
METEORITE COMMITTEE
ACADEMY OF SCIENCES
UL M ULIANOVOI 3
117313 MOSCOW
U.S.S.R.

KRISHNA SWAMY K S DR
ASTROPHYSICS GROUP
TATA INSTITUTE
COLABA
BOMBAY 400005
INDIA

KRISHNAN THIRUVENKATA DR
132-A LLOYDS ROAD
GOPALAPURAM
MADRAS 600086
INDIA

KRISTENSEN LEIF KAHL DR
INST OF PHYSICS
UNIV OF AARHUS
NY MUNKGADE
DK-8000 AARHUS C
DENMARK

KRISTENSON HENRIK DR
SLAANBAERSVAEGEN 9
S-381 00 KALMAR
SWEDEN

KRISTIAN JEROME DR
HALE OBSERVATORIES
813 SANTA BARBARA ST
PASADENA CA 91101
U.S.A.

KRISTIANSSON KRISTER
G B PROF
DEPT OF PHYSICS
SOELVEGATAN 14
S-223 62 LUND
SWEDEN

KRIVSKY L DR
ASTRONOMICAL
INSTITUTE
25165 ONDREJOV
CZECHOSLOVAKIA

KRIZ SVATOPLUK DR
DEPT.OF ASTRONOMY
CHARLES UNIVERSITY
SVEDSKA 8
150 000 PRAGUE
CZECHOSLOVAKIA

KROGDAHL W S DR
DEPT OF PHYSICS AND
ASTRONOMY
UNIVERSITY KENTUCKY
LEXINGTON KY 40506
U.S.A.

KRON GERALD E DR
PINECREST OBS
416 N BERTRAND ST
FLAGSTAFF AZ 86001
U.S.A.

KRON KATHERINE GORDON
PINECREST OBS
416 N BERTRAND ST
FLAGSTAFF AZ 86001
U.S.A.

KRON RICHARD G
YERKES OBSERVATORY
WILLIAMS BAY WI53191
U.S.A.

KRONBERG PHILIPP DR
UNIVERSITY
OF TORONTO
SCARBOROUGH COLLEGE
TORONTO M1C 1A4
CANADA

KROOK M DR
HARVARD COLLEGE OBS
CAMBRIDGE MA 02138
U.S.A.

KRUEGEL ENDRIK DR
MPI FUER
RADIOASTRONOMIE
AUF DEM HUEGEL 69
D-5300 BONN
GERMANY, F.R.

KRUEGER ALBRECHT DR
ZENTRALINSTITUT FUER
SOLAR-TERR.PHYSIK
RUDOWER CHAUSSEE 5
DDR-1199 BERLIN
G.D.R.

KRUEGER E PROF
VIA MAURO MACCHI 65
I-20100 MILANO
ITALY

KRUMM NATHAN ALLYN
PHYSICS DEPT
UNIVERSITY OF
CINCINNATI
CINCINNATI OH 45221
U.S.A.

KRUSZEWSKI ANDRZEJ PROF
ASTRONOMICAL
OBSERVATORY
AL UJAZDOWSKIE 4
00-478 WARSZAWA
POLAND

KRZEMINSKI W DR
COPERNICUS ASTRONO-
MICAL CENTER
UL BARTYCKA 18
00-716 WARSAW
POLAND

KSANFOMALITI L V DR
SPACE RESEARCH INSTI
TUTE USSR ACADEMY OF
SCIENCES
117810 MOSCOW
U.S.S.R.

KUBIAK MARCIN A DR
WARSAW UNIVERSITY
OBSERVATORY
AL UJAZDOWSKIE 4
00-478 WARSAW
POLAND

KUBICELA ALEKSANDAR DR
ASTRONOMSKA OPSERV
VOLGINA 7
11050 BEOGRAD
YUGOSLAVIA

KUBO YOSHIO
ASTRONOMICAL DIV
HYDROGRAPHIC DEPT
TSUKIJI-5 CHUO-KU
TOKYO 104
JAPAN

KUBOTA JUN DR
KASAN OBSERVATORY
YAMASHINA
KYOTO 607
JAPAN

KUDRITZKI ROLF-PETER PH D
INST F THEOR PHYSIK
UND UNIV STERNWARTE
OLSHAUSENSTRASSE
D-2300 KIEL
GERMANY, F.R.

KUEHNE CHRISTOPH F DR
CARL ZEISS COMPANY
POSTFACH 35/36
D-7082 OBERKOCHEN
GERMANY, F.R.

KUHI LEONARD V PROF
ASTRONOMY DEPT
UNIVERSITY OF CALIF
BERKELEY CA 94720
U.S.A.

KUIJPERS JAN DR
STERREKUNDIG INST
SERVAAS BOLWERK 13
3512 NK UTRECHT
THE NETHERLANDS

KUIPER THOMAS B H DR
JET PROPULSION LABO-
RATORY 183B-365
PASADENA CA 91103
U.S.A.

KUKLIN G V DR
SIBIZMIR
P B 4
664697 IRKUTSK 33
U.S.S.R.

KULIKOV K A PROF DR
STERNBERG STATE AS-
TRONOMICAL INSTITUTE
117234 MOSCOW
U.S.S.R.

KULIKOVSKIJ P G DR
STERNBERG STATE AS-
TRONOMICAL INSTITUTE
117234 MOSCOW
U.S.S.R.

KULKARNI PRABHAKAR V PROF
PHYSICAL RESEARCH
LABORATORY
AHMEDABAD 380 009
INDIA

KULSRUD R M DR
ASTROPHYSICAL SCS
PRINCETON UNIVERSITY
PRINCETON NJ 08540
U.S.A.

KULTIMA JOHANNES
GEOPHYSICAL OBS
SF-99600 SODANKYLAE
FINLAND

KUMAJGORODSKAYA RAISA DR
SPECIAL ASTROPHYS OB
NIZHNIJ ARKHYZ
357140 STAVROPOLSKIJ
KRAJ
U.S.S.R.

KUMAR CIDAMBI KRISHNA DR
DEPT PHYSICS ASTRON
HOWARD UNIV
WASHINGTON DC 20059
U.S.A.

KUMAR S S PROF
DEPT OF ASTRONOMY
UNIV OF VA POB 3818
CHARLOTTESVILLE
VA 22903
U.S.A.

KUMSIASHVILY MZIA I DR
ABASTUMANI ASTRO-
PHYSICAL OBSERVATORY
383762 GEORGIA
U.S.S.R.

KUMSISHVILI J I DR
ABASTUMANY ASTROPHY-
SICAL OBSERVATORY
383762 ABASTUMANY
GEORGIA
U.S.S.R.

KUN MARIA DR
KONKOLY OBSERVATORY
HUNG ACADEMY OF SC
BOX 67
H-1525 BUDAPEST
HUNGARY

KUNCHEV PETER DR
DEPT OF ASTRONOMY
FACULTY OF PHYSICS
ANTON IVANOV STR 5
1126 SOFIA
BULGARIA

KUNDT WOLFGANG PROF DR
INST F ASTROPHYSIK U
EXTRATERR FORSCHUNG
AUF DEM HUEGEL 71
D-5300 BONN 1
GERMANY, F.R.

KUNDU MUKUL R DR
ASTRONOMY PROGRAM
UNIVERSITY MARYLAND
COLLEGE PARK
MD 20742
U.S.A.

KUNKEL WILLIAM E DR
OBS NACIONAL
RUA GENERAL BRUCE
586 ZC 08
RIO DE JANEIRO GB
BRAZIL

KUNTH DANIEL
INST D'ASTROPHYSIQUE
98 BIS BOULEVARD
ARAGO
F-75014 PARIS
FRANCE

KUNZEL HORST
ZENTRALINSTITUT FUER
SOLARTERRESTRISCHE
PHYSIK
DDR-15 POTSDAM
G.D.R.

KUPERUS MAX PROF DR
ASTRONOMICAL
INSTITUTE
ZONNENBURG 2
3512 NL UTRECHT
THE NETHERLANDS

KUPPERIAN JAMES E DR
GODDARD SPACE FLIGHT
CENTER
CODE 410
GREENBELT MD 20771
U.S.A.

KURIL-CHIK V N DR
STERNBERG STATE AS-
TRONOMICAL INSTITUTE
117234 MOSCOW
U.S.S.R.

KUROCHKA L N DR
KIEV STATE UNIVER-
SITY ASTRONOMICAL
OBSERVATORY
252053 KIEV
U.S.S.R.

KUROKAWA HIROKI DR
HIDA OBSERVATORY
KAMITAKARA
YOSHIKI-GUN
GIFU
JAPAN

KURPINSKA-WINIARSKA M DR
ASTRON OBS OF THE
JAGELLONIAN UNIV
UL ORLA 171
30-244 KRAKOW
POLAND

KURT V G DR
SPACE RESEARCH INSTI
TUTE USSR ACADEMY OF
SCIENCES
117810 MOSCOW
U.S.S.R.

KURTH R DR
PROF OF MATHEMATICS
GEORGIA INST OF
TECHNOLOGY
ATLANTA GA 30332
U.S.A.

KURTZ DONALD WAYNY DR
ASTRONOMY DEPT
UNIV OF CAPE TOWN
RONDEBOSCH
7700 CAPE
SOUTH AFRICA

KURUCZ ROBERT L DR
SMITHSONIAN ASTRO-
PHYSICAL OBSERVATORY
60 GARDEN ST
CAMBRIDGE MA 02138
U.S.A.

KUS ANDRZEJ JAN DR
RADIOASTRONOMY OBS
COPERNICUS UNIV
CHOPINA 12/18
PL-87-100 TORUN
POLAND

KUSHWAHA R S PROF
DEPT OF MATHEMATICS
UNIV OF JODHPUR
JODHPUR RAJ
INDIA

KUSTAANHEIMO PAUL E PROF
DIA E 451
DANMARKS TEKNISKE
HOJSKOLE
DK-2800 LYNGBY
DENMARK

KUTNER MARC LESLIE DR
PHYSICS DEPT
RENSSELAER POLYTECH
INST
TROY NY 12181
U.S.A.

KUTTER G SIEGFRIED DR
THE EVERGREEN STATE
COLLEGE
OLYMPIA WA 98505
U.S.A.

KUTUZOV S A DR
LENINGRAD STATE UNI-
VERSITY ASTRONOMICAL
OBSERVATORY
199178 LENINGRAD
U.S.S.R.

KUZMANOSKI MIKE
INST OF ASTRONOMY
UNIV OF BELGRADE
STUDENTSKI TRG 16
YU-11000 BELGRADE
YUGOSLAVIA

KUZMENKO K N DR
ASTRONOMICAL OBSERVA
TORY KHARKOV STATE
UNIVERSITY
310077 KHARKOV
U.S.S.R.

KUZMIN ARKADII D
PROF DR SC
LEBEDEV PHYS INST
ACADEMY SC
117924 MOSCOW
U.S.S.R.

KUZMIN GRIGORI G PROF
IAPHA ESTONIAN
ACADEMY OF SCIENCES
TORAVERE
202444 TARTU
U.S.S.R.

KVIZ ZDENEK DR
SCHOOL OF PHYSICS
UNIV OF SOUTH WALES
P O BOX 1
KENSINGTON NSW 2033
AUSTRALIA

KWEE K K DR
STERREWACHT
POSTBUS 9513
2300 RA LEIDEN
THE NETHERLANDS

KWITTER KAREN BETH
DEPT OF PHYSICS &
ASTRONOMY
WILLIAMS COLLEGE
WILLIAMSTOWN MA02167
U.S.A.

KWOK SUN DR
HERZBERG INST ASTRO
NAT RES COUNCIL CAN
100 SUSSEX DRIVE
OTTAWA ONT K1A OR6
CANADA

KYLAFIS N D DR
INSTITUTE FOR
ADVANCED STUDY
SCHOOL OF NATURAL SC
PRINCETON NJ 08540
U.S.A.

LA BONTE BARRY JAMES
INST FOR ASTRONOMY
2680 WOODLAWN DRIVE
HONOLULU HI 96822
U.S.A.

LA PADULA CESARE
IST DI ASTROFISICA
SPAZIALE
CASELLA POSTALE 67
I-00044 FRASCATI
ITALY

LABEYRIE ANTOINE DR
CERGA
F-06460 ST VALLIER
FRANCE

LABEYRIE JACQUES DR
CENTRE DES FAIBLES
RADIOACTIVITES
LAB MIXTE CNRS-CEA
91190 GIF SUR YVETTE
FRANCE

LABORDE GEORGES DR
OBSERVATOIRE
DE MEUDON
F-92190 MEUDON
FRANCE

LABRUM NORMAN R MR
CSIRO DIVISION OF
RADIOPHYSICS
PO BOX 76
EPPING NSW 2121
AUSTRALIA

LABS DIETRICH PROF
LANDESSTERNWARTE
KOENIGSTUHL
D-6900 HEIDELBERG 1
GERMANY, F.R.

LACLARE F MR
CERGA
AVENUE COPERNIC
F-06130 GRASSE
FRANCE

LACROUTE PIERRE A PROF
2 RUE D ALISE
F-21000 DIJON
FRANCE

LACY CLAUDE H DR
DEPT OF PHYSICS
UNIV OF ARKANSAS
FAYETTEVILLE AR72701
U.S.A.

LADA CHARLES JOSEPH DR
STEWARD OBS
UNIV OF ARIZONA
TUCSON AZ 85721
U.S.A.

LAFFINEJR MARIUS MR
21 BLVD BRUNE
F-75014 PARIS
FRANCE

LAFON JEAN-PIERRE J DR
OBSERVATOIRE
DE MEUDON
F-92190 MEUDON
FRANCE

LAGERKVIST CLAES-INGVAR
ASTRONOMISKA
OBSERVATORIET
BOX 515
S-751 20 UPPSALA
SWEDEN

LAGERQVIST ALBIN PROF
INSTITUTE OF PHYSICS
VANADISVAEGEN 9
S-113 46 STOCKHOLM
SWEDEN

LAGO M T V T
GRUPO DE MATEMATICA
APLICADA
RUA DAS TAIPAS 135
4000 PORTO
PORTUGAL

LAHIRI N C
INDIAN ASTRON EPH UN
INDIAN MET DEPT
P-546 BLOCK N
NEW ALIPORE CALCUTTA
INDIA

LAHULLA J FORNICS DR
OBSERVATORIO
ASTRONOMICO
ALFONSO XII 5
MADRID 7
SPAIN

LAKE KAYLL WILLIAM DR
DEPT OF PHYSICS
QUEEN'S UNIVERSITY
KINGSTON ONT K7L 3N6
CANADA

LALA PETR DR
ASTRONOMICAL INST
CSAV
251 65 ONDREJOV
CZECHOSLOVAKIA

LAMB DONALD QUINCY JR DR
CTR FOR ASTROPHYSICS
60 GARDEN ST
CAMBRIDGE MA 02138
U.S.A.

LAMB FREDERICK K PROF
PHYSICS DEPT
UNIVERSITY OF
ILLINOIS
URBANA IL 61801
U.S.A.

LAMB SUSAN ANN DR
DEPT OF PHYSICS
UNIV OF MISSOURI
8001 NATURAL BRG RD
ST LOUIS MO 63121
U.S.A.

LAMBECK KURT PROF
AUSTRALIA NAT UNIVER
SITY RESEARCH SCHOOL
EARTH SC PO BOX 4
CANBERRA 2600
AUSTRALIA

LAMBERT DAVID L PROF
DEPT OF ASTRONOMY
UNIVERSITY OF TEXAS
R L MOORE HALL
AUSTIN TX 78712
U.S.A.

LAMBRECHT HERMANN PROF DR
SCHILLERGAESSCHEN 2
DDR-69 JENA
G.D.R.

LAMERS H J G L M DR
LAB VOOR RUIMTE-
ONDERZOEK
BENELUXLAAN 21
3527 HS UTRECHT
THE NETHERLANDS

LAMLA ERICH F DR
ASTRONOMISCHES INST
STERNW UNIV BONN
AUF DEM HUEGEL 71
D-5300 BONN
GERMANY, F.R.

LAMPTON MICHAEL
SPACE SCIENCES LAB
UNIV OF CALIFORNIA
BERKELEY CA 94729
U.S.A.

LAMY P DR
LAB D'ASTRON SPAT
TRAVERSE DU SIPHON
LES-TROIS-LUCS
F-13012 MARSEILLE
FRANCE

LANCASTER BROWN PETER
41 LINDEN CLOSE
ALDEBURGH SUFFOLK
U.K.

LANDE KENNETH PROF
PHYSICS DEPT
UNIVERSITY OF
PENNSYLVANIA
PHILADELPHIA PA19104
U.S.A.

LANDECKER PETER BRUCE DR
HUGHES AIRCRAFT CO
SP/COM GR BLDG 541
MAIL B322 POB 92219
LOS ANGELES CA 90009
U.S.A.

LANDECKER THOMAS L DR
DOMINION RADIO ASTRO
PHYSICAL OBSERVATORY
BOX 248
PENTICTON BC V2A 6K3
CANADA

LANDI DEGL INNOCENTI
EGIDIO DR
OSSERVATORIO ARCETRI
LARGO FERMI 5
I-50125 FIRENZE
ITALY

LANDI-DESSY J DR
OBSERVATORIO NACIO-
NAL ARGENTINO
CORDOBA
ARGENTINA

LANDINI MASSIMO PROF
OSSERVATORIO ARCETRI
LARGO FERMI
I-50125 FIRENZE
ITALY

LANDMAN DONALD ALAN DR
INST FOR ASTRONOMY
HALEAKALA OBS
P O BOX 157
KULA MAUI HI 96790
U.S.A.

LANDOLT ARLO U PROF
DEPT PHYS AND ASTRON
LOUISIANA STATE
UNIVERSITY
BATON ROUGE LA 70803
U.S.A.

LANDSTREET JOHN D PROF
DEPT OF ASTRONOMY
UNIVERSITY OF
WESTERN ONTARIO
LONDON ONT N6A 5B9
CANADA

LANE ARTHUR LONNE DR
JET PROPULSION LAB
4800 OAK GROVE DRIVE
PASADENA CA 91109
U.S.A.

LANG JAMES DR
RUTHERFORD APPLETON
LABORATORY
CHILTON
DIDCOT OXON OX11 0QX
U.K.

LANG KENNETH R ASST PROF
DEPT OF PHYSICS
TUFTS UNIVERSITY
MEDFORD MA 02155
U.S.A.

LANGER WILLIAM DAVID DR
PLASMA PHYSICS LAB
PRINCETON UNIV
P O BOX 451
PRINCETON NJ 08544
U.S.A.

LANTOS PIERRE
OBS DE MEUDON
F-92190 MEUDON
FRANCE

LAPAZ LINCOLN DR
INST OF METEORITICS
UNIV OF NEW MEXICO
CAMPUS PO BOX 23
ALBUQUERQUE NM 87131
U.S.A.

LAPOINTE S M DR
UNIVERSITE DE QUEBEC
2875 BOUL LAURIER
STE-FOY P Q G1V 2M3
CANADA

LAPUSHKA K K DR
LATVIAN STATE UNIVER
SITY ASTRONOMICAL
OBSERVATORY
226098 RIGA
U.S.S.R.

LAQUES PIERRE DR
OBSERVATOIRE
DU PIC DU MIDI
F-65200 BAGNERES
DE BIGORRE
FRANCE

LARGE MICHAEL I DR
SCHOOL OF PHYSICS
UNIVERSITY OF SYDNEY
SYDNEY NSW 2006
AUSTRALIA

LARI C DR
LAB NAZIONALE DI
RADIOASTRONOMIA
VIA IRNERIO 46
I-40126 BOLOGNA
ITALY

LARINK JOHANNES PROF DR
HAMBURGER STERNWARTE
BERGEDORF
GOJENBERGSWEG 78
D-2050 HAMBURG 80
GERMANY, F.R.

LARSON HAROLD P DR
LUNAR AND PLANETARY
LABORATORY
UNIVERSITY ARIZONA
TUCSON AZ 85721
U.S.A.

LARSON RICHARD B PROF
YALE UNIVERSITY
OBSERVATORY BOX 2023
YALE STATION
NEW HAVEN CT 06520
U.S.A.

LARSSON LEANDER GUNNAR
PROF
LUND OBSERVATORY
SVANEGATAN 9
S-222 24 LUND
SWEDEN

LASHER GORDON JEWETT DR
IBM WATSON RES CTR
YORKTWN HTS NY 10598
U.S.A.

LASKARIDES PAUL G ASSPROF
DEPT OF ASTRONOMY
UNIVERSITY OF ATHENS
PANEPISTIMIOPOLIS
ATHENS 621
GREECE

LASKER BARRY M DR
CERRO TOLOLO
INTERAMERICAN OBS
CASILLA 63 D
LA SERENA
CHILE

LASOTA JAN PIOTR DR
COPERNICUS ASTRONO-
MICAL CENTER
UL BARTYCKA 18
00 716 WARSAW
POLAND

LATHAM DAVID W DR
CTR F ASTROPHYSICS
60 GARDEN ST
CAMBRIDGE MA 02138
U.S.A.

LATOUR JEAN J
OBS DU PIC-DU-MIDI
ET DE TOULOUSE
14 AV EDOUARD BELIN
F-31400 TOULOUSE
FRANCE

LATTIMER JAMES M DR
ASTROPHYSICS PROGRAM
DEPT EARTH&SPACE SCS
SUNY AT STONY BROOK
STONY BROOK NY 11794
U.S.A.

LATYPOV A A DR
ASTRONOMICAL INST
UZBEKIAN ACAD OF SC
700000 TASHKENT
U.S.S.R.

LAUBERTS ANDRIS DR
ASTRONOMICAL
OBSERVATORY
BOX 515
S-751 20 UPPSALA
SWEDEN

LAUNAY J M DR
OBS DE MEUDON
F-92190 MEUDON
FRANCE

LAURENT BERTIL E PROF
INSTITUTE FOR
THEORETICAL PHYSICS
VANADISVAEGEN 9
S-113 46 STOCKHOLM
SWEDEN

LAURENT C DR
L P S P
B P NO 10
F-91370 VERRIERES
LE-BUISSON
FRANCE

LAUSBERG ANDRE DR
INST D'ASTROPHYSIQUE
AV DE COINTE 5
B-4200 OUGREE-LIEGE
BELGIUM

LAUSTSEN SVEND DR
AARHUS UNIVERSITY
DK-8000 AARHUS C
DENMARK

LAUTMAN D A DR
SMITHSONIAN ASTRO-
PHYSICAL
60 GARDEN STREET
CAMBRIDGE MA 02138
U.S.A.

LAVAL ANNIE DR
OBSERVATOIRE DE
MARSEILLE
2 PLACE LE VERRIER
F-13004 MARSEILLE
FRANCE

LAVROV M I DR
ENGELHARDT ASTRONO-
MICAL OBSERVATORY
422526 KAZAN
U.S.S.R.

LAVRUKHINA A K PROF DR
INSTITUTE OF GEOCHE-
MISTRY AND ANALYTIC
CHEMISTRY
117334 MOSCOW
U.S.S.R.

LAWRENCE G M DR
LASP
SPACE SCIENCE BLDG
UNIV OF COLORADO
BOULDER CO 80302
U.S.A.

LAYZER DAVID PROF
HARVARD COLLEGE
OBSERVATORY
60 GARDEN ST
CAMBRIDGE MA 02138
U.S.A.

LAZOVIC JOVAN P PROF
DEPT OF ASTRONOMY
FACULTY OF SCIENCES
STUDENTSKI TRG 16
11000 BEOGRAD
YUGOSLAVIA

LE CONTEL JEAN M
OBSERVATOIRE DE NICE
LE MONT GROS
B P 252
F-06007 NICE CEDEX
FRANCE

LE POOLE RUDOLF S DR
STERREWACHT
POSTBUS 9513
2300 RA LEIDEN
THE NETHERLANDS

LE SQUEREN ANNE-MARIE DR
OBSERVATOIRE
DE MEUDON
F-92190 MEUDON
FRANCE

LEA SUSAN MAUREEN DR
ASTRONOMY DEPT
UNIV OF CALIFORNIA
BERKELEY CA 94708
U.S.A.

LEACOCK ROBERT JAY
DEPT OF ASTRONOMY
UNIV OF FLORIDA
GAINESVILLE FL 32611
U.S.A.

LEBOFSKY LARRY ALLEN
LUNAR & PLANETARY
LABORATORIUM
UNIV OF ARIZONA
TUCSON AZ 85721
U.S.A.

LECKRONE DAVID S DR
GODDARD SPACE
FLIGHT CENTER
CODE 671
GREENBELT MD 20771
U.S.A.

LEE P D DR
LOUISIANA STATE UNIV
BATON ROUGE LA 70803
U.S.A.

LEER EGIL PROF
AURORAL OBSERVATORY
UNIVERSITY OF TROMSO
P O BOX 953
N-9001 TROMSO
NORWAY

LEGG THOMAS H DR
HERZBERG INSTITUTE
OF ASTROPHYSICS
NAT RES COUNC CANADA
OTTAWA ONT KIA OR6
CANADA

LEIBOWITZ ELIA M DR
DEPT OF PHYSICS
AND ASTRONOMY
TEL AVIV UNIVERSITY
TEL AVIV
ISRAEL

LEINERT CHRISTOPH DR
MPI F ASTRONOMIE
KOENIGSTUHL
D-6900 HEIDELBERG 1
GERMANY, F.R.

LEMAITRE GERARD R DR
OBSERVATOIRE
DE MARSEILLE
2 PLACE LE VERRIER
F-13004 MARSEILLE
FRANCE

LENZEN RAINER DR
MPI FUER
ASTRONOMIE
KOENIGSTUHL
D-6900 HEIDELBERG 1
GERMANY, F.R.

LEQUEUX JAMES DR
OBSERVATOIRE DE
PARIS-MEUDON
F-92190 MEUDON
FRANCE

LETFUS VOJTECH DR
ASTRONOMICAL INST
CZECH ACAD SCIENCES
251 65 ONDREJOV
OBSERVATORY
CZECHOSLOVAKIA

LEBEDINETS V N DR
ASTRONOMICAL COUNCIL
USSR ACADEMY OF SCI
PYATNITSKAYA UL 48
109017 MOSCOW
U.S.S.R.

LEBOVITZ NORMAN R PROF
MATHEMATICS DEPT
UNIV OF CHICAGO
5734 S UNIV AVE
CHICAGO IL 60637
U.S.A.

LEDERLE TRUDPERT DR
ASTRONOMISCHES
RECHEN INSTITUT
MOENCHHOFSTR 12 14
D-6900 HEIDELBERG 1
GERMANY, F.R.

LEE SEE-WOO DR
DEPT EARTH SCIENCES
KYUNGPOOK NATIONAL
UNIVERSITY
TAEGU CITY
REPUBLIC OF KOREA

LEFEBVRE MICHEL DR
CNES-GRGS
18 AVE E BELIN
F-31055 TOULOUSE CED
FRANCE

LEHNERT BO P PROF
DEPT PLASMA PHYSICS
ROYAL INSTITUTE OF
TECHNOLOGY
S-10044 STOCKHOLM 70
SWEDEN

LEIGHTON R B PROF
HALE OBSERVATORIES
CALIF INST OF TECHN
1201 E CALIFORNIA ST
PASADENA CA 91125
U.S.A.

LELIEVRE G DR
OBS DE MEUDON
F-92190 MEUDON
FRANCE

LEMKE DIETRICH DR
MPI FUER ASTRONOMIE
KOENIGSTUHL
D-6900 HEIDELBERG 1
GERMANY, F.R.

LEORAT JACQUES DR
OBSERVATOIRE
DE MEUDON
F-92190 MEUDON
FRANCE

LEROY JEAN LOUIS
OBSERVATOIRE
DE PIC DE MIDI
F-65200 BAGNERES
DE BIGORRE
FRANCE

LEUNG CHUN MING DR
DEPT OF PHYSICS
RENSSELAER POLYTECHN
INSTITUTE
TROY NY 12181
U.S.A.

LEBLANC YOLANDE DR
OBSERVATOIRE
DE MEUDON
F-92190 MEUDON
FRANCE

LECAR MYRON DR
CTR FOR ASTROPHYSICS
60 GARDEN ST
CAMBRIDGE MA 02138
U.S.A.

LEDOUX P J PROF
INST D ASTROPHYSIQUE
UNIVERSITE DE LIEGE
B-4200 COINTE OUGREE
BELGIUM

LEE TERENCE J DR
U.K.INFRARED
TELESCOPE UNIT
900 LEILANI STREET
HILO HI 96720
U.S.A.

LEFEVRE JEAN DR
OBSERVATOIRE DE NICE
B P 252
F-06007 NICE CEDEX
FRANCE

LEIBACHER JOHN DR
SPACE ASTRONONOMY GR
LOCKHEED 52-12/202
3251 HANOVER ST
PALO ALTO CA 94304
U.S.A.

LEIKIN G A DR
ASTRONOMICAL COUNCIL
USSR ACAD OF SCI
PYATNITSKAYA UL 48
109017 MOSCOW
U.S.S.R.

LEMAIRE PHILIPPE DR
L P S P CNRS
B P 10
F-91370 VERRIERES
LE BUISSON
FRANCE

LENA PIERRE J PROF
OBSERVATOIRE
DE MEUDON
F-92190 MEUDON
FRANCE

LEPINE JACQUES R D DR
CRAAM/ON/CNPQ
RUA CEARA 290
SAO PAULO
BRAZIL

LESTER JOHN B DR
ASTRONOMY DEPT
ERINDALE COLLEGE
UNIV OF TORONTO
MISSISSAUGA L5L 1C6
CANADA

LEUNG KAM CHING PROF
BEHLEN OBSERVATORY
DEPT PHYSICS ASTRON
UNIV OF NEBRASKA
LINCOLN NB 68588
U.S.A.

LEVALLOIS J J DR
INSTITUT GEOGRAPHI-
QUE NATIONAL
2 AVE PASTEUR
F-94160 SAINT-MANDE
FRANCE

LEVASSEUR-REGOURD
ANNY-CHANTAL DR
SERV AERO CNRS BP 2
F-91370 VERRIERES
LE BUISSON
FRANCE

LEVATO ORLANDO H DR
CALLE 38 NO 463
1900 LA PLATA
ARGENTINA

LEVIN B J DR
ASTRONOMICAL COUNCIL
USSR ACADEMY OF SCI
PYATNITSKAYA UL 48
109017 MOSCOW
U.S.S.R.

LEVINE RANDOLPH H DR
CTR F ASTROPHYSICS
60 GARDEN ST
CAMBRIDGE MA 02138
U.S.A.

LEVY EUGENE H DR
DEPT PLANETARY SCI
LUNAR AND PLANET LAB
UNIVERSITY ARIZONA
TUCSON AZ 85721
U.S.A.

LEVY JACQUES R DR
OBSERVATOIRE
DE PARIS
61 AV L OBSERVATOIRE
F-75014 PARIS
FRANCE

LEWIN WALTER H G PROF
MIT 37-627
CAMBRIDGE MA 02139
U.S.A.

LEWIS BRIAN MURRAY DR
CARTER OBSERVATORY
P O BOX 2909
WELLINGTON
NEW ZEALAND

LEWIS J S
L BERKELEY LAB
UNIV OF CALIFORNIA
BERKERLEY CA 94720
U.S.A.

LI CHUN-SHENG
DEPT OF ASTRONOMY
NANJING UNIVERSITY
NANJING
PEOPLE'S REPUBLIC CHINA

LI DONG-MING
BEIJING OBSERVATORY
BEIJING
PEOPLE'S REPUBLIC CHINA

LI HEN
SHANGHAI OBSERVATORY
SHANGHAI
PEOPLE'S REPUBLIC CHINA

LI J Y
PYONGYANG ASTRONOMI-
CAL OBS
ACADEMY OF SCIENCES
PYONGYANG
DEM PEOPLE'S REP OF KOREA

LI JING
BEIJING OBSERVATORY
BEIJING
PEOPLE'S REPUBLIC CHINA

LI NED C DR
CALIFORNIA UNIV
6531 WITHWORTH RD
LOS ANGELES CA 90035
U.S.A.

LI NENG-YAO
PURPLE MOUNTAIN
OBSERVATORY
NANJING
PEOPLE'S REPUBLIC CHINA

LI QI-BIN
BEIJING OBSERVATORY
BEIJING
PEOPLE'S REPUBLIC CHINA

LIDDELL U MR
NASA LUNAR AND PLAN-
ETARY PROGRAM OFF OF
SPACE SC AND APPL
WASHINGTON DC 20546
U.S.A.

LIEBSCHER D E DR
ZNTRINST ASTROPHYSIK
STERNW BABELSBERG
ROSA LUXEMBURGST 17A
DDR-1502 POTSDAM
G.D.R.

LIESKE JAY H DR
CALIF INST TECHN
JET PROPULSION LAB
4800 OAK GROVE DRIVE
PASADENA CA 91103
U.S.A.

LILLER MARTHA H DR
HARVARD COLLEGE
OBSERVATORY
CAMBRIDGE MA 02138
U.S.A.

LILLER WILLIAM PROF
CASILLA 604
VINA DEL MAR
CHILE

LILLEY EDWARD A PROF
HARVARD COLLEGE
OBSERVATORY
60 GARDEN ST
CAMBRIDGE MA 02138
U.S.A.

LILLIE C F DR
DEPT PHYSICS ASTRON
UNIV OF COLORADO
BOULDER CO 80302
U.S.A.

LIN CHIA C PROF
MIT
DEPT OF MATHEMATICS
77 MASSACHUSETTS AVE
CAMBRIDGE MA 02139
U.S.A.

LIN DOUGLAS E DR
LICK OBSERVATORY
UNIV OF CALIFORNIA
SANTA CRUZ CA 95064
U.S.A.

LIN YUANZHANG
PEKING ASTRONOMICAL
OBSERVATORY
ACADEMIA SINICA
PEKING
PEOPLE'S REPUBLIC CHINA

LINCOLN J VIRGINIA MISS
WDC-A FOR SOLAR-
TERRESTRIAL PHYSICS
NOAA D63
BOULDER CO 80303
U.S.A.

LINDBLAD BERTIL A DR
LUND OBSERVATORY
BOX 1107
S-221 04 LUND
SWEDEN

LINDBLAD PER OLOF PROF
STOCKHOLM
OBSERVATORY
S-133 00
SALTSJOEBADEN
SWEDEN

LINDEGREN LENNART DR
LUND OBSERVATORIET
BOX 1107
S-221 04 LUND
SWEDEN

LINDSEY CHARLES ALLAN
UNIVERSITY OF HAWAII
AT MANOA
2680 WOODLAWN DRIVE
HONOLULU HI 96822
U.S.A.

LING CHIH-BING DR
INST OF MATHEMATICS
ACADEMIA SINICA
P O BOX NO 143
TAIPEI
TAIWAN

LINGENFELTER RICHARD E
CTR ASTROPH & SP SCS
C-011
UN OF CALIF S DIEGO
LA JOLLA CA 92093
U.S.A.

LINK FRANCOIS PROF
INSTITUT
D ASTROPHYSIQUE
98 BIS BLVD ARAGO
F-75014 PARIS
FRANCE

LINKE RICHARD ALAN DR
BELL LABS
CRAWFORD HILL LAB
HOLMDEL NJ 07733
U.S.A.

LINNELL ALBERT P PROF
DEPT OF ASTRONOMY
AND ASTROPHYSICS
MICHIGAN STATE UNIV
E LANSING MI 48824
U.S.A.

LINNIK V P PROF DR
MAIN ASTRONOMICAL
OBSERVATORY PULKOVO
196140 LENINGRAD
U.S.S.R.

LINSKY JEFFREY L DR
JILA
UNIVERSITY OF
COLORADO
BOULDER CO 80309
U.S.A.

LINSLEY JOHN
DEPT OF PHYSICS &
ASTRONOMY
UNIV OF NEW MEXICO
ALBUQUERQUE NM 97131
U.S.A.

LIPOVETSKY V A
SPECIAL ASTROPHYS OB
ZELENCHUK
357140 STAVROPOLSKIJ
KRAJ
U.S.S.R.

LIPPINCOTT SARAH LEE DR
SPROUL OBSERVATORY
SWARTHMORE COLLEGE
SWARTHMORE PA 19081
U.S.A.

LISZT HARVEY STEVEN
NRAO
EDGEMONT ROAD
CHARLOTTESVILLE
VA 22901
U.S.A.

LITTLE ALEC G ASSOC PROF
PHYSICS DEPT
UNIVERSITY OF SYDNEY
SYDNEY NSW 2006
AUSTRALIA

LITTLE LESLIE T DR
ELECTRONICS LAB
UNIVERSITY OF KENT
CANTERBURY
KENT CT2 7NT
U.K.

LITTLETON JOHN E
DEPT OF PHYSICS
WEST VIRGINIA UNIV
MORGANTOWN WV 26506
U.S.A.

LITVAK MARVIN M DR
JPL CALTECH
MAIL STOP 168-314
4800 OAK GROVE DRIVE
PASADENA CA 91103
U.S.A.

LIU LIN
DEPT OF ASTRONOMY
NANJING UNIVERSITY
NANJING
PEOPLE'S REPUBLIC CHINA

LIU SOU-YANG DR
COMPUTER SCIENCES CO
SYSTEM SCIENCES DIV
8728 COLESVILLE ROAD
SILVER SPG MD 20910
U.S.A.

LIVINGSTON WILLIAM C
KITT PEAK NATIONAL
OBSERVATORY
P O BOX 4130
TUCSON AZ 85726
U.S.A.

LIVSHITS M A DR
INSTITUTE OF TERRE-
STRIAL MAGNETISM AND
IONOSPHERE
142092 AKADEMGORODOK
U.S.S.R.

LLOYD-EVANS THOMAS DR
S A ASTRONOMICAL OBS
P O BOX 9
OBSERVATORY
CAPE 7935
SOUTH AFRICA

LO KWOK-YUNG DR
OWENS VLY RADIO OBS
CALTECH 102-24
PASADENA CA 91125
U.S.A.

LOCANTHI DOROTHY DAVIS DR
2180 PINECREST DR
ALTADENA CA 91001
U.S.A.

LOCHTE-HOLTGREVEN W PROF
INSTITUT FUER EXPERI
MENTALPHYSIK DER
UNIVERSITAET
D-2300 KIEL
GERMANY, F.R.

LOCKE JACK L DR
HERZBERG INSTITUTE
OF ASTROPHYSICS
NAT RES COUNC CANADA
OTTAWA ONT K1A 0R6
CANADA

LOCKWOOD G WESLEY DR
LOWELL OBSERVATORY
P O BOX 1269
FLAGSTAFF AZ 86002
U.S.A.

LODEN KERSTIN R DR
STOCKHOLM
OBSERVATORY
S-133 00
SALTSJOEBADEN
SWEDEN

LODEN LARS OLOF PROF
ASTRONOMICAL OBSERVA
TORY BOX 515
S-751 20 UPPSALA
SWEDEN

LOHMANN WERNER PROF
ASTRONOMISCHES
RECHEN INSTITUT
MOENCHHOFSTR 12 14
D-6900 HEIDELBERG
GERMANY, F.R.

LOMB NICHOLAS RALPH DR
SYDNEY OBSERVATORY
OBSERVATORY HILL
SYDNEY NSW 2000
AUSTRALIA

LONGAIR M S DR
ASTRON ROYAL FOR
SCOTLAND
ROYAL OBS
EDINBURGH EH9 3JH
U.K.

LONGMORE ANDREW J
UK INFRARED TEL UNIT
900 LEILANI ST
HILO HI 96720
U.S.A.

LOPEZ GARCIA ZUELMA L DR
OBS ASTRONOMICO
FELIX AGUILAR
AV BENAVIDEZ 8175 OE
5407 MARQUESADO S J
ARGENTINA

LOPEZ JOSE A ING
OBS ASTRONOMICO
FELIX AGUILAR
AV BENAVIDEZ 8175 OE
5407 MARQUESADO S J
ARGENTINA

LOPEZ-ARROYO M
OBSERVATORIO
ASTRONOMICO
ALFONSO XII 5
MADRID 7
SPAIN

LOREN ROBERT BRUCE DR
MILLIMETER WAVE OBS
MCDONALD OBS
UNIV OF TEXAS
FT DAVIS TX 79734
U.S.A.

LORTET MARIE CLAIRE
D A F
OBSERVATOIRE
DE MEUDON
F-92190 MEUDON
FRANCE

LOSCO LUCETTE DR
FACULTE DES SCIENCES
BESANCON
F-25030 BESANCON
CEDEX
FRANCE

LOTOVA N A DR
PHYSICAL INSTITUTE
USSR ACADEMY OF
SCIENCES
117924 MOSCOW
U.S.S.R.

LOUGHHEAD RALPH E DR
NATIONAL MEASUREMENT
LABORATORY
BRADFIELD RD
LINDFIELD NSW 2070
AUSTRALIA

LOUISE RAYMOND DR
OBSERVATOIRE
DE MARSEILLE
2 PLACE LE VERRIER
F-13004 MARSEILLE
FRANCE

LOULERGUE M DR
OBS DE MEUDON
F-92190 MEUDON
FRANCE

LOVAS FRANCIS JOHN DR
MOLECULAR SPECTR
DIV 545
NAT BUREAU STANDARDS
WASHINGTON DC 20234
U.S.A.

LOVAS MIKLOS
KONKOLY OBSERVATORY
BOX 67
1525 BUDAPEST
HUNGARY

LOVELACE RICHARD V E DR
SPACE SCIENCES BLDG
CORNELL UNIV
ITHACA NY 14853
U.S.A.

LOVELL SIR BERNARD PROF
JODRELL BANK
CHESHIRE SK11 9PL
U.K.

LOW BOON CHYE
HIGH ALTITUDE OBS
NCAR
P O BOX 3000
BOULDER CO 80307
U.S.A.

LOW FRANK J DR
4940 CALLE BARRIL
TUCSON AZ 85718
U.S.A.

LOWE ROBERT P DR
DEPT OF PHYSICS
UNIV WESTERN ONTARIO
LONDON ONT N6A 3K7
CANADA

LOYOLA PATRICIO DR
OBS ASTRON NACIONAL
UNIVERSIDAD DE CHILE
CASILLA 36-D
SANTIAGO
CHILE

LOZINSKAYA T A DR
STERNBERG STATE AS-
TRONOMICAL INSTITUTE
117234 MOSCOW
U.S.S.R.

LOZINSKIJ A M DR
ASTRONOMICAL COUNCIL
USSR ACADEMY OF SCI
PYATNITSKAYA UL 48
109017 MOSCOW
U.S.S.R.

LU PHILLIP K DR
DEPT PHYSICS ASTRON
WESTERN CONN ST COL
181 WHITE ST
DANBURY CT 06810
U.S.A.

LU TAN
DEPT OF ASTRONOMY
NANJING UNIVERSITY
NANJING
PEOPLE'S REPUBLIC CHINA

LU YANG
DEPT OF ASTRONOMY
& MATHEMATICS
NANJING UNIVERSITY
NANJING
PEOPLE'S REPUBLIC CHINA

LUB JAN DR
C/O ESO
SCHLEISSHEIMERSTR 17
D-8046 GARCHING B M
GERMANY, F.R.

LUCAS R DR
GROUPE D'ASTROPHYS
UNIVERSITE
CERMO B P 53X
F-38041 GRENOBLE
FRANCE

LUCCHIN F
IST DI FISICA
G GALILEI
VIA MARZOLO 8
I-35100 PADOVA
ITALY

LUCKE PETER B DR
DEPT PHYSICS/ASTRON
MT UNION COLLEGE
ALLIANCE OH 44601
U.S.A.

LUCY LEON B PROF
DEPT OF ASTRONOMY
COLUMBIA UNIVERSITY
NEW YORK NY 10027
U.S.A.

LUEST REIMAR PROF
MAX PLANCK SOCIETY
POSTFACH 647
D-8000 MUENCHEN 1
GERMANY, F.R.

LUEST RHEA DR
MPI FUER PHYSIK UND
ASTROPHYSIK
FOEHRINGER RING 6
D-8000 MUENCHEN 40
GERMANY, F.R.

LUKACEVIC ILIJA S DR
FACULTY OF SCIENCES
DEPT OF MECHANICS
STUDENTSKI TRG 16
11000 BEOGRAD
YUGOSLAVIA

LUMME KARI A DR
OBSERVATORY
TAHTITORNINMAKI
SF-00130 HELSINKI 13
FINLAND

LUNDQUIST CHARLES A DR
SPACE SCIENCES
LABORATORY
CODE ES51
HUNTSVILLE AL 35812
U.S.A.

LUNEL MADELEINE DR
OBSERVATOIRE
DE LYON
F-69230 SAINT GENIS
LAVAL
FRANCE

LUNGU NICOLAIE DR
INST POLITEHNIC
CAT DE MATEMATICA
STR EMIL ISAC 15
3400 CLUJ-NAPOCA
ROUMANIA

LUO DING-JIANG
BEIJING OBSERVATORY
BEIJING
PEOPLE'S REPUBLIC CHINA

LUTZ BARRY L DR
LOWELL OBSERVATORY
P O BOX 1269
FLAGSTAFF AZ 86002
U.S.A.

LUTZ JULIE H DR
DEPT OF MATHEMATICS
WASHINGTON STATE
UNIVERSITY
PULLMAN WA 99164
U.S.A.

LUTZ THOMAS E DR
DEPT OF PURE AND
APPLIED MATHEMATICS
WASHINGTON ST UNIV
PULLMAN WA 99164
U.S.A.

LUUD LAURI DR
TARTU OBSERVATORY
202444 ESTONIA
U.S.S.R.

LUYTEN WILLEM J PROF
SPACE SCIENCE CENTER
UNIVERSITY OF
MINNESOTA
MINNEAPOLIS MN 55455
U.S.A.

LYNDEN-BELL DONALD PROF
INST OF ASTRONOMY
MADINGLEY ROAD
CAMBRIDGE CB3 OHA
U.K.

LYNDS BEVERLY T DR
KITT PEAK NATIONAL
OBSERVATORY
P O BOX 26732
TUCSON AZ 85726
U.S.A.

LYNDS ROGER C DR
KITT PEAK NATIONAL
OBSERVATORY
P O BOX 26732
TUCSON AZ 85726
U.S.A.

LYNE ANDREW G DR
NRAL
JODRELL BANK
MACCLESFIELD
CHESHIRE SK11 9PL
U.K.

LYNGA GOSTA DR
LUND OBSERVATORY
BOX 1107
S-221 04 LUND
SWEDEN

LYTTKENS EJNAR DR
SKOLGATAN 33 B
S-752 21 UPPSALA
SWEDEN

LYTTLETON RAYMOND A PROF
ST JOHNS COLLEGE
CAMBRIDGE
U.K.

LYUTY VICTOR M DR
CRIMEAN STATION OF
STERNBERG INSTITUTE
NAUCHNYJ
334413 CRIMEA
U.S.S.R.

MACCACARO TOMMASO DR
LAB RADIOASTRONOMIA
VIA IRNERIO 46
I-40126 BOLOGNA
ITALY

MACCALLUM M A H
DEPT OF APPLIED MATH
QUEEN MARY COLLEGE
MILE END ROAD
LONDON E1 4NS
U.K.

MACCHETTO FERDINANDO DR
EUROPEAN SP AGENCY
SPACE SCIENCE DEPT
ESTEC
NOORDWIJK
THE NETHERLANDS

MACCONNELL DARRELL J DR
CENTRO DE INVEST DE
ASTRONOMIA
APARTADO 264
MERIDA
VENEZUELA

MACDONALD GEOFFREY H DR
ELECTRONICS LAB
UNIVERSITY OF KENT
CANTERBURY
KENT CT2 7NT
U.K.

MACDONALD JAMES
DEPT OF ASTRONOMY
ASTRONOMY BUILDING
1011 WEST SPRINGFLD
URBANA IL 61801
U.S.A.

MACGILLIVRAY HARVEY T DR
ROYAL OBSERVATORY
BLACKFORD HILL
EDINBURGH EH9 3HJ
U.K.

MACHADO JOSE M A B DR
TECHNICAL UNIVERSITY
OF LISBON
AV DA IGREJA 17 1 D
LISBOA 5
PORTUGAL

MACHALSKI JERZY DR
UL MAZOWIECKA 36/33
30-019 KRAKOW
POLAND

MACIEL WALTER J DR
UNIV DE SAO PAULO
INST ASTRO/GEOFISICO
CAIXA POSTAL 30627
01000 SAO PAULO SP
BRAZIL

MACKAY CRAIG D DR
INSTITUTE ASTRONOMY
UNIVERSITY CAMBRIDGE
MADINGLEY RD
CAMBRIDGE CB3 OHA
U.K.

MACLEOD JOHM M DR
HERZBERG INSTITUTE
OF ASTROPHYSICS
NAT RES COUNC CANADA
OTTAWA ONT K1A 0R6
CANADA

MACQUEEN ROBERT M DR
HIGH ALTITUDE
OBSERVATORY
P O BOX 3000
BOULDER CO 80307
U.S.A.

MACRAE DONALD A PROF
DAVID DUNLAP OBS
P O BOX 360
RICHMOND HILL
ONT L4C 4Y6
CANADA

MACRIS CONSTANTIN J PROF
RSAAM ACADEMY ATHENS
ANAGNOSTOPOULOU 14
ATHENS 136
GREECE

MACY WILLIAM WRAY DR
LOCKHEED RES LAB
3251 HANOVER ST
BLDG 202 ORG 52-13
PALO ALTO CA 94304
U.S.A.

MADDISON RONALD CH DR
2 CHURCH PLANTATION
KEELE PARK
KEELE STAFFS
U.K.

MADORE BARRY FRANCIS DR
DAVID DUNLAP OBS
UNIV OF TORONTO
RICHMOND HL L4C 4Y6
CANADA

MAEDER ANDRE DR
OBSERVATOIRE
DE GENEVE
CH-1290 SAUVERNY GE
SWITZERLAND

MAEHARA HIDEO DR
TOKYO ASTRONOMICAL
OBSERVATORY
OSAWA MITAKA
TOKYO 181
JAPAN

MAETZLER CHRISTIAN DR
INST APPLIED PHYSICS
SIDLERSTRASSE 5
CH-3012 BERN
SWITZERLAND

MAFFEI PAOLO PROF
ISTITUTO DI FISICA
CATTEDRA DI ASTROFIS
V DELL'ELCE DI SOTTO
I-06100 PERUGIA
ITALY

MAGALASHVILI N L DR
ABASTUMANI ASTROPHY-
SICAL OBSERVATORY
383762 ABASTUMANI
GEORGIA
U.S.S.R.

MAGALHAES ANTONIO A S DR
OBS ASTRON DE BARROS
UNIV DO PORTO
MONTE DA VIRGEM
4400 VILA NOVA GAIA
PORTUGAL

MAGNAN CHRISTIAN DR
LAT
98 BIS BLVD ARAGO
F-75014 PARIS
FRANCE

MAGNARADZE N G DR
TBILISI STATE
UNIVERSITY
380043 TBILISI
U.S.S.R.

MAGNI GIANFRANCO
IST ASTROFISICA
SPAZIALE
VIALE DELL UNIV 11
I-00185 ROMA
ITALY

MAGUN ANDREAS DR
INST APPLIED PHYSICS
SIDLERSTRASSE 5
CH-3012 BERN
SWITZERLAND

MAHDY HAMED A DR
HELWAN OBSERVATORY
HELWAN-CAIRO
ARAB REP. OF EGYPT

MAHMOUD FAROUK M A B DR
HELWAN OBSERVATORY
HELWAN CAIRO
ARAB REP. OF EGYPT

MAHRA H S DR
UTTAR PRADESH ST OBS
MANORA PEAK
NAINITAL 263 129
INDIA

MAIHARA TOSHINORI DR
DEPT OF PHYSICS
KYOTO UNIVERSITY
SAKYOKU KYOTO 606
JAPAN

MAILLARD JEAN-PIERRE DR
CANADA FRANCE HAWAII
TELESCOPE CO
P O BOX 1597
KAMUELA HI 96743
U.S.A.

MAITZEN HANS M DR
UNIVERSITAETS-
STERNWARTE
TUERKENSCHANZSTR 17
A-1180 WIEN
AUSTRIA

MAKARENKO EKATERINA N DR
ASTRONOMICAL OBS
PARK SHEVCHENKO
270014 ODESSA
U.S.S.R.

MAKAROVA ELENA A DR
STERNBERG ASTRONOMI-
CAL INSTITUTE
117234 MOSCOW
U.S.S.R.

MAKINO FUMIYOSHI DR
DEPT OF ASTROPHYSICS
NAGOYA UNIVERSITY
CHIKUSA-KU
NAGOYA 464
JAPAN

MAKITA MITSUGU DR
TOKYO ASTRONOMICAL
OBSERVATORY
OSAWA MITAKA
TOKYO 181
JAPAN

MALACARA DANIEL
INAOE
AP POSTAL 216 Y51
TONANTZINTLA
PUEBLA PUE
MEXICO

MALAGNINI MARIA LUCIA
OSS ASTRONOMICO
VIA TIEPOLO 11
I-34131 TRIESTE
ITALY

MALAISE DANIEL J DR
INSTITUT
D ASTROPHYSIQUE
B-4200 COINTE-OUGREE
BELGIUM

MALAKPUR IRADJ DR
INST OF GEOPHYSICS
TEHRAN UNIVERSITY
AMIRABAD SHOMALI
TEHRAN
IRAN

MALARODA STELLA M DR
OBS ASTRONOMICO
PASEO DEL BOSQUE
1900 LA PLATA
ARGENTINA

MALIN DAVID F DR
ANGLO-AUSTRALIAN OBS
P O BOX 296
EPPING N S W 2121
AUSTRALIA

MALINOVSKY MONIQUE DR
L P S P
B P NO 10
F-91370 VERRIERES LE
BUISSON
FRANCE

MALITSON HARRIET H MS
13315 MAGELLAN AVE
ROCKVILLE MD 20853
U.S.A.

MALLIA EDWARD A DR
DEPT OF ASTROPHYSICS
SOUTH PARKS ROAD
OXFORD OX1 3RQ
U.K.

MALLIK D C V DR
INDIAN INSTITUTE OF
ASTROPHYSICS
BANGALORE 560 034
INDIA

MALTBY PER PROF
INST THEOR PHYSICS
UNIVERSITY OF OSLO
P O BOX 1029
BLINDERN OSLO 3
NORWAY

MALVILLE J MCKIM PROF
DEPT ASTROGEOPHYSICS
UNIVERSITY OF
COLORADO
BOULDER CO 80302
U.S.A.

MAMMANO A DR
OSS ASTROFISICO
I-36012 ASIAGO VIC
ITALY

MANABE SEIJI DR
INT'L LATITUDE OBS
OF MIZUSAWA
MIZUSAWA
IWATE 023
JAPAN

MANARA ALESSANDRO A DR
OSSERVATORIO ASTRONO
MICO DI MILANO
VIA BRERA 28
I-20121 MILANO
ITALY

MANCHESTER RICHARD N DR
CSIRO RADIOPHYSICS
P O BOX 76
EPPING NSW 2121
AUSTRALIA

MANCUSO-SANTI DR
CAPODIMONTE ASTRONO-
MICAL OBSERVATORY
VIA MOIARIELLO 16
I-80131 NAPOLI
ITALY

MANDELSTAM S L PROF
INST OF SPECTROSCOPY
USSR ACADEMY OF SC
142092 AKADEMIGORO-
DOK MOSCOW REGION
U.S.S.R.

MANFROID JEAN DR
INST D'ASTROPHYSIQUE
UNIVERSITE DE LIEGE
AVENUE DE COINTE 5
B-4200 COINTE-OUGREE
BELGIUM

MANGENEY ANDRE DR
OBSERVATOIRE DE
MEUDON
F-92190
FRANCE

MANNINO GIUSEPPE PROF
ISTITUTO MATEMATICO
VIA CAMPI 181
I-41100 MODENA
ITALY

MANRIQUE WALTER T PROF
OBS ASTRONOMICO
FELIX AGUILAR
AV BENAVIDEZ 8175 OE
5407 MARQUESADO S J
ARGENTINA

MANSFIELD VICTOR N PROF
COLGATE UNIVERSITY
HAMILTON NY 13346
U.S.A.

MANTEGAZZA LUCIANO
OSS ASTRONOMICO DI
MERATE
VIA BIANCHI 46
I-22055 MERATE
ITALY

MAO WEI
YUNNAN OBSERVATORY
KUNMING
YUNNAN PROVINCE
PEOPLE'S REPUBLIC CHINA

MARABINI RODOLFO JOSE
OBS ASTRONOMICO
PASEO DEL BOSQUE
1900-LA PLATA
ARGENTINA

MARAN STEPHEN P DR
NASA-GSFC CODE 680
GREENBELT MD 20771
U.S.A.

MARAR T M K
ISRO SATELLITE CTR
PEENYA
BANGALORE 560 058
INDIA

MARASCHI LAURA DR
ISTITUTO DI FISICA
VIA CELORIA 16
I-20133 MILANO
ITALY

MARCHAL C DR
DEPT D'ETUDES DE
SYNTHESE
ONERA
F-92320 CHATILLON
FRANCE

MARCUS ELLA PROF
OBSERVATOIRE
ASTRONOMIQUE
5 R CUTITUL ARGINT
BUCAREST 75212
ROUMANIA

MARDIROSSIAN FABIO
OSS ASTRONOMICO
VIA G B TIEPOLO 11
I-34131 TRIESTE
ITALY

MAREK JOHN
DEPT OF APPLIED MATH
UNIV CLG OF WALES
ABERYSTWYTH DYFED
U.K.

MARGON BRUCE H PROF
ASTRONOMY DEPT FM-20
UNIV OF WASHINGTON
SEATTLE WA 98195
U.S.A.

MARGONI R
OSS ASTROFISICO
I-36012 ASIAGO VIC
ITALY

MARGRAVE THOMAS EWING JR
1639 HELENA AVE
MISSOULA MT 59801
U.S.A.

MARIE M A E S DR
ASTRONOMY DEPT
FACULTY OF SCIENCE
CAIRO UNIVERSITY
CAIRO
ARAB REP. OF EGYPT

MARIK MIKLOS DR.
ASTRONOMICAL DEPT.
EOTVOS UNIVERSITY
KUN BELA TER 2
H-1083 BUDAPEST
HUNGARY

MARILLI ETTORE DR
OSS ASTROFISICO
CITTA UNIVERSITARIA
I-95125 CATANIA
ITALY

MARINO BRIAN F ENG
BOX 72009
NORTHCOTE
AUCKLAND 9
NEW ZEALAND

MARISKA JOHN THOMAS
CODE 4175M
NAVAL RESEARCH LAB
WASHINGTON DC 20375
U.S.A.

MARK JAMES WAI-KEE DR
L LIVERMORE LAB
L-477
UNIV OF CALIFORNIA
LIVERMORE CA 94550
U.S.A.

MARKARIAN B E DR
BYURAKAN ASTROPHYSI-
CAL OBSERVATORY
378433 ARMENIA
U.S.S.R.

MARKELLOS VASSILIS V DR
ASTRONOMY DEPT
UNIV OF GLASGOW
GLASGOW G12 8QQ
U.K.

MARKKANEN TAPIO
OBSERVATORY
TAHTITORNINMAKI
SF-00130 HELSINKI 13
FINLAND

MARKOWITZ WILLIAM DR
APT 15-B
2800 E SUNRISE BLVD
FT LAUDERDL FL 33304
U.S.A.

MARLBOROUGH J M PROF
DEPT OF ASTRONOMY
UNIVERSITY OF
WESTERN ONTARIO
LONDON ONT N6A 5B9
CANADA

MAROCHNIK L S PROF DR
USSR ACAD OF SCS
SPACE RESEARCH INST
PROFSOJUSNA A 84/32
117810 MOSCOW
U.S.S.R.

MAROV MIKHAIL YA DR
INST OF APPLIED MATH
USSR ACAD OF SCS
MIUSSKAYA SQ 4
125047 MOSCOW
U.S.S.R.

MARQUES DOS SANTOS P PROF
INST ASTRONOMICO E
GEOFISICO USP
CAIXA POSTAL 30627
01000 SAO PAULO SP
BRAZIL

MARQUES MANUEL N DR
OBSERVATORIO
ASTRONOMICO
TAPADA DA AJUDA
LISBOA 3
PORTUGAL

MARRACO HUGO G DR
OBS ASTRONOMICO
PASEO DEL BOSQUE
1900 LA PLATA
ARGENTINA

MARSCHER ALAN PATRICK
ASTRONOMY DEPT
BOSTON UNIVERSITY
725 COMMONWEALTH AVE
BOSTON MA 02215
U.S.A.

MARSDEN BRIAN G DR
SMITHSONIAN ASTROPHY
SICAL OBSERVATORY
CAMBRIDGE MA 02138
U.S.A.

MARSDEN PHILIP L PROF
DEPT OF PHYSICS
UNIVERSITY OF LEEDS
LEEDS LS2 9JT
U.K.

MARSH JULIAN C D
HATFIELD POLYTECHNIC
OBSERVATORY
BAYFORDBURY NR HERT-
FORD HERTS SG13 8LD
U.K.

MARSOGLU S DR
UNIV OBSERVATORY
UNIVERSITE
ISTANBUL
TURKEY

MARTEL MARIE-THERESE DR
OBSERVATOIRE
DE LYON
F-69230 SAINT GENIS
LAVAL
FRANCE

MARTIN DEREK H PROF
QUEEN MARY COLLEGE
MILE END ROAD
LONDON E1 4NS
U.K.

MARTIN F DR
DEPT D'ASTROPHYSIQUE
UNIV DE NICE
PARC VALROSE
F-06034 NICE CEDEX
FRANCE

MARTIN INACIO MALMONGE DR
INST PESQUISAS
ESPACIAIS
INPE C P 515
12200 S JOSE CAMPOS
BRAZIL

MARTIN NICOLE DR
OBSERVATOIRE
DE MARSEILLE
2 PLACE LE VERRIER
F-13004 MARSEILLE
FRANCE

MARTIN PETER G PROF
DEPT OF ASTRONOMY
UNIVERSITY TORONTO
TORONTO ONT M5S 1A7
CANADA

MARTIN ROBERT N DR
MPI FUER RADIOASTRON
AUF DEM HUEGEL 69
D-5300 BONN
GERMANY, F.R.

MARTIN WILLIAM C DR
A167 PHYSICS
NATIONAL BUREAU OF
STANDARDS
WASHINGTON DC 20234
U.S.A.

MARTIN WILLIAM L DR
ROYAL GREENWICH OBS
HERSTMONCEUX CASTLE
HAILSHAM
E SUSSEX BN27 1RP
U.K.

MARTIN-LORON M DR
HERMANOS MIRALLES 14
MADRID 1
SPAIN

MARTINET LOUIS PROF
OBSERVATOIRE
DE GENEVE
CH-1290 SAUVERNY GE
SWITZERLAND

MARTINEZ M DR
INST DE ASTRONOMIA
UNAM
APT POSTAL 70-264
MEXICO DF
MEXICO

MARTINI ALDO DR
CNR ASTROFISICA
SPAZIALE C P 67
I-00044 FRASCATI
ROMA
ITALY

MARTINS DONALD HENRY DR
DEPT PHYSICS ASTRON
UNIV OF GEORGIA
ATHENS GA 30602
U.S.A.

MARTRES MARIE-JOSEPHE
OBSERVATOIRE DE
PARIS-MEUDON
F-92190 MEUDON
FRANCE

MARTYNOV D YA PROF DR
STERNBERG STATE AS-
TRONOMICAL INSTITUTE
117234 MOSCOW
U.S.S.R.

MARVIN URSULA B DR
CENTER FOR
ASTROPHYSICS
60 GARDEN ST
CAMBRIDGE MA 02138
U.S.A.

MARX GYORGY PROF.
EOTVOS UNIVERSITY
PUSKIN U 5
H-1088 BUDAPEST
HUNGARY

MARX S DR
ZNTRINST ASTROPHYSIK
KARL SCHWARZSCHILD
OBSERVATORIUM
DDR-6901 TAUTENBURG
G.D.R.

MASANI A PROF
OSS ASTRONOMICO DI
BRERA
VIA BRERA 28
I-20100 MILANO
ITALY

MASLOWSKI JOZEF DR
ASTRONOMICAL
OBSERVATORY
UL ORLA 171
30-244 KRAKOW
POLAND

MASNOU FRANCOISE DR
28 ALLEE GAMBAUDERIE
F-91190 GIF S YVETTE
FRANCE

MASNOU J L DR
SECT D'ASTROPHYSIQUE
CEN SACLAY
B P NO 2
F-91120 GIF-S-YVETTE
FRANCE

MASON HELEN E DR
DEPT APPLIED MATH
AND THEORET PHYSICS
SILVER STREET
CAMBRIDGE
U.K.

MASON KEITH OWEN
MULLARD SPACE
SCIENCE LABORATORY
HOLMBURY ST MARY
DORKING SURREY
U.K.

MASSEVICH A G DR
ASTRONOMICAL COUNCIL
USSR ACADEMY OF SCI
PYATNITSKAYA UL 48
109017 MOSCOW
U.S.S.R.

MASURSKY HAROLD PROF
US GEOLOGICAL SURVEY
BR OF ASTRO STUDIES
2255 NORTH GEMINI DR
FLAGSTAFF AZ 86001
U.S.A.

MATAS VLADIMIR R DR
ASTRONOMISCHES
RECHEN INSTITUT
MOENCHHOFSTR 12-14
D-6900 HEIDELBERG
GERMANY, F.R.

MATERNE JUERGEN
INST ASTRON ASTROPH
TU BERLIN
ERNST-REUTER-PL
D-1000 BERLIN 10
GERMANY, F.R.

MATHER JOHN CROMWELL
LAB FOR EXTRATERRES-
TRIAL PHYSICS
CODE 693.2 GSFC
GREENBELT MD 20771
U.S.A.

MATHESON DAVID NICHOLAS
SERC RUTHERFORD &
APPLETON LAB
CHILTON DIDCOT
OXON OX11 0QX
U.K.

MATHEWS WILLIAM G PROF
LICK OBSERVATORY
UNIVERSITY OF CALIF
SANTA CRUZ CA 95064
U.S.A.

MATHEWSON DONALD S DR
MT STROMLO
OBSERVATORY
WODEN PO ACT 2606
AUSTRALIA

MATHEZ GUY
OBS DE PARIS
DEPT D'ASTROPHYSIQUE
FONDAMENTALE
F-92190 MEUDON
FRANCE

MATHIS JOHN S PROF
DEPT OF ASTRONOMY
UNIVERSITY OF
WISCONSIN
MADISON WI 53706
U.S.A.

MATIAGIN VALERY S DR
ASTROPHYSICAL INST
480068 ALMA-ATA
U.S.S.R.

MATSAKIS DEMETRIOS N
US NAVAL OBS
3400 MASSACHUSETTS
AVENUE NW
WASHINGTON DC 20390
U.S.A.

MATSON DENNIS L DR
JPL 183-501
4800 OAK GROVE DRIVE
PASADENA CA 91103
U.S.A.

MATSUDA TAKUYA PROF
DEPT AERONAUTIC ENG
KYOTO UNIVERSITY
YOSHIDAHONMACHI
SAKYOKU KYOTO 606
JAPAN

MATSUMOTO MASAMICHI PROF
FACULTY OF
ENGINEERING
GIFU UNIVERSITY
504 KAGAMIGAHARA
JAPAN

MATSUMOTO TOSHIO DR
DEPT OF PHYSICS
NAGOYA UNIVERSITY
FUROCHO CHIKUSAKU
NAGOYA 464
JAPAN

MATSUOKA MASARU DR
INST SPACE AERON SC
UNIVERSITY OF TOKYO
KOMABA MEGURO
TOKYO 153
JAPAN

MATSUSHIMA SATOSHI DR
DEPT ASTRO PENNA ST
UNIV 525 DAVEY LAB
UNIVERSITY PARK
PA 16802
U.S.A.

MATSUURA OSCAR T DR
DEPTO DE ASTRONOMIA
IAG-USP
CAIXA POSTAL 30627
01000 SAO PAULO SP
BRAZIL

MATTEI JANET AKYUZ DR
AAVSO
187 CONCORD AVE
CAMBRIDGE MA 02138
U.S.A.

MATTHEWS THOMAS A DR
ASTRONOMY PROGRAM
UNIVERSITY MARYLAND
COLLEGE PK MD 20742
U.S.A.

MATTIG W PROF DR
KIEPENHEUER INST
FUER SONNENPHYSIK
SCHOENECKSTRASSE 6
D-7800 FREIBURG-BR
GERMANY, F.R.

MATTILA KALEVI DR
OBSERVATORY
TAHTITORNINMAKI
SF-00130 HELSINKI 13
FINLAND

MATVEYENKO L I DR
SPACE RESEARCH
INSTITUTE
USSR ACADEMY OF SCI
117810 MOSCOW
U.S.S.R.

MATZNER RICHARD A PROF
PHYSICS DEPT
UNIVERSITY OF TEXAS
AUSTIN TX 78712
U.S.A.

MAUCHERAT-JOUBERT M MME
LAB D'ASTRON SPAT
TRAVERSE DU SIPHON
LES TROIS LUCS
F-13012 MARSEILLE
FRANCE

MAUDER HORST PROF DR
ASTRONOMISCHES INST
WALDHAUSER STR 64
D-7400 TUEBINGEN
GERMANY, F.R.

MAUGER CHANTAL DR
OBS DE PARIS
DEPT RADIOASTRONOMIE
F-92190 MEUDON
FRANCE

MAURICE ERIC N
ESO
K-SCHWARZSCHILD-ST 2
D-8046 GARCHING
GERMANY, F.R.

MAVRAGANIS A G DR
DEPT OF MECHANICS B
TECHN UNIV OF ATHENS
7 K ZOGRAFOU ST
ATHENS 624
GREECE

MAVRIDES S DR
OBS DE PARIS
DEPT RADIOASTRONOMIE
F-92190 MEUDON
FRANCE

MAVRIDIS L N PROF
DEPT GEO ASTRONOMY
UNIVERSITY
OF THESSALONIKI
THESSALONIKI
GREECE

MAVROMICHALAKI H DR
NUCLEAR PHYSICS LAB
COSMIC RAY GROUP
104 SOLONOS ST
ATHENS 144
GREECE

MAX CLAIRE E DR
L-477 LAWRENCE
LIVERMORE LABORATORY
P O BOX 808
LIVERMORE CA 94550
U.S.A.

MAXWELL ALAN DR
HARVARD RADIO
ASTRONOMY STATION
FORT DAVIS TX 79734
U.S.A.

MAY J ING
OBS RADIOASTRONOMICO
UNIV DE CHILE
CASILLA 68
MAIPU
CHILE

MAYALL MARGARET W
5 SPARKS ST
CAMBRIDGE MA 02138
U.S.A.

MAYALL NICHOLAS U ASTRON
7206 E CAMINO VEC'NO
TUCSON AZ 85715
U.S.A.

MAYANO BRUNO
OSS ASTRONOMICO
UNIVERSITARIO
CASELLA POSTALE 596
I-40100 BOLOGNA
ITALY

MAYER CORNELL H
SPACE SCIENCE DIV
NAVAL RESEARCH LAB
CODE 7130
WASHINGTON DC 20375
U.S.A.

MAYER PAVEL DR
ASTRONOMY DEPT
CHARLES UNIVERSITY
SVEDSKA 8
15000 PRAHA 5
CZECHOSLOVAKIA

MAYER URSULA DR
JUL-NACHER-STR 22
D-7530 PFORZHEIM
GERMANY, F.R.

MAYFIELD EARLE B DR
5536 MICHELLE DR
TORRANCE CA 90503
U.S.A.

MAYOR MICHEL DR
OBSERVATOIRE
DE GENEVE
CH-1290 SAUVERNY GE
SWITZERLAND

MAZA JOSE
DEPTO DE ASTRONOMIA
UNIVERSIDAD DE CHILE
CASILLA 36-D
SANTIAGO
CHILE

MAZURE ALAIN DR
OBS DE PARIS
DEPT D'ASTROPHYSIQUE
FONDAMENTALE
F-92190 MEUDON
FRANCE

MAZUREK THADDEUS JOHN DR
DEPT OF PHYSICS
SUNY AT STONY BROOK
LONG ISLAND NY 11794
U.S.A.

MAZZITELLI ITALO DR
LAB ASTROFIZICA SPAZ
P O B 67
I-00044 FRASCATI
ITALY

MAZZUCCONI FABRIZIO DR
OSSERVATORIO ARCETRI
LARGO FERMI 5
I-50125 FIRENZE
ITALY

MCADAM W BRUCE DR
SCHOOL OF PHYSICS
UNIVERSITY OF SYDNEY
SYDNEY NSW 2006
AUSTRALIA

MCALISTER HAROLD A DR
DEPT OF PHYSICS
GEORGIA STATE UNIV
UNIVERSITY PLAZA
ATLANTA GA 30303
U.S.A.

MCBREEN BRIAN PHILIP DR
PHYSICS DEPT
UNIV COLLEGE
BELFIELD
DUBLIN 4
IRELAND

MCCABE MARIE K MS
INST FOR ASTRONOMY
2680 WOODLAWN DR
HONOLULU HI 96822
U.S.A.

MCCARROLL RONALD PROF
LAB D ASTROPHYSIQUE
UNIVERSITE
DE BORDEAUX I
F-33405 TALENCE
FRANCE

MCCARTHY DENNIS D DR
US NAVAL OBSERVATORY
WASHINGTON DC 20390
U.S.A.

MCCARTHY MARTIN F DR
SPECOLA VATICANA
CITTA DEL VATICANO
VATICAN CITY STATE

MCCLAIN EDWARD F
4133 MAPLE RD
MORNINGSIDE MD 20746
U.S.A.

MCCLINTOCK JEFFREY E DR
MASSACHUSETTS INSTI-
TUTE OF TECHNOLOGY
BLDG 37-521
CAMBRIDGE MA 02139
U.S.A.

MCCLURE ROBERT D PROF
DOMINION ASTROPH OBS
5071 W SAANICH ROAD
RR 5
VICTORIA BC V8X 4M6
CANADA

MCCLUSKEY GEORGE E JR DR
DIV OF ASTRONOMY
DEPT OF MATHEMATICS
LEHIGH UNIVERSITY
BETHLEHEM PA 18015
U.S.A.

MCCORD T B DR
INST F ASTRONOMY
2680 WOODLAWN DR
HONOLULU HI 96822
U.S.A.

MCCRACKEN KENNETH G DR
CSIRO
MINERAL PHYSICS SECT
P O BOX 136
NORTH RYDE NSW 2113
AUSTRALIA

MCCRAY RICHARD DR
JILA
UNIVERSITY OF
COLORADO
BOULDER CO 80309
U.S.A.

MCCREA J DERMOTT
DEPT OF MATH PHYSICS
UNIVERSITY COLLEGE
BELFIELD
DUBLIN 4
IRELAND

MCCREA WILLIAM H PROF
ASTRONOMY CENTRE
SUSSEX UNIVERSITY
BRIGHTON BN1 9QH
U.K.

MCCROSKY RICHARD E DR
CENTER FOR
ASTROPHYSICS
CAMBRIDGE MA 02138
U.S.A.

MCCULLOCH P M DR
DEPT OF PHYSICS
UNIV OF TASMANIA
HOBART TASMANIA
AUSTRALIA

MCCUTCHEON WILLIAM H PROF
DEPT OF PHYSICS
UNIVERSITY OF B C
2075 WESBROOK MALL
VANCOUVER BC V6T 1W5
CANADA

MCDONALD FRANK B DR
CODE 660
NASA GSFC
GREENBELT MD 20771
U.S.A.

MCDONALD J K PETRIE DR
DOMINION ASTROPHYSI-
CAL OBSERVATORY
5071 W SAANICH ROAD
VICTORIA BC V8X 3X3
CANADA

MCDONNELL J A M DR
SPACE SCIENCE LAB
UNIVERSITY OF KENT
CANTERBURY
KENT CT2 7NT
U.K.

MCDONOUGH THOMAS R DR
SPACE SCIENCES DIV
JPL 264-738
4800 OAK GROVE DRIVE
PASADENA CA 91103
U.S.A.

MCELROY M B DR
KITT PEAK NAT OBS
950 N CHERRY AVE
TUCSON AZ 85726
U.S.A.

MCGEE JAMES D PROF
10 HILLTOP CRESCENT
APT 3E
FAIRLIGHT 2094
SYDNEY
AUSTRALIA

MCGEE RICHARD X DR
CSIRO DIVISION OF
RADIOPHYSICS
PO BOX 76
EPPING NSW 2121
AUSTRALIA

MCGIMSEY BEN Q JR DR
DEPT OF PHYSICS
GEORGIA STATE UNIV
UNIVERSITY PLAZA
ATLANTA GA 30303
U.S.A.

MCINTOSH BRUCE A DR
HERZBERG INSTITUTE
OF ASTROPHYSICS
NAT RES COUNCIL CAN
OTTAWA ONT K1A 0R6
CANADA

MCKEE CHRISTOPHER F PROF
PHYSICS DEPT
UNIVERSITY OF CALIF
BERKELEY CA 94720
U.S.A.

MCKEITH CONAL D DR
17 ADELAIDE PARK
BELFAST BT9 6FX
U.K.

MCKEITH NIALL ENDA DR
ST PATRICK'S COLLEGE
MAYNOOTH CO KILDARE
IRELAND

MCKENNA LAWLOR SUSAN
M P DR
ST PATRICKS COLLEGE
MAYNOOTH CO KILDARE
IRELAND

MCKINLEY DONALD W R DR
1889 FAIRMEADOW CR
OTTAWA ONT
CANADA

MCLAREN ROBERT A DR
DEPT OF ASTRONOMY
UNIV OF TORONTO
TORONTO ONT M5S 1A7
CANADA

MCLEAN DONALD J DR
DIV OF RADIOPHYSICS
CSIRO
P O BOX 76
EPPING NSW 2121
AUSTRALIA

MCLEAN IAN S DR
LUNAR AND PLANET LAB
UNIV OF ARIZONA
TUCSON AZ 85721
U.S.A.

MCMULLAN DENNIS DR
CAVENDISH LAB
MADINGLEY ROAD
CAMBRIDGE CB3 0HE
U.K.

MCNALLY DEREK DR
UNIVERSITY OF LONDON
OBSERVATORY
MILL HILL PARK
LONDON NW7 2QS
U.K.

MCNAMARA DELBERT H DR
DEPT OF PHYSICS AND
ASTRONOMY
BRIGHAM YOUNG UNIV
PROVO UT 84602
U.S.A.

MCVITTIE GEORGE C PROF
74 OLD DOVER ROAD
CANTERBURY
KENT CT1 3AY
U.K.

MCWHIRTER R W PETER DR
ASTROPHYSICS RES DIV
CULHAM LAB
ABINGDON
OXON OX14 3DB
U.K.

MEABURN J DR
DEPT OF ASTRONOMY
THE UNIVERSITY
MANCHESTER M13 9PL
U.K.

MEAD JAYLEE MONTAGUE DR
NASA-GSFC
CODE 680
GREENBELT MD 20771
U.S.A.

MEADOWS A JACK PROF
ASTRONOMY AND HISTO-
RY OF SCIENCE DEPT
UNIVERSITY LEICESTER
LEICESTER LE1 7RH
U.K.

MEBOLD U DR
MPI FUER RADIO-
ASTRONOMIE
AUF DEM HEUGEL 69
D-5300 BONN 1
GERMANY, F.R.

MEDVEDEV YURI A DR
ASTRONOMICAL OBS
ODESSA STATE UNIV
270014 ODESSA
U.S.S.R.

MEEKS M LITTLETON DR
LINCOLN LABORATORY
P O BOX 73
LEXINGTON MA 02173
U.S.A.

MEFFROY JEAN F PROF DR
INST MATHEMATIQUES
UNIV DES SCIENCES ET
TECHN DU LANGUEDOC
F-34060 MONTPELLIER
FRANCE

MEGESSIER CLAUDE DR
OBSERVATOIRE PARIS
61 AVE OBSERVATOIRE
F-75014 PARIS
FRANCE

MEGRELISHVILI T G PROF
ABASTUMANI ASTROPHY-
SICAL OBSERVATORY
383762 ABASTUMANI
GEORGIA
U.S.S.R.

MEHLTRETTER J P DR
KIEPENHEUER INST
FUER SONNENPHYSIK
SCHOENECKSTR 6
D-7800 FREIBURG-BR
GERMANY, F.R.

MEIER ROBERT R
1003 PRISCILLA LANE
ALEXANDRIA VA 22308
U.S.A.

MEIN NICOLE DR
OBS DE PARIS
DASOP
F-92190 MEUDON
FRANCE

MEIN PIERRE
OBSERVATOIRE
DE MEUDON
5 PLACE JANSSEN
F-92190 MEUDON
FRANCE

MEINEL ADEN B PROF
OPTICAL SCIENCES CTR
UNIVERSITY OF
ARIZONA
TUCSON AZ 85721
U.S.A.

MEINIG MANFRED DR
ZENTRALINSTITUT
PHYSIK DER ERDE
TELEGRAFENBERG 17 A
DDR-15 POTSDAM
G.D.R.

MEIRE RAPHAEL
WEIDERSTRAAT 11
B-9050 EVERGEM
BELGIUM

MEISEL DAVID D DR
DEPT OF PHYSICS AND
ASTRONOMY
STATE UNIV COLLEGE
GENESEO NY 14454
U.S.A.

MEKLER YURI DR
DEPT OF GEOPHYSICS
TEL-AVIV UNIVERSITY
TEL-AVIV
ISRAEL

MELBOURNE WILLIAM G DR
JPL
4800 OAK GROVE DRIVE
PASADENA CA 91103
U.S.A.

MELCHIOR PAUL J PROF
OBSERVATOIRE ROYAL
DE BELGIQUE
AVENUE CIRCULAIRE 3
B-1180 BRUXELLES
BELGIUM

MELIK-ALAVERDIAN YU DR
BYURAKAN ASTRO-
PHYSICAL OBSERVATORY
378433 ARMENIA
U.S.S.R.

MELNICK JORGE
DEPTO DE ASTRONOMIA
UNIVERSIDAD DE CHILE
CASILLA 36-D
SANTIAGO
CHILE

MELROSE DONALD B DR
DEPT OF THEORETICAL
PHYSICS
UNIVERSITY OF SYDNEY
SYDNEY N S W 2006
AUSTRALIA

MEN' A V DR
INSTITUTE OF RADIO-
PHYSICS AND ELECTRO-
NICS
310085 KHARKOV
U.S.S.R.

MENDEZ MANUEL DR
INST DE ASTRONOMIA
UNAM
APT POSTAL 70-264
MEXICO DF
MEXICO

MENDEZ ROBERTO H DR
INST ASTRON FIS ESP
CASILLA 67
SUCURSAL 28
1428 BUENOS AIRES
ARGENTINA

MENDIS D ASOKA DR
APIS UCSD
LA JOLLA CA 92093
U.S.A.

MENDOZA V EUGENIO E DR
CIDA
APARTADO 264
MERIDA
VENEZUELA

MENEGUZZI MAURICE M DR
SEC ETUDES SPATIALES
DPEH SACLAY
BP 2
F-91190 GIF S YVETTE
FRANCE

MENNESSIER MARIE-ODILE DR
LABORATOIRE
D ASTRONOMIE
U S T L
F-34060 MONTPELLIER
FRANCE

MENON T K PROF
DEPT GEOPHYS/ASTRON
UNIVERSITY OF
BRITISH COLUMBIA
VANCOUVER BC V6T 1W5
CANADA

MENTESE HUESEYIN DR
ISTANBUL UNIVERSITY
FACULTY OF SCIENCE
ISTANBUL
TURKEY

MENZIES JOHN W DR
SAAO
PO BOX 9
OBSERVATORY
CAPE TOWN 7935
SOUTH AFRICA

MERCIER CLAUDE DR
OBS DE PARIS
DASOP
F-92190 MEUDON
FRANCE

MERGENTALER JAN PROF
ASTRONOMICAL INST
KOPERNIKA 19
51-617 WROCLAW
POLAND

MERLEAU-PONTY J PROF
5 RUE DU GENERAL
DE CASTELNAU
F-75015 PARIS
FRANCE

MERMAN G A DR
INST OF THEORETICAL
ASTRONOMY
192187 LENINGRAD
U.S.S.R.

MERMAN NATALIA V DR
PULKOVO OBSERVATORY
196140 LENINGRAD
U.S.S.R.

MERMILLIOD JEAN-CLAUDE DR
INST D'ASTRONOMIE
UNIVERSITE LAUSANNE
CH-1290 CHAVANNES-DES-BOIS
SWITZERLAND

MERRILL JOHN E DR
DEPT OF ASTRONOMY
SSRB 211
UNIV OF FLORIDA
GAINESVILLE FL 32611
U.S.A.

MERTON GERALD MC MA PHD
17 HOLYWELL
OXFORD OX1 3SA
U.K.

MERTZ LAWRENCE N DR
287 FAIRFIELD COURT
PALO ALTO CA 94306
U.S.A.

MERZANIDES CONSTANTINOS DR
DEPT OF ASTRONOMY
UNIVERSITY OF THESSALONIKI
GREECE

MESSAGE PHILIP J DR
DEPT APPLIED MATHS
THEORETICAL PHYSICS
THE UNIVERSITY
LIVERPOOL L69 3BX
U.K.

MESTEL LEON PROF
ASTRONOMY CENTRE
UNIVERSITY OF SUSSEX
BRIGHTON BN1 9QH
U.K.

MESZAROS PETER DR
COD 660
NASA/GSFC
LAB HIGH EN ASTROPH
GREENBELT MD 20771
U.S.A.

METZ K DR
INSTITUT F ASTRONOMIE UND ASTROPHYSIK
SCHEINERSTR 1
D-8000 MUENCHEN 80
GERMANY, F.R.

MEURERS JOSEPH PROF DR
UNIVERSITAETS-STERNWARTE
TUERKENSCHANZSTR 17
A-1180 WIEN
AUSTRIA

MEWE R DR
LAB VOOR RUIMTE-ONDERZOEK
BENELUXLAAN 21
3527 HS UTRECHT
THE NETHERLANDS

MEYER CLAUDE DR
C E R G A
AVENUE COPERNIC
F-06130 GRASSE
FRANCE

MEYER FRIEDRICH DR
MPI FUER PHYSIK UND ASTROPHYSIK
K-SCHWARZSCHILD-ST 1
D-8046 GARCHING
GERMANY, F.R.

MEYER J P DR
SECT D'ASTROPHYSIQUE
CEN SACLAY
B P NO 2
F-91190 GIF-S-YVETTE
FRANCE

MEYER-HOFMEISTER E DR
MPI FUER PHYSIK UND ASTROPHYSIK
FOEHRINGER RING 6
D-8000 MUENCHEN 40
GERMANY, F.R.

MEZETTI MARINO
OSS ASTRONOMICO DI TRIESTE
VIA G B TIEPOLO 11
I-34131 TRIESTE
ITALY

MEZGER PETER G PROF
MPI FUER RADIO-ASTRONOMIE
AUF DEM HUEGEL 69
D-5300 BONN 1
GERMANY, F.R.

MIANES PIERRE DR
OBSERVATOIRE DE TOULOUSE
AV C FLAMMARION 1
F-31500 TOULOUSE
FRANCE

MIAO YONG-KUAN
DEPT OF ASTRONOMY
NANJING UNIVERSITY
NANJING
PEOPLE'S REPUBLIC CHINA

MIAO YONG-RUI
SHAANXI OBSERVATORY
LINTONG XIAN
SHAANXI
PEOPLE'S REPUBLIC CHINA

MICHALITSIANOS ANDREW
NASA-GSFC
CODE 685
GREENBELT MD 20771
U.S.A.

MICHARD RAYMOND DR
DIRECTOR
OBS DE NICE
B P 252
F-06007 NICE CEDEX
FRANCE

MICHAUD GEORGES J DR
250 DU FINISTERE
ST-LAMBERT J4S 1P5
CANADA

MICHEL F CURTIS PROF
RICE UNIVERSITY
HOUSTON TX 77001
U.S.A.

MICHEL K W PROF
INST PHYSIK CHEMIE
UNIVERSITAET STUTTGART
D-7000 STUTTGART 80
GERMANY, F.R.

MICZAIKA G R DR
TRW SYSTEMS R5-2291
1 SPACE PARK
REDONDO BCH CA 90278
U.S.A.

MIDDLEHURST BARBARA M MS
LUNAR AND PLANETARY INSTITUTE
3303 NASA ROAD 1
HOUSTON TX 77058
U.S.A.

MIETELSKI JAN S DR
AL MICKIEWICZA 13-5
31-120 KRAKOW
POLAND

MIGEOTTE MARCEL V PROF
INSTITUT
D ASTROPHYSIQUE
B-4200 COINTE-OUGREE
BELGIUM

MIGNARD FRANCOIS DR
CERGA
AVENUE COPERNIC
F-06130 GRASSE
FRANCE

MIHAILA IERONIM M DR
BUCHAREST UNIVERSITY
ACADEMIEI 14
70109 BUCHAREST
ROUMANIA

MIHALAS DIMITRI DR
HIGH ALTITUDE OBS
NCAR
BOX 3000
BOULDER CO 80307
U.S.A.

MIKAMI TAKAO DR
DEPT OF ASTRONOMY
UNIVERSITY OF TOKYO
BUNKYO-KU
TOKYO 113
JAPAN

MIKESELL ALFRED H MR
2109 EAST COPPER ST
TUCSON AZ 85719
U.S.A.

MIKHAEL FAHMY I PROF DR
AIN SHAMS UNIVERSITY
CAIRO
ARAB REP. OF EGYPT

MIKHAIL JOSEPH SIDKY DR
HELWAN OBSERVATORY
CAIRO
ARAB REP. OF EGYPT

MIKHAILOV ALEXANDRA PROF
PULKOVO OBSERVATORY
196140 LENINGRAD
U.S.S.R.

MIKHELSON NIKOLAJ N DR
PULKOVO OBSERVATORY
196140 LENINGRAD
U.S.S.R.

MILANI ANDREA
IST DI MATEMATICA
VIA BUONARROTI 2
I-56100 PISA
ITALY

MILANO LEOPOLDO DR
CAPODIMONTE ASTRONO-
MICAL OBSERVATORY
VIA MOIARIELLO 16
I-80131 NAPOLI
ITALY

MILES HOWARD G MR
21 BABBACOMBE RD
COVENTRY CV3 5PE
U.K.

MILET BERNARD L DR
OBSERVATOIRE DE NICE
B P 252
F-06007 NICE CEDEX
FRANCE

MILEY G K DR
STERREWACHT
POSTBUS 9513
2300 RA LEIDEN
THE NETHERLANDS

MILKEY ROBERT W DR
AURA INC
1002 N WARREN AVE
TUCSON AZ 85719
U.S.A.

MILLAR THOMAS J DR
MATHEMATICS DEPT
UMIST
P O BOX 88
MANCHESTER M60 1QD
U.K.

MILLER FREEMAN D PROF
UNIVERSITY MICHIGAN
DEPT OF ASTRONOMY
DENNISON BLDG
ANN ARBOR MI 48109
U.S.A.

MILLER HUGH R PROF
DEPT OF PHYSICS
GEORGIA STATE
UNIVERSITY
ATLANTA GA 30303
U.S.A.

MILLER JOHN C DR
DEPT OF ASTROPHYSICS
UNIV OF OXFORD
SOUTH PARKS ROAD
OXFORD
U.K.

MILLER JOSEPH S PROF
LICK OBSERVATORY
UNIVERSITY OF CALIF
SANTA CRUZ CA 95060
U.S.A.

MILLER RICHARD H DR
ASTRONOMY DEPT
UNIV OF CHICAGO
5640 ELLIS AVE
CHICAGO IL 60637
U.S.A.

MILLET JEAN DR
LAS
TRAVERSE DU SIPHON
LES TROIS LUCS
13012 MARSEILLE
FRANCE

MILLIGAN J E
ASTROPHYSICS BRANCH
GODDARD SPACE FL CTR
GREENBELT MD 20771
U.S.A.

MILLIKAN ALLAN G DR
RESEARCH LABS B-59
EASTMAN KODAK CO
ROCHESTER NY 14650
U.S.A.

MILLIS ROBERT L DR
LOWELL OBSERVATORY
P O BOX 1269
FLAGSTAFF AZ 86001
U.S.A.

MILLMAN PETER M DR
HERZBERG INSTITUTE
OF ASTROPHYSICS
NAT RES COUNC CANADA
OTTAWA ONT K1A 0R6
CANADA

MILLS ALLAN A DR
DEPT OF ASTRONOMY
THE UNIVERSITY
LEICESTER LE1 7RH
U.K.

MILLS BERNARD Y PROF
SCHOOL OF PHYSICS
UNIVERSITY OF SYDNEY
SYDNEY NSW 2006
AUSTRALIA

MILNE DOUGLAS K DR
CSIRO
DIV OF RADIOPHYSICS
P O BOX 76
EPPING NSW 2121
AUSTRALIA

MILONE EUGENE F PROF
PHYSICS DEPT
UNIVERSITY
OF CALGARY
CALGARY ALTA T2N 1N4
CANADA

MILONE LUIS A DR
OBSERVATORIO
DE CORDOBA
LAPRIDA 854
5000 CORDOBA
ARGENTINA

MILOVANOVIC VLADETA DR
INSTITUT ZA GEODE-
ZIJU BEOGRAD
BUL REVOLUCIJE 73
11000 BEOGRAD
YUGOSLAVIA

MININ I N DR
LENINGRAD STATE UNI-
VERSITY ASTRONOMICAL
OBSERVATORY
199178 LENINGRAD
U.S.S.R.

MINN YOUNG KEY DR
KOREAN NATIONAL
OBSERVATORY
KANGNAM-KU
SEOUL 134-03
REPUBLIC OF KOREA

MINNET HARRY C MR
CSIRO DIVISION OF
RADIOPHYSICS
P O BOX 76
EPPING NSW 2121
AUSTRALIA

MINTZ BLANCO MA B MR
CERRO TOLOLO INTER-
AMERICAN OBS
CASILLA 63-D
LA SERENA
CHILE

MIRABEL IGOR FELIX DR
DEPT OF PHYSICS
FAC OF NATURAL SCS
UNIV OF PUERTO RICO
RIO PIEDRAS PR 00931
U.S.A.

MIRZOYAN L N DR PROF
BYURAKAN ASTROPHYSI-
CAL OBSERVATORY
378433 ARMENIA
U.S.S.R.

MISCONI NEBIL YOUSIF DR
SPACE ASTRONOMY LAB
SUNY
EXECUTIVE PARK EAST
ALBANY NY 12203
U.S.A.

MISNER CHARLES W PROF
DEPT OF PHYSICS AND
ASTRONOMY
UNIVERSITY MARYLAND
COLLEGE PARK MD20742
U.S.A.

MISSANA MARCO DR
VIA BERGAMO
I-22058 OSNAGO COMO
ITALY

MISSANA NATALE PROF
VIA PUCCINI N.3
I-10025 PINO
TORINESE
ITALY

MITALAS ROMAS ASSOC PROF
DEPT OF ASTRONOMY
UNIVERSITY OF
WESTERN ONTARIO
LONDON ONT N6A 5B9
CANADA

MITCHELL GEORGE F DR
DEPT OF ASTRONOMY
SAINT MARY S
UNIVERSITY
HALIFAX NS B3H 3C3
CANADA

MITCHELL R MR
C/O CICESE
AV CASTELUM NO 898
ENSENADA
BAJA CALIFORNIA
MEXICO

MITCHELL WALTER E JR
ASTRONOMY DEPT
OHIO STATE UNIV
174 W 18TH AVE
COLUMBUS OH 43210
U.S.A.

MITIC LJUBISA A DR
OBSERVATOIRE
DE BELGRADE
VOLGINA 7
11050 BELGRADE
YUGOSLAVIA

MITRA A P DR
ASST DIRECTOR
NATIONAL PHYSICAL
LABORATORY
NEW DELHI
INDIA

MITROFANOVA LYUDMILA A DR
PULKOVO OBSERVATORY
196140 LENINGRAD
U.S.S.R.

MITTON SIMON DR
CAMBRIDGE UNIV PRESS
P O BOX 110
CAMBRIDGE CB2 3RL
U.K.

MIYADI MASASI DR
SENGAWA 3-6-11
CHOFU-SHI 182
JAPAN

MIYAJI SHIGEKI DR
LAB FOR ASTRONOMY &
SOLAR PHYSICS
NASA GSFC
GREENBELT MD 20771
U.S.A.

MIYAMOTO MASANORI DR
TOKYO ASTRONOMICAL
OBSERVATORY
OSAWA MITAKA
TOKYO 181
JAPAN

MIYAMOTO SIGINORI PROF
DEPT OF PHYSICS
OSAKA UNIVERSITY
MACHIKANEYAMA-CHO
TOYONAKA OSAKA 560
JAPAN

MIYAMOTO SYOTARO PROF DR
KASAN OBSERVATORY
YAMA HIGASHIYAMA
KYOTO 607
JAPAN

MNATSAKANIAN MAMIKON A DR
BYURAKAN ASTROPHYSI-
CAL OBSERVATORY
378433 ARMENIA
U.S.S.R.

MOCHNACKI S W DR
DAVID DUNLAP OBS
DEPT OF ASTRONOMY
UNIV OF TORONTO
TORONTO ONT M5S 1A7
CANADA

MOCZKO JANUSZ DR
BOROWIEC
63-120 KORNIK
POLAND

MODALI SARMA B DR
8518 NIGHTINGALE DR
LANHAM MD 20706
U.S.A.

MODISETTE JERRY L PROF
HOUSTON BAPTIST
UNIVERSITY
7502 FONDREN AVE
HOUSTON TX 77074
U.S.A.

MOELLENHOFF CLAUS DR
LANDESSTERNWARTE
KOENIGSTUHL
D-6900 HEIDELBERG 1
GERMANY, F.R.

MOELLER OLE PROF
INST OF ASTRONOMY
UNIVERSITY OF AARHUS
DK-8000 AARHUS C
DENMARK

MOERDIJK WILLY G DR
ST PIETERSAALSTR 171
B-9000 GENT
BELGIUM

MOESGAARD KRISTIAN P PH D
BYGADEN 1
TORRILD
DK-8300 ODDER
DENMARK

MOFFAT ANTHONY F J DR
DEPT DE PHYSIQUE
UNIVERSITE
DE MONTREAL
MONTREAL P Q H3C 3J7
CANADA

MOFFATT HENRY KEITH
6 BANHAMS CLOSE
CAMBRIDGE CB3
U.K.

MOFFET ALAN T PROF
OWENS VALLEY RADIO
OBSERVATORY
CALTECH 102 24
PASADENA CA 91125
U.S.A.

MOFFETT THOMAS J PROF
DEPT OF PHYSICS
PURDUE UNIVERSITY
WEST LAFAYETTE
IN 47907
U.S.A.

MOGILEVSKIJ EH I DR
INSTITUTE OF TERRE-
STRIAL MAGNETISM AND
IONOSPHERE
142092 AKADEMGORODOK
U.S.S.R.

MOHLER ORREN C PROF
405 AWIXA RD
ANN ARBOR MI 48104
U.S.A.

MOISEEV I G DR
CRIMEAN ASTROPHYSI-
CAL OBSERVATORY USSR
ACAD OF SCI NAUCHNIY
334413 CRIMEA
U.S.S.R.

MOLCHANOV A P PROF
LENINGRAD STATE UNI-
VERSITY ASTRONOMICAL
OBSERVATORY
199178 LENINGRAD
U.S.S.R.

MOLES MARIANO J DR
JOSE ANTONIO 5
BINACED HUESCA
SPAIN

MOLNAR MICHAEL R PROF
UNIVERSITY OF TOLEDO
RITTER OBSERVATORY
TOLEDO OH 43606
U.S.A.

MONAGHAN JOSEPH J DR
MATHEMATICS DEPT
MONASH UNIVERSITY
CLAYTON VICT 3168
AUSTRALIA

MONFILS ANDRE G PROF
INSTITUT
D ASTROPHYSIQUE
B-4200 COINTE-LIEGE
BELGIUM

MONNET GUY J DR
OBSERVATOIRE DE LYON
AVENUE CHARLES ANDRE
F-69230 ST GENIS
LAVAL
FRANCE

MONSIGNORI FOSSI BRUNA DR
OSSERVATORIO ARCETRI
LARGO FERMI 5
I-50125 FIRENZE
ITALY

MONTES C DR
OBS DE NICE
B P 252
F-06007 NICE CEDEX
FRANCE

MONTMERLE T DR
SECT D'ASTROPHYSIQUE
CEN SACLAY
B P NO 2
F-91190 GIF-S-YVETTE
FRANCE

MOOK DELO E PROF
DEPT OF PHYSICS AND
ASTRONOMY
DARTMOUTH COLLEGE
HANOVER NH 03755
U.S.A.

MOONS MICHELE B M M
DEPT DE MATHEMATIQUE
FAC UN ND DE LA PAIX
REMPART VIERGE 8
B-5000 NAMUR
BELGIUM

MOORE D R DR
DEPT APPLIED MATHS
AND THEORETICAL PHYS
SILVER STREET
CAMBRIDGE CB3 9EW
U.K.

MOORE ELLIOTT P PROF
JOINT OBSERVATORY
F COMETARY RESEARCH
CAMPUS STATION
SOCORRO NM 87801
U.S.A.

MOORE PATRICK DR OBE
FARTHINGS
39 WEST STREET
SELSEY SUSSEX
U.K.

MOORE RONALD L DR
SOLAR ASTRONOMY
264 33
CALIF INST TECHNOLOG
PASADENA CA 91125
U.S.A.

MOORHEAD JAMES M DR
ASTRONOMY DEPT
UNIVERSITY OF
WESTERN ONTARIO
LONDON ONT N6A 3K7
CANADA

MOOS HENRY WARREN DR
DEPT OF PHYSICS
JOHNS HOPKINS UNIV
BALTIMORE MD 21218
U.S.A.

MORAN JAMES M DR
CENTER FOR
ASTROPHYSICS
60 GARDEN ST
CAMBRIDGE MA 02138
U.S.A.

MORANDO BRUNO L DR
BUREAU D LONGITUDES
77 AVE DENFERT
ROCHEREAU
F-75014 PARIS
FRANCE

MORBEY CHRISTOPHER L
DOMINION ASTROPH OBS
5071 W SAANICH ROAD
RR 5
VICTORIA BC V8X 4M6
CANADA

MOREELS G DR
OBS DE BESANCON
41 BIS AVENUE DE
L'OBSERVATOIRE
F-25000 BESANCON
FRANCE

MOREL PIERRE JACQUES DR
OBS DE NICE
B P 252
F-06007 NICE CEDEX
FRANCE

MORENO HUGO PROF
DEP DE ASTRONOMIA
UNIVERSIDAD DE CHILE
CASILLA 36-D
SANTIAGO
CHILE

MORETON G E
15-5 THE ESPLANADE
BALMORAL BCH NSW2088
AUSTRALIA

MORGAN BRIAN LEALAN
BLACKETT LABORATORY
IMPERIAL COLLEGE
LONDON SW7 2BZ
U.K.

MORGAN DAVID H DR
ROYAL OBSERVATORY
BLACKFORD HILL
EDINBURGH EH9 3HJ
U.K.

MORGAN JOHN ADRIAN
74 HIGHDOWN ROAD
HOVE SUSSEX
U.K.

MORGAN PETER DR
DEPT EARTH PLANET SC
CODE 54-419
MIT
CAMBRIDGE MA 02139
U.S.A.

MORGAN THOMAS H DR
DEPT PHYSICS/ASTRON
SOUTHWESTERN UNIV
GEORGETOWN TX 78626
U.S.A.

MORGAN W W PROF
YERKES OBSERVATORY
WILLIAMS BAY WI53191
U.S.A.

MORGULEFF NINA ING
INST D ASTROPHYSIQUE
98 BIS BD ARAGO
F-75014 PARIS
FRANCE

MORIMOTO MASAKI DR
TOKYO ASTRONOMICAL
OBSERVATORY
OSAWA MITAKA
TOKYO 181
JAPAN

MORISON IAN MR
NRAL
JODRELL BANK
CHESHIRE SK11 9DL
U.K.

MORIYAMA FUMIO DR
TOKYO ASTRONOMICAL
OBSERVATORY
OSAWA MITAKA
TOKYO 181
JAPAN

MOROZ V I PROF DR
SPACE RESEARCH
INSTITUTE
USSR ACADEMY OF SCI
117810 MOSCOW
U.S.S.R.

MOROZHENKO A V DR
MAIN ASTRON OBS
UKRAINIAN ACAD OF SC
GOLOSEEVO
252127 KIEV
U.S.S.R.

MOROZHENKO N N DR
MAIN ASTRON OBS
UKRAINIAN ACAD OF SC
GOLOSEEVO
252127 KIEV
U.S.S.R.

MORRAS RICARDO DR.
INSTITUTO ARGENTINO
DE RADIOASTRONOMIA
CASILLA DE CORREO 5
1894-VILLA ELISA
ARGENTINA

MORRIS DAVID DR
MPI FUER RADIOASTRON
AUF DEM HUEGEL 69
D-5300 BONN 1
GERMANY, F.R.

MORRIS MARK ROOT DR
DEPT OF ASTRONOMY
COLUMBIA UNIV
NEW YORK NY 10027
U.S.A.

MORRIS S C DR
DOMINION ASTROPH OBS
5071 W SAANICH ROAD
RR 5
VICTORIA BC V8X 4M6
CANADA

MORRISON DAVID PROF
CODE SL
NASA HEADQUARTERS
WASHINGTON DC 20546
U.S.A.

MORRISON LESLIE V
ROYAL GREENWICH
OBSERVATORY
HAILSHAM
EAST SUSSEX BN27 1RP
U.K.

MORRISON NANCY DUNLAP DR
DEPT PHYSICS ASTRON
UNIV OF TOLEDO
2801 W BANCROFT ST
TOLEDO OH 43606
U.S.A.

MORRISON PHILIP PROF
PHYSICS MIT 6-205
CAMBRIDGE MA 02139
U.S.A.

MORTON DONALD C DR
ANGLO AUSTRALIAN
OBSERVATORY
P O BOX 296
EPPING N S W 2121
AUSTRALIA

MORTON G A DR
1122 SKYCREST DRIVE
APT 6
WALNUT CREEK CA94595
U.S.A.

MOSS CHRISTOPHER DR
VATICAN OBS RES GR
STEWARD OBS
UNIV OF ARIZONA
TUCSON AZ 85721
U.S.A.

MOSS DAVID L DR
MATHEMATICS DEPT
MANCHESTER
UNIVERSITY
MANCHESTER M13 9PL
U.K.

MOTTA SANTO DR
CSS ASTROFISICO
CITTA UNIVERSITARIA
I-95125 CATANIA
ITALY

MOTZ LLOYD PROF
DEPT OF ASTRONOMY
COLUMBIA UNIVERSITY
PUPIN HALL BOX 57
NEW YORK NY 10027
U.S.A.

MOULD JEREMY R
KPNO
P O BOX 26732
TUCSON AZ 85726
U.S.A.

MOURADIAN ZADIG M DR
OBSERVATOIRE DE
PARIS-MEUDON
F-92190 MEUDON
FRANCE

MOUSCHOVIAS TELEMACHOS CH
DEPT OF ASTRONOMY
UNIVERSITY ILLINOIS
URBANA IL 61801
U.S.A.

MOUSSAS XENOPHON PH D
20 SKYLITSI STREET
ATHENS 707
GREECE

MOUTSOULAS MICHAEL PROF
ASTRONOMY DEPT
UNIVERSITY OF ATHENS
PANEPISTIMIOPOLIS
ATHENS 621
GREECE

MOVAHED REZA DR
P O BOX 6
BABOLSAR
IRAN

MRKOS ANTONIN DR
DEPT OF ASTRONOMY
CHARLES UNIVERSITY
SVEDSKA 8
150 00 PRAHA 5
CZECHOSLOVAKIA

MUELLER EDITH A PROF
OBSERVATOIRE
DE GENEVE
CH-1290 SAUVERNY GE
SWITZERLAND

MUELLER HELMUT O PROF
HERZOGENMUEHLESTR 4
CH-8051 ZUERICH
SWITZERLAND

MUELLER IVAN I PROF
GEODETIC SCIENCE
OHIO ST UNIVERSITY
1958 NEIL AVE
COLUMBUS OH 43210
U.S.A.

MUENCH GUIDO PROF
MPI FUER ASTRONOMIE
KOENIGSTUHL
D-6900 HEIDELBERG 1
GERMANY, F.R.

MUFSON STUART LEE DR
ASTRONOMY DEPT
319 SWAIN WEST
INDIANA UNIV
BLOOMINGTON IN 47401
U.S.A.

MUKAI SONOYO DR
KANAZAWA INSTITUTE
OF TECHNOLOGY
ENKOJIHONMACHI
KANAZAWA 921
JAPAN

MUKAI TADASHI DR
KANAZAWA INSTITUTE
OF TECHNOLOGY
ENKOJIHONMACHI
KANAZAWA 921
JAPAN

MULDERS GERARD F W
4519 EVERETT ST
KENSINGTON MD 20795
U.S.A.

MULHOLLAND J DERRAL DR
DEPT OF ASTRONOMY
UNIVERSITY OF TEXAS
AUSTIN TX 78712
U.S.A.

MULLALY RICHARD F DR
SCHOOL OF ELECTRICAL
ENGINEERING
UNIV OF SYDNEY
SYDNEY NSW 2006
AUSTRALIA

MULLAN DERMOTT J DR
BARTOL RESEARCH
FOUNDATION
UNIV OF DELAWARE
NEWARK DE 19711
U.S.A.

MULLER A B DR
ESO
K-SCHWARZSCHILD-ST 2
D-8046 GARCHING BEI
MUENCHEN
GERMANY, F.R.

MULLER C A PROF ING
ODINKSVELD 8
7491 HD DELDEN
THE NETHERLANDS

MULLER PAUL
CERGA
AV COPERNIC
F-06130 GRASSE
FRANCE

MULLER R DR
OBS DU PIC-DU-MIDI
F-65200 BAGNERES-DE-
BIGORRE
FRANCE

MULLER RICHARD A
LAWRENCE BERKELEY
LABORATORY
BLDG 50 RM 238
BERKELEY CA 94720
U.S.A.

MUMFORD GEORGE S PROF
BALLOU HALL
TUFTS UNIVERSITY
MEDFORD MA 02155
U.S.A.

MUNRO RICHARD H DR
HIGH ALTITUDE
OBSERVATORY
P O BOX 3000
BOULDER CO 80307
U.S.A.

MURDIN PAUL G DR
ROYAL GREENWICH
OBSERVATORY
HAILSHAM
EAST SUSSEX BN27 1RP
U.K.

MURDOCH HUGH S DR
ASTROPHYSICS DEPT
UNIVERSITY OF SYDNEY
SYDNEY 2006
AUSTRALIA

MURDOCK THOMAS LEE
OPI STOP 30
AIR FORCE GEOPHYSICS
LABORATORIUM
HANSCOM AFB MA 01731
U.S.A.

MURPHY ROBERT E DR
CODE SL-4
NASA HEADQUARTERS
WASHINGTON DC 20546
U.S.A.

MURRAY C ANDREW
ROYAL GREENWICH
OBSERVATORY
HAILSHAM
EAST SUSSEX BN27 1RP
U.K.

MURRAY JOHN B DR
UNIV OF LONDON
OBSERVATORY
MILL HILL PARK
LONDON NW7 2QS
U.K.

MURRAY STEPHEN S DR
HARVARD-SMITHSONIAN
CTR FOR ASTROPHYSICS
60 GARDEN ST
CAMBRIDGE MA 02138
U.S.A.

MUSEN PETER DR
8804 ORBIT LANE
LANHAM MD 20801
U.S.A.

MUSMAN STEVEN DR
JPL
264-748
4800 OAK GROVE DRIVE
PASADENA CA 91103
U.S.A.

MUSTEL E R PROF DR
ASTRONOMICAL COUNCIL
USSR ACAD OF SCI
PYATNITSKAYA UL 48
109017 MOSCOW
U.S.S.R.

MUTEL ROBERT LUCIEN
DEPT OF PHYSICS &
ASTRONOMY
UNIVERSITY OF IOWA
IOWA CITY IA 52242
U.S.A.

MUTSCHLECNER J PAUL DR
ASTRONOMY DEPT
SWAIN HALL W
INDIANA UNIVERSITY
BLOOMINGTON IN 47405
U.S.A.

MUZZIO JUAN C DR
OBSERVATORIO
ASTRONOMICO
PASEO DEL BOSQUE
1900 LA PLATA
ARGENTINA

MYACHIN V F DR
INSTITUTE OF THEORE-
TICAL ASTRONOMY
192187 LENINGRAD
U.S.S.R.

MYERS PHILIP C PROF
PHYSICS DEPT
MIT
BLDG 26 RM 349
CAMBRIDGE MA 02139
U.S.A.

NACOZY PAUL E PROF
AEROSPACE ENGR AND
ENGR MECHANICS
UNIVERSITY OF TEXAS
AUSTIN TX 78712
U.S.A.

NADOLSCHI V PROF DR
COM.ARDEOANI
OF TESCANI JUD
BACAU
ROUMANIA

NADYOZHIN D K DR
INST THEOR EXP PHYS
CHEREMUSHKINSKAJA 25
117259 MOSCOW
U.S.S.R.

NAGASAWA SHINGO PROF
TOKYO SCIENCE UNIV
2641 YAMAZAKI
HIGASHI KAMEYAMA
NODASHI 278
JAPAN

NAGASE FUMIAKI DR
DEPT OF ASTROPHYSICS
NAGOYA UNIVERSITY
CHIKUSA-KU
NAGOYA 464
JAPAN

NAGIRNER DMITRIJ I DR
LENINGRAD UNIVERSITY
ASTRONOMICAL OBS
199178 LENINGRAD
U.S.S.R.

NAGNIBEDA VALERY G DR
LENINGRAD UNIVERSITY
ASTRONOMICAL
OBSERVATORY
199178 LENINGRAD
U.S.S.R.

NAHON FERNAND PROF
25 AVE DE L-EUROPE
F-92310 SEVRES
FRANCE

NAKADA YOSHIKAZU DR
DEPT OF ASTRONOMY
FACULTY OF SCIENCE
UNIV TOKYO BUNKYO-KU
TOKYO 113
JAPAN

NAKAGAWA NAOYA DR
UNIV OF ELECTRO-
COMMUNICATIONS
CHOFU-SHI
TOKYO 181
JAPAN

NAKAGAWA YOSHINARI DR
HIGH ALTITUDE
OBSERVATORY
P O BOX 3000
BOULDER CO 80307
U.S.A.

NAKAGAWA YOSHITSUGU DR
DEPT OF PHYSICS
FACULTY OF SCIENCE
KYOTO UNIVERSITY
KYOTO 606
JAPAN

NAKAI YOSHIHIRO
KWASAN AND HIDA
OBSERVATORIES
KYOTO UNIVERSITY
KYOTO 607
JAPAN

NAKAMURA TAKASHI DR
DEPT OF PHYSICS
KYOTO UNIVERSITY
KYOTO 606
JAPAN

NAKAMURA TSUKO DR
DODAIRA STATION
TOKYO ASTRON OBS
TOKIGAWA
SAITAMA 355-05
JAPAN

NAKANO SABURO DR
KOBINATO 1-21-7
BUNKYO KU
TOKYO 112
JAPAN

NAKANO TAKENORI DR
DEPT OF PHYSICS
KYOTO UNIVERSITY
KITASHIRAKAWA OIWAKE
SAKYOKU KYOTO 606
JAPAN

NAKAYAMA SHIGERU DR
FACULTY GENERAL EDUC
UNIVERSITY OF TOKYO
KOMABA MEGURO
TOKYO 153
JAPAN

NAKAZAWA KIYOSHI DR
DEPT OF PHYSICS
KYOTO UNIVERSITY
KITASHIRAKAWA OIWAKE
SAKYOKU KYOTO 606
JAPAN

NAMBA OSAMU DR
ASTRONOMICAL INST
ZONNENBURG 2
3512 NL UTRECHT
THE NETHERLANDS

NANDY KASHINATT DR
ROYAL OBSERVATORY
EDINBURGH
BLACKFORD EH9 3HJ
U.K.

NAPIER WILLIAM M DR
ROYAL OBSERVATORY
BLACKFORD HILL
EDINBURGH EH9 3HJ
U.K.

NAQVI S I H PROF
DEPT OF PHYS & ASTRO
UN OF REGINA
REGINA S4S 0A2
CANADA

NARANAN S PROF
TATA INSTITUTE OF
FUNDAMENTAL RESEARCH
HOMI BHABHA ROAD
BOMBAY 400005
INDIA

NARAYANA J V
REG METEOROLOGICAL
OFFICE
4 COLLEGE ROAD
MADRAS 600006
INDIA

NARIAI HIDEICHI
RESEARCH INST FOR
THEORETICAL PHYSICS
HIROSHIMA UNIVERSITY
725 TAKEHARA
JAPAN

NARIAI KYOJI DR
TOKYO ASTRONOMICAL
OBSERVATORY
MITAKA
TOKYO 181
JAPAN

NARITA SHINJI DR
DOSHISHA UNIVERSITY
KYOTO 602
JAPAN

NARLIKAR JAYANT V PROF
TATA INSTITUTE OF
FUNDAMENTAL RESEARCH
HOMI BHABHA ROAD
BOMBAY 400 005
INDIA

NATALI GIULIANO DR
LAB ASTROFISICO SPAZ
P O BOX 67
I-00044 FRASCATI
ITALY

NATHER R EDWARD
DEPT OF ASTRONOMY
UNIVERSITY OF TEXAS
AUSTIN TX 78712
U.S.A.

NATTA ANTONELLA DR
CNR ASTROFISICA
SPAZIALE C P 67
I-00044 FRASCATI
ROMA
ITALY

NAUMOV VITALIJ A DR
PULKOVO OBSERVATORY
196140 LENINGRAD
U.S.S.R.

NAWAR SAMIR DR
HELWAN OBSERVATORY
HELWAN
ARAB REP. OF EGYPT

NECKEL HEINZ DR
HAMBURGER STERNWARTE
D-2050 HAMBURG 80
GERMANY, F.R.

NECKEL TH DR
MPI FUER ASTRONOMIE
D-6900 HEIDELBERG-
KOENIGSTUHL
GERMANY, F.R.

NEE TSU-WEI DR
INST OF PHYSICS AND
ASTRONOMY
NAT CENTRAL UNIV
CHUNG-LI
TAIWAN

NEEMAN YUVAL PROF
DEPT OF PHYSICS AND
ASTRONOMY
TEL AVIV UNIVERSITY
TEL AVIV
ISRAEL

NEFEDEVA ANTONINA I DR
ENGELHARDT ASTRONO-
MICAL OBSERVATORY
OBSERVATORY STATION
422526 KAZAN
U.S.S.R.

NEFF JOHN S
605 BROOKLAND PK DR
IOWA CITY IA 52240
U.S.A.

NELSON ALISTAIR H DR
DEPT OF APPLIED MATH
AND ASTRONOMY
UNIV COLLEGE
CARDIFF
U.K.

NELSON BURT DR
DEPT OF ASTRONOMY
SAN DIEGO STATE
UNIVERSITY
SAN DIEGO CA 92115
U.S.A.

NELSON GRAHAM JOHN DR
CSIRO DIV.RADIOPHYS
SOLAR RADIO OBS
P O BOX 94
NARRABRI NSW 2390
AUSTRALIA

NELSON JERRY
LAWRENCE BERKELEY
LABORATORY
UNIV.OF CALIFORNIA
BERKELEY CA 94720
U.S.A.

NELSON ROBERT M
183/501 JET PROPUL-
SION LAB
PASADENA CA 91109
U.S.A.

NEMIRO ANDREJ A DR PROF
PULKOVO OBSERVATORY
196140 LENINGRAD
U.S.S.R.

NESS NORMAN F DR
NASA GSFC CODE 690
GREENBELT MD 20771
U.S.A.

NEUGEBAUER GERRY DR
PHYSICS DEPARTMENT
CALIF INST OF TECHN
320 DOWNS
PASADENA CA 91125
U.S.A.

NEUKUM G DR
INST FUER ALLG UND
ANGEW GEOLOGIE
LUISENSTR 37
D-8000 MUENCHEN
GERMANY, F.R.

NEUPERT WERNER M DR
NASA GODDARD SPACE
FLIGHT CENTER
CODE 682
GREENBELT MD 20771
U.S.A.

NEUZIL LUDEK DR
ASTRONOMICAL INST
CZECH ACAD SCIENCES
OBSERVATORY
251-65 ONDREJOV
CZECHOSLOVAKIA

NEVEN LUC DR
OBSERVATOIRE ROYAL
AVENUE CIRCULAIRE 3
B-1180 BRUXELLES
BELGIUM

NEVIN THOMAS E PROF
PHYSICS DEPT
UNIVERSITY COLLEGE
BELFIELD
DUBLIN 4
IRELAND

NEWBURN RAY L JR
3226 EMERALD ISLE DR
GLENDALE CA 91206
U.S.A.

NEWELL EDWARD B DR
MT STROMLO AND
SIDING SPRING OBS
PRIVATE BAG
WODEN A C T 2606
AUSTRALIA

NEWKIRK GORDON A JR DR
HIGH ALTITUDE
OBSERVATORY NCAR
P O BOX 3000
BOULDER CO 80307
U.S.A.

NEWMAN MICHAEL JOHN DR
T-DOT MS 288
LOS ALAMOS SC LAB
P O BOX 1663
LOS ALAMOS NM 87545
U.S.A.

NEWSOM GERALD H ASST PROF
ASTRONOMY DEPT
OHIO STATE UNIV
174 W 18TH AVE
COLUMBUS OH 43210
U.S.A.

NEWTON ROBERT R DR
JOHNS HOPKINS UNIV
APPLIED PHYSICS LAB
JOHNS HOPKINS RD
LAUREL MD 20810
U.S.A.

NEY EDWARD P PROF
DEPT OF ASTRONOMY
TATE LAB OF PHYSICS
UNIV OF MINNESOTA
MINNEAPOLIS MN 55455
U.S.A.

NHA II-SEONG DR
66 SINCHONDONG
SEODAEMOONKU
SEOUL 120
REPUBLIC OF KOREA

NIARCHOS PANAYIOTIS PH D
DEPT OF ASTRONOMY
UNIV OF ATHENS
PANEPISTIMIOPOLIS
ATHENS 621
GREECE

NICHOLLS RALPH W PROF
CRESS DEPT OF PHYS
YORK UNIVERSITY
4700 KEELE ST
DOWNSVIEW ONT M3J1P3
CANADA

NICHOLSON WILLIAM
ROYAL GREENWICH
OBSERVATORY
HAILSHAM
EAST SUSSEX BN27 1RP
U.K.

NICOLAS KENNETH ROBERT
CODE 4163
NAVAL RESEARCH LAB
OVERLOOK AVENUE
WASHINGTON DC 20375
U.S.A.

NICOLET MARCEL PROF
UNIVERSITE
DE BRUXELLES
30 AVE DEN DOORN
B-1180 BRUXELLES
BELGIUM

NICOLOV N S DR
UNIVERSITE DE SOFIA
FACULTE PHYSIQUE
SOFIA
BULGARIA

NICOLSON GEORGE D DR
RADIO ASTRONOMY
OBSERVATORY
P O BOX 3718
JOHANNESBURG 2000
SOUTH AFRICA

NIELL ARTHUR E DR
JPL
FHL-D 430
4800 OAK GROVE DR
PASADENA CA 91103
U.S.A.

NIEMELA VIRPI S DR
CALLE 51 ESQ 11
1894 VILLA ELISA BA
ARGENTINA

NIEUWENHUIJZEN HANS DR
ASTRONOMICAL INST
ZONNENBURG 2
3512 NL UTRECHT
THE NETHERLANDS

NIIMI HIDEYUKI DR
FACULTY ENGINEERING
UNIVERSITY OF KYOTO
YOSHIDAHONMACHI
SAKYOKU KYOTO 606
JAPAN

NIKITIN A A DR
LENINGRAD STATE UNI-
VERSITY ASTRONOMICAL
OBSERVATORY
199178 LENINGRAD
U.S.S.R.

NIKOLOFF IVAN DR
PERTH OBSERVATORY
BICKLEY WA 6076
AUSTRALIA

NIKOLOV ANDREJ DR
P O BOX 402
1000 SOFIA
BULGARIA

NIKOLSKY G M PROF DR
INSTITUTE OF TERRE-
STRIAL MAGNETISM AND
IONOSPHERE
142092 AKADEMGORODOK
U.S.S.R.

NIKONOV V B DR
CRIMEAN ASTROPHYSI-
CAL OBSERVATORY USSR
ACAD OF SCI NAUCHNIY
334413 CRIMEA
U.S.S.R.

NILSON PETER DR
ASTRONOMICAL
OBSERVATORY
BOX 515
S-751 20 UPPSALA
SWEDEN

NILSSON CARL DR
SMITHSONIAN ASTRO-
PHYSICS OBSERVATORY
60 GARDEN STREET
CAMBRIDGE MA 02138
U.S.A.

NININGER H H
BOX 420
SEDONA AZ 86336
U.S.A.

NISHI KEIZO DR
TOKYO ASTRONOMICAL
OBSERVATORY
OSAWA MITAKA
TOKYO 181
JAPAN

NISHIDA MINORU PROF
DEPT OF PHYSICS
KYOTO UNIVERSITY
KITASHIRAKAWA OIWAKE
SAKYOKU KYOTO 606
JAPAN

NISHIMURA JUN DR
INST SPACE AERON SC
UNIVERSITY OF TOKYO
KOMABA MEGURO-KU
TOKYO
JAPAN

NISHIMURA SHIRO DR
TOKYO ASTRONOMICAL
OBSERVATORY
OSAWA MITAKA
TOKYO 181
JAPAN

NISHIMURA TETSUO DR
STEWARD OBS
UNIV OF ARIZONA
TUCSON AZ 85721
U.S.A.

NISSEN POUL E PROF
INST OF ASTRONOMY
UNIVERSITY OF AARHUS
LANGELANDSKADE
DK-8000 AARHUS C
DENMARK

NITTMAN JOHANN
DEPT OF APPLIED
MATHEMATICAL STUDIES
UNIVERSITY OF LEEDS
LEEDS LS2 9JT
U.K.

NOBILI A M
ISTITUTO DI
MATEMATICA
VIA BUONARROTI 2
I-56100 PISA
ITALY

NOBILI L DR
OSS ASTRONOMICO
VICOLO DELL OSSERVA-
TORIO
PADOVA
ITALY

NOCI GIANCARLO PROF
OSSERVATORIO ARCETRI
LARGO FERMI 5
I-50125 FIRENZE
ITALY

NOEL FERNANDO
DPTO DE ASTRONOMIA
UNIVERSIDAD DE CHILE
CASILLA 36D
SANTIAGO
CHILE

NOELS ARLETTE DR
50 AVENUE DE LA PAIX
BOITE 063
B-4030 GRIVEGNEE
BELGIUM

NOERDLINGER PETER D PROF
SOLAR ENERGY RES INS
BUILDINGS SYSTEMS DV
1617 COLE BLVD
GOLDEN CO 80401
U.S.A.

NOLLEZ G DR
INST D'ASTROPHYSIQUE
98 BIS BD ARAGO
F-75014 PARIS
FRANCE

NOMOTO KEN'ICHI DR
DEPT EARTH SC/ASTRON
COLLEGE GEN EDUCAT
UNIVERSITY OF TOKYO
MEGURO-KU TOKYO 153
JAPAN

NOONAN THOMAS W PROF
PHYSICS DEPT
SUNY
BROCKPORT NY 14420
U.S.A.

NORDH H LENNART DR
STOCKHOLM
OBSERVATORY
S-133 00
SALTSJOEBADEN
SWEDEN

NORDLUND AKE DR
NORDITA
BLEGDAMSVEJ 17
DK-2100 COPENHAGEN O
DENMARK

NORDSTROM BIRGITTA DR
COPENHAGEN
OBSERVATORY
BRORFELDE
DK-4340 TOLLOSE
DENMARK

NORMAN COLIN DR
HUYGENS LAB
STERREWACHT
POSTBUS 9513
2300 RA LEIDEN
THE NETHERLANDS

NORRIS JOHN DR
MT STROMLO
OBSERVATORY
WODEN PO ACT 2606
AUSTRALIA

NORRIS RAYMOND PAUL
UNIV OF MANCHESTER
NRAL JODRELL BANK
MACCLESFIELD
CHESHIRE SK11 9DL
U.K.

NORTH JOHN D PROF
28 CHALFONT ROAD
OXFORD
U.K.

NOSKOV BORIS N DR
STERNBERG STATE
ASTRONOMICAL INST
117234 MOSCOW
U.S.S.R.

NOTNI P DR
ZNTRINST ASTROPHYSIK
STERNW BABELSBERG
ROSA LUXEMBURGST 17A
DDR-1502 POTSDAM
G.D.R.

NOTTALE LAURENT
OBS DE PARIS
DEPT D'ASTROPHYSIQUE
FONDAMENTALE
F-92190 MEUDON
FRANCE

NOTUKI MASAO DR
4-36-7 NARITAHIGASHI
SUGINAMI-KU
TOKYO 166
JAPAN

NOVICK ROBERT
DEPT OF PHYSICS
COLUMBIA UNIVERSITY
538 W 120 ST
NEW YORK NY 10027
U.S.A.

NOVIKOV I D DR
SPACE RESEARH
INSTITUTE
USSR ACADEMY OF SCI
117810 MOSCOW
U.S.S.R.

NOVIKOV SERGEJ B DR
STERNBERG STATE
ASTRONOMICAL INST
117234 MOSCOW
U.S.S.R.

NOVOSELOV V S PROF DR
LENINGRAD STATE UNI-
VERSITY ASTRONOMICAL
OBSERVATORY
199178 LENINGRAD
U.S.S.R.

NOVOTNY VACLAV
ASTRONOMICAL INST
CZECH ACADEMY OF SC
ONDREJOV OBSERVATORY
25165 ONDREJOV
CZECHOSLOVAKIA

NOYES ROBERT W PROF
CTR F ASTROPHYSICS
60 GARDEN ST
CAMBRIDGE MA 02138
U.S.A.

NUNES ROGERIO S DE SOUSA
GRUPO DE MATEMATICA
APLICADA
UNIV DO PORTO
4000 PORTO
PORTUGAL

NUNEZ JORGE DR
OBSERVATORIO FABRA
TIBIDADO
BARCELONA-22
SPAIN

NUSSBAUMER HARRY DR
ATOMIC AND ASTROPHYS
GROUP ETHZ
HUTTENSTR 34
CH-8092 ZUERICH
SWITZERLAND

O'CONNELL ROBERT F PROF
LOUISIANA STATE
UNIVERSITY
BATON ROUGE LA 70803
U.S.A.

O'CONNELL ROBERT WEST DR
ASTRO DEPT UNIV VA
PO BOX 3818 UNIV STA
CHARLOTTESVILLE
VA 22903
U.S.A.

O'CONNOR SEAMUS L DR
PHYSICS DEPT
UNIVERSITY COLLEGE
DUBLIN 4
IRELAND

O'DELL STEPHEN L
PHYSICS DEPT
V P I AND S U
BLACKSBURG VA 24061
U.S.A.

O'HANDLEY DOUGLAS A DR
JET PROPULSION
LABORATORY
BLDG 169 422
PASADENA CA 91103
U.S.A.

O'HORA NATHY P J
ROYAL GREENWICH
OBSERVATORY
HAILSHAM
EAST SUSSEX BN27 1RP
U.K.

O'KEEFE JOHN A DR
GODDARD SPACE FLIGHT
CENTER CODE 680
GREENBELT MD 20771
U.S.A.

O'LEARY BRIAN T
DEPT OF PHYSICS
PRINCETON UNIVERSITY
PRINCETON NJ 08540
U.S.A.

O'MARA BERNARD J PROF
DEPT OF PHYSICS
UNIV OF QUEENSLAND
ST LUCIA
BRISBANE QLD 4067
AUSTRALIA

O'MONGAIN EON
PHYSICS DEPT
UNIVERSITY COLLEGE
BELFIELD
DUBLIN 4
IRELAND

O'SULLIVAN DENIS F
DUBLIN INSTITUTE FOR
ADVANCED STUDIES
5 MERRION SQUARE
DUBLIN 2
IRELAND

O'SULLIVAN JOHN DAVID DR
RADIOSTERREWACHT
POSTBUS 2
7990 AA DWINGELOO
THE NETHERLANDS

OBASHEV SAKEN O DR
ASTROPHYSICAL
INSTITUTE
480068 ALMA-ATA
U.S.S.R.

OBI SHINYA PROF
FAC OF GENERAL ED.
UNIV OF TOKYO
KOMABA MEGURO
TOKYO 153
JAPAN

OBREGON DIAZ OCTAVIO J DR
UAM IZTAPALAPA
PO BOX 55 534
MEXICO 13 DF
MEXICO

OBRIDKO VLADIMIR A DR
IZMIRAN
ACADEMGORODOK
142092 MOSCOW REGION
U.S.S.R.

OCCHIONERO FRANCO PROF
CNR ASTROFISICA
SPAZIALE C P 67
I-00044 FRASCATI
ROMA
ITALY

OCHSENBEIN FRANCOIS DR
OBS DE STRASBOURG
11 RUE DE L'UNIV
F-67000 STRASBOURG
FRANCE

ODA MINORU PROF
INST SPACE AERON SC
UNIVERSITY OF TOKYO
KOMABA MEGURO KU
TOKYO 153
JAPAN

ODELL ANDREW P
DEPT OF PHYSICS &
ASTRONOMY
NORTHERN ARIZONA UN
FLAGSTAFF AZ 86011
U.S.A.

ODELL CHARLES R DR
DS30 MSFC
HUNTSVILLE AL 35812
U.S.A.

ODGERS GRAHAM J DR
DOMINION ASTROPH OBS
5071 W SAANICH ROAD
RR 5
VICTORIA BC V8X 4M6
CANADA

OEHMAN YNGVE PROF
THULEHEM 53
S-223 67 LUND
SWEDEN

OEMLER AUGUSTUS JR DR
YALE UNIV OBS
BOX 2023
YALE STATION
NEW HAVEN CT 06520
U.S.A.

OERGERLE WILLIAM R
5924 BERWYN ROAD
COLLEGE PARK MD20740
U.S.A.

OESTERWINTER CLAUS
COMMANDER
NAVAL SURFACE WEA-
PONS CENTER DK 10
DAHLGREN VA 22448
U.S.A.

OESTGAARD ERLEND
PHYSICAL INSTITUTE
UNIV OF TRONDHEIM
NLHT
N-7000 TRONDHEIM
NORWAY

OETKEN L DR
ZNTRINST ASTROPHYSIK
ASTROPHY OBSERVATORI
TELEGRAFENBERG
DDR-15 POTSDAM
G.D.R.

OGORODNIKOV KYRILL P PROF
DEPT OF ASTRONOMY
LENINGRAD UNIVERSITY
199164 LENINGRAD
U.S.S.R.

OGURA KATSUO DR
KOKUGAKUIN UNIV
CLG OF LITERATURE
HIGASHI 4-10-28
SHIBUYAKU TOKYO 150
JAPAN

OHKI KENICHIRO DR
TOKYO ASTRONOMICAL
OBSERVATORY
OSAWA
MITAKA TOKYO 181
JAPAN

OHRING GEORGE PROF
DEPT OF GEOPHYSICS
PLANETARY SCIENCES
TEL AVIV
ISRAEL

OHTANI HIROSHI DR
DEPT OF ASTRONOMY
UNIV OF KYOTO
KYOTO 606
JAPAN

OHYAMA NOBORU PROF
FAC OF ENGINEERING
SHIZUOKA UNIVERSITY
3 CHOME JYOHOKU
HAMAMATSU 432
JAPAN

OJA HEIKKI DR
UNIV OF HELSINKI
OBS & ASTROPHYS LAB
TAEHTITORNINMAEKI
SF-00130 HELSINKI 13
FINLAND

OJA TARMO PROF
KVISTABERG
OBSERVATORY
S-190 51 BRO
SWEDEN

OKA TAKESHI DR
CHEMISTRY DEPT
UNIV OF CHICAGO
5735 S ELLIS AVE
CHICAGO IL 60637
U.S.A.

OKAMOTO ISAO DR
INTERNATIONAL LATITU
DE OBSERVATORY
HOSHIGAOKA MIZUSAWA
IWATE 023
JAPAN

OKAMURA SADANORI DR
KISO BRANCH OF THE
TOKYO ASTRON OBS
MITAKEMURA KISOGUN
NAGANOKEN 397-01
JAPAN

OKAZAKI AKIRA DR
TOKYO ASTRON OBS
UNIV OF TOKYO
MITAKA
TOKYO 181
JAPAN

OKAZAKI SEICHI DR
TOKYO ASTRON OBS
UNIV OF TOKYO
MITAKA
TOKYO 181
JAPAN

OKE J BEVERLY PROF
HALE OBSERVATORIES
CALIFORNIA INSTITUTE
OF TECHNOLOGY
PASADENA CA 91125
U.S.A.

OKI TOSIO PROF DR
DEPT EARTH SC FAC OF
EDUC FUKUSHIMA UNIV
MATSUKAWA MACHI
FUKUSHIMA 960-12
JAPAN

OKOYE SAMUEL E PROF
DEPT OF PHYSICS
UNIVERSITY OF
NIGERIA
NSUKKA
NIGERIA

OKUDA HARUYUKI DR PROF
DEPT OF PHYSICS
KYOTO UNIVERSITY
KITASHIRAKAWA SAKYO
KYOTO 606
JAPAN

OKUDA TOYOZO DR
5-11-9 HANAZONO
CHIBA 281
JAPAN

OLAH KATALIN DR
KONKOLY OBS
HUNG ACADEMY OF SC
BOX 67
H-1525 BUDAPEST
HUNGARY

OLEAK H DR
ZNTRINST ASTROPHYSIK
STERNW BABELSBERG
ROSA LUXEMBURGST 17A
DDR-1502 POTSDAM
G.D.R.

OLIVER JOHN PARKER DR
DEPT PHYSICS ASTRON
UNIV OF FLORIDA
GAINESVILLE FL 32611
U.S.A.

OLLONGREN A DR
DEPT OF MATHEMATICS
WASSENAARSEWEG 80
POSTBOX 9512
2300 RA LEIDEN
THE NETHERLANDS

OLOFSSON S GOERAN DR
STOCKHOLM
OBSERVATORY
S-133 00
SALTSJOEBADEN
SWEDEN

OLSEN ERIK H
COPENHAGEN
OBSERVATORY
BRORFELDE
DK-4340 TOLLOSE
DENMARK

OLSEN KENNETH H DR
LOS ALAMOS SCIENTI-
FIC LABORATORY
BOX 1663 MS 676
LOS ALAMOS NM 87544
U.S.A.

OLSON EDWARD C PROF
OBSERVATORY
UNIVERSITY OF
ILLINOIS
URBANA IL 62801
U.S.A.

OLTHOF HINDERICUS DR
ESA
8-10 RUE MARIO NIKIS
F-75738 PARIS 15
FRANCE

OMAROV TUKEN B PROF
ASTROPHYSICAL
INSTITUTE
480068 ALMA ATA
U.S.S.R.

OMER GUY C JR PROF
1080 SW 11TH TERRACE
GAINESVILLE FL 32601
U.S.A.

OMNES ROLAND PROF
LPTHE BAT 211
UNIVERSITE DE
PARIS-SUD
F-91405 ORSAY
FRANCE

OMONT ALAIN PROF
CERMO-USMG
P B 53 X
F-38041 GRENOBLE CED
FRANCE

ONDERLICKA BEDRICH DR
DEPT OF ASTRONOMY
PURKYNE UNIVERSITY
KOTLARSKA 2
61137 BRNO
CZECHOSLOVAKIA

ONEGINA A B DR
MAIN ASTRON OBS
UKRAINIAN AC OF SCS
GOLOSEEVO
252127 KIEV
U.S.S.R.

ONO YORO PROF
DEPT OF PHYSICS
UNIVERSITY HOKKAIDO
KITAHACHIJYO NISHI 8
SAPPORO HOKKAIDO 063
JAPAN

OOE MASATSUGU DR
INTERNATIONAL LATITU
DE OBSERVATORY
HOSHIGAOKA MIZUSAWA
IWATE 023
JAPAN

OORT JAN H PROF
PRES KENNEDYLAAN 169
2343 GZ OEGSTGEEST
THE NETHERLANDS

OPALSKI WIESLAW PROF
ELEKTORALNA 4-6 M4
00-139 WARSAW
POLAND

OPHER REUVEN PROF
PHYSICS DEPT
TECHNION-ISRAEL INST
OF TECHNOLOGY
HAIFA 32000
ISRAEL

OPIK ERNST J DR
ARMAGH OBSERVATORY
ARMAGH BT61 9DG
U.K.

OPOLSKI ANTONI PROF
ASTRONOMICAL
OBSERVATORY
KOPERNIKA 11
51-622 WROCLAW
POLAND

OPROIU TIBERIU DR
STR BUCIUM 25 BLOC
R-5 SC 1 ET 11 AP 10
3400 CLUJ-NAPOCA
ROUMANIA

ORLIN HYMAN DR
NAT ACADEMY SCIENCES
2101 CONSTITUTION
AVE N W
WASHINGTON DC 20418
U.S.A.

ORLOV A A DR
STERNBERG STATE AS-
TRONOMICAL INSTITUTE
117234 MOSCOW
U.S.S.R.

ORLOV MIKHAIL DR
MAIN ASTRONOMICAL OB
252127 KIEV
U.S.S.R.

ORRALL FRANK Q PROF
INSTITUTE FOR
ASTRONOMY
2680 WOODLAWN DRIVE
HONOLULU HI 96822
U.S.A.

ORTE ALBERTO
ISTI Y OBSERVATORIO
DE MARINA
SAN FERNANDO CADIZ
SPAIN

ORTON GLENN S DR
M S 183B-365
JPL
4800 OAK GROVE DRIVE
PASADENA CA 91103
U.S.A.

ORUS JUAN J PROF
DPTO FIS TIERRA COSM
UNIVERSIDAD
DE BARCELONA
BARCELONA 28
SPAIN

OSAKI TORU DR
RYUKOKU UNIVERSITY
FUKAKUSA TSUKAMOTO
FUSHIMIKU KYOTO 612
JAPAN

OSAKI YOJI DR
DEPT OF ASTRONOMY
UNIV OF TOKYO
YAYOI BUNKYO
TOKYO 113
JAPAN

OSAWA KIYOTERU DR
TOKYO ASTRONOMICAL
OBSERVATORY
OSAWA
MITAKA TOKYO 181
JAPAN

OSBORN WAYNE DR
PHYSICS DEPT
CENTRAL MICHIGAN
UNIVERSITY
MT PLEASANT MI 48859
U.S.A.

OSBORNE JOHN L DR
DEPT OF PHYSICS
UNIV OF DURHAM
SOUTH ROAD
DURHAM DH1 3LE
U.K.

OSKANYAN V S DR
BYURAKAN ASTROPHYSI-
CAL OBSERVATORY
378433 ARMENIA
U.S.S.R.

OSMAN ANAS MOHAMED DR
HELWAN OBSERVATORY
HELWAN
ARAB REP. OF EGYPT

OSMER PATRICK S DR
CERRO TOLOLO INTER-
AMERICAN OBSERVATORY
CASILLA 63 D
LA SERENA
CHILE

OSORIO JOSE J S P DR
OBS ASTRONOMICO
UNIV DO PORTO
MONTE DA VIRGEM
VILA NOVA DE GAIA
PORTUGAL

OSTER LUDWIG F PROF DR
JILA
UNIV OF COLORADO
BOULDER CO 80309
U.S.A.

OSTERBROCK DONALD E PROF
LICK OBSERVATORY
UNIVERSITY
OF CALIFORNIA
SANTA CRUZ CA 95064
U.S.A.

OSTRIKER JEREMIAH P PROF
PRINCETON UNIVERSITY
OBSERVATORY
PEYTON HALL
PRINCETON NJ 08540
U.S.A.

OTERMA LIISI PROF
SIRKKALANKATU 31
SF-20700 TURKU
FINLAND

OTTELET I J DR
INST D ASTROPHYSIQUE
UNIVERSITE DE LIEGE
B-4200 COINTE OUGREE
BELGIUM

OVENDEN MICHAEL W PROF
DEPT OF GEOPHYSICS
AND ASTRONOMY
UNIVERSITY OF B C
VANCOUVER BC V6T 1W5
CANADA

OVERBEEK MICHIEL DANIEL
P O BOX 212
EDENVALE 1610
SOUTH AFRICA

OWAKI NAOAKI DR
TOKYO GAKUGEI
UNIVERSITY
NUKUIKITAMACHI
KOGANEI TOKYO 184
JAPAN

OWEN FRAZER NELSON DR
NATIONAL RADIO ASTR
OBSERVATORY
EDGEMONT RD
CHARLOTTESVL VA22901
U.S.A.

OWEN TOBIAS C PROF
DEPT OF EARTH AND
SPACE SCIENCES
STATE UNIVERSITY NY
STONY BROOK NY 11794
U.S.A.

OWREN LEIF DR
FYSISK INSTITUTT
UNIVERSITETET BERGEN
ALLEGATEN 53 55
N-5014 BERGEN U
NORWAY

OXENIUS JOACHIM DR
UNIV LIBRE BRUXELLES
CP 231
CAMPUS PLAINE ULB
B-1050 BRUSSELS
BELGIUM

OZEMRE KAMURAN DOCENT DR
UNIVERSITY
OBSERVATORY
UNIVERSITE
ISTANBUL
TURKEY

OZERNOY LEONID M PROF
PHYSICAL INSTITUTE
USSR ACADEMIE OF
SCIENCE
117924 MOSCOW
U.S.S.R.

OZSVATH I PROF
EARTH AND PLANET SCS
GRADUATE RES CTR
P O BOX 30365
DALLAS TX 75230
U.S.A.

PAAL GYORGY DR
KONKOLY OBSERVATORY
BOX 67
1525 BUDAPEST
HUNGARY

PACHNER JAROSLAV PROF
390 WOODSWORTH RD 11
TORONTO ONT M2L 2T9
CANADA

PACHOLCZYK ANDRZEJ G PROF
STEWARD OBSERVATORY
UNIVERSITY ARIZONA
TUCSON AZ 85721
U.S.A.

PACINI FRANCO PROF
ISTITUTO ASTRONOMIA
LARGO E FERMI 5
I-50125 FIRENZE
ITALY

PACZYNSKI B PROF
COPERNICUS ASTRONOMI
-CAL CENTER
UL BARTYCKA
00 716 WARSAW
POLAND

PADEVET VLADIMIR DR
ASTRONOMICAL INST
CZECH ACADEMY SC
251 65 ONDREJOV
OBSERVATORY
CZECHOSLOVAKIA

PADRIELLI L
LAB DI RADIOASTRON
VIA IRNERIO 46
I-40126 BOLOGNA
ITALY

PAGE CLIVE G DR
PHYSICS DEPT
UNIV OF LEICESTER
LEICESTER LE1 7RH
U.K.

PAGE DON NELSON
104 DAVEY LABORATORY
PENNSYLVANIA SU
UNIVERSITY PARK
PA 16802
U.S.A.

PAGE THORNTON L DR
18639 POINT LOOKOUT
DRIVE
HOUSTON TX 77058
U.S.A.

PAGEL BERNARD E J PROF
ROYAL GREENWICH
OBSERVATORY
HAILSHAM
EAST SUSSEX BN27 1RP
U.K.

PAL ARPAD PROF DR
STR RAKOCZI 72
3400 CLUJ NAPOCA
ROUMANIA

PALLAVICINI ROBERTO DR
OSSERVATORIO ARCETRI
LARGO FERMI 5
I-50125 FIRENZE
ITALY

PALMEIRA RICARDO A R DR
INPE CP 515
12200 SJOSE D CAMPOS
SAO PAULO
BRAZIL

PALMER PATRICK E PROF
DEPT OF ASTRONOMY
UNIVERSITY CHICAGO
1100 E 58TH ST
CHICAGO IL 60637
U.S.A.

PALUMBO GIORGIO G C DR
TE-S R E LAB CNR
VIA DE CASTAGNOLI 1
I-40100 BOLOGNA
ITALY

PAMJATN'IKH A A DR
ASTONOMICAL COUNCIL
USSR ACAD OF SCS
48 PYATNITSKAYA ST
109017 MOSCOW
U.S.S.R.

PAN JUN-HUA
NANJING ASTRONOMICAL
INSTRUMENTS FACTOR
NANJING
PEOPLE'S REPUBLIC CHINA

PAN XIAO-PEI
P.O BOX 18
LINTONG
SHAANXI
PEOPLE'S REPUBLIC CHINA

PANAGIA NINO DR
LABORATORIO DI
RADIOASTRONOMIA
VIA IRNERIO 46
I-40126 BOLOGNA
ITALY

PANDE GIRISH CHANDRA PROF
DEPT OF MECHANICS
UNIV OF PATRAS
PATRAS
GREECE

PANDE MAHESH CHANDRA DR
UP STATE OBSERVATORY
MANORA PEAK
NAINI TAL 263129
INDIA

PANEK ROBERT J DR
DEPT OF ASTRONOMY
525 DAVEY LAB
UNIV OF PENNA
UNIV PK PA 16802
U.S.A.

PANG KEVIN
JET PROPULSION LAB
48000 OAK GROVE DR
PASADENA CA 91109
U.S.A.

PANKONIN VERNON LEE DR
DIV ASTRON SCIENCES
NAT SCIENCE FOUND
1800 G ST SW
WASHINGTON DC 20550
U.S.A.

PANNUNZIO RENATO
ASTRON OBS OF TURIN
STRADA OSS 20
I-10025 PINO
TORINESE
ITALY

PANOV K DR
DEPT OF ASTRONOMY
BULGARIAN AC OF SCS
7TH NOVEMBER ST 1
1000 SOFIA
BULGARIA

PAOLICCHI PAOLO DR
OSSERVATORIO ASTRO-
NOMICO
MERATE
MILANO
ITALY

PAPAGIANNIS MICHAEL D PRO
DEPT OF ASTRONOMY
BOSTON UNIVERSITY
725 COMMONWEALTH AVE
BOSTON MA 02215
U.S.A.

PAPALIOLIOS C DR
SMITHSONIAN ASTRO-
PHYSICAL OBS
CAMBRIDGE MA 02138
U.S.A.

PAPALOIZOU JOHN C B DR
INST OF ASTRONOMY
MADINGLEY ROAD
CAMBRIDGE CB3 OHA
U.K.

PAPARO MARGIT DR
KONKOLY OBSERVATORY
HUNG ACADEMY OF SC
BOX 67
H-1525 BUDAPEST
HUNGARY

PAPATHANASOGLOU D DR
DEPT OF ASTRONOMY
UNIV OF ATHENS
PANEPISTIMIOPOLIS
ATHENS 621
GREECE

PAPAYANNOPOULOS TH DR
DEPT OF ASTRONOMY
UNIV OF ATHENS
PANEPISTIMIOPOLIS
ATHENS 621
GREECE

PAQUET PAUL EG DR
OBSERVATOIRE ROYAL
DE BELGIQUE
AVE CIRCULAIRE 3
B-1180 BRUXELLES
BELGIUM

PARCELIER PIERRE DR
OBSERVATOIRE
DE PARIS
61 AV L OBSERVATOIRE
F-75014 PARIS
FRANCE

PARESCE FRANCESCO DR
SPACE SCIENCES
LABORATORY
UNIVERSITY OF CALIF
BERKELEY CA 94720
U.S.A.

PARIJSKIJ N N DR
INSTITUTE OF PHYSICS
OF THE EARTH
USSR ACAD OF SCI
123810 MOSCOW
U.S.S.R.

PARIJSKIJ YU N DR
SPECIAL ASTROPHYS-
OBS USSR ACAD OF SCI
LENINGRAD BRANCH
196140 LENINGRAD
U.S.S.R.

PARKER EDWARD A DR
ELECTRONICS
LABORATORIES
THE UNIVERSITY
CANTERBURY CT2 7NT
U.K.

PARKER EUGENE N
LAB ASTROPHYS SP RES
UNIV OF CHICAGO
933 E 56TH ST
CHICAGO IL 60637
U.S.A.

PARKER ROBERT A R
CODE CB
NASA-JSC
HOUSTON TX 77058
U.S.A.

PARKINSON JOHN H DR
MULLARD SPACE SC LAB
UNIV COLLEGE LONDON
HOLMBURY ST MARY
DORKING SURREYRH56NS
U.K.

PARKINSON T DR
KITT PEAK NAT OBS
950 N CHERRY AVE
TUCSON AZ 85726
U.S.A.

PARKINSON WILLIAM H
HARVARD COLLEGE
OBSERVATORY
60 GARDEN ST
CAMBRIDGE MA 02138
U.S.A.

PARMA PAOLA
ISTITUTO DI
RADIOASTRONOMIA CNR
VIA IRNERIO 45
I-40126 BOLOGNA
ITALY

PARRISH ALLAN DR.
STATE UNIV.OF N.YORK
C/O 6191 GRC U MASS.
AMHERST MA 01003
U.S.A.

PARSAMYAN E S DR
BYURAKAN ASTROPHYSI-
CAL OBSERVATORY
378433 ARMENIA
U.S.S.R.

PARSONS SIDNEY B DR
DEPT OF ASTRONOMY
UNIVERSITY OF TEXAS
AUSTIN TX 78712
U.S.A.

PARTHASARATHY M DR
INDIAN INSTITUTE OF
ASTROPHYSICS
BANGALORE 560 034
INDIA

PARTRIDGE ROBERT B PROF
HAVERFORD COLLEGE
HAVERFORD PA 19041
U.S.A.

PASACHOFF JAY M PROF
WILLIAMS COLLEGE
HOPKINS OBSERVATORY
WILLIAMSTOWN MA01267
U.S.A.

PASCOAL ANTONIO J B SCI
OBS ASTRONOMICO
PROF MANUEL D BARROS
MONTE DA VIRGEM
VILA NOVA DE GAIA
PORTUGAL

PASCU DAN DR
US NAVAL OBSERVATORY
WASHINGTON DC 20390
U.S.A.

PASINETTI LAURA E PROF
OSSERVATORIO DI
MILANO MERATE
VIA E BIANCHI 46
I-22055 MERATE
ITALY

PASTORI LIVIO
OSS ASTRONOMICO
DI MERATE
VIA BIANCHI 46
I-22055 MERATE
ITALY

PASTORIZA MIRIANI G DR
INSTITUTO DE FISICA
UNIVERSIDADE FEDERAL
DO RIO GRANDE DO SUL
90000 PORTO ALEGRE
BRAZIL

PATERNO LUCIO DR
OSSERVATORIO
ASTROFISICO
CITTA UNIVERSITARIA
I-95125 CATANIA
ITALY

PATERSON-BEECKMANS F
VINCENT V GOGHLN 19
2343 RH OEGSTGEEST
THE NETHERLANDS

PATHRIA RAJ K PROF
DEPT OF PHYSICS
UNIVERSITY WATERLOO
WATERLOO ONT N2L 3G1
CANADA

PATKOS LASZLO DR
KONKOLY OBSERVATORY
HUNG ACADEMY OF SC
BOX 67
H-1525 BUDAPEST
HUNGARY

PATRIARCHI PATRIZIO
OSS ASTROFISICO DI
ARCETRI
LARGO ENRICO FERMI 5
I-50125 FIRENZE
ITALY

PAULINY TOTH IVAN K K DR
MPI FUER RADIOASTRO-
NOMIE
AUF DEM HUEGEL 69
D-5300 BONN
GERMANY, F.R.

PAULS THOMAS ALBERT DR
ERSTES PHYS INST
UNIV KOELN
UNIVERSITAETS STR 14
D-5000 KOELN 41
GERMANY, F.R.

PAVLOV NIKOLAJ N PROF DR
PULKOVO OBSERVATORY
196140 LENINGRAD
U.S.S.R.

PAVLOVSKAYA E D DR
STERNBERG STATE AS-
TRONOMICAL INSTITUTE
117234 MOSCOW
U.S.S.R.

PAXTON HAROLD J B R
ROYAL GREENWICH OBS
HERSTMONCEUX CASTLE
HAILSHAM
E SUSSEX BN27 1RP
U.K.

PEACH G DR
DEPT PHYSICS ASTRO
UNIVERSITY COLLEGE
GOWER STREET
LONDON WC1E 6BT
U.K.

PEACH JOHN V DR
DEPT OF ASTROPHYSICS
SOUTH PARKS ROAD
OXFORD OX1 3RQ
U.K.

PEALE STANTON J PROF
DEPT OF PHYSICS
UNIVERSITY OF
CALIFORNIA
SN BARBARA CA 93106
U.S.A.

PEARSE REGINALD W B DR
60 GREENLANDS ROAD
STAINES MDX TW18 4LR
U.K.

PEAT D W DR
COLLEGE OF RIPON
YORK ST JOHN
LORD MAYORS WALK
YORK YO3 7EX
U.K.

PECKER J C PROF
LAT
98 BIS BLVD ARAGO
F-75014 PARIS
FRANCE

PEDERSEN B DR
OBS DE MEUDON
F-92190 MEUDON
FRANCE

PEDERSEN HOLGER DR
E S O
CASILLA 16317
SANTIAGO 9
CHILE

PEDERSEN OLAF PROF
HISTORY SCIENCE DEPT
UNIVERSITY OF AARHUS
NY MUNKEGADE
DK-8000 AARHUS C
DENMARK

PEDLAR ALAN DR
NRAL
JODRELL BANK
CHESHIRE SK11 9DL
U.K.

PEDOUSSAUT ANDRE
OBS DE TOULOUSE
1 AVE FLAMMARION
F-31500 TOULOUSE
FRANCE

PEEBLES P JAMES E
JOSEPH HENRY LABS
JADWIN HALL
PRINCETON NJ 08540
U.S.A.

PEERY BENJAMIN F PROF
DEPT OF PHYSICS AND
ASTRONOMY
HOWARD UNIVERSITY
WASHINGTON DC 20059
U.S.A.

PEIMBERT MANUEL DR
INST DE ASTRONOMIA
APDO POSTAL 70264
04510 MEXICO DF
MEXICO

PEKERIS CHAIM LEIB PROF
DEPT APPLIED MATHS
WEIZMANN INSTITUTE
OF SCIENCE
REHOVOT
ISRAEL

PEKUENLUE E RENNAN DR
EGE UNIVERSITY
FACULTY OF SCIENCE
BORNOVA-IZMIR
TURKEY

PEL J W DR
KAPTEIJN STERREWACHT
MENSINGHEWEG 20
9301 KA RODEN
THE NETHERLANDS

PELLAS PAUL DR
61 RUE BUFFON
F-75005 PARIS
FRANCE

PELLET ANDRE
OBS DE MARSEILLE
2 PLACE LE VERRIER
F-13248 MARSEILLE C4
FRANCE

PELSENEER JEAN KON PROF
UNIV LIBRE DE BRUX
76 AV DES GRENADIERS
BOX 6
B-1150 BRUSSELS
BELGIUM

PELTIER LESLIE C
327 S BREDEICK ST
DELPHOS OH 45833
U.S.A.

PENNY ALAN JOHN DR
ROYAL GREENWICH OBS
HERSTMONCEUX CASTLE
HAILSHAM
E SUSSEX BN27 1RP
U.K.

PENSADO J DR
OBSERVATORIO
ASTRONOMICO
ALFONSO XII 5
MADRID 7
SPAIN

PENSTON MICHAEL V DR
ROYAL GREENWICH OBS
HAILSHAM
SUSSEX BN27 1RP
U.K.

PENZIAS ARNO A DR
BELL LABORATORIES
BOX 400
HOLMDEL NJ 07733
U.S.A.

PEQUIGNOT DANIEL
OBS DE PARIS
DEPT D'ASTROPHYSIQUE
FONDAMENTALE
F-92190 MEUDON
FRANCE

PERAIAH ANNAMANENI DR
INDIAN INSTITUTE OF
ASTROPHYSICS
BANGALORE 560034
INDIA

PERCY JOHN R PROF
ERINDALE COLLEGE
MISSISAUGA
ONT L5L 1C6
CANADA

PERDANG JEAN M DR
INSTITUT
D'ASTROPHYSIQUE
B-4200 COINTE-OUGREE
BELGIUM

PEREK LUBOS DR
ASTRONOMICAL INST
CZECHOSLOVAK AC SCS
BUDECSKA 6 VINOHRADY
120 23 PRAHA 2
CZECHOSLOVAKIA

PEREZ PERAZA JORGE DR
INST DE GEOFISICA
UNAM 04510 DELEGA-
CION COYOACAN C U
MEXICO 20 D F
MEXICO

PEREZ-DE-TEJADA H A DR
INST DE ASTRONOMIA
UNAM
AV GASTELUM ENSENADA
BAJA CALIFORNIA
MEXICO

PERINOTTO MARIO DR
OSSERVATORIO ASTRO-
FISICO ARCETRI
I-50125 FIRENZE
ITALY

PERKINS FRANCIS W DR
PLASMA PHYSICS
LABORATORY
P O BOX 451
PRINCETON NJ 08540
U.S.A.

PERLEY RICHARD ALAN
NRAO VLA PROGRAM
P O BOX 0
SOCORRO NM 87801
U.S.A.

PEROLA G C DR
ISTITUTO ASTRONOMICO
C/O IST DI FISICA NE
P LE A MORO 2
I-00185 ROMA
ITALY

PERRIN MARIE-NOEL DR
OBSERVATOIRE
DE PARIS
61 AVE OBSERVATOIRE
F-75014 PARIS
FRANCE

PERRY CHARLES L DR
DEPT OF PHYS/ASTRON
LOUISIANA STATE UNIV
BATON ROUGE LA 70803
U.S.A.

PERRY JUDITH J DR
C/O DR M SOHNIUS
310 WESTBOURNE PARK
AVENUE
LONDON W11
U.K.

PERSI PAOLO
ISTITUTO ASTROFISICA
SPAZIALE
CASELLA POSTALE 67
I-00044 FRASCATI
ITALY

PERSIDES SOTIRIOS C
ASST PROF
DEPT OF ASTRONOMY
UNIV OF THESSALONIKI
THESSALONIKI
GREECE

PESCH PETER DR
WARNER AND SWASEY
OBSERVATORY
CASE WESTERN RES UN
CLEVELAND OH 44106
U.S.A.

PETERS GERALDINE JOAN DR
DEPT OF ASTRONOMY
UNIV SOUTHERN CALIF
UNIVERSITY PARK
LOS ANGELES CA 90007
U.S.A.

PETERS WILLIAM L III DR
ASTRONOMY DEPT
UNIV OF TEXAS
AUSTIN TX 78712
U.S.A.

PETERSEN J O DR
UNIV OBSERVATORY
OESTER VOLDGADE 3
DK-1350 COPENHAGEN K
DENMARK

PETERSON BRADLEY MICHAEL
DEPT OF ASTRONOMY
OHIO STATE UNIV
174 W 18TH AVENUE
COLUMBUS OH 43210
U.S.A.

PETERSON BRUCE A DR
MT STROMLO & SIDING
SPRING OBS
AUSTRALIAN NAT UN
WODEN ACT 2606
AUSTRALIA

PETERSON CHARLES JOHN DR
DEPT OF PHYSICS
223 PHYSICS BLDG
UNIV OF MISSOURI
COLUMBIA MO 65211
U.S.A.

PETERSON LAURENCE E PROF
DEPT OF PHYSICS C011
UNIVERSITY OF CALIF-
AT SAN DIEGO
LA JOLLA CA 92093
U.S.A.

PETERSON RUTH CAROL DR
607 MARION PLACE
PALO ALTO CA 94301
U.S.A.

PETFORD A DAVID DR
DEPT OF ASTROPHYSICS
SOUTH PARKS ROAD
OXFORD OX1 3RQ
U.K.

PETON ALAIN DR
OBS DE MARSEILLE
2 PLACE LE VERRIER
F-13248 MARSEILLE C4
FRANCE

PETRI WINFRIED PROF DR
UNTERLEITEN 2
POSTFACH 106
8162 SCHLIERSEE OBB
GERMANY, F.R.

PETRINI DANIEL DR
OBSERVATOIRE DE NICE
B P 252
F-06007 NICE CEDEX
FRANCE

PETROPOULOS BASIL CH DR
RES CTR ASTRO AP MAT
ACADEMY OF ATHENS
14 ANAGNOSTOPOULOU
ATHENS 136
GREECE

PETROSIAN VAHE PROF
INSTITUTE FOR PLASMA
RESEARCH
STANFORD UNIVERSITY
STANFORD CA 94305
U.S.A.

PETROU M DR
DEPT OF ASTRONOMY
UNIV OF ATHENS
PANEPISTIMOPOLIS
ATHENS 621
GREECE

PETROV G I PROF DR
SPACE RESEARCH
INSTITUTE
USSR ACADEMY OF SCI
117810 MOSCOW
U.S.S.R.

PETROV G M DR
NIKOLAEV DEPT OF THE
MAIN ASTRONOMICAL
OBSERVATORY
327000 NIKOLAEV
U.S.S.R.

PETROV GEORGY TRENDAFILOV
DEPT OF ASTRONOMY
BULGARIAN ACADEMY SC
72 LENIN BLVD
1184 SOFIA
BULGARIA

PETROV NIKOLAI
ASTRONOMICAL OBS
VARNA
BULGARIA

PETROVSKAYA M S DR
INSTITUTE OF THEORE-
TICAL ASTRONOMY
192187 LENINGRAD
U.S.S.R.

PETTENGILL GORDON H PROF
RM 54-620
MASS INSTITUTE OF
TECHNOLOGY
CAMBRIDGE MA 02139
U.S.A.

PETTERSEN BJOERN RAGNVALD
INST MATH & PHYS SC
UNIV OF TROMSOE
P O BOX 953
N-9001 TROMSOE
NORWAY

PETTINI MAX
15 CHRISTIE AVE
RINGMER
EAST SUSSEX BN8 5JT
U.K.

PEYTREMANN ERIC DR
50 WATTEN VIEW
SINGAPORE 1128
SINGAPORE

PEYTURAUX ROGER H PROF
INST D ASTROPHYSIQUE
98B BLVD ARAGO
F-75014 PARIS
FRANCE

PFEIFFER RAYMOND J
8 BARBARA LANE
TITUSVILLE NJ 08560
U.S.A.

PFENNIG HANS H DR
MPI F PHYSIK UND
ASTROPHYSIK
FOEHRINGER RING 6
D-8000 MUENCHEN 40
GERMANY, F.R.

PFLEIDERER JORG PROF
INST F ASTRONOMIE
UNIVERSITAETSSTR 4
A-6020 INNSBRUCK
AUSTRIA

PFLUG KLAUS DR
ZENTRALINSTITUT FUER
SOLARTERR PHYSIK
DDR-15 POTSDAM-
TELEGRAFENBERG
G.D.R.

PHAM-VAN J MME
CERGA
AVENUE COPERNIC
F-06130 GRASSE
FRANCE

PHILIP A G DAVIS
1125 OXFORD PLACE
SCHENECTADY NY 12308
U.S.A.

PHILLIPS ANTHONY PAUL
DEPT PHYS & ASTRON
UNIV CLG LONDON
GOWER STREET
LONDON WC1E 6BT
U.K.

PHILLIPS JOHN G PROF
ASTRONOMY DEPT
UNIV OF CALIFORNIA
601 CAMPBELL HALL
BERKELEY CA 94720
U.S.A.

PHILLIPS KENNETH J H
SPACE & ASTROPH DIV
RUTHERFORD APPLETON
LAB CHILTON
DIDCOT OXON OX11 0QX
U.K.

PHILLIPS THOMAS GOULD DR
CALTECH
320-47
PASADENA CA 91125
U.S.A.

PICAT J P DR
OBS DE MEUDON
F-92190 MEUDON
FRANCE

PICCIONI ADALBERTO
ISTITUTO ASTRONOMICO
UNIVERSITARIO
CASELLA POSTALE 596
I-40100 BOLOGNA
ITALY

PICK MONIQUE DR
DASOP
OBSERVATOIRE
DE MEUDON
F-92190 MEUDON
FRANCE

PIDDINGTON JACK H RES FEL
NATIONAL MEASUREMENT
LABORATORY P O B 218
LINDFIELD
SYDNEY 2070
AUSTRALIA

PIERCE A KEITH DR
KITT PEAK NATIONAL
OBSERVATORY
PO 26732
TUCSON AZ 85726
U.S.A.

PIERCE DAVID ALLEN
PHYSICAL SCS DIV
EL CAMINO COLLEGE
TORRANCE CA 90506
U.S.A.

PIIROLA VILPPU E DR
OBSERVATORY
TAHTITORNINMAKI
SF-00130 HELSINKI 13
FINLAND

PIKE CHRISTOPHER DAVID
ROYAL GREENWICH OBS
HERSTMONCEUX CASTLE
HAILSHAM
EAST SUSSEX BN27 1RP
U.K.

PILACHOWSKI CATHERINE DR
ASTRONOMY DEPT FM-20
UNIV OF WASHINGTON
SEATTLE WA 98195
U.S.A.

PILCHER CARL BERNARD DR
INST FOR ASTRONOMY
UNIV OF HAWAII
2680 WOODLAWN DRIVE
HONOLULU HI 96822
U.S.A.

PILKINGTON JOHN D H DR
ROYAL GREENWICH
OBSERVATORY
HAILSHAM
EAST SUSSEX BN27 1RP
U.K.

PILOWSKI K PROF DR
GEODAETISCHES INST
TECHN UNIVERSITAET
NIENBURGER STR 1
D-3000 HANNOVER
GERMANY, F.R.

PINEAULT SERGE DR
DOMINION RADIO
ASTROPHYSICAL OBS
BOX 248
PENTINCTON BC V2A6K3
CANADA

PINES DAVID PROF
DEPT OF PHYSICS
UNIVERSITY ILLINOIS
URBANA IL 61801
U.S.A.

PINGREE DAVID PROF
BROWN UNIV
PROVIDENCE RI 02912
U.S.A.

PINIGIN GENNADIJ I DR
MAIN ASTRONOMICAL OB
USSR ACAD OF SCS
PULKOVO M-140
196140 LENINGRAD
U.S.S.R.

PINKAU K PROF
MPI FUER EXTRATERRE
STRISCHE PHYSIK
D-8046 GARCHING
GERMANY, F.R.

PINOTSIS A D DR
DEPT OF ASTRONOMY
UNIV OF ATHENS
PANEPISTIMIOPOLIS
ATHENS 621
GREECE

PINTO GIROLAMO PROF
OSSERVATORIO
ASTRONOMICO
I-35100 PADOVA
ITALY

PIOTROWSKI STEFAN PROF
OBSERWATORIUM
ASTRONOMICZNE UW
AL UJAZDOWSKIE 4
00478 WARSZAWA
POLAND

PIPHER JUDITH L
PHYSICS AND ASTRONO-
MY DEPT
UNIV OF ROCHESTER
ROCHESTER NY 14627
U.S.A.

PIRRONELLO VALERIO
OSS ASTROFISICO
CITTA UNIVERSITARIA
F-95125 CATANIA
ITALY

PISMIS DE RECILLAS PARIS
PROF
INST DE ASTRONOMIA
APDO POSTAL 70264
04510 MEXICO DF
MEXICO

PITTICH EDUARD M DR
ASTRONOMICAL INSTI-
TUTE SAV
DUBRAVSKA CESTA A 11
89930 BRATISLAVA
CZECHOSLOVAKIA

PITZ ECKHART DR
MPI-ASTRONOMY
KOENIGSTUHL
D-6900 HEIDELBERG
GERMANY, F.R.

PLAKIDIS STAVROS PROF
EVRYTANIAS 16
KATO HALANDRI
ATHENS
GREECE

PLAVCOVA Z DR
DEPT OF ASTRONOMY
UCLA
405 HILGARD AVE
LOS ANGELES CA 90024
U.S.A.

PODOBED V V DR
STERNBERG STATE AS-
TRONOMICAL INSTITUTE
117234 MOSCOW
U.S.S.R.

POGO ALEXANDER DR
HALE OBSERVATORIES
813 SANTA BARBARA ST
PASADENA CA 91101
U.S.A.

POLCARO V F
IST DI ASTROFISICA
SPAZIALE
CASELLA POSTALE 67
I-00044 FRASCATI
ITALY

POLLACK JAMES B DR
SPACE SC DIV MP245-3
NASA AMES RES CTR
MOFFETT FIELD
CA 94035
U.S.A.

POLUPAN P N DR
KIEV STATE UNIVERSI-
TY ASTRONOMICAL OB-
SERVATORY
252053 KIEV
U.S.S.R.

POOLEY GUY DR
CAVENDISH LABORATORY
MADINGLEY RD
CAMBRIDGE CB3 OHE
U.K.

POPOV N A DR
GRAVIMETRICAL OBSER-
VATORY OF UKRAINIAN
ACADEMY OF SCIENCES
314029 POLTAVA
U.S.S.R.

POPOVA MALINA D DR
DEPT OF ASTRONOMY
BULGARIAN ACAD SCI
7 NOVEMBER 1
SOFIA 1000
BULGARIA

POPPER DANIEL M PROF
DEPT OF ASTRONOMY
UNIVERSITY OF
CALIFORNIA
LOS ANGELES CA 90024
U.S.A.

PIZZELLA G DR
IST DI FISICA
G MARCONI
UNIVERSITA DI ROMA
ROMA
ITALY

PLASSARD J DR
KSARA OBSERVATORY
KSARA
LEBANON

PLAVEC MIROSLAV PROF
DEPT OF ASTRONOMY
UNIVERSITY OF
CALIFORNIA MS 8979
LOS ANGELES CA 90024
U.S.A.

POECKERT R H DR
2658 AVEBURY AVENUE
VICTORIA BC V8R 3W2
CANADA

POHL ECKHARD DR
STERNWARTE NUERNBERG
LUETZOWSTR 10
D-8500 NUERNBERG
GERMANY, F.R.

POLETTO GIANNINA PROF
OSSERVATORIO ARCETRI
LARGO FERMI 5
I-50125 FIRENZE
ITALY

POLNITZKY GERHARD DR
INST FUER ASTRONOMIE
UNIV WIEN
TUERKENSCHANZSTR 17
A-1180 WIEN
AUSTRIA

POMA ANGELO DR
INT ASTRONOMICAL
STATION
CARLOFORTE
CAGLIARI
ITALY

POPE JOHN D
ROYAL GREENWICH
OBSERVATORY
HAILSHAM
EAST SUSSEX BN27 1RP
U.K.

POPOV VASIL NIKOLOV
DEPT OF ASTRONOMY
BULGARIAN ACADEMY SC
72 LENIN BLVD
1184 SOFIA
BULGARIA

POPOVIC BOZIDAR PROF DR
OGNJENA PRICE 30
11000 BEOGRAD
YUGOSLAVIA

PORCAS RICHARD DR
MPI FUER
RADIOASTRONOMIE
AUF DEM HUEGEL 69
D-5300 BONN
GERMANY, F.R.

PIZZICHINI GRAZIELLA
ISTITUTO TESRE/CNR
VIA DE CASTAGNOLI 1
I-40126 BOLOGNA
ITALY

PLAUT LUCAS PROF
KAPTEYN LABORATORIUM
POSTBUS 800
9700 RV GRONINGEN
THE NETHERLANDS

PNEUMAN GERALD W
HIGH ALTITUDE
OBSERVATORY
BOULDER CO 80302
U.S.A.

POEPPEL WOLFGANG G L DR
INSTITUTO ARGENTINO
DE RADIOASTRONOMIA
CASILLA DE CORREO 5
1894 VILLA ELISA BA
ARGENTINA

POLAND ARTHUR I DR
NASA GSFC
CODE 682
GREENBELT MD 20771
U.S.A.

POLIDAN RONALD S
PRINCETON UNIV OBS
PEYTON HALL
PRINCETON NJ 08544
U.S.A.

POLOZHENTSEV DIMITRIJ DDR
PULKOVO OBSERVATORY
196140 LENINGRAD
U.S.S.R.

PONSONBY JOHN E B DR
NRAL
JODRELL BANK
CHESHIRE SK11 9DL
U.K.

POPELAR J DR
GRAVITY AND GEOD DIV
EARTH PHYSICS BRANCH
3 OBSERVATORY CRESC
OTTAWA ONT K1A 0E4
CANADA

POPOV VICTOR S DR
MAIN ASTRONOMICAL OB
USSR ACAD OF SCS
PULKOVO M-140
196140 LENINGRAD
U.S.S.R.

POPOVIC GEORGIJE DR
ASTRONOMICAL OBS
VOLGINA 7
11050 BEOGRAD
YUGOSLAVIA

PORFIR EV V V DR
KRUPSKAJA PEDAGOGIC
INSTITUTE
107846 MOSCOW
U.S.S.R.

PORTER NEIL A PROF
PHYSICS DEPT
UNIVERSITY COLLEGE
BELFIELD
DUBLIN 4
IRELAND

PORUBCAN VLADIMIR DR
ASTRONOMICAL INSTI-
TUTE SAV
DUBRAVSKA CESTA A 11
89930 BRATISLAVA
CZECHOSLOVAKIA

POTTASCH STUART R PROF
KAPTEYN LABORATORIUM
POSTBUS 800
9700 AV GRONINGEN
THE NETHERLANDS

POTTER HEINO I DR
PULKOVO OBSERVATORY
196140 LENINGRAD
U.S.S.R.

POULAKOS CONSTANTINE DR
RESEARCH CTR ASTRONO
MY AND APPLIED MATHS
ACADEMY OF ATHENS
ATHENS 136
GREECE

POULLE EMMANUEL PROF
ECOLE NATIONALE DES
CHARTES
19 RUE SORBONNE
F-75005 PARIS
FRANCE

POUMEYROL FERNAND MR
OBSERVATOIRE
DE BORDEAUX
F-33270 FLOIRAC
FRANCE

POUNDS KENNETH A PROF
DEPT OF PHYSICS
UNIVERSITY
LEICESTER LE1 7RH
U.K.

POUQUET A DR
OBS DE NICE
B P 252
F-06007 NICE CEDEX
FRANCE

POVEDA ARCADIO DR
INST DE ASTRONOMIA
UNIV NAC AUTO DE MEX
APDO POSTAL 70-264
04510 MEXICO DF
MEXICO

POYET J P DR
OBS DE TOULOUSE
1 AV C FLAMMARION
F-31500 TOULOUSE
FRANCE

PRABHU TUSHAR P
INDIAN INSTITUTE OF
ASTROPHYSICS
BANGALORE 560 034
INDIA

PRADERIE F DR
DESPA
OBSERVATOIRE MEUDON
F-92190 MEUDON
FRANCE

PRADHAN DR
NIZAMIAH OBS
HYDERABAD
INDIA

PRATAP R DR
PHYSICAL RESEARCH
LABORATORY
NAVRANGPURA
AHMEDABAD 380009
INDIA

PREITE-MARTINEZ ANDREA DR
LABORATORIO DI
ASTROFISICA SPAZIALE
CP 67
I-00044 FRASCATI
ITALY

PRENDERGAST KEVIN H PROF
DEPT OF ASTRONOMY
COLUMBIA UNIVERSITY
538 W 120 ST
NEW YORK NY 10027
U.S.A.

PRENTICE ANDREW J R DR
DEPT OF MATHEMATICS
MONASH UNIVERSITY
CLAYTON VICT 3168
AUSTRALIA

PRESS WILLIAM H DR
HARVARD COLLEGE OBS
60 GARDEN ST
CAMBRIDGE MA 02138
U.S.A.

PRESTON GEORGE W DR
HALE OBSERVATORIES
813 SANTA BARBARA ST
PASADENA CA 91101
U.S.A.

PRESTON ROBERT ARTHUR
264-720 JET PROPUL-
SION LAB
PASADENA CA 91109
U.S.A.

PREUSS EUGEN DR
MPI FUER RADIO-
ASTRONOMIE
AUF DEM HUEGEL 69
D-5300 BONN 1
GERMANY, F.R.

PREVOT BURNICHON MARIE
LOUISE DR
OBS DE MARSEILLE
2 PLACE LE VERRIER
F-13248 MARSEILLE C4
FRANCE

PREVOT LOUIS DR
OBSERVATOIRE DE
MARSEILLE
2 PLACE LE VERRIER
F-13004 MARSEILLE
FRANCE

PRICE MICHAEL J. DR.
SCIENCE APPLICATIONS
5055 E BROADWAY
SUITE A-214
TUCSON AZ 85711
U.S.A.

PRICE R MARCUS DR
CHAIRMAN DEPT OF
PHYSICS AND ASTRON
UNIV OF NEW MEXICO
ALBUQUERQUE NM 87131
U.S.A.

PRICE STEPHEN DONALD
2 POLLEY ROAD
WESTFORD MA 01886
U.S.A.

PRIEST E R DR
APPLIED MATHS DEPT
THE UNIVERSITY
ST ANDREWS FIFE
KY16 9SS
U.K.

PRIESTER WOLFGANG PROF
INST F ASTROPHYSIK
AUF DEM HUEGEL 71
D-5300 BONN
GERMANY, F.R.

PRINCE HELEN DODSON PROF
4800 FILLMORE AVE
ALEXANDRIA VA 22311
U.S.A.

PRINGLE JAMES E DR
INST OF ASTRONOMY
MADINGLEY ROAD
CAMBRIDGE CB3 0HA
U.K.

PRITCHET CHRISTOPHER J DR
DEPT GEOPHYS ASTRON
UNIVERSITY OF B C
VANCOUVER BC V6T 1W5
CANADA

PROBSTEIN R F DR
DEPT OF MECHANICAL
ENGINEERING
MIT
CAMBRIDGE MA 02139
U.S.A.

PROCHAZKA FRANZ V DR
UNIVERSITAETS-
STERNWARTE
TUERKENSCHANZSTR 17
A-1180 WIEN
AUSTRIA

PRODAN Y I DR
STERNBERG STATE AS-
TRONOMICAL INSTITUTE
117234 MOSCOW
U.S.S.R.

PROISY PAUL E DR
OBSERVATOIRE DE LYON
F-69230 ST GENIS
LAVAL
FRANCE

PROKAKIS THEODORE J DR
OBSERVATOIRE
D ATHENES
ATHENS 306
GREECE

PROKOF'EV VLADIMIR K PROF
CRIMEAN ASTROPHYSI-
CAL OBSERVATORY
PO NAUCHNY
334413 CRIMEA
U.S.S.R.

PROKOF'EVA IRINA A DR
PULKOVO OBSERVATORY
196140 LENINGRAD
U.S.S.R.

PROKOF'EVA VALENTINA V DR
CRIMEAN ASTROPHYS OB
NAUCHNYJ
334413 CRIMEA
U.S.S.R.

PRONIK I I DR
CRIMEAN ASTROPHY OBS
ACAD SCIENCES USSR
334413 P O NAUCHNY
CRIMEA
U.S.S.R.

PRONIK V I DR
CRIMEAN ASTROPHY OBS
ACAD SCIENCES USSR
334413 P O NAUCHNY
CRIMEA
U.S.S.R.

PROTHEROE WILLIAM M PROF
DEPT OF ASTRONOMY
OHIO STATE UNIV
174 W 18TH AVE
COLUMBUS OH 43210
U.S.A.

PROTICH MILORAD B
ASTRONOMICAL OBSERVA
TORY OF BELGRADE
VOLGINA 7
11050 BELGRADE
YUGOSLAVIA

PROVERBIO EDOARDO PROF
IST DI ASTRONOMIA
VIA OSPEDALE 72
I-09100 CAGLIARI
ITALY

PROVOST J DR
OBS DE NICE
B P 252
F-06007 NICE CEDEX
FRANCE

PRYCE MAURICE H L DR
DEPT OF PHYSICS
UNIVERSITY OF B C
2075 WESBROOK MALL
VANCOUVER BC V6T 1W5
CANADA

PRZYBYLSKI ANTONI DR
MOUNT STROMLO
OBSERVATORY
WODEN A C T 2606
AUSTRALIA

PSKOVSKIJ JU P DR
STERNBERG STATE AS-
TRONOMICAL INSTITUTE
117234 MOSCOW
U.S.S.R.

PUCILLO MAURO DR
OSSERVATORIO
ASTRONOMICO
VIA TIEPOLO 11
I-34131 TRIESTE
ITALY

PUGET JEAN-LOUP DR
INSTITUT
D ASTROPHYSIQUE
98 BIS BLVD ARAGO
F-75014 PARIS
FRANCE

PUNETHA LALIT MOHAN DR
UP STATE OBSERVATORY
MANORA PEAK
NAINI TAL 263129
INDIA

PURGATHOFER ALOIS PROF DR
INST FUER ASTRONOMIE
UNIV WIEN
TUERKENSCHANZSTR 17
A-1180 WIEN
AUSTRIA

PURTON CHRISTOPHER R DR
DOMINION RADIO ASTRO
PHYS OBS NRCC
P O BOX 248
PENTICTON BC V2A 6K3
CANADA

PUSCHELL JEFFREY JOHN
UNIV OF CALIFORNIA
CASS C-011
LA JOLLA CA 92093
U.S.A.

PUSHKIN SERGEY B DR
TIME AN FREQUENCY
SERVICE
GOSSTANDARD USSR
117049 MOSCOW
U.S.S.R.

PYE JOHN P DR
8 BEAUMONT WALK
LEICESTER LE4 OPQ
U.K.

PYPER D M DR
DEPT OF PHYSICS AND
ASTRONOMY
TEL-AVIV UNIVERSITY
TEL-AVIV
ISRAEL

QIAN SHAN-JIE
BEIJING OBSERVATORY
BEIJING
PEOPLE'S REPUBLIC CHINA

QU QIN-YUE
DEPT OF ASTRONOMY
NANJING UNIVERSITY
NANJING
PEOPLE'S REPUBLIC CHINA

QUAST G R
OBS NACIONAL
R GENERAL BRUCE 586
SAO CRISTOVAO
20000 RIO DE JANEIRO
BRAZIL

QUENBY JOHN J DR
BLACKETT LAB
IMPERIAL COLLEGE
PRINCE CONSORT ROAD
LONDON SW7 2BZ
U.K.

QUERCI FRANCOIS R DR
OBS DU PIC-DU-MIDI
ET DE TOULOUSE
14 AV EDOUARD BELIN
F-31400 TOULOUSE
FRANCE

QUERCI MONIQUE DR
OBS DU PIC-DU-MIDI
ET DE TOULOUSE
14 AV EDOUARD BELIN
F-31400 TOULOUSE
FRANCE

QUIJANO LUIS
INSTITUTO Y OBSERVA-
TORIO DE MARINA
SAN FERNANDO CADIZ
SPAIN

QUINTANA HERNAN DR
ASTRONOMY DEPT
UNIVERSITY CATOLICA
CASILLA 114-D
SANTIAGO
CHILE

QUINTANA JOSE M DR
INSTITUTO DE ASTROFI
SICA DE ANDALUCIA
GRANADA
SPAIN

QUIRK WILLIAM J DR
24798 JO MARY CT
HAYWARD CA 94541
U.S.A.

QVIST BERTIL PROF
MATHEMATICS DEPT
ABO AKADEMI
SF-20500 ABO 50
FINLAND

RAADU MICHAEL A DR
DEPT PLASMA PHYSICS
ROYAL INSTITUTE OF
TECHNOLOGY
S-10044 STOCKHOLM 70
SWEDEN

RACHKOVSKY D N DR
CRIMEAN ASTROPHYSI-
CAL OBSERVATORY USSR
ACAD OF SCI NAUCHNIY
334413 CRIMEA
U.S.S.R.

RACINE RENE DR
DEPT DE PHYSIQUE
UNIVERSITE DE
MONTREAL BP 6128
MONTREAL PQ H3C 3J7
CANADA

RACKHAM THOMAS W DR
JODRELL BANK
CHESHIRE SK 11 9PL
U.K.

RADHAKRISHNAN V PROF
DIRECTOR
RAMAN RESEARCH
INSTITUTE
BANGALORE 560 006
INDIA

RADLOVA L N DR
INSTITUTE OF SCIENCE
AND TECHNICS INFORMA
TION
125219 MOSCOW
U.S.S.R.

RADOSLAVOVA TZVETANA
DEPT OF ASTRONOMY
BULGARIAN ACADEMY SC
72 LENIN BLVD
1184 SOFIA
BULGARIA

RAEDLER K H DR
ZNTRINST ASTROPHYSIK
ASTROP OBSERVATORIUM
TELEGRAFENBERG
DDR-15 POTSDAM
G.D.R.

RAFANELLI PIERO DR
OSSERVATORIO ASTRO-
NOMICO
I-35100 PADOVA
ITALY

RAFERT JAMES BRUCE
DEPT PHYSICS/ASTRON
APPELACHIAN ST UNIV
BOONE NC 28608
U.S.A.

RAGHAVAN NIRUPAMA DR
2133 INDIAN INSTITU-
TE OF TECHNOLOGY
CAMPUS
NEW DELHI 110 029
INDIA

RAHE JURGEN PROF
UNIV ERL-NUERNBERG
ASTRONOMICAL INST
STERNWARTSTR 7
D-8600 BAMBERG
GERMANY, F.R.

RAIKOVA DONKA DR
DEPT OF ASTRONOMY
BULGARIAN AC OF SCS
7TH NOVEMBER ST 1
1000 SOFIA
BULGARIA

RAIMOND ERNST DR
NETHERL FOUNDATION
FOR RADIOASTRONOMY
RADIOSTERRENWACHT
7991 AC DWINGELOO
THE NETHERLANDS

RAINE DEREK J DR
DEPT OF ASTRONOMY
UNIVERSITY OF
LEICESTER
LEICESTER LEI 7RH
U.K.

RAJAMOHAN R DR
INDIAN INSTITUTE OF
ASTROPHYSICS
BANGALORE 560 034
INDIA

RAJCHL JAROSLAV DR
ASTRONOMICAL INST
CZECHOS ACAD SC
OBSERVATORY
251 65 ONDREJOV
CZECHOSLOVAKIA

RAJU P K DR
INDIAN INSTITUTE OF
ASTROPHYSICS
BANGALORE 560034
INDIA

RAKAVY GIDEON PROF
HEBREW UNIVERSITY
EINSTEIN INST PHYS
JERUSALEM
ISRAEL

RAKOS KARL D PROF
INST FUER ASTRONOMIE
UNIV WIEN
TURKENSCHANZSTR 17
A-1180 WIEN
AUSTRIA

RAKSHIT H PROF
BENGAL ENGINEERING
COLLEGE
SIBPORE
HEWRAH
INDIA

RAMADURAI S DR
TATA INST FUND RES
HOMI BHABHA ROAD
BOMBAY 400 005
INDIA

RAMATY REUVEN DR
LAB HIGH ENERGY ASTR
NASA-GODDARD SPACE
FLIGHT CTR CODE 660
GREENBELT MD 20771
U.S.A.

RAMBERG JOERAN M PROF
GENVAEGEN 4
S-133 00
SALTSJOEBADEN
SWEDEN

RAMSEY LAWRENCE W DR
DEPT OF ASTRONOMY
525 DAVEY LAB
UNIV OF PENNA
UNIV PK PA 16802
U.S.A.

RANDIC LEO PROF DR
GUNDULICEVA 54
ZAGREB
YUGOSLAVIA

RANK DAVID M PROF
LICK OBSERVATORY
UNIVERSITY OF CALIF
SANTA CRUZ CA 95064
U.S.A.

RANKIN JOANNA M DR
DEPT OF HISTORY
CORNELL UNIVERSITY
MCGRAW HALL
ITHACA NY 14853
U.S.A.

RANUHEN TIMO
OBS & ASTROPHYS LAB
UNIV OF HELSINKI
TAEHTITORNINMAEKI
SF-00130 HELSINKI 13
FINLAND

RAO A P DR
RADIO ASTRONOMY CTR
P O BOX NO 8
OOTACAMUND 643 001
INDIA

RAO N KAMESWARA
INDIAN INSTITUTE OF
ASTROPHYSICS
BANGALORE 560 034
INDIA

RAO RAMACHANDRA V PROF
ISRO SATELLITE
CENTER PEENYA
BANGALORE 560058
INDIA

RAPAPORT M DR
OBS DE BORDEAUX
F-33270 FLOIRAC
FRANCE

RAPLEY CHRISTOPHER G DR
MULLARD SPACE SC LAB
UNIV COLLEGE LONDON
LONDON
U.K.

RASMUSEN H W DR
VAERSLEVGAAREN
FRANKERUPVEI 40
DK-4400 KALUNDBORG
DENMARK

RAYMOND JOHB CHARLES
CTR FOR ASTROPHYSICS
60 GARDEN STREET
CAMBRIDGE MA 02138
U.S.A.

RAYROLE JEAN R DR
OBSERVATOIRE
DE MEUDON
F-92190 MEUDON
FRANCE

RAZIN V A DR
RADIOPHYSICAL
RESEARCH INSTITUTE
603600 GORKIJ
U.S.S.R.

READHEAD ANTHONY C S DR
RADIO ASTRONOMY DEPT
ROBINSON BLDG
CALTECH
PASADENA CA 91125
U.S.A.

REASENBERG ROBERT D DR
MIT
54-612
CAMBRIDGE MA 02139
U.S.A.

REAVES GIBSON PROF
UNIVERSITY OF
SOUTHERN CALIFORNIA
LOS ANGELES CA 90007
U.S.A.

REAY NEWRICK K DR
ASTRONOMY GROUP
BLACKETT LABORATORY
IMPERIAL COLLEGE
LONDON SW7 2BZ
U.K.

REBEIROT EDITH DR
OBS DE MARSEILLE
2 PLACE LE VERRIER
F-13004 MARSEILLE
FRANCE

REBER GROTE DR
GENERAL DELIVERY
BOTHWELL TASM 7411
AUSTRALIA

RECILLAS-CRUZ ELSA DR
INST DE ASTRONOMIA
UNAM
APDO POSTAL 70-264
04510 MEXICO DF
MEXICO

REDDISH VINCENT C PROF
ROYAL OBSERVATORY
EDINBURGH EH12 9AJ
U.K.

REES DAVID ELWYN DR
DEPT APPLIED MATH
UNIVERSITY OF SYDNEY
SYDNEY NSW 2006
AUSTRALIA

REES MARTIN J PROF
INST OF ASTRONOMY
MADINGLEY RD
CAMBRIDGE CB3 0HE
U.K.

REEVES EDMOND M DR
NASA HEADQUARTERS
SM-8
400 MARYLAND AVE
WASHINGTON DC 20546
U.S.A.

REEVES HUBERT PROF
SEP-SES BAT 28
CEN SACLAY
B P 2
F-91190 GIF S YVETTE
FRANCE

REFSDAL S PROF DR
HAMBURGER STERNWARTE
GOJENBERGSWEG 112
D-2050 HAMBURG 80
GERMANY, F.R.

REGO FERNANDES M. DR.
ASTROFISICA
FACULTA FISICA
UNIV.COMPLUTENSE
MADRID 3
SPAIN

REICH WOLFGANG
MPI FUER
RADIOASTRONOMIE
AUF DEM HUEGEL 69
D-5300 BONN
GERMANY, F.R.

REID JOHN H DR
1007 RESEDA DRIVE
HOUSTON TX 77062
U.S.A.

REID MARK JONATHAN DR
NAT RADIO ASTRON OBS
EDGEMONT RD
CHARLOTTESVL VA22901
U.S.A.

REIMERS DIETER DR
INST THEO PHY STERNW
NEUE UNIV PHYSIK ZTM
OLSHAUSENST GB N 61C
D-2300 KIEL 1
GERMANY, F.R.

REINHARDT MICHAEL PROF
ASTRONOMISCHES INST
RUHR UNIVERSITAET
POSTFACH 102148
D-4630 BOCHUM 1
GERMANY, F.R.

REINISCH G DR
OBS DE NICE
B P 252
F-06007 NICE CEDEX
FRANCE

REIZ ANDERS PROF
UNIVERSITY
OBSERVATORY
OSTERVOLDGADE 3
DK-1350 COPENHAGEN K
DENMARK

REMY BATTIAU LILIANE G A
DR
INST D ASTROPHYSIQUE
UNIVERSITY DE LIEGE
B-4200 COINTE OUGREE
BELGIUM

REN JIANG-PING
DEPT OF ASTRONOMY
NANJING UNIVERSITY
NANJING
PEOPLE'S REPUBLIC CHINA

RENGARAJAN T N DR
TATA INSTITUTE OF
FUNDAMENTAL RESEARCH
HOMI BHABBA ROAD
BOMBAY 400 005
INDIA

RENSE WILLIAM A DR
DEPT PHYS AND ASTROP
UNIV OF COLORADO
DUANE PHYSICAL LABS
BOULDER CO 80302
U.S.A.

RENSON P F M DR
INSTITUT
D ASTROPHYSIQUE
B-4200 COINTE-OUGREE
BELGIUM

RENZINI ALVIO PROF
OSSERVATORIO
ASTRONOMICO
VIA ZAMBONI 33
I-40126 BOLOGNA
ITALY

REQUIEME YVES DR
OBSERVATOIRE
DE BORDEAUX
F-33270 FLOIRAC
FRANCE

REUNING ERNEST G DR
DEPT OF PHYSICS AND
ASTRONOMY
UNIV OF GEORGIA
ATHENS GA 30602
U.S.A.

REVELLE DOUGLAS ORSON DR
DEPT OF PHYSICS
C U BOX 6010
NORTHERN ARIZ UNIV
FLAGSTAFF AZ 86011
U.S.A.

REYNOLDS JOHN H PROF
DEPT OF PHYSICS
UNIVERSITY OF CALIF
BERKELEY CA 94720
U.S.A.

REYNOLDS RONALD J DR
SPACE PHYSICS GROUP
1150 UNIVERSITY AVE
MADISON WI 53706
U.S.A.

RHODES EDWARD J JR
11801 KILLIMORE AVE
NORTHRIDGE CA 91326
U.S.A.

RIBES ELIZABETH DR
OBSERVATOIRE
DE MEUDON
F-92190 MEUDON
FRANCE

RIBES JEAN-CLAUDE DR
INAG
77 AVE DENFERT-
ROCHEREAU
F-75014 PARIS
FRANCE

RICE D A
COAST AND GEODETIC
SURVEY
ROCKVILLE MD 20852
U.S.A.

RICE JOHN B DR
DEPT PHYSICS AND
ASTRONOMY
BRANDON UNIVERSITY
BRANDON MAN R7A 6A9
CANADA

RICHARDSON E HARVEY DR
HERZBERG INST ASTRO
DOMINION ASTRO OBS
5071 W SAANICH RD
VICTORIA BC V8X 4M6
CANADA

RICHARDSON R S
GRIFFITH OBS
P O BOX 27787
LOS FELIX STATION
LOS ANGELES CA 90027
U.S.A.

RICHER HARVEY B DR
DEPT GEOPHYS-ASTRON
UNIVERSITY OF BC
2075 WESBROOK PLACE
VANCOUVER BC V6T 1W5
CANADA

RICHTER G A DR
ZENTRALINSTITUT FUER
ASTROPHYSIK
STERNWARTE SONNEBERG
DDR-64 SONNEBERG
G.D.R.

RICHTER JOHANNES PROF
INST F EXPER PHYSIK
PHYSIKZENTRUM
OLSHAUSENSTRASSE
D-2300 KIEL
GERMANY, F.R.

RICKARD JAMES JOSEPH DR
P O BOX 777
BORREGO SPRINGS
CO 92004
U.S.A.

RICKARD LEE J DR
NAT RADIO ASTRON OBS
EDGEMONT RD
CHARLOTTESVL VA22901
U.S.A.

RICKER GEORGE R DR
MIT
CTR FOR SPACE RES
77 MASS AV RM 37-527
CAMBRIDGE MA 02139
U.S.A.

RICKETT BARNABY JAMES DR
DEPT APPLIED PHYSICS
AND INFORMATION SC
UNIV CALIF SAN DIEGO
LA JOLLA CA 92093
U.S.A.

RICKMAN HANS DR
ASTRONOMISKA OBS
BOX 515
S-75120 UPPSALA
SWEDEN

RICORT GILBERT DR
DEPT D'ASTROPHYSIQUE
UNIVERSITE DE NICE
PARC VALROSE
F-06034 NICE
FRANCE

RIDDLE ANTHONY C DR
700 GRANT PL
BOULDER CO 80302
U.S.A.

RIEGEL KURT W DR
DEPT OF ASTRONOMY
UCLA
LOS ANGELES CA 90024
U.S.A.

RIEU N Q DR
OBSERVATOIRE
DE MEUDON
F-92190 MEUDON
FRANCE

RIGHINI ALBERTO PROF
OSSERVATORIO ASTRO-
FISICO ARCETRI
LARGO E FERMI 5
I-50125 FIRENZE
ITALY

RIGHINI-COHEN GIOVANNA DR
DEPT EARTH SPACE SCS
SUNY
STONY BROOK NY 11794
U.S.A.

RIGUTTI MARIO PROF
OSSERVATORIO ASTRONO
MICO DI CAPODIMONTE
MOIARIELLO 16
I-80131 NAPOLI
ITALY

RIIHIMAA JORMA J DR
AARNE KARJALAINEN
OBSERVATORY
UNIVERSITY OF OULU
SF-90100 OULU 10
FINLAND

RILEY JULIA M DR
MULLARD RADIO ASRON
LABORATORY
MADINGLEY ROAD
CAMBRIDGE
U.K.

RINDLER WOLFGANG PROF
UNIVERSITY OF TEXAS
BOX 688
RICHARDSON TX 75080
U.S.A.

RINEHART JOHN S PROF
HYPERDYNAMICS
P O BOX 392
SANTA FE NM 87501
U.S.A.

RING JAMES PROF
THE BLACKETT
LABORATORY
IMPERIAL COLLEGE
LONDON SW7 2BZ
U.K.

RINGNES TRULS S DR
INST THEOR ASTROPHYS
UNIVERSITY OF OSLO
P O BOX 1029
BLINDERN OSLO 3
NORWAY

RINGUELET ADELA E DR
49 342
1900 LA PLATA
ARGENTINA

RIPKEN HARTMUT W DR
INST F ASTROPHYS UND
EXTRATERR FORSCHUNG
AUF DEM HUEGEL 71
D-5300 BONN
GERMANY, F.R.

RITTER HANS DR
MPI FUER PHYSIK UND
ASTROPHYSIK
FOEHRINGER RING 6
D-8000 MUENCHEN
GERMANY, F.R.

RIZVANOV NAUFAL G DR
ENGELHARDT OBS
422526 KASAN
U.S.S.R.

ROACH FRANKLIN E
2969 KALAKAUA AVE
APT 605
HONOLULU HI 96815
U.S.A.

ROARK TERRY P PROF
DEPT OF ASTRONOMY
OHIO ST UNIVERSITY
174 W 18TH AVE
COLUMBUS OH 43210
U.S.A.

ROBBINS R ROBERT PROF
UNIVERSITY OF TEXAS
ASTRONOMY DEPT
AUSTIN TX 78712
U.S.A.

ROBE H A G DR
INSTITUT
D ASTROPHYSIQUE
B-4200 COINTE-OUGREE
BELGIUM

ROBERTI GIUSEPPE DR
OSSERVATORIO ASTRO-
NOMICO
NAPOLI
ITALY

ROBERTS BERNARD DR
DEPT APPLIED MATH
UNIV OF ST ANDREWS
ST ANDREWS FIFE
KY16 8QP
U.K.

ROBERTS DAVID HALL DR
RES LAB ELECTRONICS
26-321
MIT
CAMBRIDGE MA 02139
U.S.A.

ROBERTS JAMES A DR
CSIRO DIVISION OF
RADIOPHYSICS
P O BOX 76
EPPING NSW 2121
AUSTRALIA

ROBERTS MORTON S DR
NAT RADIO ASTRO OBS
EDGEMONT RD
CHARLOTTESVILLE
VA 22901
U.S.A.

ROBERTS WALTER ORR DR
UNIVERSITY CORP FOR
ATMOSPHERIC RESEARCH
P O BOX 3000
BOULDER CO 80307
U.S.A.

ROBERTS WILLIAM W JR PROF
DPT APPL MATH COM SC
THORNTON HALL UNIVVA
CHARLOTTESVILLE
VA 22901
U.S.A.

ROBERTSON DOUGLAS S
NOAA NOS
NGS OA/C12
ROCKVILLE MD 20852
U.S.A.

ROBERTSON JAMES GORDON DR
ANGLO-AUSTRALIAN
OBSERVATORY
P O BOX 296
EPPING NSW 2121
AUSTRALIA

ROBERTSON JOHN ALISTAIR
INST ASTROPHYSIK
KARL-SCHWARZSCHILD-
STRASSE 1
D-8046 GARCHING B M
GERMANY, F.R.

ROBERTSON WILLIAM H
1 DUNMORE ROAD
EPPING NSW 2121
AUSTRALIA

ROBINSON BRIAN J DR
CSIRO RADIOPHYSICS
LABORATORY
P O BOX 76
EPPING NSW 2121
AUSTRALIA

ROBINSON EDWARD LEWIS DR
DEPT OF ASTRONOMY
UNIV OF TEXAS
AUSTIN TX 78712
U.S.A.

ROBINSON I PROF
UNIVERSITY OF TEXAS
BOX 688
MAIL STATION BE 32
RICHARDSON TX 75080
U.S.A.

ROBINSON JR RICHARD D DR
CSIRO DIVISION OF
RADIOPHYSICS
P O BOX 76
EPPING NSW 2121
AUSTRALIA

ROBINSON LEIF J
14 WILLOW ROAD
WELLESLEY MA 02181
U.S.A.

ROBINSON LLOYD B DR
LICK OBSERVATORY
UNIVERSITY OF CALIF
SANTA CRUZ CA 95064
U.S.A.

ROBINSON WILLIAM J DR
DEPT OF MATHEMATICS
THE UNIVERSITY
BRADFORD BD7 1DP
U.K.

ROBLEY R DR
OBSERVATOIRE
DU PIC DU MIDI
1 AVE C FLAMMARION
F-31500 TOULOUSE
FRANCE

ROBSON IAN E DR
DIV PHYS AND ASTRON
PRESTON POLYTECHNIC
PRESTON PR1 2TQ
U.K.

ROCCA-VOLMERANGE B DR
LAB RENE BERNAS
BAT 108
F-91406 ORSAY
FRANCE

ROCHESTER MICHAEL G PROF
PHYSICS DEPT
MEMORIAL UNIVERSITY
ST JOHNS NFLD A1C5S7
CANADA

RODDIER C DR
UER DE MATHEMATIQUES
UNIV DE PROVINCE
PLACE VICTOR HUGO
F-13331 MARSEILLE
FRANCE

RODDIER FRANCOIS PROF
DEPT D'ASTROPHYSIQUE
UNIVERSITE DE NICE
PARC VALROSE
F-06034 NICE CEDEX
FRANCE

RODGERS ALEX W DR
MT STROMLO
OBSERVATORY
WODEN PO ACT 2606
AUSTRALIA

RODMAN RICHARD B DR
65 LOCUST AVE
LEXINGTON MA 02173
U.S.A.

RODONO MARCELLO DR
OSSERVATORIO
ASTROFISICO
CITTA UNIVERSITARIA
I-95125 CATANIA
ITALY

RODRIGUEZ LUIS F
INST DE ASTRONOMIA
UNAM
APDO POSTAL 70-264
04510 MEXICO DF
MEXICO

ROEDER ROBERT C PROF
DEPT OF ASTRONOMY
SCARBOROUGH COLLEGE
1265 MILITARY TRAIL
WEST HILL ONT M1C1A4
CANADA

ROEMER ELIZABETH PROF
LUNAR AND PLANETARY
LABORATORY
UNIVERSITY ARIZONA
TUCSON AZ 85721
U.S.A.

ROEMER MAX PROF
INST F ASTROPHYSIK
EXTRATERR FORSCHUNG
AUF DEM HUEGEL 71
D-5300 BONN 1
GERMANY, F.R.

ROENNAENG BERNT O DR
ONSALA SPACE
OBSERVATORY
S-430 34 ONSALA
SWEDEN

ROESER HERMANN-JOSEF DR
MPI FUER ASTRONOMIE
KOENIGSTUHL
D-6900 HEIDELBERG 1
GERMANY, F.R.

ROESER SIEGFRIED DR
ASTRONOMISCHES
RECHEN-INSTITUT
MOENCHHOFSTR 12-14
D-6900 HEIDELBERG 1
GERMANY, F.R.

ROGER ROBERTS S DR
DOMINION RADIO ASTRO
PHYSICAL OBSERVATORY
BOX 248
PENTICTON BC V2A 6K3
CANADA

ROGERS ALAN E E DR
HAYSTACK OBSERVATORY
WESTFORD MA 01886
U.S.A.

ROGERSON JOHN B PROF
PRINCETON UNIVERSITY
DEPT ASTROPHYS SCS
PEYTON HALL
PRINCETON NJ 08540
U.S.A.

ROGSTAD DAVID H DR
MAIL CODE 102-24
CALIF INST TECHN
1201 E CALIF BLVD
PASADENA CA 91125
U.S.A.

ROHLFS K PROF DR
RUHR UNIV BOCHUM
INST FUR ASTROPHYS
POSTFACH 102 148
D-4630 BOCHUM 1
GERMANY, F.R.

ROLAND GINETTE DR
INST D ASTROPHYSIQUE
UNIVERSITE DE LIEGE
B-4200 COINTE OUGREE
BELGIUM

ROLLAND ANGEL DR
JUPITER 2
HUETOR VEGA
GRANADA
SPAIN

ROMAN NANCY G DR
4620 NORTH PARK AVE
APT 306W
CHEVY CHASE MD 20015
U.S.A.

ROMANCHUK P R DR
KIEV STATE UNIVERSI-
TY ASTRONOMICAL
OBSERVATORY
252053 KIEV
U.S.S.R.

ROMANO GIULIANO PROF
VIALE S ANTONIO
DA PADOVA 7
I-31100 TREVISO
ITALY

ROMANOV YURI S DR
ODESSA ASTRONOMICAL
OBSERVATORY
SHEVCHENKO PARK
270014 ODESSA
U.S.S.R.

ROMNEY JONATHAN D DR
MPI FUER
RADIOASTRONOMIE
AUF DEM HUEGEL 69
D-5300 BONN
GERMANY, F.R.

ROMPOLT BOHDAN DR
ASTRONOMICAL INST
UL KOPERNIKA 11
51 622 WROCLAW
POLAND

RONAN COLIN A
13 ACORN AVE
BAR HILL
CAMBRIDGE CB3 8DT
U.K.

ROOD HERBERT J
52 ELIZABETH ST
SOUTH BOUND BROOK
NJ 08880
U.S.A.

ROOD ROBERT T DR
BOX 3818
UNIVERSITY STATION
CHARLOTTESVILLE
VA 22903
U.S.A.

ROOSEN ROBERT G DR
LAB FOR SOLAR PHYS
UNIV OF NEW MEXICO
ALBUQUERQUE NM 87131
U.S.A.

ROSA DOROTHEA DR
LANDESSTERNWARTE
KOENIGSTUHL
D-6900 HEIDELBERG
GERMANY, F.R.

ROSADO MARGARITA DR
INST DE ASTRONOMIA
UNAM
APDO POSTAL 70-264
04510 MEXICO DF
MEXICO

ROSCH JEAN PROF DIR
OBSERVATOIRES DU PIC
DU MIDI ET DE
TOULOUSE
F-65200 BAGNERES
FRANCE

ROSE JAMES ANTHONY
INST FOR ASTRONOMY
UNIVERSITY OF HAWAII
2680 WOODLAWN DRIVE
HONOLULU HI 96822
U.S.A.

ROSE WILLIAM K DR
ASTRONOMY PROGRAM
UNIVERSITY MARYLAND
COLLEGE PK MD 20742
U.S.A.

ROSEN E DR
DEPT OF HISTORY
CITY COLLEGE OF THE
CITY UNIV NEW YORK
NEW YORK NY 10031
U.S.A.

ROSENBERG J DR
STERRENWACHT
SERVAAS BOLWERK 13
3512 NK UTRECHT
THE NETHERLANDS

ROSENDHAL JEFFEREY D DR
NASA HEADQUARTERS
CODE SA
WASHINGTON DC 20546
U.S.A.

ROSINO LEONIDA PROF
ASTRONOMICAL OBSERVA
TORY PADOVA
PADOVA
ITALY

ROSLUND CURT DR
DEPT OF ASTRONOMY
CHALMERS UNIVERSITY
OF TECHNOLOGY FACK
S-402 20 GOETEBORG
SWEDEN

ROSNER ROBERT
HARVARD-SMITHSONIAN
CTR FOR ASTROPHYSICS
60 GARDEN STREET
CAMBRIDGE MA 02138
U.S.A.

ROSQUIST KJELL
MULTRAG 140
S-162 28 VAELLINGBY
SWEDEN

ROSS DENNIS K PROF
PHYSICS DEPT
IOWA ST UNIVERSITY
AMES IA 50010
U.S.A.

ROSS JOHN E R DR
PHYSICS DEPT
UNIV OF QUEENSLAND
ST LUCIA
BRISBANE QLD 4067
AUSTRALIA

ROSSELAND SVEIN PROF
INST THEORET ASTROPH
UNIVERSITY OF OSLO
P O BOX 1029
BLINDERN OSLO 3
NORWAY

ROSSI BRUNO B PROF
MASS INSTITUTE OF
TECHNOLOGY
RM 37-667
CAMBRIDGE MA 02139
U.S.A.

ROSSI LUCIO
IST DI ASTROFISICA
SPAZIALE
CASELLA POSTALE 67
I-00044 FRASCATI
ITALY

ROSTAS FRANCOIS DR
OBSERVATOIRE
DE MEUDON
F-92190 MEUDON
FRANCE

ROTH MARIA-LUISE PH D
HAMBURGER STERNWARTE
D-205 HAMBURG 80
GERMANY, F.R.

ROTH MIGUEL R DR
INST DE ASTRONOMIA
UNAM
APDO POSTAL 70-264
04510 MEXICO DF
MEXICO

ROTS ARNOLD H DR
NAT RADIO ASTRON OBS
P O BOX 0
SOCORRO NM 87801
U.S.A.

ROTTENBERG J A DR
2911 BAYVIEW AVE
SUITE 110C
WILLOWDALE ONTM2K1E8
CANADA

ROUEFF EVELYNE M A DR
OBSERVATOIRE MEUDON
DEPT D ASTROPHYSIQUE
FOTLE
F-92190 MEUDON
FRANCE

ROUNTREE JANET DR
AFE TECHN SVCS STAFF
BOLLING AFB
WASHINGTON DC 20332
U.S.A.

ROUSE CARL A DR
627 15TH ST
DEL MAR CA 92014
U.S.A.

ROUSSEAU JEAN-MICHEL MR
OBSERVATOIRE
DE BORDEAUX
F-33270 FLOIRAC
FRANCE

ROUSSEAU JEANINE DR
OBSERVATOIRE DE LYON
F-69230 ST GENIS
LAVAL
FRANCE

ROUTLEDGE DAVID DR
ELECTRICAL ENGG DEPT
UNIVERSITY OF
ALBERTA
EDMONTON T6G 2J1
CANADA

ROUTLY PAUL M DR
US NAVAL OBSERVATORY
34TH AND MASS AVE NW
WASHINGTON DC 20390
U.S.A.

ROVITHIS PETER EMM DR
NATIONAL OBSERVATORY
OF ATHENS
ATHENS 306
GREECE

ROVITHIS-LIVANIOU H PH D
ASTRONOMY DEPT
UNIV OF ATHENS
PANEPISTIMIOPOLIS
ATHENS 621
GREECE

ROWAN-ROBINSON MICHAEL DR
DEPT OF APPLIED MATH
QUEEN MARY COLLEGE
MILE END RD
LONDON E1 4NS
U.K.

ROWSON BARRIE DR
NRAL
JODRELL BANK
MACCLESFIELD
CHESHIRE SK11 9DL
U.K.

ROXBURGH IAN W PROF
DEPT APPLIED MATHS
QUEEN MARY COLLEGE
MILE END ROAD
LONDON E1
U.K.

ROY ARCHIE E PROF
DEPT OF ASTRONOMY
GLASGOW UNIVERSITY
GLASGOW G12 8QQ
U.K.

ROY JEAN-RENE
DEPT DE PHYSIQUE
UNIVERSITE LAVAL
CITE UNIVERSITAIRE
QUEBEC G1K 7P4
CANADA

ROZELOT JEAN P
150 BLVD DIDEROT
IMMEUBLE F 128
F-75012 PARIS
FRANCE

ROZHKOVSKIJ DIMITRIJ A
PROF
ASTROPHYSICAL
INSTITUTE
48068 ALMA ATA
U.S.S.R.

RUBASHEV BORIS M DR
PULKOVO OBSERVATORY
196140 LENINGRAD
U.S.S.R.

RUBEN G PROF DR
ZNTRINST ASTROPHYSIK
ASTROPHYS OBS
TELEGRAFENBERG
DDR-15 POTSDAM
G.D.R.

RUBIN ROBERT HOWARD
NASA AMES RESEARCH
CENTER MS 245-6
MOFFETT FIELD
CA 94035
U.S.A.

RUBIN VERA C DR
DEPT TERRESTRIAL MAG
CARNEGIE INST WASH
5241 BROAD BR RD N W
WASHINGTON DC 20015
U.S.A.

RUCINSKI SLAWOMIR M DR
ASTRONOMICAL
OBSERVATORY
AL UJAZDOWSKIE 4
00 478 WARSZAWA
POLAND

RUDDY VINCENT P DR
REGIONAL TECHNICAL
COLLEGE
ROSSA AVE
CORK
IRELAND

RUDERMAN MALVIN A
PHYSICS DEPT
COLUMBIA UNIVERSITY
NEW YORK NY 10027
U.S.A.

RUDKJOBING MOGENS PROF
ASTRONOMISK INSTITUT
LANGELANDSGADE
DK-8000 AARHUS C
DENMARK

RUDNICK LAWRENCE DR
NAT RADIO ASTRON LAB
EDGEMONT RD
CHARLOTTESVL VA22901
U.S.A.

RUDNICKI KONRAD PROF
ASTRONOMICAL OBS
UL ORLA 171
30 244 KRAKOW
POLAND

RUFENER FREDY G PROF
OBSERVATOIRE
DE GENEVE
CH-1290 SAUVERNY GE
SWITZERLAND

RUGGE HUGO R DR
THE AEROSPACE
CORPORATION
P O BOX 92957
LOS ANGELES CA 90009
U.S.A.

RUIZ MARIA TERESA DR
OBS ASTRON NACIONAL
UNIVERSIDAD DE CHILE
CASILLA 36-D
SANTIAGO
CHILE

RULE BRUCE H STAFF RETIRE
HALE OBSERVATORIES
2205 MONTE VISTA ST
PASADENA CA 91107
U.S.A.

RUMSEY NORMAN J
PHYSICS AND ENGINEE-
RING LABORATORY
D S I R
LOWER HUTT
NEW ZEALAND

RUNCORN S K PROF
SCHOOL OF PHYSICS
THE UNIVERSITY
NEWCASTLE UPON TYNE
NE1 7RU
U.K.

RUPRECHT JAROSLAV DR
ASTRONOMICAL INST OF
CZECHOSLOVAK ACAD SC
BUDECSKA 6
120 23 PRAHA 2
CZECHOSLOVAKIA

RUSCONI LUIGIA DR
VIA DEL VELTRO 77
I-34100 TRIESTE
ITALY

RUSIN VOJTECH
ASTRONOMICAL INST
SLOVAK ACADEMY OF SC
05960 TATRANSKA
LOMNICA
CZECHOSLOVAKIA

RUSKOL EUGENIA L DR
OJSCHMIDT INSTITUTE
OF PHYSICS OF EARTH
AC SCI USSR
123810 MOSCOW
U.S.S.R.

RUSSELL CHRISTOPHER T
INST OF GEOPHYSICS
& PLANETARY PHYSICS
UNIV OF CALIFORNIA
LOS ANGELES CA 90024
U.S.A.

RUSSELL JANE L DR
DEPT OF PHYSICS
IOWA STATE UNIV
AMES IA 50011
U.S.A.

RUSSELL JOHN A PROF
DEPT OF ASTRONOMY
UNIV OF SOUTHERN CAL
UNIVERSITY PARK
LOS ANGELES CA 90007
U.S.A.

RUSSEV RUSCHO DR
DEPT OF ASTRONOMY
ANTON IVANOV ST 5
1126 SOFIA
BULGARIA

RUSSO GUIDO DR
OSSERVATORIO ASTRO-
NOMICO CAPODIMONTE
VIA MOIARELLO 16
I-80131 NAPOLI
ITALY

RUST DAVID M DR
AMERICAN SCIENCE AND
ENGINEERING CO
955 MASS AVE
CAMBRIDGE MA 02139
U.S.A.

RUSU I DR
ASTRONOMICAL OBS
5 CUTITUL DE ARGINT
BUCAREST 28
ROUMANIA

RUSU L DR
ASTRONOMICAL OBS
5 CUTITUL DE ARGINT
BUCAREST 28
ROUMANIA

RUTTEN ROBERT JELLE DR
OBS SONNENBORGH
ZONNENBURG 2
3512 NL UTRECHT
THE NETHERLANDS

RUZDJAK VLADIMIR DR
INST OF PHYSICS
UNIVERSITY
P O BOX 304
41001 ZAGREB
YUGOSLAVIA

RUZICKOVA TOPOLOVA
BLAZENA DR
ASTRONOMICAL INST
CZECHOSLOVAK ACAD SC
251 65 ONDREJOV OBS
CZECHOSLOVAKIA

RYABOV YU A PROF DR
MATHEMATICS DEPT OF
MADI
125319 MOSCOW
U.S.S.R.

RYBANSKY MILAN
ASTRONOMICAL INST
SLOVAK ACADEMY OF SC
05960 TATRANSKA
LOMNICA
CZECHOSLOVAKIA

RYBICKI GEORGE B DR
HARVARD SMITHSONIAN
CTR F ASTROPHYSICS
60 GARDEN ST
CAMBRIDGE MA 02138
U.S.A.

RYBKA EUGENIUSZ PROF DR
UL SZOPENA 23
30047 KRAKOW
POLAND

RYBKA PRZEMYSLAW DR
INSTITUTE OF HISTORY
OF SCIENCE
NOWY SWIAT 72
00-330 WARSZAWA
POLAND

RYDBECK G H B DR
PL 6730
ALGARDA
S-434 00 KUNGSBACKA
SWEDEN

RYDBECK OLOF E H PROF
ONSALA SPACE
OBSERVATORY
S-430 34 ONSALA
SWEDEN

RYDGREN ALFRED ERIC JR DR
PHYSICS DEPT
RENSSELAER POLYTECHN
INSTITUTE
TROY NY 12180
U.S.A.

RYKHLOVA LIDIJA V DR
ASTRONOMICAL COUNCIL
USSR ACADEMY OF SC
PYATNITSKAYA 48
109017 MOSCOW
U.S.S.R.

RYLE MARTIN PROF SIR
CAVENDISH LABORATORY
MADINGLEY ROAD
CAMBRIDGE CB3 OHE
U.K.

RYLOV V S DR
SPECIAL ASTROPHYSI-
CAL OBSERVATORY
USSR ACADEMY OF SCI
357140 N ARKHYZ
U.S.S.R.

RYTER CHARLES E DR
CEN SACLAY
DPH EP A
BP 2
F-91190 GIF S YVETTE
FRANCE

RYZHKOV N F DR
SPECIAL ASTROPHYS-
OBS USSR ACADEMY SCI
LENINGRAD BRANCH
196140 LENINGRAD
U.S.S.R.

RZHIGA O N DR
INSTITUTE OF RADIO
AND ELECTRONICS
USSR ACADEMY OF SCI
103907 MOSCOW
U.S.S.R.

SAAR ENN DR
OBSERVATORY 5-19
202444 TORAVERE
ESTONIA
U.S.S.R.

SABANO YUTAKA DR
ASTRONOMICAL INST
TOHOKU UNIVERSITY
ARAMAKI
SENDAI 980
JAPAN

SABBADIN FRANCO DR
OSSERVATORIO ASTRO-
FISICO
I-36012 ASIAGO
ITALY

SACK NOAM DR
DEPT OF THEORETICAL
PHYSICS
HEBREW UNIVERSITY
JERUSALEM
ISRAEL

SACKMANN I JULIANA DR
KELLOGG RADIATION
LABORATORY
CALIF INST OF TECHN
PASADENA CA 91125
U.S.A.

SADAKANE KOZO DR
ASTRONOMICAL INST
OSAKA KYOIKU UNIV
TENNOJI-KU
OSAKA 543
JAPAN

SADEH D DR
DEPT OF PHYSICS AND
ASTRONOMY
TEL AVIV UNIVERSITY
TEL AVIV
ISRAEL

SADIK AZIZ R DR
ASTR & SPACE RES CTR
COUNCIL OF SC RES
P O BOX 255
JADIRIYAH BAGHDAD
REP. OF IRAQ

SADLER DONALD H DR
8 COLLINGTON RISE
BEXHILL ON SEA
SUSSEX TN39 3RT
U.K.

SADZAKOV SOFIJA DR
ASTRONOMICAL OBS
VOLGINA 7
11050 BEOGRAD
YUGOSLAVIA

SAFKO JOHN L
DEPT OF PHYSICS
& ASTRONOMY
UNIV OF S CAROLINA
COLUMBIA SC 29208
U.S.A.

SAFRONOV VICTOR S DR
INSTITUTE OF PHYSICS
OF THE EARTH
B GRUZINSKAYA 10
123242 MOSCOW
U.S.S.R.

SAGAN CARL DR
302 SPACE SCIENCES
CORNELL UNIVERSITY
ITHACA NY 14853
U.S.A.

SAGGION ANTONIO PROF
ISTITUTO DI FISICA
G GALILEI
VIA MARZOLO 8
I-35100 PADOVA
ITALY

SAGITOV M U DR
STERNBERG STATE AS-
TRONOMICAL INSTITUTE
117234 MOSCOW
U.S.S.R.

SAGNIER JEAN LOUIS MR
BUREAU D LONGITUDES
77 AVENUE DENFERT-
ROCHEREAU
F-75014 PARIS
FRANCE

SAHADE JORGE PROF
49 342
1900 LA PLATA
ARGENTINA

SAHAL-BRECHOT SYLVIE DR
OBSERVATOIRE DE
PARIS-MEUDON
F-92190 MEUDON
FRANCE

SAIO HIDEYUKI DR
DEPT OF ASTRONOMY
UNIVERSITY OF TEXAS
AT AUSTIN
AUSTIN TX 78712
U.S.A.

SAISSAC JOSEPH DR
OBSERVATOIRE DU
PIC-DU-MIDI-TOULOUSE
F-65200 BAGNERES-DE-
BIGORRE
FRANCE

SAITO KUNIJI PROF
TOKYO ASTRONOMICAL
OBSERVATORY
OSAWA MITAKA
TOKYO 181
JAPAN

SAITO MAMORU DR
DEPT OF ASTRONOMY
UNIVERSITY OF KYOTO
SAKYOKU
KYOTO 606
JAPAN

SAITO SUMISABURO DR
KASAN OBSERVATORY
YAMASH HIGASHIYAMA
KYOTO 607
JAPAN

SAKAI JUN-ICHI
HIGH ALTITUDE
OBSERVATORY
BOULDER CO 80307
U.S.A.

SAKASHITA SHIRO PROF
DEPT OF PHYSICS
HOKKAIDO UNIVERSITY
KITAJYUJO NISHI 8
SAPPORO HOKKAIDO 063
JAPAN

SAKHAROV VLADIMIR I DR
PULKOVO OBSERVATORY
196140 LENINGRAD
U.S.S.R.

SAKHIBULLIN NAIL A DR
DEPT OF ASTRONOMY
KAZAN STATE UNIV
420008 KAZAN
U.S.S.R.

SAKURAI KUNITOMO PROF
DEPT OF PHYSICS
KANAGAWA UNIVERSITY
KANAGAWAKU
YOKOHAMA 221
JAPAN

SAKURAI TAKASHI DR
DEPT OF ASTRONOMY
UNIV OF TOKYO
BUNKYO-KU
TOKYO 113
JAPAN

SAKURAI TAKEO T PROF
FACULTY OF
ENGINEERING
KYOTO UNIVERSITY
SAKYO-KU KYOTO 606
JAPAN

SALETIC DUSAN
ASTRONOMICAL
OBSERVATORY
VOLGINA 7
11050 BEOGRAD
YUGOSLAVIA

SALISBURY J W DR
AFCRL
BLDG 1121
L G HANSCOM FIELD
BEDFORD MA 01730
U.S.A.

SALOMONOVICH A E DR
PHYSICAL INSTITUTE
USSR ACADEMY OF SCI
117924 MOSCOW
U.S.S.R.

SALPETER EDWIN E PROF
NEWMAN LAB OF NS
CORNELL UNIVERSITY
ITHACA NY 14853
U.S.A.

SALUKVADZE G N DR
ABASTUMANY ASTROPHY-
SICAL OBSERVATORY
383762 ABASTUMANY
U.S.S.R.

SALVATI MARCO
IST DI ASTROFISICA
SPAZIALE
CASELLA POSTALE 67
I-00044 FRASCATI
ITALY

SAMAIN D DR
LPSP
B P NO 10
F-91370 VERRIERES LE
BUISSON
FRANCE

SAMPSON DOUGLAS H PROF
DEPT OF ASTRONOMY
PENN STATE UNIV
525 DAVEY LAB
UNIV PARK PA 16802
U.S.A.

SAMUS NIKOLAI N DR
ASTRONOMICAL COUNCIL
USSR AC OF SCIENCES
PYATNITSKAYA 48
109017 MOSCOW
U.S.S.R.

SANAMIAN V A DR
BYURAKAN ASTROPHYSI-
CAL OBSERVATORY
378433 ARMENIA
U.S.S.R.

SANCHEZ FRANCISCO PROF
INSTITUTO ASTROFISI-
CA CANARIAS
UNIVERSIDAD LAGUNA
LA LAGUNA TENERIFE
SPAIN

SANCHEZ MAGRO C DR
IAC INSTITUTO DE
ASTROFISICA CANARIAS
UNIV DE LA LAGUNA
TENERIFE
SPAIN

SANCHEZ MANUEL
INSTITUTO Y OBSERVA-
TORIO DE MARINA
SAN FERNANDO CADIZ
SPAIN

SANCISI RENZO DR
KAPTEYN LABORATORIUM
POSTBUS 800
9700 RV GRONINGEN
THE NETHERLANDS

SANDAGE ALLAN
MT WILSON & LAS CAM-
PANAS OBSERVATORIES
813 SN BARBARA ST
PASADENA CA 91101
U.S.A.

SANDAKOVA E V DR
KIEV STATE UNIVERSI-
TY ASTRONOMICAL
OBSERVATORY
252053 KIEV
U.S.S.R.

SANDELL GORAN HANS L DR
STOCKHOLM OBS
S-13300 SALTSJOBADEN
SWEDEN

SANDERS R DR
KAPTEYN LABORATORY
P O BOX 800
9700 RV GRONINGEN
THE NETHERLANDS

SANDERS W L PROF
NEW MEXICO STATE
UNIVERSITY
BOX 4500
LAS CRUCES NM 88003
U.S.A.

SANDERS WILTON TURNER III
DEPT OF PHYSICS
UNIV OF WISCONSIN
MADISON WI 53706
U.S.A.

SANDFORD MAXWELL T II
LOS ALAMOS SCIENTI-
FIC LABORATORY
LOS ALAMOS NM 87545
U.S.A.

SANDMANN WILLIAM HENRY
PHYSICS DEPT
HARVEY MUDD COLLEGE
CLAREMONT CA 91711
U.S.A.

SANDQVIST AAGE DR
STOCKHOLM
OBSERVATORY
S-133 00
SALTSJOEBADEN
SWEDEN

SANDULEAK N DR
WARNER AND SWASEY
OBSERVATORY
CASE WESTERN RES UN
CLEVELAND OH 44106
U.S.A.

SANFORD P W MR
MULLARD SPACE SC LAB
HOLMBURY ST MARY
DORKING
SURREY RH5 6NS
U.K.

SANG-GAK LEE
DEPT OF ASTRONOMY
CLG OF NATURAL SC
SEOUL NATIONAL UNIV
SEOUL 151-00
REPUBLIC OF KOREA

SANTIN PAOLO DR
VIA DUDOVICH 3
I-34100 TRIESTE
ITALY

SANWAL N B DR
DEPT OF ASTRONOMY
OSMANIA UNIVERSITY
HYDERABAD 500 007
INDIA

SANYAL ASHIT DR
DEPT OF ASTRONOMY
PENNA STATE UNIV
UNIV PK PA 16802
U.S.A.

SAPAR ARVED DR
IAPHA ESTONIAN AC OF
SCIENCE TORAVERE
202444 TARTU
U.S.S.R.

SARAZIN CRAIG L DR
DEPT OF ASTRONOMY
UNIV OF VIRGINIA
POB 3818 UNIV STA
CHARLOTTESVL VA22903
U.S.A.

SAREYAN J P DR
OBS DE NICE
B P 252
F-06007 NICE CEDEX
FRANCE

SARGENT WALLACE L W DR
HALE OBSERVATORIES
CALTECH
1201 E CALIFORNIA ST
PASADENA CA 91125
U.S.A.

SARMA M B K PROF
DEPT OF ASTRONOMY
OSMANIA UNIVERSITY
HYDERABAD 500 007
INDIA

SARRIS ELEFTHERIOS PH D
NAT OBS OF ATHENS
ASTRONOMICAL INST
ATHENS 306
GREECE

SARRIS EMMANUEL T PH D
DEMOCRITUS UNIV OF
THRACE
XANTHI
GREECE

SARTORI LEO PROF
BEHLEN LABORATORY OF
PHYSICS
UNIVERSITY NEBRASKA
LINCOLN NB 68588
U.S.A.

SASAO TETSUO DR
INTERNATIONAL
LATITUDE OBSERVATORY
OF MIZUSAWA
MIZUSAWA IWATE 023
JAPAN

SASLAW WILLIAM C PROF
ASTRONOMY DEPT
UNIVERSITY VIRGINIA
BOX 3818 UNIVSTATION
CHARLOTTESV VA 22903
U.S.A.

SASTRY CH V
INDIAN INSTITUTE OF
ASTROPHYSICS
BANGALORE 560 034
INDIA

SASTRY SHANKARA K
DEPT OF ASTRONOMY
OSMANIA UNIVERSITY
HYDERABAD 500 007
INDIA

SATO FUMIO DR
HYOGO UNIVERSITY OF
TEACHER EDUCATION
YASHIRO
HYOGO 673-14
JAPAN

SATO FUMITAKA PROF
INSTITUTE OF FUNDA-
MENTAL PHYSICS
UNIVERSITY OF KYOTO
SAKYOKU KYOTO 606
JAPAN

SATO KATSUHIKO DR
RESEARCH INST FOR
FUNDAMENTAL PHYSICS
KYOTO UNIVERSITY
SAKYOKU KYOTO 606
JAPAN

SATO KOICHI DR
INTERNATIONAL
LATITUDE OBSERVATORY
OF MIZUSAWA
MIZUSAWA IWATE 023
JAPAN

SATO NAONOBU DR
AKITA UNIVERSITY
AKITA 010
JAPAN

SATO SHUJI DR
DEPT OF PHYSICS
UNIVERSITY OF KYOTO
KITASHIRAKAWA
SAKYOKU KYOTO 606
JAPAN

SATO YUZO DR
OSAWA 4-8-19
MITAKA
TOKYO 181
JAPAN

SAUVAL JACQUES DR
OBS ROYAL DE BELG.
3 AVE CIRCULAIRE
B-1180 BRUXELLES
BELGIUM

SAVAGE ANN DR
ROYAL OBSERVATORY
BLACKFORD HILL
EDINBURGH EH9 3HJ
U.K.

SAVAGE BLAIR D DR
DEPT OF ASTRONOMY
UNIVERSITY WISCONSIN
475 N.CHARTER STR
MADISON WI 53706
U.S.A.

SAVEDOFF MALCOLM P PROF
PHYS AND ASTRON DEPT
UNIVERSITY ROCHESTER
BAUSCH AND LOMB BLDG
ROCHESTER NY 14627
U.S.A.

SAVONIJE G J DR
ASTRONOMISCH INST
UNIV OF AMSTERDAM
ROETERSSTRAAT 15
1018 WB AMSTERDAM
THE NETHERLANDS

SAWYER CONSTANCE S DR
HIGH ALTITUDE OBS
NCAR
P O BOX 3000
BOULDER CO 80301
U.S.A.

SAWYER HOGG HELEN B DR
DAVID DUNLAP OBS
P O BOX 360
RICHMOND HILL
ONT L4C 4Y6
CANADA

SAXENA P P DR
UTTAR PRADESH
STATE OBSERVATORY
MANORA PEAK
NAINI TAL
INDIA

SCALISE JR EUGENIO DR
INPE/CRAAM
AV ASTRONAUTAS 1758
C P 515
12200 S JOSE CAMPOS
BRAZIL

SCALO JOHNMICHAEL DR
DEPT OF ASTRONOMY
UNIV OF TEXAS
AUSTIN TX 78712
U.S.A.

SCALTRITI FRANCO DR
OSSERVATORIO ASTRO-
NOMICO
PINO TORINESE
TORINO
ITALY

SCARFE COLIN D DR
DEPT OF PHYSICS
UNIV OF VICTORIA
P O BOX 1700
VICTORIA BC V8W 2Y2
CANADA

SCARGLE JEFFREY D DR
NASA-AMES RESEARCH
CENTER MS245-3
MOFFETT FIELD
CA 94035
U.S.A.

SCARROTT STANLEY M DR
PHYSICS DEPT
UNIV OF DURHAM
SOUTH ROAD
DURHAM DH1 3LE
U.K.

SCHADEE AERT DR
STERREWACHT
SONNENBORGH
ZONNENBURG 2
3512 NL UTRECHT
THE NETHERLANDS

SCHAIFERS KARL DR
LANDESSTERNWARTE
KOENIGSTUHL
D-6900 HEIDELBERG 1
GERMANY, F.R.

SCHALEN CARL PROF
LUND OBSERVATORY
SVANEGATAN 9
S-222 24 LUND
SWEDEN

SCHANDA ERWIN PROF
INST OF APPLIED
PHYSICS
SIDLERSTR 5
CH-3012 BERN
SWITZERLAND

SCHARLEMANN ERNST T DR
DEPT OF ASTRONOMY
COLUMBIA UNIV
BOX 98 PUPIN
NEW YORK NY 10027
U.S.A.

SCHARMER GOERAN BJARNE
STOCKHOLM OBS
S-133 00
SALTSJOEBADEN
SWEDEN

SCHATTEN K H DR
PHYSICS DEPT
VICTORIA UNIVERSITY
PRIVATE BAG
WELLINGTON
NEW ZEALAND

SCHATZMAN EVRY PROF
OBSERVATOIRE DE NICE
BP 252
F-06007 NICE CEDEX
FRANCE

SCHEEPMAKER ANTON DR
COSM RAY WORK GROUP
HUYGENS LABORATORY
WASSENAARSEWEG 78
2300 RA LEIDEN
THE NETHERLANDS

SCHEFFLER HELMUT PROF
LANDESSTERNWARTE
KOENIGSTUHL
D-6900 HEIDELBERG 1
GERMANY, F.R.

SCHEID PAOLO LEITE DR
OBS NACIONAL
R GENERAL BRUCE 586
20000 RIO DE JANEIRO
BRAZIL

SCHEIDECKER JEAN-PAUL DR
OBSERVATOIRE DE NICE
BP 252
F-06007 NICE CEDEX
FRANCE

SCHERB FRANK PROF
PHYSICS DEPT
UNIVERSITY WISCONSIN
MADISON WI 53706
U.S.A.

SCHERRER PHILIP H DR
INST OF PLASMA RES
STANFORD UNIV
VIA CRESPI
STANFORD CA 94305
U.S.A.

SCHEUER PETER A G DR
CAVENDISH LABORATORY
MADINGLEY RD
CAMBRIDGE CB3 0HE
U.K.

SCHILD RUDOLPH E DR
CTR F ASTROPHYSICS
60 GARDEN STREET
CAMBRIDGE MA 02138
U.S.A.

SCHILIZZI RICHARD T DR
RADIOSTERRENWACHT
POSTBUS 2
7990 AA DWINGELOO
THE NETHERLANDS

SCHILLER KARL PROF DR
PIRSCH-WEG 6
D-6072 DREIEICH-
BUCHSCHLAG
GERMANY, F.R.

SCHINDLER KARL PROF DR
LEHRSTUHL FUER
THEORETISCHE PHYSIK
RUHR-UNIVERSITAET
D-4630 BOCHUM
GERMANY, F.R.

SCHKODROV V G DR
DEPT OF ASTRONOMY
7TH NOVEMBER ST 1
SOFIA
BULGARIA

SCHLESINGER BARRY M DR
SYSTEMS AND APPLIED
SCIENCES CORPORATION
6811 KENILWORTH AVE
RIVERDALE MD 20840
U.S.A.

SCHLICKEISER REINHARD DR
MPI FUER
RADIOASTRONOMIE
AUF DEM HUEGEL 69
D-5300 BONN
GERMANY, F.R.

SCHLOSSER WOLFHARD PROF
ASTRONOMISCHES
INSTITUT
POSTFACH 102148
D-4630 BOCHUM
GERMANY, F.R.

SCHLUETER A PROF DR
MPI F PLASMAPHYSIK
D-8046 GARCHING
GERMANY, F.R.

SCHLUETER DIETER PROF
INST THEO PHY STERNW
NEUE UNIV PHYSIK ZTM
OLSHAUSENST GB N 61C
D-2300 KIEL 1
GERMANY, F.R.

SCHMADEL LUTZ D DR
ASTRONOMISCHES
RECHEN-INSTITUT
MOENCHHOFSTR 12-14
D-6900 HEIDELBERG 1
GERMANY, F.R.

SCHMAHL EDWARD J DR
ASTRONOMY PROGRAM
UNIV OF MARYLAND
COLLEGE PK MD 20742
U.S.A.

SCHMAHL GUENTER PROF
UNIVERSITAETS-
STERNWARTE
GEISMARLANDSTR 11
D-3400 GOETTINGEN
GERMANY, F.R.

SCHMALBERGER DONALD C DR
DUDLEY OBSERVATORY
1202 TROY-
SCHENECTADY RD
LATHAM NY 12110
U.S.A.

SCHMEIDLER F PROF DR
INSTITUT F ASTRONO-
MIE UND ASTROPHYSIK
SCHEINERSTR 1
D-8000 MUENCHEN 80
GERMANY, F.R.

SCHMID-BURGK J DR
MPI RADIOASTRONOMIE
AUF DEM HUEGEL 69
D-5300 BONN 1
GERMANY, F.R.

SCHMIDT EDWARD G
DEPT OF PHYSICS
UNIVERSITY OF
NEBRASKA
LINCOLN NB 68588
U.S.A.

SCHMIDT H U DR
MPI FUER PHYSIK UND
ASTROPHYSIK
FOEHRINGER RING 6
D-8000 MUENCHEN 40
GERMANY, F.R.

SCHMIDT HANS PROF
UNIVERSITAETS-
STERNWARTE BONN
AUF DEM HUEGEL 71
D-5300 BONN 1
GERMANY, F.R.

SCHMIDT K H DR
ZNTRINST ASTROPHYSIK
STERNW BABELSBERG
ROSA LUXEMBURGST 17A
DDR-1502 POTSDAM
G.D.R.

SCHMIDT MAARTEN PROF
CALTECH
ASTRONOMY 105-24
PASADENA CA 91125
U.S.A.

SCHMIDT THOMAS DR
MPI FUER ASTRONOMIE
KOENIGSTUHL
D-6900 HEIDELBERG
GERMANY, F.R.

SCHMIDT-KALER TH PROF
STEINHUEGEL 105
D-5810 WITTEN
GERMANY, F.R.

SCHMIEDER B DR
OBS DE MEUDON
F-92190 MEUDON
FRANCE

SCHMITTER EDWARD F DR
DEPT OF PHYSICS
UNIVERSITY OF LAGOS
AKOKA LAGOS
NIGERIA

SCHNELL ANNELISE DR
INST FUER ASTRONOMIE
UNIVERSITAT WIEN
TUERKENSCHANZST 17
A-1180 WIEN
AUSTRIA

SCHNOPPER HERBERT W DR
CTR FOR ASTROPHYSICS
60 GARDEN ST
CAMBRIDGE MA 02138
U.S.A.

SCHNUR G F O
ESO ADMINISTRATION
ALONSO CORDOBA 3107
VITACURA CAS 11 P
SANTIAGO DE CHILE
CHILE

SCHOBER HANS J DR
ASTRONOMISCHES INST
DER UNIVERSITAET
UNIVERSITAETSPLATZ 5
A-8010 GRAZ
AUSTRIA

SCHOEFFEL EBERHARD F DR
MERIANESTRASSE 42
D-86 BAMBERG
GERMANY, F.R.

SCHOEMBS ROLF DR
INST FUER ASTRONOMIE
ASTROPHYSIK
SCHEINERSTR 1
D-8000 MUENCHEN 81
GERMANY, F.R.

SCHOENBERNER DETLEF PH D
INST F THEOR PHYSIK
UND UNIV STERNWARTE
OLSHAUSENSTRASSE
D-2300 KIEL
GERMANY, F.R.

SCHOENEICH W DR
ZNTRINST ASTROPHYSIK
ASTROP OBSERVATORIUM
TELEGRAFENBERG
DDR-15 POTSDAM
G.D.R.

SCHOENFELDER VOLKER DR
MPI F EXTRATERRE-
STRISCHE PHYSIK
D-8046 GARCHING
GERMANY, F.R.

SCHOLL HANS DR
ASTRONOMISCHES
RECHEN-INSTITUT
MOENCHHOFSTR 12-14
D-6900 HEIDELBERG
GERMANY, F.R.

SCHOLZ GERHARD DR
ZENTR INST ASTROPHYS
DER ADW DER DDR
ROSA-LUXEMBURGST 17A
DDR-1502 POTSDAM
G.D.R.

SCHOLZ M PROF
INST F THEORETISCHE
ASTROPHYS DER UNIV
NEUENHEIMER FELD 294
D-6900 HEIDELBERG
GERMANY, F.R.

SCHOMBERT JOHN L MR
21920 DAVIS MILL RD
GERMANTOWN MD 20767
U.S.A.

SCHOOLMAN STEPHEN A DR
LOCKHEED PA RES LAB
3251 HANOVER ST
PALO ALTO CA 94304
U.S.A.

SCHRAMM DAVID N PROF
UNIVERSITY CHICAGO
EFI-LASR-205
933 E 55TH ST
CHICAGO IL 60637
U.S.A.

SCHRIJVER J DR
LABORATORIUM VOOR
RUIMTEONDERZOEK
BENELUXLAAN 21
3527 HS UTRECHT
THE NETHERLANDS

SCHROEDER DANIEL J PROF
DEPT PHYS AND ASTRON
BELOIT COLLEGE
BELOIT WI 53511
U.S.A.

SCHROEDER ROLF DR
HAMBURGER STERNWARTE
GOJENBERGSWEG 112
D-2050 HAMBURG 80
GERMANY, F.R.

SCHROETER EGON H PROF
KIEPENHEUER INST
FUER SONNENPHYSIK
SCHOENECKSTRASSE 6
D-7800 FREIBURG I BR
GERMANY, F.R.

SCHROLL ALFRED DR
SONNENOBSERVATORIUM
KANZELHOEHE
A-9520 SATTENDORF
AUSTRIA

SCHRUEFFER EBERHARD DR
INST F ASTROPHYSIK
UNIVERSITAET BONN
AUF DEM HUEGEL 71
D-5300 BONN 1
GERMANY, F.R.

SCHRUTKA RECHTENSTAMM
GUNTRAM PROF DR
WILLERGASSE 27/4/7
A-1238 WIEN
AUSTRIA

SCHUBART JOACHIM DR
ASTRONOMISCHES
RECHEN INSTITUT
MOENCHHOFSTR 12-14
D-6900 HEIDELBERG
GERMANY, F.R.

SCHUECKING E L DR
DEPT OF PHYSICS
NEW YORK UNIVERSITY
NEW YORK NY 10012
U.S.A.

SCHUERER MAX PROF
ASTRONOMISCHES INST
UNIVERSITAET BERN
SIDLERSTR 5
CH-3012 BERN
SWITZERLAND

SCHUESSLER MANFRED DR
UNIVERSITAETSSTERN-
WARTE
GEISMARLANDSTR 11
D-3400 GOETTINGEN
GERMANY, F.R.

SCHULER WALTER DR
STERNWARTE DER
KANTONSSCHULE
CH-4500 SOLOTHURN
SWITZERLAND

SCHULTE D H DR
ITEK CORPORATION
10 MAGUIRE ROAD
LEXINGTON MA 02173
U.S.A.

SCHULZ G V DR
MPI FUER
RADIOASTRONOMIE
AUF DEM HUEGEL 69
D-5300 BONN
GERMANY, F.R.

SCHULZ HARTMUT DR
MPI FUER ASTRONOMIE
KOENIGSTUHL
D-6900 HEIDELBERG 1
GERMANY, F.R.

SCHUMANN JOERG DIETER DR
OBS HOHER LIST
UNIV STERNWARTE BONN
D-5568 DAUN
GERMANY, F.R.

SCHUSTER HANS-EMIL PH D
EUROPEAN SOUTHERN OB
CAS 16317 CORREO 9
ALONSO CORDOVA 3107
SANTIAGO
CHILE

SCHUSTER WILLIAM JOHN DR
INSTITUTO DE
ASTRONOMIA UNAM
APDO POSTAL 877
ENSENADA BC
MEXICO

SCHUTZ BERNARD F DR
APPLIED MATHEMATICS
AND ASTRONOMY
UNIVERSITY COLLEGE
CARDIFF CF1 1XL
U.K.

SCHUTZ BOB EWALD
DEPT OF AEROSPACE
ENG & ENG MECHANICS
UNIVERSITY OF TEXAS
AUSTIN TX 78712
U.S.A.

SCHWAN HEINER DR
ASTRONOMISCHES
RECHEN INSTITUT
MOENCHHOFSTR 12-14
D-6900 HEIDELBERG
GERMANY, F.R.

SCHWARTZ DANIEL A DR
CENTER FOR ASTROPHYS
60 GARDEN ST
CAMBRIDGE MA 02138
U.S.A.

SCHWARTZ PHILIP R DR
NAVAL RESEARCH LAB
CODE 7138
WASHINGTON DC 20375
U.S.A.

SCHWARTZ RICHARD D
ASST PROF DEPT PHYS
UNIVERSITY MISSOURI
8001 NAT BRIDGE RD
ST LOUIS MO 63121
U.S.A.

SCHWARTZ ROLF PH D
MPI FUER
RADIOASTRONOMIE
AUF DEM HUEGEL 69
D-5300 BONN 1
GERMANY, F.R.

SCHWARTZ STEVEN JAY
DEPT OF APPLIED MATH
QUEEN MARY COLLEGE
MILE END ROAD
LONDON E1 4NS
U.K.

SCHWARZ ULRICH J DR
KAPTEYN LABORATORIUM
POSTBUS 800
9700 AV GRONINGEN
THE NETHERLANDS

SCHWARZSCHILD MARTIN PROF
PRINCETON UNIVERSITY
OBSERVATORY
PEYTON HALL
PRINCETON NJ 08540
U.S.A.

SCHWEIZER FRANCOIS DR
DEPT OF TERRESTRIAL
MAGNETISM
5241 BROAD BR RD NW
WASHINGTON DC 20015
U.S.A.

SCHWERDTFEGER H DR
ASTRONOMISCHES
RECHEN-INSTITUT
MOENCHHOFSTR 12-14
D-6900 HEIDELBERG 1
GERMANY, F.R.

SCIAMA DENNIS W DR
DEPT OF ASTROPHYSICS
SOUTH PARKS ROAD
OXFORD OX1 3RQ
U.K.

SCONZO PASQUALE DR
29 OLD MYSTIC ST
ARLINGTON MA 02174
U.S.A.

SCOTT ELIZABETH L PROF
STATISTICS LAB
UNIVERSITY OF CALIF
367 EVANS HALL
BERKELEY CA 94720
U.S.A.

SCOTT EUGENE HOWARD
NASA-AMES RESEARCH
CTR MAIL STOP 245-3
MOFFETT FIELD
CA 94035
U.S.A.

SCOTT JOHN S DR
ASTRONOMY PROGRAM
UNIV OF MARYLAND
COLLEGE PK MD 20742
U.S.A.

SCOTT PAUL F DR
CAVENDISH LABORATORY
MADINGLEY ROAD
CAMBRIDGE CB3 OHE
U.K.

SCOVILLE NICHOLAS Z
ASTRO-PHYSICS DEPT
UNIVERSITY OF MASS
GRADUATE RES CENTER
AMHERST MA 01003
U.S.A.

SCUFLAIRE RICHARD DR
INST D'ASTROPHYSIQUE
UNIVERSITE DE LIEGE
5 AVENUE DE COINTE
B-4200 COINTE-OUGREE
BELGIUM

SEAQUIST ERNEST R PROF
DEPT OF ASTRONOMY
UNIVERSITY
TORONTO ONT M5S 1A7
CANADA

SEARLE L DR
HALE OBSERVATORIES
813 SANTA BARBARA ST
PASADENA CA 91101
U.S.A.

SEARS RICHARD LANGLEY DR
DEPT OF ASTRONOMY
UNIV OF MICHIGAN
PHYSICS-ASTRO BLDG
ANN ARBOR MI 48109
U.S.A.

SEATON MICHAEL J PROF
DEPT PHYS AND ASTRON
UNIV COLLEGE LONDON
GOWER ST
LONDON WC1E 6BT
U.K.

SEDDON HARRY MR
ROYAL OBSERVATORY
EDINBURGH EH9 3HJ
U.K.

SEDLMAYER ERWIN DR
INST FUER ASTRON UND
ASTROPHYK DER TU
ERNST-REUTER-PLATZ 7
D-1000 BERLIN 10
GERMANY, F.R.

SEDMAK GIORGIO DR
OSSERVATORIO ASTRONO
MICO DI TRIESTE
VIA TIEPOLO 11
I-34131 TRIESTE
ITALY

SEEGER CHARLES LOUIS III
SAN FRANCISCO STATE
UNIVERSITY
473 JAMES ROAD
PALO ALTO CA 94306
U.S.A.

SEEGER PHILIP A DR
LOS ALAMOS SCIENTI-
FIC LABORATORY
LOS ALAMOS NM 87545
U.S.A.

SEGAL IRVING E DR
MIT
2-224
CAMBRIDGE MA 02139
U.S.A.

SEGALUVITZ ALEXANDER DR
11 HABANIM ST
KEFAR SAVA
ISRAEL

SEGGEWISS WILHELM DR
OBSERVATORIUM
HOHER LIST
D-5568 DAUN EIFEL
GERMANY, F.R.

SEHNAL LADISLAV DR
ASTRONOMICAL INST
CSAV
25165 ONDREJOV
CZECHOSLOVAKIA

SEIDELMANN P KENNETH DR
US NAVAL OBSERVATORY
WASHINGTON DC 20390
U.S.A.

SEIDOV ZAKIR F DR
SHEMAKHA ASTRO-
PHYSICAL OBSERVATORY
373243 SHEMAKHA AZER
U.S.S.R.

SEIELSTAD GEORGE A
OWENS VALLEY RADIO
OBSERVATORY
PO BOX 387
BIG PINE CA 93513
U.S.A.

SEIRADAKIS J H DR
MPI FUER RADIOASTRON
AUF DEM HUEGEL 69
D-5300 BONN 1
GERMANY, F.R.

SEITTER WALTRAUT C PROF
ASTRONOMISHES INST
STEINFURTER STR 107
D-4400 MUENSTER/W
GERMANY, F.R.

SEKANINA ZDENEK DR
EARTH SPACE SCS DIV
JPL
4800 OAK GROVE DRIVE
PASADENA CA 91103
U.S.A.

SEKI MUENZO DR
ASTRONOMICAL INST
TOHOKU UNIVERSITY
KAWAUCHI
SENDAI 980
JAPAN

SEKIGUCHI NAOSUKE DR
TOKYO ASTRONOMICAL
OBSERVATORY
OSAWA MITAKA
TOKYO 181
JAPAN

SELLWOOD JEREMY ARTHUR
INSTITUTE OF
ASTRONOMY
MADINGLEY ROAD
CAMBRIDGE CB3 OHA
U.K.

SELVELLI PIERLUIGI DR
OSS ASTRONOMICO
VIA TIEPOLO 11
I-34131 TRIESTE
ITALY

SEMEL MEIR DR
OBSERVATOIRE
DE MEUDON
F-92190 MEUDON
FRANCE

SEMENIUK IRENA DR
ASTRONOMICAL OBS
UL UJAZDOWSKIE 4
00478 WARZAWA
POLAND

SEN S N DR
INDIAN ASSN FOR THE
CULTIVATION OF SC
JADAVPUR
INDIA

SENGBUSCH KURT V DR
MPI FUER PHYSIK UND
ASTROPHYSIK
FOEHRINGER RING 40
D-8000 MUENCHEN 40
GERMANY, F.R.

SERRANO ALFONSO DR
INST DE ASTRONOMIA
UNAM
APDO POSTAL 70-264
MEXICO D F
MEXICO

SERSIC J L DR
OBSERVATORIO
ASTRONOMICO
5000 CORDOBA
ARGENTINA

SERVAJEAN ROGER DR
OBSERVATOIRE
DE MEUDON
F-92190 MEUDON
FRANCE

SERVAN BERNARD
OBS DE PARIS
61 AV DE L'OBS
F-75014 PARIS
FRANCE

SESSIN WAGNER DR
INST TECNOLOGICA DE
AERONAUTICA
DEPTO DE ASTRONOMIA
12200 S JOSE CAMPOS
BRAZIL

SETTI GIANCARLO PROF
LABORATORIO RADIOAST
ISTITUTO DI FISICA
VIA IRNERIO
I-40126 BOLOGNA
ITALY

SEVARLIC BRANISLAV M PROF
DEPT OF ASTRONOMY
UNIVERSITY BEOGRAD
VOLGINA 7
11050 BEOGRAD
YUGOSLAVIA

SEVERNYJ A B PROF DR
CRIMEAN ASTROPHYSI-
CAL OBSERVATORY USSR
ACAD OF SCI NAUCHNIY
334413 CRIMEA
U.S.S.R.

SEVILLA MIGUEL J DR
ALMANSA 76
MADRID 20
SPAIN

SEYMOUR P A H
57 HERMITAGE ROAD
PLYMOUTH DEVON
U.K.

SEZER GENGIZ DR
EGE UNIVERSITY
FACULTY OF SCIENCE
BORNOVA-IZMIR
TURKEY

SHAFFER DAVID B DR
CODE 974
NASA/GFSC
GREENBELT MD 20771
U.S.A.

SHAH GHANSHYAM A DR
INDIAN INST ASTROPHY
SARJAPUR RD
KORAMANGALA
BANGALORE 560 034
INDIA

SHAHAM JACOB PROF
RACAH INSTITUTE OF
PHYSICS
HEBREW UNIVERSITY
JERUSALEM 91000
ISRAEL

SHAHBAZIAN ROMELIA K DR
BYURAKAN ASTRO-
PHYSICAL OBSERVATORY
378433 BYURAKAN ARM
U.S.S.R.

SHAKESHAFT JOHN R DR
CAVENDISH LABORATORY
MADINGLEY ROAD
CAMBRIDGE CB3 OHE
U.K.

SHAKHBAZYAN YURIJ L DR
PULKOVO OBSERVATORY
196140 LENINGRAD
U.S.S.R.

SHAKHOVSKOJ N M DR
CRIMEAN ASTROPHYSI-
CAL OBSERVATORY USSR
ACAD OF SCI NAUCHNIY
334413 CRIMEA
U.S.S.R.

SHAKURA N I DR
STERNBERG STATE AS-
TRONOMICAL INSTITUTE
117234 MOSCOW
U.S.S.R.

SHALLIS MICHAEL J DR
DEPT OF ASTROPHYSICS
SOUTH PARKS ROAD
OXFORD OX1 3RQ
U.K.

SHALTOUT MESALAM A M DR
HELWAN OBSERVATORY
HELWAN
ARAB REP. OF EGYPT

SHANE C D ASTRONOMER
BOX 582
SANTA CRUZ CA 95061
U.S.A.

SHANE W W DR
STERRENKUNDIG INST
KATH UNIV
TOERNOOIVELD
6525 ED NIJMEGEN
THE NETHERLANDS

SHAO CHING-YUAN
HARVARD COLLEGE
OBSERVATORY
60 GARDEN STR
CAMBRIDGE MA 02138
U.S.A.

SHAPIRO IRWIN I PROF
54-620 MIT
CAMBRIDGE MA 02139
U.S.A.

SHAPIRO MAURICE M
LAB FOR COSMIC RAY
PHYSICS CODE 7020
NAVAL RESEARCH LAB
WASHINGTON DC 20375
U.S.A.

SHAPIRO STUART L
ASST PROF
CRSR
CORNELL UNIVERSITY
ITHACA NY 14853
U.S.A.

SHAPLEY ALAN H
NOAA
BOULDER CO 80302
U.S.A.

SHARA MICHAEL DR
DEPT OF PHYSICS
ARIZONA STATE UNIV
TEMPE AZ 85281
U.S.A.

SHARAF SH G DR
INSTITUTE OF THEORE-
TICAL ASTRONOMY
192187 LENINGRAD
U.S.S.R.

SHAROV A S DR
STERNBERG STATE AS-
TRONOMICAL INSTITUTE
117234 MOSCOW
U.S.S.R.

SHARPLESS STEWART PROF
DEPT OF PHYSICS AND
ASTRONOMY
UNIVERSITY ROCHESTER
ROCHESTER NY 14627
U.S.A.

SHAVER PETER A DR
ESO
K-SCHWARZSCHILD-ST 2
D-8046 GARCHING
GERMANY, F.R.

SHAVIV GIORA DR
DEPT OF PHYSICS
TEL-AVIV UNIVERSITY
RAMAT-AVIV
TEL-AVIV
ISRAEL

SHAW J SCOTT DR
DEPT OF PHYSICS AND
ASTRONOMY
UNIVERSITY GEORGIA
ATHENS GA 30602
U.S.A.

SHAW JOHN H PROF
174 W 18TH AVE
COLUMBUS OH 43210
U.S.A.

SHAW R WILLIAM PROF
CORNELL UNIVERSITY
105 HALCYON HILL
ITHACA NY 14850
U.S.A.

SHAWHAN STANLEY D DR
DEPT PHYSICS ASTRON
UNIV OF IOWA
IOWA CITY IA 52242
U.S.A.

SHAWL STEPHEN J DR
DEPT OF PHYSICS AND
ASTRONOMY
UNIVERSITY OF KANSAS
LAWRENCE KS 66045
U.S.A.

SHCHEGLOV P V DR
STERNBERG STATE AS-
TRONOMICAL INSTITUTE
117234 MOSCOW
U.S.S.R.

SHCHEGLOV V P PROF DR
ASTRONOMICAL INSTITU
TE UZBEKIAN ACADEMY
OF SCIENCES
700000 TASHKENT
U.S.S.R.

SHCHEGOLEV DIMITRIJ E DR
PULKOVO OBSERVATORY
196140 LENINGRAD
U.S.S.R.

SHCHERBINA-SAMOJLOVA I S
INSTITUTE OF SCIENCE
AND TECHNICS INFORMA
TION
125219 MOSCOW
U.S.S.R.

SHEA MARGARET A DR
AIR FORCE GEOPHYSICS
LAB PHC
HANSCOM AFB MA 01732
U.S.A.

SHEELEY NEIL R DR
CODE 7140NS
NAVAL RESEARCH LAB
WASHINGTON DC 20375
U.S.A.

SHEFFER ERGENIJ K DR
STERNBERG STATE
ASTRONOMICAL INST
117234 MOSCOW
U.S.S.R.

SHELUS PETER J DR
ASTRONOMY DEPT
UNIVERSITY OF TEXAS
R L M 15-316
AUSTIN TX 78712
U.S.A.

SHEN BENJAMIN S P PROF
DEPT ASTRONOMY E1
UNIVERSITY OF
PENNSYLVANIA
PHILADELPHIA PA19104
U.S.A.

SHEN CHUN-SHAN PRESIDENT
ASTRO SOCIETY CHINA
DEAN SCHOOL OF SC
NAT TSING-HUA UNIV
HSIN-CHU 300
TAIWAN

SHEN LIANG-ZHAO
BEIJING OBSERVATORY
BEIJING
PEOPLE'S REPUBLIC CHINA

SHEN LONG-XIANG
BEIJING OBSERVATORY
ACADEMIA SINICA
BEIJING
PEOPLE'S REPUBLIC CHINA

SHEN ZEE DR
DEPT OF MATHEMATICS
NATIONAL TAIWAN UNIV
TAIPEI
TAIWAN

SHER DAVID DR
2837 MINTO DR APT 2
CINCINNATI OH 45208
U.S.A.

SHERIDAN K V DR
C S I R O
DIV OF RADIOPHYSICS
P O BOX 76
EPPING NSW 2121
AUSTRALIA

SHERWOOD WILLIAM A DR
MPI FUER RADIO
ASTRONOMIE
AUF DEM HUEGEL 69
D-5300 BONN
GERMANY, F.R.

SHEVCHENKO V V DR
STERNBERG STATE AS-
TRONOMICAL INSTITUTE
117234 MOSCOW
U.S.S.R.

SHI GUANG-CHEN
PURPLE MOUNTAIN
OBSERVATORY
NANJING
PEOPLE'S REPUBLIC CHINA

SHIBAHASHI HIROMOTO DR
DEPT OF ASTRONOMY
UNIV OF TOKYO
BUNKYO-KU
TOKYO 113
JAPAN

SHIBATA YUKIO DR
RESEARCH INST SCIEN-
TIFIC MEASUREMENTS
TOHOKU UNIVERSITY
SENDAI 980
JAPAN

SHIELDS GREGORY A DR
DEPT OF ASTRONOMY
RLM 15 212
UNIV OF TEXAS
AUSTIN TX 78712
U.S.A.

SHIM WOON-TAIK PROF
236-53 SINDAN-DONG
JOONG-KU
SEOUL 100
REPUBLIC OF KOREA

SHIMIZU MIKIO ASST PROF
INST SPACE AERON SC
UNIVERSITY OF TOKYO
KOMABA MEGURO KU
TOKYO 153
JAPAN

SHIMIZU TSUTOMU PROF
DEPT OF ASTRONOMY
UNIVERSITY OF KYOTO
SAKYOKU KYOTO 606
JAPAN

SHIMMINS A J
AUSTRALIAN NAT RADIO
ASTRONOMY OBS
P O BOX 276
PARKES NSW 2870
AUSTRALIA

SHINE RICHARD A DR
NASA GSFC
CODE 685
GREENBELT MD 20771
U.S.A.

SHIPMAN HENRY L DR
PHYSICS DEPT
UNIV OF DELAWARE
NEWARK DE 19711
U.S.A.

SHIRYAEV A V DR
LENINGRAD STATE UNI-
VERSITY ASTRONOMICAL
OBSERVATORY
199178 LENINGRAD
U.S.S.R.

SHIVANANDAN KANDIAH DR
NAVAL RESEARCH LAB
CODE 7122 1
WASHINGTON DC 20375
U.S.A.

SHKLOVSKY I S DR
SPACE RESEARCH
INSTITUTE
USSR ACADEMY OF SCI
117810 MOSCOW
U.S.S.R.

SHOBBROOK ROBERT R DR
CHATTERTON DEPT OF
ASTRONOMY
UNIVERSITY OF SYDNEY
SYDNEY NSW 2006
AUSTRALIA

SHOEMAKER E
BRANCH ASTROGEOL ST
US GEOLOGICAL SURVEY
2255 N GEMINI DRIVE
FLAGSTAFF AZ 86001
U.S.A.

SHOLOMITSKY G B DR
SPACE RESEARCH
INSTITUTE
USSR ACADEMY OF SC
117810 MOSCOW
U.S.S.R.

SHOR VIKTOR A DR
INST THEOR ASTRONOMY
NABEREZHNAYA
KUTUZOVA 10
192187 LENINGRAD
U.S.S.R.

SHORE BRUCE W
LAWRENCE LIVERMORE
LABORATORY
LIVERMORE CA 94550
U.S.A.

SHORE STEVEN N
WARNER & SWASEY OBS
CASE WESTERN RESERVE
UNIVERSITY
CLEVELAND OH 44106
U.S.A.

SHOSTAK G S DR
KAPTEYN ASTRON INST
POSTBUS 800
9700 AV GRONINGEN
THE NETHERLANDS

SHTEINS K A DR
LATVIAN STATE UNIVER
SITY ASTRONOMICAL
OBSERVATORY
226098 RIGA
U.S.S.R.

SHU FRANK H PROF
ASTRONOMY DEPT
UNIV OF CALIFORNIA
CAMPBELL HALL
BERKELEY CA 94720
U.S.A.

SHUKLA K
DEPT OF MATHEMATICS
AND ASTRONOMY
LUCKNOW UNIVERSITY
LUCKNOW U P
INDIA

SHUKRE C S DR
CSIRO
DIV OF RADIOPHYSICS
P O BOX 76
EPPING NSW 2121
AUSTRALIA

SHUL MAN L M DR
MAIN ASTRONOMICAL OB
SERVATORY UKRAINIAN
ACADEMY OF SCIENCES
252127 KIEV
U.S.S.R.

SHUL'BERG A M DR
ODESSA STATE UNIVER-
SITY ASTRONOMICAL
OBSERVATORY
270014 ODESSA
U.S.S.R.

SHULL JOHN MICHAEL
UNIV OF COLORADO
JILA
BOULDER CO 80309
U.S.A.

SHULOV O S DR
LENINGRAD STATE UNI-
VERSITY ASTRONOMICAL
OBSERVATORY
199178 LENINGRAD
U.S.S.R.

SHUSTOV BORIS M DR
ASTRONOMICAL COUNCIL
USSR ACADEMY OF SC
PYATNITSKAYA 48
109017 MOSCOW
U.S.S.R.

SHUTER WILLIAM L H DR
DEPT OF PHYSICS
UNIVERSITY OF B C
2075 WESTBROOK MALL
VANCOUVER BC V6T 1W5
CANADA

SIBILLE FRANCOIS
OBSERVATOIRE DE LYON
F-69230 ST-GENIS
LAVAL
FRANCE

SIDA DERRICK W PROF
DEPT OF PHYSICS
CARLETON UNIVERSITY
OTTAWA K1S 5B6
CANADA

SIDLICHOVSKY MILOS DR
ASTRONOMICAL INST
CSAV
BUDECSKA 6
120 23 PRAHA 2
CZECHOSLOVAKIA

SIEBER WOLFGANG PH D
MPI FUER
RADIOASTRONOMIE
AUF DEM HUEGEL 69
D-5300 BONN 1
GERMANY, F.R.

SIEGEL C L
UNIVERSITAET
GOETTINGEN
BUNSENSTR 3-5
D-3400 GOETTINGEN
GERMANY, F.R.

SIENKIEWICZ RYSZARD DR
N COPERNICUS ASTRON
CENTER
BARTYCKA 18
00-716 WARSAW
POLAND

SIGNORE M DR
OBS DE PARIS
61,AV DE L'OBS
F-75014 PARIS
FRANCE

SILBERBERG REIN DR
NAVAL RESEARCH LAB
CODE 7028
WASHINGTON DC 20375
U.S.A.

SILK JOSEPH I PROF
ASTRONOMY DEPT
UNIV OF CALIFORNIA
BERKELEY CA 94720
U.S.A.

SILVERBERG ERIC C DR
MCDONALD OBSERVATORY
UNIVERSITY OF TEXAS
BOX 1337
FORT DAVIS TX 79734
U.S.A.

SILVESTRO GIOVANNI
ISTITUTO DI FISICA
GENERALE DELL UNIV
CR M D AZEGLIO 46
I-10125 TORINO
ITALY

SIM MARY E DR
ROYAL OBSERVATORY
BLACKFORD HILL
EDINBURGH EH9 3HJ
U.K.

SIMA ZDISLAV
ASTRONOMICAL INST
CZECH ACADEMY OF SC
BUDECSKA 6
12023 PRAHA 2
CZECHOSLOVAKIA

SIMEK MILOS DR
ASTRONOMICAL INST
CZECHO ACAD SCIENCES
OBSERVATORY
251 65 ONDREJOV
CZECHOSLOVAKIA

SIMIEN FRANCOIS DR
OBS DE MARSEILLE
2 PLACE LE VERRIER
F-13248 MARSEILLE C4
FRANCE

SIMKIN SUSAN M DR
MT STROMLO OBSERVA-
TORY
CANBERRA ACT 2606
AUSTRALIA

SIMMONEAU E DR
INST D'ASTROPHYSIQUE
98 BIS BD ARAGO
F-75014 PARIS
FRANCE

SIMMONS J F L
31 HAVELOCK STREET
GLASGOW G11 5HA
U.K.

SIMO CHARLES DR
PJE MARIMON 11
BARCELONA 21
SPAIN

SIMODA MAHIRO PROF
DEPT ASTRON/EARTH SC
TOKYO GAKUGEI UNIV
KOGANEI
TOKYO 184
JAPAN

SIMON GEORGE W DR
SACRAMENTO PEAK
OBSERVATORY
SUNSPOT NM 88349
U.S.A.

SIMON J L MR
BUREAU D LONGITUDES
77 AVENUE DENFERT-
ROCHEREAU
F-75014 PARIS
FRANCE

SIMON MICHAL PROF
DEPT EARTH SPACE SC
STATE UNIVERSITY OF
NEW YORK
STONY BROOK NY 11794
U.S.A.

SIMON NORMAN R ASSOC PROF
BEHLEN LAB PHYSICS
UNIVERSITY NEBRASKA
LINCOLN NB 68588
U.S.A.

SIMON PAUL A DR
OBSERVATOIRE
DE MEUDON
F-92190 MEUDON
FRANCE

SIMON PAUL C DR
AV MANOIR D'ANJOU 14
B-1150 BRUXELLES
BELGIUM

SIMON RENE L E PROF
INSTITUT
D ASTROPHYSIQUE
B-4200 COINTE-LIEGE
BELGIUM

SIMON THEODORE
INST FOR ASTRONOMY
UNIV OF HAWAII
2680 WOODLAWN DRIVE
HONOLULU HI 96822
U.S.A.

SIMONENKO ALLA N DR
ASTRON COUNCIL
USSR ACADEMY OF SCS
PYATNITSKAYA 48
109017 MOSCOW
U.S.S.R.

SIMONS STUART DR
170 HOLMLEIGH ROAD
LONDON N16 5PY
U.K.

SIMONSON S CHRISTIAN DR
1061 RUSSELL AVE
LOS ALTOS CA 94022
U.S.A.

SIMOVLJEVITCH JOVAN L DR
DEPT OF ASTRONOMY
FACULTY OF SCIENCES
STUDENTSKI TRG 16
11000 BEOGRAD
YUGOSLAVIA

SIMS KENNETH P DR
SYDNEY OBSERVATORY
SYDNEY N S W 2000
AUSTRALIA

SINCLAIR ANDREW T DR
ROYAL GREENWICH
OBSERVATORY
HAILSHAM
EAST SUSSEX BN27 1RP
U.K.

SINGH PATAN DEEN DR
UNIV DE SAO PAULO
INST ASTRO/GEOFISICO
CAIXA POSTAL 30627
01000 SAO PAULO SP
BRAZIL

SINHA RAMESHWAR P
5512 S WICK ST
BETHESDA MD 20034
U.S.A.

SINNERSTAD ULF E PROF
STOCKHOLMS
OBSERVATORIUM
S-13300 SALTSJOBADEN
SWEDEN

SINTON WILLIAM M
UNIV OF HAWAII
INST FOR ASTRONOMY
2680 WOODLAWN DR
HONOLULU HI 96822
U.S.A.

SINVHAL SHAMBHU DAYAL DR.
KUMAUN UNIVERSITY
NAINI TAL 263001
INDIA

SINZI AKIRA M DR
HYDROGRAPHIC DEPT
TSUKIJI 5 CHUO KU
TOKYO 104
JAPAN

SION EDWARD MICHAEL
DEPT OF ASTRONOMY
VILLANOVA UNIV
VILLANOVA PA 19085
U.S.A.

SIRY JOSEPH W
4438 42ND ST N W
WASHINGTON DC 20016
U.S.A.

SISSON GEORGE M MR
PLANETREES WALL
HEXHAM NE46 4EQ
U.K.

SISTERO ROBERTO F DR
OBSERVATORIO
ASTRONOMICO
LAPRIDA 854
5000 CORDOBA
ARGENTINA

SITARSKI GRZEGORZ DR
CTR FOR SPACE RES
UL BARTYCKA 18
00716 WARSAW
POLAND

SITNIK G F DR
STERNBERG STATE AS-
TRONOMICAL INSTITUTE
117234 MOSCOW
U.S.S.R.

SITTERLY CHARLOTTE M DR
3711 BRANDYWINE STNW
WASHINGTON DC 20016
U.S.A.

SIVAN J P DR
LAB D'ASTRON SPAT
TRAVERSE DU SIPHON
LES TROIS LUCS
F-13012 MARSEILLE
FRANCE

SIVARAMAN K R DR
INDIAN INSTITUTE OF
ASTROPHYSICS
BANGALORE 560034
INDIA

SJOGREN WILLIAM L DR
JPL
MS 264-738
4800 OAK GROVE DRIVE
PASADENA CA 91103
U.S.A.

SKALAFURIS ANGELO J
MATH RESEARCH CTR
NAVAL RESEARCH LAB
C7840
WASHINGTON DC 20375
U.S.A.

SKILLING JOHN DR
DEPT APPLIED
MATHS AND THEO PHYS
SILVER STREET
CAMBRIDGE CB3 9EW
U.K.

SKUMANICH ANDRE PROF
HIGH ALTITUDE
OBSERVATORY
PO BOX 3000
BOULDER CO 80307
U.S.A.

SLADE MARTIN A III DR
JPL
264-720
4800 OAK GROVE DRIVE
PASADENA CA 91103
U.S.A.

SLEBARSKI TADEUSZ DR
UNIVERSITY OBS
BUCHANAN GARDENS
ST ANDREWS FIFE
KY16 9LZ
U.K.

SLEE O B DR
RADIOPHYSICS
CSIRO
BOX 76
EPPING NSW 2121
AUSTRALIA

SLETTEBAK ARNE PROF
PERKINS OBSERVATORY
DELAWARE OH 43015
U.S.A.

SLONIM E M DR
ASTRONOMICAL INSTITU
TE UZBEKIAN ACADEMY
OF SCIENCES
700000 TASHKENT
U.S.S.R.

SLYSH V I DR
SPACE RESEARCH
INSTITUTE
USSR ACADEMY OF SCI
117810 MOSCOW
U.S.S.R.

SMAK JOSEPH I PROF
COPERNICUS ASTRONOMI
CAL CENTER
UL BARTYCKA 18
00 716 WARSAW
POLAND

SMEYERS PAUL PROF
ASTRONOMISCH INST
NAAMSESTRAAT 61
B-3000 LEUVEN
BELGIUM

SMIT J A PROF
PHYSICAL LAB
UNIVERSITY
SORBONNELAAN 4
3584 CA UTRECHT
THE NETHERLANDS

SMITH ALEX G PROF
DEPT PHYSICS AND
ASTRONOMY
231 SPACE SC BLDG
GAINESVILLE FL 32611
U.S.A.

SMITH ANDREW M DR
CODE 672
GODDARD SPACE FLIGHT
CENTER
GREENBELT MD 20771
U.S.A.

SMITH BARHAM W DR
G-6 MS 565
LASL
LOS ALAMOS NM 87544
U.S.A.

SMITH BRADFORD A PROF
DEPT OF PLANETARY
SCIENCES
UNIVERSITY ARIZONA
TUCSON AZ 85721
U.S.A.

SMITH BRUCE F DR
THEORET & PLAN STUD
BR 245-3
NASA-AMES RES CTR
MOFFETT FLD CA 94035
U.S.A.

SMITH CLAYTON A JR DR
U S NAVAL OBS
WASHINGTON DC 20390
U.S.A.

SMITH DEAN F DR
BERKELEY RESEARCH
ASSOCIATES
BOX 983
BERKELEY CA 94701
U.S.A.

SMITH ELSKE V P DR
DEAN,SCHOOL ARTS&SC
VIRG COMMONW UNIV
901 W FRANKLIN ST
RICHMOND VA 23284
U.S.A.

SMITH F GRAHAM PROF
DIRECTOR NRAL
JODRELL BANK
MACCLESFIELD
CHESHIRE SK11 9DL
U.K.

SMITH GEOFFREY DR
DEPT OF ASTROPHYSICS
SOUTH PARKS ROAD
OXFORD OX1 3RQ
U.K.

SMITH HARDING E JR DR
PHYSICS DEPT C-011
UNIV OF CALIFORNIA
SAN DIEGO
LA JOLLA CA 92093
U.S.A.

SMITH HARLAN J PROF
ASTRONOMY DEPT
UNIVERSITY OF TEXAS
RL MOORE HALL 15,206
AUSTIN TX 78712
U.S.A.

SMITH HAYWOOD C DR
DEPT OF ASTRONOMY
UNIVERSITY OF
SOUTH FLORIDA
TAMPA FL 33620
U.S.A.

SMITH HENRY J DR
NASA HEADQUARTERS
CODE SG
WASHINGTON DC 20546
U.S.A.

SMITH HOWARD ALAN
CODE 4139 5
SPACE SC DIVISION
NAVAL RESEARCH LAB
WASHINGTON DC 20375
U.S.A.

SMITH HUMPHRY M
23 NORMANDALE
BEXHILL ON SEA
EAST SUSSEX TN39 3LU
U.K.

SMITH LINDSEY F DR
DEPT OF PHYSICS
UNIV OF WOLLONGONG
P O BOX 1144
WOLLONGONG NSW 2500
AUSTRALIA

SMITH MALCOLM G DR
ROYAL OBSERVATORY
BLACKFORD HILL
EDINBURGH EH9 3HJ
U.K.

SMITH MYRON A ASST PROF
SACRAMENTO PEAK OBS
SUNSPOT NM 88349
U.S.A.

SMITH PETER L DR
CTR FOR ASTROPHYSICS
60 GARDEN ST
CAMBRIDGE MA 02138
U.S.A.

SMITH ROBERT CONNON DR
ASTRONOMY CENTRE
UNIVERSITY OF SUSSEX
BRIGHTON BN1 9QH
U.K.

SMITH WM HAYDEN PROF
MCDONNELL CENTER FOR
SPACE SCIENCES
WASHINGTON UNIVERSIT
ST LOUIS MO 63130
U.S.A.

SMOLINSKI JAN DR
COPERNICUS ASTRONO-
MICAL CENTER
CHOPINA 12-18
87-100 TORUN
POLAND

SMOLKOV G YA DR
SIBIZMIR
P B 4
664697 IRKUTSK 33
U.S.S.R.

SMOLUCHOWSKI ROMAN PROF
ASTRONOMY DEPT
UNIVERSITY OF TEXAS
RLM HALL 15 220
AUSTIN TX 78712
U.S.A.

SMOOT III GEORGE F.
BLDG 50-230
L BERKELEY LAB
UNIV OF CALIFORNIA
BERKELEY CA 94720
U.S.A.

SMRIGLIO FILIPPO PROF
OSSERVATORIO
ASTRONOMICO
VIA PARCO MELLINI 84
I-00136 ROMA
ITALY

SMYLIE DOUGLAS E DR
DEPT OF PHYSICS
YORK UNIVERSITY
4700 KEELE STREET
DOWNSVIEW ONT M3J1P3
CANADA

SMYTH MICHAEL J DR
DEPT OF ASTRONOMY
ROYAL OBSERVATORY
EDINBURGH EH9 3HJ
U.K.

SNELL RONALD L
FIVE COLLEGE RAO
GRC TOWER B
UNIV MASSACHUSETTS
AMHERST MA 01003
U.S.A.

SNIJDERS M A J
DEPT OF PHYSICS AND
ASTRONOMY
UNIV CLG LONDON
LONDON WC1E 6BT
U.K.

SNOW THEODORE P PROF
DEPT OF PHYSICS AND
ASTROPHYSICS
UNIVERSITY COLORADO
BOULDER CO 80309
U.S.A.

SNYDER LEWIS E
DEPT OF ASTRONOMY
UNIV OF ILLINOIS
URBANA IL 61801
U.S.A.

SOBERMAN ROBERT K DR
FRANKLIN RES CENTER
THE FRANKLIN INST
20+B.FRANKLIN PARKW
PHILADELPHIA PA19103
U.S.A.

SOBIESKI STANLEY DR
CODE 673
GODDARD SPACE FLIGHT
CENTER
GREENBELT MD 20771
U.S.A.

SOBOLEV V V DR
LENINGRAD STATE UNIV
ASTRONOMICAL OBS
199178 LENINGRAD
U.S.S.R.

SOBOLEV VLADISLAV M DR
MAIN ASTRONOMICAL
OBSERVATORY PULKOVO
196140 LENINGRAD
U.S.S.R.

SOBOLEVA N S DR
SPECIAL ASTROPHYS
OBS USSR ACAD OF SCI
LENINGRAD BRANCH
196140 LENINGRAD
U.S.S.R.

SOBOUTI YOUSEF PROF
BIRUNI OBSERVATORY
AND PHYSICS DEPT
UNIVERSITY
SHIRAZ
IRAN

SOCHILINA ALLA S DR
INSTITUTE FOR THEORE
TICAL ASTRONOMY
10 KUTUZOV QUAY
192187 LENINGRAD
U.S.S.R.

SODERBLOM LARRY DR
BR ASTROGEOLOGIC ST
US GEOLOGICAL SURVEY
2555 N GEMINI DRIVE
FLAGSTAFF AZ 86001
U.S.A.

SOEDERHJELM STAFFAN DR
LUND OBSERVATORIET
BOX 1107
S-221 04 LUND
SWEDEN

SOFIA SABATINO PROF
GODDARD SPACE FLIGHT
CENTER
CODE 681
GREENBELT MD 20771
U.S.A.

SOFUE YOSHIAKI DR
NOBEYAMA RADIO ASTRO
NOMY OBSERVATORY
MINAMISAKU-GUN
NAGANO-KEN 384-13
JAPAN

SOIFER BARUCH T DR
320-47 DOWNS LAB
PHYSICS DEPT
CALTECH
PASADENA CA 91125
U.S.A.

SOLAZZO CLAUDIO
OSS ASTRONOMICO
DI CAPODIMONTE
VIA MOIARIELLO 16
I-80131 NAPOLI
ITALY

SOLC IVAN DR
ASTRON INST CSAV
GROUP OF OPTICS
DVORAKOWA 298
511 01 TURNOV
CZECHOSLOVAKIA

SOLF JOSEF DR
MPI FUER ASTRONOMIE
KOENIGSTUHL
D-6900 HEIDELBERG 1
GERMANY, F.R.

SOLHEIM JAN ERIK
IMR P O BOX 953
N-9001 TROMSO
NORWAY

SOLIMAN MOHAMED AHMED
HELWAN OBSERVATORY
HELWAN NEAR CAIRO
ARAB REP. OF EGYPT

SOLOMON PHILIP M DR
ASTROPHYSICS PROGRAM
DEPT EARTH SPACE SCS
SUNY AT STONY BROOK
STONY BROOK NY 11794
U.S.A.

SOLTAN ANDRZEJ MARIA DR
N COPERNICUS ASTRON
CENTER
BARTYCKA 18
00-716 WARSAW
POLAND

SOMERVILLE WILLIAM B DR
UNIVERSITY COLLEGE
LONDON WC1E 6BT
U.K.

SONETT CHARLES P PROF
DEPT PLANETARY SCI
UNIVERSITY OF
ARIZONA
TUCSON AZ 85721
U.S.A.

SORENSEN GUNNAR DR
INSTITUTE OF PHYSICS
LANGELANDSGADE
DK-8000 AARHUS C
DENMARK

SORENSEN SOREN-AKSEL DR
UNIV OF LONDON
OBSERVATORY
MILL HILL PARK
LONDON NW7 2QS
U.K.

SOROCHENKO R L DR
PHYSICAL INSTITUTE
USSR ACADEMY OF
SCIENCES
117924 MOSCOW
U.S.S.R.

SORU-ESCAUT IRINA MRS
OBSERVATOIRE
DE MEUDON
F-92190 MEUDON
FRANCE

SOTIROVSKI PASCAL DR
OBSERVATOIRE
DE MEUDON
F-92190 MEUDON
FRANCE

SOUFFRIN PIERRE B DR
OBSERVATOIRE DE NICE
B P 252
F-06007 NICE CEDEX
FRANCE

SOULIE GUY
OBS DE L'UNIVERSITE
DE BORDEAUX
F-33270 FLOIRAC
FRANCE

SOUTHWORTH R B DR
HARVARD COLLEGE OBS
60 GARDEN STREET
CAMBRIDGE MA 02138
U.S.A.

SPADA GIANFRANCO DR
TE S R E
CNR
VIA DE CASTAGNOLI 1
I-40126 BOLOGNA
ITALY

SPAENHAUER ANDREAS MARTIN
ASTRON INST BASEL
VENUSSTRASSE 7
CH-4102 BINNINGEN
SWITZERLAND

SPARKS WARREN M DR
CODE 671
GODDARD SPACE FLIGHT
CENTER
GREENBELT MD 20771
U.S.A.

SPARROW JAMES G DR
AERONAUTICAL RES
LABORATORIES
BOX 4331
MELBOURNE 3001
AUSTRALIA

SPASOVA NEDKA MARINOVA
DEPT OF ASTRONOMY
BULGARIAN ACADEMY SC
72 LENIN BLVD
1184 SOFIA
BULGARIA

SPEER R J DR
DEPT OF PHYSICS
IMPERIAL COLLEGE
LONDON SW7 2BZ
U.K.

SPENCER JOHN HOWARD
NAVAL RESEARCH LAB
CODE 4134 2
WASHINGTON DC 20375
U.S.A.

SPENCER RALPH E DR
NUFFIELD RADIO ASTRO
NOMY LABS
JODRELL BANK MACCLES
FIELD CHES SK11 9DL
U.K.

SPICER DANIEL SHIELDS DR
INST FOR PHYSICAL SC
AND TECHNOLOGY
UNIV OF MARYLAND
COLLEGE PK MD 20742
U.S.A.

SPIEGEL E DR
COURANT INST OF
MATHEMATICAL SCS
251 MERCER STREET
NEW YORK NY 10003
U.S.A.

SPINRAD HYRON PROF
DEPT OF ASTRONOMY
UNIVERSITY
OF CALIFORNIA
BERKELEY CA 94720
U.S.A.

SPITE FRANCOIS M DR
OBSERVATOIRE
DE PARIS
F-92190 MEUDON
FRANCE

SPITE MONIQUE DR
OBSERVATOIRE
DE PARIS
F-92190 MEUDON
FRANCE

SPITHAS ELEFTERIOS N DR
DEPT OF ASTRONOMY
UNIVERSITY OF ATHENS
PANEPISTIMIOPOLIS
ATHENS 621
GREECE

SPITZER LYMAN JR DR
PRINCETON UNIVERSITY
OBSERVATORY
PEYTON HALL
PRINCETON NJ 08540
U.S.A.

SPOELSTRA T A TH DR
STERREWACHT
POSTBUS 9513
2300 RA LEIDEN
THE NETHERLANDS

SPRUIT HENK C DR
MPI FUER ASTROPHYSIK
KARL-SCHWARZSCHILD-
STR 1
D-8046 GARCHING B M
GERMANY, F.R.

SPYROU NICOLAOS DR
DEPT OF ASTRONOMY
UNIVERSITY OF
THESSALONIKI
THESSALONIKI
GREECE

SRAMEK RICHARD A DR
NAT RADIO ASTRON OBS
P O BOX O
SOCORRO NM 87801
U.S.A.

SREEKANTAN B V DR
TATA INSTITUTE OF
FUNDAMENTAL RESEARCH
HOMI BHABHA ROAD
BOMBAY 400005
INDIA

SREENIVASAN S RANGA PROF
PHYSICS DEPT
THE UNIVERSITY
2920-24 AVE N W
CALGARY ALTA T2N 1N4
CANADA

SRINIVASAN G
RAMAN RESEARCH INST
BANGALORE 560 080
INDIA

STABELL ROLF DR
INSTITUTE OF THEORE-
TICAL ASTROPHYSICS
UNIVERSITY OF OSLO
BLINDERN OSLO 3
NORWAY

STAGNI RUGGERO
OSS ASTROFISICO
DI ASIAGO
I-36012 ASIAGO
ITALY

STAHR-CARPENTER M DR
LEANDER MCCORMICK OB
BOX 3818
UNIVERSITY STATION
CHARLOTTESVL VA22903
U.S.A.

STALIO ROBERTO DR
OSSERVATORIO ASTRONO
MICO DI TRIESTE
VIA TREPOLO 11
I-34131 TRIESTE
ITALY

STANDISH E MYLES DR
JET PROPULSION
LABORATORY
PASADENA CA 91103
U.S.A.

STANGE LOTHAR
TECHNICAL UNIVERSITY
DRESDEN
MOMMSENSTR.13
DDR-8027 DRESDEN
G.D.R.

STANILA GEORGE DR
ASTRONOMICAL OBS
5 CUTITUL DE ARGINT
BUCAREST 28
ROUMANIA

STANKEVICH K S DR
RADIOPHYSICAL
RESEARCH INSTITUTE
603600 GORKIJ
U.S.S.R.

STANLEY G J
CALTECH
PASADENA CA 91109
U.S.A.

STANNARD DAVID DR
JODRELL BANK
MACCLESFIELD
CHESHIRE SK11 9DL
U.K.

STARK ANTONY A
HOH L-231
BELL LABORATORIES
HOLMDELL NJ 07733
U.S.A.

STARRFIELD SUMNER PROF
DEPT OF PHYSICS
ARIZONA STATE
UNIVERSITY
TEMPE AZ 85281
U.S.A.

STARYTSIN GENNADIJ V DR
PULKOVO OBSERVATORY
196140 LENINGRAD
U.S.S.R.

STASINSKA G DR
OBS DE MEUDON
F-92190 MEUDON
FRANCE

STAUBERT RUDIGER DR
ASTRONOMISCHES INST.
DER UNIV TUEBINGEN
WALDHAUSERSTR 64
D-7400 TUEBINGEN
GERMANY, F.R.

STAUDE HANS JAKOB PH D
MPI FUER ASTRONOMIE
KOENIGSTUHL
D-6900 HEIDELBERG 1
GERMANY, F.R.

STAUDE JUERGEN DR
ZENTRALINST FUER
SOLAR-TERR PHYSIK
TELEGRAFENBERG
DDR-15 POTSDAM
G.D.R.

STAWIKOWSKI ANTONI DR
ASTRONOMICAL CENTER
UL CHOPINA 12-18
87-100 TORUN
POLAND

STEARNS C L DR
VAN VLECK OBS
MIDDLETOWN CT 06457
U.S.A.

STEBBINS ROBIN
SACRAMENTO PEAK OBS
SUNSPOT NM 88349
U.S.A.

STECHER THEODORE P
CODE 672
GODDARD SPACE FLIGHT
CENTER
GREENBELT MD 20771
U.S.A.

STECKER FLOYD W DR
LAB FOR HIGH ENERGY
ASTROPHYS 660 NASA
GODDARD SPACE CENTER
GREENBELT MD 20771
U.S.A.

STEENMAN-CLARK L DR
OBS DE NICE
B P 252
F-06007 NICE CEDEX
FRANCE

STEFANOVITCH-GOMEZ A E DR
OBSERVATOIRE
DE MEUDON
F-92190 MEUDON
FRANCE

STEIGER W R PROF
HAWAII INST GEOPHYS
UNIV OF HAWAII
2525 CORREA ROAD
HONOLULU HI 96822
U.S.A.

STEIGMAN GARY PROF
BARTOL RESEARCH
FOUNDATION
UNIV OF DELAWARE
NEWARK DE 19711
U.S.A.

STEIN JOHN WILLIAM
555 HILL STREET
SEWICKLEY PA 15143
U.S.A.

STEIN ROBERT F ASSOC PROF
ASTRONOMY DEPT
MICHIGAN STATE
UNIVERSITY
EAST LANSING MI48824
U.S.A.

STEIN WAYNE A PROF
SCHOOL OF PHYSICS
AND ASTRONOMY
UNIVERSITY MINNESOTA
MINNEAPOLIS MN 55455
U.S.A.

STEINBERG JEAN L DR
OBSERVATOIRE
DE MEUDON
F-92190 MEUDON
FRANCE

STEINITZ RAPHAEL PROF
PHYSICS DEPT
BG UNIVERSITY OF THE
NEGEV
BEER SHEVA
ISRAEL

STEINLIN ULI PROF
ASTRONOMISCHES INST
UNIVERSITAET BASEL
VENUSSTR 7
CH-4102 BINNINGEN
SWITZERLAND

STELLINGWERF ROBERT F DR
DEPT PHYSICS ASTRON
RUTGERS UNIV
PISCATAWAY NJ 08904
U.S.A.

STELLMACHER GOETZ
INST D ASTROPHYSIQUE
98 BIS BD ARAGO
F-75014 PARIS
FRANCE

STELLMACHER IRENE DR
BUREAU D LONGITUDES
77 AVENUE DENFERT-
ROCHEREAU
F-75014 PARIS
FRANCE

STENCEL ROBERT EDWARD
JOINT INST FOR
LAB ASTROPHYSICS
UNIV OF COLORADO
BOULDER CO 80309
U.S.A.

STENFLO JAN O DR
INST FUER ASTRONOMIE
ETH-ZENTRUM
CH-8092 ZUERICH
SWITZERLAND

STEPANIAN A A DR
CRIMEAN ASTROPHYSI-
CAL OBSERVATORY USSR
ACAD OF SCI NAUCHNIY
334413 CRIMEA
U.S.S.R.

STEPANIAN N N DR
CRIMEAN ASTROPHYSI-
CAL OBSERVATORY USSR
ACAD OF SCI NAUCHNIY
334413 CRIMEA
U.S.S.R.

STEPANOV V E PROF
SIBIZMIR
P B 4
664697 IRKUTSK 33
U.S.S.R.

STEPHENSON C BRUCE PROF
WARNER AND SWASEY
OBSERVATORY
CASE WESTERN RES UN
CLEVELAND OH 44106
U.S.A.

STEPHENSON F RICHARD DR
SCHOOL OF PHYSICS
UNIVERSITY
NEWCASTLE UPON TYNE
NE1 7RU
U.K.

STEPIEN KAZIMIERZ DR
ASTRONOMICAL
OBSERVATORY
AL UJAZDOWSKIE 4
00478 WARSZAWA
POLAND

STEPPE HANS DR
ASTRON INST BASEL
VENUSSTRASSE 7
CH-4102 BINNINGEN
SWITZERLAND

STERKEN CHRISTIAAN LEO DR
ASTROPHYSICAL INST
V U B
PLEINLAAN 2
B-1050 BRUSSELS
BELGIUM

STERN ROBERT ALLAN
MS 169-327
JET PROPULSION DRIVE
4800 OAK GROVE DRIVE
PASADENA CA 91109
U.S.A.

STERNBERK BOHUMIL
ASTRON INST CSAV
BUDECSKA 6
120 23 PRAHA 2
CZECHOSLOVAKIA

STESHENKO N V DR
CRIMEAN ASTROPHYSI-
CAL OBSERVATORY USSR
ACAD OF SCI NAUCHNIY
334413 CRIMEA
U.S.S.R.

STEVENS GERARD A DR
LABORATORIUM VOOR
RUIMTEONDERZOEK
BENELUXLAAN 21
3527 HS UTRECHT
THE NETHERLANDS

STEWART J M DR
INSTITUTE OF
ASTRONOMY
MADINGLEY ROAD
CAMBRIDGE CB3 0HA
U.K.

STEWART PAUL DR
MATHEMATICS DEPT
THE UNIVERSITY
MANCHESTER M13 9PL
U.K.

STEWART RONALD T MR
CSIRO DIVISION OF
RADIOPHYSICS
P O BOX 76
EPPING NSW 2121
AUSTRALIA

STEYAERT HERMAN PROF DR
STERRENKUNDIG INST
KRYGSLAAN 271 S9
B-9000 GENT
BELGIUM

STIBBS DOUGLAS W N PROF
UNIV OBSERVATORY
BUCHANAN GARDENS
ST ANDREWS
FIFE KY16 9L2
U.K.

STICKLAND DAVID J DR
ROYAL GREENWICH OBS
HAILSHAM
E SUSSEX BN27 1RP
U.K.

STIFT M J DR
INST FUER ASTRONOMIE
TUERKENSCHANZSTR 17
A-1180 WIEN
AUSTRIA

STIX MICHAEL DR
KIEPENHEUER INST
FUER SONNENPHYSIK
SCHOENECKSTR 6
D-7800 FREIBURG
GERMANY, F.R.

STOBIE ROBERT S DR
ROYAL OBS
BLACKFORD HILL
EDINBURGH EH9 3HJ
U.K.

STOCK JURGEN
CENTRO DE INVESTIGA-
CION DE ASTRONOMIA
APARTADO 264
MERIDA
VENEZUELA

STOCKMAN HERVEY S JR DR
STEWARD OBS
UNIV OF ARIZONA
TUCSON AZ 85721
U.S.A.

STOCKTON ALAN N DR
INSTITUTE FOR
ASTRONOMY
2680 WOODLAWN DR
HONOLULU HI 96822
U.S.A.

STODOLKIEWICZ JERZY S DR
COPERNICUS ASTRONOMI
CAL CENTER
UL BARTYCKA 18
00 716 WARSAW
POLAND

STOEGER W R DR
SPECOLA VATICANA
I-00120
CITTA DEL VATICANO
VATICAN CITY STATE

STOHL JAN DR
ASTRONOMICAL INST
SAV
DUBRAVSKA CESTA A 11
89930 BRATISLAVA
CZECHOSLOVAKIA

STONE EDWARD C DR
CALTECH
PASADENA CA 91125
U.S.A.

STONE R G DR
CODE 693
LAB FOR EXTRATERRES
PHYSICS NASA-GSFC
GREENBELT MD 20771
U.S.A.

STONE REMINGTON P S DR
LICK OBSERVATORY
MT HAMILTON CA 95140
U.S.A.

STONE RONALD CECIL
US NAVAL OBSERVATORY
TRANSIT CIRCLE
DIVISION
WASHINGTON DC 20390
U.S.A.

STOREY JOHN W V DR
DEPT OF PHYSICS
UNIV NEW SOUTH WALES
P O BOX 1
KENSINGTON NSW 2033
AUSTRALIA

STOYKO ANNA
OBSERVATOIRE
DE PARIS
61 AVE OBSERVATOIRE
F-75014 PARIS
FRANCE

STRAFELLA FEDERICO
DIPARTIMENTO DI
FISICA
UNIVERSITY OF LECCE
I-73100 LECCE
ITALY

STRAIZYS V PROF DR
ASTRONOMIJOS OBS
VILNIUS 31
CIURLIONIO 29
LITHUANIA
U.S.S.R.

STRAND KAJ AA DR
3200 ROWLAND PL NW
WASHINGTON DC 20008
U.S.A.

STRASSL HANS L PROF
OCHTRUPWEG 39
D-4400 MUENSTER
GERMANY, F.R.

STRAZZULLA GIOVANNI
OSS ASTROFISICO
CITTA UNIVERSITARIA
I-95125 CATANIA
ITALY

STREL'NITSKIJ VLADIMIR DR
ASTRONOMICAL COUNCIL
USSR ACAD OF SCS
48 PYATNITSKAYA ST
109017 MOSCOW
U.S.S.R.

STRITTMATTER PETER A PROF
STEWARD OBSERVATORY
TUCSON AZ 85721
U.S.A.

STROBEL ANDRZEJ DR
INST OF ASTRONOMY
N COPERNICUS UNIV
CHOPINA 12/18
87-100 TORUN
POLAND

STROBEL DARRELL F
CODE 4780
NAVAL RESEARCH LAB
WASHINGTON DC 20375
U.S.A.

STROBEL WILLI DR
ASTRONOMISCHES
RECHEN INSTITUUT
MOENCHHOFSTR 12 14
D-6900 HEIDELBERG 1
GERMANY, F.R.

STROHMEIER WOLFGANG PROF
VOLKFELDSTRASSE 5
D-8600 BAMBERG
GERMANY, F.R.

STROM KAREN M
KITT PEAK NATIONAL
OBSERVATORY
P O BOX 26732
TUCSON AZ 85726
U.S.A.

STROM RICHARD G DR
RADIOSTERRENWACHT
POSTBUS 2
7990 AA DWINGELOO
THE NETHERLANDS

STROM ROBERT G PROF
DEPT OF PLANETARY SC
UNIVERSITY
OF ARIZONA
TUCSON AZ 85721
U.S.A.

STROM STEPHEN E
GALACTIC EXTRAG PROG
KITT PEAK NAT OBS
P O BOX 26732
TUCSON AZ 85726
U.S.A.

STROMGREN BENGT PROF
GAMLE CARLSBERGV 15
DK-2500 COPENHAGEN
VALBY
DENMARK

STRONG IAN B DR
LOS ALAMOS SCI LAB
MAIL STOP 436
P O BOX 1663
LOS ALAMOS NM 87545
U.S.A.

STRONG JOHN D PROF
ASTRONOMY RESEARCH
FACILITY
UNIVERSITY OF MASS
AMHERST MA 01003
U.S.A.

STRUBLE MITCHELL F
DEPT OF ASTRONOMY
& ASTROPHYSICS E-1
UNIV OF PENNSYLVANIA
PHILADELPHIA PA19104
U.S.A.

STUMPFF PETER DR
MPI FUER RADIO
ASTRONOMIE
AUF DEM HUEGEL 69
D-5300 BONN
GERMANY, F.R.

STURCH CONRAD R DR
1942 LEWIS AVE
ROCKVILLE MD 20851
U.S.A.

STURROCK PETER A PROF
INST PLASMA RESEARCH
VIA CRESPI
STANFORD UNIVERSITY
STANFORD CA 94305
U.S.A.

SU DING-QIANG
NANJING ASTRONOMICAL
INSTRUMENTS FACTORY
NANJING
JIANGSU PROVINCE
PEOPLE'S REPUBLIC CHINA

SU HONG-JUN
PURPLE MOUNTAIN
OBSERVATORY
NANJING
PEOPLE'S REPUBLIC CHINA

SUBRAHMANYA C R
TATA INSTITUTE OF
FUNDAMENTAL RESEARCH
INDIAN INST SC CAMP
BANGALORE 560 012
INDIA

SUDA JAN
ASTRONOMICAL INST
CZECH ACADEMY OF SC
ONDREJOV OBSERVATORY
25165 ONDREJOV
CZECHOSLOVAKIA

SUDA KAZUO ASST PROF
DEPT OF ASTRONOMY
TOHOKU UNIVERSITY
ARAMAKI SENDAI 980
JAPAN

SUEMOTO ZENZABURO PROF DR
TOKYO ASTRONOMICAL
OBSERVATORY
OSAWA MITAKA
TOKYO 181
JAPAN

SUESS STEVEN T DR
SPACE ENVIRONMENT
LABORATORY
NOAA ERL
BOULDER CO 80302
U.S.A.

SUGAWA CHIKARA DR
HANANOI 1586-25
KASHIWA-SHI
CHIBA-KEN 277
JAPAN

SUGIMOTO DAIICHIRO PROF
DEPT EARTH SC ASTRO
COLL GENERAL EDUC
UNIVERSITY OF TOKYO
MEGURO-KU TOKYO 153
JAPAN

SULENTIC JACK W DR
DEPT PHYS & ASTRON
UNIV OF ALABAMA
P O BOX 1921
UNIVERSITY AL 35486
U.S.A.

SULLIVAN DENIS JOHN DR
VICTORIA UNIV
WELLINGTON
NEW ZEALAND

SULLIVAN WOODRUFF T IIIDR
DEPT ASTRONOMY FM 20
UNIVERSITY OF
WASHINGTON
SEATTLE WA 98195
U.S.A.

SULTANOV G F ACAD
SHEMAKHA ASTROPHYSI-
CAL OBSERVATORY
373243 AZERBAIDZAN
U.S.S.R.

SUMMERS HUGH P DR
DEPT OF APPLIED MATH
AND THEORETICAL PHYS
SILVER ST
CAMBRIDGE CB3 9EW
U.K.

SUN KAI
ASTROPHYSICS DIV
GEOPHYSICS DEPT
BEIJING UNIVERSITY
BEIJING
PEOPLE'S REPUBLIC CHINA

SUNDMAN ANITA DR
STOCKHOLM
OBSERVATORIET
S-133 00
SALTSJOEBADEN
SWEDEN

SUNYAEV R A DR
SPACE RESEARCH
INSTITUTE
USSR ACADEMY OF SCI
117810 MOSCOW
U.S.S.R.

SURDEJ JEAN M G
INST D'ASTROPHYSIQUE
UNIVERSITE DE LIEGE
AV DE COINTE 5
B-4200 COINTE-OUGREE
BELGIUM

SUTANTYO WINARDI
BOSSCHA OBS
LEMBANG JAVA
INDONESIA

SUTHERLAND P G DR
PHYSICS DEPT
MCMASTER UNIVERSITY
HAMILTON ONT L8S 4M1
CANADA

SUZUKI YOSHIMASA PROF
23-1 NAKAJIMA
HIRONOMACHI
UJI SHI 611
JAPAN

SVALGAARD LEIF DR
INST FOR PLASMA RES
STANFORD UNIV
VIA CRESPI
STANFORD CA 94305
U.S.A.

SVATOS JAN DR
DEPT OF ASTRONOMY
CHARLES UNIVERSITY
SVEDSKA 8
150 00 PRAHA 5
CZECHOSLOVAKIA

SVECHNIKOV M A DR
ASTRONOMICAL DEPT OF
URALSKIJ
STATE UNIVERSITY
620083 SVERDLOVSK
U.S.S.R.

SVESTKA ZDENEK DR
SPACE RESEARCH LAB
BENELUXLAAN 21
3527 HS UTRECHT
THE NETHERLANDS

SVOLOPOULOS SOTIRIOS PROF
DEPT OF ASTROPHYSICS
UNIVERSITY OF ATHENS
PANEPISTIMIOPOLIS
ATHENS 621
GREECE

SWANENBURG B N DR
COSMIC RAY WORKING
GROUP HUYGENS LAB
WASSENAARSEWEG 78
2300 RA LEIDEN
THE NETHERLANDS

SWANK JEAN HEBB
CODE 661
NASA GSFC
GREENBELT MD 20771
U.S.A.

SWARUP GOVIND PROF
RADIO ASTRONOMY
CENTRE OF TIFR
POST BOX 8
OOTACAMUND 643001
INDIA

SWEET PETER A PROF
DEPT OF ASTRONOMY
THE UNIVERSITY
GLASGOW G12 8QW
U.K.

SWEIGART ALLEN V DR
CODE 681
NASA GODDARD SPACE
FLIGHT CENTER
GREENBELT MD 20771
U.S.A.

SWENSON GEORGE W JR PROF
ELECTRICAL ENG DEPT
UNIV OF ILLINOIS U C
1406 WEST GREEN ST
URBANA IL 61801
U.S.A.

SWENSSON JOHN W DR
DEPT OF PHYSICS
SOELVEGATAN 14
S-223 62 LUND
SWEDEN

SWERDLOV NOEL PROF
UNIV OF CHICAGO
CHICAGO IL 60637
U.S.A.

SWIHART THOMAS L DR
STEWARD OBSERVATORY
UNIVERSITY OF
ARIZONA
TUCSON AZ 85721
U.S.A.

SWINGS JEAN-PIERRE DR
INST D ASTROPHYSIQUE
UNIVERSITE DE LIEGE
5 AVENUE DE COINTE
B-4200 COINTE-OUGREE
BELGIUM

SWINGS P DR
AVE L SOUGUENET 23
B-4050 ESNEUX
BELGIUM

SYKES-HART AVRIL B DR
DEPT OF ASTROPHYSICS
SOUTH PARKS ROAD
OXFORD OX1 3RQ
U.K.

SYKORA JULIUS DR
SKALNATE PLESO
OBSERVATORY
059 60 TATRANSKA
LOMNICA
CZECHOSLOVAKIA

SZABADOS LASZLO PH D
KONKOLY OBSERVATORY
P O BOX 67
H-1525 BUDAPEST
HUNGARY

SZAFRANIEC ROZALIA DR
ASTRONOMICAL OBS
UL ORLA 171
30 244 KRAKOW
POLAND

SZEBEHELY VICTOR G PROF
DEPT AEROSPACE ENG
UNIVERSITY OF TEXAS
AUSTIN TX 78712
U.S.A.

SZECSENYI-NAGY GABOR DR
DEPT OF ASTRONOMY
ROALND EOETVOES UNIV
KUN BELA TER 2
H-1083 BUDAPEST
HUNGARY

SZEIDL BELA DR
KONKOLY OBSERVATORY
BOX 67
1525 BUDAPEST
HUNGARY

SZKODY PAULA DR
DEPT OF ASTRONOMY
UNIV OF WASHINGTON
SEATTLE WA 98195
U.S.A.

TABARA HIROTO DR
FACULTY OF EDUCATION
UTSUNOMIYA UNIV
MINEMACHI
UTSUNOMIYA 320
JAPAN

TADEMARU E DR
UNIV OF MASS
AMHERST MA 01002
U.S.A.

TAFFARA SALVATORE PROF
VIA CALZA 5 BIS
I-35100 PADOVA
ITALY

TAGLIAFERRI GIUSEPPE PROF
OSSERVATORIO ARCETRI
LARGO FERMI 5
I-50125 FIRENZE
ITALY

TAI YUIN-KWEI PROF
NO 7 CHAO-CHOU ST
TAIPEI
TAIWAN

TAKADA MASAHIDE DR
DEPT OF ASTRONOMY
UNIV OF TOKYO
BUNKYO-KU
TOKYO 113
JAPAN

TAKAGI KOSHIRO PROF
DEPT OF PHYSICS
TOYAMA UNIVERSITY
3190 GOFUKU
TOYAMA 920
JAPAN

TAKAGI SHIGETSUGU DR
INTERNATIONAL LATITU
DE OBSERVATORY
MIZUSAWA IWATE 023
JAPAN

TAKAHARA FUMIO DR
TOKYO ASTRONOMICAL
OBSERVATORY
UNIV OF TOKYO
MITAKA 181
JAPAN

TAKAKUBO KEIYA PROF
DEPT OF ASTRONOMY
TOHOKU UNIVERSITY
ARAMAKI SENDAI 980
JAPAN

TAKAKURA TATSUO PROF
DEPT OF ASTRONOMY
UNIVERSITY OF TOKYO
BUNKYO-KU
TOKYO 113
JAPAN

TAKAYANAGI KAZUO PROF
INST SPACE AERON SC
UNIVERSITY OF TOKYO
KOMABA MEGURO KU
TOKYO 153
JAPAN

TAKENS R DRS
ASTRONOMISCH INST
UNIV OF AMSTERDAM
ROETERSSTRAAT 15
1018 WB AMSTERDAM
THE NETHERLANDS

TALON R DR
CESR
9 AV DU COL ROCHE
B P 4335
F-31029 TOULOUSE CED
FRANCE

TAMMANN G ANDREAS PROF DR
ASTRONOMISCHES INST
UNIVERSITAET BASEL
VENUSST 7
CH-4102 BINNINGEN
SWITZERLAND

TANAKA HARUO PROF
TOKYO ASTRONOMICAL
OBSERVATORY
OSAWA MITAKA
TOKYO 181
JAPAN

TANAKA WATARU DR
DEPT OF ASTRONOMY
UNIV OF TOKYO
BUNKYO-KU
TOKYO 113
JAPAN

TANDBERG HANSSEN EINAR A
DR
NASA MARSHALL SPACE
FLIGHT CENTER ES51
HUNTSVILLE AL 35812
U.S.A.

TAPLEY BYRON D DR
DEPT AEROSPACE ENGR
AND ENGR MECHANICS
UNIV TX-AUST WRW 402
AUSTIN TX 78712
U.S.A.

TARENGHI MASSIMO DR
ESO
K SCHWARZSCHILD ST 2
D-8046 GARCHING B M
GERMANY, F.R.

TARTER JILL C DR
NASA-AMES RES CTR
MS 245-3
MOFFETT FLD CA 94035
U.S.A.

TATEVYAN S K DR
ASTRONOMICAL COUNCIL
USSR ACADEMY OF SCI
PYATNITSKAYA UL 48
109017 MOSCOW
U.S.S.R.

TAKARADA KATSUO DR
KYOTO UNIVERSITY IND
ARTS TEXTILE FIBERS
MATSUGASAKI SAKYOKU
KYOTO 606
JAPAN

TAKEDA HIDENORI DR
DEPT OF AERONAUTICAL
ENGINEERING
KYOTO UNIVERSITY
SAKYOKU KYOTO 606
JAPAN

TAKEUTI MINE DR
DEPT OF ASTRONOMY
TOHOKU UNIVERSITY
ARAMAKI SENDAI 980
JAPAN

TALWAR SATYA P DR
PHYSICS DEPT
DELHI UNIVERSITY
DELHI 7
INDIA

TAMURA SHINICHI DR
DEPT OF ASTRONOMY
TOHOKU UNIVERSITY
ARAMAKI SENDAI 980
JAPAN

TANAKA KATSUO DR
TOKYO ASTRONOMICAL
OBSERVATORY
OSAWA MITAKA
TOKYO 181
JAPAN

TANAKA YASUO DR
FACULTY OF EDUCATION
IBARAKI UNIVERSITY
BUNKYO MITO 310
JAPAN

TANDON JAGDISH NARAIN DR
DEPT OF PHYSICS AND
ASTROPHYSICS
UNIVERSITY OF DELHI
DELHI 110007
INDIA

TARADIJ VLADIMIR K DR
MAIN ASTRON OBS
UKRAINIAN AC OF SC
GOLOSEEVO
252127 KIEV
U.S.S.R.

TARNSTROM GUY DR
MIT LINCOLN LAB
P O BOX 73
LEXINGTON MA 02173
U.S.A.

TASSOUL JEAN-LOUIS PROF
DEPT DE PHYSIQUE
UNIVERSITE MONTREAL
CP 6128
MONTREAL PQ H3C 3J7
CANADA

TATON RENE PROF
CENTRE ALEXANDRE
KOYRE
12 RUE COLBERT
F-75002 PARIS
FRANCE

TAKASE BUNSHIRO PROF
TOKYO ASTRONOMICAL
OBSERVATORY
OSAWA MITAKA
TOKYO 181
JAPAN

TAKENOUCHI TADAO DR
1-28-30 KICHIJYOJI
KITA MACHI
MUSASHINO TOKYO 180
JAPAN

TALBOT RAYMOND J JR PROF
SPACE PHYSICS AND
ASTRONOMY
RICE UNIVERSITY
HOUSTON TX 77001
U.S.A.

TAMENAGA TATSUO DR
FACULTY OF EDUCATION
MIE UNIVERSITY
TSU SHI MIE 514
JAPAN

TANABE HIROYOSHI DR
TOKYO ASTRONOMICAL
OBSERVATORY
OSAWA MITAKA
TOKYO 181
JAPAN

TANAKA RIICHIRO PROF
FACULTY GENERAL EDUC
NIIGATA UNIVERSITY
ASAHIMACHIDORI
NIIGATA 951
JAPAN

TANAKA YASURO PROF
INST SPACE AERON SC
UNIVERSITY OF TOKYO
KOMABA MEGURO-KU
TOKYO 153
JAPAN

TAPIA MAURICIO DR
INSTITUTO DE
ASTRONOMIA UNAM
APDO POSTAL 877
ENSENADA BC
MEXICO

TARAFDAR S P DR
TATA INST FUND RES
HOMI BHABHA ROAD
BOMBAY 400 005
INDIA

TARTER C BRUCE DR
LAWRENCE LIVERMORE
LABORATORY L-71
UNIVERSITY OF CALIF
LIVERMORE CA 94550
U.S.A.

TASSOUL MONIQUE DR
C/O DEPT DE PHYSIQUE
UNIV DE MONTREAL
C P 6128
MONTREAL P Q H3C 3J7
CANADA

TATUM JEREMY B DR
UNIVERSITY OF VICTO-
RIA OBSERVATORY
VICTORIA BC V8W 2Y2
CANADA

TAUBER GERALD E PROF
DEPT OF PHYSICS
TEL AVIV UNIVERSITY
TEL AVIV
ISRAEL

TAWADROS MAHET JACOUB DR
HELWAN OBSERVATORY
HELWAN-CAIRO
ARAB REP. OF EGYPT

TAYLOR DONALD J DR
DEPT OF PHYSICS AND
ASTRONOMY
UN OF NEBRASKA
LINCOLN NB 68588
U.S.A.

TAYLOR KEITH DR
ROYAL GREENWICH
OBSERVATORY
HAILSHAM
EAST SUSSEX BN27 1RP
U.K.

TEDESCO EDWARD F
JET PROPULSION LAB
MS 183-501
4800 OAK GROVE DRIVE
PASADENA CA 91109
U.S.A.

TEJFEL VIKTOR G DR
LAB OF LUNAR AND
PLANETARY PHYSICS
ASTROPHYSICAL INST
480068 ALMA-ATA
U.S.S.R.

TEMESVARY S DR
R STRAUSSSTRASSE 127
D-8000 MUENCHEN 80
GERMANY, F.R.

TEPLITSKAYA R B DR
SIBIZMIR
P B 4
664697 IRKUTSK 33
U.S.S.R.

TEREBIZH VALERY YU DR
CRIMEAN STATION OF
STERNBERG
ASTRONOMICAL INST
NAUCHNY CRIMEA334413
U.S.S.R.

TERRAZAS MR L.R.
INST NAC DE ASTROFIS
OPT Y ELECTRONICA
APT POSTALES 216 51
PUEBLA
MEXICO

TERRILE RICHARD JOHN
MS 183-30
JET PROPULSION LAB
PASADENA CA 91109
U.S.A.

TERZIDES CH DR
DEPT OF ASTRONOMY
UNIV OF THESSALONIKI
THESSALONIKI
GREECE

TAVAKOL REZA K
DEPT OF APPLIED MATH
QUEEN MARY COLLEGE
MILE END ROAD
LONDON E1 4NS
U.K.

TAYLER ROGER J PROF
ASTRONOMY CENTRE
UNIVERSITY OF SUSSEX
BRIGHTON BN1 9QH
U.K.

TAYLOR GORDON E
ROYAL GREENWICH
OBSERVATORY
HAILSHAM
EAST SUSSEX BN27 1RP
U.K.

TAYLOR KENNETH N R PROF
105A COPELAND ROAD
BEECROFT NSW 2119
AUSTRALIA

TEERIKORPI VELI PEKKA DR
OBS AND ASTROPH LAB
UNIV OF HELSINKI
TAHTITORNINMAKI
SF-00130 HELSINKI 13
FINLAND

TEKTULANI H GOEKMEN DR
ISTANBUL UNIVERSITY
FACULTY OF SCIENCE
ISTANBUL
TURKEY

TEMPESTI PIERO PROF
OSSERVATORIO ASTRONO
MICO DI COLLURANIA
I-64100 TERAMO
ITALY

TER HAAR DIRK
MAGDALEN COLLEGE
OXFORD OX1 4AU
U.K.

TERENJEVA ALEXANDRA K DR
ASTRONOMICAL COUNCIL
USSR ACADEMY OF SC
PYATNITSKAYA 48
109017 MOSCOW
U.S.S.R.

TERRELL NELSON JAMES JR
ESS-9 MS/436
LOS ALAMOS NAT LAB
BOX 1663
LOS ALAMOS NM 87545
U.S.A.

TERZAN AGOP DR
OBSERVATOIRE DE LYON
F-69230 SAINT GENIS-
LAVAL
FRANCE

TESKE RICHARD G PROF
DEPT OF ASTRONOMY
UNIVERSITY OF
MICHIGAN
ANN ARBOR MI 48109
U.S.A.

TAVARES J T L DR
AV DIAS DA SILVA
173 R/C ESQ
COIMBRA
PORTUGAL

TAYLOR DONALD BOGGIA
2 ROSE COTTAGE
WARTLING NEAR
HAILSHAM EAST SUSSEX
U.K.

TAYLOR JOSEPH H PROF
UNIV MASSACHUSETTS
GRADUATE RES CTR
TOWER B
AMHERST MA 01003
U.S.A.

TECH JACK L DR
B 306 METROLOGY BLDG
NATIONAL BUREAU OF
STANDARDS
WASHINGTON DC 20234
U.S.A.

TEHERANY D
83 AVENUE REY
TEHERAN
IRAN

TELEKI GEORGE DR
ASTRONOMSKA
OPSERVATORIJA
VOLGINA 7
11050 BEOGRAD
YUGOSLAVIA

TENORIO-TAGLE G DR
MPI F PHYSIK & ASTRO
PHYSIK
D-8046 GARCHING B M
GERMANY, F.R.

TERASHITA YOICHI PROF
KANAZAWA INST TECHN
NONOICH CHO
MINAMIKYOKU
KANAZAWA 921
JAPAN

TERLEVICH ROBERTO JUAN
INST OF ASTRONOMY
MADINGLEY ROAD
CAMBRIDGE CB3 OHA
U.K.

TERRIEN JEAN DIRECTEUR
B I P M
PAVILLON DE BRETEUIL
F-92310 SEVRES
FRANCE

TERZIAN YERVANT PROF
SPACE SCIENCES BLDG
CORNELL UNIVERSITY
ITHACA NY 14853
U.S.A.

TEXEREAU JEAN M
CERGA
AVENUE COPERNIC
F-06130 GRASSE
FRANCE

THADDEUS PATRICK PROF
INSTITUTE FOR
SPACE STUDIES
2880 BROADWAY
NEW YORK NY 10025
U.S.A.

THE PIK SIN PROF
ASTRONOMICAL INST
ROETERSSTRAAT 15
1018 WB AMSTERDAM
THE NETHERLANDS

THERNOE KARL-AUGUST
LEKTOR
VINKELVEJ 36
DK-2800 LYNGBY
DENMARK

THIELHEIM KLAUS O DR
INST REINE & A KERNP
UNIVERSITAET KIEL
OLSHAUSENSTR 40/60
D-2300 KIEL
GERMANY, F.R.

THIRY YVES R PROF
UNIVERSITE DE PARIS
VI TOUR 66
4 PLACE JUSSIEU
F-75230 PARIS CDX 05
FRANCE

THOMAS DAVID V DR
SCIENCE RES COUNCIL
CENTRAL OFFICE
P O BOX 18
SWINDON SN2 1ET WILT
U.K.

THOMAS HANS-CHRISTOPH DR
MPI FUER PHYSIK UND
ASTROPHYSIK
FOEHRINGER RING 6
D-8000 MUENCHEN 40
GERMANY, F.R.

THOMAS JOHN A DR
PHYSICS RAAF DEPT
UNIV OF MELBOURNE
PARKVILLE VIC 3052
AUSTRALIA

THOMAS JOHN H PROF
DEPT MECH AERO SC
UNIVERSITY ROCHESTER
ROCHESTER NY 14627
U.S.A.

THOMAS RICHARD N DR
LAT
98 BIS BLVD ARAGO
F-75014 PARIS
FRANCE

THOMAS ROGER J DR
CODE 682
NASA GODDARD SPACE
FLIGHT CENTER
GREENBELT MD 20771
U.S.A.

THOMASSON PETER DR
N R A L
JODRELL BANK
MACCLESFIELD
CHESHIRE SK11 9DL
U.K.

THOMPSON A RICHARD DR
NAT RADIO ASTRO OBS
VLA PROJECT
P O BOX O
SOCORRO NM 87801
U.S.A.

THOMPSON G I DR
ROYAL OBSERVATORY
EDINBURGH EH9 3HJ
U.K.

THOMPSON LAIRD A DR
INST FOR ASTRONOMY
2680 WOODLAWN DRIVE
HONOLULU HI 96822
U.S.A.

THOMPSON ROGER I PROF
STEWARD OBSERVATORY
UNIVERSITY
OF ARIZONA
TUCSON AZ 85721
U.S.A.

THOMPSON THOMAS WILLIAM
2043 CLOUD CREST AVE
LA CRESCENTA
CA 91214
U.S.A.

THOMSEN BJARNE B PROF
INST OF ASTRONOMY
UNIVERSITY OF AARHUS
DK-8000 AARHUS C
DENMARK

THONNARD NORBERT DR
CARNEGIE INST WASH
DTM
5241 BROAD BR RD NW
WASHINGTON DC 20015
U.S.A.

THOREN VICTOR E PROF
130 GOODBODY HALL
INDIANA UNIVERSITY
BLOOMINGTON IN 47401
U.S.A.

THORNE KIP S PROF
130-33
CALTECH
PASADENA CA 91125
U.S.A.

THRONSON HARLEY ANDREW JR
DEPT OF PHYSICS AND
ASTRONOMY
UNIV OF WYOMING
LARAMIE WY 82071
U.S.A.

THUAN TRINH X DE
DEPT OF ASTRONOMY
UNIV OF VIRGINIA
BOX 3818 UNIV STA
CHARLOTTESVL VA22903
U.S.A.

THUM CLEMENS DR
MPI FUER
RADIOASTRONOMIE
AUF DEM HUEGEL 69
D-5300 BONN
GERMANY, F.R.

TIFFT WILLIAM G PROF
STEWARD OBSERVATORY
UNIVERSITY
OF ARIZONA
TUCSON AZ 85721
U.S.A.

TIFREA EMILIA DR
OBS DE BUCAREST
STR CUTITUL DE
ARGINT 5
75212 BUCAREST
ROUMANIA

TIMOTHY J GETHYN DR
DEPT ASTRO-GEOPHYS
CAMPUS BOX 391
UNIV OF COLORADO
BOULDER CO 80309
U.S.A.

TINBERGEN JAAP DR
KAPTEYN STERREWACHT
WERKGROEP
MENSINGHEWEG 20
9301 KA RODEN
THE NETHERLANDS

TING YEOU-TSWEN
ASTRONOMY SECTION
CENTRAL WEATHER BUR
64 KUNG YUEN ROAD
TAIPEI 100
TAIWAN

TIURI MARTTI PROF
HELSINKI UNIV TECHN
RADIO LABORATORY
OTAKAARI 5 A
SF-02150 ESPOO 15
FINLAND

TJIN-A-DJIE HERMAN R E DR
ASTRONOMICAL INST
UNIVERSITY AMSTERDAM
ROETERSSTRAAT 15
1018 WB AMSTERDAM
THE NETHERLANDS

TLAMICHA ANTONIN DR
ASTRONOMICAL
OBSERVATORY
25165 ONDREJOV
CZECHOSLOVAKIA

TOBIN WILLIAM
UNIVERSITY OBS
BUCHANAN GARDENS
ST ANDREWS
FIFE KY16 9LZ
U.K.

TODORAN IOAN DR
ASTRONOMICAL
OBSERVATORY
STR REPUBLICII 109
3400 CLUJ-NAPOCA
ROUMANIA

TOFANI GIANNI PROF
OSSERVATORIO ARCETRI
LARGO FERMI 5
I-50125 FIRENZE
ITALY

TOHLINE JOEL EDWARD
MS 288 GROUP T-6
LOS ALAMOS NAT LAB
P O BOX 1663
LOS ALAMOS NM 87545
U.S.A.

TOLBERT CHARLES R DR
LEANDER MCCORMICK OB
BOX 3818
UNIV STATION
CHARLOTTESVL VA22901
U.S.A.

TOMASI PAOLO DR
LAB RADIOASTRONOMIA
VIA IRNERIO 46
I-40126 BOLOGNA
ITALY

TOMASKO MARTIN G DR
LUNAR PLANETARY LAB
UNIV OF ARIZONA
SPACE SCIENCES BLDG
TUCSON AZ 85721
U.S.A.

TOMBAUGH CLYDE W PROF
DEPT OF ASTRONOMY
BOX 4500
NM STATE UNIVERSITY
LAS CRUCES NM 88003
U.S.A.

TOMIMATSU AKIRA DR
RES INST THEOR PHYS
HIROSHIMA UNIVERSITY
TAKEHARA-MACHI
TAKEHARA 725
JAPAN

TOMITA KENJI PROF
RESEARCH INSTITUTE F
THEORETICAL PHYSICS
HIROSHIMA UNIVERSITY
TAKEHARA 725
JAPAN

TOMITA KOICHIRO MR
TOKYO ASTRONOMICAL
OBSERVATORY
OSAWA MITAKA
TOKYO 181
JAPAN

TOMOV ALEXANDER NIKOLOV
DEPT OF ASTRONOMY
BULGARIAN ACADEMY SC
72 LENIN BLVD
1184 SOFIA
BULGARIA

TONG FU
PURPLE MOUNTAIN
OBSERVATORY
NANJING
PEOPLE'S REPUBLIC CHINA

TOOMRE ALAN DR
ROOM 2-371
MIT
77 MASSACHUSETTS AVE
CAMBRIDGE MA 02139
U.S.A.

TOOMRE JURI
DEPT ASTRO-GEOPHY-
SICS AND JILA
UNIV OF COLORADO
BOULDER CO 80309
U.S.A.

TOPAKTAS LATIF A DR
UNIVERSITY
OBSERVATORY
UNIVERSITY ISTANBUL
TURKEY

TORAO MASAHISA PROF
5-7-9-401 YOYOGI
SHIBUYA
TOKYO 151
JAPAN

TORELLI M DR
OSS ASTRONOMICO
VIA DEL P MELLINI 84
ROMA
ITALY

TOROSHLIDZE TEIMURAZ I DR
ABASTUMANI ASTRO-
PHYSICAL OBSERVATORY
383762 GEORGIA
U.S.S.R.

TORRES CARLOS ALBERTO DR
OBS NACIONAL
R GENERAL BRUCE 586
20000 RIO DE JANEIRO
BRAZIL

TORRES CARLOS DR
OBS ASTRON NACIONAL
UNIVERSIDAD DE CHILE
CASILLA 36-D
SANTIAGO
CHILE

TORRES PEIMBERT SILVIA DR
INST DE ASTRONOMIA
APDO POSTAL 70264
04510 MEXICO DF
MEXICO

TORROJA J PROF
CATEDRA DE ASTRON
FACULTAD DE CIENCIAS
UNIV COMPLUTENSE
MADRID 3
SPAIN

TOSA MAKOTO DR
DEPT OF PHYSICS
NAGOYA UNIVERSITY
FURUCHO CHIGUSA
NAGOYA 464
JAPAN

TOUSEY RICHARD DR
CODE 7140
NAVAL RESEARCH LAB
WASHINGTON DC 20375
U.S.A.

TOVMASSIAN H M DR
BYURAKAN ASTROPHYSI-
CAL OBSERVATORY
378433 ARMENIA
U.S.S.R.

TOWNES CHARLES HARD DR
DEPT OF PHYSICS
UNIVERSITY OF CALIF
RM 557 BIRGE HALL
BERKELEY CA 94720
U.S.A.

TOZER DAVID C DR
SCHOOL OF PHYSICS
UNIVERSITY NEWCASTLE
NEWCASTLE UPON TYNE
NE1 7RU
U.K.

TRAFTON LAURENCE M DR
ASTRONOMY DEPT
UNIVERSITY OF TEXAS
AT AUSTIN
AUSTIN TX 78712
U.S.A.

TRAN MINH N DR
OBS DE MEUDON
F-92190 MEUDON
FRANCE

TRAUB WESLEY ARTHUR
CTR FOR ASTROPHYSICS
60 GARDEN STREET
CAMBRIDGE MA 02138
U.S.A.

TRAVING GERHARD PROF
INSTITUT FUER THEORE
TISCHE ASTROPHYSIK
NEUENHEIMER FELD 294
D-6900 HEIDELBERG
GERMANY, F.R.

TREDER H J PROF DR
ZNTRINST ASTROPHYSIK
STERNW BABELSBERG
ROSA LUXEMBURGST 17A
DDR-1502 POTSDAM
G.D.R.

TREFFERS RICHARD R
ASTRONOMY DEPT
UNIV OF CALIFORNIA
BERKELEY CA 94720
U.S.A.

TREFFTZ ELEONORE E DR
MPI F PHYSIK UND
ASTROPHYSIK
FOEHRINGER RING 6
D-8000 MUENCHEN 40
GERMANY, F.R.

TREFZGER CHARLES F DR
ASTRON INST BASEL
VENUSSTRASSE 7
CH-4102 BINNINGEN
SWITZERLAND

TREHAN SURINDAR K PROF
DEPT OF MATHEMATICS
PANJAB UNIVERSITY
CHANDIGARH 160014
INDIA

TRELLIS MICHEL DR
OBSERVATOIRE DE NICE
B P 252
F-06007 NICE CEDEX
FRANCE

TREMAINE SCOTT DUNCAN
RM 6-211
DEPT OF PHYSICS MIT
77 MASSACHUSETTS AVE
CAMBRIDGE MA 02139
U.S.A.

TREMKO JOZEF DR
SKALNATE PLESO
OBSERVATORY
059 60 TATRANSKA
LOMNICA
CZECHOSLOVAKIA

TREUMANN RUDOLF DR
HOCHSTRASSE 58
CH-8044 ZUERICH
SWITZERLAND

TREVESE DARIO
OSS ASTRONOMICO
VIALE DEL PARCO
MELLINI 84
I-00136 ROMA
ITALY

TREXLER JAMES H MR
5609 OTTAWA STREET
LABORATORY
OXON HILL MD 20021
U.S.A.

TRIMBLE VIRGINIA L DR
DEPT OF PHYSICS
UNIVERSITY OF CALIF
IRVINE CA 92717
U.S.A.

TRITAKIS BASIL P DR
ASTRONOMY APPL MATHS
ACADEMY OF ATHENS
14 ANAGNOSTOPOULOU
ATHENS 136
GREECE

TRITTON KEITH P DR
ROYAL OBS EDINBURGH
BLACKFORD HILL
EDINBURGH EH9 3HJ
U.K.

TRITTON SUSAN BARBARA
ROYAL OBSERVATORY
BLACKFORD HILL
EDINBURGH EH9 3HJ
U.K.

TRODAHL HARRY JOSEPH DR
VICTORIA UNIVERSITY
WELLINGTON
NEW ZEALAND

TROITSKY V S PROF DR
RADIOPHYSICAL
RESEARCH INSTITUTE
603600 GORKIJ
U.S.S.R.

TROLAND THOMAS HUGH
PHYSICS ASTRONOMY
DEPT
UNIV OF KENTUCKY
LEXINGTON KY 40506
U.S.A.

TROTTET GERARD DR
OBS DE PARIS
DASOP
F-92190 MEUDON
FRANCE

TRUEMPER JOACHIM PROF
MPI FUER EXTRATERRE-
STRISCHE PHYSIK
D-8046 GARCHING BEI
MUENCHEN
GERMANY, F.R.

TRULSEN JAN K PROF
P O BOX 953
UNIVERSITY OF TROMSO
N-9001 TROMSO
NORWAY

TRURAN JAMES W JR
DEPT OF ASTRONOMY
UNIV OF ILLINOIS
OBSERVATORY
URBANA IL 61801
U.S.A.

TRUTSE YU L DR
INSTITUTE OF PHYSICS
OF ATMOSPHERE
USSR ACADEMY OF SCI
109017 MOSCOW
U.S.S.R.

TSAI CHANG-HSIEN DIRECTOR
GENERAL SECRETARY
ASTRO SOCIETY CHINA
TAIPEI OBSERVATORY
TAIPEI 104
TAIWAN

TSAO M PROF
NO 47 SEC 3
HSIN-I ROAD
TAIPEI 106
TAIWAN

TSAP T T DR
CRIMEAN ASTROPHYSI-
CAL OBSERVATORY USSR
ACAD OF SCI NAUCHNIY
334413 CRIMEA
U.S.S.R.

TSCHARNUTER WERNER M DR
MPI FUER ASTROPHYSIK
FOEHRINGER RING 6
D-8000 MUENCHEN 40
GERMANY, F.R.

TSESEVICH V P PROF DR
ODESSA STATE UNIVER-
SITY ASTRONOMICAL
OBSERVATORY
270014 ODESSA
U.S.S.R.

TSIKOUDI VASSILIKI PH D
DEPT OF ASTRONOMY
UNIV OF IOANNINA
IOANNINA
GREECE

TSIOUMIS ALEXANDROS DR
DEPT OF GEODETIC
ASTRONOMY
UNIV OF THESSALONIKI
THESSALONIKI
GREECE

TSUBAKI TOKIO PROF
FACULTY OF EDUCATION
SHIGA UNIVERSITY
ISHIYAMAHIRATSUCHO
OHTSU 520
JAPAN

TSUBOKAWA IETSUNE DR
INTERNATIONAL LATITU
DE OBSERVATORY
MIZUSAWA IWATE 023
JAPAN

TSUCHIYA ATSUSHI DR PROF
TOKYO ASTRONOMICAL
OBSERVATORY
OSAWA
OSAWA MITAKA TOK 181
JAPAN

TSUJI TAKASHI
TOKYO ASTRONOMICAL
OBSERVATORY
MITAKA
TOKYO 181
JAPAN

TSURUTA SACHIKO DR
MPI PHYS UND ASTROPH
FOEHRINGER RING 6
POSTFACH 401212
D-8000 MUENCHEN 40
GERMANY, F.R.

TSVETANOV ZLATAN IVANOV
DEPT OF ASTRONOMY
BULGARIAN ACADEMY SC
72 LENIN BLVD
1184 SOFIA
BULGARIA

TSVETKOV TSVETAN DR
DEPT OF ASTRONOMY
FACULTY OF PHYSICS
ANTON IVANOV STR 5
1126 SOFIA
BULGARIA

TUCKER ROY H
ROYAL GREENWICH
OBSERVATORY
HAILSHAM
EAST SUSSEX BN27 1RP
U.K.

TUCKER WALLACE H DR
CTR FOR ASTROPHYSICS
60 GARDEN STREET
CAMBRIDGE MA 02138
U.S.A.

TUEG HELMUT DR
ASTRONOMISCHES INST
UNIVERSITAET BOCHUM
POSTFACH 102148
D-4630 BOCHUM 1
GERMANY, F.R.

TUFEKCIOGLU ZEKI DR
ASTRONOMY DEPT
UNIV OF ANKARA
FEN FAKULTESI
ANKARA
TURKEY

TULL ROBERT G
DEPT OF ASTRONOMY
UNIVERSITY OF TEXAS
AT AUSTIN RLM 15 212
AUSTIN TX 78712
U.S.A.

TULLY JOHN A DR
OBSERVATOIRE DE NICE
BP 252
F-06007 NICE CEDEX
FRANCE

TULLY RICHARD BRENT DR
UNIVERSITY OF HAWAII
INST OF ASTRONOMY
2680 WOODLAWN DR
HONOLULU HI 96822
U.S.A.

TUNCA ZEYNEL DR
EGE UNIVERSITY
FACULTY OF SCIENCE
BORNOVA-IZMIR
TURKEY

TURLO ZYGMUNT DR
ASTRONOMICAL CENTER
UL CHOPINA 12-18
87-100 TORUN
POLAND

TURNER EDWIN L DR
PRINCETON UNIV OBS
PEYTON HALL
PRINCETON NJ 08540
U.S.A.

TURNER MICHAEL S
ASTRONOMY & ASTRO-
PHYSICS CENTER
5460 S ELLIS AVE
CHICAGO IL 60637
U.S.A.

TURTLE A J DR
DEPT OF PHYSICS
UNIVERSITY OF SYDNEY
SYDNEY NSW 2006
AUSTRALIA

TUZI KONOSUKE DR
2-2-20 OSAWA
MITAKA
TOKYO 181
JAPAN

TYLENDA ROMUALD DR
N COPERNICUS ASTRON
CENTER
CHOPINA 12/18
87-100 TORUN
POLAND

TYSON JOHN A DR
BELL LABS RM 1D-316
600 MOUNTAIN AVE
MURRAY HILL NJ 07974
U.S.A.

UCHIDA JUICHI DR
TOHOKU GAKUEN
UNIVERSITY
TAGAJYOMACHI
MIYAGI 985
JAPAN

UENO SUEO PROF
KANAZAWA INSTITUTE
OF TECHNOLOGY
NONOICHIMACHI
ISHIKAWA 921
JAPAN

ULICH BOBBY LEE
MULTIPLE MIRROR
TELESCOPE OBS
UNIV OF ARIZONA
TUCSON AZ 85721
U.S.A.

ULRICH BRUCE T PROF
ESO
C/O CERN
CH-1211 GENEVA 23
SWITZERLAND

TUOMINEN ILKKA V DR
OBSERVATORY
TAHTITORNINMAKI
SF-00130 HELSINKI 13
FINLAND

TURNER BARRY E DR
NRAO
EDGEMONT RD
CHARLOTTESVL VA22901
U.S.A.

TURNER KENNETH C DR
ARECIBO OBSERVATORY
P O BOX 995
ARECIBO PR 00613
U.S.A.

TURON PIERRE
INST D'ASTROPHYSIQUE
98 BIS BD ARAGO
F-75014 PARIS
FRANCE

TUTUKOV A V DR
ASTRONOMICAL COUNCIL
USSR ACADEMY OF SCI
PYATNITSKAYA UL 48
109017 MOSCOW
U.S.S.R.

TWISS R Q DR
C/O A R BOSCHI
96A HOLLAND ROAD
LONDON W14 8BD
U.K.

TYLER JR G L DR
CTR FOR RADAR ASTRON
STANFORD UNIV
STANFORD CA 94305
U.S.A.

TZVETKOV MILCHO K DR
BULGARIAN ACADEMY
OF SCIENCES
DEPT OF ASTRONOMY
1000 SOFIA
BULGARIA

UCHIDA YUTAKA PROF
TOKYO ASTRONOMICAL
OBSERVATORY
OSAWA
MITAKA TOKYO 181
JAPAN

UESUGI AKIRA DR
DEPT OF ASTRONOMY
UNIVERSITY OF KYOTO
SAKYOKU KYOTO 606
JAPAN

ULMER MELVILLE P PROF
DEARBORN OBSERVATORY
NORTHWESTERN
UNIVERSITY
EVANSTON IL 60201
U.S.A.

ULRICH MARIE HELENE D DR
ESO
K SCHWARZSCHILD ST 2
D-8046 GARCHING B M
GERMANY, F.R.

TUOMINEN JAAKKO V PROF
PIHLAJATIE 49 B 20
SF-00270 HELSINKI 27
FINLAND

TURNER DAVID G DR
DEPT PHYSICS&ASTRON
LAURENTIAN UNIV
SUDBURY ONT P3E 2C6
CANADA

TURNER MARTIN J L DR
X-RAY ASTRONOMY GR
PHYSICS DEPT
UNIV OF LEICESTER
LEICESTER LE1 7RH
U.K.

TURON-LACARRIEU C DR
OBS DE PARIS
DEPEG
F-92190 MEUDON
FRANCE

TUVE MERLE A DR
CARNEGIE INSTITUTION
5241 BROAD BRANCH RD
WASHINGTON DC 20015
U.S.A.

TWOKOWSKI A S
DEPT OF APPLIED MATH
QUEEN MARY COLLEGE
MILE END ROAD
LONDON E1 4NS
U.K.

TYPHOON LEE
C/O ACADEMIA SINICA
NANKUNG TAIPEI
TAIWAN

UBERTINI PIETRO
ISTITUTO ASTROFISICA
SPAZIALE
CASELLA POSTALE 67
I-00044 FRASCATI
ITALY

UDAL'TSOV V A DR
PHYSICAL INSTITUTE
USSR ACADEMY OF
SCIENCES
117924 MOSCOW
U.S.S.R.

ULFBECK OLE DR
NIELS BOHR INST
BLEGDAMSVEJ 17
DK-2100 COPENHAGEN-O
DENMARK

ULMSCHNEIDER PETER DR
INST THEOR ASTROPHYS
IM NEUENHEIMER F 294
D-6900 HEIDELBERG
GERMANY, F.R.

ULRICH ROGER K PROF
DEPT OF ASTRONOMY
UNIV OF CALIFORNIA
8931 MSB
LOS ANGELES CA 90024
U.S.A.

UNDERHILL ANNE B DR
CODE 680
GODDARD SPACE FL CTR
GREENBELT MD 20771
U.S.A.

UNDERWOOD JAMES H DR
INSTITUTE F PLASMA
RESEARCH
STANFORD UNIVERSITY
STANFORD CA 94305
U.S.A.

UNNO WASABURO PROF
DEPT OF ASTRONOMY
UNIVERSITY OF TOKYO
BUNKYO-KU
TOKYO 113
JAPAN

UNSOELD ALBRECHT PROF
INST THEO PHY STERNW
NEUE UNIV PHYSIK ZNT
OLSHAUSENST GB N 61C
D-2300 KIEL 1
GERMANY, F.R.

UPGREN ARTHUR R DR
VAN VLECK
OBSERVATORY
WESLEYAN UNIVERSITY
MIDDLETOWN CT 06457
U.S.A.

UPSON WALTER L II DR
PRINCETON UNIV OBS
PEYTON HALL
PRINCETON NJ 08540
U.S.A.

UPTON E K L DR
ASTRONOMY DEPT
UCLA
LOS ANGELES CA 90024
U.S.A.

URAS S DR
ISTITUTO DI
ASTRONOMIA
VIA OSPEDALE 72
I-09100 CAGLIARI
ITALY

URASIN L A DR
ENGELGARDT ASTRONOMI
CAL OBSERVATORY
422526 KAZAN
U.S.S.R.

URBANIK MAREK
ASTRONOMICAL OBS
JAGIELLONIAN UNIV
UL ORLA 171
30-244 KRAKOW
POLAND

URBARZ H DR
ASTRONOMISCHES INST
DER UNIVERSITAET
WALDHAUSER STR 64
D-7400 TUEBINGEN
GERMANY, F.R.

URECHE VASILE DR
UNIVERSITY OF CLUJ
NAPOCA
3400 CLUJ NAPOCA
ROUMANIA

USHER PETER D DR
DEPT ASTRONOMY
PENNSYLVANIA ST UNIV
507 DAVEY LAB
UNIV PARK PA 16802
U.S.A.

UTSUMI KAZUHIKO DR
FAC OF GEN EDUCATION
HIROSHIMA UNIVERSITY
HIGASHI-SENDACHO
HIROSHIMA 730
JAPAN

UUS UNDO DR
IAPHA ESTONIAN ACAD
OF SCIENCE
TORAVERE
202444 TARTU
U.S.S.R.

VAGER ZEEV DR
DEPT OF PHYSICS
WEIZMANN INSTITUTE
REHOVOT
ISRAEL

VAGHI SERGIO DR
ESTEC/PHA
POSTBUS 299
2200 AG NOORDWIJK
THE NETHERLANDS

VAIANA GIUSEPPE S DR
HARVARD COLLEGE OBS
HARVARD UNIVERSITY
60 GARDEN ST
CAMBRIDGE MA 02138
U.S.A.

VAIDYA P C PROF
VICE CHANCELLOR
GUJARAT UNIVERSITY
AHMEDABAD 380009
INDIA

VAINSTEIN L A DR
PHYSICAL INSTITUTE
USSR ACADEMY OF
SCIENCES
117924 MOSCOW
U.S.S.R.

VALBOUSQUET ARMAND DR
OBSERVATOIRE
DE STRASBOURG
11 RUE L UNIVERSITE
F-67000 STRASBOURG
FRANCE

VALENTIJN EDWIN A DR
ESO
KARL-SCHWARZSCHILD-
STR 2
D-8046 GARGING B M
GERMANY, F.R.

VALLEE JACQUES P DR
HERZBERG INST CNRC
100 SUSSEX DRIVE
OTTAWA ONT K1A OR6
CANADA

VALNICEK BORIS DR
ASTRONOMICAL INST
CZECHO ACADEMY SC
OBSERVATORY
251 65 ONDREJOV
CZECHOSLOVAKIA

VALTIER J C DR
OBS DE NICE
B P 252
F-06007 NICE CEDEX
FRANCE

VALTONEN MAURI J DR
DEPT PHYSICAL SCS
UNIVERSITY OF TURKU
SF-20500 TURKU 50
FINLAND

VAN AGT S L TH J DR
STERRENKUNDIG INST
TOERNOOIVELD
6525 ED NIJMEGEN
THE NETHERLANDS

VAN ALBADA TJEERD S DR
KAPTEYN LABORATORIUM
POSTBUS 800
9700 AV GRONINGEN
THE NETHERLANDS

VAN ALLEN JAMES A PROF
DEPT OF PHYSICS AND
ASTRONOMY
UNIVERSITY OF IOWA
IOWA CITY IA 52242
U.S.A.

VAN ALTENA WILLIAM F PROF
YALE UNIVERSITY
OBSERVATORY BOX 2023
YALE STATION
NEW HAVEN CT 06520
U.S.A.

VAN BEEK F DR
LAB VOOR RUIMTE-
ONDERZOEK
BENELUXLAAN 21
3527 HS UTRECHT
THE NETHERLANDS

VAN BLERKOM DAVID J PROF
ASTRONOMY DEPT
UNIVERSITY OF MASS
AMHERST MA 01002
U.S.A.

VAN BREDA IAN G DR
ROYAL GREENWICH
OBSERVATORY
HAILSHAM
EAST SUSSEX BN27 1RP
U.K.

VAN BUEREN HENDRIK G PROF
ADVISORY COUNCIL FOR
SCIENCE POLICY
P O BOX 18524
2502 EM THE HAGUE
THE NETHERLANDS

VAN CITTERS GORDON W DR
DEPT OF ASTRONOMY
UNIV OF TEXAS
AUSTIN TX 78712
U.S.A.

VAN DE HULST H C PROF DR
STERREWACHT
POSTBUS 9513
2300 RA LEIDEN
THE NETHERLANDS

VAN DE KAMP PETER
15 WELLESLEY ROAD
SWARTHMORE PA 19081
U.S.A.

VAN DE STADT HERMAN DR
ASTRONOMICAL
INSTITUTE
ZONNENBURG 2
3512 NL UTRECHT
THE NETHERLANDS

VAN DEN BERGH SIDNEY PROF
HERZBERGH INST ASTRO
DOMINION ASTRO OBS
5071 W SAANICH ROAD
VICTORIA BC V8X 4M6
CANADA

VAN DEN HEUVEL EDWARD P J
ASTRONOMICAL INST
ROETERSSTRAAT 15
1018 WB AMSTERDAM
THE NETHERLANDS

VAN DER BORGHT RENE PROF
DEPT OF MATHEMATICS
MONASH UNIVERSITY
CLAYTON VICT 3168
AUSTRALIA

VAN DER HUCHT KAREL A DR
SPACE RESEARCH LAB
BENELUXLAAN 21
3527 HS UTRECHT
THE NETHERLANDS

VAN DER HULST J M DR
NETHERLANDS FOUNDA-
TION RADIOASTRONOMY
POSTBUS 2
7990 AA DWINGELOO
THE NETHERLANDS

VAN DER HULST THIJS
3235 1ST AVENUE
S MINNEAPOLIS
MN 55408
U.S.A.

VAN DER KRUIT PIETER C DR
KAPTEYN LABORATORIUM
POSTBUS 800
9700 AV GRONINGEN
THE NETHERLANDS

VAN DER LAAN H PROF DR
STERREWACHT
POSTBUS 9513
2300 RA LEIDEN
THE NETHERLANDS

VAN DER RAAY HERMAN B
DEPT OF PHYSICS
UNIV OF BIRMINGHAM
P O BOX 363
BIRMINGHAM B15 2TT
U.K.

VAN DESSEL EDWIN LUDO DR
KON STERRENWACHT
RINGLAAN 3
B-1180 BRUSSELS
BELGIUM

VAN DIGGELEN J DR
STERREWACHT
ZONNENBURG 2
3512 NL UTRECHT
THE NETHERLANDS

VAN DORN BRADT H DR
MIT RM 37-581
CTR FOR SPACE RES
CAMBRIDGE MA 02139
U.S.A.

VAN DUINEN R J DR
FOKKER BV
P O BOX 7600
1117 ZJ SCHIPHOL
THE NETHERLANDS

VAN FLANDERN THOMAS DR
US NAVAL OBSERVATORY
WASHINGTON DC 20390
U.S.A.

VAN GENDEREN A.M. DR.
STERRENWACHT LEIDEN
POSTBUS 9513
2300 RA LEIDEN
THE NETHERLANDS

VAN HAMME WALTER
STERREKUNDIG OBSER-
VATORIUM RUG
KRIJGSLAAN 281 S9
B-9000 GENT
BELGIUM

VAN HERK G
STERREWACHT VAN
LEIDEN
POSTBUS 9513
2300 RA LEIDEN
THE NETHERLANDS

VAN HERK-KLUYVER H A DR
KORTE WAARDER 58
2415 AV NIEUWERBRUG
THE NETHERLANDS

VAN HOOF A PROF EM
ASTRONOMISCH
INSTITUUT KU LEUVEN
NAAMSESTRAAT 61
B-3000 LEUVEN
BELGIUM

VAN HORN HUGH M PROF
DEPT PHYSICS AND
ASTRONOMY
UNIVERSITY ROCHESTER
ROCHESTER NY 14627
U.S.A.

VAN HOUTEN C J DR
STERREWACHT
POSTBUS 9513
2300 RA LEIDEN
THE NETHERLANDS

VAN HOUTEN-GROENEVELD
INGRID DR
STERREWACHT LEIDEN
POSTBUS 9513
2300 RA LEIDEN
THE NETHERLANDS

VAN HOVEN GERARD DR
DEPT OF PHYSICS
UNIV OF CALIFORNIA
IRVINE CA 92717
U.S.A.

VAN LEER B DR
STERREWACHT
POSTBUS 9513
2300 RA LEIDEN
THE NETHERLANDS

VAN NIEUWKOOP J DR IR
STERREWACHT
SONNENBORGH
ZONNENBURG 2
3512 WK UTRECHT
THE NETHERLANDS

VAN PARADIJS J DR
ASTRONOMICAL INST
ROETERSSTRAAT 15
1018 WB AMSTERDAM
THE NETHERLANDS

VAN REGEMORTER HENRI DR
OBSERVATOIRE
DE MEUDON
F-92190 MEUDON
FRANCE

VAN RENSBERGEN WALTER DR
ASTROPHYSISCH INST
VRIJE UNIVERSITEIT
PLEINLAAN 2
B-1050 BRUSSEL
BELGIUM

VAN RIPER KENNETH A DR
MS 625 X-7
LOS ALAMOS NAT LAB
P O BOX 608
LOS ALAMOS NM 87545
U.S.A.

VAN SPEYBROECK LEON P DR
CTR FOR ASTROPHYSICS
60 GARDEN ST
CAMBRIDGE MA 02138
U.S.A.

VAN T VEER FRANK DR
INST D ASTROPHYSIQUE
98 BIS BLVD ARAGO
F-75014 PARIS
FRANCE

VAN T VEER-MENNERET
CLAUDE DR
INST D ASTROPHYSIQUE
98 BIS BLVD ARAGO
F-75014 PARIS
FRANCE

VAN WOERDEN HUGO PROF DR
KAPTEYN LABORATORIUM
POSTBUS 800
9700 AV GRONINGEN
THE NETHERLANDS

VANDEN BOUT PAUL A
ASTRONOMY DEPT
RLM 15 212
UNIVERSITY OF TEXAS
AUSTIN TX 78712
U.S.A.

VANDENBERG DON DR
C/O PHYSICS DEPT
UNIV OF VICTORIA
P O BOX 1700
VICTORIA BC V8W 2Y2
CANADA

VANDERLINDEN HENRI L PROF
UNIVERSITEIT GENT
HERTOGENDREEF 15
B-1170 BRUSSEL
BELGIUM

VANDERVOORT PETER O DR
DEPT ASTRON ASTROPH
1100-14 E 58TH ST
CHICAGO IL 60637
U.S.A.

VANYSEK VLADIMIR PROF
ASTRONOMY DEPT
CHARLES UNIVERSITY
SVEDSKA 8
150 00 PRAHA 5
CZECHOSLOVAKIA

VAPILLON LOIC J DR
OBSERVATOIRE DE
PARIS-MEUDON
F-92190 MEUDON
FRANCE

VARDANIAN R A DR
BYURAKAN ASTROPHYSI-
CAL OBSERVATORY
378433 ARMENIA
U.S.S.R.

VARDAVAS ILIAS MIKHAIL
RESEARCH SCHOOL OF
PHYSICAL SCIENCES
P O BOX 4
CANBERRA ACT 2600
AUSTRALIA

VARDYA M S DR
TATA INSTITUTE OF
FUNDAMENTAL RESEARCH
HOMI BHABHA RD
BOMBAY 400 005
INDIA

VARMA RAM KUMAR PROF
PHYSICAL RESEARCH
LAB
AHMEDABAD 380 009
INDIA

VARSAVSKY C M DR
LAS HERAS 1975 6 A
BUENOS AIRES
ARGENTINA

VARSHALOVICH D A DR
PHYSIKO-TECHNICAL
INSTITUTE USSR ACADE
MY OF SCIENCES
194021 LENINGRAD
U.S.S.R.

VARVOGLIS CH DR
DEPT OF ASTRONOMY
UNIV OF THESSALONIKI
THESSALONIKI
GREECE

VASHKOV'YAK S N DR
STERNBERG STATE AS-
TRONOMICAL INSTITUTE
117234 MOSCOW
U.S.S.R.

VASILEV VLADIMIR M DR
PULKOVO OBSERVATORY
196140 LENINGRAD
U.S.S.R.

VASILEVA GALINA J DR
PULKOVO OBSERVATORY
196140 LENINGRAD
U.S.S.R.

VASILEVSKIS STANISLAUS
PROF
455 GRAND AVE NO 16
PALO ALTO CA 94306
U.S.A.

VAUCLAIR GERARD P DR
OBSERVATOIRE DE
PARIS-MEUDON
F-92190 MEUDON
FRANCE

VAUCLAIR SYLVIE D DR
OBSERVATOIRE DE
PARIS-MEUDON
F-92190 MEUDON
FRANCE

VAUGHAN ARTHUR H DR
HALE OBSERVATORIES
813 SANTA BARBARA ST
PASADENA CA 91101
U.S.A.

VAZQUEZ MANUEL DR
INST ASTROFISICA
DE CANARIAS
UNIV DE LA LAGUNA
TENERIFE
SPAIN

VEEDER GLENN J DR
JPL
4800 OAK GROVE DRIVE
PASADENA CA 91103
U.S.A.

VEIS GEORGE PH D
GEODESY LAB
NAT TECHNICAL UNIV
ATHENS
GREECE

VEISMANN UNO DR
TARTU OBSERVATORY
202444 ESTONIA
U.S.S.R.

VELGHE ALBERT G PROF DR
OBSERVATOIRE ROYAL
DE BELGIQUE
AVENUE CIRCULAIRE 3
B-1180 BRUSSELS
BELGIUM

VELTMANN ULO ILMAR DR
TARTU OBSERVATORY
202444 ESTONIA
U.S.S.R.

VELUSAMY T DR
RADIO ASTRONOMY CTR
P O BOX NO 8
OOTACAMUND 643 001
INDIA

VENKATESAN DORASWAMY DR
PHYSICS DEPT
UNIV OF CALGARY
CALGARY ALB T2N 1N4
CANADA

VENUGOPAL V R DR
RADIO ASTRONOMY
CENTRE OF TIFR
POST BOX 8
OOTACAMUND 643001
INDIA

VERBEEK PAUL DR
GEORGE MINNELAAN 50
9830 S MARTENS-LATEM
BELGIUM

VERDET JEAN-PIERRE DR
OBSERVATOIRE
DE PARIS
61 AVE OBSERVATOIRE
F-75014 PARIS
FRANCE

VERGNANO A PROF
OSS ASTRONOMICO
10025 PINO TORINESE
ITALY

VERNIANI FRANCO PROF
CNR METEOR PHYSICS
VIA DEI CASTAGNOLI 1
I-40126 BOLOGNA
ITALY

VERON MARIE-PAULE DR
OBSERVATOIRE DE
MEUDON
F-92190 MEUDON
FRANCE

VERON PHILIPPE DR
OBSERVATOIRE
DE MEUDON
F-92190 MEUDON
FRANCE

VERSCHUUR GERRIT L PROF
FISKE PLANETARIUM
UNIVERSITY OF
COLORADO
BOULDER CO 80302
U.S.A.

VESECKY J F DR
DEPT OF ASTRONOMY
THE UNIVERSITY
LEICESTER LE1 7RH
U.K.

VETESNIK MIROSLAV DR
DEPT OF ASTRONOMY
PURKYNE UNIVERSITY
KOTLARSKA 2
61137 BRNO
CZECHOSLOVAKIA

VETTOLANI GIAMPAOLO
ISTITUTO DI
RADIOASTRONOMIA CNR
VIA IRNERIO 46
I-40126 BOLOGNA
ITALY

VEVERKA JOSEPH DR
312 SPACE SC BLDG
CORNELL UNIVERSITY
ITHACA NY 14853
U.S.A.

VIAL JEAN-CLAUDE
LPSP
B P 10
F-91370 VERRIERES-LE
BUISSON
FRANCE

VICENTE RAIMUNDO O PROF
R MESTRE AVIZ 30 R C
P 1495 LISBOA
PORTUGAL

VIDAL J L DR
OBS DU PIC-DU-MIDI
F-65200 BAGNERES DE
BIGORRE
FRANCE

VIDAL N V DR
DEPT OF PHYSICS AND
ASTRONOMY
TEL-AVIV UNIVERSITY
TEL-AVIV
ISRAEL

VIDAL-MADJAR ALFRED DR
L P S P
BP 10
F-91370 VERRIERES LE
BUISSON
FRANCE

VIGIER JEAN-PIERRE DR.
INSTITUT H.POINCARE
11 RUE P.ET M.CURIE
F-75005 PARIS
FRANCE

VIGOTTI MARIO
LABORATORIO DI
RADIOASTRONOMIA
VIA IRNERIO 46
I-40126 BOLOGNA
ITALY

VIIK TONU DR
TORAVERE OBS
202444 TARTU RAJ EST
U.S.S.R.

VILA SAMUEL C PROF
DEPT OF ASTRONOMY
UNIVERSITY OF PENNA
33RD AND WALNUT STR
PHILADELPHIA PA19104
U.S.A.

VILHENA DE MORAES R DR
DEPTO DE ASTRONOMIA
ITA-CTA
12 200 SAO JOSE DOS
CAMPOS SAO PAULO
BRAZIL

VILHU OSMI DR
OBS AND ASTROPHY LAB
UNIV OF HELSINKI
TAHTITORNINMAKI
SF-00130 HELSINKI 13
FINLAND

VINER MELVYN R DR
ASTRONOMY GROUP
DEPT OF PHYSICS
QUEEN'S UNIVERSITY
KINGSTON ONT K7L 3N6
CANADA

VINOD S KRISHNAN MRS DR
INDIAN INST OF
ASTROPHYSICS
BANGALORE 560 034
INDIA

VINTI JOHN P DR
RM W91-202
M I T
MEASUREMENT SYST LAB
CAMBRIDGE MA 02139
U.S.A.

VIOTTI ROBERTO DR
CNR ASTROFISICA
SPAZIALE C P 67
I-00044 FRASCATI
ROMA
ITALY

VIRGOPIA N DR
OSS ASTRONOMICO
VIA TRIONFALE 204
I-00136 ROMA
ITALY

VISHVESWARA C V PROF
RAMAN RESEARCH
INST
BANGALORE 560 080
INDIA

VISVANATHAN NATARAJAN DR
MT STROMLO
OBSERVATORY
WODEN PO ACT 2606
AUSTRALIA

VITINSKIJ YURIJ I DR
PULKOVO OBSERVATORY
196140 LENINGRAD
U.S.S.R.

VITON MAURICE DR
L A S
ALLEE PEIRESC
F-13012 MARSEILLE
FRANCE

VITTONE ALBERTO ANGELO
OSS ASTRONOMICO
DI CAPODIMONTE
VIA MOIARIELLO 16
I-80131 NAPOLI
ITALY

VIVES TEODORO JOSE DR
CTR ASTRON H ALEMAN
REINA 50 90B
APT CORREOS 511
ALMERIA
SPAIN

VLACHOS DEMETRIUS G PROF
UNIV THESSALONIKI
DEPT OF GEODESY
THESSALONIKI
GREECE

VOELK HEINRICH J PROF
MPI FUER KERNPHYSIK
POSTFACH 103 980
D-6900 HEIDELBERG
GERMANY, F.R.

VOGT NIKOLAUS DR
UNIV STERNWARTE
SCHEINERSTR 1
D-8000 MUENCHEN 80
GERMANY, F.R.

VOGT STEVEN SCOTT
LICK OBSERVATORY
UNIV OF CALIFORNIA
SANTA CRUZ CA 95064
U.S.A.

VOIGT HANS H PROF
UNIVERSITAETS-
STERNWARTE
GEISMARLANDSTR 11
D-3400 GOETTINGEN
GERMANY, F.R.

VOLLAND H DR
ASTRONOMISCHES INST
DER UNIVERSITAET
AUF DEM HUEGEL 71
D-5300 BONN
GERMANY, F.R.

VOLONTE SERGE DR
19 RUE CAPORAL
TRESIGNIES
B-7020 HYON
BELGIUM

VON BORZESZKOWSKI H H DR
ZENTR INST ASTROPHYS
DER ADW DER DDR
ROSA-LUXEMBURGST 17A
DDR-1502 POTSDAM
G.D.R.

VON DER HEIDE JOHANN DR
ALARDUSSTR 12
D-2000 HAMBURG 19
GERMANY, F.R.

VON HOERNER SEBASTIAN DR
NATIONAL RADIO ASTRO
NOMY OBSERVATORY
P O BOX 2
GREEN BANK WV 24944
U.S.A.

VON SOCHER HERMANN DR
INST FUER ASTRONOMIE
DER UNIV WIEN
TUERKENSCHANZSTR 17
A-1180 WIEN
AUSTRIA

VON WEIZSAECKER C F
PROF DR
MAX-PLANCK-INSTITUT
RIEMERSCHMID STR 7
D-8130 STARNBERG
GERMANY, F.R.

VONDRAK JAN DR
ASTRONOMICAL INST
CSAV
BUDECSKA 6
120 23 PRAHA 2
CZECHOSLOVAKIA

VORONTSOV-VEL'YAMINOV B A
PROF DR
STERNBERG STATE AS-
TRONOMICAL INSTITUTE
117234 MOSCOW
U.S.S.R.

VOROSHILOV V I DR
MAIN ASTRONOMICAL
OBSERVATORY UKRAINI-
AN ACAD OF SCIENCES
252127 KIEV
U.S.S.R.

VORPAHL JOAN A DR
AEROSPACE
CORPORATION
PO BOX 92957
LOS ANGELES CA 90009
U.S.A.

VRBA FREDERICK J DR
U S NAVAL OBS
P O BOX 1149
FLAGSTAFF AZ 86002
U.S.A.

VREUX JEAN MARIE DR
INST D ASTROPHYSIQUE
UNIVERSITE DE LIEGE
B-4200 COINTE OUGREE
BELGIUM

VSEKHSVYATSKIJ SERGIUS K
PROF
KIEV ASTRONOMICAL
OBSERVATORY
252053 KIEV PER 3
U.S.S.R.

VU D DR
BUREAU D LONGITUDES
77 AVENUE DENFERT
ROCHEREAU
F-75014 PARIS
FRANCE

VUJNOVIC VLADIS DR
INSTITUTE OF PHYSICS
OF THE UNIVERSITY
P O B 304
41001 ZAGREB
YUGOSLAVIA

VUKICEVIC K M PROF DR
DEPT OF ASTRONOMY
FACULTY OF SCIENCES
STUDENTSKI TRG 16
11000 BEOGRAD
YUGOSLAVIA

VYALSHIN GENNADIJ F DR
MAIN ASTRONOMICAL OB
PULKOVO M-140
196140 LENINGRAD
U.S.S.R.

WACHMANN A A PROF DR
SCHMIEDESBERG 2 B
D-2057 REINBEK
GERMANY, F.R.

WACKERNAGEL H BEAT DR
2939 COUNTRY CLUB DR
COLORADO SPRINGS
CO 80909
U.S.A.

WADDINGTON C JAKE PROF
SCHOOL PHYS ASTRON
UNIV MINNESOTA
116 CHURCH ST S.E.
MINNEAPOLIS MN 55455
U.S.A.

WADE CAMPBELL M DR
NAT RADIO ASTRON OBS
P O BOX O
SOCORRO NM 87801
U.S.A.

WAGNER RAYMOND L DR
FORD AEROSPACE
2880 E.FOUNTAIN BLVD
F-10
COLO SPGS CO 80910
U.S.A.

WAGNER WILLIAM J DR
HIGH ALTITUDE
OBSERVATORY
P O BOX 3000
BOULDER CO 80307
U.S.A.

WAGONER ROBERT V PROF
DEPT OF PHYSICS
STANFORD UNIVERSITY
STANFORD CA 94305
U.S.A.

WAKAMATSU KEN-ICHI DR
JUNIOR TECHN COLLEGE
GIFU UNIVERSITY
KAGAMIGAHARA
GIFU 504
JAPAN

WAKO KOJIRO DR
INTERNATIONAL LATITU
DE OBSERVATORY
MIZUSAWA IWATE 023
JAPAN

WALBORN NOLAN R DR
CERRO TOLOLO INTER-
AMERICAN OBSERVATORY
CASILLA 63 D
LA SERENA
CHILE

WALDMEIER MAX PROF
SWISS FEDERAL OBS
SCHMELZBERGSTR 25
CH-8006 ZUERICH
SWITZERLAND

WALKER ALISTAIR ROBIN DR
SAAO
P O BOX 9
OBSERVATORY
CAPE
SOUTH AFRICA

WALKER ARTHUR B C JR PROF
INST FOR PLASMA
RESEARCH
STANFORD UNIVERSITY
STANFORD CA 94305
U.S.A.

WALKER EDWARD N MR
ROYAL GREENWICH OBS
HERSTMONCEUX CASTLE
HAILSHAM
E SUSSEX BN27 1RP
U.K.

WALKER GORDON A H PROF
INST ASTRON SPACE SC
UNIVERSITY OF B C
2075 WESBROOK PLACE
VANCOUVER BC V6T 1W5
CANADA

WALKER MERLE F PROF
LICK OBSERVATORY
UNIVERSITY OF CALIF
SANTA CRUZ CA 95064
U.S.A.

WALKER R L
NAVAL OBSERVATORY
FLAGSTAFF STATION
BOX 1149
FLAGSTAFF AZ 86002
U.S.A.

WALKER ROBERT M A PROF
PHYSICS DEPT
BOX 1105
WASHINGTON UNIV
ST LOUIS MO 63130
U.S.A.

WALKER WILLIAM S G
14 APPLEYARD CRES
AUCKLAND 5
NEW ZEALAND

WALL JASPER V DR
MULLARD RADIO ASTRO-
OBS CAVENDISH LAB
MADINGLEY RD
CAMBRIDGE CB3 OHE
U.K.

WALLACE LLOYD V DR
KITT PEAK NATIONAL
OBSERVATORY
P O BOX 26732
TUCSON AZ 85726
U.S.A.

WALLACE PATRICK T MR
STARLINK RUTHERFORD
& APPLETON LABS
CHILTON DIDCOT
OXON OX11 OOX
U.K.

WALLENQUIST AAKE A E PROF
NORRLANDSGATAN 34 D
S-752 29 UPPSALA
SWEDEN

WALLERSTEIN GEORGE PROF
ASTRONOMY DEPT FM 20
UNIVERSITY OF
WASHINGTON
SEATTLE WA 98195
U.S.A.

WALLIS MAX K. DR
DEPT APP MATH ASTRON
UNIVERSITY COLLEGE
P O BOX 78
CARDIFF CF1 1XL
U.K.

WALMSLEY C MALCOLM DR
MPI FUER RADIO-
ASTRONOMIE
AUF DEM HUEGEL 69
D-5300 BONN 1
GERMANY, F.R.

WALRAVEN TH DR
P O BOX 98
CORNELIA
ORANGE FREESTATE
9850
SOUTH AFRICA

WALSH DENNIS DR
NRAL
JODDRELL BANK
MACCLESFIELD
CHESHIRE SK11 9DL
U.K.

WALTER HANS G DR
ASTRONOMISCHES
RECHEN-INSTITUT
MONCHHOFSTER 12-14
D-6900 HEIDELBERG 1
GERMANY, F.R.

WALTER KURT PROF DR
ASTRONOMISCHES INSTI
TUT DER UNIVERSITAET
WALDHAUSERSTR 64
D-7400 TUEBINGEN
GERMANY, F.R.

WAMPLER E JOSEPH PROF
LICK OBSERVATORY
UNIVERSITY OF CALIF
SANTA CRUZ CA 95064
U.S.A.

WAMSTEKER W DR
EUROPEAN SOUTHERN
OBSERVATORY
SCHLEISSHEIMERSTR 17
D-8046 GARCHING
GERMANY, F.R.

WAN LAI
ZO-SE SECTION
SHANGHAI OBSERVATORY
SHANGHAI
PEOPLE'S REPUBLIC CHINA

WAN TONG-SHAN
SHANGHAI OBSERVATORY
SHANGHAI
PEOPLE'S REPUBLIC CHINA

WANAS M I DR
FACULTY OF SCIENCE
CAIRO UNIVERSITY
GEZA ORMAN
ARAB REP. OF EGYPT

WANG SHOU-GUAN
BEIJING OBSERVATORY
BEIJING
PEOPLE'S REPUBLIC CHINA

WANNIER PETER GREGORY DR
CALTECH
OWENS VLY RADIO OBS
PASADENA CA 91125
U.S.A.

WAPSTRA AALDERT H PROF
VAN WOENSEL KOOIJ-
LAAN 48
1411 JZ NAARDEN
THE NETHERLANDS

WARD MARTIN JOHN
INST OF ASTRONOMY
MADINGLEY ROAD
CAMBRIDGE CB3 OHA
U.K.

WARD RICHARD A DR
MPI FUER ASTROPHYSIK
K-SCHWARZSCHILD-ST 1
D-8046 GARCHING
GERMANY, F.R.

WARD WILLIAM R DR
JPL
4800 OAK GROVE DRIVE
PASADENA CA 91103
U.S.A.

WARDLE JOHN F C PROF
PHYSICS DEPT
BRANDEIS UNIVERSITY
WALTHAM MA 02154
U.S.A.

WARES GORDON W DR
73 PERKINS ST
WEST NEWTON MA 02165
U.S.A.

WARMAN J DR
INST DI ASTRONOMIA
UNAM
APT POSTAL 70-264
04510 MEXICO DF
MEXICO

WARNER BRIAN PROF
ASTRONOMY DEPT
UNIVERSITY OF CAPE
TOWN RONDEBOSCH
CAPE TOWN 7700
SOUTH AFRICA

WARNER JOHN W DR
NASA CODE TA-01
MARSHALL SFC AL35812
U.S.A.

WARNER PETER J DR
CAVENDISH LABS
MADINGLEY ROAD
CAMBRIDGE CB3 OHE
U.K.

WARREN WAYNE H JR DR
NASA-GSFC
CODE 601
GREENBELT MD 20771
U.S.A.

WARWICK JAMES W DR
ASTRO-GEOPHYSICS
DEPT
UNIV OF COLORADO
BOULDER CO 80309
U.S.A.

WARWICK ROBERT S DR
PHYSICS DEPT
UNIV OF LEICESTER
LEICESTER LE1 7RH
U.K.

WARZEE J DR
115 RUE MATTOT
B-1410 WATERLOO
BELGIUM

WASHIMI HARUICHI DR
RESEARCH INSTITUTE
OF ATMOSPHERICS
NAGOYA UNIVERSITY
TOYOKAWA AICHI 442
JAPAN

WASSERMAN LAWRENCE H DR
LOWELL OBS
P O BOX 1269
FLAGSTAFF AZ 86002
U.S.A.

WASSON JOHN T
DEPT CHEMISTRY
UNIVERSITY OF CALIF
LOS ANGELES CA 90024
U.S.A.

WATANABE TAKASHI DR
RESEARCH INSTITUTE
OF ATMOPHERICS
NAGOYA UNIVERSITY
TOYOKAWA 442
JAPAN

WATERFIELD REGINALD L DR
WOOLSTON OBSERVATORY
NORTH CADBURY NR
YEOVIL SOMERSET
U.K.

WATERWORTH MICHAEL DR
UNIVERSITY TASMANIA
GPO BOX 252C
HOBART 7001
AUSTRALIA

WATSON FREDERICK GARNETT
ROYAL OBSERVATORY
BLACKFORD HILL
EDINBURGH EH9 3HJ
U.K.

WATSON WILLIAM D PROF
PHYSICS DEPT
UNIVERSITY
OF ILLINOIS
URBANA IL 61801
U.S.A.

WATT GRAEME DAVID
11 HIGHFIELD AVE
SALE CHESHIRE M333DW
U.K.

WATTENBERG D PROF
LINDERHOFSTR 57
DDR 1150 BERLIN
G.D.R.

WAYMAN PATRICK A PROF
DUNSINK OBSERVATORY
CASTLEKNOCK
CO DUBLIN
IRELAND

WEAVER HAROLD F PROF
DEPT OF ASTRONOMY
UNIVERSITY OF
CALIFORNIA
BERKELEY CA 94720
U.S.A.

WEAVER THOMAS A DR
PHYSICS DEPT L-372
LAWRENCE LIVERM LAB
P O BOX 808
LIVERMORE CA 94550
U.S.A.

WEBBER JOHN C DR
UNIVERSITY OF ILLI-
NOIS OBSERVATORY
URBANA IL 61801
U.S.A.

WEBBINK RONALD F DR
DEPT OF ASTRONOMY
UNIV OF ILLINOIS
URBANA IL 61801
U.S.A.

WEBER STEPHEN VANCE
DEPT OF PHYSICS &
ASTRONOMY
DARTMOUTH COLLEGE
HANOVER NH 03755
U.S.A.

WEBROVA LUDMILA DR
ASTRONOMICAL INST
CZECHO ACADEMY SC
BUDECSKA 6
120 23 PRAHA
CZECHOSLOVAKIA

WEBSTER ADRIAN S DR
14 CLARE HALL
HERSCHEL ROAD
CAMBRIDGE
U.K.

WEBSTER B LOUISE DR
ANGLO AUSTRALIAN
OBSERVATORY
P O BOX 296
EPPING NSW 2121
AUSTRALIA

WEEDMAN DANIEL W ASS PROF
DYER OBSERVATORY
VANDERBILT UNIV
BOX 1803
NASHVILLE TN 37235
U.S.A.

WEEKES TREVOR C DR
MOUNT HOPKINS
OBSERVATORY
P O BOX 97
AMADO AZ 85640
U.S.A.

WEGNER GARY ALAN
DEPT OF ASTRONOMY
525 DAVEY LAB
UNIVERSITY PARK
PA 16802
U.S.A.

WEHINGER PETER A DR
1606 E WEST WIND WAY
TEMPE AZ 85283
U.S.A.

WEHLAU AMELIA DR
ASTRONOMY DEPT
UNIVERSITY OF
WESTERN ONTARIO
LONDON ONT N6A 5B9
CANADA

WEHLAU WILLIAM H PROF
ASTRONOMY DEPT
UNIVERSITY OF
WESTERN ONTARIO
LONDON ONT N6A 5B9
CANADA

WEHRSE RAINER DR
INSTITUT F THEORETI-
SCHE ASTROPHYSIK
IM NEUENHEIMER F 294
D-6900 HEIDELBERG
GERMANY, F.R.

WEIDEMANN VOLKER PROF
INST THEO PHY STERNW
NEUE UNIV PHYSIK ZNT
OLSHAUSENST GB N 61C
D-2300 KIEL 1
GERMANY, F.R.

WEIGERT ALFRED PROF
HAMBURGER STERNWARTE
GOJENBERGSWEG 112
D-2050 HAMBURG 80
GERMANY, F.R.

WEILER EDWARD J DR
PRINCETON UNIV OBS
PEYTON HALL
PRINCETON NJ 08540
U.S.A.

WEILER KURT W DR
DIV ASTRONOMICAL SCS
NSF
1800 G ST NW
WASHINGTON DC 20550
U.S.A.

WEILL GILBERT M DR
CNES
129 RUE L'UNIVERSITE
F-75327 PARIS
FRANCE

WEIMER THEOPHILE P F DR
OBSERVATOIRE
DE PARIS
61 AV L'OBSERVATOIRE
F-75014 PARIS
FRANCE

WEINBERG J L DR
SPACE ASTRONOMY LAB
UNIV OF FLORIDA
1810 NW 6TH STREET
GAINESVILLE FL 32601
U.S.A.

WEINBERG STEVEN DR.
DEPT.OF PHYSICS
HARVARD UNIVERSITY
CAMBRIDGE MA 02138
U.S.A.

WEISHEIT JON C DR
THEORETICAL PHYSICS
DIV L-71 LAWRENCE
LIVERMOR NAT LAB
LIVERMORE CA 94550
U.S.A.

WEISS EDWARD W DR
VAN VLECK OBS
WESLEYAN UNIV
MIDDLETOWN CT 06457
U.S.A.

WEISS NIGEL O DR
DAMPT SILVER STREET
CAMBRIDGE CB3 9EW
U.K.

WEISS WERNER W DR
INST FUR ASTRON DER
UNIV WIEN
TUERKENSCHANZSTR 17
A-1180 WIEN
AUSTRIA

WEISSKOPF MARTIN CH DR
NASA/MSFC CODE ES-62
MARSHALL SFC AL35812
U.S.A.

WEISSMAN PAUL ROBERT
JET PROPULSION
LAB 183-301
4800 OAK GROVE DRIVE
PASADENA CA 91109
U.S.A.

WEISTROP DONNA W DR
LAB FOR ASTRONOMY
AND SOLAR PHYSICS
CODE 681 NASA GSFC
GREENBELT MD 20771
U.S.A.

WELCH GARY A DR
DEPT OF ASTRONOMY
ST MARY'S UNIVERSITY
HALIFAX NS B3H 3C3
CANADA

WELCH WILLIAM J PROF
RADIO ASTRONOMY LAB
UNIV OF CALIFORNIA
617 CAMPBELL HALL
BERKELEY CA 94720
U.S.A.

WELIACHEW LEONID DR
IRAM
53 RUE DES MARTYRS
F-38026 GRENOBLE
FRANCE

WELLER CHARLES S DR
NAVAL RESEARCH
LAB CODE 4144
WASHINGTON DC 20375
U.S.A.

WELLGATE G BERNARD MR
CANEHEATH HOUSE
ARLINGTON
POLEGATE
EAST SUSSEX BN26 6SJ
U.K.

WELLMANN PETER PROF DR
INST F ASTRON UND
ASTROPHYSIK
SCHEINER-STR 1
D-8000 MUENCHEN 80
GERMANY, F.R.

WELLS DONALD C III DR
KITT PEAK NAT OBS
P O BOX 26732
TUCSON AZ 85726
U.S.A.

WENDKER HEINRICH J PROF
HAMBURGER STERNWARTE
GOJENBERGSWEG 112
D-2050 HAMBURG 80
GERMANY, F.R.

WENIGER SCHAME DR
OBSERVATOIRE
DE MEUDON
F-92190 MEUDON
FRANCE

WENTZEL DONAT G DR
ASTRONOMY PROGRAM
UNIVERSITY MARYLAND
COLLEGE PARK MD20742
U.S.A.

WENZEL W DR
ZENTRALINSTITUT FUER
ASTROPHYSIK
STERNWARTE SONNEBERG
DDR-64 SONNEBERG
G.D.R.

WESEMAEL FRANCOIS DR
DEPT DE PHYSIQUE
UNIV DE MONTREAL
C P 6128 SUCC A
MONTREAL QUE H3C 3J7
CANADA

WESSELINK ADRIAAN J DR
143 FALLS ROAD
BETHANY CT 06525
U.S.A.

WESSELIUS PAUL R DR
DEPT SPACE RESEARCH
P O BOX 800
9700 AV GRONINGEN
THE NETHERLANDS

WESSON PAUL S DR
DEPT OF PHYSICS
UNIV OF ALBERTA
EDMONTON ALB T6G 2J1
CANADA

WEST RICHARD M DR
ESO
K-SCHWARZSCHILD-ST 2
D-8046 GARCHING
GERMANY, F.R.

WESTERHOUT GART DR
SCIENTIFIC DIRECTOR
US NAVAL OBSERVATORY
WASHINGTON DC 20390
U.S.A.

WESTERLUND BENGT E PROF
ASTRONOMICAL
OBSERVATORY
BOX 515
S-751 20 UPPSALA
SWEDEN

WESTFOLD KEVIN C PROF
DEPT OF MATHEMATICS
MONASH UNIVERSITY
CLAYTON VICT 3168
AUSTRALIA

WESTPHAL JAMES A DR
CALTECH
170-25
169 SOUTH MUDD
PASADENA CA 91125
U.S.A.

WETHERILL GEORGE W
DEPT TERRESTR MAGNET
CARNEGIE INST WASH
5241 BROAD BR RD N W
WASHINGTON DC 20015
U.S.A.

WEYMANN RAY J PROF
STEWARD OBSERVATORY
UNIVERSITY OF
ARIZONA
TUCSON AZ 85721
U.S.A.

WHEELER J CRAIG
ASSOC PROF
DEPT OF ASTRONOMY
UNIVERSITY OF TEXAS
AUSTIN TX 78712
U.S.A.

WHEELER JOHN A DR
DEPT OF PHYSICS
UNIVERSITY OF TEXAS
AUSTIN TX 78712
U.S.A.

WHIPPLE FRED L DR
CTR FOR ASTROPHYSICS
60 GARDEN ST
CAMBRIDGE MA 02138
U.S.A.

WHITAKER EWEN A
4332 E SIXTH STREET
TUCSON AZ 85711
U.S.A.

WHITE GLENN J
PHYSICS DEPT
QUEEN MARY COLLEGE
MILE END ROAD
LONDON E1 4NS
U.K.

WHITE NATHANIEL M DR
LOWELL OBS
P O BOX 1269
FLAGSTAFF AZ 86002
U.S.A.

WHITE ORAN R DR
HIGH ALTITUDE
OBSERVATORY
BOX 3000
BOULDER CO 80307
U.S.A.

WHITE R STEPHEN PROF
IGPP
UNIVERSITY OF
CALIFORNIA
RIVERSIDE CA 92521
U.S.A.

WHITE RAYMOND E DR
STEWARD OBSERVATORY
UNIVERSITY OF
ARIZONA
TUCSON AZ 85721
U.S.A.

WHITE SIMON DAVID MANION
ASTRONOMY DEPT
UNIV OF CALIFORNIA
BERKELEY CA 94720
U.S.A.

WHITELOCK PATRICIA ANN DR
SAAAO
P O BOX 9
OBSERVATORY 7935
CAPE
SOUTH AFRICA

WHITEOAK J B DR
CSIRO
DIV OF RADIOPHYSICS
P O BOX 76
EPPING NSW 2121
AUSTRALIA

WHITFORD ALBERT E PROF
LICK OBSERVATORY
UNIVERSITY OF CALIF
SANTA CRUZ CA 95064
U.S.A.

WHITNEY BALFOUR S
1102 E MISSOURI
NORMAN OK 73069
U.S.A.

WHITNEY CHARLES A PROF
CENTER FOR ASTRO-
PHYSICS
60 GARDEN STREET
CAMBRIDGE MA 02138
U.S.A.

WHITROW GERALD JAMES PROF
41 HOME PARK RD
WIMBLEDON
LONDON SW19 7HS
U.K.

WHITTET DOUGLAS C B DR
DIV OF ASTRONOMY
PRESTON POLYTECHNIC
CORPORATION ST
PRESTON PR1 2TQ
U.K.

WHITWORTH ANTHONY PETER
DEPT OF APPLIED MATH
AND ASTRONOMY
UNIVERSITY COLLEGE
CARDIFF CF1 1XL
U.K.

WICKRAMASINGHE D T DR
INSTITUTE OF
ASTRONOMY
MADINGLEY ROAD
CAMBRIDGE CB3 OHA
U.K.

WICKRAMASINGHE N C PROF
UNIVERSITY COLLEGE
DEPT OF APPLIED MATH
AND THEO PHYS
CARDIFF CF1 1XL
U.K.

WIDING KENNETH G DR
CODE 7144
US NAVAL RESEARCH
LABORATORY
WASHINGTON DC 20375
U.S.A.

WIDORN THOMAS DR
GRINZINGERSTR 93 6
A-1190 WIEN
AUSTRIA

WIEDLING TORD DR
VILLAVAEGEN 15
S-611 00 NYKOEPING
SWEDEN

WIEHR EBERHARD DR
UNIV STERNWARTE
GEISMARLAND STR 11
D-3400 GOETTINGEN
GERMANY, F.R.

WIELEBINSKI RICHARD PROF
MPI FUER RADIO-
ASTRONOMIE
AUF DEM HUEGEL 69
D-5300 BONN
GERMANY, F.R.

WIELEN ROLAND PROF DR
TECH UNIV BERLIN
INST ASTRON/ASTROPH
ERNST REUTER PLATZ 7
D 1000 BERLIN 10
GERMANY, F.R.

WIESE WOLFGANG L DR
NATIONAL BUREAU OF
STANDARDS
WASHINGTON DC 20234
U.S.A.

WIETH KNUDSEN NIELS P DR
SVEND TROSTSVEJ 12
DK-1912 COPENHAGEN V
DENMARK

WIITA JOSEPH PAUL
DEPT OF ASTRONOMY &
ASTROPHYSICS E1
UNIV OF PENNSYLVANIA
PHILADELPHIA PA19104
U.S.A.

WIJNBERGEN JOHANNES DRS
LAB VOOR
RUIMTEONDERZOEK
HOOGBOUW 800
9700 AV GRONINGEN
THE NETHERLANDS

WILCOCK WILLIAM L PROF
SCHOOL PHY MOLEC SC
UNIV COLLEGE N WALES
BANGOR GWYNEDD
LL57 2UW WALES
U.K.

WILCOX JOHN M PROF
INSTITUTE FOR PLASMA
RESEARCH
STANFORD UNIVERSITY
STANFORD CA 94305
U.S.A.

WILD JOHN PAUL DR
CSIRO RADIOPHYSICS
PO BOX 76
EPPING NSW 2021
AUSTRALIA

WILD PAUL DR
ASTRONOMISCHES INST
UNIVERSITAET BERN
SIDLERSTR 5
CH-3012 BERN
SWITZERLAND

WILD PETER A T ASSOC PROF
ASTRONOMY DEPT
UNIVERSITY OF
CAPE TOWN
RONDEBOSCH 7700
SOUTH AFRICA

WILDEY ROBERT L PROF DR
NORTHERN ARIZONA
UNIVERSITY
ASTROPHYSICAL OBS
FLAGSTAFF AZ 86011
U.S.A.

WILKENING LAUREL L DR
DEPT PLANETARY SCS
UNIV OF ARIZONA
TUCSON AZ 85721
U.S.A.

WILKINS GEORGE A DR
ROYAL GREENWICH
OBSERVATORY
HAILSHAM
EAST SUSSEX BN27 1RP
U.K.

WILKINSON ALTHEA
6 HAWTHORN GROVE
WILMSLOW
CHESHIRE SK9 5DE
U.K.

WILKINSON DAVID T
PRINCETON UNIVERSITY
JADWIN HALL
PRINCETON NJ 08540
U.S.A.

WILKINSON PETER N DR
N R A L
JODRELL BANK
MACCLESFIELD
CHESHIRE SK11 9DL
U.K.

WILL CLIFFORD M DR
DEPT OF PHYSICS
WASHINGTON UNIV
ST LOUIS MO 63130
U.S.A.

WILLIAMON RICHARD M
FERNBANK SCIENCE CTR
156 HEATON PARK DR
ATLANTA GA 30307
U.S.A.

WILLIAMS CAROL A
ASSOC PROF
DEPT OF ASTRONOMY
UNIVERSITY S FLORIDA
TAMPA FL 33620
U.S.A.

WILLIAMS D DR
RADIO-ASTRONOMY LAB
UNIV OF CALIFORNIA
BERKELEY CA 94720
U.S.A.

WILLIAMS DAVID A DR
MATHEMATICS DEPT
UMIST
MANCHESTER M60 1QD
U.K.

WILLIAMS IWAN P DR
APPLIED MATH DEPT
QUEEN MARY COLLEGE
MILE END RD
LONDON E1 4NS
U.K.

WILLIAMS JAMES G DR
JPL 264-720
4800 OAK GROVE DRIVE
PASADENA CA 91103
U.S.A.

WILLIAMS JOHN A DR
PHYSICS DEPT
ALBION COLLEGE
ALBION MI 49224
U.S.A.

WILLIAMS PEREDUR M DR
UK INFRARED
TELESCOPE UNIT
900 LEILANI ST
HILO HI 96720
U.S.A.

WILLIAMS ROBERT E PROF
STEWARD OBSERVATORY
UNIVERSITY OF
ARIZONA
TUCSON AZ 85721
U.S.A.

WILLIAMS THEODORE B DR
PRINCETON UNIV OBS
129 PEYTON HALL
PRINCETON NJ 08540
U.S.A.

WILLIS A G DR
DEPT OF PHYSICS
BRANDEIS UNIVERSITY
WALTHAM MA 02154
U.S.A.

WILLIS ALAN J DR
DEPT OF PHYSICS
AND ASTRONOMY
UNIV COLLEGE LONDON
LONDON
U.K.

WILLMORE A PETER PROF
SPACE RESEARCH DEPT
UNIV OF BIRMINGHAM
PO BOX 363
BIRMINGHAM B15 2TT
U.K.

WILLNER STEVEN PAUL DR
CTR FOR ASTROPHYSICS
60 GARDEN STREET
CAMBRIDGE MA 02138
U.S.A.

WILLS BEVERLEY J DR
DEPT OF ASTRONOMY
RLM 15 212
UNIVERSITY OF TEXAS
AUSTIN TX 78712
U.S.A.

WILLS DEREK DR
ASTRONOMY DEPT
UNIVERSITY OF TEXAS
AUSTIN TX 78712
U.S.A.

WILLSON LEE ANNE DR
E W FICK OBSERVATORY
PHYSICS
IOWA ST UNIVERSITY
AMES IA 50011
U.S.A.

WILLSON ROBERT FREDERICK
DEPT OF PHYSICS
TUFTS UNIVERSITY
MEDFORD MD 02155
U.S.A.

WILLSTROP RODERICK V DR
INST OF ASTRONOMY
MADINGLEY ROAD
CAMBRIDGE CB3 0HA
U.K.

WILSON ALBERT G DR
RESEARCH PROGRAM
STUDIES
P O BOX 113
TOPANGA CA 90290
U.S.A.

WILSON ANDREW S DR
ASTRONOMY CENTRE
UNIVERSITY OF SUSSEX
FALMER
BRIGHTON BN1 9QH
U.K.

WILSON BRIAN G PROF
UN OF QUEENSLAND
55 WALCOTT STREET
ST LUCIA QLD 4067
AUSTRALIA

WILSON CURTIS A
ST.JOHN'S COLLEGE
ANNAPOLIS MD 21404
U.S.A.

WILSON JAMES R DR
LAWR LIVERMORE LAB
L-35
LIVERMORE CA 94550
U.S.A.

WILSON LIONEL DR
LUNAR AND PLANETARY
UNIT E S DEPT
LANCASTER UNIVERSITY
LANCASTER LA1 4YR
U.K.

WILSON OLIN C
HALE OBSERVATORIES
813 SANTA BARBARA ST
PASADENA CA 91106
U.S.A.

WILSON P DR
INST ANGEW GEODAESIE
R-STRAUSS-ALLEE 11
D-6000 FRANKFURT M
GERMANY, F.R.

WILSON PETER R PROF
DEPT OF APPLIED
MATHEMATICS
UNIVERSITY OF SYDNEY
SYDNEY 2007
AUSTRALIA

WILSON R N DR
ESO
K SCHWARZSCHILD ST 2
D-8046 GARCHING B M
GERMANY, F.R.

WILSON RAYMOND H DR
5325 GAINSBOROUGH DR
FAIRFAX VA 22032
U.S.A.

WILSON ROBERT E PROF
DEPT OF ASTRONOMY
UNIV OF FLORIDA
GAINESVILLE FL 32611
U.S.A.

WILSON ROBERT PROF
DEPT PHYS ASTRONOMY
UNIV COLLEGE LONDON
GOWER ST
LONDON WC1E 6BT
U.K.

WILSON ROBERT W DR
BELL LABORATORIES
HOH L229
BOX 400
HOLMDEL NJ 07733
U.S.A.

WILSON S J
DEPT OF APPLIED PHYS
& INFORMATION SC
UNIV OF CALIFORNIA
LA JOLLA CA 92093
U.S.A.

WILSON THOMAS L DR
MPI FUER RADIO-ASTRO
NOMIE
AUF DEM HUEGEL 69
D-5300 BONN
GERMANY, F.R.

WILSON WILLIAM J DR
JPL
BLDG 168-327
4800 OAK GROVE DRIVE
PASADENA CA 91109
U.S.A.

WINCKLER JOHN R PROF
SCHOOL OF PHYSICS
AND ASTRONOMY
UNIV OF MINNESOTA
MINNEAPOLIS MN 55455
U.S.A.

WING ROBERT F PROF
ASTRONOMY DEPT
OHIO ST UNIVERSITY
174 WEST 18TH AVE
COLUMBUS OH 43210
U.S.A.

WINK JOERN ERHARD DR
MPI FUER
RADIOASTRONOMIE
AUF DEM HUEGEL 69
D-5300 BONN
GERMANY, F.R.

WINKLER FRANK P DR
DEPT OF PHYSICS
MIDDLEBURY COLLEGE
MIDDLEBURY VT 05753
U.S.A.

WINKLER GERNOT M R DR
US NAVAL OBSERVATORY
WASHINGTON DC 20390
U.S.A.

WINKLER KARL-HEINZ A DR
MPI FUER ASTROPHYSIK
KARL-SCHWARZSCHILD-
STR 1
D-8046 GARCHING B M
GERMANY, F.R.

WINNBERG ANDERS DR
ONSALA SPACE OBS
S-43900 ONSALA
SWEDEN

WINNEWISSER GISBERT DR
MPI FUER RADIO-
ASTRONOMIE
AUF DEM HUEGEL 69
D-5300 BONN
GERMANY, F.R.

WISNIEWSKI WIESLAW Z
UNIVERSITY ARIZONA
LUNAR AND PLANETARY
LABORATORY
TUCSON AZ 85721
U.S.A.

WITHBROE GEORGE L DR
CTR F ASTROPHYSICS
HARVARD COLLEGE OBS
60 GARDEN ST
CAMBRIDGE MA 02138
U.S.A.

WITT ADOLF N DR
RITTER ASTROPHYSICAL
RESEARCH CENTER
UNIVERSITY OF TOLEDO
TOLEDO OH 43606
U.S.A.

WITTEN LOUIS PROF
DEPT OF PHYSICS
UNIVERSITY OF
CINCINNATI
CINCINNATI OH 45221
U.S.A.

WITTMANN AXEL DIETER PH D
UNIV STERNWARTE
GEISMARLANDSTR 11
D-3400 GOETTINGEN
GERMANY, F.R.

WITZEL ARNO DR
MPI FUER RADIO-
ASTRONOMIE
AUF DEM HUEGEL 69
D-5300 BONN
GERMANY, F.R.

WLERICK GERARD DR
OBSERVATOIRE PARIS
F-92190 MEUDON
FRANCE

WOEHL HUBERTUS DR
KIEPENHEUER INSTITUT
FUER SONNENPHYSIK
SCHOENECKSTRASSE 6
D-7800 FREIBURG
GERMANY, F.R.

WOLF BERNHARD PH D
MOMBERTSTRASSE 5
D-6900 HEIDELBERG
GERMANY, F.R.

WOLF RAINER E A DR
MPI FUER ASTRONOMIE
KOENIGSTUHL
D-6900 HEIDELBERG 1
GERMANY, F.R.

WOLFE ARTHUR M PROF
DEPT OF PHYSICS AND
ASTRONOMY
UNIV OF PITTSBURGH
PITTSBURGH PA 15260
U.S.A.

WOLFENDALE ARNOLD W PROF
PHYSICS DEPT
UNIVERSITY
SOUTH ROAD
DURHAM DH1 3LE
U.K.

WOLFF SIDNEY C DR
INSTITUTE FOR
ASTRONOMY
2680 WOODLAWN DRIVE
HONOLULU HI 96822
U.S.A.

WOLSTENCROFT RAMON D DR
ROYAL OBSERVATORY
BLACKFORD HILL
EDINBURGH EH9 3HJ
U.K.

WOLSZCZAN ALEKSANDER DR
N COPERNICUS ASTRON
CENTER
CHOPINA 12/18
87-100 TORUN
POLAND

WOLTJER LODEWIJK PROF
ESO
K SCHWARZSCHILD ST 2
D-8046 GARCHING B M
GERMANY, F.R.

WOOD DAVID B DR
WOOD ASSOCIATES
5108 VIKING RD
BETHESDA MD 20014
U.S.A.

WOOD F BRADSHAW PROF
DEPT OF ASTRONOMY
SSRB 211
UNIVERSITY FLORIDA
GAINESVILLE FL 32611
U.S.A.

WOOD HARLEY W DR
178 KENTHURST RD
KENTHURST 2154
AUSTRALIA

WOOD III H J DR
DEPT OF ASTRONOMY
UNIV OF VIRGINIA
CHARLOTTESVL VA22903
U.S.A.

WOOD JOHN A DR
CTR FOR ASTROPHYSICS
60 GARDEN ST
CAMBRIDGE MA 02138
U.S.A.

WOOD PETER R DR
MT STROMLO AND
SIDING SPRING OBS
PRIVATE BAG
WODEN A C T 2606
AUSTRALIA

WOOD ROGER DR
ROYAL GREENWICH OBS
HERSTMONCEUX CASTLE
HAILSHAM
E SUSSEX BN27 1RP
U.K.

WOODSWORTH ANDREW W.DR
HERZBERG INSTITUTE
OF ASTROPHYSICS
NRC
OTTAWA ONT K1A 0R6
CANADA

WOODWARD P R DR
T DIV L-71 UN OF CAL
LAWRENCE LIVERM LAB
P O BOX 808
LIVERMORE CA 94550
U.S.A.

WOOLEY SIR RICHARD
4 MYRTLE STREET
SOMERSET WEST
CAPE 7130
SOUTH AFRICA

WOOLF NEVILLE J
STEWARD OBSERVATORY
UNIV OF ARIZONA
TUCSON AZ 85721
U.S.A.

WOOLFSON MICHAEL M PROF
UNIVERSITY OF YORK
YORK YO1 5DD
U.K.

WOOLSEY E G
695 BRIERWOOD
OTTAWA ONT
CANADA

WOOSLEY S E PROF
LICK OBSERVATORY
UNIVERSITY OF CALIF
SANTA CRUZ CA 95064
U.S.A.

WOOTTEN HENRY ALWYN
NRAO
EDGEMONT ROAD
CHARLOTTESVL VA22901
U.S.A.

WORDEN SIMON P DR
HQ SD/YBE
P O BOX 92960
L A AIR FORCE STA
LOS ANGELES CA 90009
U.S.A.

WORLEY CHARLES E
ASTROMETRY AND ASTRO
PHYSICS DIVISION
US NAVAL OBSERVATORY
WASHINGTON DC 20390
U.S.A.

WORRALL DIANA MARY
CASS CODE C-011
UNIV OF CALIFORNIA
AT SAN DIEGO
LA JOLLA CA 92093
U.S.A.

WORRALL GORDON DR
BIRDSWOOD
EARDISLEY HEREFDS
U.K.

WOSZCZYK ANDRZEJ DR
INST OF ASTRONOMY
CHOPINA 12-18
87100 TORUN
POLAND

WRAMDEMARK STIG S O DR
LUND OBSERVATORIET
BOX 1107
S-221 04 LUND
SWEDEN

WRAY JAMES D DR
DEPT OF ASTRONOMY
UNIVERSITY OF TEXAS
RL MOORE HALL 15-212
AUSTIN TX 78712
U.S.A.

WRIGHT ALAN E DR
AUSTRALIAN NATIONAL
RADIO ASTRONOMY OBS
P O BOX 276
PARKES NSW 2870
AUSTRALIA

WRIGHT EDWARD L DR
DEPT OF PHYSICS
RM 26-345
MIT
CAMBRIDGE MA 02139
U.S.A.

WRIGHT FRANCES W DR
ASTRONOMY DEPT
HARVARD COLLEGE
60 GARDEN ST
CAMBRIDGE MA 02138
U.S.A.

WRIGHT GEOFFREY A E DR
INST F ASTROPHYS UND
EXTRATERR FORSCHUNG
AUF DEM HUEGEL 71
D-5300 BONN
GERMANY, F.R.

WRIGHT HELEN
173 FOREST LAKE DR
ANDOVER NJ 07821
U.S.A.

WRIGHT JAMES P DR
DIV ASTRONOMICAL SC
NAT SC FOUND RM 615
1800 G STREET N W
WASHINGTON DC 20550
U.S.A.

WRIGHT KENNETH O DR
DOMINION ASTROPH OBS
5071 W SAANICH ROAD
RR 5
VICTORIA BC V8X 4M6
CANADA

WRIGHT MELVYN C H DR
RADIO ASTRONOMY
LABORATORY
UNIVERSITY OF CALIF
BERKELEY CA 94720
U.S.A.

WRIXON GEORGE T DR
DEPT OF ELECTRICAL
ENGINEERING
UNIVERSITY COLLEGE
CORK
IRELAND

WU CHI CHAO DR
COMPUTER SCIENCES CO
NASA GODDARD SPACE
FLIGT CNTR CODE 685
GREENBELT MD 20771
U.S.A.

WU HSIN-HENG DR
DEPT OF PHYSICS
NAT CENTAL UNIV
CHUNG-LI
TAIWAN

WU LIN-XIANG
DEPT OF GEOPHYSICS
BEIJING UNIVERSITY
BEIJING
PEOPLE'S REPUBLIC CHINA

WU MING-CHAN
YUNNAN OBSERVATORY
KUNMING
YUNNAN PROVINCE
PEOPLE'S REPUBLIC CHINA

WU SHI TSAN DR
SCHOOL SCIENCE ENGIN
UNIV OF ALABAMA
P O BOX 1247
HUNTSVILLE AL 35801
U.S.A.

WU SHOU-XIAN
SHAANXI OBSERVATORY
LINTONG XIAN
SHAANXI
PEOPLE'S REPUBLIC CHINA

WU XUE-LIN
NANJING ASTRONOMICAL
INSTRUMENT FACTORY
NANJING
JIANGSU PROVINCE
PEOPLE'S REPUBLIC CHINA

WYCKOFF SUSAN DR
PHYSICS DEPT
ARIZONA STATE UNIV
TEMPE AZ 85281
U.S.A.

WYLLER ARNE A PROF
STOCKHOLM
OBSERVATORY
S-133 00
SALTSJOEBADEN
SWEDEN

WYNN-WILLIAMS C G DR
INST FOR ASTRONOMY
UNIVERSITY OF HAWAII
2680 WOODLAWN DRIVE
HONOLULU HI 96822
U.S.A.

WYNNE CHARLES G PROF
ROYAL GREENWICH
OBSERVATORY
HAILSHAM
EAST SUSSEX BN27 1RP
U.K.

XANTHAKIS JOHN N PROF
R C A A M
ACADEMY OF ATHENS
14 ANAGNOSTOPOULOU
ATHENS 136
GREECE

XANTHOPOULOS B C DR
DEPT OF ASTRONOMY
UNIV OF THESSALONIKI
THESSALONIKI
GREECE

XI ZE-ZONG
INSTITUTE OF THE
HISTORY OF NATURAL
SCIENCE
BEIJING
PEOPLE'S REPUBLIC CHINA

XING JUN
DEPT OF GEOPHYSICS
BEIJING UNIVERSITY
BEIJING
PEOPLE'S REPUBLIC CHINA

XIONG DA-RUN
PURPLE MOUNTAIN
OBSERVATORY
NANJING
PEOPLE'S REPUBLIC CHINA

XU BANG-XIN
DEPT OF ASTRONOMY
NANJING
PEOPLE'S REPUBLIC CHINA

XU JIA-YAN
P O BOX 18
LINTONG
SHAAXI
PEOPLE'S REPUBLIC CHINA

XU PEI-YUAN
PURPLE MOUTAIN
OBSERVATORY
NANJING
PEOPLE'S REPUBLIC CHINA

XU TONG-QI
SHANGHAI OBSERVATORY
SHANGHAI
PEOPLE'S REPUBLIC CHINA

YABUSHITA SHIN A PROF
DEPT OF APPLIED MATH
AND PHYSIC
KYOTO UNIVERSITY
SAKYOKU KYOTO 606
JAPAN

YABUUTI KIYOSHI PROF
20 TANAKA HIGASKI
HINOKUCH MACHI
SAKYOKU 606 KYOTO
JAPAN

YAHIL AMOS DR.
ASTRONOMY PROGRAM
ESS BUILDING
SUNY AT STONY BROOK
STONY BROOK NY 11794
U.S.A.

YAKOVKIN N A DR
KIEV STATE UNIV
ASTRONOMICAL OBS
252053 KIEV
U.S.S.R.

YALLOP BERNARD D DR
ROYAL GREENWICH
OBSERVATORY
HAILSHAM
EAST SUSSEX BN27 1RP
U.K.

YAMAMOTO TETSUO DR
INST OF SPACE &
ASTRONAUTICAL SC
KOMABA MEGURO-KU
TOKYO 153
JAPAN

YAMASAKI ATSUMA DR
DEPT EARTH SC ASTRON
UNIVERSITY OF TOKYO
KOMABA MEGURO-KU
TOKYO 153
JAPAN

YAMASHITA KOJUN DR
DEPT OF PHYSICS
NAGOYA UNIVERSITY
FUROCHO CHIKUSAKU
NAGOYA 464
JAPAN

YAMASHITA YASUMASA PROF
TOKYO ASTRONOMICAL
OBSERVATORY
OSAWA
MITAKA TOKYO 181
JAPAN

YAMAZAKI AKIRA DR
HYDROGRAPHIC DEPT
TSUKIJI 5 CHUO KU
TOKYO 104
JAPAN

YANG HAI SHOU
ASTROPHYSICS DIV
GEOPHYSICS DEPT
PEKING UNIVERSITY
BEIJING
PEOPLE'S REPUBLIC CHINA

YANG SHI JIE
PURPLE MOUNTAIN
OBSERVATORY
NANJING
PEOPLE'S REPUBLIC CHINA

YANKULOVA IVANKA DR
DEPT OF ASTRONOMY
FACULTY OF PHYSICS
ANTON IVANOV STR 5
1126 SOFIA
BULGARIA

YANOVITSKIJ EDGAROT G DR
MAIN ASTRONOMICAL OB
UKRAINIAN
252127 KIEV
U.S.S.R.

YAPLEE B S
8 CREST VIEW COURT
ROCKVILLE MD 20854
U.S.A.

YAROV-YAROVOJ M S DR
MATHEMATICS DEPT
OF MVTU
107005 MOSCOW
U.S.S.R.

YASUDA HARUO PROF DR
TOKYO ASTRONOMICAL
OBSERVATORY
OSAWA
MITAKA TOKYO 181
JAPAN

YATSKIV YA S DR
MAIN ASTRONOMICAL
OBS UKRAINIAN ACADE-
MY OF SCIENCES
252127 KIEV
U.S.S.R.

YAVNEL ALEXANDER A DR
METEORITE COMMITTEE
ACADEMY OF SCIENCES
UL M ULIANOVOJ 3 K 1
117313 MOSCOW
U.S.S.R.

YE SHI-HUI
PURPLE MOUNTAIN
OBSERVATORY
NANJING
PEOPLE'S REPUBLIC CHINA

YE SHU-HUA
SHANGHAI OBSERVATORY
SHANGHAI
PEOPLE'S REPUBLIC CHINA

YEIVIN Y PROF
TEL-AVIV UNIVERSITY
TEL-AVIV
ISRAEL

YEN YUI-LIN PROF
DEPT OF ELECTRICAL
ENGINEERING
UNIVERSITY TORONTO
ONT M5S 1A4
CANADA

YEOMANS DONALD K DR
JET PROPULSION LAB
4800 OAK GROVE DR
PASADENA CA 91103
U.S.A.

YI ZHAO-HUA
DEPT OF ASTRONOMY
NANJING UNIVERSITY
NANJING
PEOPLE'S REPUBLIC CHINA

YILMAZ FATMA DR
OBS DE L'UNIVERSITE
ISTANBUL
TURKEY

YILMAZ NIHAL DR
FACULTY OF SCIENCE
DEPT OF ASTRONOMY
ANKARA
TURKEY

YIN QI-FENG
DEPT GEOPHYSICS
PEKING UNIVERSITY
PEKING
PEOPLE'S REPUBLIC CHINA

YODER CHARLES F
JET PROPULSION LAB
4800 OAK GROVE DRIVE
PASADENA CA 91109
U.S.A.

YOKOYAMA KOICHI DR
INTERNATIONAL LATITU
DE OBSERVATORY
MIZUSAWA IWATE 023
JAPAN

YONEYAMA TADAOKI DR
3-14-3-802
OHKUBO
SHINJUKU-KU
TOKYO 112
JAPAN

YORK DONALD G DR
PEYTON HALL
PRINCETON UNIVERSITY
PRINCETON NJ 08540
U.S.A.

YORKE HAROLD W DR
UNIV STERNWARTE
GEISMARLANDSTR 11
D-3400 GOETTINGEN
GERMANY, F.R.

YOSHIDA JUNZO PROF
DEPT OF PHYSICS
KYOTO SANGYO UNIV
KAMIGAMO KITA-KU
KYOTO 603
JAPAN

YOSHII YUZURU DR
TOKYO ASTRON OBS
UNIV OF TOKYO
MITAKA 181
JAPAN

YOSHIMURA HIROKAZU DR
DEPT OF ASTRONOMY
UNIVERSITY OF TOKYO
YAYOI BUNKYO
TOKYO 113
JAPAN

YOSHIZAWA MASANORI DR
TOKYO ASTRON OBS
OSAWA 2-21-1
MITAKA
TOKYO 181
JAPAN

YOSS KENNETH M DR
ASTRONOMY DEPT
UNIV OF ILLINOIS
OBSERVATORY
URBANA IL 61801
U.S.A.

YOU JIAN-QI
PURPLE MOUNTAIN
OBSERVATORY
NANJING
PEOPLE'S REPUBLIC CHINA

YOU JUNHAN
ASTROPHYS RES DIV
UNIV SC & TECH CHINA
HEIFI
ANHUI PROVINCE
PEOPLE'S REPUBLIC CHINA

YOUNG ANDREW T DR
DEPT OF ASTRONOMY
SAN DIEGO STATE UN
SAN DIEGO CA 92182
U.S.A.

YOUNG ARTHUR DR
ASTRONOMY DEPT
SAN DIEGO STATE UNIV
SAN DIEGO CA 92180
U.S.A.

YOUNG JUDITH SHORN
FIVE CLG RAD
619 I GRC TOWER B
UNIV MASSACHUSETTS
AMHERST MA 01003
U.S.A.

YOUNG LOUISE GRAY DR
DEPT OF ASTRONOMY
SAN DIEGO STATE UN
SAN DIEGO CA 92182
U.S.A.

YOUSEF SHAHINAZ M DR
ASTRONOMY DEPT
FACULTY OF SCIENCE
CAIRO UNIVERSITY
CAIRO
ARAB REP. OF EGYPT

YOUSSEF NAHED H DR
ASTRONOMY DEPT
FACULTY OF SCIENCE
CAIRO UNIVERSITY
CAIRO
ARAB REP. OF EGYPT

YU KYUNG-LOH PROF
SEOUL NAT UNIVERSITY
SINLIM-DONG
KWANAK-KU
SEOUL 151
REPUBLIC OF KOREA

YUAN CHI PROF
DEPT OF PHYSICS
CITY COLLEGE OF N Y
138 ST CONVENE AVE
NEW YORK NY 10031
U.S.A.

YUASA MANABU DR
DEPT OF MATH & PHYS
KINKI UNIVERSITY
HIGASHI-OSAKA
OSAKA 577
JAPAN

YUE ZENG-YUAN
DEPT OF GEOPHYSICS
PEKING UNIVERSITY
PEKING
PEOPLE'S REPUBLIC CHINA

YUMI S DR
TOYO UNIVERSITY
HAKUSAN 5-28-20
BUNKYO-KU
TOKYO 112
JAPAN

YUN HONG-SIK ASSOC PROF
SEOUL NAT UNIVERSITY
SINLIM-DONG
KWANAK-KU
SEOUL 151
REPUBLIC OF KOREA

ZABRISKIE F R PROF
R D 1
ALEXANDRIA PA 16611
U.S.A.

ZACHAROV IGOR DR
ASTRONOMICAL INST
CZECHO ACADEMY SC
OBSERVATORY
251 65 ONDREJOV
CZECHOSLOVAKIA

ZADUNAISKY PEDRO E PROF
OBS NAC FISICA COSM
AV MITRE 3100
1663 SAN MIGUEL
BUENOS AIRES
ARGENTINA

ZAHN JEAN-PAUL DR
OBSERVATOIRE DE NICE
BP 252
F-06007 NICE CEDEX
FRANCE

ZAITSEV VALERII V DR
INST OF APPIED PHYS
ULYANOVA ST 46
603600 GORKY
U.S.S.R.

ZAMORANI GIOVANNI
ISTITUTO DI
RADIOASTRONOMIA
VIA IRNERIO 46
I-40126 BOLOGNA
ITALY

ZANDER RODOLPHE DR
INST D ASTROPHYSIQUE
UNIVERSITE DE LIEGE
B-4200 COINTE OUGREE
BELGIUM

ZAPPALA ROSARIO ALDO DR
OSS ASTROFISICO
CITTA UNIVERSITARIA
I-95125 CATANIA
ITALY

ZAPPALA VINCENZO DR
OSSERVATORIO ASTRO-
NOMICO
PINO TORINESE
TORINO
ITALY

ZARE K DR
DEPT OF MATHEMATICS
TECHNICAL UNIVERSITY
P O BOX 6934
ISFAHAN
IRAN

ZARNECKI JAN CHARLES DR
GIOTTO PROJECT OFF.
PHYSICS LABORATORY
UNIV.KENT AT CANTERB
CANTERBURY CT2 7NT
U.K.

ZASOV ANATOLE V DR
STERNBERG ASTRONOMI-
CAL INSTITUTE
117234 MOSCOW
U.S.S.R.

ZDANAVICIUS KAZIMERAS DR
ASTRONOMIJOS OBS
VILNIUS 31
CIURLIONIO 29
LITHUNIA
U.S.S.R.

ZEALEY WILLIAM J DR
ROYAL OBS EDINBURGH
BLACKFORD HILL
EDINBURGH EH9 3HJ
U.K.

ZEILIK MICHAEL II DR
DEPT PHYSICS ASTRON
UNIV OF NEW MEXICO
800 YALE BLVD BE
ALBUQUERQUE NM 87131
U.S.A.

ZEIPPEN C DR
OBS DE MEUDON
F-92190 MEUDON
FRANCE

ZEL'DOVICH YA B ACAD
SPACE RESEARCH
INSTITUTE
USSR ACADEMY OF SCI
117810 MOSCOW
U.S.S.R.

ZEL'MANOV A L DR
STERNBERG STATE AS-
TRONOMICAL INSTITUTE
117234 MOSCOW
U.S.S.R.

ZELENKA ANTOINE DR
DACHSLENBERGSTR
CH 8180 BUELACH
SWITZERLAND

ZELLNER BENJAMIN H DR
LUNAR AND PLANETARY
LABORATORY
UNIVERSITY ARIZONA
TUCSON AZ 85721
U.S.A.

ZERULL REINER H DR
RUHR-UNIV BOCHUM
BEREICH EXTRATERRES-
TRISCHE PHYSIK
D-4630 BOCHUM
GERMANY, F.R.

ZHANG BAI-RONG
YUNNAN OBSERVATORY
KUNMING
YUNNAN PROVINCE
PEOPLE'S REPUBLIC CHINA

ZHANG HE-QI
PURPLE MOUNTAIN
OBSERVATORY
NANJING
JIANGSU PROVINCE
PEOPLE'S REPUBLIC CHINA

ZHANG JIE-LIU
ASTROPHYSICS RES DIV
UNIV OF SCIENCE &
TECHNOLOGY OF CHINA
HEFEI ANHUI PROVINCE
PEOPLE'S REPUBLIC CHINA

ZHANG JIE-XIANG
PURPLE MOUNTAIN
OBSERVATORY
ACADEMIA SINICA
NANJING
PEOPLE'S REPUBLIC CHINA

ZHANG YU-ZHE
PURPLE MOUNTAIN
OBSERVATORY
NANJING
PEOPLE'S REPUBLIC CHINA

ZHANG ZHEN-DA
DEPT OF ASTRONOMY
NANJING UNIVERSITY
NANJING
PEOPLE'S REPUBLIC CHINA

ZHAO JUN-LIANG
SHANGHAI OBSERVATORY
SHANGHAI
PEOPLE'S REPUBLIC CHINA

ZHAO QUE-MIN
DEPT OF ASTRONOMY
NANJING UNIVERSITY
NANJING
PEOPLE'S REPUBLIC CHINA

ZHAO XIAN-ZI
PURPLE MOUNTAIN
OBSERVATORY
NANJING
PEOPLE'S REPUBLIC CHINA

ZHELEZNIAKOV VLADIMIR V
PROF APPLIED PHYSICS
INST ACADEMY SCIENCE
ULYANOV STREET 46
603600 GORKIJ
U.S.S.R.

ZHENG WENGUANG
BEIJING ASTRONOMICAL
OBSERVATORY
BEIJING
PEOPLE'S REPUBLIC CHINA

ZHEVAKIN S A PROF DR
RADIOPHYSICAL
RESEARCH INSTITUTE
603600 GORKIJ
U.S.S.R.

ZHU NENGHONG
SHANGHEI OBSERVATORY
ACADEMIA SINICA
SHANGHAI
PEOPLE'S REPUBLIC CHINA

ZHUGZHDA YUZEF D DR
INST OF TERRESTRIAL
MAGNETISM &
IONOSPHERE
142092 ACADEMGOROK
U.S.S.R.

ZIEBA ANDRZEJ PROF
ASTRONOMICAL
OBSERVATORY
UL ORLA 171
30-244 KRAKOW
POLAND

ZIEBA STANISLAW DR
OBSERVATORIUM
ASTRONOMICZNE UJ
UL ORLA 171
30-244 KRAKOW
POLAND

ZIKIDES MICHAEL C DR
DEPT OF ASTRONOMY
PANEPISTIMIOPOLIS
ATHENS 621
GREECE

ZIMMERMAN HELMUT DR
UNIVERSITAETS
STERNWARTE JENA
SCHILLERGAESSCHEN 2
DDR-69 JENA
G.D.R.

ZINN ROBERT J DR
DEPT OF ASTRONOMY
YALE UNIVERSITY
BOX 2023 YALE STAT
NEW HAVEN CT 06520
U.S.A.

ZIOLKOWSKI J DR
COPERNICUS ASTRONOMI
CAL CENTER
UL BARTYCKA 18
00 716 WARSAW
POLAND

ZIOLKOWSKI KRZYSZTOF DR
MICKIEWICZA 4 M 17
01-517 WARSZAWA
POLAND

ZIRIN HAROLD DR
CAL TECH 264-33
PASADENA CA 91125
U.S.A.

ZIRKER JACK B DR
SACRAMENTO PEAK
OBSERVATORY
SUNSPOT NM 88349
U.S.A.

ZLOBEC PAOLO DR
S FLORIANA 1/A
I-34015 MUGGIA
ITALY

ZOMBECK MARTIN V DR
CTR FOR ASTROPHYSICS
60 GARDEN ST
CAMBRIDGE MA 02138
U.S.A.

ZOU HUI-CHENG
SHANGHAI OBSERVATORY
SHANGHAI
PEOPLE'S REPUBLIC CHINA

ZOU YI-XIN
BEIJING OBSERVATORY
BEIJING
PEOPLE'S REPUBLIC CHINA

ZOU ZHEN-LONG
BEIJING OBSERVATORY
BEIJING
PEOPLE'S REPUBLIC CHINA

ZUCKERMAN BEN M DR
ASTRONOMY PROGRAM
UNIVERSITY MARYLAND
COLLEGE PARK MD20742
U.S.A.

ZUIDERWIJK E J
BEECHES
JOES LANE
WINDMILL HILL
HAILSHAM E SUSSEX
U.K.

ZVEREV MITROFAN S PROF DR
PULKOVO OBSERVATORY
196140 LENINGRAD
U.S.S.R.

ZWAAN CORNELIS DR
OBSERVATORY
ZONNENBURG 2
3512 NL UTRECHT
THE NETHERLANDS

GPSR Compliance

*The European Union's (EU) General Product Safety Regulation (GPSR) is a set of rules that requires consumer products to be safe and our obligations to ensure this.*

*If you have any concerns about our products, you can contact us on ProductSafety@springernature.com*

In case Publisher is established outside the EU, the EU authorized representative is:

Springer Nature Customer Service Center GmbH
Europaplatz 3
69115 Heidelberg, Germany

**Batch number: 10800581**

Printed by Printforce, the Netherlands